THE
INTERNATIONAL SERIES
OF
MONOGRAPHS ON PHYSICS

NUCLEAR MAGNETISM:
Order and Disorder

A. ABRAGAM
AND
M. GOLDMAN
Collège de France and Centre d'Etudes Nucléaires de Saclay

CLARENDON PRESS · OXFORD
1982

Oxford University Press, Walton Street, Oxford OX2 6DP

OXFORD LONDON GLASGOW
NEW YORK TORONTO MELBOURNE WELLINGTON
KUALA LUMPUR SINGAPORE JAKARTA HONG KONG TOKYO
DELHI BOMBAY CALCUTTA MADRAS KARACHI
NAIROBI DAR ES SALAAM CAPE TOWN

Published in the United States by Oxford University Press, New York

British Library Cataloguing in Publication Data

Abragam, A.
Nuclear magnetism.—(International series of monographs in physics)
1. Nuclear magnetic resonance
I. Title II. Goldman, M. III. Series 538′.3 QC762

ISBN 0–19–851294–5

Printed in Great Britain by J. W. Arrowsmith Ltd., Bristol

PREFACE

TWENTY years ago one of the authors wrote a comprehensive book on *The principles of nuclear magnetism* ('*The Principles*' for brevity). Ten years later the other author wrote a more specialized monograph on *Spin temperature and nuclear magnetic resonance in solids* ('*Spin Temperature*' for brevity).

Although the present book does deal with the principles of nuclear magnetism and although spin temperature is one of its key concepts, it should be made emphatically clear at the outset that this is not a new edition of either one of the earlier books.

The motivations for preparing a new edition of a book are manifold but can be summed up as a wish to add and a wish to suppress: to add, because in a field which remains alive, new directions of research, new theories, new experimental results keep cropping up and want coverage; to suppress, because these new experiments and theories make some of the former descriptions and explanations incomplete or even incorrect, and rob some others of much of their interest.

Through care, foresight, caution and sheer luck, it has turned out that there is very little to suppress in our former books, or so we think. On the other hand at least for the more comprehensive of the two, *The Principles*, there is far too much to add. It is our considered opinion that today no single author (nor even a couple of authors) could cover in a single volume (or even in a couple of volumes) the whole field of nuclear magnetism. This is why in spite of its size this new book covers only a limited part of the field, which does not overlap with either *The Principles* or *Spin Temperature*.

Nuclear magnetism is both a tool and a field that is an object of study for its own sake. With respect to its first aspect, to quote from the preface to *The Principles*: 'it has become a major tool in the study of the finer properties of matter in bulk. Structure of molecules, reaction rates and chemical equilibria, chemical bonding, crystal structures, internal motions in solids and liquids, electronic densities in metals, alloys, and semiconductors, internal fields in ferromagnetic and antiferromagnetic substances, density of states in superconductors, properties of quantum liquids, are some of the topics where nuclear magnetism has so far provided specific and detailed information.' Twenty years later, with biologists and biochemists taking now the lion's part among the users, the role of nuclear magnetism as a scientific tool is greater than ever. The amount of factual information gathered is staggering. The techniques and instrumentation of nuclear magnetic resonance (NMR) and above all of high resolution in liquids have reached new heights of sophistication and efficiency, thanks in particular to the tremendous progress in the art of computers.

In this book however we have deliberately chosen to turn our backs on this aspect of nuclear magnetism. Rather than a tool, nuclear magnetism is here viewed as a field where the nuclear spins are actors rather than spectators. It is true that the two approaches have much in common. The best way to turn a field into a useful tool is to understand it well. Thus, systematic studies of nuclear relaxation in liquids, as described for instance in *The Principles* have paved the way to many physicochemical studies and one of the purposes of *Spin Temperature* was to provide solid state physicists with a better understanding of improved NMR signals.

This book is a study in depth of what appears to us as the most interesting advances of the last ten to fifteen years in nuclear magnetism. This includes the study of methods, new or improved, such as coherent manipulation of nuclear spins in solids, dynamic nuclear polarization at low temperatures or pseudomagnetic neutron precession within polarized nuclear targets but also nuclear spin systems of particular interest such as solid ^{3}He, superfluid ^{3}He, dipolar ferromagnets and antiferromagnets. In spite of this variety of subjects there is, we believe, an internal unity among them, summed up in the title. The order and disorder mentioned there express the correlations between the orientations of the nuclear spins. It may be a Zeeman order where nuclear spins are correlated with each other through their common orientation along the same direction of space. It is usually the direction of an applied magnetic field but it can be another direction along which the spin orientation is maintained by cleverly applied r.f. pulses. It may be a short range order for which dipolar or exchange interactions are responsible for sufficiently low temperatures. It may be a more subtle order of a collective nature such as exists in superfluid ^{3}He. Finally for very low spin temperatures a transition to a phase with a long range spin order may occur.

Before describing the contents of the book a few words about the philosophy which has presided over its writing are in order.

As for our former books, and in accordance with the very nature of nuclear magnetism, this is neither a theoretical nor an experimental book, or rather it is both. The large number of equations is balanced by a selection of numerical results and experimental curves. To quote once more from the preface to *The Principles*: 'there is [in nuclear magnetism] little room for a theory that could not be tested by a suitable experiment or for an experiment that does not admit of a well-defined theoretical interpretation'. Twenty years later, in spite of the increased sophistication of some theories and the increased difficulty and complexity of the experiments this is still what we believe.

In incorporating recent results we have tried to steer a course between the twin snags of excessive caution and excessive haste which both lead to early obsolescence. Our golden rule in the matter, which has proven reasonably successful in the past, is to include only results that we think we understand, irrespective of their date.

We have strived hard to make this book self contained. This does not mean of course that the reader who wants to *know* more should not look for it elsewhere in the literature. What it does mean is that he should not *have* to look elsewhere for a better *understanding* of what *is* in the book. As a consequence of this endeavour a theoretically-minded reader may sometimes find our presentation pedestrian and too detailed. He ought then to skip what he already knows, as we invite him to do in a few places.

On the other hand some previous knowledge of the principles of nuclear magnetism is useful.

The contents of the book are listed in detail in a table and we limit ourselves to their brief description.

The subject matter is organized into eight chapters as follows.

CHAPTER 1. *Dynamics and thermodynamics of systems of interacting spins* is introductory in nature. It contains the basic formalism of interacting spins dynamics. With respect to *The Principles* and *Spin Temperature* this formalism has been reformulated, modernized and enlarged by the introduction of Liouville operators and Mori's memory functions. It is illustrated by applications to specific experimental problems. Last the validity of the concept of spin temperature, on which some doubts had been cast is re-examined and the conditions of its applicability redefined.

CHAPTER 2. *Coherent manipulation of nuclear spins and high resolution in solids* is a nice example of the duality field-tool in nuclear magnetism. We have included it in this book because it is a beautiful study in nuclear spin dynamics. Yet these spin manipulations were invented for the specific purpose of endowing chemists with a technique for high resolution NMR in solids, and in writing this chapter we have greatly benefited from two excellent monographs written for that very purpose.

CHAPTER 3. *Nuclear magnetic resonance in solid ^{3}He.* Solid ^{3}He is an ideal field of study in nuclear magnetism: experimentally because of its large magnetization and narrow NMR line; theoretically because of the nice illustration that it provides for all the concepts and methods of a quantum mechanical theory of nuclear magnetism: this with the very simple assumption of a Heisenberg exchange coupling between nearest neighbours and down to temperatures of a few times the strength of the exchange coupling. For lower temperatures the model begins to fail and breaks down completely in the millikelvin range where a long range order appears. Chapter 3 is almost entirely devoted to the temperature range where the Heisenberg model is satisfactory. The ordered state is discussed briefly in the last chapter.

CHAPTER 4. *Nuclear magnetism of superfluid ^{3}He.* The discovery in the early seventies of the superfluid phases of ^{3}He has led to observations of nuclear magnetic behaviour which has no counterpart anywhere else in nuclear magnetism. In trying to understand the theoretical explanations of this behaviour we realized how ill-prepared for this task, through their background, were most students of nuclear magnetism, including ourselves. This chapter is an earnest attempt to spare the reader some of the difficulties that the authors experienced in going through the original papers and the review articles. It is a painstaking, somewhat pedestrian account of the magnetic properties of superfluid ^{3}He, which does not assume any previous knowledge of the theory of Fermi liquids let alone that of the BCS theory of superconductivity.

With this chapter ends what may be considered as the first part of the book, although it is not so formalized in the table of contents. It differs from the second part, formed by the last four chapters, at two levels.

The spin temperatures encountered in the first part are high: whenever Boltzmann statistics applies to the spins, the first order expansion of the Boltzmann exponentials is an adequate approximation. This is not so in the second part and in particular in the chapter on the long range ordering where all expansions in powers of the inverse temperature break down.

The second distinction between the two parts is of a personal nature. In contrast with the first, the second part covers subjects on which the authors have worked actively for a long time. To use an image that we applied earlier to nuclear spins: spectators in the first part, actors in the second.

CHAPTER 5. *Spin systems at low temperatures* plays for part II the role that chapter 1 played for part I. The theoretical methods necessary for dealing with spin systems at very low temperatures are developed. The expansion of expectation values of various physical observables in powers of inverse temperature, valid in the paramagnetic range is performed by the diagrammatic method, with what the *cognoscenti* may deem excessive detail but others may find useful. General relations, valid also in the ordered states, are established and experimental illustrations are given.

CHAPTER 6. *Thermal contact between nuclear spins and paramagnetic impurities: nuclear relaxation and dynamic polarization.* These studies are not new since they are already dealt with in both *The Principles* and *Spin Temperature*. However the advances in this field, resulting both in a better understanding of the various processes involved and in a large harvest of new and spectacular results, warrant a new chapter in a new book. Indeed this is one of the very few topics where *The Principles*, and even the more recent *Spin Temperature*, could do with some little suppressions. The duality field-tool appears again. Dynamic nuclear polarization, a field of study in its own right is a tool for building polarized targets for nuclear, neutron, and elementary particle physics. It is also a tool for another field of nuclear magnetism, namely dipolar magnetic ordering.

CHAPTER 7. *Nuclear magnetism and neutrons: nuclear pseudomagnetism.* The nuclear forces between slow neutrons and nuclei have a spin-dependent part which can be conveniently described by assigning to each nuclear species a fictitious pseudomagnetic moment. The various aspects of the interaction of the neutron spin with a system of nuclear spins are likewise referred to as nuclear pseudomagnetism. Nuclear pseudomagnetism exhibits some very interesting features of its own such as pseudomagnetic resonance and pseudomagnetic precession. It is however chiefly a tool for the study of long range nuclear magnetic ordering, its structures, and its imperfections.

CHAPTER 8. *Nuclear dipolar magnetic ordering.* This is the key chapter of the second part of this book. All the methods and results described in the three previous chapters converge towards the realization of a goal pursued over the years namely the prediction, production, observation, and study by various techniques, of magnetically ordered structures of nuclear spins produced by their dipolar magnetic interactions.

The ordering of nuclear spins of ^{3}He is described in far less detail in view of the preliminary character of the experiments and of the tentative character of the theory. The so-called enhanced nuclear magnetism with its potentialities for ordering nuclear spins is not considered at all. It is felt that because of its strong dependence on the magnetism of the electronic clouds surrounding the nuclei, which give rise to it, its description belongs in a book on electron magnetism.

A few more topics such as the nuclear magnetism of solid hydrogen and solid methane could arguably have been included in this book. Among other arguments against their inclusion, such as the few uncertainties still present in the experimental results and in their interpretation, was the wish to complete in a finite time a book of finite size.

Like its predecessors (*The Principles* 1961, *Spin Temperature* 1970) this book reflects the spirit of our laboratory, dedicated for more than twenty-five years now to the study and progress of nuclear magnetism. To all our coworkers past and present, too numerous to be cited here but not too many to be all very present in our mind, our heartfelt thanks.

We are grateful to Professor Bleaney for suggesting the title of this book which is a concise and accurate statement of its contents.

This manuscript could never have been prepared in the allotted time without the kind and competent assistance of Madame Porneuf and her staff. Special thanks are due to Madame Neveux who typed most of the preliminary draft and to Madame Parent who produced the final typescript.

We thank the authors and the publishers of the scientific journals who kindly gave us permission to reproduce the figures.

Orme des Merisiers A.A.
August 1981 M.G.

JOHNSON: Make a large book; a folio.

BOSWELL: But of what use will it be, sir?

JOHNSON: Never mind the use; do it.

Boswell's Life of Johnson

CONTENTS

1. DYNAMICS AND THERMODYNAMICS OF SYSTEMS OF INTERACTING SPINS 1

A. Epitome of the theory of spin temperature 2
- (*a*) The Hamiltonian 2
- (*b*) The concept of spin temperature 4
  1. Low fields 5
  2. High fields 6
  3. High fields in the presence of r.f. 7
- (*c*) NMR signals and Provotorov equations 9
  1. Free precession signal 9
  2. Absorption and dispersion at low r.f. level 10
  3. The Provotorov equations 12

B. The Liouville formalism and memory functions 15
- (*a*) Definition 15
- (*b*) 'Operators of interest' and memory functions 17
- (*c*) The generalized Provotorov equations: an illustration of the Liouville formalism 19
  1. Long term trend toward thermal equilibrium 19
  2. Short term transient oscillations 22
- (*d*) Resonance line shape and memory functions 25
  1. Introduction of the memory function 25
  2. Gaussian approximation for the memory function 28
- (*e*) An illustration: Coupling between unlike spin systems 32
  1. The basic equations 33
  2. The experimental sequence 36
  3. Analysis of the experimental results 38

C. On the validity of the concept of spin temperature 41
- (*a*) The random phase assumption 41
- (*b*) The trend toward spin temperature: Production and observation of dipolar order in high field 43
- (*c*) Phase refocussing. The magic sandwich 45
  1. The argument for irreversibility 45
  2. The magic sandwich 47

D. Spin–lattice relaxation 49

2. COHERENT MANIPULATION OF NUCLEAR SPINS AND HIGH RESOLUTION IN SOLIDS 53

A. The spin Hamiltonians 54

(*a*) The quadrupole interaction 55
(*b*) The chemical shift 56
(*c*) The indirect couplings 56
B. Fast rotation of the sample 57
C. Homonuclear narrowing through spin manipulation 58
(*a*) Large effective field in the rotating frame 58
(*b*) Multipulse methods 64
1. An illustrative example 64
2. α The effective periodic Hamiltonian 66
β The Magnus expansion 69
3. Two efficient multipulse cycles 73
α The WHH–4 sequence 73
β The MREV sequence 75
4. Off-resonance irradiation and second averaging 76
5. Pulses of finite width 77
6. Limitations due to pulse imperfections 82
D. Heteronuclear narrowing for spins of rare isotopes 84
(*a*) Measurement of chemical shifts 86
(*b*) Measurement of dipolar interactions 89
E. An illustrative example of heteronuclear spin manipulation 93
F. Multipulses and spin temperature: two experiments 95
G. The Magnus paradox 103

3. NUCLEAR MAGNETIC RESONANCE OF SOLID ^{3}He 108
A. Introduction 108
B. The various 'motions' in solid ^{3}He 112
(*a*) Exchange 112
1. The exchange Hamiltonian 112
2. The exchange integral 114
(*b*) Vacancies 117
1. The formation energy 117
2. The tunnelling frequency 118
3. Vacancy waves 120
(*c*) Impurities of ^{4}He 121
(*d*) Phonons 122
C. The effects of exchange on NMR in solid ^{3}He 122
(*a*) The theory of the spin–lattice relaxation T_{Ze} 123
(*b*) The theory of line width and transverse relaxation time 127
1. Adiabatic linewidth 127
2. Transverse relaxation: general theory 132
(*c*) Exchange and diffusion of nuclear magnetization 134
(*d*) NMR and exchange: experimental results 138
D. The effects of vacancies on NMR in solid ^{3}He 144

(*a*) Introduction 144
(*b*) Vacancies and diffusion of nuclear magnetization 146
(*c*) Theory of the relaxation and narrowing by vacancies 147
(*d*) Diffusion and Zeeman relaxation caused by vacancies: experimental results 153
1. Diffusion 153
2. Zeeman-vacancies relaxation 154
E. Non-Zeeman relaxation mechanisms 156
(*a*) Exchange–defects coupling 156
1. Exchange–vacancies coupling 156
2. Exchange–impurities coupling 159
(*b*) Defects–phonons coupling 161
1. Relaxation 161
2. Vacancy waves and phonons 164
F. Multiple echoes in solid ^{3}He 166
(*a*) Simplified theory of multiple echoes 167
(*b*) General case 172
1. The effects of relaxation 172
2. Sample shape and field gradient 173
3. The effects of diffusion 174
(*c*) Experimental results 175
G. Low temperature properties of solid ^{3}He 180
(*a*) Inadequacy of the Heisenberg Hamiltonian in the paramagnetic state 180
(*b*) Physical origin of multiple–spin exchange 182

4. NUCLEAR MAGNETISM OF SUPERFLUID ^{3}He 186
A. Introduction: The new phases of liquid ^{3}He 186
B. Normal ^{3}He and the Landau interactions 188
(*a*) Molecular fields 190
(*b*) Free energy of a Landau fluid 191
(*c*) Static magnetic susceptibility 192
(*d*) Some numerical data on liquid ^{3}He 193
C. The Cooper pairs and the BCS theory applied to ^{3}He 193
D. The order parameter 195
E. Elementary excitations and energy gap in superfluid ^{3}He 200
(*a*) Linearization of the interaction Hamiltonian 200
(*b*) Diagonalization of the energy matrix 202
(*c*) The self-consistent equation 204
F. Solutions of the BCS equations and the Ginzburg–Landau approximation 206
(*a*) The critical temperature 206
(*b*) Solutions below the critical temperature 208

(*c*) The free energy and the Ginzburg–Landau approximation 210
(*d*) The $l = 1$ solutions of the BCS equation: the ABM and BW phases 212
G. Magnetic susceptibility 214
(*a*) Susceptibility of a single pair $d(n)$ 214
(*b*) Susceptibility of a superfluid phase 220
(*c*) Landau corrections to the susceptibility 221
(*d*) Experimental results 222
(*e*) The adiabatic susceptibility 223
H. The magnetic dipole–dipole interactions in superfluid ^{3}He 226
(*a*) The local field 226
(*b*) The dipolar interactions in superfluid ^{3}He 227
(*c*) Estimate of the constant g_D 229
I. Magnetic resonance: The various energies, the commutation relations, the equations of motion 232
(*a*) Magnetic and dipolar energy: equilibrium states 232
(*b*) The adiabatic approximation 234
(*c*) The commutation relations 236
(*d*) The equations of motion 238
J. Magnetic resonance in the ABM and BW phases 240
(*a*) Equations of motion 240
1. The ABM phase 240
2. The BW phase 240
(*b*) Motions of small amplitude 242
1. The ABM phase 242
2. The BW phase 243
(*c*) Experimental results 244
(*d*) Motions of large amplitude 247
1. Longitudinal ringing in the ABM phase 247
2. Free precession in the ABM phase 248
3. Free precession in the BW phase 251
(*e*) Experimental results 254
K. Walls 255
(*a*) General 255
(*b*) Nuclear magnetism of the BW phase in restricted geometry 258
(*c*) The wall-pinned mode 261
L. Relaxation 263
(*a*) The theory 263
1. Superfluid and normal magnetizations 263
2. The equations of motion 266
3. Relaxation of energy 268
(*b*) Relaxation of the wall-pinned mode 268
(*c*) Relaxation of oscillations of small amplitude 269

5. SPIN SYSTEMS AT LOW TEMPERATURE 273
A. Formal expression of the resonance signal 274
B. The Zeeman resonance signal 276
(*a*) The moments of the absorption signal 279
(*b*) The first moment 281
1. One spin species 281
2. Several spin species 285
(*c*) The second moment 288
(*d*) Higher moments 291
C. Rigorous NMR results 293
(*a*) Area of the absorption signal 293
(*b*) Rate of change of the polarization 293
(*c*) Slightly saturating linear passages 294
1. Saturation of the polarization 294
2. Saturation of the dipolar energy 295
(*d*) First moment of the absorption signal 299
(*e*) Qualitative shape of the absorption signal 301
(*f*) Linear response and transverse spin susceptibility 302
(*g*) Cotanh transform of the absorption signal. Measurement of the dipolar temperature 306
D. Non-linear effects in spin temperature 313
(*a*) The diagrammatic method 315
1. The longitudinal case 316
α The semi-invariants 317
β Contractions and diagrams 318
2. The general case 322
(*b*) Thermodynamic properties of the spin system 327
1. Dipole–dipole energy 328
2. Spin polarization 329
3. Free energy 330
4. Entropy 331
5. Transverse susceptibility 332
(*c*) Experimental illustration 332
1. Variation of dipolar energy with polarization in high effective field 332
2. Susceptibilities in zero effective field 336

6. THERMAL CONTACT BETWEEN NUCLEAR SPINS AND PARAMAGNETIC IMPURITIES: NUCLEAR RELAXATION AND DYNAMIC POLARIZATION 339
Introduction 339
A. Structure and relaxation of paramagnetic centres 341
(*a*) Spin Hamiltonian and hyperfine structure 342

(*b*) Electronic spin–lattice relaxation: spins and phonons 344
1. Pure electronic relaxation 344
2. Nuclear hyperfine relaxation 346
B. Nuclear relaxation by isolated paramagnetic impurities 349
(*a*) An isolated electron–nucleus pair 349
1. The random field approach 349
2. The scrambled states approach 351
(*b*) The relaxation of an assembly of nuclear spins by non-interacting paramagnetic centres 354
C. The well-resolved solid effect 355
(*a*) The rate equations 355
(*b*) Experimental results 359
D. Electronic spin–spin interactions and electronic spin–spin temperature 362
E. Electronic spin–spin reservoir and nuclear relaxation 365
(*a*) The coupling between the nuclear Zeeman and the electronic non-Zeeman reservoirs 365
(*b*) Nuclear Zeeman relaxation 368
(*c*) Nuclear dipolar relaxation 371
F. Electronic spin–spin reservoir and nuclear dynamic polarization 373
(*a*) Historical background 373
(*b*) High temperature case 377
(*c*) Low temperature case 379
1. Inhomogeneous broadening of the first type 379
2. Inhomogeneous broadening of the second type 383
(*d*) Numerical results 388
(*e*) Experimental results 393
G. Applications of DNP 402
(*a*) The Nedor method 402
(*b*) Polarized targets for nuclear and particle physics 405
H. Other examples of coupling between the electronic spin–spin reservoir and the nuclear Zeeman reservoir 408
(*a*) The rotating crystal 408
(*b*) The bootstrap effect: cooling of electron spins by nuclei in low fields 411
1. The triplet 411
2. The singlet 415
(*c*) Cooling of a gas by laser irradiation: an analogue of DNP 417

7. NUCLEAR MAGNETISM AND NEUTRONS: NUCLEAR PSEUDOMAGNETISM 419
A. Slow neutron scattering by an isolated nucleus 420
(*a*) Scattering on a spinless nucleus 420

(*b*) Slow neutron scattering by a nucleus with a spin *I* 424
B. Slow neutron scattering by a macroscopic target: Bragg scattering 426
(*a*) Spinless nuclei 427
(*b*) Nuclei with spin 430
C. Pseudomagnetic nuclear field and pseudomagnetic resonance 433
(*a*) A simple scheme 434
(*b*) Rotating pseudomagnetic field 435
(*c*) Experimental procedure and results 436
(*d*) Methodological digressions 438
D. Systematic measurements of nuclear pseudomagnetic moments 440
(*a*) Bragg scattering on polarized targets 441
(*b*) Pseudomagnetic precession and the two-coils method 443
1. The principle 443
2. The results 447
3. Fall-out for nuclear magnetism 450
E. Wave-like description of the neutron 452
(*a*) Refractive index 452
(*b*) Transmission and absorption 453
(*c*) Pseudomagnetic precession 456
F. Neutron scattering and spatial correlations between nuclear spins; domain size 459
(*a*) The correlation function 459
(*b*) Mapping of the correlation function and domain size 462
(*c*) Short-range spin correlations in a paramagnetic dipolar state 466

8. NUCLEAR DIPOLAR MAGNETIC ORDERING 470
A. Introduction 470
(*a*) Production of nuclear magnetic ordering: the principle 470
(*b*) General features of nuclear magnetic ordering 473
B. Prediction of the ordered structures 476
(*a*) The method of Villain 477
1. General form of the solutions 480
2. Systems with two spin species 483
(*b*) The method of Luttinger and Tisza 485
(*c*) The stable structures in CaF_2 487
(*d*) The stable structures in LiF and LiH 493
(*e*) The stable structures in $Ca(OH)_2$ 495
(*f*) General character of the nuclear ordered structures 498
C. Approximate theories of ordering 500
(*a*) The Weiss-field approximation 500
1. The Weiss-field equations 500
2. Weiss-field and spin temperature 503
(*b*) The restricted-trace approximation 505

D. Antiferromagnetism in zero field 510
(*a*) Transverse susceptibility 511
1. Calcium fluoride 511
2. Lithium fluoride and hydride 516
(*b*) Longitudinal susceptibility 520
(*c*) Non-uniform longitudinal susceptibility of the ^{19}F spins in CaF_2 524
1. Quadrupole alignment of the rare spins 529
(*d*) Spin–lattice relaxation 531
1. Short correlation time: $\tau_c \ll T_2$ 532
2. Long correlation time: $\tau_c \gg T_2$ 533
(*e*) Neutron diffraction in lithium hydride 535
E. Ferromagnetism 539
(*a*) Characteristic properties of ferromagnets with domains 539
(*b*) Resonance of ^{43}Ca in ferromagnetic CaF_2 545
1. Evidence for the production of ferromagnetism 545
2. Dipolar field as a function of energy 549
(*c*) Ferromagnetic–antiferromagnetic transition in CaF_2 551
(*d*) Transverse susceptibilities 553
1. Ferromagnetic calcium fluoride 554
2. $Ca(OH)_2$ at negative temperature 555
3. $Ca(OH)_2$ at positive temperature 556
(*e*) Neutron diffraction in LiH 558
F. Antiferromagnetism in non-zero field 560
(*a*) Second order transition 561
(*b*) The mixed state 564
1. Qualitative discussion 564
2. Quantitative description 566
(*c*) Experimental results 570
(*d*) Comparison with other antiferromagnetic systems 571
G. Investigation of the ordered domains 575
(*a*) An estimate of the thickness of ferromagnetic domains 576
(*b*) Neutron diffraction and domains in LiH 577
(*c*) NMR investigation of ferromagnetic domains in CaF_2 580
1. Reproducibility of the domains 580
2. Thickness of the domains 583
(*d*) Control of paramagnetic impurities 587
H. Some further aspects of the theory of nuclear antiferromagnetism 589
(*a*) The spin wave approximation 589
1. Elementary excitations 590
2. Sublattice polarization 591
3. Energy 594
(*b*) The random phase approximation (RPA) 596

(c) Monte Carlo calculations 598
I. Magnetic ordering in solid ^{3}He 599
(a) Ordering properties of solid ^{3}He 599
(b) NMR study of the ordered phase in low field 601
(c) Multiple-spin exchange model: a summary 607

References 610

Name Index 617

Subject Index 620

1

DYNAMICS AND THERMODYNAMICS OF SYSTEMS OF INTERACTING SPINS

Memento Mori

THE first few years after the discovery of nuclear magnetic resonance in late 1945 were devoted mainly to studies on liquid samples. The fast relative motions of the atomic nuclei in such samples annihilate, at least in first order, the effects of the dipolar interactions between the nuclear spins: as a consequence both the experiments and their theoretical interpretation were made far easier than in solids. The averaging out of the local fields, known as the 'motion narrowing' of the resonance lines, yielded large signals, easy to observe with primitive detection devices. At the same time the effective uncoupling of the spins from each other made very simple the description of their resonant behaviour by means of the Bloch equations, guessed with great insight by their author, long before their microscopic derivation.

In the studies of solids, where spin–spin interactions play a vital role, a similar progress was made a few years later through the assumption of the existence of a spin temperature.

The great majority of these studies were performed under conditions where the approximation of so-called 'high spin temperatures', whose meaning will be recalled shortly, was justified. This approximation leads to a tremendous simplification of the underlying theory. There are many descriptions of such studies in books and review articles among which are those by Abragam (1961) and Goldman (1970).

This chapter, restricted to the domain of high temperatures, is devoted to a concise description of the dynamics and thermodynamics of interacting spins.

Some results are stated without proof. The derivations can then be found at various places in Abragam (1961) or Goldman (1970). On the other hand some novel derivations have been introduced for the sake of compactness. Then, still in the high temperature limit, a few more recent results, experimental and theoretical, are presented.

The low temperature domain where the linear expansion in powers of the inverse spin temperature is invalid will be dealt with in chapter 5.

A. Epitome of the theory of spin temperature

(*a*) *The Hamiltonian*

Our main concern will be spin systems interacting with an applied magnetic field through a Zeeman coupling and with each other through dipolar couplings. The dipolar Hamiltonian of two spins $\mathbf{I}^1$ and $\mathbf{I}^2$ is (in frequency units)

$$\mathcal{H}_{I^1I^2} = \hbar\gamma_1\gamma_2\{\mathbf{I}^1 . \mathbf{I}^2 - 3(\mathbf{I}^1 . \hat{\mathbf{r}})(\mathbf{I}^2 . \hat{\mathbf{r}})\}r^{-3}, \tag{1.1}$$

where $\mathbf{r}$ is the separation $\mathbf{r}_1 - \mathbf{r}_2$ of the two spins and $\hat{\mathbf{r}} = \mathbf{r}/|\mathbf{r}| = \mathbf{r}/r$, a unit vector. γ_1 and γ_2 are the gyromagnetic factors of the two spins. When $\gamma_1 = \gamma_2$ (like spins) it is convenient to rewrite (1.1) by means of spherical harmonics and irreducible tensor operators of order two.

We define a set of orbital functions F_q as:

$$F_q(\Omega) = F_q(\theta, \varphi) = (\tfrac{6}{5})^{1/2}\sqrt{(4\pi)}Y_2^q(\theta, \varphi), \tag{1.2}$$

whence, from the orthogonality properties of spherical harmonics:

$$\langle F_q F_{q'}^* \rangle = \int F_q F_{q'}^* \frac{d\Omega}{4\pi} = \tfrac{6}{5}\delta_{qq'}.$$

Explicitly the F_q are:

$$F_0 = \sqrt{(\tfrac{3}{2})}(3\cos^2\theta - 1); \tag{1.3a}$$

$$F_{\pm 1} = \mp 3 \sin\theta \cos\theta \exp(\pm i\varphi); \tag{1.3b}$$

$$F_{\pm 2} = \tfrac{3}{2}\sin^2\theta \exp(\pm 2i\varphi). \tag{1.3c}$$

We also introduce the tensor spin operators $T_q = T_q^{(2)}$ which transform under a rotation $\mathcal{R}$ of the spin axes, like spherical harmonics, that is:

$$\tilde{T}_q = \mathcal{R}T_q\mathcal{R}^{-1} = \sum_{q'} T_{q'}\mathcal{D}_{q'q}^{(2)}. \tag{1.4}$$

The operators T_q are orthogonal in the sense that

$$\mathrm{Tr}\{T_q^{(2)}T_{q'}^{(2)\dagger}\} = 0 \quad \text{for } q' \neq q. \tag{1.5}$$

We renormalize them to the condition:

$$\mathrm{Tr}\{T_q^{(2)}T_{q'}^{(2)\dagger}\} = \left(\frac{I(I+1)}{3}\right)^2 \delta_{qq'}. \tag{1.6}$$

Thus normalized, the T_q are:

$$T_0 = \frac{1}{\sqrt{6}}\{3I_z^1I_z^2 - \mathbf{I}^1 . \mathbf{I}^2\} = T_0^\dagger; \tag{1.7a}$$

$$T_{\pm 1} = \mp\tfrac{1}{2}\{I_\pm^1 I_z^2 + I_z^1 I_\pm^2\} = -T_{\mp 1}^\dagger; \tag{1.7b}$$

$$T_{\pm 2} = \tfrac{1}{2}I_\pm^1 I_\pm^2 = T_{\mp 2}^\dagger. \tag{1.7c}$$

The dipolar Hamiltonian (1.1) can be expressed as:

$$\mathcal{H}_{I^1I^2} = -\gamma^2\hbar r^{-3} \sum_{-2\leq q\leq 2} F_q^*(\Omega)T_q(\mathbf{I}). \tag{1.8}$$

For infinitesimal rotations $\mathcal{R}$ along the x, y, z axes, the eqns (1.4) reduce to the commutation relations:

$$[I_z, T_q^{(2)}] = qT_q^{(2)}, \qquad [I_\pm, T_q^{(2)}] = T_{q\pm1}^{(2)}[(2\mp q)(2\pm q+1)]^{1/2}, \tag{1.9a}$$

whence, for $q=0$, a relation we shall use later:

$$[I_x, T_0] = \tfrac{1}{2}[I_+ + I_-, T_0] = \sqrt{(\tfrac{3}{2})}[T_1 + T_{-1}]. \tag{1.9b}$$

The most general rotation $\mathcal{R}(\alpha, \beta, \gamma)$ where α, β, γ are the Euler angles and can be written in operator form:

$$\mathcal{R}(\alpha, \beta, \gamma) = \exp(-i\alpha I_z)\exp(-i\beta I_y)\exp(-i\gamma I_z), \tag{1.10}$$

where $\mathbf{I} = \mathbf{I}^1 + \mathbf{I}^2$. Any rotation can thus be expressed as a product of rotation around $0z$ and $0y$.

The general relation (1.4) specialized to a rotation around $0z$, yields:

$$\tilde{T}_q = \exp(-i\alpha I_z)T_q\exp(i\alpha I_z) = \exp(-i\alpha q)T_q.$$

For a rotation around $0y$, the formulae are more complicated. Writing for brevity:

$$\cos\beta = c \qquad \sin\beta = s;$$

$$\tilde{T}_0 = \exp(-i\beta I_y)T_0\exp(i\beta I_y) = \frac{3c^2-1}{2}T_0 - \sqrt{(\tfrac{3}{2})}cs(T_1 - T_{-1})$$

$$+\sqrt{(\tfrac{3}{8})}s^2(T_2 + T_{-2}); \tag{1.11a}$$

$$\tilde{T}_1 = \sqrt{(\tfrac{3}{2})}csT_0 + \tfrac{1}{2}(2c^2+c-1)T_1 - \tfrac{1}{2}(2c^2-c-1)T_{-1} - \tfrac{1}{2}s(1+c)T_2$$

$$+\tfrac{1}{2}s(1-c)T_{-2}; \tag{1.11b}$$

$$\tilde{T}_{-1} = -\tilde{T}_1^\dagger;$$

$$\tilde{T}_2 = \sqrt{(\tfrac{3}{8})}s^2T_0 + \tfrac{1}{2}s(1+c)T_1 + \tfrac{1}{2}s(1-c)T_{-1} + \tfrac{1}{4}(1+c)^2T_2$$

$$+\tfrac{1}{4}(1-c)^2T_{-2}; \tag{1.11c}$$

$$\tilde{T}_{-2} = \tilde{T}_2^\dagger.$$

It is also convenient to know the transformation of T_q through a rotation around the x-axis.

If

$$\tilde{T}_q = \exp(-i\beta I_y)T_q\exp(i\beta I_y) = \sum_{q'} T_{q'}d_{q'q},$$

we have,

$$\tilde{\tilde{T}}_q = \exp(-i\beta I_x) T_q \exp(i\beta I_x),$$

$$= \exp\left(i\frac{\pi}{2}I_z\right)\exp(-i\beta I_y)\exp\left(-i\frac{\pi}{2}I_z\right)T_q\exp\left(i\frac{\pi}{2}I_z\right)\exp(i\beta I_y)\exp\left(-i\frac{\pi}{2}I_z\right)$$

$$= \sum_{q'}\exp\left(i(q'-q)\frac{\pi}{2}\right)T_{q'}d_{q'q}(\beta).$$

In particular:

$$\tilde{\tilde{T}}_0 = \frac{3c^2-1}{2}T_0 - i\surd(\tfrac{3}{2})cs(T_1+T_{-1}) - \surd(\tfrac{3}{8})s^2(T_2+T_{-2}). \qquad (1.11d)$$

(b) The concept of spin temperature

The assumption (for it *is* an assumption) of a spin temperature can be stated as follows:

A system of interacting spins, isolated from its surroundings, will evolve toward an equilibrium state characterized by a temperature. Its density matrix is then of the form:

$$\sigma_{\text{eq}} = \exp(-\beta\mathcal{H})/\text{Tr}\{\exp(-\beta\mathcal{H})\}. \qquad (1.12)$$

$\mathcal{H}$ is expressed in frequency units and β is related to the spin temperature T by:

$$\beta = \hbar/k_{\text{B}}T.$$

For brevity we shall call β, which has the dimensions of a time, the inverse spin temperature. The density matrix (1.12) maximizes the entropy of the system

$$S = -k_{\text{B}}\,\text{Tr}(\sigma\ln\sigma). \qquad (1.13)$$

The maximum value of the entropy is:

$$\frac{S}{k_{\text{B}}} = -\beta\langle\mathcal{H}\rangle - \ln\{\text{Tr}(\exp - \beta\mathcal{H})\}, \qquad (1.14)$$

where

$$\langle\mathcal{H}\rangle = \text{Tr}(\sigma_{\text{eq}}\mathcal{H}).$$

The sign of the spin temperature depends on the value of the energy. We shall choose the origin of the energies of the spin system so as to have $\text{Tr}(\mathcal{H}) = 0$. Then:

$$\langle\mathcal{H}\rangle < 0 \quad \text{for } T > 0, \qquad \langle\mathcal{H}\rangle > 0 \quad \text{for } T < 0. \qquad (1.15)$$

A key feature is the rate at which the system evolves toward equilibrium. This depends on the Hamiltonian.

1. *Low fields*

'Low fields' means fields such that the Zeeman 'splitting' ω_0 is comparable to the span $\Delta\omega$ of the dipolar spectrum of $\mathscr{H}_D$. The time constant for reaching equilibrium is then of the order of $(\Delta\omega)^{-1}$ which is also of the order of T_2, the lifetime of the free precession signal observed in high fields. A typical value is 100 μsec.

If $\hbar\omega_0, \hbar\Delta\omega \ll k_B T$ one may use for σ in (1.12) an expansion in powers of β, limited to the linear term. This is the 'high temperature' approximation. Equation (1.12) reduces to:

$$\sigma = A(1-\beta\mathscr{H}), \tag{1.16}$$

where the constant $A = (\text{Tr } 1)^{-1}$ in order that $\text{Tr}\,\sigma = 1$. For a system of N identical spins I: $\text{Tr } 1 = (2I+1)^N$.

For a system of non-interacting spins, the Hamiltonian reduces to the Zeeman term: $\mathscr{H} = -\gamma H I_z = \sum_i -\gamma H I_z^i = \sum_i \omega_0 I_z^i$ and the density matrix (1.12) factors in the product: $\sigma = \prod_i \sigma_i$ with

$$\sigma_i = \exp(-\beta\omega_0 I_z^i)/\text{Tr}_i\{\exp - \beta\omega_0 I_z^i\},$$

where the symbol Tr_i means trace with respect to the variables $\mathbf{I}^i$ only. It is easy to see that in this special case the condition for the validity of (1.16) is $|\beta\omega_0| \leqslant 1$ which can be written as:

$$\frac{1}{N}\beta^2 \,\text{Tr}\{\mathscr{H}^2\} \ll 1. \tag{1.17}$$

We shall *assume* that this criterion remains valid for *interacting* spins. We shall see in chapter 5 that the first few terms in the expansion of (1.12) in powers of β are in agreement with this criterion.

In the high temperature approximation the energy is given by:

$$\langle\mathscr{H}\rangle = -A\beta \,\text{Tr}\{\mathscr{H}^2\}, \tag{1.18}$$

and the entropy is:

$$\frac{S}{k_B} = N \ln(2I+1) - \tfrac{1}{2}A\beta^2 \,\text{Tr}\{\mathscr{H}^2\}. \tag{1.19}$$

For a Hamiltonian, sum of Zeeman and dipolar interactions:

$$\mathscr{H} = -\gamma H I_z + \mathscr{H}_D \tag{1.20}$$

it is easily found:

$$\text{Tr}\{\mathscr{H}^2\} = \gamma^2 H^2 \,\text{Tr}\{I_z^2\} + \text{Tr}\{\mathscr{H}_D^2\} = \gamma^2 \,\text{Tr}\{I_z^2\}(H^2 + H_L^2), \tag{1.21}$$

where the local field H_L is defined by:

$$H_L^2 = \mathrm{Tr}\{\mathcal{H}_D^2\}/\gamma^2\,\mathrm{Tr}\{I_z^2\}. \tag{1.22}$$

We recall below some results relative (i) to a sudden change of $\mathcal{H}$ or σ; (ii) to an adiabatic variation of $\mathcal{H}$; (iii) to spin-lattice relaxation.

(i) Assume that at time $t=0$ the density matrix σ_{in} of the spin system is different from its equilibrium form $1-\beta\mathcal{H}$ (we drop for brevity the normalization constant A of (1.16)). This may occur for instance if just prior to $t=0$ either the density matrix or the Hamiltonian have been changed *suddenly*, that is in a time short compared to T_2. After an interval τ equal to a few times T_2, the spin system will reach an equilibrium state with a density matrix: $\sigma_f = 1-\beta\mathcal{H}$. Since the system is isolated its energy $\langle\mathcal{H}\rangle = \mathrm{Tr}(\sigma\mathcal{H})$ remains constant during this evolution whence:

$$\begin{aligned}\langle\mathcal{H}\rangle = \mathrm{Tr}(\sigma_{in}\mathcal{H}) = \mathrm{Tr}\{\sigma_f\mathcal{H}\} = -\beta\,\mathrm{Tr}\{\mathcal{H}^2\},\\ \beta = -\mathrm{Tr}(\sigma_{in}\mathcal{H})/\mathrm{Tr}\{\mathcal{H}^2\}.\end{aligned} \tag{1.23}$$

(ii) If the Hamiltonian $\mathcal{H}$ is varied so slowly that the spin system is at all times very near equilibrium its entropy remains constant whence, according to (1.19),

$$\beta^2\,\mathrm{Tr}\{\mathcal{H}^2\} = \mathrm{const.} \tag{1.24}$$

For a combination of Zeeman and dipolar interactions (1.24) reduces to:

$$\beta \propto (H^2 + H_L^2)^{-1/2}. \tag{1.24$'$}$$

(iii) Let T_{1z} and T_{1D} be the respective relaxation times for I_z and $\mathcal{H}_D$. If both are much longer than T_2 the system will at all times have an inverse spin-temperature β evolving toward the inverse lattice temperature β_L at a rate $1/T_1$ given by:

$$\frac{1}{T_1} = [H^2 + H_L^2]^{-1}\left[\frac{H^2}{T_{1z}} + \frac{H_L^2}{T_{1D}}\right]. \tag{1.25}$$

2. *High fields*

When $H \gg H_L$ the time τ for reaching equilibrium becomes very long. This is due to the fact that the energy spectra of the Zeeman energy and the dipolar energy are not on 'speaking terms': the Zeeman energy can only change by quanta $\pm\omega_0$, much larger than the total span of the dipolar spectrum.

However a time of the order of T_2 is sufficient for the establishment of a quasi-equilibrium state characterized by more than one spin-temperature.

Consider first a system of like spins. It is convenient to single out in the dipolar Hamiltonian $\mathcal{H}_D$, much smaller than the Zeeman Hamiltonian $Z = \omega_0 I_z$, the secular part $\mathcal{H}'_D$, that is the part that commutes with Z. The non-secular part $\mathcal{H}''_D = \mathcal{H}_D - \mathcal{H}'_D$ does not affect in first order the energy levels of the spin system and can be disregarded in high field, leaving an approximate Hamiltonian, sum of two commuting terms

$$\mathcal{H} \simeq -\gamma H I_z + \mathcal{H}'_D = \omega_0 I_z + \mathcal{H}'_D. \tag{1.26}$$

From (1.8) and the properties under rotation of the F_q and T_q we get for $\mathcal{H}'_D$ the expression

$$\mathcal{H}'_D = -\gamma^2 \hbar \sum_{i<j} r_{ij}^{-3} F_0^{ij} T_0^{ij}. \tag{1.27}$$

$\mathcal{H}'_D$ is often called the 'truncated' dipolar Hamiltonian.

It is reasonable to assume and well verified by experiment, that a quasi-equilibrium state of the spin system can be described by a density matrix of the form:

$$\sigma = \exp(-\beta_Z \omega_0 I_z - \beta \mathcal{H}'_D)/\mathrm{Tr}\{\exp(-\beta_Z \omega_0 I_z - \beta \mathcal{H}'_D)\}. \tag{1.28}$$

The quantities β_Z and β are called the inverse Zeeman and dipolar temperatures. In the high temperature limit:

$$\sigma \simeq 1 - \beta_Z \omega_0 I_z - \beta \mathcal{H}'_D, \tag{1.29}$$

whence noticing that $\mathrm{Tr}\{I_z \mathcal{H}'_D\} = 0$:

$$\langle Z \rangle = \omega_0 \langle I_z \rangle = -\beta_Z \omega_0^2 \,\mathrm{Tr}\{I_z^2\} = -\beta_Z \gamma^2 H^2 \,\mathrm{Tr}\{I_z^2\}; \tag{1.30}$$

$$\langle \mathcal{H}'_D \rangle = -\beta \,\mathrm{Tr}(\mathcal{H}'^2_D) = -\beta \gamma^2 H'^2_L \,\mathrm{Tr}\{I_z^2\}; \tag{1.31}$$

where the local field H'_L (different from H_L, defined in (1.22)) is given by:

$$\gamma^2 H'^2_L = D^2 = \mathrm{Tr}\{\mathcal{H}'^2_D\}/\mathrm{Tr}\{I_z^2\}. \tag{1.32}$$

According to (1.30) and (1.31), $\langle Z \rangle$ depends only on β_Z and $\langle \mathcal{H}'_D \rangle$ on β. We shall see in chapter 5 that this is not true anymore when β_Z and β are not small, that is for low temperatures.

The separation of the dipolar Hamiltonian into a secular and a nonsecular part applies also when there are two spin species I and S (or more). However then the relation (1.27) does not apply anymore. The secular part $\mathcal{H}'_{IS}$ of the dipolar Hamiltonian is given by:

$$\mathcal{H}'_{IS} = \hbar \gamma_I \gamma_S \sum_{i,\mu} \frac{(1 - 3\cos^2 \theta_{i\mu})}{r_{i\mu}^3} \mathrm{I}_z^i \mathrm{S}_z^\mu. \tag{1.27$'$}$$

3. *High fields in the presence of r.f.*

In the presence of an r.f. field of frequency ω in the neighbourhood of ω_0, the Zeeman and dipolar energies *are* on 'speaking terms' because then the span of the dipolar energies is sufficient to take up or yield the energy $\pm|\omega - \omega_0|$, difference between the energy of a photon, absorbed or emitted by the spins, and the energy ω_0 absorbed or emitted by the Zeeman system.

The Hamiltonian of the system is

$$\mathcal{H} = \omega_0 I_z + \mathcal{H}'_D + \omega_1(I_x \cos \omega t + I_y \sin \omega t) \tag{1.33}$$

if we assume that the r.f. field of amplitude $H_1 = -\omega_1/\gamma$ is rotating in the plane perpendicular to the d.c. field H.

In a frame rotating with velocity ω around $0z$, the effective Hamiltonian is:

$$\mathcal{H}_{\text{eff}} = \exp(i\omega I_z t)\mathcal{H} \exp(-i\omega I_z t) - \omega \mathrm{I}_z = \Delta \mathrm{I}_z + \mathcal{H}'_D + \omega_1 I_x, \tag{1.34}$$

where $\Delta = \omega_0 - \omega$.

If Δ is comparable to $\gamma H'_L$, all terms of (1.34) are on speaking terms and the system will evolve toward a state describable in the rotating frame by a single spin temperature, that is by a density matrix of the form:

$$\sigma_{\text{r.f.}} = \exp(-\beta\mathcal{H}_{\text{eff}})/\text{Tr}\{\exp(-\beta\mathcal{H}_{\text{eff}})\}. \tag{1.35}$$

If H_1 is of the order of H'_L the time constant for reaching equilibrium is of the order of T_2. When $H_1 \ll H'_L$ this time becomes much longer. It can then be calculated by the formula due to Provotorov (see below **A**(c)3).

The effects of sudden or adiabatic changes in the system are calculated as in the low field case, replacing simply $\mathcal{H}$ by $\mathcal{H}_{\text{eff}}$ in (1.23), (1.24), (1.24′) and $\text{Tr}\{\mathcal{H}^2\}$ by:

$$\text{Tr}\{\mathcal{H}^2_{\text{eff}}\} = \text{Tr}(I_z^2)(\Delta^2 + \omega_1^2 + D^2) = \gamma^2\,\text{Tr}(I_z^2)(h^2 + H_1^2 + H_L'^2), \tag{1.36}$$

where $D = \gamma H'_L$, $h = -\Delta/\gamma$.

In an adiabatic change of the field (or frequency), the inverse temperature changes as:

$$\beta \propto (h^2 + H_1^2 + H_L'^2)^{-1/2}. \tag{1.37}$$

An adiabatic sweep of h through resonance is the well known adiabatic 'fast passage': it reverses the magnetization. An adiabatic passage stopped at $h = 0$ followed by the adiabatic suppression of H_1, is the well known ADRF (adiabatic demagnetization in the rotating frame): it transfers order from the Zeeman system into the dipolar system by cooling $\langle\mathcal{H}'_D\rangle$ and warming up $\langle Z\rangle$. In the presence of spin lattice relaxation the relaxation rate $1/T_1$ of the inverse spin temperature β is related to the respective relaxation rates $1/T_{1z}$, $1/T_{1x}$, $1/T_{1D}$ for ΔI_z, $\omega_1 I_x$ and $\mathcal{H}'_D$ by:

$$\frac{1}{T_1} = (h^2 + H_1^2 + H_L'^2]^{-1}\left\{\frac{h^2}{T_{1z}} + \frac{H_1^2}{T_{1x}} + \frac{H_L'^2}{T_{1D}}\right\}. \tag{1.38}$$

Under steady state conditions the inverse spin temperature β_{eq} is related to the inverse lattice temperature β_L by:

$$\beta_{eq} = \beta_L \frac{\omega_0 \Delta}{\Delta^2 + \dfrac{T_{1z}}{T_{1x}}\omega_1^2 + \dfrac{T_{1z}}{T_{1D}}D^2}. \tag{1.39}$$

(*c*) *NMR signals and Provotorov equations*

1. *Free precession signal*

In high field the Hamiltonian of the system is $\mathcal{H} = \omega_0 I_z + \mathcal{H}'_D$ and its density matrix is:

$$\sigma = 1 - \beta_Z \omega_0 I_z - \beta \mathcal{H}'_D. \tag{1.40}$$

It is invariant by a rotation around $0z$ and thus has the same expression in a frame rotating around $0z$ as in the laboratory frame. A pulse of duration τ produced by a field of amplitude H_1 applied along the axis $0x$ of a frame rotating with velocity ω, induces a rotation of the spins around $0x$ by an angle $\theta = -\gamma H_1 \tau$. The expression, in the rotating frame, of the density matrix immediately after the pulse can be obtained from the expression (1.27) of $\mathcal{H}'_D$ and from the transformation under rotation of the spin operators T_q, as given in (1.11*d*) and (1.9*b*). This expression is:

$$\sigma_0 = 1 - \beta_Z \omega_0 (cI_z - sI_y) - \beta[\tfrac{1}{2}(3c^2 - 1)\mathcal{H}'_D - ics[I_x, \mathcal{H}'_D] + \mathcal{H}_2], \tag{1.41}$$

where $c = \cos\theta$, $s = \sin\theta$ and $\mathcal{H}_2$ is a combination of tensor operators T_2 and T_{-2}. The evolution of the density matrix after the end of the pulse is given by:

$$\sigma(t) = \exp(-i\mathcal{H}'_D t)\sigma_0 \exp(i\mathcal{H}'_D t). \tag{1.42}$$

It is associated with a precessing magnetization whose components $\langle I_x \rangle$ and $\langle I_y \rangle$ in the rotating frame are given by:

$$\begin{aligned}\langle I_y \rangle = \mathrm{Tr}\{\sigma(t) I_y\} &= \beta_Z \omega_0 s \,\mathrm{Tr}\{\exp(-i\mathcal{H}'_D t) I_y \exp(i\mathcal{H}'_D t) I_y\} \\ &= \beta_Z \omega_0 s \,\mathrm{Tr}\{I_y^2\} G(t),\end{aligned} \tag{1.43}$$

with

$$G(t) = \mathrm{Tr}\{\exp(-i\mathcal{H}'_D t) I_y \exp(i\mathcal{H}'_D t) I_y\} / \mathrm{Tr}\{I_y^2\}. \tag{1.43$'$}$$

In writing (1.43) we have made use of the fact that in the product $\sigma(t) I_y$ where $\sigma(t)$ is defined by (1.42) and (1.41) there is only one term with a non-vanishing trace. Similarly one finds:

$$\begin{aligned}\langle I_x \rangle = \mathrm{Tr}\{\sigma(t) I_x\} &= \beta cs \,\mathrm{Tr}\{\exp(-i\mathcal{H}'_D t)[-i\mathcal{H}'_D, I_x] \exp(i\mathcal{H}'_D t) I_x\} \\ &= \beta cs \,\mathrm{Tr}(I_x^2) \frac{d}{dt} G(t).\end{aligned} \tag{1.44}$$

Use is made here of the fact that:

$$\exp(-i\mathcal{H}'_D t)[-i\mathcal{H}'_D, I_x]\exp(i\mathcal{H}'_D t) = \frac{d}{dt}\{\exp(-i\mathcal{H}'_D t)I_x \exp(i\mathcal{H}'_D t)\}. \quad (1.44')$$

The function $G(t)$ defined by (1.43′) is called the free induction decay (FID). It can be calculated in principle (in principle only) from the knowledge of its successive derivatives for $t = 0$:

$$\left(\frac{d^{(n)}G}{dt^n}\right)_{t=0} = \frac{(-i)^n}{\mathrm{Tr}\{I_x^2\}}\underbrace{\mathrm{Tr}\{[\mathcal{H}'_D, [\mathcal{H}'_D \cdots [\mathcal{H}'_D, I_x]] \cdots]I_x\}}_{n \text{ times}}. \quad (1.45)$$

All the odd derivatives (1.45) vanish. At low temperatures the FID has a form different from (1.43′) and its odd derivatives do not vanish at the origin. The separate measurements of $\langle I_x\rangle$ and $\langle I_y\rangle$, that is of the components of the precessing magnetization in phase and in quadrature with the rotating field H_1, yield according to (1.43) and (1.44) separate measurements of β and β_Z.

If the system contains two spin species I and S only one of which, say I is affected by the pulse, the results are somewhat different. In a rotation of angle θ of the spins I, the secular interaction $\mathcal{H}'_{IS}$ given by (1.27′) becomes:

$$\tilde{\mathcal{H}}'_{IS} = c\mathcal{H}'_{IS} - is[I_x, \mathcal{H}'_{IS}], \quad (1.46)$$

and the expression (1.44) for $\langle I_x\rangle$ is replaced by:

$$\langle I_x\rangle = \beta s\ \mathrm{Tr}\{\exp(-i\mathcal{H}'_D t)(-ic[\mathcal{H}'_{II}, I_x] - i[\mathcal{H}'_{IS}, I_x])\exp(i\mathcal{H}'_D t)I_x\}. \quad (1.44'')$$

The form of the time dependence of $\langle I_x\rangle$ in (1.44″) depends on the angle θ. Only for infinitely small rotations θ is it proportional to dG/dt as in (1.44). On the other hand the time dependence of $\langle I_y\rangle$ is unchanged.

2. *Absorption and dispersion at low r.f. level*

We calculate the steady state linear response of the spin system to a small excitation produced by a field perpendicular to the field H_0 and rotating at a frequency $\omega = \omega_0 - \Delta$. In a frame rotating at the Larmor frequency ω_0, this field is a vector rotating at the frequency $-\Delta$, with components: $H_x = H_1 \cos \Delta t$, $H_y = -H_1 \sin \Delta t$ and can be written in the usual complex representation for two-dimensional vectors as:

$$H(t) = H_x + iH_y = H_1 \exp(-i\Delta t). \quad (1.47)$$

The excitation is in the same notation:

$$\mathcal{E}(t) = -\gamma H(t) = -\gamma H_1 \exp(-i\Delta t) = \omega_1 \exp(-i\Delta t). \quad (1.47')$$

The linear response $\mathcal{R}(t)$ to this excitation can be related to the linear response $\mathcal{R}_0(t)$ to a pulse of angle θ, written as an excitation:

$$\mathcal{E}_0(t) = \theta\delta(t). \quad (1.47'')$$

The response to a pulse of arbitrary angle θ is calculated in **A**(c)1. It is $\langle I_x(t)\rangle + \mathrm{i}\langle I_y(t)\rangle = \langle I_+(t)\rangle$ for $t>0$ and 0 for $t<0$, where $\langle I_x(t)\rangle$ and $\langle I_y(t)\rangle$ are given by (1.44) and (1.43). It is only when the angle θ is infinitely small that this response is linear:

$$\mathcal{R}_0(t) = \theta\ \mathrm{Tr}\{I_x^2\}\left\{\beta\frac{\mathrm{d}G}{\mathrm{d}t} + \mathrm{i}\beta_Z\omega_0 G\right\}. \qquad (1.47''')$$

The excitation $\mathcal{E}(t)$ of (1.47′) can be written:

$$\mathcal{E}(t) = \frac{\omega_1}{\theta}\exp(-\mathrm{i}\Delta t)\int_{-\infty}^{\infty}\mathcal{E}_0(t')\exp(\mathrm{i}\Delta t')\,\mathrm{d}t'. \qquad (1.48)$$

The linear response to it, $\mathcal{R}(t)$ will be:

$$\begin{aligned}\mathcal{R}(t) &= \frac{\omega_1}{\theta}\exp(-\mathrm{i}\Delta t)\int_{-\infty}^{+\infty}\mathcal{R}_0(t')\exp(\mathrm{i}\Delta t')\,\mathrm{d}t' \\ &= \frac{\omega_1}{\theta}\exp(-\mathrm{i}\Delta t)\int_0^{\infty}\{\langle I_x(t')\rangle + \mathrm{i}\langle I_y(t')\rangle\}\exp(\mathrm{i}\Delta t')\,\mathrm{d}t'. \qquad (1.49)\end{aligned}$$

This response can be rewritten:

$$\mathcal{R}(t) = \omega_1\chi(-\Delta)\exp(-\mathrm{i}\Delta t), \qquad (1.49')$$

where according to (1.49):

$$\chi(-\Delta) = \frac{1}{\omega_1}\{\langle I_x(\Delta)\rangle + \mathrm{i}\langle I_y(\Delta)\rangle\}; \qquad (1.50)$$

where:

$$\langle I_x(\Delta)\rangle = -\mathrm{Tr}\{I_x^2\}\omega_1\{\beta + \pi(\beta_Z\omega_0 - \Delta\beta)g'(\Delta)\}; \qquad (1.51a)$$

$$\langle I_y(\Delta)\rangle = \mathrm{Tr}\{I_x^2\}\pi\omega_1\{\beta_Z\omega_0 - \beta\Delta\}g(\Delta); \qquad (1.51b)$$

with:

$$g(\Delta) = \frac{1}{\pi}\int_0^{\infty}G(t)\cos(\Delta t)\,\mathrm{d}t; \qquad (1.52)$$

whence:

$$g(\Delta) = g(-\Delta), \qquad \int_{-\infty}^{+\infty}g(\Delta)\,\mathrm{d}\Delta = 1;$$

$$g'(\Delta) = \frac{1}{\pi}\int_0^{\infty}G(t)\sin(\Delta t)\,\mathrm{d}t. \qquad (1.53)$$

The present derivation of $\chi(-\Delta)$ is different from the one usually adopted (cf. Abragam (1961), chapter IV and Goldman (1970), chapter 4). The more conventional derivation will be recalled in chapter 5, section **C**(f). The functions $g(\Delta)$ and $g'(\Delta)$ are Kramers–Kronig transforms of each other with the following well-known properties: The successive moments of $g(\Delta)$ are related to the successive derivatives of $G(t)$:

$$M_{2n} = (-1)^n \left(\frac{\mathrm{d}^{2n} G}{\mathrm{d}t^{2n}} \right)_{t=0}, \qquad M_{2n+1} = 0. \tag{1.54}$$

The expansion of $g'(\Delta)$ in powers of $1/\Delta$ is:

$$g'(\Delta) = \frac{1}{\pi \Delta} \sum_{n=0}^{\infty} \frac{M_{2n}}{\Delta^{2n}}. \tag{1.55}$$

$\langle I_x(\Delta) \rangle$ and $\langle I_y(\Delta) \rangle$ given by (1.51a) and (1.51b) are the dispersion and absorption signals represented in the literature by the symbols u and v.

The function $G(t)$ decays in a time T_2, defined somewhat loosely since the shape of $G(t)$ is complex and certainly not exponential.

The conditions for the validity of the linear approximation will be stated in the next section.

3. *The Provotorov equations*

Under the conditions of a linear response to a vanishingly small excitation ω_1, the quantities $\langle I_z \rangle = -\beta_Z \omega_0 \,\mathrm{Tr}\{I_z^2\}$ and $\langle \mathcal{H}'_D \rangle = -\beta \,\mathrm{Tr}\{\mathcal{H}'^2_D\}$ are unaffected by the presence of the r.f. field.

The Provotorov equations describe the evolution of $\langle I_z \rangle$ and β when H_1 is not vanishingly small anymore, whilst still much smaller than the local field H'_L defined in (1.32):

$$|\omega_1| = |-\gamma H_1| \ll D = |\gamma H'_L|. \tag{1.56}$$

In the frame rotating at the frequency $\omega = \omega_0 - \Delta$, of the r.f. field the effective Hamiltonian is:

$$\mathcal{H}_{\mathrm{eff}} = \Delta I_z + \mathcal{H}'_D + \omega_1 I_x. \tag{1.34}$$

The quasi-equilibrium density matrix defined in high field in the absence of r.f.,

$$\sigma = 1 - \beta_Z \omega_0 I_z - \beta \mathcal{H}'_D, \tag{1.29}$$

is invariant through rotation around $0z$ and has the same form in the rotating frame ω. It is convenient to define a Zeeman temperature in the rotating frame such that:

$$\beta_Z \omega_0 = \Delta \alpha.$$

The density matrix (1.29) can then be rewritten as:

$$\sigma = 1 - \alpha \Delta I_z - \beta \mathcal{H}'_D. \tag{1.57}$$

In the absence of the r.f. field the energy in the rotating frame is:

$$\langle \mathcal{H} \rangle = \langle \Delta I_z + \mathcal{H}'_{\mathrm{D}} \rangle = -(\alpha \Delta^2 + \beta D^2)\, \mathrm{Tr}\{I_z^2\}. \tag{1.58}$$

The presence of the r.f. field modifies to a certain extent the form of the density matrix (1.57) and the Hamiltonian, $(\Delta I_z + \mathcal{H}'_{\mathrm{D}})$, to which the term $\omega_1 I_x$ must be added, but, if $|\omega_1| \ll D$, the corresponding change in the total energy $\langle \mathcal{H} \rangle$ given by (1.58) can be neglected.

Within the framework of this approximation $\langle \mathcal{H} \rangle$ given by (1.58) is the total energy of the spin system (in the rotating frame) and therefore time-independent, whence:

$$\Delta^2 \frac{\mathrm{d}\alpha}{\mathrm{d}t} + D^2 \frac{\mathrm{d}\beta}{\mathrm{d}t} = 0. \tag{1.59}$$

The time dependence of $\langle I_z \rangle$ is given by:

$$\begin{aligned} \frac{\mathrm{d}}{\mathrm{d}t}\langle I_z \rangle &= \mathrm{Tr}\left[I_z \frac{\mathrm{d}\sigma}{\mathrm{d}t} \right] = \mathrm{Tr}\{I_z(-\mathrm{i}[\mathcal{H}, \sigma]\} \\ &= \mathrm{Tr}\{-\mathrm{i}[I_z, \mathcal{H}]\sigma\} = \langle -\mathrm{i}[I_z, \mathcal{H}] \rangle, \end{aligned} \tag{1.60}$$

where $\mathcal{H}$ is given by (1.34) from which it follows,

$$\frac{\mathrm{d}}{\mathrm{d}t}\langle I_z \rangle = \omega_1 \langle I_y \rangle. \tag{1.61}$$

In contrast with the other formulae of this section (1.61) is quite general. It is independent of the assumption $|\omega_1| \ll D$ and even of the high temperature approximation. It is true for any Hamiltonian where the only part that does not commute with I_z is $\omega_1 I_x$. If we use for $\langle I_y \rangle$ the expression (1.51*b*) of the linear approximation we find:

$$\frac{\mathrm{d}}{\mathrm{d}t}\langle I_z \rangle = -\Delta\, \mathrm{Tr}\{I_z^2\} \frac{\mathrm{d}\alpha}{\mathrm{d}t} = \mathrm{Tr}\{I_x^2\} \pi \omega_1^2 (\Delta\alpha - \Delta\beta) g(\Delta)$$

or:

$$\frac{\mathrm{d}\alpha}{\mathrm{d}t} = -\pi \omega_1^2 (\alpha - \beta) g(\Delta), \tag{1.62a}$$

and from (1.59) we get:

$$\frac{\mathrm{d}\beta}{\mathrm{d}t} = -\pi \omega_1^2 \frac{\Delta^2}{D^2} (\beta - \alpha) g(\Delta). \tag{1.62b}$$

The eqns (1.62) are the Provotorov equations which describe the evolution of the Zeeman and dipolar temperatures in the rotating frame. Instead of α it is convenient to use i_z (a c-number): $i_z = \Delta\alpha = -\langle I_z \rangle / \mathrm{Tr}\{I_z^2\}$ and to rewrite the eqns (1.62) as:

$$\frac{\mathrm{d}i_z}{\mathrm{d}t} = -W(i_z - \Delta\beta); \tag{1.63a}$$

$$\frac{\mathrm{d}\beta}{\mathrm{d}t} = W\frac{\Delta}{D^2}(i_z - \Delta\beta); \tag{1.63b}$$

where:

$$W = \pi\omega_1^2 g(\Delta). \tag{1.64}$$

The eqns (1.63) are in fact more general than the eqns (1.62). It can be shown (Goldman (1970), p. 103) that they are valid even if the departure from resonance Δ is time-dependent provided it changes little during a time T_2.

These equations provide the criterion for the validity of the linear response approximation of **A**(c)2 namely that i_z and β be unaffected by the irradiation.

The time constants $[W(\Delta)]^{-1}$, for the change of i_z due to irradiation and $[W(\Delta)(\Delta^2/D^2)]^{-1}$ for the change of β, must be long compared to the duration, τ, of the irradiation.

The criterion for the validity of the Provotorov equations themselves follows from the assumption that the spins are at all times in internal quasi-equilibrium. Since the relevant time constant is T_2, i_z and β should change little during T_2, which can be expressed by the condition $WT_2 \ll 1$. Since the time τ of the irradiation is always much longer than T_2 it is seen that this condition is much less stringent than the condition $W\tau \ll 1$ required for the validity of the linear approximation **A**(c)2.

From the expression (1.64) for W and the fact that the three quantities $1/T_2$, $\gamma H'_L$, $g(0)^{-1}$ are all of the same order of magnitude we see that our initial assumption that $|\omega_1| \ll D$, in (1.56) provides an adequate criterion for the validity of Provotorov's equations.

If on the other hand ω_1 is not small compared to D we still expect the density matrix of the spin system in the rotating frame to reach the equilibrium form predicted by (1.34) and (1.35):

$$\sigma = 1 - \beta(\Delta I_z + \mathscr{H}'_D + \omega_1 I_x). \tag{1.65}$$

However the trend toward that form cannot be described by the Provotorov equations anymore.

It is possible to include in the Provotorov equations the effects of spin-lattice relaxation by adding to the eqns (1.63) the rates of change of i_z

and β originating in the coupling with the lattice.

$$\frac{di_z}{dt} = -W(i_z - \Delta\beta) - \frac{1}{T_{1z}}(i_z - i_L); \tag{1.66a}$$

$$\frac{d\beta}{dt} = W\frac{\Delta}{D^2}(i_z - \Delta\beta) - \frac{1}{T_{1D}}\beta, \tag{1.66b}$$

where $i_L = \beta_L\omega_0$ and β_L is the inverse lattice temperature (omitted in (1.66*b*) because of its smallness).

B. The Liouville formalism and memory functions

(*a*) *Definition*

In studies of irreversible evolutions of spin systems it is sometimes possible to derive for the density matrix, what is known as a master equation (see for instance Abragam (1961), chapter VIII), of the following type.

$$\frac{d\sigma_{\alpha\beta}}{dt} = -i[\mathcal{H}, \sigma]_{\alpha\beta} - \sum_{\gamma\varepsilon} R_{\alpha\beta,\gamma\varepsilon}\sigma_{\gamma\varepsilon}, \tag{1.67}$$

where $\mathcal{H}$ is some unperturbed spin Hamiltonian and $R_{\alpha\beta,\gamma\varepsilon}$ is a supermatrix operating in a supermanifold of dimension n^2, square of the dimension n of the Hilbert space where operate $\mathcal{H}$ and σ.

It may be convenient to write the combinations of indices $(\alpha\beta)$ or $(\gamma\varepsilon)$ as single indices ν and ν' and rewrite:

$$\sum_{\gamma\varepsilon} R_{\alpha\beta,\gamma\varepsilon}\sigma_{\gamma\varepsilon} = \sum_{\nu'} R_{\nu\nu'}\sigma_{\nu'}. \tag{1.68}$$

We can then treat the quantities $\sigma_{\nu'}$ as components of a ket $|\sigma\rangle$ and $R_{\nu\nu'}$ as matrix elements of a super-operator R acting on the ket $|\sigma\rangle$ in the same way as an ordinary quantum-mechanical operator acts on an ordinary ket. The super-operator R is called a Liouville operator, the supermanifold of dimension n^2 where it operates, a Liouville space, and an ordinary quantum-mechanical operator becomes a Liouville ket. The first term $-i[\mathcal{H}, \sigma]_{\alpha\beta}$ of (1.67) can also be written in the Liouville notation with a Liouville operator $\hat{\mathcal{H}}$ defined by:

$$\begin{aligned} -i[\mathcal{H}, \sigma]_{\alpha\beta} &\equiv -i\sum_{\gamma}(\mathcal{H}_{\alpha\gamma}\sigma_{\gamma\beta} - \sigma_{\alpha\gamma}\mathcal{H}_{\gamma\beta}) \\ &\equiv -i(\hat{\mathcal{H}}|\sigma\rangle)_{\alpha\beta} = -i\sum_{\gamma\varepsilon}\hat{\mathcal{H}}_{\alpha\beta,\gamma\varepsilon}\sigma_{\gamma\varepsilon}, \end{aligned} \tag{1.69}$$

whence:

$$\hat{\mathcal{H}}_{\alpha\beta,\gamma\varepsilon} = \mathcal{H}_{\alpha\gamma}\,\delta_{\beta\varepsilon} - \mathcal{H}_{\varepsilon\beta}\,\delta_{\alpha\gamma}. \tag{1.70}$$

Using (1.69) and (1.70) it is possible to write and to solve systematically the quantum mechanical equation:

$$\frac{d\sigma}{dt} = -i[\mathcal{H}, \sigma] \quad \text{as} \quad \frac{d|\sigma\rangle}{dt} = -i\hat{\mathcal{H}}|\sigma\rangle. \tag{1.71}$$

This is known as the Liouville formalism.

We shall from now on speak of Hilbert operators and Liouville operators, Hilbert operators being kets in Liouville space, or Liouville kets. It should be abundantly clear that there are no results obtainable with this formalism that could not be derived with the usual formalism of quantum mechanics. We devote however the present section to this formalism for the following reasons: (i) it is often found in the literature; (ii) its compactness leads to more concise derivations and formulations; (iii) it introduces naturally the so-called memory-functions, useful for certain applications (Mori, 1965).

Our notations follow closely those of Mehring (1976), Appendix E.

If $\mathcal{H}$ (and $\hat{\mathcal{H}}$) in (1.71) is time-independent its solution can be written in Liouville space:

$$|\sigma(t)\rangle = \exp - \{it\hat{\mathcal{H}}\}|\sigma(0)\rangle, \tag{1.72}$$

equivalent in Hilbert space to:

$$\sigma(t) = \exp(-it\mathcal{H})\sigma(0)\exp(it\mathcal{H}). \tag{1.72$'$}$$

In the general case when $\mathcal{H}$ is time-dependent, (1.72) must be replaced by the more general formula:

$$|\sigma(t)\rangle = T\exp\left\{-i\int_0^t dt'\hat{\mathcal{H}}(t')\right\}|\sigma(0)\rangle. \tag{1.73}$$

$T\exp\{-i\int_0^t \hat{\mathcal{H}}(t')\,dt'\}$ is Feynman's time-ordered operator (a concept by no means restricted to the Liouville formalism) which has the property that for any $t_2 > t_1 > 0$:

$$T\exp\left\{-i\int_0^{t_2}\hat{\mathcal{H}}(t')\,dt'\right\} = T\exp\left\{-i\int_{t_1}^{t_2}\hat{\mathcal{H}}(t')\,dt'\right\}T\exp\left\{-i\int_0^{t_1}\hat{\mathcal{H}}(t')\,dt'\right\}. \tag{1.73$'$}$$

Pursuing the analogy between Hilbert operators A, B and Liouville vectors $|A\rangle$, $|B\rangle$ we define the scalar product of two Liouville vectors as:

$$\langle A|B\rangle \equiv \text{Tr}\{A^{\dagger}B\} = \langle B|A\rangle^*. \tag{1.74}$$

The expectation value $\langle Q\rangle = \text{Tr}\{Q\sigma\}$ of a Hermitian operator Q becomes in Liouville space:

$$\langle Q\rangle = \text{Tr}\{Q\sigma\} = \text{Tr}\{Q^{\dagger}\sigma\} = \langle Q|\sigma\rangle, \tag{1.75}$$

whence from (1.73):

$$\langle Q(t)\rangle = \langle Q|T\exp\left\{-\mathrm{i}\int_0^t \hat{\mathscr{H}}(t')\,\mathrm{d}t'\right\}|\sigma(0)\rangle. \tag{1.76}$$

(b) 'Operators of interest' and memory functions

In the Liouville space where Hilbert operators (and in particular the density matrix) are kets, we can expand any Liouville ket $|\xi\rangle$ into a linear combination of Liouville kets (Hilbert operators):

$$|\xi\rangle = \beta_0|1\rangle - \sum_i \beta_i|Q_i\rangle,$$

where $|1\rangle$ is the Liouville ket for the unit Hilbert operator.

We can always assume the $|Q_i\rangle$ orthogonal to each other (and to the ket $|1\rangle$)

$$\langle Q_i|Q_j\rangle = \langle Q_i|Q_i\rangle\,\delta_{ij} \qquad \langle 1\,|\,Q_i\rangle = \mathrm{Tr}\{Q_i\} = 0. \tag{1.77}$$

If they form a complete set:

$$\sum_i |Q_i\rangle\langle Q_i|/\langle Q_i|Q_i\rangle = \hat{1} \tag{1.78}$$

where $\hat{1}$ is the unit Liouville operator (not to be confused with the unit operator in the Hilbert space, which is a ket $|1\rangle$ in Liouville space).

The density matrix σ is expanded as:

$$|\sigma\rangle = \beta_0|1\rangle - \sum_i \beta_i|Q_i\rangle, \tag{1.79}$$

where

$$\beta_i = -\frac{\langle Q_i|\sigma\rangle}{\langle Q_i|Q_i\rangle} = -\frac{\langle Q_i\rangle}{\langle Q_i|Q_i\rangle}.$$

In the following we shall always assume $\beta_0 = 1$ and the β_i small. This is coherent with the high temperature approximation within the spin temperature assumption. From (1.71), (1.78), and (1.79) we get:

$$\frac{\mathrm{d}\langle Q_i\rangle}{\mathrm{d}t} = -\mathrm{i}\langle Q_i|\hat{\mathscr{H}}|\sigma\rangle = -\mathrm{i}\sum_j \frac{\langle Q_i|\hat{\mathscr{H}}|Q_j\rangle\langle Q_j|\sigma\rangle}{\langle Q_j|Q_j\rangle}$$
$$\frac{\mathrm{d}\langle Q_i\rangle}{\mathrm{d}t} = -\mathrm{i}\sum_j \frac{\langle Q_i|\hat{\mathscr{H}}|Q_j\rangle}{\langle Q_j|Q_j\rangle}\langle Q_j(t)\rangle. \tag{1.80}$$

It should be noted that so far the information content of (1.80) is identical to that of (1.71).

However it will often happen that the expectation values of only a few of the complete set of operators Q_i are of interest. It is then desirable to obtain a set of equations which contain the expectation values $\langle Q_i \rangle$ of these operators only. They form a manifold in the Liouville space, spanned by the kets $|Q_1\rangle, |Q_2\rangle, \ldots |Q_m\rangle$. We define the Liouville operator projection on this manifold:

$$\hat{P} = \sum_{k=1}^{m} \frac{|Q_k\rangle\langle Q_k|}{\langle Q_k|Q_k\rangle}. \tag{1.81}$$

The expectation value $\langle Q_k \rangle = \langle Q_k|\sigma\rangle$, where $k \leqslant m$, obeys the equation:

$$\frac{\mathrm{d}}{\mathrm{d}t}\langle Q_k \rangle = -\mathrm{i}\langle Q_k|\hat{\mathcal{H}}\hat{P}|\sigma\rangle - \mathrm{i}\langle Q_k|\hat{\mathcal{H}}(\hat{1}-\hat{P})|\sigma\rangle, \tag{1.82}$$

where we have split $|\sigma\rangle$ into its so-called 'diagonal' part $\hat{P}|\sigma\rangle$ and the 'off-diagonal' part $(\hat{1}-\hat{P})|\sigma\rangle$.

These two parts are related to each other by the equation:

$$\frac{\mathrm{d}}{\mathrm{d}t}\{\hat{1}-\hat{P})|\sigma\rangle + \mathrm{i}(\hat{1}-\hat{P})\hat{\mathcal{H}}(\hat{1}-\hat{P})|\sigma\rangle = -\mathrm{i}(\hat{1}-\hat{P})\hat{\mathcal{H}}\hat{P}|\sigma\rangle, \tag{1.83}$$

which can be integrated by the standard formula:

$$(\hat{1}-\hat{P})|\sigma(t)\rangle = \hat{S}(t,0)(\hat{1}-\hat{P})|\sigma(0) - \mathrm{i}\int_0^t \hat{S}(t,t')(\hat{1}-\hat{P})\hat{\mathcal{H}}(t')\hat{P}|\sigma(t')\rangle\,\mathrm{d}t', \tag{1.84}$$

where

$$\hat{S}(t,t') = T\exp\left[-\mathrm{i}\int_{t'}^{t}(\hat{1}-\hat{P})\hat{\mathcal{H}}(\tau)\,\mathrm{d}\tau\right]. \tag{1.85}$$

Carrying (1.84) and (1.85) into (1.82) we get:

$$\frac{\mathrm{d}}{\mathrm{d}t}\langle Q_k(t)\rangle = K + L + M, \tag{1.86}$$

where:

$$K = -\mathrm{i}\langle Q_k|\hat{\mathcal{H}}\hat{P}|\sigma\rangle = -\mathrm{i}\sum_{j=1}^{m}\frac{\langle Q_k|\hat{\mathcal{H}}(t)|Q_j\rangle}{\langle Q_j|Q_j\rangle}\langle Q_j(t)\rangle; \tag{1.87}$$

$$L = -\mathrm{i}\langle Q_k|\hat{\mathcal{H}}(t)\hat{S}(t,0)(\hat{1}-\hat{P})|\sigma(0)\rangle; \tag{1.88}$$

$$M = -\sum_{j=1}^{m}\int_0^t \mathrm{d}t'\frac{\langle Q_k|\hat{\mathcal{H}}(t)\hat{S}(t,t')(\hat{1}-\hat{P})\hat{\mathcal{H}}(t')|Q_j\rangle}{\langle Q_j|Q_j\rangle}\langle Q_j(t')\rangle. \tag{1.89}$$

We shall make two restrictions on the properties of the 'operators of interest' and on the initial stae of the spin system which simplify considerably the eqns (1.86).

(i) All commutators $[Q_i^\dagger, Q_j]$ vanish for the 'operators of interest'. Then for these operators:

$$\langle Q_k|\hat{\mathcal{H}}(t)|Q_j\rangle = \mathrm{Tr}\{Q_k^\dagger[\mathcal{H}(t), Q_j]\} = \mathrm{Tr}\{[Q_j, Q_k^\dagger]\mathcal{H}(t)\} = 0 \qquad (1.90)$$

and the term K in (1.87) vanishes.

(ii) We assume that at time $t = 0$, σ is 'diagonal';

$$(\hat{1} - \hat{P})|\sigma(0)\rangle = 0. \qquad (1.91)$$

This suppresses the term L in (1.88). We are left with the system:

$$\frac{\mathrm{d}}{\mathrm{d}t}\langle Q_k(t)\rangle = -\sum_{j=1}^{m}\int_0^t K_{kj}(t, t')\langle Q_j(t')\rangle\,\mathrm{d}t'. \qquad (1.92)$$

The functions $K_{kj}(t, t')$,

$$K_{kj}(t, t') = \frac{\langle Q_k|\hat{\mathcal{H}}(t)\hat{S}(t, t')(\hat{1} - \hat{P})\hat{\mathcal{H}}(t')|Q_j\rangle}{\langle Q_j|Q_j\rangle}, \qquad (1.93)$$

are called memory functions.

If the Hamiltonian $\hat{\mathcal{H}}(t)$ is time-independent, $K_{kj}(t, t')$ becomes $K_{kj}(t - t')$ with:

$$K_{kj}(\tau) = \frac{\langle Q_k|\hat{\mathcal{H}}\exp -\{\mathrm{i}\tau(\hat{1} - \hat{P})\hat{\mathcal{H}}\}(\hat{1} - \hat{P})\hat{\mathcal{H}}|Q_j\rangle}{\langle Q_j|Q_j\rangle}; \qquad (1.94)$$

$$K_{kj}(0) = \frac{\langle Q_k|\hat{\mathcal{H}}(\hat{1} - \hat{P})\hat{\mathcal{H}}|Q_j\rangle}{\langle Q_j|Q_j\rangle} = \frac{\langle Q_k|\hat{\mathcal{H}}^2|Q_j\rangle}{\langle Q_j|Q_j\rangle}; \qquad (1.95)$$

for

$$\hat{P}\hat{\mathcal{H}}|Q_j\rangle = \sum_{i=1}^{m}|Q_i\rangle\langle Q_i|\hat{\mathcal{H}}|Q_j\rangle/\langle Q_i|Q_i\rangle,$$

vanishes according to (1.90).

It is useful to emphasize again that the eqns (1.92), (1.93) are direct consequences of (1.71) with the restrictions (1.90), (1.91) and have no new physical content such as irreversibility. Such a content is introduced if the nature of the spin system allows some specific assumptions on the nature of the memory functions (1.93) or (1.94). Such assumptions will be introduced in the next section.

(c) *The generalized Provotorov equations: an illustration of the Liouville formalism*

1. *Long term trend toward thermal equilibrium*

We assume that the Hamiltonian of the spin system can be written $\mathcal{H} = \mathcal{H}_0 + V$ where $\mathcal{H}_0$ is the sum of two orthogonal and commuting operators $\mathcal{H}_0=\mathcal{H}_1+\mathcal{H}_2$ with:

$$\mathrm{Tr}\{\mathcal{H}_1\}=\mathrm{Tr}\{\mathcal{H}_2\}=\mathrm{Tr}\{\mathcal{H}_1\mathcal{H}_2\}=0, \qquad [\mathcal{H}_1, \mathcal{H}_2]=0, \tag{1.96}$$

whereas V is a small perturbation which commutes with neither $\mathcal{H}_1$ nor $\mathcal{H}_2$. With the notations of the section **B**(b) we select $\mathcal{H}_1$ and $\mathcal{H}_2$ as our 'operators of interest' and assume that at time $t=0$ the density matrix is 'diagonal';

$$|\sigma(0)\rangle=|1\rangle-\beta_1|\mathcal{H}_1\rangle-\beta_2|\mathcal{H}_2\rangle, \qquad (\hat{1}-\hat{P})|\sigma(0)\rangle=0. \tag{1.97}$$

The new physical assumption that we make now is that $K_{jk}(\tau)$, given by (1.95) for $\tau=0$, has a very short correlation time τ_c, that is, decays very rapidly toward zero as τ increases, at a rate much faster than the rate of change of the $\langle Q_j(t')\rangle$ in the equation (1.92). Once more, this is an *assumption* on the physical nature of our spin system. We could not give a general proof of its validity, for the very good reason that it is not true in general.

With this assumption we can replace in (1.92) $\langle Q_j(t')\rangle$ by $\langle Q_j(t)\rangle$. If we take $t\gg\tau_c$, the system (1.92), with the notations (1.96) becomes the very simple system:

$$\frac{\mathrm{d}\langle\mathcal{H}_1\rangle}{\mathrm{d}t}=-A_{11}\langle\mathcal{H}_1\rangle-A_{12}\langle\mathcal{H}_2\rangle; \tag{1.98a}$$

$$\frac{\mathrm{d}\langle\mathcal{H}_2\rangle}{\mathrm{d}t}=-A_{21}\langle\mathcal{H}_1\rangle-A_{22}\langle\mathcal{H}_2\rangle; \tag{1.98b}$$

with

$$A_{kj}=\int_0^\infty K_{kj}(\tau)\,\mathrm{d}\tau. \tag{1.99}$$

We note that in the expression (1.94) of K_{kj}, because of the commutation relations (1.96) we can replace $\langle\mathcal{H}_k|\hat{\mathcal{H}}$ by $\langle\mathcal{H}_k|\hat{V}$ and $(\hat{1}-\hat{P})\hat{\mathcal{H}}|\mathcal{H}_j\rangle$ by $(\hat{1}-\hat{P})\hat{V}|\mathcal{H}_j\rangle$ or, since $\hat{P}\hat{V}|\mathcal{H}_j\rangle=0$, by $\hat{V}|\mathcal{H}_j\rangle$ whence:

$$K_{kj}(\tau)\simeq\frac{\langle\mathcal{H}_k|\hat{V}\exp\{-\mathrm{i}\tau(\hat{1}-\hat{P})(\hat{\mathcal{H}}_0+\hat{V})\}\hat{V}|\mathcal{H}_j\rangle}{\langle\mathcal{H}_j|\mathcal{H}_j\rangle}. \tag{1.100}$$

Since V is a small perturbation, $K_{kj}(\tau)$ is of order V^2 and it is a reasonable approximation to replace in the exponential $(\hat{\mathcal{H}}_0+\hat{V})$ by $\hat{\mathcal{H}}_0$, and since $\hat{P}\hat{\mathcal{H}}_0=0$, $K_{kj}(\tau)$ reduces to:

$$K_{kj}(\tau)\simeq\frac{\langle\mathcal{H}_k|\hat{V}\exp(-\mathrm{i}\tau\hat{\mathcal{H}}_0)\hat{V}|\mathcal{H}_j\rangle}{\langle\mathcal{H}_j|\mathcal{H}_j\rangle}. \tag{1.101}$$

Transcribed in the traditional Hilbert space notation, (1.101) takes the more familiar form:

$$K_{kj}(\tau)=\frac{\mathrm{Tr}\{[\mathcal{H}_k, V]\exp(-\mathrm{i}\tau\mathcal{H}_0)[V,\mathcal{H}_j]\exp(\mathrm{i}\tau\mathcal{H}_0)\}}{\mathrm{Tr}\{\mathcal{H}_j^2\}}. \tag{1.102}$$

The coefficients A_{kj} in (1.98) have important symmetry properties. Let us define coefficients $a_{ij} = A_{ij}\langle \mathcal{H}_j | \mathcal{H}_j \rangle$. It can be shown easily that:

$$a_{11} = a_{21} = -a_{12} = -a_{22} = a. \tag{1.103}$$

As an example we show that $a_{11} + a_{12} = 0$. Using (1.102) we have:

$$\begin{aligned} a_{11} + a_{12} &= \int_0^\infty \mathrm{Tr}\{[\mathcal{H}_1, V]\exp(-\mathrm{i}\tau\mathcal{H}_0)[V, \mathcal{H}_0]\exp(\mathrm{i}\tau\mathcal{H}_0)\}\,\mathrm{d}t \\ &= \mathrm{i}\frac{\mathrm{d}}{\mathrm{d}t}\int_0^\infty \mathrm{Tr}\{[\mathcal{H}_1, V]\exp(-\mathrm{i}\tau\mathcal{H}_0)V\exp(\mathrm{i}\tau\mathcal{H}_0)\}\,\mathrm{d}t \\ &= |-\mathrm{i}\,\mathrm{Tr}\{[\mathcal{H}_1, V]\exp(-\mathrm{i}\tau\mathcal{H}_0)V\exp(\mathrm{i}\tau\mathcal{H}_0)\}|_{\tau=0}^{\infty} = 0, \end{aligned} \tag{1.104}$$

$|\mathcal{H}_1\rangle$ and $|\mathcal{H}_2\rangle$ being the 'operators of interest', the density matrix of the spin system can be written:

$$|\sigma\rangle = \hat{P}|\sigma(t)\rangle + (\hat{1} - \hat{P})|\sigma(t)\rangle = |1\rangle - \beta_1|\mathcal{H}_1\rangle - \beta_2|\mathcal{H}_2\rangle + (\hat{1} - P)|\sigma(t)\rangle;$$

$$\langle \mathcal{H}_1 \rangle = -\beta_1\,\mathrm{Tr}\{\mathcal{H}_1^2\}, \qquad \langle \mathcal{H}_2 \rangle = -\beta_2\,\mathrm{Tr}\{\mathcal{H}_2^2\},$$

whence from the symmetry relations (1.103) the eqn (1.98) can be rewritten:

$$\begin{aligned} \frac{\mathrm{d}\beta_1}{\mathrm{d}t} &= -a(\beta_1 - \beta_2)/\mathrm{Tr}\{\mathcal{H}_1^2\}; \\ \frac{\mathrm{d}\beta_2}{\mathrm{d}t} &= -a(\beta_2 - \beta_1)/\mathrm{Tr}\{\mathcal{H}_2^2\}. \end{aligned} \tag{1.105}$$

These are the generalized Provotorov equations, sometimes written as:

$$\frac{\mathrm{d}\beta_1}{\mathrm{d}t} = -W(\beta_1 - \beta_2), \qquad \frac{\mathrm{d}\beta_2}{\mathrm{d}t} = -\varepsilon W(\beta_2 - \beta_1), \tag{1.105$'$}$$

where,

$$\varepsilon = \mathrm{Tr}\{\mathcal{H}_1^2\}/\mathrm{Tr}\{\mathcal{H}_2^2\}.$$

The Provotorov eqns (1.62) relative to saturation by an r.f. field are a special case of (1.105) with:

$$\begin{aligned} &\mathcal{H}_1 = \Delta I_z; \quad \mathcal{H}_2 = \mathcal{H}'_\mathrm{D}; \qquad V = \omega_1 I_x; \\ &\qquad \beta_1 = \alpha; \qquad \beta_2 = \beta. \end{aligned} \tag{1.106}$$

A little algebra shows that

$$W = \frac{a}{\mathrm{Tr}\{\mathscr{H}_1^2\}} = \frac{\int_0^\infty \mathrm{Tr}\{[\mathscr{H}_1, V]\exp(-\mathrm{i}\mathscr{H}_0 t)[V, \mathscr{H}_1]\exp(\mathrm{i}\mathscr{H}_0 t)\}\,\mathrm{d}t}{\mathrm{Tr}\{\mathscr{H}_1^2\}}$$

$$= \omega_1^2 \int_0^\infty \mathrm{Tr}\{I_y \exp(-\mathrm{i}\mathscr{H}'_{\mathrm{D}}t) I_y \exp(\mathrm{i}\mathscr{H}'_{\mathrm{D}}t)\}\cos(\Delta t)\,\mathrm{d}t/\mathrm{Tr}\{I_y^2\}$$

$$= \omega_1^2 \int_0^\infty G(t)\cos(\Delta t)\,\mathrm{d}t = \pi\omega_1^2 g(\Delta), \qquad (1.107)$$

in agreement with (1.43′), (1.52), and (1.64).

It is interesting to note that in the Liouville derivation of the generalized Provotorov equations, in contrast to the earlier derivation, no explicit assumption on the density matrix was necessary, besides $(\hat{1}-\hat{P})|\sigma(0)\rangle = 0$; that is besides the assumption of spin temperature (different for $\mathscr{H}_1$, *and* $\mathscr{H}_2$) for $t = 0$, rather than at all times as in the earlier derivation **A**(c)3. On the other hand we had to make the far reaching assumption of very short correlation times for the memory functions $K_{kj}(\tau)$. Such an assumption is not always warranted and can then lead to incorrect conclusions. Consider for instance the well known problem of cross relaxation between two spin systems, of neighbouring Larmor frequencies ω_{I} and ω_{S}. If we take for $|\mathscr{H}_1\rangle$ and $|\mathscr{H}_2\rangle$ the Zeeman energies $\omega_{\mathrm{I}} I_z$ and $\omega_{\mathrm{S}} S_z$ of the spin systems I and S, with a total dipolar energy $\mathscr{H}'_{\mathrm{D}}$, and assume short correlation times for the memory functions $K_{11}, K_{12}, K_{21}, K_{22}$ we are led to the system of eqns (1.98) which in this case is known to be incorrect unless $\omega_{\mathrm{I}} = \omega_{\mathrm{S}}$ (Goldman (1970), p. 160).

Furthermore, even when the long term behaviour of the spin system is correctly described by equations similar to (1.98) its short time behaviour due to transient phenomena occurring immediately after an abrupt change in the spin system requires a special treatment given in the next section.

2. *Short term transient oscillations*

In this section we examine the behaviour of a spin system immediately after its Hamiltonian has been abruptly modified. As in the previous section the Hamiltonian is made of three parts:

$$\mathscr{H} = \mathscr{H}_1 + \mathscr{H}_2 + V, \qquad (1.108)$$

where $\mathscr{H}_1$ and $\mathscr{H}_2$ are orthogonal and commuting whereas V commutes with neither. We assume that the Hamiltonian $\mathscr{H}_1$ is suddenly created at time $t = 0$. In practice it will be the Zeeman coupling of a spin species with a magnetic field applied suddenly, in the laboratory frame (a d.c. field) or in a rotating frame (an r.f. field). Prior to its application the system is in thermal equilibrium with a density matrix.

$$\sigma(0) = 1 - \beta_0(\mathscr{H}_2 + V). \tag{1.109}$$

If in the spirit of the previous section we select as operators 'of interest' $\mathscr{H}_1$ and $\mathscr{H}_2$ we must remark that because of the presence of V in (1.109) the initial condition (1.97): $(\hat{1} - \hat{P})|\sigma(0)\rangle$ is not verified and that the rate of change of $\langle\mathscr{H}_1\rangle$ contains a term $L(t)$ proportional to $(\hat{1} - \hat{P})$ and given by (1.88). Defining the inverse temperatures $\alpha(t)$ and $\beta(t)$ of $\mathscr{H}_1$ and $\mathscr{H}_2$

$$\alpha(t) = -\frac{\langle\sigma(t)|\mathscr{H}_1\rangle}{\langle\mathscr{H}_1|\mathscr{H}_1\rangle} \qquad \beta(t) = -\frac{\langle\sigma(t)|\mathscr{H}_2\rangle}{\langle\mathscr{H}_2|\mathscr{H}_2\rangle}, \tag{1.110}$$

the rate of change of $\alpha(t)$ is given by:

$$\frac{\mathrm{d}\alpha}{\mathrm{d}t} = -L(t)/\langle\mathscr{H}_1|\mathscr{H}_1\rangle - \int_0^t \mathrm{d}t'\left\{K_{11}(t-t')\alpha(t') + K_{12}(t-t')\frac{\langle\mathscr{H}_2|\mathscr{H}_2\rangle}{\langle\mathscr{H}_1|\mathscr{H}_1\rangle}\beta(t')\right\}. \tag{1.111}$$

Immediately after the introduction of $\mathscr{H}_1$ the density matrix is still (1.109) and $\alpha(0) = 0$, $\beta(0) = \beta_0$. For very short times t, we have approximately:

$$\frac{\mathrm{d}\alpha}{\mathrm{d}t} \simeq -L(t)/\langle\mathscr{H}_1|\mathscr{H}_1\rangle - \beta_0\frac{\langle\mathscr{H}_2|\mathscr{H}_2\rangle}{\langle\mathscr{H}_1|\mathscr{H}_1\rangle}\int_0^t K_{12}(t')\,\mathrm{d}t'. \tag{1.112}$$

We can get rid of the cumbersome term $L(t)$ in (1.112) by the following argument: if after the introduction of $\mathscr{H}_1$ we had $\alpha(0) = \beta_0$ that is a density matrix:

$$\sigma(0) = 1 - \beta_0(\mathscr{H}_1 + \mathscr{H}_2 + V) = 1 - \beta_0\mathscr{H}, \tag{1.113}$$

this density matrix would commute with the total Hamiltonian $\mathscr{H}$ of the system and remain unchanged. For short times this would give:

$$0 = \frac{\mathrm{d}\alpha}{\mathrm{d}t} = -L(t)/\langle\mathscr{H}_1|\mathscr{H}_1\rangle - \beta_0\int_0^t \mathrm{d}t'\left\{K_{11}(t') + \frac{\langle\mathscr{H}_2|\mathscr{H}_2\rangle}{\langle\mathscr{H}_1|\mathscr{H}_1\rangle}K_{12}(t')\right\}. \tag{1.114}$$

Subtracting (1.114) from (1.112) we get for short times t:

$$\frac{d\alpha}{\mathrm{d}t} \simeq \beta_0\int_0^t K_{11}(t')\,\mathrm{d}t'; \tag{1.115}$$

$$\alpha(t) \simeq \beta_0\int_0^t \mathrm{d}t'\int_0^{t'} \mathrm{d}t''\,K_{11}(t'') = \beta_0\int_0^t (t-\tau)K_{11}(\tau)\,\mathrm{d}t. \tag{1.116}$$

It is interesting to remark that in the derivation of (1.116) we did not assume that, compared to $\mathscr{H}_1$ and $\mathscr{H}_2$, V was a small perturbation. Even if V is comparable to $\mathscr{H}_1$ or $\mathscr{H}_2$ (1.116) is still valid for sufficiently short times. We must be careful however to use in that case for K_{11} the exact expression (1.100) rather than the approximate one (1.101).

An example of this situation is provided by the experiment of Strombotne and Hahn (1964), where, the spin system being in thermal equilibrium in zero field, a d.c. magnetic field H_0 is applied suddenly. In this example $\mathcal{H}_1$ is the Zeeman Hamiltonian $Z = \omega_0 I_z$ in that field, $\mathcal{H}_2$ is the secular part $\mathcal{H}'_D$ of the dipolar Hamiltonian which commutes with Z, and V is its non-secular part $\mathcal{H}''_D$. Although $\mathcal{H}''_D = V$ may be much smaller than $Z = \mathcal{H}_1$, it is not smaller than $\mathcal{H}'_D = \mathcal{H}_2$.

This problem is discussed in Goldman (1970) (chapter 6) in considerable detail, although with a different approach, and we shall not consider it in this book.

We shall assume from now on that V is much smaller than $\mathcal{H}_1$ and $\mathcal{H}_2$ and can be treated as a perturbation. We can then use for $K_{kj}(\tau)$ the approximate eqn (1.101), which makes life much easier.

The eqn (1.116) which describes the early evolution of the spin temperature $\alpha(t)$ for small t is valid as long as it is permissible to replace on the right-hand side of the *exact* eqn (1.111) $\alpha(t)$ by $\alpha(0)=0$ and $\beta(t)$ by $\beta(0)=\beta_0$.

The behaviour of $\alpha(t)$ as given by (1.116) is very different depending on whether t is comparable to the correlation time τ_c of the memory function $K_{11}(\tau)$, or, while sufficiently small for (1.116) to be valid, very much larger than τ_c. When as often happens, the memory function is an oscillating function of τ with a damping constant of the order of $1/\tau_c$, it follows that if in (1.116) t is not large compared to τ_c, $\alpha(t)$ will exhibit the same type of damped oscillation.

On the other hand if $t \gg \tau_c$, (1.116) can be written approximately as:

$$\alpha(t) \simeq \beta_0 t \int_0^\infty K_{11}(\tau)\,\mathrm{d}\tau - \beta_0 \int_0^\infty \tau K_{11}(\tau)\,\mathrm{d}\tau. \tag{1.117}$$

The first integral $\int_0^\infty K_{11}(\tau)\,\mathrm{d}t$ is according to (1.98) and (1.99) the transition probability W in the first generalized Provotorov eqn (1.105′) and the first term in (1.117) is $\beta_0 Wt$ with $Wt, \ll 1$, in order for (1.117) to be valid. The second term in (1.117) is a constant that we shall call α'_0. Equation (1.117) then reads:

$$\alpha(t) \simeq \beta_0 Wt + \alpha'_0, \tag{1.118}$$

or, since $Wt \ll 1$, with the same accuracy:

$$\alpha(t) \simeq Wt(\beta_0 - \alpha'_0) + \alpha'_0. \tag{1.119}$$

Equation (1.119) is the solution for small t of the Provotorov equation:

$$\frac{\mathrm{d}\alpha}{\mathrm{d}t} = -W(\alpha - \beta) \qquad \text{with} \qquad \beta(0) = \beta_0, \alpha(0) = \alpha'_0; \tag{1.120}$$

$$\alpha'_0 = \beta_0 \int_0^\infty \tau K_{11}(\tau)\,\mathrm{d}\tau. \tag{1.121}$$

We see now that the evolution of the inverse spin temperature $\alpha(t)$ goes through two different regimes.

After the sudden introduction of $\mathcal{H}_1$, $\alpha(t)$ undergoes a damped oscillation and reaches a quasi-equilibrium value α'_0 given by (1.121), after a time of the order of several times τ_c, the correlation time of $K_{11}(\tau)$.

The density matrix has then reached a quasi-equilibrium form,

$$\sigma \simeq 1 - \alpha\mathcal{H}_1 - \beta\mathcal{H}_2, \tag{1.122}$$

with $\alpha = \alpha'_0$, $\beta = \beta_0$ after which it proceeds towards an equilibrium state where $\alpha = \beta$, at a rate described by the generalized Provotorov equations. α'_0 is very much smaller than β_0. According to (1.121):

$$\alpha'_0 \simeq \beta_0\tau_c \int_0^\infty K_{11}(\tau)\,\mathrm{d}\tau = \beta_0\tau_c W \ll \beta_0.$$

(The inequality $W\tau_c \ll 1$ is a necessary condition for the validity of the Provotorov equations.)

In section **B**(e) we shall discuss in some detail experiments where extensive measurements have been made on both the short-time and the long-time evolution of a spin system.

(d) *Resonance line shape and memory functions*

1. *Introduction of the memory function*

In this section we use the memory function as an approach to the well known problem of the shape of the absorption and dispersion signals at low r.f. level. The relevant formulae were given in **A**(c)2, with the absorption signal $\langle I_y(\Delta)\rangle$ and the dispersion signal $\langle I_x(\Delta)\rangle$ given by (1.51), (1.52), (1.53), and (1.43′). If the spin system is at or near equilibrium the Zeeman and dipolar temperatures β_Z and β are equal or nearly equal and since, in high field: $\omega_0 \gg \Delta$, and $\omega_0 g'(\Delta) \approx \omega_0 T_2 \gg 1$, we neglect in (1.51) all terms proportional to β and keep only those proportional to β_Z,

$$\langle I_x(\Delta)\rangle + \mathrm{i}\langle I_y(\Delta)\rangle = -\mathrm{i}\beta_Z\omega_0\omega_1\,\mathrm{Tr}\{I_x^2\}\int_0^\infty G(t')\exp(\mathrm{i}\Delta t')\,\mathrm{d}t'$$

$$= \mathrm{i}\pi\omega_1\beta_Z\omega_0\{g(\Delta) + \mathrm{i}g'(\Delta)\}. \tag{1.123}$$

The FID function,

$$G(t) = \mathrm{Tr}\{I_x\exp(-\mathrm{i}\mathcal{H}'_D t)I_x\exp(\mathrm{i}\mathcal{H}'_D t)\}/\mathrm{Tr}\{I_x^2\} = \frac{\langle I_x(t)\rangle}{\mathrm{Tr}\{I_x^2\}}, \tag{1.124}$$

is (normalized to unity for $t = 0$) the expectation value at time t of the operator I_x if at time $t = 0$ the density matrix $\sigma(0)$ is $1 - \omega_0\beta_Z I_x$.

This is the problem we have tackled in the last section, but now with a single 'operator of interest', namely I_x, and a single memory function $K(t)$

for the calculation of:

$$\langle I_x(t)\rangle \propto G(t).$$

The system (1.92) becomes a single equation:

$$\frac{\mathrm{d}\langle I_x\rangle}{\mathrm{d}t} \propto \frac{\mathrm{d}G}{\mathrm{d}t} = -\int_0^t K(t-t')G(t')\,\mathrm{d}t', \tag{1.125}$$

with K given by (1.94):

$$K(\tau) = \frac{\langle I_x|\mathcal{H}\exp\{-\mathrm{i}\tau(\hat{1}-\hat{P})\mathcal{H}\}(\hat{1}-\hat{P})\mathcal{H}|I_x\rangle}{\langle I_x|I_x\rangle}; \tag{1.126a}$$

$$\hat{P} = \frac{|I_x\rangle\langle I_x|}{\langle I_x|I_x\rangle}. \tag{1.126b}$$

We had assumed so far that the only Hamiltonian responsible for the evolution of $\langle I_x(t)\rangle$ in (1.124) was the truncated dipolar Hamiltonian $\mathcal{H}'_{\mathrm{D}}$ given by (1.27) for a system of identical spins.

There may exist other bilinear interactions between spins, the so-called indirect interactions, $\mathcal{H}_{\mathrm{ind}}$ due to the magnetic polarizability of the electron clouds surrounding the nuclear magnetic moments (Abragam (1961), p. 183). There may also be spin–spin interactions of yet another origin the so-called exchange interactions which we shall represent by the symbol $\mathcal{H}_{\mathrm{e}}$. These interactions exist only in solid ^{3}He and we shall discuss them in chapter 3 which is devoted to that subject. Suffice it to say at present that $\mathcal{H}_{\mathrm{e}}$ has two characteristics: (i) it commutes with I_x (but not with $\mathcal{H}'_{\mathrm{D}}$); (ii) it is often very much larger than $\mathcal{H}'_{\mathrm{D}}$. The existence of such a Hamiltonian modifies profoundly the resonance lineshapes.

It should be understood that in all cases where indirect or exchange interactions exist, the corresponding Hamiltonians $\mathcal{H}_{\mathrm{ind}}$ or $\mathcal{H}_{\mathrm{e}}$ must be included in the Hamiltonian $\mathcal{H}$ in (1.126*a*) or added to $\mathcal{H}'_{\mathrm{D}}$ in (1.124). If there are two spin systems I and S, the Hamiltonians $\mathcal{H}_{\mathrm{IS}}$ and also $\mathcal{H}_{\mathrm{SS}}$ should be included.

If the exchange interaction $\mathcal{H}_{\mathrm{e}}$ is so large as to be comparable with the Zeeman energy there are further complications. The simple addition of $\mathcal{H}_{\mathrm{e}}$ to $\mathcal{H}'_{\mathrm{D}}$ in (1.124) is not correct. This situation which may occur in solid ^{3}He will be examined in chapter 3.

After this aside we return to the problem of the determination of $K(t)$, $G(t)$, and the shapes $g(\Delta)$ and $g'(\Delta)$ of the absorption and dispersion signals. Introducing $K(t)$ to determine $G(t)$ through (1.125) may appear as a strange idea since the expression (1.124) for $G(t)$ looks rather simpler than that for $K(t)$, (1.126*a*). The opposite procedure, namely that of using $G(t)$ to determine $K(t)$ by inverting (1.125) would if anything seem more logical. The heart of the matter is that we are unable to calculate exactly either $G(t)$

through (1.124) or $K(t)$ through (1.126*a*). We are thus led to try out, for one or the other, analytical expressions, Gaussian, exponential, Lorentzian, or others, with parameters adjusted to fit the first few terms of a limited expansion of $G(t)$ from (1.124) or $K(t)$ from (1.126*a*).

As we shall see shortly the justification for the introduction of the memory function K is that a single, very simple approximation for K, namely a Gaussian shape, happens to lead through (1.125) to rather good (and sometimes surprisingly good) approximations for G. We now solve the eqn (1.125) by means of a Laplace transform. Let us define:

$$\mathcal{G}(z) = \int_0^\infty G(t)\exp(-zt)\,\mathrm{d}t \tag{1.127a}$$

$$\mathcal{K}(z) = \int_0^\infty K(t)\exp(-zt)\,\mathrm{d}t \tag{1.127b}$$

Equation (1.125) is transformed into:

$$\begin{cases} z\mathcal{G}(z) - 1 = -\mathcal{G}(z)\mathcal{K}(z) \\ \quad \mathcal{G}(z) = \dfrac{1}{z+\mathcal{K}(z)}. \end{cases} \tag{1.128}$$

The shapes $g(\Delta)$ and $g'(\Delta)$ of the absorption and dispersion signals are, according to (1.123) and (1.127) given by:

$$g(\Delta) - \mathrm{i}g'(\Delta) = \frac{1}{\pi}\mathcal{G}(\mathrm{i}\Delta).$$

If we write,

$$\mathcal{K}(\mathrm{i}\Delta) = \mathcal{K}'(\Delta) - \mathrm{i}\mathcal{K}''(\Delta),$$

we get:

$$g(\Delta) = \frac{1}{\pi}\,\frac{\mathcal{K}'(\Delta)}{\mathcal{K}'^2(\Delta) + [\Delta - \mathcal{K}''(\Delta)]^2}; \tag{1.129a}$$

$$g'(\Delta) = \frac{1}{\pi}\,\frac{\Delta - \mathcal{K}''(\Delta)}{\mathcal{K}'^2(\Delta) + [\Delta - \mathcal{K}''(\Delta)]^2}. \tag{1.129b}$$

Inverting (1.128) we get:

$$\mathcal{K}(z) = [\mathcal{G}(z)]^{-1} - z; \tag{1.130a}$$

$$\mathcal{K}'(\Delta) = \frac{1}{\pi}\,\frac{g}{g^2 + g'^2}; \tag{1.130b}$$

$$\mathcal{K}''(\Delta) = \Delta - \frac{1}{\pi}\,\frac{g'}{g^2 + g'^2}. \tag{1.130c}$$

We can see that, even granted the usefulness of the memory function for the study of line shapes, its introduction through (1.126*a*) by the Liouville formalism is not necessary since it can be defined directly through (1.130*a*).

2. *Gaussian approximation for the memory function*

The only rigorous information available on $G(t)$ is an expansion in powers of t originating in (1.45):

$$G(t)=1-\frac{M_2}{2!}t^2+\frac{M_4}{4!}t^4-\frac{M_6}{6!}t^6+\cdots, \tag{1.131}$$

which through (1.127*a*) leads to an expansion in powers of $1/z$ for $\mathcal{G}(z)$:

$$\mathcal{G}(z)=\frac{1}{z}\left(1-\frac{M_2}{z^2}+\frac{M_4}{z^4}+\cdots\right). \tag{1.132}$$

This through (1.130*a*) leads to a similar expansion for $\mathcal{K}(z)$:

$$\mathcal{K}(z)=\frac{M_2}{z}\left(1-\frac{N_2}{z^2}+\frac{N_4}{z^4}+\cdots\right), \tag{1.133}$$

where N_2 and N_4 are identified as:

$$N_2=M_2(\mu-1); \tag{1.134a}$$

$$N_4=M_2^2(\nu-2\mu+1); \tag{1.134b}$$

with

$$\mu=M_4/(M_2)^2; \qquad \nu=M_6/(M_2)^3. \tag{1.134c}$$

From (1.133) we get for $K(t)$ an expansion in powers of t:

$$K(t)=M_2\left(1-\frac{N_2t^2}{2!}+\frac{N_4}{4!}t^4+\cdots\right). \tag{1.135}$$

A very simple approximation (Lado, Memory, and Parker, 1971) is to take for $K(t)$ a Gaussian form,

$$K(t)=M_2\exp-\left(\frac{N_2t^2}{2}\right), \tag{1.136}$$

where M_2 and N_2 are defined by (1.131) and (1.134).

To this very simple expression for $K(t)$ there are corresponding forms for $\mathcal{K}'(\Delta)$ and $\mathcal{K}''(\Delta)$ through (1.127*b*):

$$\mathcal{K}'(\Delta)=\sqrt{\left(\frac{\pi}{2}\right)}\frac{M_2}{\sqrt{N_2}}\exp(-\Delta^2/2N_2); \tag{1.137a}$$

$$\mathcal{K}''(\Delta)=\frac{M_2\Delta}{N_2}\exp(-\Delta^2/2N_2)F(\tfrac{1}{2},\tfrac{3}{2};\Delta^2/2N_2); \tag{1.137b}$$

where $F(\frac{1}{2},\frac{3}{2};x^2)$ is the confluent hypergeometrical function.

The Gaussian approximation (1.136) for $K(t)$ yields through (1.129) an expression for the line shape $g(\Delta)$ with the right 2nd and 4th moments. For brevity we shall call this shape the 'Gaussian memory shape' (not to be confused with a Gaussian shape!).

The 'Gaussian memory shape' $g(\Delta)$ can be considered as one step beyond the Gaussian approximation for $G(t)$ which corresponds to a Gaussian shape $g(\Delta)$:

$$g(\Delta)=(2\pi M_2)^{-1/2}\exp-(\Delta^2/2M_2).$$

This Gaussian shape, which has the right second moment is expected to be a reasonable approximation to the exact theoretical shape, (unknown in practice) when the theoretical value of $\mu=M_4/M_2^2$ (eqn (1.134c)) is in the neighbourhood of 3.

On the other hand, the general form of the 'Gaussian memory shape' depends considerably on the magnitude of μ.

For large μ, $N_2=M_2(\mu-1)$ is large and $K(t)$ given by (1.136) decays rapidly, far more rapidly than $G(t)$.

We can then replace the eqn (1.125) by:

$$\frac{dG}{dt}=-G(t)\int_0^\infty K(\tau)\,d\tau=-\sqrt{\left(\frac{\pi}{2}\right)}\,\frac{M_2}{\sqrt{N_2}}G(t). \qquad (1.125')$$

$G(t)$ becomes an exponential and its cosine Fourier transform $g(\Delta)$ becomes a Lorentzian obtained by replacing $\mathcal{K}'(\Delta)$ and $\mathcal{K}''(\Delta)$ in (1.129a) by

$$\mathcal{K}'(0)=\sqrt{\left(\frac{\pi}{2}\right)}\,\frac{M_2}{\sqrt{N_2}} \quad\text{and}\quad \mathcal{K}''(0)=0.$$

The half-intensity half-width of $g(\Delta)$ is then:

$$\delta=\mathcal{K}'(0)=\sqrt{\left(\frac{\pi}{2}\right)}\left(\frac{M_2}{\mu-1}\right)^{1/2}\simeq\sqrt{\left(\frac{\pi}{2}\right)}\left(\frac{M_2}{\mu}\right)^{1/2}. \qquad (1.138)$$

This is comparable to the width predicted by the model of the truncated Lorentzian (Abragam (1961), p. 107):

$$\delta'=\frac{\pi}{2\sqrt{3}}\left(\frac{M_2}{\mu}\right)^{1/2}\simeq 1.35\delta.$$

Not too much importance should be attached to the numerical constant of order unity in front of $(M_2/\mu)^{1/2}$.

To illustrate the situation when μ is not a large number, Fig. 1.1 shows the shapes relative to the resonance of ^{19}F in CaF_2 with the field along [100] when $\mu=2.07$. The three curves show:

(a) the experimental shape;
(b) a Gaussian shape with the correct M_2;

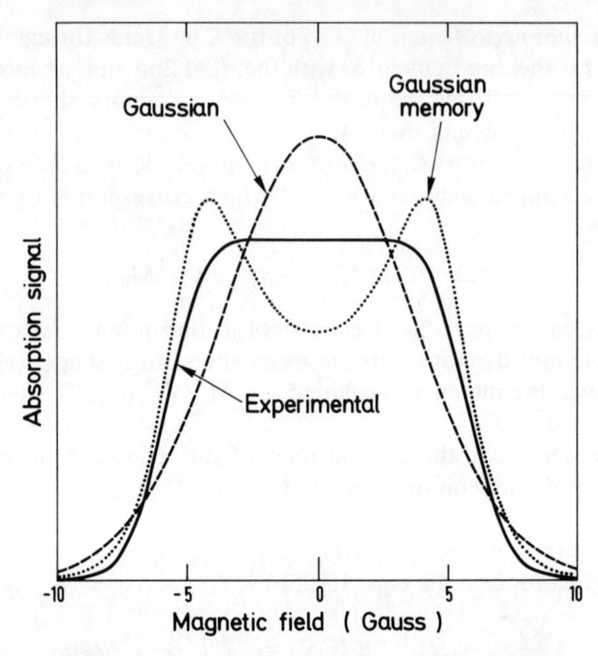

FIG. 1.1. Normalized absorption signal of ^{19}F in CaF_2, with H_0 parallel to [100]. Comparison of the experimental shape with the Gaussian and Gaussian memory approximations.

(c) a 'Gaussian memory shape' with the correct M_2 and M_4 which differs appreciably from the experimental curve in shape if not in width.

The main merit of the 'Gaussian memory shape' is to give an expression for the line width, approximately valid over a considerable range of values of μ and of line widths. From the eqns (1.129*a*) we find that the half-intensity half-width is defined implicitly by:

$$\delta = \mathcal{K}''(\delta) + [2\mathcal{K}'(0)\mathcal{K}'(\delta) - \mathcal{K}'^2(\delta)]^{1/2}.$$

Replacing $\mathcal{K}'$ and $\mathcal{K}''$ by their expressions (1.137) and introducing the variable,

$$x = \delta(2N_2)^{-1/2}, \tag{1.139a}$$

yields:

$$\delta = \sqrt{\left(\frac{\pi}{2}\right)}\frac{M_2}{\sqrt{N_2}}f(x), \tag{1.139b}$$

where:

$$f(x) = [2 - \exp(-x^2)]^{1/2}\exp(-x^2/2) + \frac{2x}{\sqrt{\pi}}\exp(-x^2)F(\tfrac{1}{2}, \tfrac{3}{2}; x^2). \tag{1.139c}$$

If, on the other hand we replace δ on the left-hand side of (1.139*b*) by $(2N_2)^{1/2}x$, it yields:

$$\frac{N_2}{M_2} = (\mu - 1) = \sqrt{\left(\frac{\pi}{2}\right)}\frac{f(x)}{x},$$

from which we can extract $x = x_0(\mu)$ which upon being fed into $f(x)$ given by (1.139*c*) yields a function:

$$f[x_0(\mu)] \equiv h(\mu).$$

$h(\mu)$ is plotted in Fig. 1.2. It tends toward unity when $\mu \to \infty$ and has a maximum value of 1.5 for $\mu \simeq 3$. It is then not a bad approximation to take $f[x_0(\mu)] \equiv h(\mu) = \text{const.}$ whence from (1.139*b*):

$$\delta \simeq \sqrt{\left(\frac{\pi}{2}\right)}[M_2/(\mu - 1)]^{1/2}. \tag{1.140}$$

An even better approximation is introduced by defining

$$g(\mu) = \left(\frac{\mu - 2}{\mu - 1}\right)^{1/2} h(\mu), \tag{1.141}$$

also plotted on Fig. 1.2 and very nearly unity, whatever μ. This leads to an approximate δ by

$$\delta = \sqrt{\left(\frac{\pi}{2}\right)}[M_2/(\mu - 2)]^{1/2}. \tag{1.142}$$

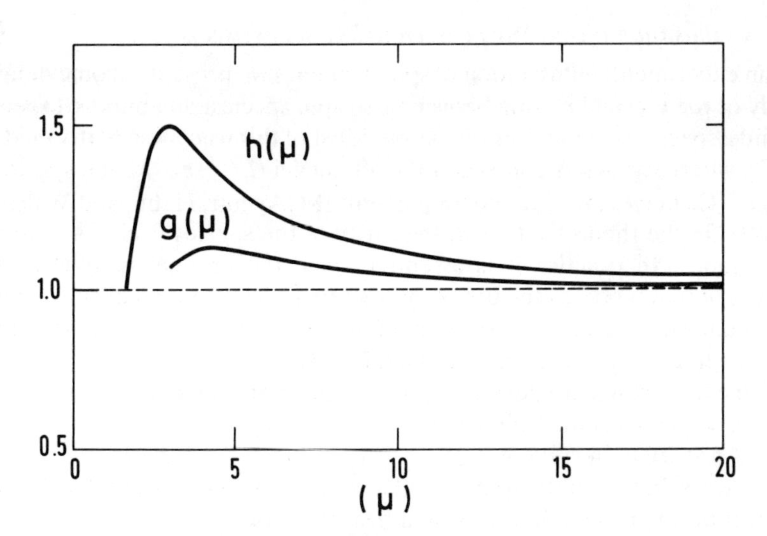

FIG. 1.2. Line width correction functions $h(\mu)$ and $g(\mu)$ versus moment ratio $\mu = M_4/M_2^2$ (see text). (After Mehring, 1976.)

Considering the arbitrariness of the 'Gaussian memory shape' assumption it seems safer to limit one's ambitions to the prediction of the line width by means of (1.142).

It is possible to pursue further the approximation process for the line shape. Instead of $\mathscr{K}(z)$ introduced by $\mathscr{G}(z)=[z+\mathscr{K}(z)]^{-1}$ one can introduce a new function $\mathscr{L}(z)$ through:

$$\mathscr{G}(z)=[z+M_2/[z+\mathscr{L}(z)]]^{-1}, \tag{1.143}$$

where $\mathscr{L}(z)$ is the Laplace transform of a function $L(t)$ whose expansion in powers of t turns out to be:

$$L(t)=N_2\left(1-\frac{R_2}{2}t^2+\cdots\right),$$

with

$$R_2=N_2\left(\frac{N_4}{N_2^2}-1\right)=M_2(\mu-1)(\nu-\mu^2).$$

By using for $L(t)$ the Gaussian approximation,

$$L(t)=N_2\exp-\left(\frac{R_2t^2}{2}\right),$$

one obtains for $g(\Delta)$ a rather complicated analytical expression with correct values for the 2nd, 4th, and 6th moments. We shall make use of this approximation in chapter 3.

(*e*) *An illustration: coupling between unlike spin systems*

As an experimental illustration of spin dynamics we present in some detail a study of the thermal mixing between two spin species: an abundant species A and a species B of low concentration. Most of this study was performed on CaF_2, where species A consists of the ^{19}F nuclei ($I=\frac{1}{2}$, $c=1$) and species B of the ^{43}Ca nuclei ($S=\frac{7}{2}$, $c\simeq 0{\cdot}13$ percent) (McArthur, Hahn, and Walstedt 1969). On the theoretical level, the rarity of the species B is of secondary importance. In practice it is essential, since the original motivation for studies of that type was the use of indirect (trigger) detection of spins of low abundance through the observation of abundant spins, a method imagined and developed by Hartmann and Hahn (1962).

The experiments are performed in high magnetic field. The basic experimental sequence is the following:

1. An ADRF on spins A.
2. An excitation of the spins B under conditions which establish a thermal contact between spins B and A, and heat the latter.
3. A measurement of the inverse temperature β of the secular dipolar reservoir by a $\pi/4$ pulse on the spins A as explained in **A**(*c*)1, eqn (1.44).

Any decrease of β larger than that due to spin-lattice relaxation is caused by the processes of step 2.

The excitation used in these experiments is a series of r.f. pulses at a frequency close to the Larmor frequency of spins B.

First we analyse the evolution of the system during a pulse. Then we describe the experimental sequence. Finally we discuss the experimental results.

1. *The basic equations*

In the presence of an r.f. field H_1 of frequency ω close to the Larmor frequency $\omega_B = -\gamma_B H_0$ of spins B, we use for the spins B a frame rotating at frequency ω, and for spins A a frame rotating at the frequency $\omega_A = -\gamma_A H_0$. The effective Hamiltonian in these frames is:

$$\mathcal{H} = \mathcal{H}'_D + Z_B;$$

where:

$$Z_B = \omega_e S_Z,$$

is the effective Zeeman interaction of the spins B in the effective field H_e. This field has components $(\omega - \omega_B)/\gamma_B$, along H_0, and H_1 at right angles to it. Its direction $0Z$ makes with H_0 and angle: $\theta = \tan^{-1}[\gamma_B H_1/(\omega - \omega_B)]$. $\mathcal{H}'_D$ is the spin–spin interaction:

$$\mathcal{H}'_D = \mathcal{H}'_{AA} + \mathcal{H}'_{AB} + \mathcal{H}'_{BB}. \tag{1.144}$$

The spins B being very diluted, we have:

$$\mathrm{Tr}\{\mathcal{H}'^2_{BB}\} \ll \mathrm{Tr}\{\mathcal{H}'^2_{AB}\} \ll \mathrm{Tr}\{\mathcal{H}'^2_{AA}\}, \tag{1.145}$$

and we can neglect altogether the dipolar coupling $\mathcal{H}'_{BB}$ between spins B:

$$\mathcal{H}'_D \simeq \mathcal{H}'_{AA} + \mathcal{H}'_{AB}. \tag{1.146}$$

Furthermore we choose experimental conditions for the pulse such that:

$$Z_B \gg \mathcal{H}'_{AB} \tag{1.147}$$

The Hamiltonian can then be split into a main Hamiltonian $\mathcal{H}_0$ and a perturbation V:

$$\mathcal{H}_0 = \mathcal{H}'_d + Z_B = \mathcal{H}'_d + \omega_e S_Z. \tag{1.148}$$

$\mathcal{H}'_d$ is the secular part of $\mathcal{H}'_D$: it is the sum of $\mathcal{H}'_{AA}$ and the part of $\mathcal{H}'_{AB}$ that commutes with S_Z. According to (1.145) $\mathcal{H}'_d$ differs little from $\mathcal{H}'_{AA}$.

The perturbation $V = \mathcal{H}'_D - \mathcal{H}'_d$ is the part of $\mathcal{H}'_{AB}$ that does not commute with S_Z. It is of the form:

$$V = \hbar\gamma_A\gamma_B \sum_{i,\mu} \frac{1 - 3\cos^2\theta_{i\mu}}{r_{i\mu}{}^3}\, I^i_z S^\mu_X \sin\theta. \tag{1.149}$$

According to (1.145) and (1.146), the perturbation V is much smaller than $\mathscr{H}'_d$ and Z_B. We can then use the formalism developed in section **B**(c), with the choice:

$$\mathscr{H}_1 = Z_B;$$

and:

$$\mathscr{H}_2 = \mathscr{H}'_d.$$

Before the application of the r.f. field, the density matrix is:

$$\sigma_i = 1 - \beta_A(\mathscr{H}'_{AA} + \mathscr{H}'_{AB}) = 1 - \beta_A(\mathscr{H}'_d + V). \tag{1.150}$$

When the field H_1 is suddenly switched on, the magnetization $\langle S_Z \rangle$ of the spins B undergoes an evolution in two steps: short-term transient behaviour which, as will be seen, involves oscillations damped in a time of the order of T_2, followed by a much slower evolution described by Provotorov equations.

The function $K_{11}(t)$ which determines the dynamics of $\langle S_Z \rangle$ is according to (1.102):

$$K_{11}(t) = \mathrm{Tr}\{[S_Z, V] \exp(-i\mathscr{H}_0 t)[V, S_Z] \exp(i\mathscr{H}_o t\}/\mathrm{Tr}\{S_Z^2\}. \tag{1.151}$$

With the help of (1.148) and (1.149) and a little algebra, one finds:

$$K_{11}(t) = \frac{\mathrm{Tr}\{V^2\}}{\mathrm{Tr}\{S_Z^2\}} G(t) \cos(\omega_e t), \tag{1.152}$$

with,

$$G(t) = \mathrm{Tr}\{\exp(-\mathscr{H}'_d t) V \exp(i\mathscr{H}'_d t) V\}/\mathrm{Tr}\{V^2\}. \tag{1.153}$$

This function is damped in a time of the order of T_2.

The short-term behaviour of $\langle S_Z \rangle$ is, according to (1.115) given by:

$$\frac{d\beta_B}{dt} = \beta_A \frac{\mathrm{Tr}\{V^2\}}{\mathrm{Tr}\{S_Z^2\}} \int_0^t G(t) \cos(\omega_e t)\, dt, \tag{1.154}$$

which exhibits the existence for both $d\beta_B/dt$ and β_B, of an oscillation of frequency ω_e whose damping, due to that of $G(t)$, occurs in a time of the order of T_2. The value of $\beta_B(t)$ when $T_2 \ll t \ll W^{-1}$ is, according to (1.117):

$$\beta_B(t) = \beta_A W t - \beta_A \int_0^\infty \tau K_{11}(\tau)\, d\tau, \tag{1.155}$$

where:

$$W = \frac{\mathrm{Tr}\{V^2\}}{\mathrm{Tr}\{S_Z^2\}} \int_0^\infty G(t) \cos(\omega_e t)\, dt. \tag{1.156}$$

It appears from (1.149) that this probability is of the form:

$$W(\omega_e) = \sin^2 \theta f(\omega_e). \tag{1.157}$$

The constant term of (1.155) is:

$$\beta_0' = -\beta_A \frac{\mathrm{Tr}\{V^2\}}{\mathrm{Tr}\{S_Z^2\}} \int_0^\infty \tau G(\tau) \cos(\omega_e \tau)\, d\tau$$

$$= -\beta_A \frac{\mathrm{Tr}\{V^2\}}{\mathrm{Tr}\{S_Z^2\}} \frac{d}{d\omega_e} \int_0^\infty G(\tau) \sin(\omega_e \tau)\, d\tau. \qquad (1.158)$$

We consider only the case when $\omega_e T_2 \gg 1$. We can then use, for the integral on the right-hand side of (1.158), an expansion limited to the first order in ω_e^{-1}:

$$\int_0^\infty G(\tau) \sin(\omega_e \tau)\, d\tau \simeq 1/\omega_e, \qquad (1.159)$$

as obtained by integration by parts.

Equation (1.158) then becomes:

$$\beta_0' = \beta_A \,\mathrm{Tr}\{V^2\}/[\omega_e^2 \,\mathrm{Tr}\{S_Z^2\}] = \beta_A \,\mathrm{Tr}\{V^2\}/\mathrm{Tr}\{Z_B^2\}. \qquad (1.160)$$

This result can be interpreted as corresponding to a thermal equilibrium between Z_B and V (Goldman (1970), chapter 6).

The long term behaviour of β_B is governed by the Provotorov equations:

$$\frac{d\beta_B}{dt} = -W(\beta_B - \beta_A); \qquad (1.161a)$$

$$\frac{d\beta_A}{dt} = -\varepsilon W(\beta_A - \beta_B) - \frac{1}{T_{1A}}\beta_A; \qquad (1.161b)$$

where we suppose that the relaxation rate of spins B is negligible.

The coefficient ε, which is the ratio of the heat capacities of Z_B and $\mathcal{H}_d'$, ensures the conservation of the energy, $\langle \mathcal{H}_0 \rangle$, in the course of the thermal mixing:

$$\varepsilon \simeq \mathrm{TR}\{Z_B^2\}/\mathrm{Tr}\{\mathcal{H}_{AA}'^2\} = \frac{N_B}{N_A} \frac{\omega_e^2}{\gamma_A^2 H_{LA}'^2} \frac{S(S+1)}{I(I+1)}, \qquad (1.162)$$

where N_B/N_A is the ratio of spins B and A, and H_{LA}' is the local field of spins A in the rotating frame. Since $N_B/N_A \ll 1$, we have $\varepsilon \ll 1$ unless,

$$(\omega_e/\gamma_A H_{LA}')^2 \sim N_A/N_B,$$

a situation we dismiss here.

If the time t is much shorter than T_{1A}, the change in the value of β_A is very small and we may replace β_A in (1.161a) by its initial value. We have then, following the initial transient behaviour:

$$\beta_B \simeq \beta_0' + (\beta_A - \beta_0')[1 - \exp(-Wt)]. \qquad (1.163)$$

According to (1.160) we have $\beta_0' \ll \beta_A$ that is, most of the variation of β_B is exponential and described by (1.163). The short-term behaviour corresponds to $\beta_B \approx \beta_0'$, that is to very small values of $\langle S_Z \rangle$. They can however be measured experimentally and yield a detailed check of the theory.

A calculation similar to the previous one yields the variations of $\langle V \rangle$ and $\langle \mathcal{H}_d' \rangle$ in the course of the pulse. This calculation is found in Goldman (1970, chapter 6) and will not be reproduced here. A qualitative result of this calculation is that, although we have at all times $\langle \mathcal{H}_d' \rangle \gg \langle V \rangle$, we must distinguish two regimes:

(i) the short-term regime ($t \lesssim T_2$) where the *variation* of $\langle V \rangle$ is comparable to, or larger than, that of $\langle \mathcal{H}_d' \rangle$,

(ii) the long-term regime ($t \gg T_2$) where the variation of $\langle V \rangle$ is much less than that of $\langle \mathcal{H}_d' \rangle$ and where it is legitimate to neglect $\langle V \rangle$ and its variation, which is the procedure used for deriving the Provotorov eqns (1.161), with the coefficient ε given by eqn (1.162).

All we need in the present case is the fact that the *total* effective energy,

$$\langle \mathcal{H}_D' + Z_B \rangle = \langle \mathcal{H}_d' + V + Z_B \rangle,$$

remains constant during the pulse. The initial density matrix being of the form (1.150) we have $\langle Z_B \rangle(0) = 0$, whence:

$$\langle \mathcal{H}_D' \rangle(0) = \langle \mathcal{H}_D' \rangle(\tau) + \langle Z_B \rangle(\tau). \tag{1.164}$$

It is the variation of $\langle \mathcal{H}_D' \rangle$ which is measured experimentally, and it yields the value of $\langle Z_B(\tau) \rangle$ according to (1.164).

2. *The experimental sequence*

After the ADRF on the spins A, the heating sequence consists in a series of N pulses of duration τ, separated by time intervals $\tau_0 \gg T_2$.

We suppose first that the pulse frequency is equal to ω_B, so that the direction $0Z$ coincides with $0x$. We neglect for the moment the spin-lattice relaxation of the spins A.

At the beginning of a pulse, the density matrix is of the form (1.150). At the end of the pulse of duration τ, the magnetization $\langle S_Z \rangle \equiv \langle S_x \rangle$ of the spins B is proportional to β_A. This proportionality can be expressed by the relation:

$$\beta_B(\tau) = \beta_A \mu(\tau). \tag{1.165}$$

When the r.f. field is turned off, the Hamiltonian reduces to $\mathcal{H}_D' = \mathcal{H}_d' + V$ and two things happen during the time $\tau_0 \gg T_2$ preceding the next pulse:

(i) the magnetization $\langle S_x \rangle$ decays to zero;

(ii) the dipolar interactions $\mathcal{H}_D'$ reaches a state of thermal equilibrium at constant energy:

$$\langle \mathcal{H}_D' \rangle(\tau + \tau_0) = \langle \mathcal{H}_D' \rangle(\tau).$$

The density matrix is then again of the form (1.150), but with an inverse temperature β'_A slightly less than β_A. This decrease is deduced from (1.164) expressed in terms of β_A, β'_A and β_B:

$$\beta_A \operatorname{Tr}\{\mathcal{H}'^2_D\} = \beta'_A \operatorname{Tr}\{\mathcal{H}'^2_D\} + \beta_B \operatorname{Tr}\{Z^2_B\};$$

or else, according to (1.145) and (1.162):

$$\beta'_A \simeq \beta_A - \varepsilon\beta_B = \beta_A(1-\varepsilon\mu), \tag{1.166}$$

where we have used (1.165).

After N pulses we have then:

$$\beta_A(N) = \beta_A(0)(1-\varepsilon\mu)^N \simeq \beta_A(0)\exp(-N\varepsilon\mu).$$

Taking into account spin-lattice relaxation, if t is the time elapsed between the ADRF and the measurement of β_A, we now have:

$$\beta_A(t) = \beta_A(0)\exp(-N\varepsilon\mu - t/T_{1A}), \tag{1.167}$$

from which we obtain the value of $\varepsilon\mu$. Although $\varepsilon\mu$ may be very small, it can be measured by using enough pulses to have $N\varepsilon\mu \simeq 1$.

Experimentally, it is necessary to choose:

$$N(\tau+\tau_0) \lesssim T_{1A}.$$

There are three distinct ranges of values of τ:

(i) When $\tau \gg W^{-1}$, the inverse temperatures β_A and β_B become equal, and we have according to (1.163), $\mu = 1$ and we obtain the value of ε.

(ii) The values of τ such that $T_2 \ll \tau \lesssim W^{-1}$ allow the measurement of the mixing rate W.

(iii) The values of τ comparable to T_2 make it possible to study the transient behaviour.

The experiment is more complicated when the r.f. frequency ω is not equal to ω_B. The magnetization $\langle S_Z \rangle$ has then along the external field a component $\langle S_z \rangle = \cos\theta \langle S_Z \rangle$, that will not decay between the pulses. It is then necessary to saturate this component by suitable irradiations (McArthur *et al.* 1969).

The internal consistency of the method was checked in the experiment on CaF_2 by verifying that, in accordance with (1.162) and (1.157) one finds indeed:

$$\varepsilon \propto \omega_e^2,$$

and, at constant ω_e:

$$W \propto \sin^2\theta.$$

3. *Analysis of the experimental results*

The most direct access to the function $G(t)$ is obtained by determining the mixing rate $W(\omega_e)$, which is proportional to its Fourier transform. The rate $W(\omega_e)$ was measured in CaF_2 for two orientations of the external field: $H_0\|[110]$ and $H_0\|[111]$ ($\|$ stands for 'parallel to'). They yield in both cases a rather surprising result: the function $f(\omega_e)$ of (1.157) is, to an excellent approximation, an exponential function of the frequency ω_e:

$$f(\omega_e) = \xi \exp(-\lambda\omega_e), \tag{1.168}$$

as observed over more than two orders of magnitude of $f(\omega_e)$.

According to (1.156), this implies that $G(t)$ is a Lorentzian function of time:

$$G(t) = \lambda^2/(\lambda^2 + t^2). \tag{1.169}$$

This is a very remarkable behaviour for a correlation function, and very unusual in nuclear magnetism. The consistency of this result is verified by adjusting λ to the second derivative at the origin of $G(t)$, and then ξ in (1.168) to the expression (1.156):

$$\left.\frac{d^2G}{dt^2}\right|_{t=0} = \frac{-\mathrm{Tr}\{[V, \mathscr{H}'_{AA}]^2\}}{\mathrm{Tr}\{V^2\}} = \frac{-\mathrm{Tr}\{[\mathscr{H}'_{AB}, \mathscr{H}'_{AA}]^2\}}{\mathrm{Tr}\{\mathscr{H}'^2_{AB}\}} = 2/\lambda^2, \tag{1.170}$$

and, according to (1.156), (1.157), and (1.169):

$$\xi = \frac{\pi}{2}\lambda\, \mathrm{Tr}\{\mathscr{H}'^2_{AB}\}/\mathrm{Tr}\{S_z^2\}. \tag{1.171}$$

The result of this adjustment is shown in Fig. 1.3 for the orientation $H_0\|[111]$, together with the experimental points. The agreement is satisfactory.

This exponential variation with frequency, of the rate of thermal mixing between the effective Zeeman interaction of a rare spin species and the dipolar interaction of an abundant one, is not limited to the system ^{43}Ca–^{19}F in CaF_2. The same behaviour has been observed in several substances:

system ^{13}C–^{19}F in CF_3COOAg (Mehring, Raber, and Sinning, 1974);
system ^{13}C–^{1}H in adamantane (Pines and Shattuck, 1974);
system ^{6}Li–^{7}Li in metallic lithium (Stokes and Ailion, 1977).

These last authors also obtain a fair account for the rate of thermal mixing between the effective Zeeman interaction of ^{6}Li, and the dipolar reservoir of ^{7}Li and ^{19}F in LiF (Lang and Moran, 1970), by adjusting separately to Lorentzian functions the auto-correlation functions of (^{6}Li–^{7}Li) and (^{6}Li–^{19}F).

In an attempt to go beyond the mere adjustment of the *experimental* curve to a single theoretical parameter of $G(t)$, namely its second derivative at

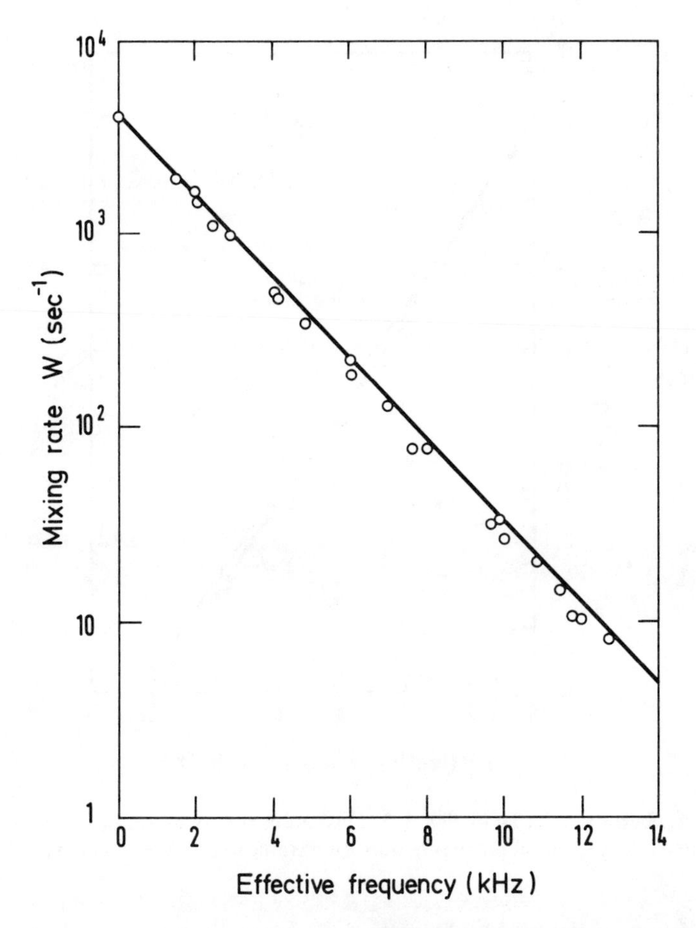

FIG. 1.3. Mixing rate between ^{43}Ca effective Zeeman reservoir and dipole–dipole reservoir versus ^{43}Ca effective Zeeman frequency in CaF_2 with H_0 parallel to [111]. The solid curve is an exponential fit (see text). (After McArthur *et al.*, 1969.)

$t = 0$, Demco, Tegenfeld, and Waugh (1975) have used the Gaussian approximation for the memory function of $G(t)$, adjusted to fit two theoretical parameters of $G(t)$, its second and fourth derivatives at $t = 0$. We saw in **B**(d)2 that a Gaussian shape for the memory function can give rise to a wide variety of shapes for the associated correlation function. In the present case, what distinguishes this procedure from the former is that it does not make any *a priori* reference to the experimental shape. The calculation was performed for the two orientations $H_0 \| [110]$ and $H_0 \| [111]$ in CaF_2. The agreement with the experimental results is remarkable. The calculated and experimental variations are shown in Fig. 1.4 for the orientation $H_0 \| [111]$.

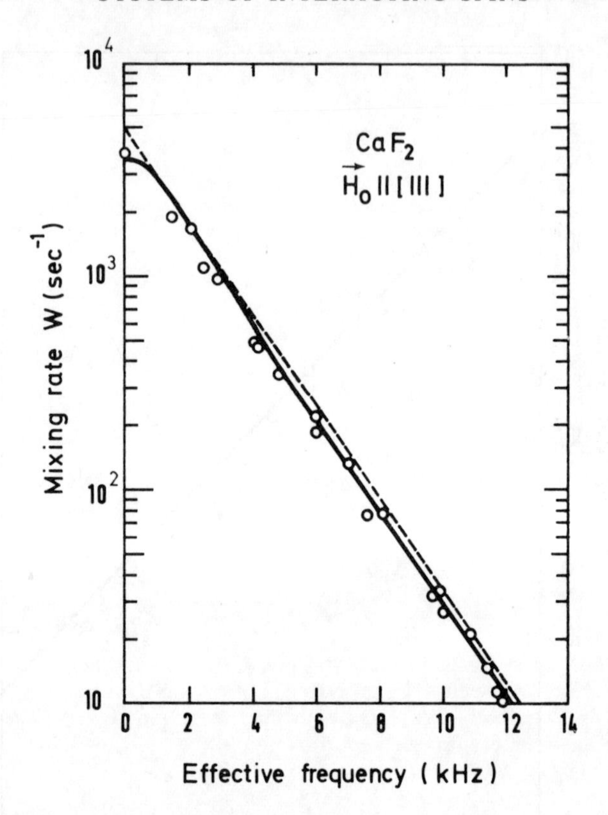

FIG. 1.4. Same results as in Fig. 1.3. Dashed curve: exponential approximation. Solid curve: Gaussian memory function approximation. (After Demco *et al.*, 1975.)

The function derived by Demco *et al. is not* an exponential, but is very close to it in the frequency interval experimentally studied. This agreement could be considered as a fortuitous consequence of the value of $M_4/(M_2^2)$, where M_2 and M_4 are the second and fourth derivatives of $G(t)$ at $t = 0$, (that is, the 2nd and 4th moments of $f(\omega)$). This ratio, calculated from lattice sums, is respectively 6·6 and 5·9 for H_0 parallel to [110] and H_0 parallel to [111] in CaF_2, whereas for an exponential function $f(\omega)$ it is equal to 6.

A further check of these approximations is provided by a comparison of the experimental short-term transient variation of $\langle S_Z \rangle$, with that computed from (1.154), using for $G(t)$ either the Lorentzian shape (1.169) or that derived by Demco *et al.* The results are shown in Fig. 1.5. The agreement is very reasonable. The approximation of Demco *et al.* is slightly better than the Lorentzian approximation.

What this agreement reveals is that in the present case, the memory function of $G(t)$ is actually close to a Gaussian.

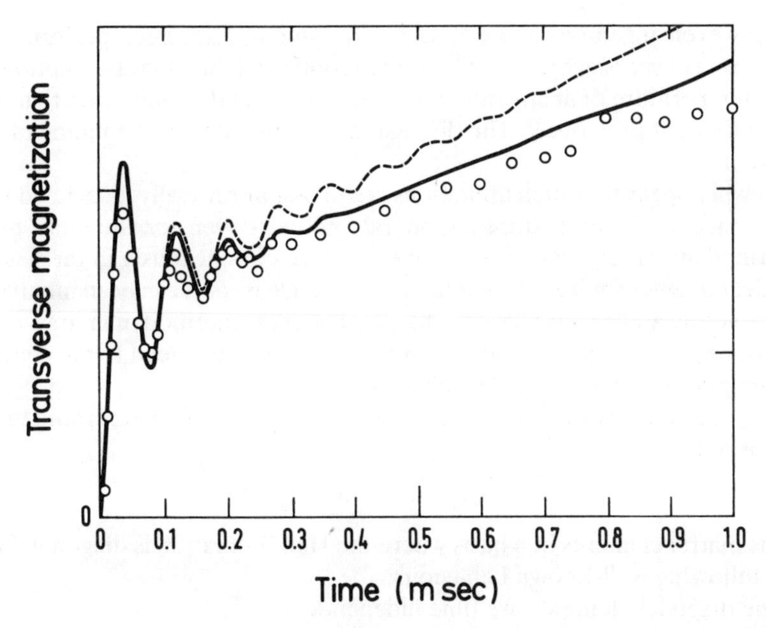

FIG. 1.5. Short-term transient variation of ^{43}Ca transverse magnetization during thermal coupling between ^{43}Ca effective Zeeman reservoir and dipole–dipole reservoir in CaF_2 with $H_0 \| [111]$. Dashed curve: exponential approximation. Solid curve: Gaussian memory function approximation. (After Demco *et al.*, 1975.)

However striking the present success of the Gaussian approximation for the memory function of $G(t)$, one should remember that this approximation does not arise from any real theory. There is no fundamental argument as to why this memory function should be Gaussian, and no way to tell how good this approximation would be in another case.

In nuclear magnetism where as a rule correlation functions have rather complicated shapes, it is a noticeable fact that in the present problem such simple mathematical functions as the Lorentzian and the Gaussian fit so well, respectively the shape of $G(t)$ and that of its memory function.

C. On the validity of the concept of spin temperature

(*a*) *The random phase assumption*

The concept of spin temperature has over the years proved extremely beneficial to the development of nuclear magnetism in solids. The validity of the underlying assumptions has been verified in many instances not only in the domain of high temperature considered so far, but also at very low spin temperatures as will appear in chapters 5 and 8.

However, ingenious and elaborate experiments have been performed in more recent years, whose results seem to contradict the usual assumptions of spin temperature or at any rate to set limits to its validity, narrower than had been thought previously. The discussion of these limits is the theme of this section.

It will appear that such limitations are in fact intrinsically associated with the concept of temperature in general and exist even for the more usual thermodynamic systems. If they emerge more conspicuously in the field of nuclear magnetism it is because interacting nuclear spins can be manipulated more subtly and observed more sharply than most thermodynamic systems.

We begin by recalling briefly some of the steps which lead to the concept of temperature in statistical mechanics.

(i) The density matrix of a macroscopic system evolves according to the equation:

$$\sigma(t) = \exp(-\mathrm{i}\mathcal{H}t)\sigma(0)\exp(\mathrm{i}\mathcal{H}t). \tag{1.172}$$

Its matrix elements in a basis where the Hamiltonian $\mathcal{H}$ is diagonal, have the following well-known behaviour:

the diagonal elements are time independent,

$$\langle i|\sigma(t)|i\rangle = \langle i|\sigma(0)|i\rangle; \tag{1.173}$$

the off-diagonal elements oscillate without damping,

$$\langle i|\sigma(t)|j\rangle = \langle i|\sigma(0)|j\rangle \exp\{-\mathrm{i}(\mathcal{H}_i - \mathcal{H}_j)t\}. \tag{1.174}$$

(ii) The expectation value of an operator Q with *no* diagonal matrix elements is a sum of oscillating terms:

$$\langle Q\rangle = \sum \langle i|\sigma(0)|j\rangle\langle j|Q|i\rangle \exp\{-\mathrm{i}(\mathcal{H}_i - \mathcal{H}_j)t\}. \tag{1.175}$$

When the frequencies $\mathcal{H}_i - \mathcal{H}_j$ form a quasi-continuous spectrum the destructive interference of the various terms leads to a decay of $\langle Q\rangle$ towards zero in a time T_2 of the order of the inverse width of the frequency spectrum.

Thus, if at time $t = 0$ the system was prepared in a state where the expectation values $\langle Q\rangle$ of off-diagonal operators Q are different from zero, the *assumption* is made that after a time of the order of several times T_2 it is permissible to 'forget' the off-diagonal part of the density matrix $\sigma(t)$ and to replace it by a 'truncated' density matrix σ_d reduced to the diagonal (and time-independent part) of $\sigma(t)$.

(iii) For macroscopic operators with diagonal matrix elements only, it is then assumed that the truncated matrix σ_d can be represented with good accuracy, (the better the larger the number of degrees of freedom of the system), by the canonical density matrix,

$$\sigma(\beta) = \exp(-\beta\mathcal{H})/\mathrm{Tr}\{\exp(-\beta\mathcal{H})\}, \tag{1.176}$$

where β is chosen to give the same value for the total energy E of the system as the original density matrix σ_d (or $\sigma(0)$, or $\sigma(t)$ which all give the same result).

$$\mathrm{Tr}\{\mathcal{H}\sigma(t)\} = \mathrm{Tr}\{\mathcal{H}\sigma_d\} = \mathrm{Tr}\{\mathcal{H}\sigma(0)\} = E. \quad (1.177)$$

We shall not examine here the validity of the replacement of $\sigma(t)$ by σ_d. This problem, well known in statistical mechanics has been discussed in the literature at great length and over many years and there is nothing in the nature of interacting spin systems to warrant a more specialized discussion.

On the other hand the random phase approximation where the off-diagonal part of the density matrix is 'forgotten' has special aspects for spin systems and deserves more careful consideration.

A first point is that in the usual thermodynamic systems the time required to perform a measurement is in general far longer than the time constant T_2 for the randomization of the phases. Thus for every physical quantity one usually measures a time-average to which the oscillating off-diagonal matrix elements of $\sigma(t)$ bring a negligible contribution.

The originality of the nuclear spin systems resides in part in the fact that the measurements can be performed on a time scale far shorter than T_2: oscillatory behaviour of longitudinal or precessing magnetization and of other quantities such as dipolar energy can be observed *without* the time averaging which in most thermodynamical systems wipes out the oscillations. This puts more stringent conditions on the validity of the random phase approximation for nuclear spin systems than for conventional systems.

We have already discussed in detail in section **B**(e) an experiment which illustrates the decay of the oscillations in the trend toward an equilibrium or quasi-equilibrium regime described by a spin temperature (or temperatures).

A second example, of a somewhat different nature, which demonstrates that the establishment of a spin temperature is not instantaneous is given below. We then describe a very remarkable experiment which shows how the effects of the off-diagonal matrix elements of $\sigma(t)$, apparently buried in the graveyard of spin temperature can be brought to life again.

(b) *The trend toward spin temperature: production and observation of dipolar order in high field: An example*

We had stated in section **A**(c)1 that in high field whatever the initial conditions a quasi-equilibrium state of the spin system was expected to occur after a time interval of the order of a few times T_2, described by the density matrix (1.29):

$$\sigma = 1 - \beta_Z \omega_0 I_z - \beta \mathcal{H}'_D. \quad (1.178)$$

In such a state the expectation values $\langle Z\rangle=\langle\omega_0 I_z\rangle$ and $\langle\mathcal{H}'_{\mathrm{D}}\rangle$ are given respectively by:

$$\langle Z\rangle=-\beta_{\mathrm{Z}}\omega_0^2\,\mathrm{Tr}\{I_z^2\}\qquad\langle\mathcal{H}'_{\mathrm{D}}\rangle=-\beta\,\mathrm{Tr}\{\mathcal{H}'^2_{\mathrm{D}}\}.\tag{1.179}$$

β_{Z} and β can be measured from the signals $\langle I_x(t)\rangle$ and $\langle I_y(t)\rangle$, following a pulse of angle θ around 0x (in the rotating frame) given by (1.43) and (1.44). The ratio of these signals is:

$$\frac{\langle I_x(t)\rangle}{\langle I_y(t)\rangle}=\frac{\beta}{\beta_{\mathrm{Z}}}\cos\theta\frac{1}{\omega_0}\frac{1}{G}\frac{\mathrm{d}G}{\mathrm{d}t}\approx\frac{1}{\omega_0 T_2}\frac{\beta}{\beta_{\mathrm{Z}}}.\tag{1.180}$$

Since $\omega_0 T_2\gg 1$ in high field, a dipolar order can only be observed through $\langle I_x(t)\rangle$ if $|\beta/\beta_{\mathrm{Z}}|\gg 1$, that is if the dipolar energy is much colder than the Zeeman energy, hardly a surprising result, since $\mathcal{H}'_{\mathrm{D}}\ll Z$. One of the methods for achieving the dipolar cooling is the ADRF described in **A**(b)3. We recall here another method due to Jeener and Broekaert (1967), (and described in Goldman, 1970), because it serves as an illustration of the trend of the system to the quasi-equilibrium state (1.29). The procedure is the following.

Starting from $\sigma_0=1-\beta_{\mathrm{Z}}^0(\omega_0 I_z+\mathcal{H}'_{\mathrm{D}})\approx 1-\beta_{\mathrm{Z}}^0\omega_0 I_z$, a $(\pi/2)$ pulse around 0y is applied, followed τ seconds later by a $(\pi/4)$ pulse around 0x. t seconds after the second pulse the density matrix is:

$$\begin{aligned}\sigma(\tau,t)=1-\beta_{\mathrm{Z}}^0\omega_0\exp(-\mathrm{i}\mathcal{H}'_{\mathrm{D}}t)\exp(-\mathrm{i}(\pi/4)I_x)\exp(-\mathrm{i}\mathcal{H}'_{\mathrm{D}}\tau)I_x\\ \times\exp(\mathrm{i}\mathcal{H}'_{\mathrm{D}}\tau)\exp(\mathrm{i}(\pi/4)I_x)\exp(\mathrm{i}\mathcal{H}'_{\mathrm{D}}t).\end{aligned}\tag{1.181}$$

The expectation values of Z and $\mathcal{H}'_{\mathrm{D}}$ at time $(t+\tau)$,

$$Z(\tau,t)=\omega_0\,\mathrm{Tr}\{I_z\sigma(\tau,t)\},\qquad\langle\mathcal{H}'_{\mathrm{D}}(\tau,t)\rangle=\mathrm{Tr}\{\mathcal{H}'_{\mathrm{D}}\sigma(\tau,t)\},\tag{1.182}$$

are easily calculated using the transformation properties of $\mathcal{H}'_{\mathrm{D}}$ following from (1.11a) and (1.9b). They are:

$$\langle Z(\tau,t)\rangle=0;\tag{1.183a}$$

$$\langle\mathcal{H}'_{\mathrm{D}}(\tau,t)\rangle=\beta_{\mathrm{Z}}^0\frac{\omega_0}{2}\mathrm{Tr}\{I_x^2\}\left(\frac{\mathrm{d}G}{\mathrm{d}t}\right)_{t=\tau}.\tag{1.183b}$$

We see that at *any time* after the second pulse these expectation values, independent of t, are according to (1.179) and (1.183) rigorously the same as if the density matrix at that time were the spin temperature matrix (1.178) with:

$$\beta_{\mathrm{Z}}=0;\qquad\beta=-\beta_{\mathrm{Z}}^0\frac{\omega_0}{2D^2}\left(\frac{\mathrm{d}G}{\mathrm{d}t}\right)_{t=\tau}.\tag{1.184}$$

Can we conclude that at a time t after the second pulse we can replace the *real* density matrix (1.181) by the spin temperature matrix (1.178), with

$\beta_Z = 0$, which gives for *any* t the same expectation values of Z and $\mathcal{H}'_D$? The answer is yes if the time t is long compared to T_2, no if t is much shorter than T_2.

The first answer is based solely on experiment. After *preparing* the spin system by a pulse $(\pi/2, 0y)$ and a pulse $(\pi/4, 0x)$ separated by a time τ it is possible to *look* at it by a pulse $(\theta, 0x)$ t seconds later. When $t \gg T_2$ it has indeed been found experimentally (Jeener and Broekaert, 1967) that $\langle I_y \rangle = 0$ and that the time-dependence of $\langle I_x(t) \rangle$ is given by (1.44) where β has the value given by (1.184):

$$\langle I_x(t') \rangle = -\beta_Z^0 \frac{\omega_0}{2D^2} \sin\theta \cos\theta \operatorname{Tr}\{I_x^2\} \left(\frac{dG}{dt}\right)_{t=\tau} \left(\frac{dG}{dt}\right)_{t=t'}. \tag{1.185}$$

Its dependence on τ and t' is $(dG/d\tau)(dG/dt')$.

The second answer is a straightforward consequence of quantum mechanics. If the observation pulse $(\theta, 0x)$ occurs shortly after the pulse $(\pi/4, 0x)$, then $\sigma(\tau, t)$ does *not* behave at all as a spin temperature matrix. As an example assume that the observation pulse $(\theta, 0x)$ occurs immediately after the pulse $(\pi/4, 0x)$ and that θ is given the value $-\pi/4$. The two pulses cancel each other and t' seconds later:

$$\begin{aligned}\langle I_x(t') \rangle &= -\beta_Z^0 \omega_0 \operatorname{Tr}\{I_x \exp(-i\mathcal{H}'_D t') \exp(-i\mathcal{H}'_D \tau) I_x \exp(i\mathcal{H}'_D \tau) \exp(i\mathcal{H}'_D t')\} \\ &= -\beta_Z^0 \omega_0 \operatorname{Tr}\{I_x^2\} G(\tau + t'),\end{aligned} \tag{1.186}$$

a dependence on τ and t' quite different from (1.185). This discrepancy is due to the fact that while $\langle Z \rangle$ and $\langle \mathcal{H}'_D \rangle$ are unaffected by the off-diagonal matrix elements of $\sigma(\tau, t)$ in (1.181), the transverse component $\langle I_x \rangle$ is, and one must therefore wait for the randomization of the phases before attempting a measurement of β_Z and β.

(c) *Phase refocussing: the magic sandwich*

1. *The argument for irreversibility*

In the early years of magnetic resonance Erwin Hahn had performed an experiment, the spin echo, which to the uninitiated, had appeared little short of miraculous. The transverse nuclear magnetization of a liquid sample, precessing in an inhomogeneous magnetic field, once it had decayed to zero, could be restored to its full, or almost full value, by an appropriate r.f. pulse, and this after a time far longer than the decay time.

The principle of Hahn's spin echo is too well known to be discussed here in any detail. The gist of it is that the Larmor precession frequencies of the different parts of the sample, although spread over an interval $\Delta\omega$, whence their destructive interference and the magnetization decay in a time of the order of $(\Delta\omega)^{-1}$, are constant in *time*. A π pulse, by reversing at a time

$\tau \gg (\Delta\omega)^{-1}$ after the beginning of the precession all the phase angles accumulated by the various spins, brings them automatically into phase again at time 2τ, if the precession velocities have remained constant; whence comes the 'resurrection' of the magnetization.

On the contrary, in a solid where the decay of the nuclear magnetization is due to the dipolar interaction $\mathcal{H}'_{\mathrm{D}}$, this decay has been for a long time considered as completely irreversible.

The case for this irreversibility has rested on various arguments, some more convincing than others.

(i) It has been argued that in the solid each spin precesses in the local field of its neighbours. Because of flip-flops between the spins this field is not constant in time but fluctuates in a random and hence, irreversible manner: the accumulated phase lags or advances cannot be retrieved. One must beware of arguments of that type. It was a similar argument which, before the historical spin temperature experiment of Pound, had led to the belief that the loss of magnetization after a demagnetization into zero field was always irreversible: it was argued that in zero field fast random reversals of one spin, or of two spins in the same direction, (processes forbidden by energy conservation in high field but allowed in zero field), would destroy for ever the magnetization in a time T_2, a conclusion that contradicts both the experiment and the spin temperature theory.

(ii) A second argument is to state that because of the bilinear nature of the dipolar Hamiltonian $\mathcal{H}'_{\mathrm{D}}$ the refocussing effect of a pulse *à la* Hahn would be nil. This is quite true but only shows that the decayed precessing magnetization cannot be retrieved by *that* particular method.

(iii) By far the most cogent argument for irreversibility was the spin temperature assumption.

Immediately after a $\pi/2$ pulse around $0x$ the density matrix of the spin system is, in the rotating frame: $\sigma(0) = 1 - \alpha\omega_0 I_y$ and the transverse magnetization:

$$\langle I_y \rangle_0 = \mathrm{Tr}\{I_y \sigma(0)\} = -\alpha\omega_0 \,\mathrm{Tr}\{I_y^2\}. \qquad (1.187)$$

After a time $t \gg T_2$ the transverse magnetization has completely disappeared and according to the spin temperature assumption and to all the experimental evidence backing this assumption we expect $\sigma(t)$ to be represented with good accuracy by the expression: $\sigma = 1 - \beta_Z \omega_0 I_z - \beta\mathcal{H}'_{\mathrm{D}}$ where the quasi-invariants $\mathcal{H}'_{\mathrm{D}}$ and $Z = \omega_0 I_z$ have the same expectation values:

$$\langle Z \rangle = -\beta_Z \omega_0^2 \,\mathrm{Tr}\{I_z^2\}, \qquad \langle \mathcal{H}'_{\mathrm{D}} \rangle = -\beta \,\mathrm{Tr}\{\mathcal{H}'^2_{\mathrm{D}}\} \qquad (1.188)$$

as at the time $t = 0$:

$$\langle Z \rangle_0 = \omega_0 \,\mathrm{Tr}\{I_z \sigma(0)\} \qquad \langle \mathcal{H}'_{\mathrm{D}} \rangle = \mathrm{Tr}\{\mathcal{H}'_{\mathrm{D}} \sigma(0)\}. \qquad (1.189)$$

The latter however are clearly zero and so therefore are β_Z and β. The spin temperature density matrix should thus be the unit matrix and no magnetic signal could seemingly be extracted from it.

'Nothing will come out of nothing.' And yet $\cdots$.

2. *The magic sandwich*

After a pulse $(\pi/2, 0x)$ at time $t=0$ and once the precessing magnetization has disappeared presumably for ever, the following sequence of pulses called 'magic sandwich' is applied (Rhim *et al.* 1971).

(i) At time $T_A \gg T_2$ a pulse $(\pi/2, 0y)$ followed by the sudden application for a duration $T_B \gg T_2$ of a strong r.f. field $H_1 = -\omega_1/\gamma$ along $0x$.

(ii) At time $(T_A + T_B)$ the field H_1 is cut off and a pulse $(-\pi/2, 0y)$ applied.

After the magic sandwich sequence one looks for the appearance of a signal. Indeed if $T_B \gg T_2$, has been chosen greater than $2T_A$ a signal shaped like an echo of the initial decay, does appear at a time $T_C = \frac{1}{2}T_B - T_A$ after the magic sandwich!

The theory of this echo is actually quite straightforward. The unit operator $U_M(T_B)$ which corresponds to the magic sandwich is:

$$\begin{aligned} U_M(T_B) &= R_y^{-1}\left(\frac{\pi}{2}\right)\exp\{-iT_B(\omega_1 I_x + \mathscr{H}'_D)\}R_y\left(\frac{\pi}{2}\right) \\ &= \exp\left\{-iT_B R_y^{-1}\left(\frac{\pi}{2}\right)[\omega_1 I_x + \mathscr{H}'_D]R_y\left(\frac{\pi}{2}\right)\right\} \\ &= \exp\left\{-iT_B\left[\omega_1 I_z + R_y^{-1}\left(\frac{\pi}{2}\right)\mathscr{H}'_D R_y\left(\frac{\pi}{2}\right)\right]\right\}. \end{aligned} \quad (1.190)$$

According to (1.10*a*),

$$R_y^{-1}\left(\frac{\pi}{2}\right)\mathscr{H}'_D R_y\left(\frac{\pi}{2}\right) = -\tfrac{1}{2}\mathscr{H}'_D + G_2 + G_{-2}, \quad (1.191)$$

where G_2 and G_{-2} transform under rotation like the tensor operators T_2 and T_{-2}. If $\omega_1 I_z \gg G_2, G_{-2}, \mathscr{H}'_D$, that is if $H_1 \gg H'_L$ we can within first order perturbation theory disregard the effects of the off-diagonal terms $G_2 + G_{-2}$ whence:

$$U_M(T_B) \simeq \exp\{-i(-\tfrac{1}{2}\mathscr{H}'_D + \omega_1 I_z)T_B\}. \quad (1.192)$$

If furthermore we choose T_B such that $\omega_1 T_B = 2n\pi$ (which, as a more elaborate discussion shows, is not really necessary) $U_M(T_B)$ becomes:

$$U_M(T_B) \simeq \exp\{-i[-\tfrac{1}{2}\mathscr{H}'_D T_B]\}. \quad (1.193)$$

The magic sandwich has the uncanny property of reversing the effective sign of $\mathscr{H}'_D$, which is equivalent to making the time flow backwards (at half the normal speed). The evolution operator of the system at time $T_A+T_B+T_C$ becomes:

$$U(T_A+T_B+T_C)=\exp(-i\mathcal{H}'_D T_C)U_M(T_B)\exp(-i\mathcal{H}'_D T_A)$$
$$=\exp(-i\mathcal{H}'_D(T_C-\tfrac{1}{2}T_B+T_A)). \qquad (1.194)$$

The magic sandwich is thus capable of cancelling the sum of the phase $-i\mathcal{H}'_D T_A$ accumulated before it and of the phase $-i\mathcal{H}'_D T_C$ accumulated after it if:

$$T_A+T_C=T_B/2. \qquad (1.195)$$

It thus restores the initial density matrix $\sigma(0)$ and the initial signal $\langle I_y(0)\rangle$ (Fig. 1.6).

It would thus appear that a decayed precessing magnetization can always be retrieved whatever the time elapsed after its decay (within the limitations caused by spin-lattice relaxation). There are actually other limitations. The formula (1.192) for the effect of the magic sandwich is only approximate. The effective Hamiltonian for U_M is:

$$\mathcal{H}_{\text{eff}}=\omega_1 I_z-\tfrac{1}{2}\mathcal{H}'_D+G_2+G_{-2}. \qquad (1.196)$$

The off-diagonal term G_2+G_{-2} does induce a departure of $U_M(T_B)$ from the approximate expression (1.192).

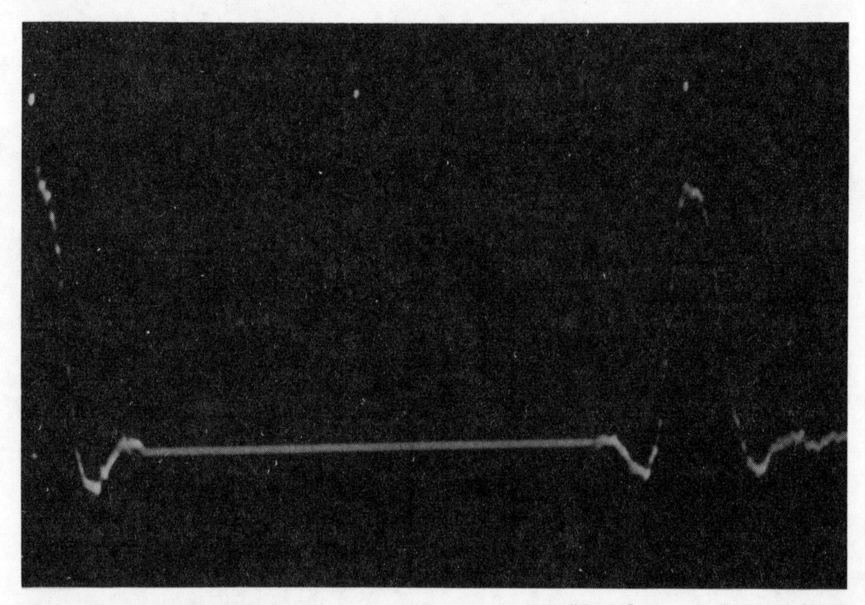

FIG. 1.6. 'Magic sandwich' experiment in CaF_2 with $H_0\|[111]$. Free induction decay and 'magic' echo of ^{19}F spins. (After Rhim *et al.*, 1971.)

The longer the time T_B during which it acts, the larger the departure of $U_M(T_B)$ from (1.192) and either, the poorer the phase cancellation by the magic sandwich, or conversely the larger the value required for ω_1 to maintain this cancellation.

Eventually, the magnetization would be lost because its retrieval would require impossibly high values of the r.f. field H_1 and irreversibility wins.

Still this experiment shows that phase coherence in the spin system may persist for times much longer than was thought previously.

Does it mean that one should give up the concept of spin temperature, such as it has been used successfully in innumerable experiments? The answer in the opinion of the authors is no. The point is not whether after a time of the order of a few T_2 the off-diagonal matrix elements of the density matrix have 'disappeared', a philosophical question, but whether their existence can have observable consequences in a given experimental situation. It turns out that unless an elaborate experiment, like the magic sandwich, has been designed with the express purpose of tracking down these off-diagonal matrix elements, the usual criterions for the validity of the spin temperature assumption keep their usefulness.

The new contribution of the magic sandwich experiment is to deepen our understanding of the meaning of the spin temperature assumption and to call for a careful assessment of the experimental conditions lest an accidental refocussing of the phases occurs.

D. Spin-lattice relaxation

At various places in the foregoing, eqn (1.25), eqn (1.38) we have introduced as phenomenological parameters spin-lattice relaxation times for various spin operators, such as T_{1z} and T_{1D} for the Zeeman and dipolar energies and T_{1x} for the relaxation of the transverse magnetization. A detailed microscopic theory of spin-lattice relaxation can be found in Abragam (1961), chapters VIII and IX and, with special reference to spin temperature, in Goldman (1970), chapter 3.

In this section we shall show how the formalism of spin-lattice relaxation can be derived as a consequence of the generalized Provotorov equations of **B**(c).

The principle is very simple. In order to study the spin-lattice relaxation of an operator Q of the spin system such as, say, I_z or I_x (in the rotating frame) or $\mathscr{H}'_D$ we choose this operator as an 'operator of interest' in the sense of the sections **B**(b) and **B**(c). As a second 'operator of interest' we choose the Hamiltonian $\mathscr{F}$ of the 'lattice' which is a physical system whose variables commute with those of the nuclear spin system.

The lattice is assumed to be at all times in internal thermal equilibrium with an inverse 'lattice' temperature β_L. In the course of the coupling between Q and $\mathscr{F}$ this temperature β_L (unless explicitly stated otherwise) is expected to remain constant. This will happen if the lattice has a 'heat capacity' $\mathrm{Tr}\{\mathscr{F}^2\}$ much higher than the 'heat capacity' $\mathrm{Tr}\{Q^2\}$ of the operator to be relaxed so that the parameter ε in the generalized Provotorov equation (1.105′) is negligible.

It may still happen if ε is not very small provided $\mathscr{F}$ is in turn closely coupled to a thermal bath of much higher heat capacity which maintains its temperature constant.

The generalized Provotorov equation becomes the spin-lattice equation:

$$\left(\frac{\mathrm{d}\beta}{\mathrm{d}t}\right)_{\mathrm{SL}} = -\frac{1}{T_1}(\beta - \beta_L); \tag{1.197}$$

or

$$\left(\frac{\mathrm{d}\langle Q\rangle}{\mathrm{d}t}\right)_{\mathrm{SL}} = -\frac{1}{T_1}(\langle Q\rangle - \langle Q_L\rangle),$$

with

$$\langle Q_L\rangle = -\beta_L\,\mathrm{Tr}\{Q^2\}.$$

It remains to write down an expression for $1/T_1 = W$.

This can be taken directly from a transcription (with slightly different notations) of (1.107)

$$W = \frac{1}{\mathrm{Tr}\{Q^2\}}\int_0^\infty \mathrm{Tr}\{[Q, V]\exp[-\mathrm{i}(\mathscr{H} + \mathscr{F})t][V, Q]\exp[\mathrm{i}(\mathscr{H} + \mathscr{F})t]\,\mathrm{d}t. \tag{1.198}$$

In (1.198) $\mathscr{H}$ is the Hamiltonian of the spin system (which in particular can itself be the operator Q); $\mathscr{F}$ is the lattice Hamiltonian; V, much smaller than $\mathscr{H}$ and $\mathscr{F}$, is the coupling Hamiltonian between them.

Consider the case when V is the product of a lattice operator F by a spin operator A with $[F, A] = 0$; eqn (1.198) can be rewritten:

$$W = \frac{1}{\mathrm{Tr}\{Q^2\}}\int_0^\infty \mathrm{tr}\{FF(t)\}\,\mathrm{Tr}\{[A, Q]\exp(-\mathrm{i}\mathscr{H}t)[Q, A]\exp(\mathrm{i}\mathscr{H}t)\}\,\mathrm{d}t. \tag{1.199}$$

In this equation, we distinguish between the symbols Tr, which means trace with respect to spin variables and tr which means trace with respect to lattice variables. $F(t)$ stands for $\exp-(\mathrm{i}\mathscr{F}t)F\exp(\mathrm{i}\mathscr{F}t)$. The function,

$$G(t) = \frac{\mathrm{tr}\{FF(t)\}}{\mathrm{tr}\{F^2\}} \quad \text{with} \quad G(0) = 1, \tag{1.200}$$

is called the lattice correlation function of the variable F. In all physical situations it decays to zero when t increases. The rate at which it decays (not necessarily exponentially) is described qualitatively by a constant τ_c called the lattice correlation time.

The case when $(1/\tau_c)$ is much larger than all the relevant frequencies of the spin system is known in the literature as 'extreme narrowing'. In that case we give as a quantitative definition of τ_c:

$$\tau_c = \int_0^\infty \mathrm{tr}\{FF(t)\}\,\mathrm{d}t/\mathrm{tr}\{F^2\}, \tag{1.201}$$

and W can be written:

$$W = \frac{-\mathrm{tr}\{F^2\}}{\mathrm{Tr}\{Q^2\}}\tau_c \mathrm{Tr}\{[A, Q]^2\}. \tag{1.202}$$

If the narrowing is not extreme we have the more general expression:

$$W = \frac{-\mathrm{tr}\{F^2\}}{\mathrm{Tr}\{Q^2\}}\int_0^\infty \mathrm{d}t \sum_{m,n} |\langle m|[Q, A]|n\rangle|^2 G(t)\exp\{-\mathrm{i}(E_m - E_n)t\}\,\mathrm{d}t. \tag{1.203}$$

If we assume for $G(t)$ the exponential form $\exp(-\tau/\tau_c)$, (1.203) becomes:

$$W = \frac{-\mathrm{tr}\{F^2\}}{\mathrm{Tr}\{Q^2\}}\sum_{m,n} |\langle m|[Q, A|n\rangle|^2 \frac{\tau_c}{1+(E_m - E_n)^2\tau_c^2}. \tag{1.204}$$

We have omitted in (1.204) a small imaginary part which corresponds to a usually unobservable frequency shift.

The case when the spin-lattice coupling Hamiltonian V is a single product FA is very exceptional.

In general it will be of the form of a sum $V = \sum_k F_k A_k$. Each spin operator A_k can refer to a single spin i or to two spins i and j (or even more, in principle if not in practice). In the case of a single spin it can be linear or quadratic with respect to the components of the spin (respectively magnetic or quadrupole relaxation). In the case of two spins it will be bilinear with respect to the components of the two spins. All this is analysed in detail in Abragam (1961) and Goldman (1970) to which the reader is referred.

Needless to say the foregoing is valid under the high temperature approximation. The low temperature case will receive special consideration in chapter 8.

Even so the use of the expression 'high temperature' approximation in the present section must be clarified.

Broadly speaking it means that a thermal density matrix $\rho = \exp(-\beta\mathcal{H})/\mathrm{Tr}\exp(-\beta\mathcal{H})$ can be approximated by $\rho = (1-\beta\mathcal{H})/\mathrm{Tr}\{1\}$.

Whereas it is perfectly legitimate in most cases for the spin system it is not so for the lattice, and expressions such as (1.199) for the relaxation rate, which are independent of the lattice temperature are unjustified and contradicted by experiment.

This point is discussed in some detail in Abragam (1961, chapter VIII) and will not be repeated here.

The upshot of it is that in the eqns (1.199) and (1.200) expressions such as $\mathrm{tr}\{FF(t)\}$ or

$$\frac{\mathrm{tr}\{FF(t)\}}{\mathrm{tr}\{F^2\}} = G(t)$$

must be replaced by the thermal averages:

$$\langle FF(t)\rangle = \mathrm{tr}\{\rho FF(t)\}; \qquad G(t) = \frac{\langle FF(t)\rangle}{\langle F^2\rangle} = \frac{\mathrm{tr}\{\rho FF(t)\}}{\mathrm{tr}\{\rho F^2\}}, \tag{1.205}$$

where $\rho(\mathcal{F})$ is the Boltzmann density matrix:

$$\rho(\mathcal{F}) = \exp-(\beta_{\mathrm{L}}\mathcal{F})/\mathrm{Tr}\{\exp-(\beta_{\mathrm{L}}\mathcal{F})\}. \tag{1.206}$$

2

COHERENT MANIPULATION OF NUCLEAR SPINS AND HIGH RESOLUTION IN SOLIDS

Double, double toil and trouble.
Fire burn and cauldron bubble.

SHAKESPEARE (*Macbeth IV, 1*)

IT is well known that in diamagnetic substances an applied magnetic field H_0 induces a small electronic magnetization by polarizing the electronic shells. The shells, thus polarized, produce at the site of an atomic nucleus a small magnetic field proportional to H_0. As a consequence the NMR frequency of the nucleus is shifted by a small amount also proportional to H_0. In a molecule, nuclei which belong to the same species I but occupy chemical sites with different electronic environments, exhibit different frequency shifts which for that reason have received the name of chemical shifts. The NMR spectrum of the nuclear species I breaks down into as many lines as there are chemical shifts. The spectrum is further complicated by the existence of so-called indirect couplings between nuclear spins, which are mediated by the electronic clouds surrounding the nuclei and which give rise to still more lines. With the possible exception of heavy atoms where the electronic density at the nuclei is high, chemical shifts and indirect couplings are, as a rule, much smaller than the dipolar nuclear couplings. As a consequence, in solid samples these spectral structures are hidden within the dipolar linewidth, resulting in an almost complete loss of important chemical information. For spins $I > \frac{1}{2}$ a potential quadrupole broadening is an aggravating circumstance.

The situation is completely different in liquids where fast molecular motions average out to zero dipolar couplings and quadrupole splittings. The chemical shifts and indirect interactions are also averaged by the molecular motions but (in contrast to dipole and quadrupole interactions) to values which are *not* zero and can be observed and measured. There is *some* loss of information in the averaging process but the remaining information is still of paramount importance. High resolution NMR in liquids, devoted to the collection of this information, has grown into an enormous field which at present occupies an overwhelming proportion of the practitioners of NMR. This field is outside of the scope of this book. A brief introduction can be found in Abragam (1961), chapter XI.

The potential extension of high resolution NMR to solid samples by suppressing the dipolar and quadrupolar broadening is clearly of great interest. It extends structural studies to compounds not available in liquid form. Furthermore by making observable the *angular* dependence of chemical shifts and indirect interactions, suppressed in liquids by the motional averaging, valuable new information on chemical structures can be obtained. Several methods of high resolution in solids have been made to work. All involve a fast modulation, periodic rather than stochastic as in liquids, of the dipolar and quadrupole Hamiltonians. This modulation must reduce to zero their average values while leaving non-vanishing averages for chemical shifts and indirect interactions.

All the Hamiltonians to be averaged are sums of products of two tensors one of which depends on the orbital co-ordinates and the other on the spin variables. The averaging can thus be produced either by a fast rotation of the sample which implies an averaging over the orbital variables or by a fast rotation of the spins, or sometimes by a combination of the two.

Spin rotation can be produced either by a strong CW r.f. field or by a sequence of suitably programmed pulses, the multipulse method. The latter has the advantage of allowing a sampling of the state of the spin system in the intervals between the pulses. Besides their usefulness for high resolution in solids the multipulse methods raise some interesting problems in nuclear magnetism proper, relative to the dynamics and thermodynamics of interacting spins. Some of these problems are considered in this chapter. An extensive description of multipulse methods is given in monographs by Mehring (1976) and Haeberlen (1976).

The spin rotation methods are somewhat different depending on whether the dipolar interactions to be averaged out are between like or unlike spins. We shall refer to these methods respectively as homonuclear and heteronuclear averaging. On the other hand the sample rotation yields the same results for the two types of broadening.

A. The spin Hamiltonians

The expansion of the homonuclear dipolar interaction into irreducible tensors and its transformation under rotation has been written out in detail in chapter 1, eqns (1.1) to (1.11). In the presence of a high magnetic field, only the secular part $\mathscr{H}'_{\mathrm{D}}$ of the dipolar interaction, given by eqn (1.27) need be considered. In this equation the orbital part F^0_{ij} is proportional to $(1–3\cos^2\theta_{ij})$ where θ_{ij} is the angle between the magnetic field and the vector $\mathbf{r}_{ij}$ linking the two interacting spins. For heteronuclear dipolar broadening the same orbital factor F^0_{ij} is present.

In introducing the spin Hamiltonians relative to the other interactions listed in the introduction it is convenient to define suitable frames of reference.

The first is a frame $0xyz$ *linked to the applied magnetic field.* The axis $0z$ is along the applied field. The axes $0x$ and $0y$ may be either fixed (laboratory frame) or rotate around $0z$ at the Larmor frequency (rotating frame).

The second is a frame $0x'y'z'$ whose axis $0z'$ is fixed with respect to both the laboratory *and* the sample. This necessarily implies that if the sample is rotated, the axis of rotation is $0z'$. It makes with $0z$ a *fixed* angle β. The axes $0x'$ and $0y'$ are fixed with respect to the sample and thus can rotate with it around $0z'$ with an angular velocity Ω.

Then for each interaction, be it quadrupole coupling, chemical shift, or indirect interaction we choose a frame $0XYZ$, *fixed with respect to the sample*, whose axes are the principal axes of a tensor related to the interaction under consideration. It is clear that there may be more than one such frame even in a single crystal and that there is an infinity of them in a polycrystalline sample.

(*a*) *The quadrupole interaction* (Abragam (1961), chapter VI)

In a frame $0XYZ$ where, X, Y, Z are the principal axes of the electric field gradient seen by a spin $I > \frac{1}{2}$, the quadrupole interaction can be written:

$$\mathcal{H}_Q = aI_X^2 + bI_Y^2 + cI_Z^2, \tag{2.1}$$

with $a + b + c = 0$.

This can be rewritten as:

$$\mathcal{H}_Q = (2c + a)[I_Z^2 - \tfrac{1}{3}I(I+1)] + (2a + c)[I_X^2 - \tfrac{1}{3}I(I+1)]. \tag{2.2}$$

We must now take into account the existence of a Zeeman energy $Z = -\gamma H_0 I_z = \omega_0 I_z \gg \mathcal{H}_Q$ and select in (2.2) the part that commutes with Z. This is easily achieved by noticing that:

$$[I_Z^2 - \tfrac{1}{3}I(I+1)] = \sqrt{(\tfrac{2}{3})}T_{0Z}^{(2)}, \tag{2.3}$$

where $T_{0Z}^{(2)}$ represents the component $m = 0$ of the tensor operator $T_m^{(2)}(I)$, the axis of quantization being along $0Z$. Using the transformation relations (1.4) we get:

$$\sqrt{(\tfrac{2}{3})}T_{0Z}^{(2)} = \sqrt{(\tfrac{2}{3})}\sum_q T_{qz}^{(2)}\mathcal{D}_{0q}^{(2)}(R), \tag{2.4}$$

where R is the rotation which transforms the frame $0xyz$ into $0XYZ$ and the $T_{qz}^{(2)}$ are quantized along $0z$. In the summation (2.4) only the operator $\sqrt{(\frac{2}{3})}T_{0z}^{(2)} = [I_z^2 - \frac{1}{3}I(I+1)]$ commutes with $Z = \omega_0 I_z$ and thus represents the secular part of (2.3) which can be written as:

$$\sqrt{(\tfrac{2}{3})}T_{0z}^{(2)}\mathcal{D}_{00}^{(2)}(R) = \tfrac{1}{2}(3\cos^2\theta_{zZ} - 1)[I_z^2 - \tfrac{1}{3}I(I+1)], \tag{2.5}$$

where θ_{zZ} is the angle between $0z$ and $0Z$.

The secular part of $\mathcal{H}_Q$ given by (2.2) is then:

$$\mathcal{H}'_Q = \tfrac{1}{2}\{(2c+a)(3\cos^2\theta_{zZ}-1)+(2a+c)(3\cos^2\theta_{zX}-1)\}[I_z^2-\tfrac{1}{3}I(I+1)]. \tag{2.6}$$

In a polycrystalline sample, even if the constants a and c have the same values over the whole sample, the distribution of the angle θ_{zZ} and θ_{zX} smears out the contributions of $\mathcal{H}'_Q$ into a continuous broadening.

(*b*) *The chemical shift* (Abragam (1961), chapter VI)

The internal field H_i responsible for the chemical shift can be written as $\mathbf{H}_i = -\boldsymbol{\sigma}\,.\,\mathbf{H}$ where $\boldsymbol{\sigma}$ is a small dimensionless tensor which depends on the electronic environment of the nucleus, and the chemical shift Hamiltonian is:

$$\mathcal{H}_c = \gamma\mathbf{I}\,.\,\boldsymbol{\sigma}\,.\,\mathbf{H}. \tag{2.7}$$

It is easy to see that an antisymmetrical tensor $\boldsymbol{\sigma}_a$ contributes no secular part to $\mathcal{H}_c$ and is not observed: $\gamma\mathbf{I}\,.\,\boldsymbol{\sigma}_a\,.\,\mathbf{H}$ can be written as $\gamma\mathbf{I}\,.\,(\boldsymbol{\Sigma}_a \wedge \mathbf{H})$ where $\boldsymbol{\Sigma}_a$ is an axial vector with components: $\frac{1}{2}\boldsymbol{\sigma}_{ayz}$, $\frac{1}{2}\boldsymbol{\sigma}_{azx}$, $\frac{1}{2}\boldsymbol{\sigma}_{axy}$. The vector $(\boldsymbol{\Sigma}_a \wedge \mathbf{H})$ is orthogonal to $\mathbf{H}$ which proves the point. Without loss of generality we can then assume $\boldsymbol{\sigma}$ symmetrical. The interaction (2.7) can be rewritten as:

$$\mathcal{H}_c = \gamma\sigma_0(\mathbf{I}\,.\,\mathbf{H}) + \gamma\mathbf{I}\,.\,\boldsymbol{\sigma}'\,.\,\mathbf{H} \tag{2.8}$$

where $3\sigma_0$ is the trace of $\boldsymbol{\sigma}$ and $\boldsymbol{\sigma}' = \boldsymbol{\sigma} - \sigma_0\mathbf{1}$ is a symmetrical traceless tensor.

Let $0XYZ$ be the principal axes of the tensor $\boldsymbol{\sigma}'$ and σ'_X, σ'_Y, σ'_Z its principal values:

$$\gamma\mathbf{I}\,.\,\boldsymbol{\sigma}'\,.\,\mathbf{H} = \gamma(\sigma'_X H_X I_X + \sigma'_Y H_Y I_Y + \sigma'_Z H_Z I_Z). \tag{2.9}$$

Let $\lambda_X = \cos\theta_{zX}$, $\lambda_Y = \cos\theta_{zY}$, $\lambda_Z = \cos\theta_{zZ}$ be the cosines of the magnetic field $\mathbf{H}$, in the frame $0XYZ$. The secular part of (2.9) is $\gamma\sigma_0 HI_z + \gamma HI_z(\sigma'_X\lambda_X^2 + \sigma'_Y\lambda_Y^2 + \sigma'_Z\lambda_Z^2)$ which with a little algebra and taking into account $\sigma'_X + \sigma'_Y + \sigma'_Z = 0$ can be given a form similar to (2.6):

$$\begin{aligned}\mathcal{H}_c = \gamma\sigma_0 HI_z + \tfrac{1}{3}\gamma HI_z[(2\sigma'_Z + \sigma'_X)(3\cos^2\theta_{zZ} - 1)\\ + (2\sigma'_X + \sigma'_Z)(3\cos^2\theta_{zX} - 1).\end{aligned} \tag{2.10}$$

(*c*) *The indirect couplings* (Abragam (1961), chapter VI)

The indirect coupling between two spins I_1 and I_2 can be written quite generally as:

$$\mathcal{H}_{12} = A_0\mathbf{I}_1\,.\,\mathbf{I}_2 + \mathbf{I}_1\,.\,\mathbf{A}\,.\,\mathbf{I}_2 \tag{2.11}$$

where A_0 is a scalar and $\mathbf{A}$ a traceless tensor.

There is no symmetry argument to show that $\mathbf{A}$ is a symmetrical tensor (in contradiction with a statement found in early editions of Abragam (1961), chapter VI). Furthermore (in contrast to the chemical shift tensor) the antisymmetrical part of (2.11), of the form:

$$\mathbf{I}_1 \,.\, \mathbf{A}_{\mathrm{a}} \,.\, \mathbf{I}_2 = \mathbf{A}_{\mathrm{a}} \,.\, (\mathbf{I}_2 \wedge \mathbf{I}_1) \tag{2.12}$$

where $\mathbf{A}_{\mathrm{a}}$ is an axial vector, does have a secular part for like nuclei, namely $A_{az}(\mathbf{I}_2 \wedge \mathbf{I}_1)_z$ which commutes with $\omega_0 I_z$.

There are however reasons to believe that the antisymmetrical part $\mathbf{A}_{\mathrm{a}}$ of the tensor $\mathbf{A}$, which originates in the difference between the electronic environments of the interacting nuclei I_1 and I_2 is, for like nuclei, much smaller than its symmetrical part, and we shall make the simplifying assumption that $\mathbf{A}_{\mathrm{a}} = 0$ and that $\mathbf{A}$ is symmetrical (this assumption might be less justified for unlike nuclei but then $\mathbf{A}_{\mathrm{a}} \,.\, (\mathbf{I}_1 \wedge \mathbf{I}_2)$ has no secular part). If A_X, A_Y, A_Z are its principal values, with $A_X + A_Y + A_Z = 0$ and $0XYZ$ its principal axes:

$$\mathcal{H}_{12} = A_0 \mathbf{I}_1 \,.\, \mathbf{I}_2 + A_X I_X^1 I_X^2 + A_Y I_Y^1 I_Y^2 + A_Z I_Z^1 I_Z^2. \tag{2.13}$$

A calculation in every way similar to that found in $\mathbf{A}(b)1$ and $(b)2$ shows that the secular part of $\mathcal{H}_{12}$ is given by:

$$\mathcal{H}'_{12} = A_0(\mathbf{I}_1 \,.\, \mathbf{I}_2) + \tfrac{1}{6}[(2A_Z + A_X)(3\cos^2\theta_{zZ} - 1) + (2A_X + A_Z)(3\cos^2\theta_{zX} - 1)](3I_z^1 I_z^2 - \mathbf{I}_1 \,.\, \mathbf{I}_2). \tag{2.14}$$

We shall for brevity call the second term in (2.14) the pseudo-dipolar interaction. Its spin part is the same as that of a true dipolar interaction but not its orbital part.

For a heteronuclear indirect coupling (2.14) must be modified, for then in $(\mathbf{I}^1 \,.\, \mathbf{I}^2)$ only the part $I_z^1 I_z^2$ is secular.

Replacing $\mathbf{I}_1$ and $\mathbf{I}_2$ by $\mathbf{I}$ and $\mathbf{S}$, (2.14) is replaced by:

$$\mathcal{H}'_{\mathrm{IS}} = I_z S_z[A_0 + \tfrac{1}{3}(2A_Z + A_X)(3\cos^2\theta_{zZ} - 1) + \tfrac{1}{3}(2A_X + A_Z)(3\cos^2\theta_{zX} - 1)]. \tag{2.15}$$

B. Fast rotation of the sample

In $\mathbf{A}$. we have considered four different interactions: dipolar, electric quadrupole, chemical shifts, and indirect.

Their secular parts are given respectively by (1.27), (2.6), (2.10) and (2.14) or (2.15). The first two, dipolar and quadrupolar, are traceless and have no part invariant through arbitrary rotation of the sample whereas the last two do have such a part respectively $\gamma\sigma_0(\mathbf{I} \,.\, \mathbf{H})$ and $A_0(\mathbf{I}^1 \,.\, \mathbf{I}^2)$ or $A_0 I_z S_z$.

We are going to show that a fast rotation of the sample around a suitably chosen axis averages out to zero the four traceless interactions listed above leaving, as in liquids, a discrete spectrum originating in the scalar terms $\gamma\sigma_0(\mathbf{I} \,.\, \mathbf{H})$ and $A_0(\mathbf{I}^1 \,.\, \mathbf{I}^2)$ (or $A_0 I_z S_z$).

We notice that the traceless parts of all the interactions contain a factor $(3\cos^2\theta_{zn}-1)\propto Y^{(2)}_{0z}(\mathbf{n})$. $\mathbf{n}$ is a unit vector of one of the principal axes of quadrupole interaction, chemical shift, indirect interaction, or yet $\mathbf{r}_{12}/|\mathbf{r}_{12}|$ for a pure dipolar interaction between two nuclear spins $\mathbf{I}_1$, $\mathbf{I}_2$. In all cases $\mathbf{n}$ is a vector fixed with respect to the sample, $Y^{(2)}_{0z}(\mathbf{n})$ is the spherical harmonic $m=0$ in the frame $0xyz$ where the z axis is along the applied field. If the sample is rotated, $\mathbf{n}$, linked to the sample, changes its orientation with respect to $0z$ and $Y^{(2)}_{0z}(\mathbf{n})$ becomes time-dependent. Its time dependence can be obtained from the relation:

$$Y^{(2)}_{0z}(\mathbf{n})=\sum_q Y^{(2)}_{qz'}(\mathbf{n})\mathscr{D}_{q0}(\alpha,\beta,\gamma). \tag{2.16}$$

In (2.16) $Y^{(2)}_{qz'}(\mathbf{n})$ is a spherical harmonic relative to the axes $x'y'z'$, a frame which has been defined as linked to the sample and since $\mathbf{n}$ is also linked to the sample $Y^{(2)}_{qz'}(\mathbf{n})$ is time-independent. (α,β,γ) are the Euler angles of the rotation which takes the frame $0x'y'z'$ into $0xyz$, that is the rotating sample frame into the laboratory frame. β is the angle between the axes z and z', that is between the magnetic field and the axis of rotation, γ is the angle of rotation of the sample around $0z'$ equal to Ωt. From the well-known formula:

$$\mathscr{D}_{q0}(\alpha,\beta,\gamma)=\exp(-iq\gamma)\,d_{q0}(\beta)=\exp(-iq\Omega t)\,d_{q0}(\beta),$$

we see that the only time-independent term on the right-hand side of (2.16) is:

$$Y^{(2)}_{0z'}(\mathbf{n})d_{00}(\beta)=\tfrac{1}{2}(3\cos^2\beta-1)Y^{(2)}_{0z'}(\mathbf{n}) \tag{2.17}$$

If we choose for $\cos\beta$ the magic value $1/\sqrt{3}$ the time-independent term (2.17) vanishes and all the traceless interactions have a fast time dependence of frequency $q\Omega$. If Ω is much larger than the line width of the static sample, this line is considerably narrowed and the remaining terms: $A_0\mathbf{I}^1\,.\,\mathbf{I}^2$, $A_0I_zS_z$ and $\sigma_0(\mathbf{I}\,.\,\mathbf{H})$ can be observed.

The condition $\Omega\gg\Delta\omega$ is a stringent one since it involves sample rotations of many thousands of turns per second.

We shall not discuss any further the various theoretical and experimental aspects of this method of narrowing, described in a review article by its pioneer E. R. Andrew (Andrew, 1971).

Figures 2.1 and 2.2. illustrate the kind of high resolution spectra that can be obtained by this method.

C. Homonuclear narrowing through spin manipulation

(a) Large effective field in the rotating frame

We have remarked in the last section on the technical difficulty of rotating a sample at angular velocities of many thousands of turns per second. On the

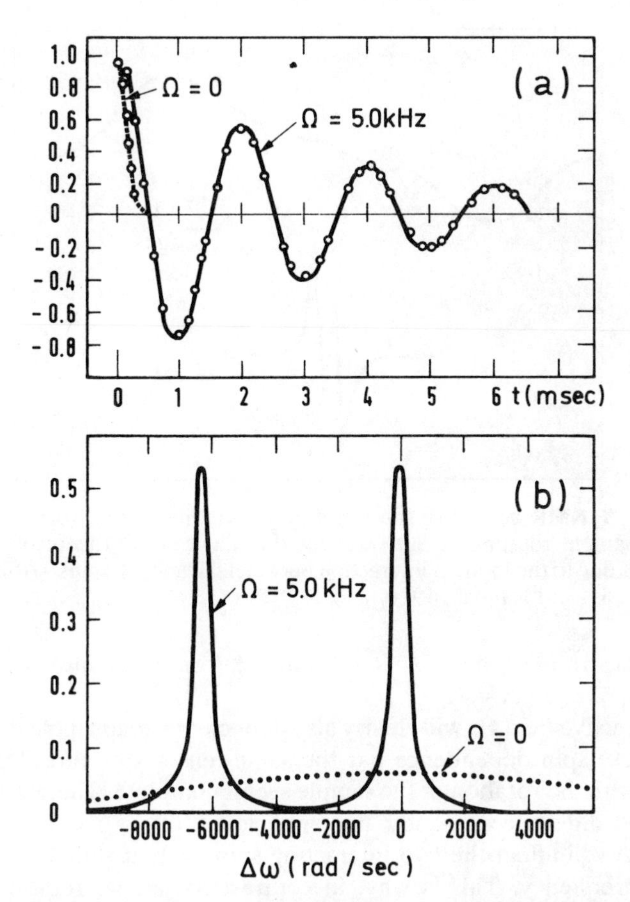

FIG. 2.1. (a) Normalized ^{31}P free induction decay for a polycrystalline specimen of zinc phosphide. Dotted curve: static sample. Solid curve: sample rotating at 5 kHz about the magic axis. (b) Fourier transforms of the FID signals. (After Kessemeier and Norberg, 1967.)

other hand spin precessions of such angular velocities exist in fields of a few gauss. Fast spin precession in the applied magnetic field $\mathbf{H}$ is actually the mechanism which reduces the dipolar and indirect interactions to their secular parts. We notice that the spin dependence of these secular parts is, for like spins, given by the tensor $T_{0z}^{(2)}(\mathbf{I}_1, \mathbf{I}_2) = 1/\sqrt{(6)}[3I_z^1 I_z^2 - \mathbf{I}^1 . \mathbf{I}^2]$ (eqns (1.127) and (2.14)). In full analogy with sample rotation a fast precession of the spins around an axis making the magic angle β with the applied field $\mathbf{H}$ will reduce considerably the dipolar and pseudo-dipolar interactions.

Such a precession can be produced by creating in the rotating frame an effective field $\mathbf{H}_e$ making an angle $\beta = \cos^{-1}(1/\sqrt{3})$ with $0z$, that is by

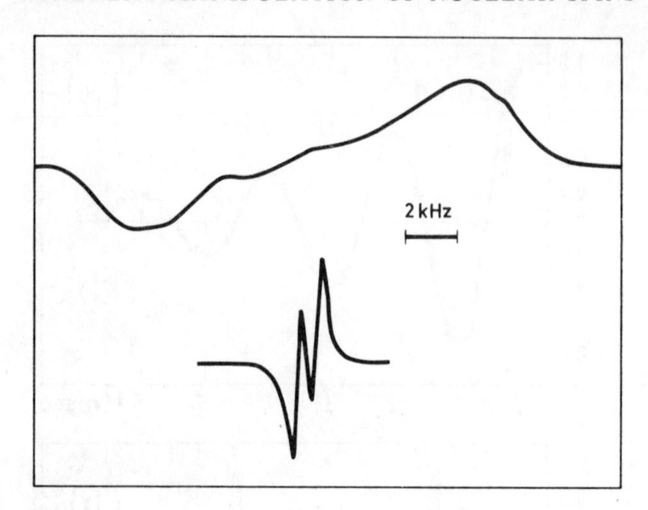

FIG. 2.2. ^{19}F NMR derivative spectra of polycrystalline KPF_6. Top: static sample. Bottom: sample rotating at 8 kHz about the magic axis. The resolved doublet structure is due to the indirect interaction between ^{19}F and ^{31}P spins. (After Andrew, Firth, Jasinski, and Randall, 1970.)

applying a strong r.f. field H_1 at a distance Δ from resonance, such that tan $\beta = -\gamma H_1/\Delta = \omega_1/\Delta = \sqrt{2}$.

The effective field $\mathbf{H}_e$ will clearly also quench the quadrupole interactions (2.6) whose spin-dependence has the same tensor structure $T^{(2)}_{0z}(I)$. The analogy with the rotation of the sample seems complete. There are however important differences. Firstly the application of an r.f. field near the resonance will affect the two interacting spins only if they have the same Larmor frequency. This is why, in contrast to sample rotation, hetero-nuclear couplings require a different treatment and will be considered in a later section. Secondly the dependence of the anisotropic chemical shift interaction $\gamma \mathbf{I} \cdot \boldsymbol{\sigma}' \cdot \mathbf{H}$ on the spin variables is a first order tensor rather than a second order one as for orbital variables. Rotation of the spins will *not* affect the chemical shift Hamiltonian in the same way as rotation of the sample.

After these remarks we return to the study of the quenching of the dipolar interaction by the effective field (neglecting for the moment other interactions). The time evolution of the spin system in the rotating frame is given by the operator:

$$\begin{aligned} U(t) &= \exp[-i(\omega_e I_Z + \mathcal{H}'_D)t] \\ &= \exp(-i\omega_e I_Z t) T \exp\left[-i \int_0^t \tilde{\mathcal{H}}'_D(t')\,dt'\right], \end{aligned} \tag{2.18}$$

where $0Z$ is the direction of the effective field and

$$\tilde{\mathcal{H}}'_D(t') = \exp(i\omega_e I_Z t')\mathcal{H}'_D \exp(-i\omega_e I_Z t'). \tag{2.19}$$

Since $\mathcal{H}'_D$ transforms under rotation of the spins as $T_{0z}^{(2)}(I)$, we have:

$$\tilde{\mathcal{H}}'_D(t') = \tfrac{1}{2}(3\cos^2\beta - 1)\mathcal{H}'_D + \mathcal{H}''_D(t'), \tag{2.20}$$

where $\mathcal{H}''_D(t')$ contains terms proportional to $\exp(\pm i\omega_e t')$ and $\exp(\pm 2i\omega_e t')$. If $\omega_e \gg \mathcal{H}'_D$ we may consider that these varying terms of $\mathcal{H}''_D(t')$ average out to zero and we are left with:

$$U(t) \simeq \exp(-i\omega_e I_Z t)\exp\left(-\frac{i}{2}(3\cos^2\beta - 1)\mathcal{H}'_D t\right), \tag{2.21}$$

which for $\cos\beta = 1/\sqrt{3}$ reduces to:

$$U(t) \simeq \exp(-i\omega_e I_Z t) \tag{2.22}$$

At time $t = n\tau_c = n2\pi/\omega_e$, $U(n\tau_c) = \exp(-i2\pi n I_Z)$ represents a rotation by an angle $2\pi n$ of all the spins of the system, which leads to the same values for all observable quantities as for $t = 0$. (Strictly speaking, $U(\tau_c) = \pm 1$, depending on whether the total spin of the system is integer or half-integer but for every observable A, $A(\tau_c) = U(\tau_c)AU^\dagger(\tau_c) = A$, and there is no loss of generality in making the convention, used henceforth, that:

$$\exp(-i2\pi I_z) = 1. \tag{2.22'}$$

In the language of the Liouville formalism it is the operator $\hat{U}$, such that $\hat{U}|A\rangle = UAU^{-1}$, which is equal to $\hat{1}$.)

The secular part of the pseudo-dipolar indirect interaction has the same spin dependence as $\mathcal{H}'_D$ (eqn 2.14) and is averaged out in the same way. The spin dependence $[I_z^2 - \frac{1}{3}I(I+1)]$ of the secular quadrupoles interaction (eqn 2.6) which transforms as $T_{0z}^{(2)}$ is also averaged out. Having thus quenched the dipolar, pseudodipolar and quadrupole interactions we turn to the chemical shift.

In the expression (2.10) we have to replace I_z by its secular part, that is its projection on the effective field: $I_Z \cos\beta = I_Z/\sqrt{3}$. The narrowed chemical shifts appear as reduced in the ratio $1/\sqrt{3}$.

The final effective Hamiltonian becomes:

$$\bar{\mathcal{H}}_0 = \frac{1}{\sqrt{3}}\gamma H \sum_i I_Z^i\{\sigma_0^i + \tfrac{1}{3}(2\sigma'^i_{Z_i} + \sigma'^i_{X_i})(3\cos^2\theta_{zZ_i} - 1)$$

$$+ \tfrac{1}{3}(2\sigma'^i_{X_i} + \sigma'^i_{Z_i})(3\cos^2\theta_{zX_i} - 1)\} + \sum_{i<j} A_0^{ij}\mathbf{I}^i \cdot \mathbf{I}^j. \tag{2.23}$$

Assume first that the sample is a single crystal. The index i refers to the various non-equivalent positions of the nuclei in the sample. The differences between these positions are relative to the principal values $\sigma_0^i, \sigma'^i_{Z_i}, \sigma'^i_{X_i}$ of the chemical shift tensor as well as to the orientations of the principal axes of this tensor in the crystal. There is a finite number of non-equivalent positions in the crystals and the high resolution spectrum of the Hamiltonian (2.23) is a line spectrum. A study of this spectrum for various orientations of the magnetic field yields far more information than the isotropic spectrum obtained by rotation of the sample, where only the σ_0^i and the A_0^{ij} could be observed.

If the sample is polycrystalline this wealth of information may become a mixed blessing. There are now a very large number of orientations of the principal axes X_i, Y_i, Z_i with respect to the magnetic field and a quasi-continuous distribution of the factors $(3\cos^2\theta_{zZ_i}-1)$ and $(3\cos^2\theta_{zX_i}-1)$.

When all spins are chemically equivalent, the spectrum acquires a characteristic shape exhibited in Fig. 2.3, with a width of the order of $\gamma H\sigma'_X$. It is possible to extract from this spectrum the principal values σ'_X, σ'_Y, and σ'_Z of the chemical shift tensor. The same is true when spins with different chemical environments have values of σ_{0i} so different that their powder

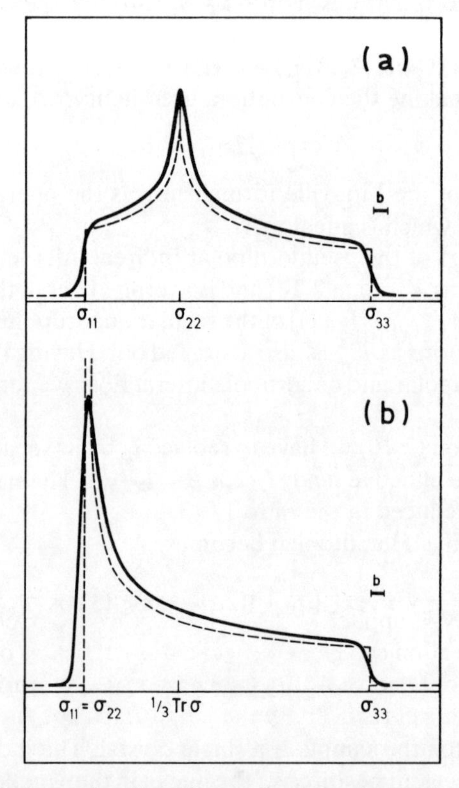

FIG. 2.3. Theoretical powder line shapes for an anisotropic chemical shift. (a) Arbitrary chemical shift tensor. (b) Axially symmetric chemical shift tensor. Dashed curves: the resonance lines of each microcrystal are infinitely sharp. Solid curves: the resonance lines of each microcrystal are Lorentzian, width with b.

spectra do not overlap. However, in the case of overlapping spectra, it may be difficult or even impossible to extract the parameters of the Hamiltonian (2.23).

It is sometimes advisable to sacrifice some of the information contained in (2.23) namely the anisotropic part of the chemical shift in order to retrieve more easily the isotropic part of the spectrum. This is achieved by rotating the sample around an axis making the magic angle $\beta = \cos^{-1}(1/\sqrt{3})$ with the magnetic field. One ends up with a doubly averaged Hamiltonian:

$$\bar{\bar{\mathcal{H}}}^0 = \frac{1}{\sqrt{3}} \gamma H \sum_i I_z^i \sigma_{0i} + \sum_{i,j} A_0^{ij} \mathbf{I}^i \cdot \mathbf{I}^j, \tag{2.24}$$

which has an isotropic line spectrum. An example of a double averaging is described in section **D**(a). Since the final result is the same as with a single sample rotation (apart from the $1/\sqrt{3}$ reduction of the chemical shift) one may question the usefulness of combining spin and sample rotation. If the overall width due to dipolar, pseudodipolar and quadrupole interactions is much larger than that due to the spread of anisotropic chemical shifts the double averaging is advantageous: the very large angular velocity required for the averaging of the former is easier to produce by rotating the spins rather than the sample. To average out the residual width due to the anisotropy of the chemical shifts, a much slower sample rotation will suffice.

As will appear in the next section continuous precession of the spins around a CW effective field at magic angle to the dc field H is by no means the only method of averaging out second order spin tensors $T_{0z}^{(2)}(I)$. All the statements relative to the effective Hamiltonian (2.23) based on this averaging remain valid for these new methods.

The spin precession around a CW effective field ω_e/γ was suggested by Barnaal and Lowe (1963) and experimented by Lee and Goldburg (1965). Much earlier (Redfield, 1955) it had been noticed that the line width observed in a rotary saturation experiment (Abragam (1961), chapter XII; Goldman (1970), chapter 4) narrowed considerably when the effective field was at a magic angle to the d.c. field.

In practice the method experimented by Lee and Goldburg is cumbersome, for it involves a large number of lengthy sequences. In each sequence a strong r.f. field H_1 is applied at a distance $\Delta = \gamma H_1/\sqrt{2}$ from resonance for a time τ. No observation can be made while this r.f. field is on. The r.f. field is then cut off sharply and the subsequent free precession is observed. One must then wait for full recovery of the longitudinal magnetization, that is a few times T_1, before repeating the sequence with a different duration τ.

The amplitude of the free precession signal plotted as a function of τ, appears as a damped oscillation of frequency ω_e, whose envelope is the Fourier transform of the high resolution spectrum for the average Hamiltonian (2.23).

The main interest of this method of narrowing is pedagogical for it exhibits very clearly the principle of narrowing through spin rotation. Another of its weaknesses is that, as will be shown in section **C**(*b*)2, the reduction of the dipolar width D is of the order of (D/ω_e) that is of first order with respect to the precession velocity, in contrast to some other methods to be described later.

The method was tested on the resonance of ^{19}F in $Ca\,F_2$ with $H\|$ to a [111] axis where the resonance is narrowest. With an r.f. field of 5·2 gauss the free precession lifetime was indeed increased a few times.

(*b*) *Multipulse methods*

The multipulse methods of line narrowing represent a major breakthrough in the field of high resolution NMR in solids. In these methods the continuous (CW) precession of the spins described in the last section, is replaced by a succession of discrete quasi-instantaneous rotations produced by strong r.f. pulses.

The main advantage over the CW spin rotation is that in the intervals between the pulses a precession signal can be observed and thus monitored all through the decay of this precession. If the repetition rate of the pulse sequence is fast, many experimental points can be obtained during a single decay.

The enormous gain in time with respect to the cumbersome procedure described in the last section for CW precession, is what eventually turned an ingenious game with spins, into a practical method for high resolution in solids.

1. *An illustrative example*

In order to illustrate the mechanism of multipulse narrowing we describe first a pulse sequence (Haeberlen and Waugh, 1968) which is of pedagogical interest mainly, because it provides a direct link with the continuous spin precession around a 'magic' axis, of last section.

In this sequence, $2\pi/3$ rotations around the 'magic' axis are produced by short and intense pulses of an r.f. field ω_1, applied at a distance $\Delta=\omega_1/\sqrt{2}$ from resonance for times t_W such that:

$$\theta=\omega_e t_W=\omega_1\sqrt{(3)}t_W=2\pi/3. \tag{2.25}$$

Let

$$R_Z=\exp\left(-i\frac{2\pi}{3}I_Z\right)$$

be the unitary operator for such a rotation, where $0Z$ is along the effective field during the pulses, that is also along the magic axis in the rotating frame.

We can always select the xy axes of that frame so that:

$$I_Z = \frac{1}{\sqrt{3}}(I_x + I_y + I_z). \tag{2.26}$$

It is clear that the rotation R_Z produces on the spins (not on the orbital variables of course) a cyclical permutation $(xyz) = x \to y \to z \to x$ of the axes, and $R^\dagger = R^{-1}$ the inverse permutation $(xzy) = x \to z \to y \to x$, whence:

$$R_Z I_z R_Z^\dagger = I_x; \qquad R_Z^\dagger I_z R_Z = I_y, \tag{2.27}$$

and if, in the sense of eqn (2.22′), we write: $\exp(-\mathrm{i}2\pi I_Z) = 1$,

$$R_Z = \exp\left(-\mathrm{i}\frac{2\pi}{3}I_Z\right) = \exp(-\mathrm{i}2\pi I_Z)\left[\exp\left(\mathrm{i}\frac{2\pi}{3}I_Z\right)\right]^2.$$
$$= R_Z^\dagger R_Z^\dagger. \tag{2.28}$$

A continuous train of pulses R_Z separated by intervals 2τ is applied to the spin system. We define as a cycle a period $\tau_c = 6\tau$ represented in Fig. 2.4. Between pulses, in the rotating frame the evolution of the system is given by the unitary operator $U_z(t) = \exp(-\mathrm{i}\mathcal{H}_{0z}t)$ where $\mathcal{H}_{0z}$ is the dipolar Hamiltonian truncated with respect to the z axis. During a period this evolution is:

$$U(\tau_c) = U_z(\tau)R_Z U_z(2\tau)R_Z U_z(2\tau)R_Z U_z(\tau) \tag{2.29}$$

We define the Hamiltonians $\mathcal{H}_{0x}$ and $\mathcal{H}_{0y}$ which result from $\mathcal{H}_{0z}$ by the cyclical permutations (xyz) and (xzy) of the spin co-ordinates, that is such that:

$$R_Z \mathcal{H}_{0z} R_Z^\dagger = \mathcal{H}_{0x}; \qquad R_Z^\dagger \mathcal{H}_{0z} R_Z = \mathcal{H}_{0y}. \tag{2.30}$$

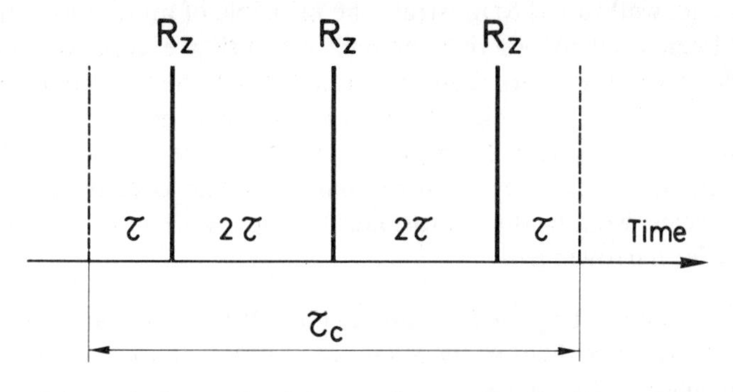

FIG. 2.4. High resolution cycle of three pulses causing $2\pi/3$ rotations around the magic axis.

Using (2.28), $U(\tau_c)$ can be rewritten as:

$$\begin{aligned} U(\tau_c) &= U_z(\tau)\underbrace{R_Z U_z(2\tau) R_Z^\dagger}\,\underbrace{R_Z^\dagger U_z(2\tau) R_Z}\,U_z(\tau) \\ &= U_z(\tau)U_x(2\tau)U_y(2\tau)U_z(\tau), \end{aligned} \tag{2.31}$$

where $U_x(t) = \exp(-\mathrm{i}\mathcal{H}_{0x}t)$, $U_y(t) = \exp(-\mathrm{i}\mathcal{H}_{0y}t)$. Representing for brevity $U_x(\tau)$ by the symbol X, etc. (2.30) becomes:

$$U(\tau_c) = [ZXXYYZ]. \tag{2.32}$$

We introduce now the assumption that the cycle $\tau_c = 6\tau$ is short, meaning thus that, say, $U_z(\tau) = \exp(-\mathrm{i}\mathcal{H}_{0z}\tau)$ differs little from $(1 - \mathrm{i}\mathcal{H}_{0z}\tau)$. To the extent that this condition is verified we get from (2.31):

$$U(\tau_c) \simeq 1 - \mathrm{i}\frac{\tau_c}{3}(\mathcal{H}_{0x} + \mathcal{H}_{0y} + \mathcal{H}_{0z}), \tag{2.33}$$

instead of $(1 - \mathrm{i}\tau_c\mathcal{H}_{0z})$ in the absence of pulses. It is easy to verify that $\mathcal{H}_{0z} + \mathcal{H}_{0x} + \mathcal{H}_{0y}$ is a sum of terms such as:

$$A_{12}\{(3I_z^1 I_z^2 - \mathbf{I}^1 . \mathbf{I}^2) + (3I_x^1 I_x^2 - \mathbf{I}^1 . \mathbf{I}^2) + (3I_y^1 I_y^2 - \mathbf{I}^1 . \mathbf{I}^2)\} = 0$$

and to the approximation of (2.33), $U(\tau_c) = 1$.

The attenuation of the free precession signal by the dipolar interaction is in that approximation completely quenched. To appreciate the actual narrowing one must calculate higher order terms in the expansion of (2.32) in powers of τ_c. The method for calculating these terms will be outlined in section **C**(b)2. Anticipating the results of this section we state that the next term of order $(\mathcal{H}_{0z}\tau_c)^2$ does *not* vanish in contrast to other more efficient cycles.

One should notice that the monitoring of the decay should be done by observing the signal always at the same position in the window between the pulses, say at times $n\tau_c$. If the observation time changes its position from cycle to cycle, the pulses cause the signal to oscillate wildly.

This cycle, well suited to illustrate the principle of multipulse narrowing, has not been used in practice for high resolution except for illustrative purposes. One of the reasons, already stated, is that the quenching of dipolar terms can be achieved to a higher approximation with better designed cycles. The other is that perfect $2\pi/3$ pulses off resonance are more difficult to produce than, say $\pi/2$ pulses on resonance. Before describing such more efficient cycles, we introduce a formalism which will enable us to calculate higher order narrowing.

2. (α) *The effective periodic Hamiltonian*. When the spins are manipulated, be it by a CW field or by sequence of pulses the Hamiltonian of the system is made of two parts:

$$\mathcal{H} = \mathcal{H}_e(t) + \mathcal{H}_i. \tag{2.34}$$

In (2.34), the 'external' Hamiltonian $\mathcal{H}_e(t)$ represents the coupling of the spins with the applied r.f. fields. The 'internal' Hamiltonian $\mathcal{H}_i$ will be the sum of all the Hamiltonians listed in **A**, among which we propose to average out the dipolar interaction and more generally all the parts which transforms like $T^{(2)}_{0z}(I)$. With respect to the external Hamiltonian $\mathcal{H}_e(t)$ we make the following assumptions:

(a) it is a periodical function of time;

(b) after a time τ_c the external Hamiltonian leaves the spin system in the same state.

It is important to realize that the two conditions are not equivalent. Thus for instance in **C**(b)1 the time dependence of the Hamiltonian representing an infinite train of $2\pi/3$ pulses around the magic axis, has as its period the interval 2τ between pulses, but the system returns to its initial state only after a time $\tau_c = 6\tau$. For brevity we shall say that if condition (b) is satisfied the Hamiltonian $\mathcal{H}_e(t)$ is 'cyclical'.

Let $U_e(t)$ be the unitary evolution operator such that:

$$\begin{cases} \dfrac{\mathrm{d}}{\mathrm{d}t} U_e(t) = -\mathrm{i}\mathcal{H}_e(t)U_e(t), \qquad U_e(0) = 1 \\[2ex] U_e(t) = U_e(0, t) = T \exp\left[-\mathrm{i}\displaystyle\int_0^t \mathcal{H}_e(t')\,\mathrm{d}t\right]. \end{cases} \tag{2.35}$$

If $\mathcal{H}_e(t)$ is periodical, for $t > \tau_c$:

$$\begin{aligned} U_e(t) &= U_e(0, t) = U_e(\tau_c, t)U_e(0, \tau_c) \\ &= U_e(0, t-\tau_c)U_e(0, \tau_c) = U_e(t-\tau_c)U_e(\tau_c), \end{aligned}$$

whence by induction:

$$U_e(n\tau_c) = [U_e(\tau_c)]^n. \tag{2.36}$$

If furthermore the system is cyclical:

$$U_e(\tau_c) = 1; \qquad U_e(t) = U_e(t-\tau_c),$$

and the evolution operator U_e itself is periodical.

The evolution operator $U(t)$ of the spin system in the presence of *both* the external Hamiltonian $\mathcal{H}_e(t)$ and of $\mathcal{H}_i$ is given by the equation:

$$\frac{\mathrm{d}}{\mathrm{d}t} U(t) = -\mathrm{i}[\mathcal{H}_e(t) + \mathcal{H}_i]U(t). \tag{2.37}$$

In the interaction representation we define $\tilde{U}_i(t)$ by:

$$U(t) = U_e(t)\tilde{U}_i(t) \tag{2.37'}$$

$\tilde{U}_i(t)$ obey the equation:

$$\begin{cases} \dfrac{d}{dt}\tilde{U}_i = -i\tilde{\mathcal{H}}_i(t)\tilde{U}_i \\ \tilde{\mathcal{H}}_i(t) = U_e^{\dagger}(t)\mathcal{H}_i U_e(t). \end{cases} \tag{2.38}$$

Since $U_e(t)$ is periodical and $\mathcal{H}_i$ is time-independent, $\tilde{\mathcal{H}}_i(t)$ is also periodical with the same period τ_c (but not cyclical anymore). If the frequency τ_c^{-1} is fast compared to the dipolar frequency, $\tilde{\mathcal{H}}_i(t)$ is to a good approximation reduced to its time average over one period and if the external Hamiltonian $\mathcal{H}_e(t)$ is chosen adequately this time average vanishes. The rate of evolution of $\tilde{U}_i(t)$ is then strongly reduced compared to the 'natural' evolution $U_i(t) = \exp(-i\mathcal{H}_i t)$. However the evolution of the spin system is not given by $\tilde{U}_i(t)$ but rather by the full evolution operator $U(t) = U_e(t)\tilde{U}_i(t)$. Within a period, $U_e(t)$ oscillates strongly. As a consequence, a slow regular decay of $U(t)$, which by Fourier transform yields a high resolution spectrum, can only be obtained if between pulses, the signal is always observed in the same 'window' that is at the same position in the cycle.

In the multipulse methods the external Hamiltonian $\mathcal{H}_e(t)$ acts only during a small part of the total period, or cycle, τ_c, the r.f. pulses being applied during time intervals t_W, in general much shorter than τ_c. The rest of the time, the spin system evolves freely under the influence of its internal, time independent Hamiltonian $\mathcal{H}_i$. Let R_j be the unitary operator describing the evolution of the spin system during an r.f. pulse j of duration t_W^j. If, as is usually the case, the r.f. field is very much stronger than the dipolar local field it is permissible to neglect during the time t_W^j the effect of the 'internal' Hamiltonian $\mathcal{H}_i$, in which case R_j is a pure spin rotation operator. This is the short pulse approximation. Let $R_1, \ldots, R_j, \ldots R_p$ be the spin rotation operators during a cycle. The assumption of a cyclical external Hamiltonian $\mathcal{H}_e(t)$ implies that:

$$R_p R_{p-1} \cdots R_j \cdots R_2 R_1 = 1. \tag{2.39}$$

Let $\tau_0, \tau_1, \ldots \tau_p$ be the intervals between the pulses when the spin system evolves freely and $Z_0 = \exp(-i\mathcal{H}_i\tau_0), \ldots Z_q = \exp(-i\mathcal{H}_i\tau_q) \cdots$ the evolution operators during these intervals. The evolution operator for the whole cycle τ_c will be:

$$U(\tau_c) = Z_p R_p Z_{p-1} \cdots R_2 Z_1 R_1 Z_0. \tag{2.40}$$

Let us assume for argument's sake that there are 3 spin rotations R_1, R_2, R_3 in the cycle with $R_3R_2R_1 = 1$. $U(\tau_c) = Z_3R_3Z_2R_2Z_1R_1Z_0$ can be rewritten:

$$U(\tau_c) = Z_3R_3R_2R_1[(R_2R_1)^{-1}Z_2(R_2R_1)]R_1^{-1}Z_1R_1Z_0, \tag{2.41}$$

or, since $R_3R_2R_1=1$

$$U(\tau_c)=Z_3\tilde{Z}_2\tilde{Z}_1Z_0, \tag{2.42}$$

where:

$$\tilde{Z}_1=R_1^{-1}Z_1R_1, \qquad \tilde{Z}_2=(R_2R_1)^{-1}Z_2(R_2R_1), \tag{2.43a}$$

or yet:

$$\tilde{Z}_1=\exp(-i\tilde{\mathcal{H}}_i^1\tau_1), \qquad \tilde{Z}_2=\exp(-i\tilde{\mathcal{H}}_i^2\tau_2), \tag{2.43b}$$

with:

$$\tilde{\mathcal{H}}_i^1=R_1^{-1}\mathcal{H}_iR_1, \qquad \tilde{\mathcal{H}}_i^2=(R_2R_1)^{-1}\mathcal{H}_i(R_2R_1). \tag{2.43c}$$

According to eqns (2.42) and (2.43), the evolution of the spin system through the cycle τ_c can be described by an internal Hamiltonian which is $\mathcal{H}_i$ during the interval τ_0, then $\tilde{\mathcal{H}}_i^1$ derived from $\mathcal{H}_i$ by the rotation $P_1^{-1}=R_1^{-1}$ during τ_1, $\tilde{\mathcal{H}}_i^2$ derived from $\mathcal{H}_i$ by the rotation $P_2^{-1}=(R_2R_1)^{-1}$ during τ_2 and finally during the last interval τ_3, again $\mathcal{H}_i$.

The generalization to more than three rotations R_j is obvious. The successive internal Hamiltonians are:

$$\begin{cases} \tilde{\mathcal{H}}_i^0=\mathcal{H}_i, \ldots \tilde{\mathcal{H}}_i^q=P_q^{-1}\mathcal{H}_iP_q, \ldots \tilde{\mathcal{H}}_i^n=\mathcal{H}_i \\ P_1=R_1, P_2=R_2R_1, \ldots P_q=R_q\cdots R_2R_1. \end{cases} \tag{2.44}$$

The multipulse cycle described in **C**(b)1 is an example of the above.

(β) *The Magnus expansion.* The Magnus expansion provides approximate expressions for the evolution operator,

$$U(\tau_c)=T\exp\left[-i\int_0^{\tau_c}\mathcal{H}(t')\,dt'\right], \tag{2.45}$$

associated with a periodic Hamiltonian $\mathcal{H}(t)$ of period τ_c. The standard expansion of (2.45),

$$U(\tau_c)=1-i\int_0^{\tau_c}\mathcal{H}(t')\,dt'-\int_0^{\tau_c}dt_1\int_0^{t_1}dt_2\,\mathcal{H}(t_1)\mathcal{H}(t_2)\ldots, \tag{2.46}$$

suffers from the defect that at each stage n of the expansion the approximate operator $U^{(n)}(\tau_c)$ is *not* unitary. In the Magnus expansion a static hermitian Hamiltonian $\bar{\mathcal{H}}(\tau_c)$ is *defined* by the relation:

$$U(\tau_c)=\exp\{-i\tau_c\bar{\mathcal{H}}(\tau_c)\}. \tag{2.47}$$

$\bar{\mathcal{H}}(\tau_c)$ is then expanded into a series of decreasing terms:

$$\bar{\mathcal{H}}(\tau_c)=\bar{\mathcal{H}}^{(0)}(\tau_c)+\bar{\mathcal{H}}^{(1)}(\tau_c)+\bar{\mathcal{H}}^{(2)}(\tau_c)+\ldots. \tag{2.48}$$

The advantage of this expansion is that at all stages of the approximation the corresponding evolution operator,

$$U_n(\tau_c) = \exp\left\{-i\tau_c \sum_1^n \mathcal{H}^{(i)}(\tau_c)\right\}, \tag{2.49}$$

is unitary.

It is convenient to introduce a frequency Ω which expresses the magnitude, that is the spread of the eigenvalues, of the Hamiltonian $\mathcal{H}(t)$. When $\mathcal{H}(t)$ is the effective Hamiltonian $\tilde{\mathcal{H}}_i(t)$ of eqn (2.38), it has the same eigenvalues as the internal Hamiltonian $\mathcal{H}_i$ and Ω is of the order of $D = \gamma H'_L$ where H'_L is the dipolar local field. We assume that $\Omega\tau_c = x$ is a small number and the nth term of the expansion (2.46) is then of order $x^n = (\Omega\tau_c)^n$. Similarly in the Magnus expansion (2.48) $\tau_c\mathcal{H}^{(0)}(\tau_c)$ is taken of order x and $\tau_c\mathcal{H}^{(n)}(\tau_c)$ of order x^{n+1}. We calculate the various terms of (2.48) by equating terms of the same order in the expansions (2.46) and (2.49). Equation (2.49) yields:

$$U_n(\tau_c) = 1 - i\tau_c\mathcal{H}^{(0)} - [i\tau_c\mathcal{H}^{(1)} + \tfrac{1}{2}\tau_c^2(\mathcal{H}^{(0)})^2]$$
$$- \left[-i\tau_c\mathcal{H}^{(2)} + \tfrac{1}{2}\tau_c^2(\mathcal{H}^{(0)}\mathcal{H}^{(1)} + \mathcal{H}^{(1)}\mathcal{H}^{(0)}) - i\frac{\tau_c^3}{3!}(\mathcal{H}^{(0)})^3\right]. \tag{2.50}$$

The identification of terms of order x^n in (2.46) and (2.50) leads after a little algebra to:

$$\mathcal{H}^{(0)} = \frac{1}{\tau_c}\int_0^{\tau_c} \mathcal{H}(t')\,dt';$$
$$\mathcal{H}^{(1)} = \frac{i}{2\tau_c}\int_0^{\tau_c} dt_1 \int_0^{t_1} dt_2[\mathcal{H}(t_1), \mathcal{H}(t_2)];$$
$$\mathcal{H}^{(2)} = -\frac{1}{6\tau_c}\int_0^{\tau_c} dt_1 \int_0^{t_1} dt_2 \int_0^{t_2} dt_3 \times \{[\mathcal{H}(t_1), [\mathcal{H}(t_2), \mathcal{H}(t_3)]]$$
$$+ [\mathcal{H}(t_3), [\mathcal{H}(t_2), \mathcal{H}(t_1)]]\}. \tag{2.51}$$

One may notice that all terms in (2.51) apart from $\mathcal{H}^{(0)}$ contain commutators of $\mathcal{H}(t)$ at various times. This is to be expected: if for all times t and t', $\mathcal{H}(t)$ and $\mathcal{H}(t')$ did commute, the time ordered operator (2.45) would become the ordinary operator $\exp(-i\int_0^{\tau_c}\mathcal{H}(t')\,dt') = \exp(-i\tau_c\mathcal{H}^{(0)})$ and all terms $\mathcal{H}^{(n)}$ with $n > 0$ would vanish.

The expressions for $\mathcal{H}^{(n)}$ for $n > 2$ become quite complicated and are seldom used.

If $x = \Omega\tau_c$ is a small number the first few terms of (2.48) provide a good approximation for $\bar{\mathcal{H}}(\tau_c)$ and $U(\tau_c)$ and this independently of whether the *infinite* series (2.48) converges or not. If it does not, the approximation might lead to values of $U(N\tau_c)$ seriously in error for very large N. However in practice the spin signal will have decayed much earlier than $N\tau_c$ for other causes not taken into account, such as instrumental faults or spin-lattice relaxation, and so the convergence of the series is of little practical importance. We shall come back to some of the problems raised by the definition (2.47) of $\bar{\mathcal{H}}(\tau_c)$ at the end of this chapter.

Whatever the cycle chosen, in order to be of any use, it must at least annul the first order term $\mathcal{H}^{(0)}$ in (2.51) (more precisely, the contribution to $\mathcal{H}^{(0)}$ of the dipolar and quadrupole couplings). In the short-pulse approximation this is obtained by the condition:

$$\tau_0\mathcal{H}_i+\tau_1\mathcal{H}_i^{(1)}+\cdots+\tau_{n-1}\mathcal{H}_i^{(n-1)}+\tau_n\mathcal{H}_i=0, \qquad (2.52)$$

where as defined in (2.44):

$$\mathcal{H}_i^{(p)}=P_p^{-1}\mathcal{H}_iP_p; \qquad P_p=R_p\cdots R_2R_1. \qquad (2.53)$$

To select more efficient cycles where the next average Hamiltonian $\mathcal{H}^{(1)}$ given by (2.51) also vanishes, we must consider more carefully the dependence of the various terms $\mathcal{H}^{(n)}$ of the expansion (2.48) on τ_c. The question of what becomes of these terms when the period τ_c of the cycle is changed, has a meaning only if we state unambiguously what is left unchanged. In the short-pulse approximation this is straightforward: all the rotations R_i (or P_i) are left unchanged and all the time intervals $\tau_0, \ldots \tau_p$ are multiplied by the same factor λ as τ_c. Two cycles which differ in this manner are called homothetical. If the pulses are not infinitely short we have to introduce the durations t_W of the pulses and a simple time *dilation* of the τ_j *and* the t_W would not be correct. Indeed the spin rotation angles $\theta_W=\omega_1t_W$ (where ω_1/γ is the amplitude of the r.f. field) would be changed and the evolutions of two Hamiltonians $\mathcal{H}(t)$ differing only by a time *dilation*, would be qualitatively different. We can circumvent these difficulties by reasoning on the 'interaction representation' Hamiltonian $\tilde{\mathcal{H}}(t)=U^{\dagger}(t)\mathcal{H}_iU_e(t)$ defined in (2.38) rather than on the total Hamiltonian $\mathcal{H}_e(t)+\mathcal{H}_i$. We then compare cycles where $\tilde{\mathcal{H}}(t)$ depends on the time only through the ratio $(t/\tau_c)=\lambda$. For pulses of finite duration this involves adjustments in the strength of the r.f. fields, the details of which we need not go into. If we make this assumption a formula such as that giving say $\mathcal{H}^{(1)}$ becomes:

$$\mathcal{H}^{(1)}=i\frac{\tau_c}{2}\int_0^1 d\lambda_1\int_0^{\lambda_1} d\lambda_2[\mathcal{H}(\lambda_1), \mathcal{H}(\lambda_2)], \qquad (2.51')$$

and thus can be written $\mathcal{H}^{(1)}=\tau_cF^{(1)}$, where $F^{(1)}$ is an operator that does not explicitly contain τ_c. With this convention,

$$\bar{\mathcal{H}}(\tau_c)=\sum_{n=1}^{\infty}\tau_c^nF^{(n)} \qquad (2.54)$$

is a *power* series of τ_c.

We now prove the following theorem. If the cycle is symmetrical, that is if for $0 \leq t \leq \tau_c$, $\tilde{\mathscr{H}}(t) = \tilde{\mathscr{H}}(\tau_c - t)$, all the $F^{(n)}$ with n odd vanish. For the theorem to be valid it is sufficient to prove that $\bar{\mathscr{H}}(-\tau_c) = \bar{\mathscr{H}}(\tau_c)$ where $\bar{\mathscr{H}}(-\tau_c)$ is *defined* through:

$$U(-\tau_c) = \exp\{-i(-\tau_c)\bar{\mathscr{H}}(-\tau_c)\}. \tag{2.55}$$

There is no ambiguity in the definition of $U(-\tau_c)$ for $\tau_c > 0$. It relates the state of the system at time $t = 0$ to the time $-\tau_c < 0$. The only obvious change from (2.45) is the replacement of the time ordered symbol T by $\tilde{T}$, defined by:

$$\tilde{T}A(t_1)A(t_2) = \begin{cases} A(t_1)A(t_2) & \text{if } t_1 < t_2; \\ A(t_2)A(t_1) & \text{if } t_1 > t_2, \end{cases}$$
$$U(-\tau_c) = \tilde{T} \exp\left\{-i \int_0^{-\tau_c} \mathscr{H}(t')\, dt'\right\}. \tag{2.56}$$

Another obvious remark is that a change of variable $t' \to t''$ such as say $t'' = -t'$ or $t'' = \tau_c - t'$ which reverses the order of two values t'_1 and t'_2 leads to the interchange of the symbols $\tilde{T}$ and T.

Changing t' into $-t'$ yields in (2.56):

$$U(-\tau_c) = T \exp\left\{i \int_0^{\tau_c} \mathscr{H}(-t')\, dt'\right\}. \tag{2.57}$$

Since $\mathscr{H}$ is periodic of period τ_c *and* symmetrical, $\mathscr{H}(-t') = \mathscr{H}(\tau_c - t') = \mathscr{H}(t')$ and

$$U(-\tau_c) = T \exp\left\{i \int_0^{\tau_c} \mathscr{H}(t')\, dt\right\}. \tag{2.58}$$

On the other hand the operator $U^\dagger(\tau_c)$, because of the relation $(AB)^\dagger = B^\dagger A^\dagger$ is clearly:

$$U^\dagger(\tau_c) = \exp[i\tau_c \bar{\mathscr{H}}(\tau_c)] = \tilde{T} \exp\left\{i \int_0^{\tau_c} \mathscr{H}(t')\, dt'\right\}, \tag{2.59}$$

or since $\mathscr{H}(t') = \mathscr{H}(\tau_c - t')$:

$$U^\dagger(\tau_c) = \tilde{T} \exp\left\{i \int_0^{\tau_c} \mathscr{H}(\tau_c - t')\, dt'\right\}. \tag{2.60}$$

Making the substitution $t' \to \tau_c - t'$:

$$U^\dagger(\tau_c) = \exp[i\tau_c \bar{\mathscr{H}}(\tau_c)] = T \exp\left\{i \int_0^{\tau_c} \mathscr{H}(t')\, dt'\right\} = U(-\tau_c). \tag{2.61}$$

Comparing (2.55) and (2.61) we get: $\bar{\mathscr{H}}(\tau_c) = \bar{\mathscr{H}}(-\tau_c)$.

Symmetrical cycles have thus the great advantage that if the cycle has been adjusted for $\mathscr{H}^{(0)}(\tau_c)=0$, the first non-vanishing term in the Magnus expansion is $\mathscr{H}^{(2)}(\tau_c)$. It can be shown (Mehring and Waugh, 1972) that in that case,

$$\mathscr{H}^{(2)}(\tau_c)=\mathscr{H}^{(2)}(\tau_c/2), \tag{2.61$'$}$$

which simplifies the calculation of $\mathscr{H}^{(2)}$ by reducing the integration range. It is easy to verify that the cycle of section **C**(b)1 made of three $2\pi/3$ pulses is not symmetrical and therefore is rather inefficient for quenching the dipolar interaction.

3. *Two efficient multipulse cycles*

We now describe two cycles among the most widely used for high resolution NMR in solids.

(α) *The WHH–4 sequence.* The first multipulse sequence to yield genuine high resolution spectra in solids was the WHH–4 sequence (Waugh, Huber, and Haeberlen, 1968). This sequence is a series of four $\pi/2$ pulses at the Larmor frequency ω_0 successively along the axes $0x$, $-0y$, $0y$, $-0x$ (Fig. 2.5(a)) separated by intervals τ or 2τ, with $\tau_c=6\tau$. In the short-pulse approximation the unitary operator $U(\tau_c)$ could be written straightaway using (2.44), as:

$$U(\tau_c)=Z_4\tilde{Z}_3\tilde{Z}_2\tilde{Z}_1Z_0, \tag{2.62}$$

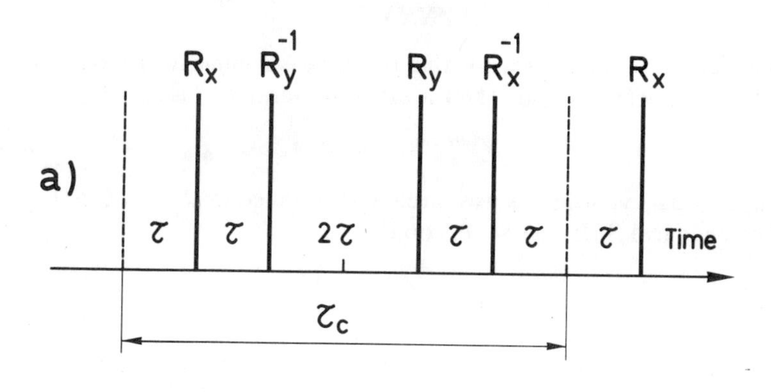

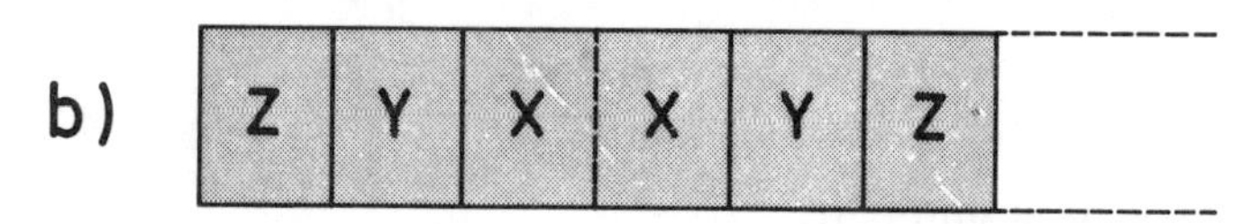

FIG. 2.5. The WHH–4 sequence. (a) Four cycle corresponding to $\pi/2$ rotations around $0x$, $-0y$, $0y$ and $-0x$. (b) Effective unitary operators in the various time intervals of the cycle.

with:

$$\begin{cases} Z_j = \exp(-\mathrm{i}\tau_j\mathcal{H}_\mathrm{i}), \quad \tilde{Z}_j = P_j^{-1}Z_jP_j, \quad P_j = R_j \cdots R_2R_1; \\ R_1 = R_x = R\left(\frac{\pi}{2}, \hat{x}\right), \quad R_2 = R_{\bar{y}} = R\left(\frac{\pi}{2}, -\hat{y}\right) = R_y^{-1}; \\ R_3 = R_y, \quad R_4 = R_{\bar{x}} = R_x^{-1}; \\ \tau_0 = \tau_1 = \tau_3 = \tau_4 = \tau, \quad \tau_2 = 2\tau, \quad \tau_c = 6\tau. \end{cases} \tag{2.63}$$

In practice it turns out to be simpler not to compute the various P_j but rather to write directly $U(\tau_c)$ as it follows from Fig. 2.5(a) and to determine its value by inspection of the Fig. 2.5(b). Let us first define the Hamiltonians $\mathcal{H}_z$, $\mathcal{H}_x$, $\mathcal{H}_y$ as:

$$\mathcal{H}_z = \mathcal{H}_\mathrm{i}; \quad \mathcal{H}_x = R_y\mathcal{H}_zR_y^{-1}; \quad \mathcal{H}_y = R_x^{-1}\mathcal{H}_zR_x; \tag{2.64}$$

$$Z = \exp(-\mathrm{i}\mathcal{H}_z\tau); \quad X = \exp(-\mathrm{i}\mathcal{H}_x\tau), \quad Y = \exp(-\mathrm{i}\mathcal{H}_y\tau); \tag{2.65}$$

whence:

$$\begin{aligned} U(\tau_c) &= ZR_x^{-1}ZR_yZZR_y^{-1}ZR_xZ \\ &= ZR_x^{-1}ZXXZR_xZ \\ &= ZYXXYZ. \end{aligned} \tag{2.66}$$

It is clear from this expression that the cycle is symmetrical and that all the odd terms of $\bar{\mathcal{H}}(\tau_c)$ vanish. The lowest order term $\mathcal{H}^{(0)}$ is clearly:

$$\mathcal{H}^{(0)} = \mathcal{H}_z + \mathcal{H}_x + \mathcal{H}_y. \tag{2.67}$$

Its dipolar part vanishes as was already mentioned in section **C**(b)1.

The chemical shift part δI_z becomes:

$$\tfrac{1}{3}\delta(I_x + I_y + I_z) = \frac{1}{\sqrt{3}}\,\delta I_Z, \tag{2.68}$$

where $0Z$ is along the axis [111]. The chemical shift is reduced by $\sqrt{3}$. The next term $\mathcal{H}^{(2)}$ can be calculated using (2.51). The calculation, straightforward although somewhat lengthy, yields for the dipolar part of $\mathcal{H}_\mathrm{i}$:

$$\mathcal{H}_\mathrm{D}^{(2)} = \frac{\tau^2}{18}[(\mathcal{H}_{\mathrm{D}x} - \mathcal{H}_{\mathrm{D}z}), [\mathcal{H}_{\mathrm{D}y}, \mathcal{H}_{\mathrm{D}x}]] = \frac{\tau_c^2}{648}[\cdots]; \tag{2.69}$$

and for the chemical shift part:

$$\mathcal{H}_\mathrm{S}^{(2)} = -\frac{\tau^2}{18}\delta^3(I_y - 2I_x + 4I_z). \tag{2.70}$$

One may notice in $\mathcal{H}_D^{(2)}$ that $\tau_c\mathcal{H}_D^{(2)}$ is indeed of order $(\tau_c\mathcal{H}_D)^3$ but with, as a bonus, a small numerical factor (1/648) which increases the efficiency of the narrowing.

We do not write out in $\mathcal{H}^{(2)}$ the cross-term between $\mathcal{H}_D$ and $\mathcal{H}_S$ which can also be calculated from (2.51).

The WHH–4 sequence is greatly superior to that described in **C**(b)1. Besides the absence of $\mathcal{H}^{(1)}$ there is also on the technical side the greater simplicity of the pulse sequence: $\pi/2$ pulses at resonance rather than $2\pi/3$ pulses at a prescribed distance from resonance.

(β) *The MREV sequence.* Another sequence of considerable practical interest is the so-called MREV sequence discovered independently by Mansfield (1971) and Rhim, Elleman, and Vaughan (1973a, b). The sequence is represented in Fig. 2.6.

It involves two successive WHH–4 sequences where the first differs from the second by the interchange of R_x and $R_{\bar{x}} = R_x^{-1}$.

Writing out explicitly the cycle and using the same inspection method which yielded (2.66), one finds:

$$U(\tau_c) = [ZYXXYZ, Z\bar{Y}XX\bar{Y}Z]. \qquad (2.71)$$

In (2.71) $\bar{Y} = \exp(-i\tau\mathcal{H}_{\bar{y}})$ and $\mathcal{H}_{\bar{y}} = R_x\mathcal{H}_z R_x^{-1}$ is the transform of $\mathcal{H}_i = \mathcal{H}_z$ by a $\pi/2$ rotation around $-0x$ rather than $0x$. For the dipolar Hamiltonian $\mathcal{H}_{D\bar{y}} = \mathcal{H}_{Dy}$ but for the chemical shift term the sign is reversed: $I_{\bar{y}} =$

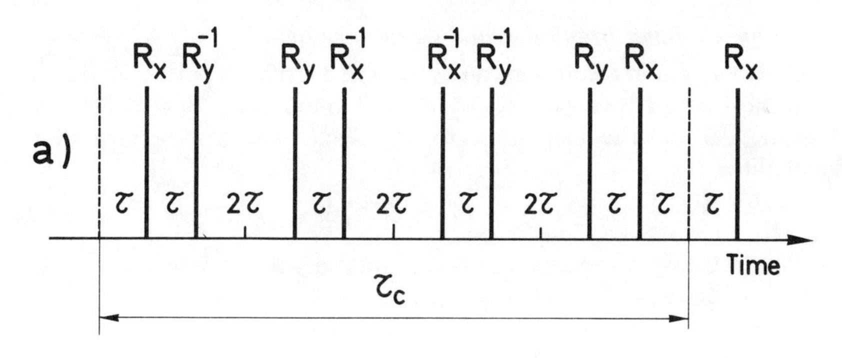

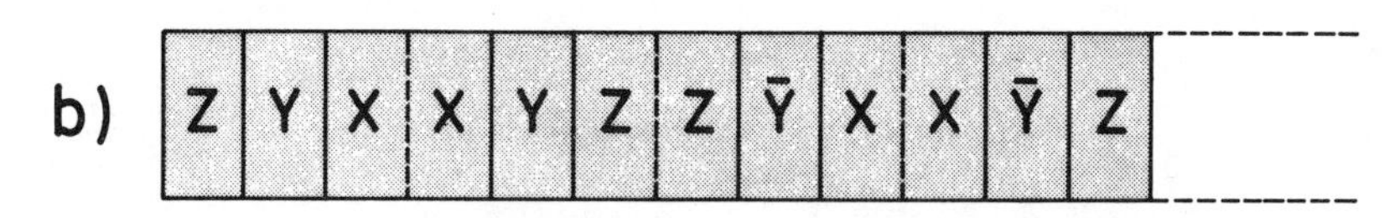

FIG. 2.6. The MREV sequence. (a) Eight-pulse cycle, corresponding to $\pi/2$ rotations around $\pm 0x$ or $\pm 0y$. (b) Effective unitary operators in the various time intervals of the cycle.

$R_x^{-1}I_zR_x = -R_xI_zR_x^{-1} = -I_y$. As far as the dipolar interaction is concerned the cycle (2.71) is then symmetrical and the terms $\mathcal{H}_D^{(0)}$ and $\mathcal{H}_D^{(1)}$ are missing as in WHH–4 and $\mathcal{H}_D^{(2)}$ has the same expression as in (2.69).

For the chemical shift Hamiltonian one finds:

$$(\delta I_z)^{(0)} = \frac{\delta}{3}(I_z + I_x) = \delta\frac{\sqrt{2}}{3}\frac{1}{\sqrt{2}}(I_z + I_x)$$
$$= (\sqrt{2}/3)\delta I_{Z'},$$

where $0Z'$ is the axis [101].

The next order term $(\delta I_z)^{(1)}$ does *not* vanish, because, as explained above the cycle (2.71) is symmetrical for the dipolar interaction but *not* for the chemical shift, whence a Hamiltonian $\mathcal{H}_S^{(1)}$:

$$\mathcal{H}_S^{(1)} = \frac{\tau}{3}\delta^2(I_z - I_x) \tag{2.72}$$

One may question the usefulness of a sequence, twice as long and more complicated than WHH–4, no better for the dipolar quenching and with a zero order chemical shift, smaller by a factor $\sqrt{\frac{3}{2}}$, and perturbed by a first order correction (2.72).

The main superiority of MREV over WHH–4 is the possibility of correcting much more efficiently the effects of pulse imperfections such as in particular the inhomogeneity of the r.f. field H_1 as will appear in **C**(b)6.

4. Off-resonance irradiation and second averaging

A further gain in resolution can be obtained if the r.f. pulses are detuned from the Larmor frequency ω_0 by a small amount $\omega_0 - \omega = \Delta$ (Pines and Waugh, 1972c). The magnitude to be selected for Δ is defined by the inequalities

$$1/\tau_c \gg |\Delta| \gg \bar{\mathcal{H}}_D(\tau_c)|. \tag{2.73}$$

We study the spin motion in a frame rotating at the frequency ω. This introduces into the spin Hamiltonian an extra Zeeman term ΔI_z which, provided the condition $|\Delta\tau_c| \ll 1$ is verified, brings an extra contribution to the average Hamiltonian $\bar{\mathcal{H}}(\tau_c)$. This contribution according to (2.68) and (2.71) is:

$$\begin{aligned}\delta\mathcal{H}^{(0)} &= \frac{1}{\sqrt{3}}\Delta I_Z \quad \text{for WHH–4;} \\ &= \frac{\sqrt{2}}{3}\Delta I'_Z \quad \text{for MREV.}\end{aligned} \tag{2.74}$$

If the second condition (2.73) is satisfied the contribution of $\delta\mathcal{H}^{(0)}$ to $\bar{\mathcal{H}}(\tau_c)$ is much greater than that of $\bar{\mathcal{H}}_D$ and in the latter, only the part that commutes with $\delta\mathcal{H}^{(0)}$ must be retained. The chemical shift part $\mathcal{H}_S^{(0)}$ commutes naturally with $\delta\mathcal{H}^{(0)}$ and is not effected by it.

In WHH–4, the second order term $\mathcal{H}_D^{(2)}$ of (2.69) (apart from the factor $\tau^2/18$) can be rewritten:

$$[\mathcal{H}_{Dx}, [\mathcal{H}_{Dy}, \mathcal{H}_{Dx}]] - R[\mathcal{H}_{Dx}, [\mathcal{H}_{Dz}, \mathcal{H}_{Dy}]]R^{\dagger}, \qquad (2.75)$$

where R is the operator for a permutation $z \to y \to x$ that is a rotation of angle $-2\pi/3$ around the axis $0Z$ parallel to [111]. Using the relation $\mathcal{H}_{Dx} + \mathcal{H}_{Dy} + \mathcal{H}_{Dz} = 0$, eqn (2.75) can be rewritten:

$$[\mathcal{H}_{Dx}, [\mathcal{H}_{Dy}, \mathcal{H}_{Dx}]] - R[\mathcal{H}_{Dx}, [\mathcal{H}_{Dy}, \mathcal{H}_{Dx}]]R^{\dagger}. \qquad (2.76)$$

Equation (2.76) is the difference of two expressions which can be deduced from each other by a rotation $\pm 2\pi/3$ around $0Z$. It is clear that the secular part of (2.76) with respect to I_Z that is the part invariant with respect to a rotation around $0Z$ vanishes. Thus the second averaging combined with a WHH–4 cycle annuls $\mathcal{H}_D^{(2)}$ as well.

It can be shown that a similar complete cancellation of $\mathcal{H}_D^{(2)}$ does not occur in the MREV cycle where the axis of quantization $0Z'$ is [101]. One should not draw conclusions about the relative superiority of WHH–4 over MREV for, in general, other causes such as pulse imperfections outweigh the effect of $\mathcal{H}_D^{(2)}$.

5. Pulses of finite width

In practice the r.f. pulses have a duration t_w which is not negligibly small compared to the period τ_c. Thus for instance for protons a $\pi/2$ pulse with a fairly large rotating amplitude of 60 gauss has a duration of 1 μsec. Since the period τ_c must be as short as possible: $\tau_c \ll T_2$, it follows that the ratio t_W/τ_c cannot be very small.

The evolution of the spin system is described by the general equations (2.35) and (2.38). The Magnus expansion (2.51) requires the knowledge of the effective Hamiltonian $\tilde{\mathcal{H}}_i(t)$ and therefore also of the unitary operator $U_e(t)$ which describes the evolution of the spin system under the sole influence of the r.f. field. As an illustration we take first two r.f. pulses R_y and R_y^{-1} of angle β around y and $-y$. If the pulses have a finite width the r.f. field has a time-dependent amplitude $\omega_{1y}(t)$ along y. If the pulses have a square shape, $\omega_{1y}(t) = \omega_1$ during the first pulse of duration t_W and $\omega_{1y}(t) = -\omega_1$ during the second pulse, with $\omega_1 t_W = \beta$. The cycle is shown in Fig. 2.7(a).

We define an angle $\alpha_y(t)$ through $\alpha_y(t) = \int_0^t \omega_{1y}(t')\, dt$. The time dependence of $\alpha_y(t)$ is shown in Fig. 2.7(b).

It is clear that $U_e(t)$, solution of $dU_e/dt = -i\omega_{1y}(t) I_y U_e(t)$ can be written $U_e(t) = \exp[-i\alpha_y(t) I_y]$ and $\tilde{\mathcal{H}}_i(t)$ is

$$\tilde{\mathcal{H}}_i(t) = \exp[i\alpha_y(t) I_y]\mathcal{H}_i \exp[-i\alpha_y(t) I_y]. \qquad (2.77)$$

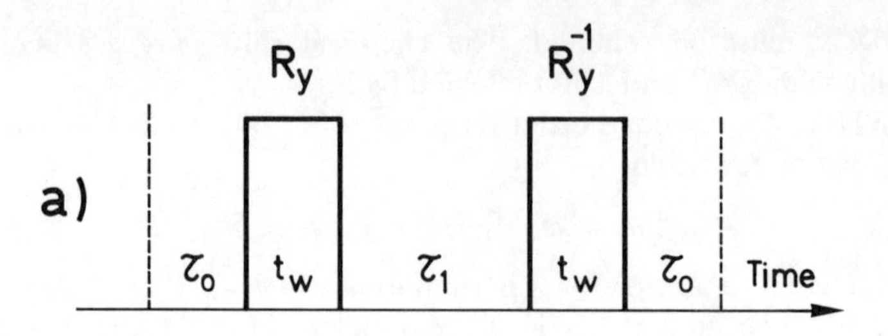

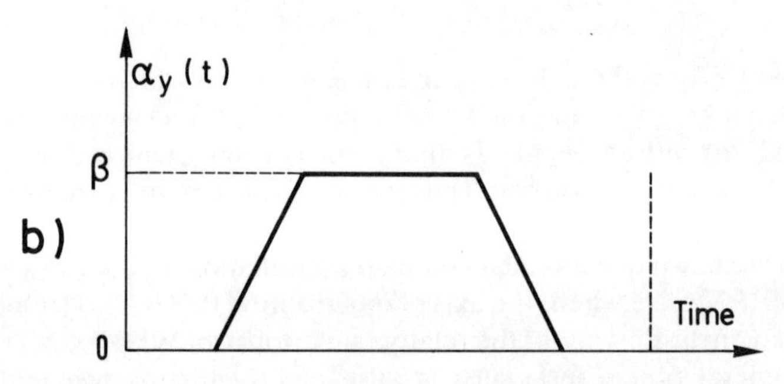

FIG. 2.7. Cycle of two pulses of finite width. (a) Pulse sequence. (b) Variation of the angle $\alpha_y(t)$ during the cycle.

The dipolar Hamiltonian $\mathcal{H}_\mathrm{i}$ which is the truncated Hamiltonian $\mathcal{H}_{\mathrm{D}z}$ transforms under rotation like $T_0^{(2)}(I)$ and using the eqns (2.77) and (1.11a) we have:

$$\begin{aligned}\tilde{\mathcal{H}}_\mathrm{i}(t) = \tilde{T}_0^{(2)}(t) = \tfrac{1}{2}(3\cos^2\alpha_y(t)-1)T_0^{(2)} \\ -\sqrt{(\tfrac{3}{2})}\sin\alpha_y(t)\cos\alpha_y(t)(T_{-1}^{(2)}-T_1^{(2)}) \\ +\sqrt{(\tfrac{3}{8})}\sin^2\alpha_y(t)(T_2^{(2)}+T_{-2}^{(2)}).\end{aligned} \tag{2.78}$$

We can use (2.78) to calculate the first term $\tau_\mathrm{c}\mathcal{H}^0(\tau_\mathrm{c})$ of the Magnus expansion: $\tau_\mathrm{c}\mathcal{H}^{(0)}(\tau_\mathrm{c}) = \int_0^{\tau_\mathrm{c}} \tilde{\mathcal{H}}_\mathrm{i}(t)\,\mathrm{d}t$. From the variation of $\alpha_y(t)$ given in Fig. 2.7(b) we find:

$$\begin{aligned}\tau_\mathrm{c}\mathcal{H}^{(0)}(\tau_\mathrm{c}) = T_0^{(2)}\left\{\tau_0+\tau_0+\tau_1\frac{3\cos^2\beta-1}{2}+2\frac{t_\mathrm{w}}{\beta}\int_0^\beta\frac{3\cos^2\alpha-1}{2}\,\mathrm{d}\alpha\right\} \\ -\sqrt{(\tfrac{3}{2})}(T_{-1}^{(2)}-T_1^{(2)})\left\{\tau_1\sin\beta\cos\beta+2\frac{t_\mathrm{w}}{\beta}\int_0^\beta\sin\alpha\cos\alpha\,\mathrm{d}\alpha\right\} \\ +\sqrt{(\tfrac{3}{8})}(T_2^{(2)}+T_{-2}^{(2)})\left\{\tau_1\sin^2\beta+2\frac{t_\mathrm{w}}{\beta}\int_0^\beta\sin^2\alpha\,\mathrm{d}\alpha\right\}.\end{aligned} \tag{2.79}$$

Let us now consider a WHH–4 cycle with pulses of duration t_W (Fig. 2.8). The unitary operator $U_e(t)$ is now of the form:

$$U_e(t) = \exp[-i\alpha_y(t)I_y]\exp[-i\alpha_x(t)I_x],$$

with:

$$\alpha_x(t) = \int_0^t \omega_{1x}(t')\,dt'.$$

The time dependences of $\alpha_x(t)$ and $\alpha_y(t)$ are shown in Fig. 2.9. The maximum values of $|\alpha_x(t)|$ and $|\alpha_y(t)|$ are both equal to β. It is clear on this figure that the cycle has reflection symmetry, so that $\mathcal{H}^{(1)}(\tau_c)$ vanishes. The calculation of:

$$\mathcal{H}_D^{(0)}(\tau_c) = \frac{1}{\tau_c}\int_0^{\tau_c} \tilde{\mathcal{H}}_i(t)\,dt,$$

where,

$$\tilde{\mathcal{H}}_i(t) = U_e^{\dagger}(t)\mathcal{H}_{Dz}U_e(t),$$

is a little complicated by the existence of a time interval where both $\alpha_x(t)$ and $\alpha_y(t)$ differ from zero. This complication can be short-circuited as follows.

We are looking for the possibility of making $\mathcal{H}_D^{(0)}(\tau_c) = 0$. It is equivalent to annul:

$$\mathcal{H}_D'^{(0)}(\tau_c) = \exp(i\beta I_x)\mathcal{H}_D^{(0)}(\tau_c)\exp(-i\beta I_x).$$

The right-hand side is the average of:

$$\tilde{\mathcal{H}}_i'(t) = U_e'^{\dagger}(t)\mathcal{H}_{Dz}U_e'(t),$$

where:

$$U_e'(t) = \exp[-i\alpha_y(t)I_y]\exp[-i\alpha_x'(t)I_x],$$

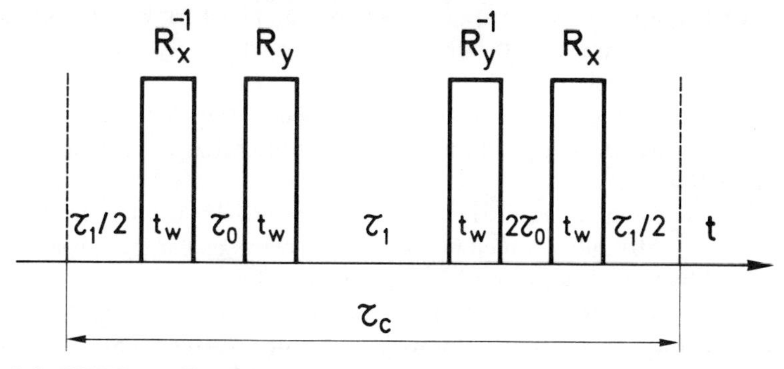

FIG. 2.8. WHH–4 cycle with pulses of finite width.

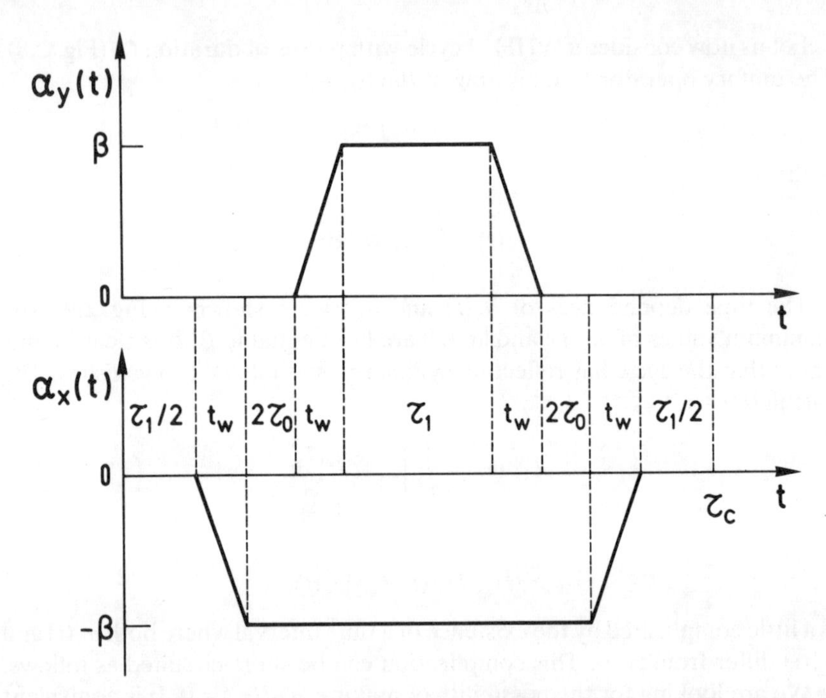

FIG. 2.9. Variation of the angles $\alpha_y(t)$ and $\alpha_x(t)$ during the WHH–4 cycle of Fig. 2.8.

and:

$$\alpha'_x(t) = \alpha_x(t) + \beta.$$

The variations of $\alpha'_x(t)$ and $\alpha'_y(t)$ are shown in Fig. 2.10. There is no overlap between the time intervals where $\alpha'_x(t) \neq 0$ and $\alpha_y(t) \neq 0$.

The integral of $\tilde{\mathscr{H}}_i(t)$ during the central half of the cycle (where $\alpha'_x(t) = 0$) is given by (2.79). In order to determine this integral over the remaining half cycle (first quarter plus last quarter where $\alpha_y(t) = 0$) we use a calculation similar to that giving (2.78) but using (1.11*d*) which gives the result of a rotation 0*x* rather than 0*y*.

It appears in these equations that with respect to the coefficients in (2.78), those of $T_0^{(2)}$ are unchanged, those of $T_{\pm 1}^{(2)}$ multiplied by $\pm$i and those of $T_{\pm 2}^{(2)}$ multiplied by -1.

If we now calculate $\tau_c \mathscr{H}_D^{(0)}(\tau_c)$ for the whole cycle, it appears that with respect to (2.79) the coefficient of $T_0^{(2)}$ is twice as large, those of $T_{\pm 1}^{(2)}$ multiplied by $(1 \pm \mathrm{i})$ and those of $T_{\pm 2}^{(2)}$ vanish.

The problem of interest is to find out whether in the generalized cycle WHH–4 with pulses of finite duration it is still possible to annul $\mathscr{H}_D^{(0)}(\tau_c)$. This is achieved if the first two curly brackets in (2.79) vanish. It is easily

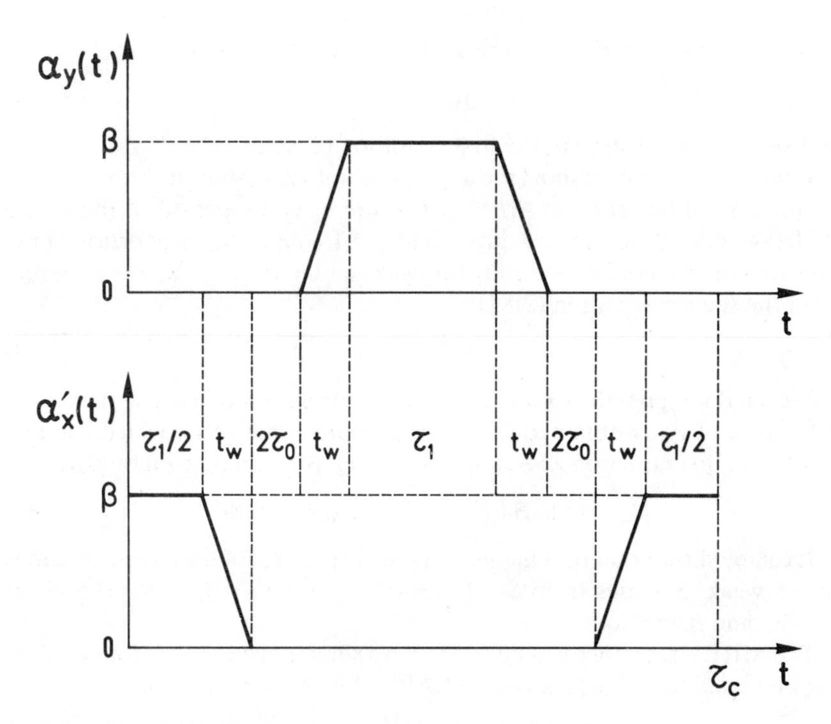

FIG. 2.10. Variation of the angles $\alpha_y(t)$ and $\alpha'_x(t)$ during the WHH–4 cycle of Fig. 2.8 (see text).

found that these two brackets are respectively:

$$2\tau_0-\frac{\tau_1}{2}+\frac{t_W}{2}+\frac{3}{2}\cos\beta\left\{\tau_1\cos\beta+t_W\frac{\sin\beta}{\beta}\right\} \tag{2.80}$$

and

$$\sin\beta\left\{\tau_1\cos\beta+t_W\frac{\sin\beta}{\beta}\right\}. \tag{2.80'}$$

It is possible to annul both of these terms by taking $\beta=\beta_0\neq\pi/2$ and $\tau_1\neq 4\tau_0$:

$$2\tau_0-\frac{\tau_1}{2}+\frac{t_W}{2}=0; \qquad \frac{t_W}{\tau_1}=-\beta_0\operatorname{cotan}\beta_0. \tag{2.81}$$

The two conditions are satisfied, as they should be, for infinitely short pulses by taking $\beta_0=\pi/2$, $\tau_1=4\tau_0$, which are the parameters of the ideal WHH–4 cycle described in **C**(b)2: the long time interval τ_1 between pulses is twice as long as the short time interval $2\tau_0$. This remains true in the case when t_w is finite. As seen from Fig. 2.8, the short and long time intervals are:

$$t_s = 2\tau_0 + t_W; \qquad t_l = \tau_1 + t_W$$

and we have, according to the first condition (2.81), $t_l = 2t_s$.

If the pulses are very short the angle β_0 is not very different from $\pi/2$ and τ_1 not very different from $4\tau_0 = \tau_c/3$ where τ_c is the period of the whole WHH–4 cycle. If we define a duty-cycle $\eta = 4t_W/\tau_c$ as the proportion of the time that the r.f. field is on during the cycle, and write $\beta_0 = (\pi/2) + \varepsilon$ we find from the second equation (2.81):

$$\varepsilon \simeq 3\eta/(2\pi). \tag{2.82}$$

We shall not give the exact numerical solutions of (2.81). The maximum value β_{0M} of β_0 is obtained for the maximum value of t_W which can still satisfy the first equation (2.81) that is $t_W = \tau_1$. β_0 is then given by β_{0M}:

$$\beta_{0M} \operatorname{cotan} \beta_{0M} = -1, \qquad \beta_{0M} = 116{\cdot}84°. \tag{2.83}$$

It can be shown that the change in the scale factor S of the chemical shift is rather weak. S varies from $S = 1/\sqrt{3} = 0{\cdot}579$ for $t_W = 0$, to $S = 0{\cdot}566$ for $t_W = \tau_{11}$ and $\beta_0 = \beta_{0M}$.

The MREV sequence can be treated in a similar fashion. It is found that if the first condition (2.81) is satisfied $\mathcal{H}^{(0)}(\tau_c)$ is proportional to:

$$\mathcal{H}^{(0)}(\tau_c)[\text{MREV}] \propto \cos \beta \left\{ \tau_1 \cos \beta + t_W \frac{\sin \beta}{\beta} \right\}. \tag{2.84}$$

Thus it vanishes not only for $\beta = \beta_0$ given by (2.81) but also for $\beta = \pi/2$. The importance of this fact will appear in the next section.

6. *Limitations due to pulse imperfections*

The high resolution in solids achieved with the multipulse techniques is affected by the various experimental imperfections absent from the ideal sequences described so far.

The first of these imperfections is the inhomogeneity of the longitudinal field H. It is reduced by the scaling factor S ($S = 1/\sqrt{3}$ for WHH–4 and $\sqrt{2/3}$ for MREV) but since the chemical shifts are reduced by the same amount, the relative resolution is independent of S. This inhomogeneity is not a serious limitation for modern high resolution magnets. A more important cause of broadening resides in the diamagnetic electronic susceptibility of the sample. The induced electronic magnetization produces inside the sample a local field which can be calculated by classical magnetostatic theory and which is inhomogeneous for samples which are not of ellipsoïdal shape. As an example of this broadening, in an experiment performed on CaF_2 the linewidth of the ^{19}F resonance was reduced by a factor 3·5 through the replacement of a rectangular sample by a spherical one (Rhim *et al.*, 1973*b*).

A second type of imperfection is an error in the pulse angle which affects equally all the pulses of a cycle. It can be caused either by a faulty adjustment or by a variation of the transmitter power during a long series of pulses (power droop) or by the inhomogeneity of the r.f. field H_1. The latter is the most serious for it affects differently the various points of the sample and is the most difficult to correct.

All these imperfections do not affect the symmetrical character of a cycle. The errors in the pulse angles have a double effect:

1. the average of $\mathscr{H}_{\mathrm{D}}^{(0)}$ does not vanish anymore;
2. the scaling factor S of the terms proportional to I_Z is modified.

The change in the scaling factor S is linear with respect to the departure $\varepsilon = \beta - \beta_0$ of the pulse angle from its theoretical value. When the field H_1 is inhomogeneous the distribution of the angles β and thus of the departures ε results in a broadening of the lines. This broadening is enhanced if one works off-resonance, as in the second averaging method of section **C**(b)4. The variation of the scaling factor S affects the effective Zeeman Hamiltonian:

$$\mathscr{H}_Z = \sum_k (\delta_k + \Delta) I_z^k.$$

Beyond a certain value of Δ this broadening more than offsets the extra narrowing due to second averaging (Rhim *et al.*, 1973*b*; Garroway, Mansfield, and Storker, 1975). As for the residual dipolar term $\mathscr{H}_{\mathrm{D}}^{(0)}$ it is linear with respect to the pulse angle error ε in WHH–4 but quadratic in MREV, at least in the limit of very short pulses.

This result is a consequence of the eqn (2.84). $\mathscr{H}_{\mathrm{D}}^{(0)}$ is proportional to a function $f(\beta)$ which vanishes for the two values $\beta = \pi/2$ and $\beta = \beta_0$ and can be written:

$$f(\beta) = \left(\beta - \frac{\pi}{2}\right)(\beta - \beta_0) h(\beta). \tag{2.85}$$

When $t_{\mathrm{W}} = 0$, $\beta_0 = \pi/2$ and $f(\beta)$ is a quadratic function of $\varepsilon = (\beta - \pi/2)$. When the duty cycle is small ($t_{\mathrm{W}}/\tau = 0.2$ is a typical value) β_0 differs little from $\pi/2$ and $f(\beta)$ although not strictly quadratic has a dependence on ε slower than linear. It is this relative insensitivity of the MREV sequence to the inhomogeneity of H_1 which confers it a certain advantage over the simpler sequence WHH–4.

There are a few other types of imperfections that we shall not discuss here. Their effects can be amended at least in part by complicating further the pulse sequences and also by second averaging. The reader is referred to the specialized monographs of Mehring and Haeberlen for discussion and references. In the present state of the art the experimental quality of the shapes and phases of the pulses, of the methods of compensation of errors and of the adjustment techniques are such that the resolution is insensitive to small *intentional* misalignments.

As an example, in a study of the ^{19}F resonance in CaF_2, for an optimum value of the offset Δ, the line width measured as a function of τ_c exhibited a τ_c^4 dependence which could be attributed to a term $\mathscr{H}^{(4)}$ in the Magnus expansion (Rhim *et al.*, 1973*b*).

Figures 2.11 and 2.12 are examples of high resolution spectra observed respectively on a single crystal and on a powder.

D. Heteronuclear narrowing for spins of rare isotopes

The importance of the study of rare spins stems from the fact that a large majority of observations of chemical shifts come from organic substances. In these substances by far the largest nuclear signal comes from protons whose

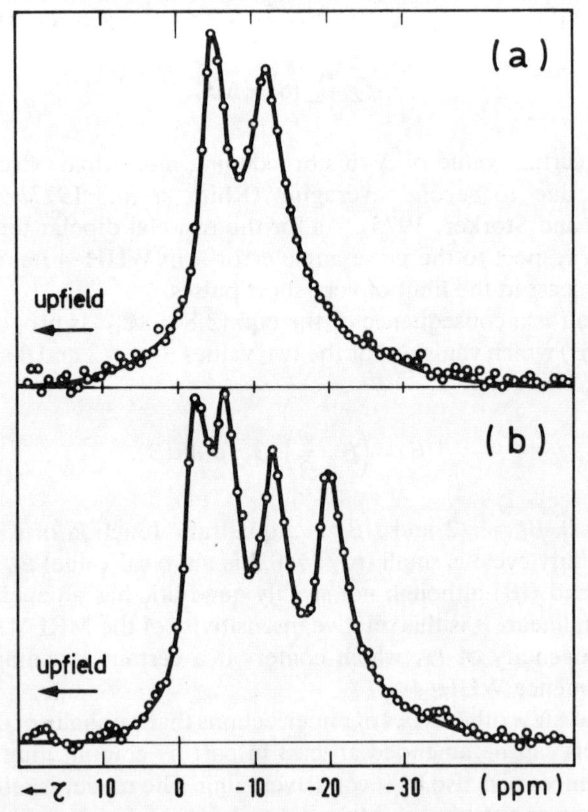

FIG. 2.11. Multipulse spectra of protons in a single crystal of malonic acid, for two orientations of the sample. (After Haeberlen, Kohlshutter, Kempf, Spiess, and Zimmermann, 1974.)

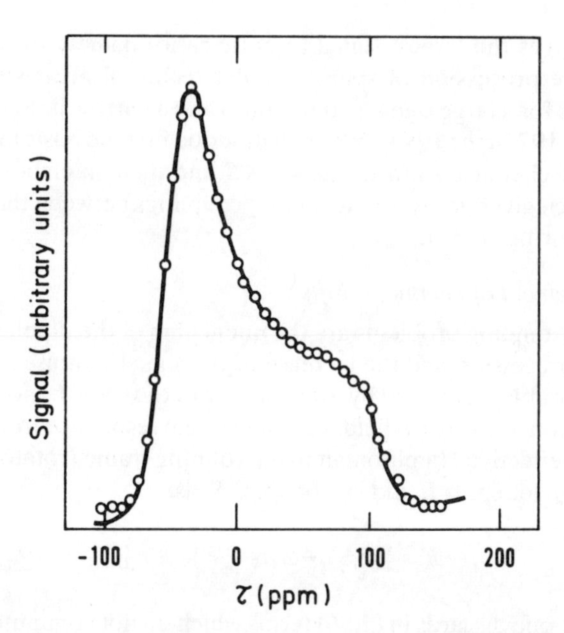

FIG. 2.12. Multipulse spectrum of ^{19}F in C_6F_6 at 40 K. Solid line: experimental spectrum. Open circles: calculated for an axially symmetric chemical shift tensor and Gaussian microcrystal resonance lines. (After Mehring, Griffin, and Waugh, 1971.)

chemical shifts in solids are studied using the homonuclear narrowing described in the previous sections. It is however no less interesting to study the electronic structure of these substances in the neighbourhood of carbon nuclei by high resolution on the isotope ^{13}C of spins $\frac{1}{2}$. Its isotopic abundance is 1·1 per cent and for many compounds of average molecular composition $(CH_2)_n$ the ratio of concentrations $N(^{13}C)/N(^1H)$, termed from now on (N_S/N_I), is of the order of $\frac{1}{200}$. The problem on hand is the observation of a high resolution spectrum of the rare isotope of ^{13}C in a sea of protons. In view of the small concentration of ^{13}C and of its relatively weak magnetic moment $(\gamma_S/\gamma_I \sim 1/4)$ dipolar interactions between the spins S can be neglected, leaving as dipolar broadening agents for the resonance of spins S the truncated Hamiltonian: $\mathcal{H}'_{IS} + \mathcal{H}'_{II}$.

Apart from the considerable dipolar width due to the presence of the spins I, the observation of the free precession of the spins S suffers from a weakness of the signal which has a triple origin: small abundance, small γ, and long spin-lattice relaxation time T_{1S} (much longer than T_{1I}). The last forbids a fast repetition rate and thus prevents an improvement in the signal to noise ratio by accumulation of successive signals.

Among the several methods used for the high resolution NMR of rare spins we shall describe briefly the most practical one (if not necessarily the one which gives the largest signal to noise ratio) namely the *direct* observation of the precession of spins S, under technical arrangements which provide both for a large signal to noise ratio and a narrow line (Pines, Gibby, and Waugh, 1972*a*, *b*; 1973). We shall describe first the basic method giving access to the chemical shifts of the spins S and then an elaboration on this method which gives access to the dipolar couplings between the spins S and the neighbouring spins I.

(*a*) *Measurement of chemical shifts*

The two outstanding problems are the quenching of the dipolar broadening during the precession and the increase of the signal to noise ratio.

The first problem is solved by a decoupling of the spins I and S: during the free precession a strong r.f. field is applied at the resonance frequency of the spins I. The effective Hamiltonian in the rotating frame (rotating as usual at frequency ω_I for spins I, and ω_S for spins S) is:

$$\mathcal{H} = \sum_i \delta_i S_z^i + \omega_1 I_x + \mathcal{H}'_{IS} + \mathcal{H}'_{II}. \tag{2.86}$$

If ω_1 is large one discards in (2.86) terms which do not commute with I_x. $\mathcal{H}'_{IS}$ which is of the form $\sum_{i\mu} B_{i\mu} I_z^i S_z^\mu$ disappears completely and we are left with:

$$\mathcal{H}' = \sum_i \delta_i S_z^i + \omega_1 I_x + (\mathcal{H}'_{II})', \tag{2.87}$$

where:

$$(\mathcal{H}'_{II})' = -\tfrac{1}{2}(\mathcal{H}'_{IIx}). \tag{2.87'}$$

The last two terms of (2.87) which depend only on the variables I do not affect the precession of the spins S, which in this approximation is entirely determined by the first term of (2.87). Its Fourier transform provides a high resolution spectrum of the spins S.

The increase in the signal to noise ratio is achieved by a succession of dynamic polarizations of spins S by thermal contact in the rotating frame with the spins I. The procedure is as follows.

The spins I, having reached thermal equilibrium with a value $\langle I_z \rangle = I_{\text{eq}}$, are spin-locked along $0x$ by the standard procedure of a $\pi/2$ pulse along $0y$ followed by a $\pi/2$ phase shift of the r.f. field. The value $\langle I_x \rangle$ along the r.f. field is then $\langle I_x \rangle = I_{\text{eq}}$. An r.f. field H_1^S is then applied at the Larmor frequency of the spins S. The magnitude of H_1^S is chosen so as to match the effective Larmor frequencies of the two spin species in the rotating frame:

$$\omega_1^S = -\gamma_S H_1^S = \omega_1^I = -\gamma_I H_1^I. \tag{2.88}$$

Under the influence of the thermal coupling produced by the dipolar interaction the two Zeeman energies $\omega_1^S S_x$ and $\omega_1^I I_x$ reach a common final temperature under which, the spin species being both 1/2, we have:

$$\langle S_x\rangle_f/N_S = \langle I_x\rangle_f/N_I. \tag{2.89}$$

If initially $\langle S_x\rangle = 0$ conservation of energy, coupled with (2.88) yields:

$$\langle S_x\rangle_f + \langle I_x\rangle_f = I_{\text{eq}}, \tag{2.90}$$

whence using (2.89):

$$\langle I_x\rangle_f = I_{\text{eq}}/(1+\varepsilon); \qquad \langle S_x\rangle_f = \varepsilon I_{\text{eq}}/(1+\varepsilon), \tag{2.91}$$

with:

$$\varepsilon = N_S/N_I \ll 1.$$

As soon as thermal equilibrium is reached H_1^S is cut off while H_1^I is kept on: the latter is necessary to keep the dipolar coupling $\mathcal{H}'_{IS}$ quenched. A free precession of the spins S, decoupled from the influence of the spins I is then observed.

It is worth insisting on the fact that the r.f. field H_1^I serves a *double* purpose: to polarize rapidly the spins S, *and* then to decouple their free precession from the influence of the spins I. In chapter 1 we saw that the first of these objects could also be achieved by demagnetizing the spins I in the rotating frame rather than by spin-locking them as in the present method, but not the second.

The thermal equilibrium values S_{eq} and I_{eq} being in the ratio,

$$S_{\text{eq}}/I_{\text{eq}} = (N_S/N_I)(\gamma_S/\gamma_I) = \varepsilon(\gamma_S/\gamma_I), \tag{2.92}$$

we get from (2.91):

$$\langle S_x\rangle_f \simeq (\gamma_I/\gamma_S)S_{\text{eq}} \simeq 4S_{\text{eq}} \quad (\text{for } ^{13}\text{C and } ^1\text{H}),$$

that is a first increase of the signal by a factor 4, from 'natural' thermal equilibrium polarization.

There is more to the method. It appears from (2.91) that the thermal mixing has reduced $\langle I_x\rangle_f$ only by a small amount $\sim\varepsilon$ from I_{eq}. When the free precession of S has decayed to zero it is possible to resume the cross-polarization of the spins S by a second application of the r.f. field H_1^S, and so on.

The successive free precession signals of the spins S can be accumulated and after n polarization observation sequences the total signal is:

$$\begin{aligned} S_{\text{T}} &= \varepsilon I_{\text{eq}}\,.\,\{(1+\varepsilon)^{-1} + \cdots + (1+\varepsilon)^{-n}\} \\ &= \varepsilon I_{\text{eq}}\{1-(1+\varepsilon)^{-n}\}/\varepsilon \\ &\simeq \varepsilon I_{\text{eq}}\{1-\exp(-n\varepsilon)\}/\varepsilon. \end{aligned} \tag{2.93}$$

The total noise B_T is $B = b\sqrt{n}$ where b is the noise for a single sequence, whence for the total signal to noise ratio:

$$\frac{S_T}{B_T} = \frac{\gamma_I}{\gamma_S}\left(\frac{S_{eq}}{b}\right)\frac{1}{\sqrt{\varepsilon}}\frac{1-\exp(-n\varepsilon)}{\sqrt{(n\varepsilon)}}. \tag{2.94}$$

The last factor in (2.94) is maximum for $n\varepsilon \simeq 1{\cdot}25$ and is equal to 0.64, whence:

$$S_T/B_T = G(S_{eq}/b), \qquad G = 0{\cdot}64\frac{\gamma_I}{\gamma_S}\sqrt{(N_I/N_s)}, \tag{2.95}$$

for:

$$\gamma_I/\gamma_S = 4, \qquad N_I/N_S = 200, \qquad G = 36.$$

The sequence is represented on Fig. 2.13. After the spin-locking of the spins I the field H_1^I is applied continuously whereas the field H_1^S is applied for brief intervals t_W devoted to thermal mixing between the two spin species and followed by periods τ of free precession.

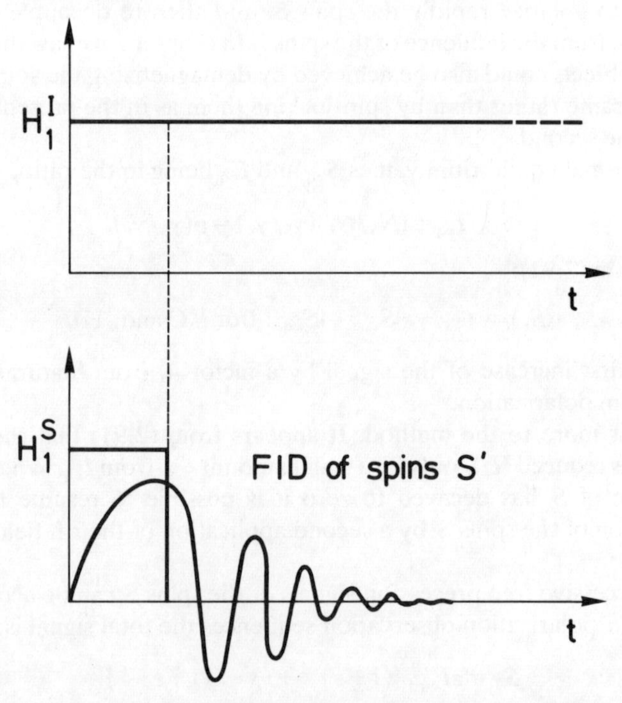

FIG. 2.13. High resolution of rare spins S in the presence of abundant spins I. The spins S are first polarized by thermal mixing with the spins I. Their FID is then observed while they are decoupled from the spins I.

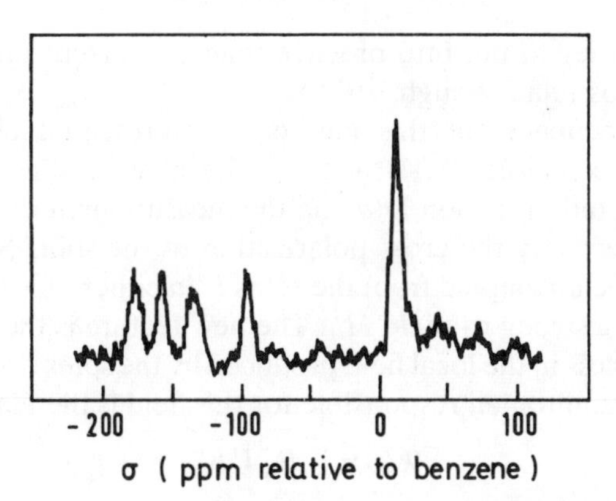

FIG. 2.14. High-resolution spectrum of ^{13}C in a single crystal of hydrogen malonate. (After Chang, Griffin, and Pines, 1975.)

We have neglected so far the effect of the spin-lattice relaxation rate $(T_{1\rho I})^{-1}$ of the spin-locked spins I, assuming implicitly $n(t_w+\tau) \ll T_{1\rho I}$. If the condition is not satisfied or if, in spite of its being satisfied, the signal to noise ratio obtained after $n = 1{\cdot}25/\varepsilon$ sequences is deemed insufficient, it may be advisable to repolarize thermally the spins I in order to have more than one set of polarization-precession sequences for the spins S. The problem then is to obtain the optimum signal to noise ratio for the spins S in an experiment of given total duration t. By suitably modifying the various pulse sequences, in a manner that we shall not discuss here, it is possible to obtain gains G higher than (2.95). A high resolution spectrum of ^{13}C obtained by this method is shown in Fig. 2.14.

An example of the combination of this high-resolution technique with 'magic angle' mechanical spinning can be found in Shaeffer and Stejskal (1976). As explained in section **C**(a), the spinning averages out the anisotropic chemical shift and, in a powder sample, yields a line spectrum determined solely by the scalar chemical shifts of the various chemically different spins S.

(b) *Measurement of dipolar interactions*

The purpose of the method to be outlined in this section is to study in detail the coupling for each type of spins S (having a given chemical shift) with the neighbouring spins I. For instance in organic substances each carbon nucleus has in general a few proton neighbours much nearer to it than the rest of the protons. The study of the dipolar couplings between a ^{13}C spin and these neighbouring protons provides some very direct information on the length and orientation of C—H bonds.

We shall describe only one of the methods which have been used for that purpose (Hester, Ackerman, Neff, and Waugh, 1976). It is both simpler to analyse and easier to put into practice than concurrent methods (Hester, Ackerman, Cross, and Waugh, 1975).

The basic sequence for this method is represented on Fig. 2.15. It comprises three phases: A, B and C. The phases A and C have been described in detail in section **D**(a) on the measurement of chemical shifts. They are respectively the cross polarization of the spins S and their free precession while uncoupled from the spins I. In either phase the spins I are spin-locked to a strong r.f. field H_1^I. The new feature is the phase B where the spins S precess in the local field produced by the spins I and 'seen' by the spins S. The Hamiltonian responsible for this field is the Hamiltonian $\mathcal{H}'_{IS}$:

$$\mathcal{H}'_{IS} = \sum_{i,\mu} B_{i\mu} I^i_z S^\mu_z. \tag{2.96}$$

The observation of the coupling $B_{i\mu}$ being the purpose of the phase B it is clear that the spins I should *not* be spin-locked during this phase for this spin-locking is precisely the procedure devised to quench the interaction (2.96). This being understood two situations have to be considered.

The simplest one arises when the spins I, bound to a spin S are sufficiently removed from each other to render negligible the frequency of their mutual flip-flop due to the dipolar Hamiltonian $\mathcal{H}'_{II}$. The local field 'seen' by the spins S whilst they precess in the phase B is then essentially static and their precession provides a direct measurement of the $B_{i\mu}$.

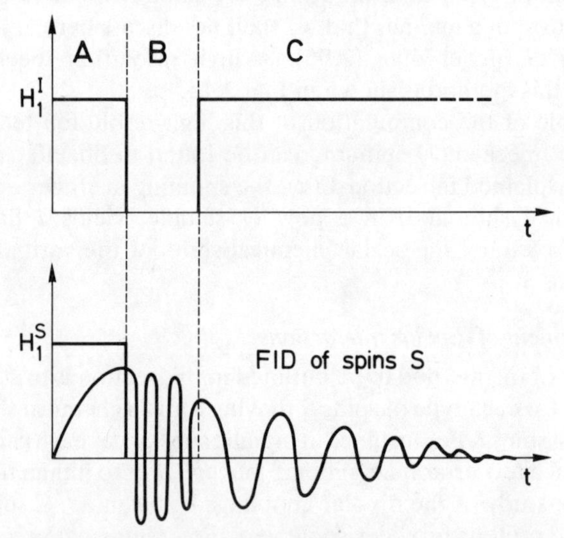

FIG. 2.15. Schematic sequence for measuring the chemical shifts of rare spins S and their dipolar coupling with the abundant spins I. A: Polarization of the spins S by thermal mixing with the spins I. B: Free precession of the spins S coupled to the spins I. C: Free induction decay of the spins S decoupled from the spins I.

On the other hand if flip-flops between the spins I occur at a fast rate they perturb the simple picture of the precession of S in a static local field. To restore this situation one resorts to the homonuclear quenching of the Hamiltonian $\mathcal{H}'_{II}$ by a WHH–4 sequence. This has however as a consequence (as shown in section **C**(b)2(β)) the reduction of all terms linear with respect to I_z by a factor $\sqrt{3}$ and thus a reduction by the same factor of the Hamiltonian (2.96).

We have omitted in the above description technical details about the ways of maintaining intact the magnetization of spins I in the phase B where it is not protected by spin-locking.

The Hamiltonian (2.96) (possibly reduced by $\sqrt{3}$) produces an oscillation in the transverse magnetization of the spins S which depends on the duration t_1 of phase B.

The method whereby the results of the experiment are collected and displayed is a special case of a recent technique widely used in conventional Fourier transform (Aue, Bartholdi, and Ernst, 1976). Each experiment involves a series of n sequences of same duration t_1 for the B phase where the signals are accumulated. The experiment is then repeated with several different values of t_1. Let $S(t_1, t_2)$ be the signal observed at time t_2 after the beginning of phase C, the duration of phase B being t_1. The double Fourier transform,

$$f(\omega_1, \omega_2) = \iint S(t_1, t_2) \exp[i(\omega_1 t_1 + \omega_2 t_2)] \, dt_1 \, dt_2, \tag{2.97}$$

exhibits peaks from which one can extract values for the chemical shifts and coupling coefficients $B_{i\mu}$ and establish correlations between them.

Consider the case of a single type of spin S, with a shift δ_0, connected to a single spin I. The Hamiltonian of the system is

$$\mathcal{H} = \delta_0 S_z + B I_z S_z \tag{2.98}$$

during the B phase and,

$$\mathcal{H} = \delta_0 S_z \tag{2.98$'$}$$

during the C phase. Let $S_+ = S_x + iS_y = S_0$ be the transverse magnetization at the beginning of phase B:

$$\begin{aligned} S(t_1, t_2) &= S_+(t_1, t_2) \\ &= \tfrac{1}{2} S_0 \left\{ \exp\left[i\left(\delta_0 + \frac{B}{2}\right) t_1 \right] + \exp\left[i\left(\delta_0 - \frac{B}{2}\right) t_1 \right] \right\} \exp(i\delta_0 t_2). \end{aligned} \tag{2.99}$$

Its Fourier transform is:

$$f(\omega_1, \omega_2) = \tfrac{1}{2} S_0 \left\{ \delta\left(\omega_1 - \delta_0 - \frac{B}{2}\right) + \delta\left(\omega_1 - \delta_0 + \frac{B}{2}\right) \right\} \delta(\omega_2 - \delta_0). \tag{2.99$'$}$$

In practice $f(\omega_1, \omega_2)$ will have peaks of finite width centred around:

$$\omega_1 = \delta_0 \pm \frac{B}{2}; \qquad \omega_2 = \delta_0.$$

As stated earlier B is to be replaced by $B/\sqrt{3}$ if a WHH–4 sequence is necessary during the B phase.

As a first example we take solid benzene (Hester, Ackerman, Cross, and Waugh, 1975). All the carbons are equivalent and each one is linked to a single proton. The experiment is performed on a powder. The chemical shift tensor has axial symmetry and the resonance frequency of each crystallite depends only on its angle with respect to the magnetic field. The determination of the $\mathcal{H}'_{IS}$ interactions requires the uncoupling of $\mathcal{H}'_{II}$ (the method used is *not* that described here but a different one (Hester, Ackerman, Cross and Waugh, 1975) but this is immaterial). The dependence of the dipolar frequency on the chemical shift is shown in Fig. 2.16. Within experimental uncertainties these results establish that the axis of the chemical shift tensor is along the C—H bond.

The second example refers to ammonium hydrogen malonate $COOH.CH_2.COONH_4$ (Rybaczewski, Neff, Waugh, and Sherfinski, 1977). The experiment was performed on a single crystal with the binary axis (this crystal is monoclinic) along the magnetic field. No decoupling $\mathcal{H}'_{II}$ was necessary in phase B. The relation between chemical shift and dipolar frequency is shown on Fig. 2.17. Only the ^{13}C of the CH_2 group experiences dipolar interactions. The experimental values of the dipolar frequencies: 20·2 and 61·5 kHz are in good agreement with theoretical values (21·0 and

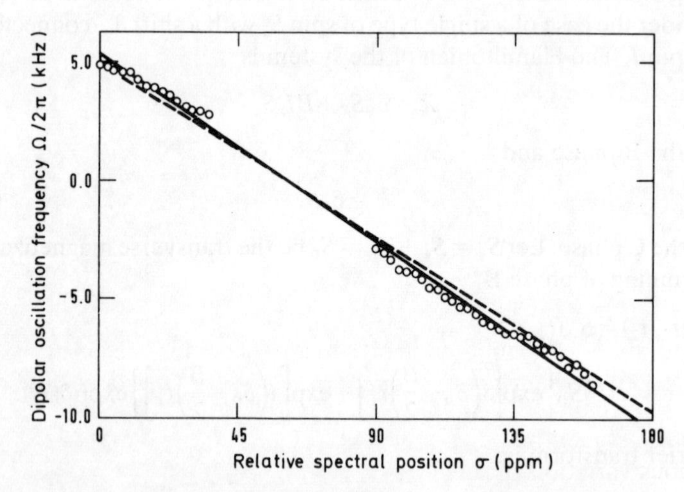

FIG. 2.16. Double Fourier transform of ^{13}C in polycrystalline benzene. ^{1}H–^{13}C dipolar frequency versus ^{13}C chemical shift. (After Hester *et al.*, 1975.)

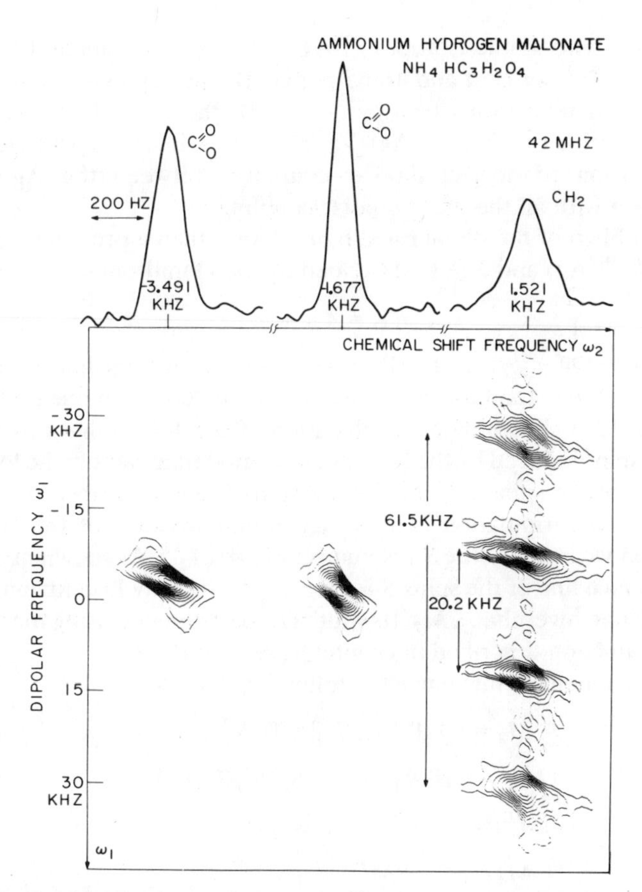

FIG. 2.17. Double Fourier transform of ^{13}C in a single crystal of ammonium hydrogen malonate. ^{1}H–^{13}C dipolar frequency versus ^{13}C chemical shift. (After Rybaczewski *et al.*, 1977.)

62·5) which gives a confirmation of the structure attributed to this compound.

E. An illustrative example of heteronuclear spin manipulation

All the examples of manipulation of spin Hamiltonians cited so far had as a practical motivation high resolution NMR in solids. The example described in this section is of a different nature: it is a study of the evolution of an NMR linewidth (actually broadening of a naturally narrow line) produced by a continuous variation of the Hamiltonian. The sample under study is AgF where the resonance of ^{109}Ag is observed (Mehring, Sinning, and Pines, 1976).

Silver has two stable isotopes ^{107}Ag and ^{109}Ag of comparable concentrations, (51·35 per cent and 48·65 per cent) equal spins 1/2 and comparable gyromagnetic ratios ($\gamma_{109}/\gamma_{107} \simeq 1{\cdot}15$), the latter being very much smaller than that of ^{19}F ($\gamma(^{109}\text{Ag})/\gamma(^{19}\text{F}) \simeq 0{\cdot}05$). It is then possible in a first approximation to neglect dipolar couplings between the Ag spins in comparison with all the other dipolar couplings.

The problem of the dipolar width of ^{109}Ag is then a problem of two spin species: $S(^{109}\text{Ag})$ and I (^{19}F) described by the Hamiltonian

$$\mathscr{H}'_{\text{D}} = \mathscr{H}'_{IS} + \mathscr{H}'_{II}. \tag{2.100}$$

The smallness of γ_S/γ_I has a well-known consequence: the linewidth of the S spins is much less than the square-root of their second moment (Abragam (1961), p. 123). The physical explanation of this fact is that fast flip-flops among I spins induced by the $\mathscr{H}'_{II}$ coupling, modulate rapidly the local fields produced by the spins I, 'seen' by the spins S and described by $\mathscr{H}'_{IS}$. The mathematical formulation of this fact is that because of the strong I–I interaction we have for the S resonance: $M_4 \gg (M_2)^2$. As shown in chapter 1 the resonance line of the spins S is then approximately Lorentzian in shape and much narrower than $\sqrt{M_2}$. Its width can be estimated using the Gaussian memory function described in chapter 1, section **B**(d)2.

It is convenient to introduce the following definitions:

$$M_2 = -\text{Tr}\{[\mathscr{H}'_{IS}, S_x]\}^2/\text{Tr}(S_x^2); \tag{2.101a}$$

$$M_4^{IS} = \text{Tr}\{[\mathscr{H}'_{IS}, [\mathscr{H}'_{IS}, S_x]]\}^2/\text{Tr}(S_x^2); \tag{2.101b}$$

$$M_4^{II} = \text{Tr}\{[\mathscr{H}'_{II}, [\mathscr{H}'_{IS}, S_x]]\}^2/\text{Tr}(S_x^2); \tag{2.101c}$$

$$M_4 = M_4^{IS} + M_4^{II};$$

$$\mu_1 = M_4^{IS}/(M_2)^2; \tag{2.102a}$$

$$\mu_2 = M_4^{II}/(M_2)^2. \tag{2.102b}$$

According to (2.101b) and (2.101c), whilst μ_1 is of the order of a few units at most, μ_2 is a large number:

$$\mu_2 \propto \gamma_I^6 \gamma_S^2/(\gamma_I \gamma_S)^4 = (\gamma_I/\gamma_S)^2 \gg 1. \tag{2.103}$$

In the expansion (1.135) of the memory function $K(t)$ relative to the FID function $G(t)$ for the spins S, the coefficient $-\frac{1}{2}N_2$ of t^2 is given by,

$$N_2 = M_2(\mu_2 + \mu_1 - 1), \tag{2.104}$$

and the Gaussian approximation for $K(t)$ leads through (1.140) to a linewidth of the spins S:

$$\delta_0 = \sqrt{\left(\frac{\pi}{2}\right)}\,[M_2/(\mu_2 + \mu_1 - 1)]^{1/2}. \tag{2.105}$$

This is confirmed by a conventional observation of the NMR of ^{109}Ag (Abragam and Winter, 1959). In the experiment described in this section the S resonance is observed whilst a strong r.f. field ω_1 is applied in the neighbourhood Δ of the I resonance. The Hamiltonian (2.100) has to be truncated further so as to keep only the part which commutes with the effective Zeeman coupling of the I spins. Let $\theta = \tan^{-1}(\omega_1/\Delta)$, be the angle of the effective field acting on the I spins, with the z-axis. The twice truncated dipolar Hamiltonian becomes:

$$(\mathcal{H}'_D)' = \cos\theta \mathcal{H}'_{IS} + \tfrac{1}{2}(3\cos^2\theta - 1)\mathcal{H}'_{II}. \tag{2.106}$$

The change thus brought to the Hamiltonian (2.100) is more than a simple scaling: it is a change in the relative importance of the couplings between like and unlike spins. In particular at the magic angle, the former disappears altogether. In the absence of flip-flops between the I spins one then expects a significant *broadening* of the S resonance. It is actually observed as described presently. Using (2.106) instead of (2.100) in the eqn (2.101) we find new (dashed) values for M_2, μ_1 and μ_2:

$$M'_2 = \cos^2\theta M_2; \tag{2.107}$$

$$\mu'_1 = \mu_1; \tag{2.108a}$$

$$\mu'_2 = \left(\frac{3\cos^2\theta - 1}{2\cos\theta}\right)^2 \mu_2; \tag{2.108b}$$

and (2.105) becomes:

$$\delta(\theta) = \sqrt{\left(\frac{\pi}{2}\right)} \left[\frac{\cos^4\theta M_2}{\frac{1}{4}(3\cos^2\theta - 1)^2\mu_2 + \cos^2\theta(\mu_1 - 1)}\right]^{1/2}. \tag{2.109}$$

$\delta(\theta)$ is plotted on Fig. 2.18 together with the experimental results. There is qualitative agreement. In particular the broadening of the line at the magic angle is unmistakable in spite of the decrease in the second moment. A numerical calculation (Mehring *et al.*, 1976) predicts for this angle a ratio $\delta/\delta_0 \simeq 4$. This value is not reached, presumably for the following reasons: the insufficient magnitude of the effective field which quenches imperfectly $\mathcal{H}'_{II}$, and above all the inhomogeneity of the r.f. amplitude H_1, which results in a distribution of angles θ through the sample around the magic value, thus also contributing to the failure to quench $\mathcal{H}'_{II}$.

This example of pedagogical interest only, provides a nice illustration of the ways in which Hamiltonian manipulation can modify the dynamics of the spins.

F. Multipulses and spin temperature: two experiments

In dealing with the behaviour of a spin system subjected to multipulse sequences it is convenient to introduce the concept of a model system S_M

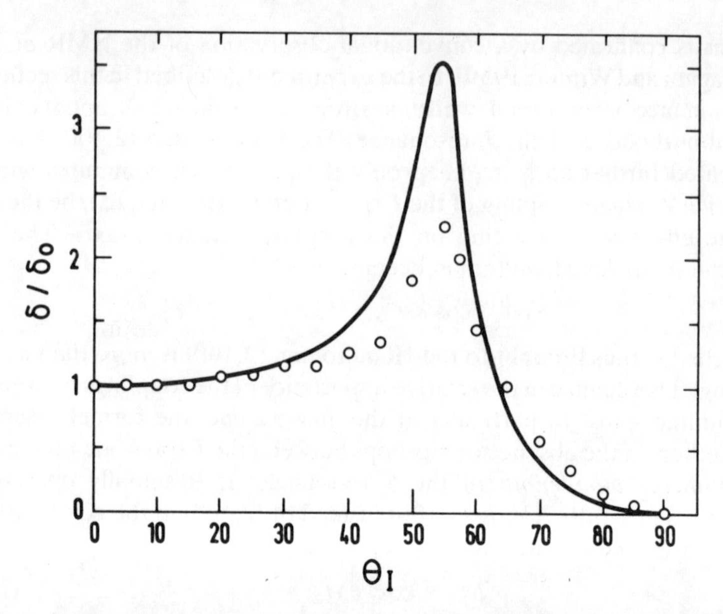

FIG. 2.18. Resonance of ^{109}Ag in AgF in the presence of strong r.f. irradiation near the resonance of the ^{19}F spins. The effective field experienced by the ^{19}F spins in the rotating frame makes an angle θ with H_0. Reduced linewidth of the ^{109}Ag signal versus the angle θ. (After Mehring *et al.*, 1976.)

with a model Hamiltonian. A model Hamiltonian is determined by the condition that the state of S_M coincides with that of S at all the times 0, $\tau_c, \ldots n\tau_c$ when the system S is observed (or as we say 'sampled'). This requires that at $t=0$, S and S_M have the same density matrix, and that the unitary operators $U(t)$ and $U_M(t)$ which describe their evolution, satisfy the condition: $U(n\tau_c) = U_M(n\tau_c)$.

An obvious candidate as a model Hamiltonian is the time-independent Magnus Hamiltonian $\bar{\mathscr{H}}(\tau_c)$ defined by:

$$U(\tau_c) = \exp[-i\tau_c\bar{\mathscr{H}}(\tau_c)].$$

From the theory of spin temperature we can expect that after a time $t \gg T_2$ a model system, which has a time-independent Hamiltonian $\mathscr{H}_M$ will reach thermal equilibrium with a density matrix $\sigma_M^{eq} = 1 - \beta\mathscr{H}_M$. This means that the expectation value of any variable Q will be given by:

$$\langle Q \rangle = -\beta \operatorname{Tr}\{\mathscr{H}_M Q\},$$

for the model system, and therefore also for the real system, if observed at the sampling times $n\tau_c$.

We describe two experiments designed to verify this extension of the validity of the spin temperature theory.

In the first experiment (Pines and Waugh, 1974) the multipulse sequence represented on Fig. 2.19 is an alternating succession of pulses θ along $0x$ and $-0x$, separated by intervals τ.

Prior to the application of the θ pulses the spin system is cooled by an ADRF: spin-locking along a strong r.f. field H_1, followed by an adiabatic removal of H_1. At the end of the ADRF the density matrix of the system is:

$$\sigma_0 = 1 - \beta_0 \mathscr{H}'_{\mathrm{D}}.$$

As a model Hamiltonian for a cycle $(\theta, -\theta)$ of duration 2τ we use a time-dependent Hamiltonian $\mathscr{H}_{\mathrm{M}}(t)$ equal to $\mathscr{H}_1 = \mathscr{H}'_{\mathrm{D}}$ during one half of the cycle and $\mathscr{H}_2 = \tilde{\mathscr{H}}'_{\mathrm{D}} = \exp(\mathrm{i}\theta I_x)\mathscr{H}'_{\mathrm{D}}\exp(-\mathrm{i}\theta I_x)$ during the other half, where:

$$\begin{aligned}\tilde{\mathscr{H}}'_{\mathrm{D}} &= \exp(\mathrm{i}\theta I_x)\mathscr{H}'_{\mathrm{D}}\exp(-\mathrm{i}\theta I_x)\\ &= \frac{3c^2-1}{2}\mathscr{H}'_{\mathrm{D}} - \mathrm{i}\sqrt{(\tfrac{3}{2})}cs(\mathrm{G}_1+\mathrm{G}_{-1}) - \sqrt{(\tfrac{3}{8})}s^2(G_2+G_{-2}),\end{aligned} \quad (2.110)$$

with $c = \cos\theta$, $s = \sin\theta$ and

$$G_m = -\gamma^2\hbar \sum_{i<j} r_{ij}^{-3} F_0^{ij} T_m^{(2)ij}. \quad (2.111)$$

As an approximation for the Magnus Hamiltonian $\bar{\mathscr{H}}(\tau_c)$ we take the lowest order term in the Magnus expansion:

$$\begin{aligned}\mathscr{H}^{(0)} = \mathscr{H}_{\mathrm{M}}^{(0)} &= \frac{1}{2\tau}\int_0^{2\tau}\mathscr{H}_{\mathrm{M}}(t)\,\mathrm{d}t = \tfrac{1}{2}(\mathscr{H}'_{\mathrm{D}} + \tilde{\mathscr{H}}'_{\mathrm{D}})\\ &= \frac{3c^2+1}{4}\mathscr{H}'_{\mathrm{D}} - \frac{\mathrm{i}}{2}\sqrt{(\tfrac{3}{2})}cs(G_1+G_{-1}) - \tfrac{1}{2}\sqrt{(\tfrac{3}{8})}s^2(G_2+G_{-2}).\end{aligned} \quad (2.112)$$

With this approximation the density matrix σ_{M} of the model system should eventually reach the equilibrium form $\sigma_{\mathrm{M}}^{\mathrm{eq}} = 1 - \beta_1\mathscr{H}_{\mathrm{M}}^{(0)}$ where the inverse

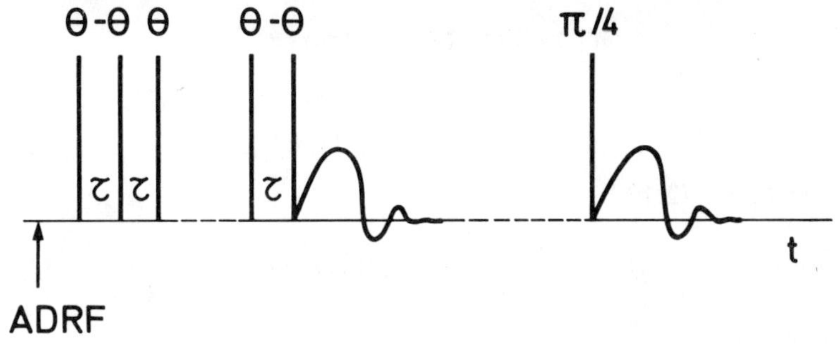

FIG. 2.19. Multipulses and spin temperature. Pulse sequence in the experiment by Pines and Waugh (1974).

temperature β_1 is related to the initial value β_0 by the relation:

$$\langle \mathscr{H}_M^{(0)} \rangle = \mathrm{Tr}\{\sigma_0 \mathscr{H}_M^{(0)}\} = \mathrm{Tr}\{\sigma_M^{eq} \mathscr{H}_M^{(0)}\},$$

or:

$$-\beta_0 \,\mathrm{Tr}\{\mathscr{H}'_D \mathscr{H}_M^{(0)}\} = -\beta_1 \,\mathrm{Tr}\{(\mathscr{H}_M^{(0)})^2\}; \tag{2.113}$$

$$\frac{\beta_1}{\beta_0} = \mathrm{Tr}\{\tfrac{1}{2}\mathscr{H}'_D(\mathscr{H}'_D + \tilde{\mathscr{H}}'_D)\}/\mathrm{Tr}\{\tfrac{1}{4}(\mathscr{H}'_D + \tilde{\mathscr{H}}'_D)^2\}. \tag{2.114}$$

Since $\tilde{\mathscr{H}}'_D$ and $\mathscr{H}'_D$ which differ by a canonical transformation clearly have the same trace, it follows from (2.114) that $\beta_1 = \beta_0$.

At the end of the last pulse θ one observes a free precession signal which should be proportional to:

$$S_1 = -\beta_0 \,\mathrm{Tr}\{I_x \mathscr{H}_M^{(0)}\} = -\tfrac{1}{2}\beta_0 cs(\mathrm{d}G/\mathrm{d}t), \tag{2.115}$$

where $G(t)$ is the FID shape (see the derivation of (1.44)). The signal S_1 can be calibrated by comparing it to a free precession signal produced by a $\pi/4$ $0x$ pulse following the ADRF. This signal S_D is according to (1.44):

$$S_D \propto \tfrac{1}{2}\beta_0 \quad (\mathrm{d}G/\mathrm{d}t), \tag{2.116}$$

whence for the ratio S_1/S_D:

$$S_1/S_D = -cs = -\tfrac{1}{2}\sin 2\theta.$$

The first part of Fig. 2.20 shows a plot of the experimental values of S_1/S_D together with theoretical curve (corrected for the finite duration of the pulses θ). The agreement is reasonable.

Another check involves the observation of the precession signal S_2 after a $\pi/4$ pulse applied a long time $t \gg T_2$, after the end of the sequence of θ pulses. At that time the system (model or real, they are the same) has reached an equilibrium with a density matrix $\sigma_2 = 1 - \beta_2 \mathscr{H}'_D$ where β_2 is given by:

$$\langle \mathscr{H}'_D \rangle = \mathrm{Tr}\{\sigma_M^{eq} \mathscr{H}'_D\} = \mathrm{Tr}\{\sigma_2 \mathscr{H}'_D\},$$

whence:

$$\beta_2 = \beta_0 \,\mathrm{Tr}\{\mathscr{H}'_D \mathscr{H}_M^{(0)}\}/\mathrm{Tr}\{\mathscr{H}'^2_D\}, \tag{2.117}$$

or according to (2.112):

$$\beta_2 = \beta_0 \frac{1 + 3c^2}{4} = \beta_0 \times \tfrac{1}{4}(3\cos^2\theta + 1), \tag{2.118}$$

whence:

$$S_2/S_D = \beta_2/\beta_0 = \tfrac{1}{4}(3\cos^2\theta + 1). \tag{2.119}$$

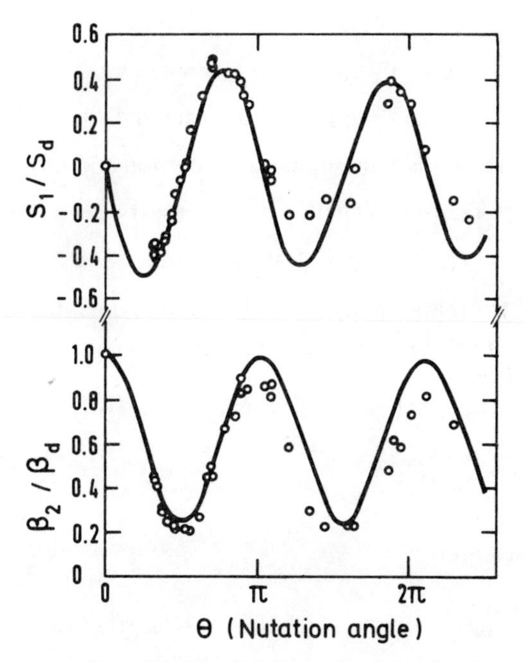

FIG. 2.20. Dipolar signal amplitudes after the train of $\theta, -\theta$ pulses of Fig. 2.19 (top), and dipolar inverse temperature several times T_2 later (bottom) versus the nutation angle θ of the pulses. The solid curves are theoretical. ^{19}F resonance in CaF_2. (After Pines and Waugh, 1974.)

This is shown in the second part of Fig. 2.20. Again the agreement is acceptable.

A second experiment (Rhim, Burum, and Elleman, 1976) affords a more detailed investigation of the relationship between multipulses and spin temperature. Again there is a long sequence of pulses of angle θ (separated by intervals that we call 2τ rather than τ). However instead of alternating as in the last experiment they are all along the same axis $0x$. The system is prepared by a $\pi/2$ pulse along $0y$ applied a time 2τ before the series of the θ pulses. The evolution operator for a cycle 2τ can be written:

$$U(2\tau) = \exp(-i\mathscr{H}'_D\tau)\exp(-i\theta I_x)\exp(-i\mathscr{H}'_D\tau). \qquad (2.120)$$

In a search for a convenient model system we define an average r.f. field $\bar{\omega}_1$ by $\bar{\omega}_1\tau = \theta/2$ and a rotated dipolar Hamiltonian $\tilde{\mathscr{H}}'_D$ by:

$$\tilde{\mathscr{H}}'_D = \exp\left(i\frac{\theta}{2}I_x\right)\mathscr{H}'_D\exp\left(-i\frac{\theta}{2}I_x\right).$$

It is easy to check that (2.120) can be rewritten as a product,

$$U(2\tau) = U_2(\tau)U_1(\tau),$$

with,

$$\begin{cases} U_1(\tau) = \exp(-\mathrm{i}\bar{\omega}_1\tau I_x)\exp(-\mathrm{i}\mathscr{H}'_{\mathrm{D}}\tau) \\ U_2(\tau) = \exp(-\mathrm{i}\bar{\omega}_1\tau I_x)\exp(-\mathrm{i}\tilde{\mathscr{H}}'_{\mathrm{D}}\tau). \end{cases} \tag{2.121}$$

We now define a time-dependent model evolution operator $U_{\mathrm{M}}(t)$ through:

$$\begin{cases} U_{\mathrm{M}}(t) = U_1(t) & \text{for } 0 \leqslant t \leqslant \tau \\ U_{\mathrm{M}}(t) = U_2(t-\tau)U_1(\tau) & \text{for } \tau \leqslant t \leqslant 2\tau, \end{cases} \tag{2.122}$$

and $U_{\mathrm{M}}(t)$ qualifies therefore as a model evolution operator. Taking the time derivative of $U_1(t)$ and $U_2(t)$ it is easily found:

$$\begin{cases} \mathrm{i}\dfrac{\mathrm{d}}{\mathrm{d}t}U_1(t) = \{\bar{\omega}_1 I_x + \exp(-\mathrm{i}\bar{\omega}_1 t I_x)\mathscr{H}'_{\mathrm{D}}\exp(\mathrm{i}\bar{\omega}_1 t I_x\}U_1(t) \\ \mathrm{i}\dfrac{\mathrm{d}}{\mathrm{d}t}U_2(t) = \{\bar{\omega}_1 I_x + \exp[-\mathrm{i}\bar{\omega}_1(t-\tau)I_x]\mathscr{H}'_{\mathrm{D}}\exp\{\mathrm{i}\bar{\omega}_1(t-\tau)I_x]\}U_2(t). \end{cases} \tag{2.123}$$

Using (2.122) we find:

$$\begin{cases} \mathrm{i}\dfrac{\mathrm{d}}{\mathrm{d}t}U_{\mathrm{M}}(t) = \{\bar{\omega}_1 I_x + \exp(-\mathrm{i}\bar{\omega}_1 t I_x)\mathscr{H}'_{\mathrm{D}}\exp(\mathrm{i}\bar{\omega}_1 t I_x)\}U_{\mathrm{M}}(t), \\ \qquad \text{for } 0 \leqslant t \leqslant \tau, \\ \qquad = \{\bar{\omega}_1 I_x + \exp[-\mathrm{i}\bar{\omega}_1(t-2\tau)I_x]\mathscr{H}'_{\mathrm{D}}\exp[\mathrm{i}\bar{\omega}_1(t-2\tau)I_x]\}U_{\mathrm{M}}(t), \\ \qquad \text{for } \tau \leqslant t \leqslant 2\tau. \end{cases} \tag{2.124}$$

Equation (2.124) is equivalent to saying that the time-dependent model Hamiltonian $\mathscr{H}_{\mathrm{M}}(t)$ which describes the evolution of the model system is given by the quantity enclosed in curly brackets in (2.124), and can be rewritten as:

$$\begin{cases} \mathscr{H}_{\mathrm{M}}(t) = \bar{\omega}_1 I_x + \exp[-\mathrm{i}\beta(t)I_x]\mathscr{H}'_{\mathrm{D}}\exp[\mathrm{i}\beta(t)I_x]; \\ \beta(t) = \dfrac{\theta}{2}\dfrac{t}{\tau} \quad \text{for } 0 \leqslant t \leqslant \tau; \\ \beta(t) = \dfrac{\theta}{2}\left(\dfrac{t}{\tau} - 2\right) \quad \text{for } \tau \leqslant t \leqslant 2\tau. \end{cases} \tag{2.125}$$

The model Hamiltonian (2.125) can be split into an average time-independent part $\mathscr{H}_{\mathrm{M}}^{(0)}$ and a sum of oscillating terms which average to zero over the period 2τ. The constant part $\mathscr{H}_{\mathrm{M}}^{(0)}$ would by itself produce an evolution of the model spin system, (and thus also of the real spin system at the proper times), toward an equilibrium state described by a density matrix:

$$\sigma_{\mathrm{M}}^{\mathrm{eq}} = 1 - \beta\mathscr{H}_{\mathrm{M}}^{(0)}.$$

The oscillating terms produce a saturation in the model system and thus a slow reduction of the inverse spin-temperature β (see Goldman (1970), chapter 5). The rate of saturation will be the greater, the larger the amplitudes of the oscillating terms and the greater their saturating efficiency, that is the smaller their frequency $\Omega = \pi/\tau$.

The simplest way to calculate these terms is to express $\mathcal{H}'_{\mathrm{D}} = \mathcal{H}'_{\mathrm{D}z}$ as a sum of irreducible second order tensors with respect to the $0x$ axis. Using (1.11) and (2.110) we get:

$$\mathcal{H}'_{\mathrm{D}z} = -\tfrac{1}{2}\mathcal{H}'_{\mathrm{D}x} - \sqrt{(\tfrac{3}{8})}(G_{2x} + G_{-2x}), \tag{2.126}$$

where:

$$\begin{cases} [I_x, \mathcal{H}'_{\mathrm{D}x}] = 0 \\ \exp(-\mathrm{i}\beta I_x) G_{\pm 2x} \exp(\mathrm{i}\beta I_x) = \exp(\mp \mathrm{i}2\beta) G_{\pm 2x}. \end{cases} \tag{2.127}$$

It is then a matter of straightforward algebra to calculate the Fourier expansion of the Hamiltonian (2.125):

$$\mathcal{H}_{\mathrm{M}}(t) = \bar{\omega}_1 I_x - \tfrac{1}{2}\mathcal{H}'_{\mathrm{D}x} - \sqrt{(\tfrac{3}{8})}\left\{ G_{2x} \sum_{-\infty}^{+\infty} (-1)^n \frac{\sin\theta}{\theta + n\pi} \exp(\mathrm{i}n\Omega t) + \mathrm{h.c.} \right\}. \tag{2.128}$$

The term $n = 0$ in the curly brackets is time-independent and thus contributes to $\mathcal{H}_{\mathrm{M}}^{(0)}$. Using (2.126) we rewrite $\mathcal{H}_{\mathrm{M}}(t)$ as:

$$\begin{aligned} \mathcal{H}_{\mathrm{M}}(t) &= \mathcal{H}_{\mathrm{M}}^{(0)} + \mathcal{H}_{\mathrm{Mosc.}}(t) \\ &= \left\{ \bar{\omega}_1 I_x + \mathcal{H}'_{\mathrm{D}z} + \sqrt{(\tfrac{3}{8})}\left(1 - \frac{\sin\theta}{\theta}\right)(G_{2x} + G_{-2x}) \right\} \\ &\quad - \sqrt{(\tfrac{3}{8})}\left\{ G_{2x} \sum_{-\infty}^{+\infty}{}' (-1)^n \frac{\sin\theta}{\theta + n\pi} \exp(\mathrm{i}n\Omega t) + \mathrm{h.c.} \right\}, \end{aligned} \tag{2.129}$$

where the summation $\sum'$ excludes $n = 0$. Equation (2.129) shows that the amplitudes of the oscillating terms and therefore the saturation rate decrease with the pulse angle θ. Furthermore if the decrease of θ is done at constant $\bar{\omega}_1 = \theta/2\tau$ it involves an increase in the frequency $\Omega = \pi/\tau$ and thus a further decrease in the saturation rate. This is illustrated in Fig. 2.21 where the saturation time is plotted against $\bar{H}_1 = \bar{\omega}_1/\gamma$ for various pulse angles θ. For a given $\bar{H}_1$ the increase in saturation time with decreasing θ is spectacular.

If θ is small, that is if the saturation is weak, one may neglect the effect of the oscillating terms for times short compared to the saturating times and take $\mathcal{H}_{\mathrm{M}}^{(0)}$ as the total Hamiltonian of the model system. The predictions of the theory of spin temperature should then coincide with observations made

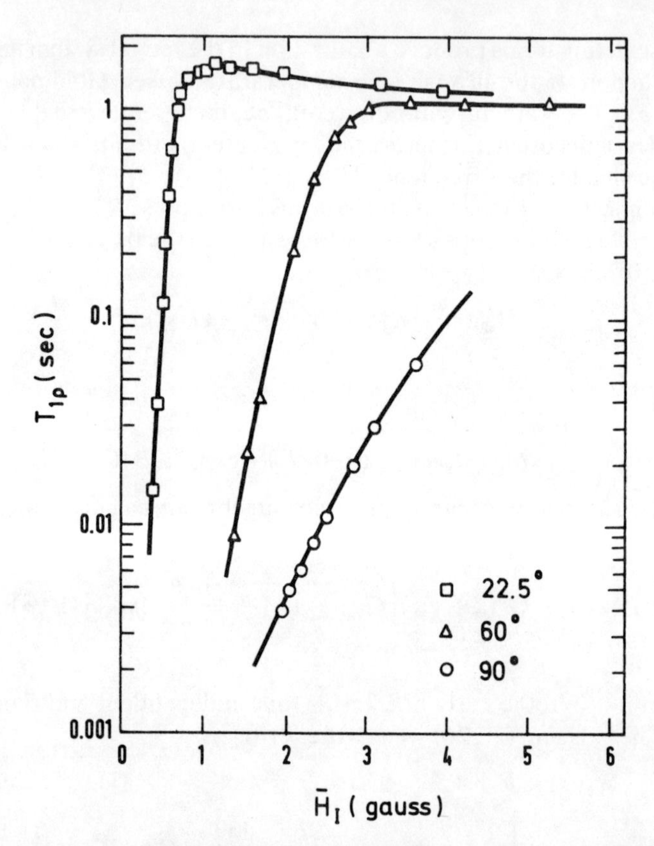

FIG. 2.21. Train of pulses of angle θ at ^{19}F resonance in CaF_2 with $H_0 \| [111]$. Decay time constant $T_{1\rho}$ of the transverse magnetization versus average r.f. field for different pulse angles. (After Rhim *et al.* (1976). Provided through the courtesy of the National Aeronautics and Space Administration, California Institute of Technology, Jet Propulsion Laboratory, Pasadena, California, U.S.A.)

on the real system at sampling times. Two experiments performed on the spins of ^{19}F in CaF_2 confirm this prediction.

The first experiment involves a pulse of $\pi/2$ along $0y$, followed by a series of θ pulses along $0x$, applied at intervals 2τ, until the magnetization $\langle I_x \rangle$ reaches a steady state value. According to (1.23) of the spin temperature theory, this value is given by:

$$\langle I_x \rangle / I_0 = \bar{H}_1^2 / (\bar{H}_1^2 + H_L'^2), \tag{2.130}$$

where $\bar{H}_1 = \theta/2\gamma\tau$ and $H_L'^2 = \mathrm{Tr}\{(\mathcal{H}_M^{(0)})^2\}/\gamma^2 \mathrm{Tr}\{I_x^2\}$.

Figure 2.22 which is a plot of $\langle I_x \rangle$ against $\bar{H}_1$ for three different values of θ exhibits an excellent agreement with (2.130).

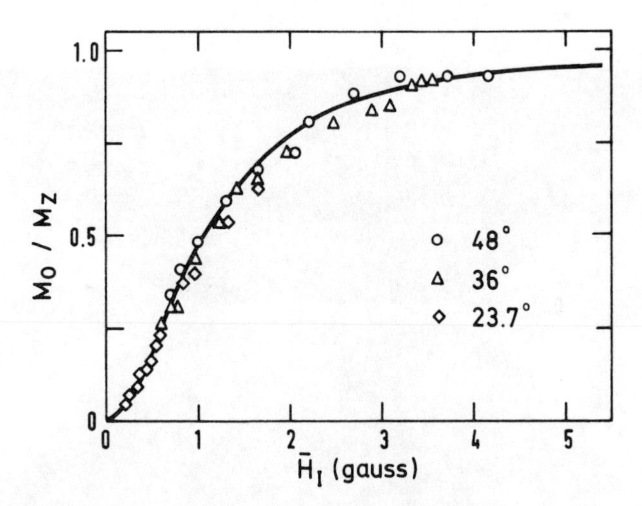

FIG. 2.22. Train of pulses of angle θ at ^{19}F resonance in CaF_2 with $H_0 \| [111]$. Equilibrium transverse magnetization versus average r.f. field for different pulse angles. The solid curve is theoretical. (After Rhim *et al.* (1976). Provided through the courtesy of the National Aeronautics and Space Administration, California Institute of Technology, Jet Propulsion Laboratory, Pasadena, California, U.S.A.)

The second experiment goes one step further. In the usual continuous wave irradiation, adiabatic demagnetization and magnetization in the rotating frame are obtained by slow variation of the CW field H_1. In the present experiment the average amplitude $\bar{H}_1$ is changed by a slow variation of amplitude of the pulses θ. Figure 2.23 shows the variation of $\langle I_x \rangle$ measured in the sampling windows during this process. During remagnetization the phase of the pulses has been shifted by π. This corresponds to a change of sign of H_1 in the rotating frame and therefore, in agreement with experiment, to a change in the sign of $\langle I_x \rangle$.

The experiments described in this section establish the validity of the assumption of a spin temperature relative to the average Hamiltonian $\mathcal{H}_M^{(0)}$ over times short compared to the time of the saturation induced by the oscillating part $\mathcal{H}_{M\,\text{osc.}}(t)$ of the Hamiltonian.

G. The Magnus paradox

We saw in the last section how the effects of a long series of pulses could be described in a satisfactory manner by separating a model Hamiltonian $\mathcal{H}_M(t)$ into an average part $\mathcal{H}_M^{(0)}$ and an oscillating part with zero mean value $\mathcal{H}_{M\,\text{osc.}}(t)$. If the effects of the oscillating terms are small because of the small values of their amplitudes and/or the large values of their frequencies, the behaviour of the system is the following: an inverse spin temperature β is

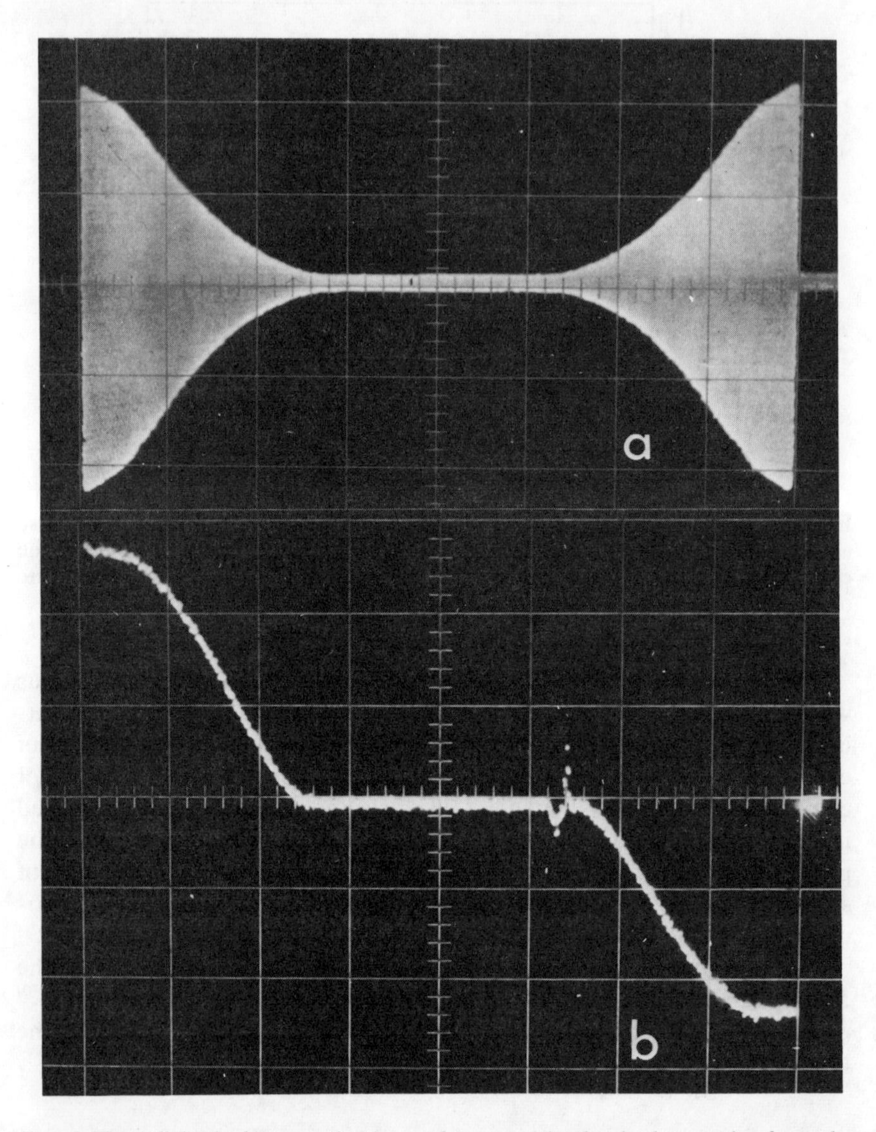

FIG. 2.23. Adiabatic demagnetization and remagnetization in the rotating frame by a train of r.f. pulses at ^{19}F resonance in CaF_2 with $H_0 \| [111]$. (a) Modulation envelope of the r.f. burst. (b) Resonance signal sampled between r.f. pulses. In the remagnetization process, the r.f. phase is shifted by π, producing inverse remagnetization. (After Rhim *et al.* (1976). Provided through the courtesy of the National Aeronautics and Space Administration, California Institute of Technology, Jet Propulsion Laboratory, Pasadena, California, U.S.A.)

reached in a time $T_{eq} \gg T_2$ and then decreases slowly towards zero through saturation by the oscillating terms at a rate $W \ll T_{eq}^{-1}$. Much has been made of the following paradox (Ivanov, Provotorov, and Fel'dman, 1978): if one selects as a model Hamiltonian, the time-independent Magnus Hamiltonian $\bar{\mathcal{H}}(\tau_c)$ defined by:

$$U(\tau_c) = \exp[-i\tau_c \bar{\mathcal{H}}(\tau_c)], \tag{2.131}$$

one expects in the framework of the spin temperature hypothesis that the model system reaches an equilibrium described by a density matrix $\sigma_M^{eq} = 1 - \beta\bar{\mathcal{H}}(\tau_c)$; since (2.131) is by definition of $\bar{\mathcal{H}}(\tau_c)$ an *exact* equation there are no oscillating terms left in the model Hamiltonian and no saturation should occur in contradiction to the theory and the experimental results of section F. At this stage one should remark that the paradox, if paradox there be, is in no way special to the multipulse problem. Consider *any* spin system with a time-independent Hamiltonian such as, say, the truncated dipolar Hamiltonian $\mathcal{H}'_D$, for which the spin temperature hypothesis is known to apply, and add to it a small periodical perturbation of frequency $\Omega = 2\pi/\tau_c$, say for simplicity:

$$V = \omega_1 I_x \cos \Omega t.$$

The evolution operator of such a system can be written:

$$U(\tau_c) = T \exp\left[-i \int_0^{\tau_c} \{\omega_1 I_x \cos \Omega t + \mathcal{H}'_D\} \, dt\right]. \tag{2.132}$$

We know (Goldman (1970), chapter 5) that the inverse spin temperature β, or the expectation value $\langle \mathcal{H}'_D \rangle$, will be saturated according to the rate equation:

$$\begin{cases} \dfrac{d}{dt}\langle \mathcal{H}'_D \rangle = -2W(\Omega^2/D^2)\langle \mathcal{H}'_D \rangle \\ W = \pi \left(\dfrac{\omega_1}{2}\right)^2 g(\Omega), \end{cases} \tag{2.133}$$

where $D^2 = \mathrm{Tr}\{\mathcal{H}'^2_D\}/\mathrm{Tr}\{I_z^2\}$ and $g(\Omega)$ is the shape function of the Zeeman NMR of the spin system. $\langle \mathcal{H}'_D \rangle$ will then reduce to zero in several saturation times W^{-1}. On the other hand suppose that the periodic perturbation is applied at time $t = 0$, when the system is known to be in a state: $\sigma_0 = 1 - \beta_0 \mathcal{H}'_D$ and the expectation value $\langle \mathcal{H}'_D \rangle_0$ is $-\beta_0 \mathrm{Tr}\{\mathcal{H}'^2_D\}$. If the spin temperature hypothesis is applicable to the time-independent Hamiltonian $\bar{\mathcal{H}}(\tau_c)$ one may expect after a certain time T_{eq} a density matrix of the form $\sigma = 1 - \beta_{eq}\bar{\mathcal{H}}(\tau_c)$ whence:

$$\begin{cases} \beta_{eq} = \beta_0 \mathrm{Tr}\{\mathcal{H}'_D \bar{\mathcal{H}}(\tau_c)\}/\mathrm{Tr}\{(\bar{\mathcal{H}}(\tau_c))^2\} \\ \langle \mathcal{H}'_D \rangle_{eq.} = -\beta_{eq} \mathrm{Tr}\{\mathcal{H}'_D \bar{\mathcal{H}}(\tau_c)\}, \end{cases} \tag{2.134}$$

or:

$$\langle \mathcal{H}'_{\mathrm{D}} \rangle_{\mathrm{eq.}} = \langle \mathcal{H}'_{\mathrm{D}} \rangle_0 \frac{[\mathrm{Tr}\{\mathcal{H}'_{\mathrm{D}} \bar{\mathcal{H}}(\tau_{\mathrm{c}})\}]^2}{\mathrm{Tr}\{\mathcal{H}'^2_{\mathrm{D}}\}\,\mathrm{Tr}\{\bar{\mathcal{H}}(\tau_{\mathrm{c}})^2\}}. \tag{2.135}$$

The paradox arises if one assumes that because of the smallness of the periodic perturbation $\omega_1 I_x \cos \Omega t$ the Magnus Hamiltonian $\bar{\mathcal{H}}(\tau_c)$ differs little from $\mathcal{H}'_{\mathrm{D}}$. Equation (2.135) then predicts that $\langle \mathcal{H}'_{\mathrm{D}} \rangle_{\mathrm{eq.}}$ differs little from $\langle \mathcal{H}'_{\mathrm{D}} \rangle_0$ in contradiction with both the well-established theory of saturation and with well-established experimental facts. The inescapable conclusion is that however small the perturbation, and provided W defined by (2.133) does not vanish identically the operators $\mathcal{H}'_{\mathrm{D}}$ and $\bar{\mathcal{H}}(\tau_c)$ must be orthogonal in Liouville space. We can use yet another approach to the properties of the mysterious $\bar{\mathcal{H}}(\tau_c)$. At any sampling time $t = n\tau_c$ after the switching on of the periodic perturbation, the expectation value $\langle \mathcal{H}'_{\mathrm{D}} \rangle(t)$ evolves according to the rigorous equation:

$$\begin{aligned}\frac{\langle \mathcal{H}'_{\mathrm{D}} \rangle(t)}{\langle \mathcal{H}'_{\mathrm{D}} \rangle_0} &= \frac{1}{\langle \mathcal{H}'_{\mathrm{D}} \rangle_0} \mathrm{Tr}\{\exp[-\mathrm{i}t\bar{\mathcal{H}}(\tau_c)]\sigma_0 \exp[\mathrm{i}t\bar{\mathcal{H}}(\tau_c)]\mathcal{H}'_{\mathrm{D}}\} \\ &= \frac{1}{\mathrm{Tr}\{\mathcal{H}'^2_{\mathrm{D}}\}} \iint \mathrm{d}E\,\mathrm{d}E' |\langle E|\mathcal{H}'_{\mathrm{D}}|E'\rangle|^2 \exp[\mathrm{i}(E-E')t],\end{aligned} \tag{2.136}$$

where $|E\rangle$ and E are the (unknown) eigenstates and eigenvalues of $\bar{\mathcal{H}}(\tau_c)$. In order for the ratio (2.136) to decay irreversibly toward zero with time we make the usual assumption that the contribution the off-diagonal matrix elements disappears for large t by destructive interference of the exponentials $\exp[\mathrm{i}(E-E')t]$, and therefore that the diagonal term,

$$\int |\langle E|\mathcal{H}'_{\mathrm{D}}|E\rangle|^2 \,\mathrm{d}E = \iint |\langle E|\mathcal{H}'_{\mathrm{D}}|E'\rangle|^2 \delta(E-E')\,\mathrm{d}E\,\mathrm{d}E', \tag{2.137}$$

is zero.

Equation (2.137) expresses the fact that $\mathcal{H}'_{\mathrm{D}}$ has no diagonal elements in the basis where $\bar{\mathcal{H}}(\tau_c)$ is diagonal, that is again the fact that $\mathcal{H}'_{\mathrm{D}}$ and $\bar{\mathcal{H}}(\tau_c)$ are orthogonal. On the other hand as long as $t \ll W^{-1}$, the saturation time, we expect (2.136) to be nearly unity. This can be expressed by a relation of the following type:

$$\iint |\langle E|\mathcal{H}'_{\mathrm{D}}|E'\rangle|^2 \,\mathrm{d}E\,\mathrm{d}E'/\mathrm{Tr}\{\mathcal{H}'^2_{\mathrm{D}}\} \approx 1; \qquad |E-E'| < W. \tag{2.138}$$

Equation (2.138) shows that although $\mathcal{H}'_{\mathrm{D}}$ is orthogonal to $\bar{\mathcal{H}}(\tau_c)$, it is 'almost diagonal' in the representation $|E\rangle$ in the sense that off-diagonal matrix elements with energy differences $|E-E'| < W$, that is very much smaller than the total width of the spectrum, are sufficient to calculate $\mathrm{Tr}\{\mathcal{H}_D'^2\}$.

We shall not pursue further our investigation of the properties of $\bar{\mathcal{H}}(\tau_c)$ which is clearly a difficult mathematical problem, as are most problems concerned with the appearance of irreversibility in systems with a large number of degrees of freedom.

Suffice it to say that the Magnus paradox rests on the unwarranted assumption that the unperturbed Hamiltonian $\mathcal{H}_D'$ and the Magnus Hamiltonian $\bar{\mathcal{H}}(\tau_c)$ differ very little whereas in fact they are orthogonal as shown by eqn (2.137).

3

NUCLEAR MAGNETIC RESONANCE OF SOLID ^{3}He

O! that this too too solid flesh would melt.

SHAKESPEARE (*Hamlet 1, 2*)

A. Introduction

SOLID ^{3}He is an ideal object to be studied by nuclear magnetic resonance. On the experimental side it has a large magnetic moment $\gamma(^3\text{He})/2\pi = 3{\cdot}24 \times 10^3$ Hz G^{-1} and a spin $\frac{1}{2}$. Furthermore in contrast to all other solids at low temperature, and for reasons to be discussed later in great detail, it offers a narrow NMR line.

These two features, together with the large Boltzmann factor associated with low temperatures provides for strong signals, which make detailed quantitative NMR studies relatively easy. But it is on the theoretical side that solid ^{3}He offers the best illustration of the basic concepts of NMR: exchange narrowing, spin-lattice relaxation, thermal mixing, spin diffusion, etc. . . . .

The motion of the atoms in ^{3}He is quantum mechanical and requires a quantum mechanical treatment of the various parameters pertaining to a theoretical description of NMR such as they are introduced for instance in Abragam (1961): correlation time, hopping time, mixing time, spin diffusion constant, etc. . . . They can be related to various moments of spectral densities of correlation functions, which can be written down unambiguously, and sometimes calculated, from first principles.

It has been surmised long ago and proven since (Halperin, Archie, Rasmussen, Buhrman, and Richardson, 1974; Halperin, Rasmussen, Archie, and Richardson, 1978; Osheroff, Cross, and Fisher, 1980) that, because of their exchange coupling, at sufficiently low temperatures (approximately 1 mK) the nuclear spins of ^{3}He undergo a phase transition to an ordered state. The exact nature of this transition and of the ordered state is being actively investigated at present (1980), both theoretically and experimentally.

On the other hand the nuclear magnetism of ^{3}He, observed by NMR at temperatures well above the transition, can be given a very coherent description. It is to this description that this chapter is devoted. A summary of the present attempts to describe the low temperature situation will however be given at the end of the chapter and also at the end of chapter 8.

The paramagnetic phases of solid ^{3}He

The phase diagram of solid ^{3}He is shown in Fig. 3.1. There are two phases, b.c.c. between 29 atm and roughly 100 atm, and h.c.p. above. A third phase, not shown on the figure is observed above 1000 atm. Two remarkable features appear on this diagram. The first is that ^{3}He does *not* solidify below a minimum pressure of 28·9 atm. The second is that below a temperature T_m of 0·32 K the slope dp/dT of the fusion curve is negative.

The absence of solidification at zero pressure is to be ascribed to two features.

(i) Weakness of the attractive forces between the atoms. These are van der Waals forces related to the polarizability of the helium atoms which is known to be small.

(ii) Smallness of the atomic mass $m = 3$ which entails a zero-point motion of large amplitude and thus large zero-point kinetic energy. There is little difference in that respect between ^{3}He and ^{4}He, which does not solidify either below a minimum pressure of 25 atm. As a consequence the cohesive energy $(-E_0)$ of solid ^{3}He is small and negative below the pressure of 29 atm., small and positive above. The large zero-point motion has however important consequences for the behaviour of the nuclear spins of ^{3}He, a phenomenon obviously without analogue in ^{4}He.

The negative slope of the fusion curve has the following consequence. Because of the well-known Clapeyron relation,

$$\frac{dp}{dT} = (S_S - S_L)/(V_S - V_L), \tag{3.1}$$

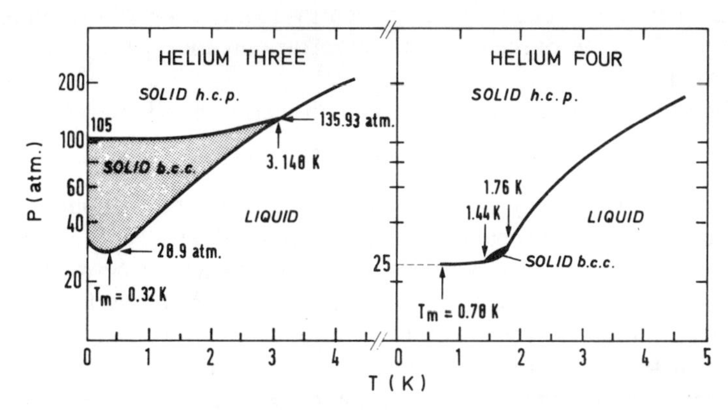

FIG. 3.1. Solid-liquid phase diagram of ^{3}He as compared to that of ^{4}He.

where S_S, S_L and V_S, V_L are the molar entropies and volumes of the solid and the liquid, and since $V_S < V_L$, a negative slope (dp/dT) corresponds to a higher entropy, that is more disorder in the solid than in the liquid, an unusual feature.

This disorder is that of the nuclear spins. In liquid ^{3}He, a Fermi liquid, the nuclear spins are paired off at low temperature and the entropy is a linear function of temperature with, per spin, the value:

$$S_L/Nk_B = bT \quad \text{with } b \approx 2.$$

On the other hand in the solid at temperatures well above the ordering, and in fact as low as 5 mK, the spins are almost completely disordered and the entropy per spin is:

$$S/(Nk_B) \simeq \ln 2 \approx 0{\cdot}35b. \tag{3.2}$$

This shows that temperatures of the liquid–solid mixture well below 5 mK can be obtained by adiabatic compression of the liquid. As the pressure is increased the temperature must decrease according to (3.1) and more solid is formed. The limiting temperature thus reached cannot be lower than that for which the entropy curve of the solid crosses that of the liquid. This phenomenon has resulted in a very efficient cooling method by adiabatic compression, first proposed by Pomerantchuk. Temperatures below 2 mK have been obtained by this method leading to the discovery of the superfluid phases of ^{3}He whose nuclear magnetism is described in chapter 4.

Motions in solid ^{3}He: vacancies, impurities, exchange, phonons

The large amplitude of the zero-point motion of the atoms in solid ^{3}He has far reaching consequences for the magnetic resonance of these nuclear spins. This amplitude δ can be related in order of magnitude to the Debye temperature θ_D (or the Debye frequency $\Omega_D = k_B\theta_D/\hbar$) by realizing that the average quantum mechanical kinetic energy connected with the amplitude δ, $T \simeq 3\hbar^2/2m\,\delta^2$, is of the order of the zero-point energy $\hbar\Omega_D/2$; whence: $\delta^2 \simeq 3\hbar/m\,\Omega_D$.

A slightly more careful estimate given in section **B**(a)1 suggests

$$\delta^2 = 9\hbar/4m\,\Omega_D. \tag{3.3}$$

The Debye frequency $\Omega_D = k_B\theta_D/\hbar$ can be extracted without excessive ambiguity from specific heat measurements of solid ^{3}He. Table 3.1 shows in both b.c.c. and h.c.p. phases the values of the distance a between nearest neighbours, θ_D, and δ, where δ^2 is extracted from (3.3), for various values of the molar volume. It appears that the ratio (δ/a) is very high, 30 per cent and, for instance, ten times that estimated for solid xenon.

We consider now the types of internal motions connected with the large value of δ/a.

Table 3.1

Molar volume V (cm^3)	n–n distance a(Å)	Debye temperature θ_D(K)	Zero point amplitude δ(Å)
b.c.c.			
23·80	3·72	27	1·17
22·86	3·67	28	1·14
21·46	3·59	31	1·09
20·18	3·52	34	1·04
h.c.p.			
19·05	3·55	39	0·97
17·13	3·42	52	0·84
15·29	3·30	68	0·73
13·71	3·18	87	0·65

Vacancies

In every solid at a finite temperature there is a finite number of vacancies, that is atoms missing from lattices sites. This is due to the fact that although transferring an atom from a lattice site to the surface of the crystal (or to an interstitial position) increases the internal energy U, it also augments the disorder, that is the entropy S of the crystal, which may lead to a lower value for the free energy $F = U - TS$. The concentration x of the vacancies is, as we shall see, given by $x = \exp - (\Phi/k_B T)$ where Φ is the energy for creating a vacancy. (There have been suggestions of the existence in ^{3}He of vacancies at zero temperature, which would mean that a crystal with fewer atoms than lattice sites would have a lower energy than a crystal with as many atoms as lattice sites. We shall not consider these suggestions here.)

The large zero-point motion of the atom near a vacancy leads to a spilling of its wave function into the vacancy and to an appreciable probability of the atom moving into this vacancy, leaving a new vacancy behind. This can clearly be visualized as a displacement of the vacancy to the former position of that atom. Successive jumps of that type result in a mobility of the vacancies, which is very much higher in solid helium than in other solids. We shall examine the nature of the motion of the vacancies and their influence on the NMR of ^{3}He later in section **B**(b) of this chapter.

Impurities of ^{4}He

Solid ^{3}He contains, unless very special care is taken, a certain proportion of atoms of ^{4}He. An atom of ^{4}He can exchange its position with a neighbouring atom of ^{3}He and through successive jumps move through the crystal. This motion too can influence the NMR of ^{3}He and will have to be considered.

Exchange

This is probably the most remarkable feature of the nuclear magnetism of solid ^{3}He. Its nature can be grasped qualitatively as follows. In ^{4}He the interchange of two atoms at neighbouring lattice sites is physically meaningless. Two such atoms are identical and their interchange is unobservable.

By contrast the interchange of two neighbouring atoms of ^{3}He is observable if their spin states are different.

The large zero-point motion of the atoms clearly favours such interchanges. Permutations of more than two atoms, 3, 4, or more, are also physically meaningful operations. We shall see in section **C** how the effects of all such permutations can be expressed as spin–spin couplings.

Phonons

Because of the large zero-point motion, solid ^{3}He is a highly anharmonic crystal and the definition of the phonons which are quantized vibrations of small amplitude is not straightforward. Still the existence in ^{3}He of a specific contribution proportional to T^3 and characteristic of phonons lends some validity to the use of this concept.

We have thus introduced a certain number of energy reservoirs or sets of degrees of freedom: vacancies, exchange, impurities, phonons. To these we have to add the Zeeman coupling of the nuclear moments with an applied magnetic field and the dipolar interactions between the nuclear moments. We shall have to examine how the couplings, via the dipolar Hamiltonian, between the Zeeman energy and the other energy reservoirs, influence the NMR of solid ^{3}He.

B. The various 'motions' in solid ^{3}He

(*a*) *Exchange*

1. *The exchange Hamiltonian*

The exact wave function describing the positions of the atoms in solid ^{3}He is naturally unknown.

As a trial-function convenient for a variational calculation of the energy the following expression is generally used (Landesman, 1973):

$$\psi(\mathbf{r}_1, \ldots, \mathbf{r}_n) = \prod_{i<j} \varphi_{\sigma_i}(\mathbf{r}_i - \mathbf{R}_i) g(\mathbf{r}_i - \mathbf{r}_j). \tag{3.4}$$

In (3.4), the $\mathbf{r}_i$ are the spatial co-ordinates of the ^{3}He nuclei and $\mathbf{R}_i$ those of the lattice sites, $g(r)$ is a pair function which expresses the correlation between the positions of neighbouring atoms in particular their repulsion when they come close to each other. The index σ_i refers to the value $\pm\frac{1}{2}$ of the nuclear spin I_z^i. The use of the trial function (3.4) for the calculation of the energy assumes, (legitimately) that the forces between the helium atoms do *not* depend on the spins ($\varphi_+(\mathbf{r}_i - \mathbf{R}_i) = \varphi_-(\mathbf{r}_i - \mathbf{R}_i)$).

A convenient expression for $\varphi(\mathbf{r})$ is a Gaussian:

$$\varphi(\mathbf{r}) = \left(\frac{A}{\pi}\right)^{3/4} \exp\left(-\frac{Ar^2}{2}\right), \tag{3.5}$$

which is normalized to:

$$\langle|\varphi|^2\rangle = \int |\varphi(\mathbf{r})|^2 \, \mathrm{d}^3 r = 1 \quad \text{and} \quad \langle r^2\rangle = \delta^2 = 3/(2A). \tag{3.6}$$

The kinetic energy per particle is given by:

$$T = \left\langle \varphi \left| -\frac{\hbar^2}{2m}\nabla^2 \right| \varphi \right\rangle = \frac{3}{4}\frac{\hbar^2 A}{m} = \frac{9}{8}\frac{\hbar^2}{m\,\delta^2}. \tag{3.7}$$

Equated to the zero-point Debye energy $\hbar\Omega_{\mathrm{D}}/2$, it yields:

$$\delta^2 = 9\hbar/4m\,\Omega_{\mathrm{D}}. \tag{3.3}$$

Naturally the wave function (3.4) cannot describe correctly an eigenstate of solid ^{3}He because it is not properly antisymmetrized with respect to the interchange of the spatial and spin co-ordinates of the nuclei. When this antisymmetrization is done the spin degeneracy of the state (3.4), which is of order 2^N, is lifted to some extent. As Dirac has shown (Dirac, 1930), the energy levels which correspond to this lifting of the degeneracy can be obtained from an effective Hamiltonian which operates only on *the spin variables* (of the ^{3}He nuclei in the present case), and which is called the exchange Hamiltonian $\mathcal{H}_{\mathrm{e}}$.

The most general expression of the exchange Hamiltonian is:

$$\mathcal{H}_{\mathrm{e}} = -\sum_p J_p P_p, \tag{3.8}$$

where P_p is a given permutation of the spin variables and the summation is over all such permutations.

The simplest of the permutations is the interchange of two spins say 1 and 2.

The corresponding spin operator P_{12} can be expressed in the well-known form

$$P_{12} = \tfrac{1}{2} + 2(\mathbf{I}_1 \cdot \mathbf{I}_2). \tag{3.9}$$

Since every permutation can be expressed as a product of interchanges the exchange Hamiltonian can be expressed (apart from a constant) as a sum of products of scalar products $(\mathbf{I}_i \cdot \mathbf{I}_j)$.

It should be quite clear that the statement made earlier about the spin-independence of the forces, between the ^{3}He atoms, and the existence of an exchange Hamiltonian $\mathcal{H}_{\mathrm{e}}$ explicitly dependent on the spin variables I_i, are *not* contradictory. Eigenstates of the system of nuclei of ^{3}He which have different *orbital* symmetries, and, as a consequence, different energies, have also different nuclear spin symmetries because of the Pauli exclusion principle. The spin-dependent exchange Hamiltonian $\mathcal{H}_e$ is an expression of this fact.

The two-body spin-independent interactions $U_{ij}(\mathbf{r}_i - \mathbf{r}_j)$, between the nuclei i and j which lift the degeneracy of the state (3.4), treated as perturbations in the first order, lead as shown by Dirac (1930) to an exchange spin Hamiltonian $\mathcal{H}_e$ linear with respect to the interchange operators (3.9).

If this approximation is made the exchange Hamiltonian can be written:

$$\mathcal{H}_e = -2 \sum_{i<j} J_{ij} \mathbf{I}_i \cdot \mathbf{I}_j. \tag{3.10}$$

The actual behaviour of solid ^{3}He at temperatures that are not large compared to the transition temperature T_c to an ordered state, leads to the belief that the bilinear Hamiltonian (3.10) must be augmented by higher order terms (M. Roger in thesis No. 2297, Orsay, France, 1980) in order to account for the experimental results. Furthermore there are calculations (Delrieu and Roger, 1978) which indicate that some 4th order terms in the exchange Hamiltonian could be comparable to or even larger than the bilinear terms (3.10). On the other hand at temperatures much higher than T_c, where practically all of the studies of NMR in solid ^{3}He have been performed, it turns out that within the accuracy of the measurements, the Hamiltonian (3.10) describes adequately the experimental results. It is even sufficient to limit the summation (3.10) to nearest neighbours with a single exchange constant J and to write $\mathcal{H}_e$ as:

$$\mathcal{H}_e = -2J \sum{}' \mathbf{I}_i \cdot \mathbf{I}_j, \tag{3.11}$$

the summation $\sum'$ being performed over nearest neighbours only.

2. *The exchange integral*

The principle of the evaluation of the exchange integral J_{12} is well known. It involves the calculation of the difference in energy: $(E(1) - E(3))$, between the singlet and the triplet spin states of the total spin $\mathbf{I} = \mathbf{I}_1 + \mathbf{I}_2$, which is also the difference in energy $(E_S - E_A)$ between the symmetrical and the antisymmetrical orbital wave functions $\psi_S(\mathbf{r}_1, \mathbf{r}_2)$ and $\psi_A(\mathbf{r}_1, \mathbf{r}_2)$ constructed from the wave functions (3.4):

$$\psi_{S,A}(\mathbf{r}_1, \mathbf{r}_2) \approx \frac{1}{\sqrt{2}} \{\varphi(\mathbf{r}_1 - \mathbf{R}_1)\varphi(\mathbf{r}_2 - \mathbf{R}_2) \pm \varphi(\mathbf{r}_1 - \mathbf{R}_2)\varphi(\mathbf{r}_2 - \mathbf{R}_1)\} \times g(\mathbf{r}_1 - \mathbf{r}_2). \tag{3.12}$$

From the definition (3.11) of J we get:

$$\begin{aligned} -2J\{\langle \mathbf{I}_1 \cdot \mathbf{I}_2 \rangle(\text{singlet}) - \langle \mathbf{I}_1 \cdot \mathbf{I}_2 \rangle(\text{triplet})\} &= 2J = E_S - E_A \\ &= \langle \psi_S(1,2) | \mathcal{H}(1,2) | \psi_S(1,2) \rangle - \langle \psi_A(1,2) | \mathcal{H}(1,2) | \psi_A(1,2) \rangle, \end{aligned} \tag{3.13}$$

where $\mathcal{H}(1, 2)$ is the orbital Hamiltonian,

$$\mathcal{H}(1,2) = -\frac{\hbar^2}{2m}(\nabla_1^2 + \nabla_2^2) + U(1) + U(2) + v(1,2),$$

describing the motion of two neighbouring nuclei in the self consistent field of all the others.

The node (or nodes), necessarily present in an antisymmetrical wave function ψ_A, make for larger spatial gradients $\nabla\psi_A$ and larger kinetic energies than in the symmetrical wave function ψ_S. As a result the coefficient J as defined in (3.13) turns out to be negative, a rather general result (Herring, 1966). This sign favours a lower energy for an antiferromagnetic arrangement of the nuclear spins, whence the expression 'antiferromagnetic exchange' used for solid ^{3}He. The reader is referred to the literature (Landesman, 1973) for the actual calculation of J, straightforward but cumbersome, using for the wave-function (3.12) the approximation (3.5) and for $g(r)$ the hard spheres approximation: $g(r) = 0, r < \lambda$; $g(r) = 1, r > \lambda$ where λ is of the order of the range of the interaction potential between two ^{3}He atoms ($\lambda \simeq 2{\cdot}5$ Å).

The result is:

$$\hbar J = -\frac{\hbar^2 A}{m}\left(\frac{Aa^2}{2\pi}\right)^{1/2} \exp -\left\{\frac{A}{2}(a^2 + \lambda^2)\right\}, \tag{3.14}$$

where $a = |R_1 - R_2|$. This expression is only valid if $A\lambda^2 \gg 1$ or $\lambda^2 \gg \delta^2$. Using $A = 3/(2\delta^2)$; $\delta^2 = 9\hbar/4m\,\Omega_D$ where Ω_D is the Debye frequency, (3.14) can be rewritten:

$$J \simeq -\frac{\Omega_D}{\sqrt{(3\pi)}}\left(\frac{a}{\delta}\right) \exp -\left\{\frac{3}{4}\frac{a^2 + \lambda^2}{\delta^2}\right\}. \tag{3.15}$$

The calculation leading to (3.14) has been rightfully criticized, on the following grounds. The wave functions (3.5) used in the calculation of J are trial functions selected to minimize the energy. They may be good approximations in regions of space where they are large and bring appreciable contributions to the energy. On the other hand the main contribution to J as given by (3.13) comes from integrals containing overlap products such as $\varphi(\mathbf{r} - \mathbf{R}_1)\varphi(\mathbf{r} - \mathbf{R}_2)$. Since the functions φ are peaked around the lattice sites R_1 and R_2 and fall off sharply, away from these sites, the main contribution to exchange comes from regions where the functions are small and bring negligible contributions to the total energy. They are therefore likely to be rather unreliable, precisely in the regions where they are needed for the calculation of exchange.

The expressions (3.14) and (3.15) should therefore be considered as being qualitatively correct at best.

Still, imperfect as they are, they are enlightening in two respects.

(i) They give correctly an order of magnitude for J. Using for a, θ_D, and δ values taken from Table 3.1, (3.15) yields values for J agreeing in order of magnitude (if not better) with values extracted from NMR measurements or from static measurements to be mentioned later.

(ii) They describe correctly over more than three decades the tremendous variation of exchange with the molar volume.

At first sight it looks surprising that when the volume increases the exchange should also go up. A simple-minded view would be that as the nuclear sites move apart from each other, the overlap between the nuclear wave functions, and thus also the exchange integral, should decrease. This overlooks the fact that the *spread* of the wave functions φ around the nuclear sites, measured by δ, and thus also their overlap, also increase with increasing molar volume and this more than compensates for the increasing lattice spacings.

The increase in exchange with increasing molar volume appears natural if one thinks of exchange as actual permutation of helium atoms themselves. The strong repulsion at short distances between helium atoms appears then as a steric hindrance to the motion of atoms considered as hard spheres. This hindrance is alleviated and exchange made easier by an increase in molar volume.

The increase in exchange with molar volume can be put on a semiquantitative basis as follows (Landesman, 1973). The dependence of the Debye frequency Ω_D on the volume is known. The dimensionless constant,

$$\Gamma = -\frac{V}{\Omega_D} \times \frac{\partial \Omega_D}{\partial V} = -\frac{\partial \ln \Omega_D}{\partial \ln V}, \tag{3.16}$$

known as the Grüneisen constant has been measured for ^{3}He (Trickey, Kirk, and Adams, 1972) and is approximately 2·2.

Similarly a Grüneisen constant $\Gamma_J = -\partial \ln |J| / \partial \ln V$ can be extracted from (3.15) using the expression (3.3) for δ^2.

Neglecting in (3.15) the volume dependence of the pre-exponential factor and taking $\lambda = 2.5$ Å, independent of the volume, we get:

$$\Gamma_J = \frac{\partial}{\partial \ln V}\left\{\frac{3(a^2+\lambda^2)}{4\,\delta^2}\right\} = -\frac{\hbar\Omega_D}{3(\hbar^2/ma^2)}\left\{\Gamma\left(1+\frac{\lambda^2}{a^2}\right)-\frac{2}{3}\right\}. \tag{3.17}$$

For: $V = 23{\cdot}8\ \text{cm}^3$, $a = 3{\cdot}72$ Å, $\theta_D = \hbar\Omega_D/k_B = 27$ K (3.17) yields the very high value $\Gamma_J \simeq -19$, which exhibits the very stiff positive dependence of $|J|$ on the molar volume, in reasonable agreement with experiment, as will appear later.

This concludes the survey of the properties of exchange as required by the NMR studies of ^{3}He at high temperature. We shall come back to exchange at the end of the chapter.

(b) Vacancies

The behaviour of vacancies in solid ^{3}He is characterized by two energies: the energy Φ for the creation of a vacancy and the tunnelling energy $\hbar\omega_V$ which describes its motion.

1. *The formation energy*

Φ can be written approximately as

$$\Phi = Pv - \langle V \rangle,$$

where $\langle V \rangle$ is the potential energy of an atom in the average field of its neighbours and v the atomic volume. The creation of a vacancy in a sample at pressure P, through transport of an atom to the surface of the sample, involves an increase v in volume and requires the work Pv; the removal of an atom from the mean field of its neighbours corresponds to the term $-\langle V \rangle$. Φ can then be related to the kinetic energy $T = \langle -(\hbar^2/2m)\nabla^2 \rangle$ of an atom and to its cohesive energy $-E_0 = -(T + \langle V \rangle)$ by the relation:

$$\Phi = Pv - (E_0 - T) = Pv + T - E_0. \tag{3.18}$$

If we keep for T the approximate relation $T = \hbar\Omega_D/2$ we get:

$$\Phi = Pv - \left(E_0 - \frac{\hbar\Omega_D}{2}\right), \tag{3.18$'$}$$

where all the parameters on the right hand side can be extracted from experiment. The eqns (3.18) do not take into account the fact that when an atom is removed, the positions and the wave functions of the neighbouring atoms are modified. In particular if a Gaussian approximation is used it is reasonable to replace the isotropic wave function (3.5) by an anisotropic function $\varphi_D(\mathbf{r})$,

$$\varphi_D(\mathbf{r}) = \varphi(\mathbf{r})\left(\frac{A_D}{A}\right)^{1/4} \exp\left(\frac{A - A_D}{2} . x^2\right) \tag{3.19}$$

where the choice of $A_D < A$ expresses the expansion of the atomic wave function along the direction $0x$ towards the vacancy. To this expansion corresponds a decrease $|\delta T|$ in the kinetic energy of the z atoms surrounding the vacancy,

$$|\delta T| = \left\langle \varphi(\mathbf{r}) \right| -\frac{\hbar^2}{2m}\nabla^2 \left| \varphi(r) \right\rangle - \left\langle \varphi_D(\mathbf{r}) \right| -\frac{\hbar^2}{m}\nabla^2 \left| \varphi_D(\mathbf{r}) \right\rangle = \frac{3}{8}\frac{\hbar^2}{m}(A - A_D), \tag{3.20}$$

and a decrease $|\delta\Phi|$ in the formation energy of the vacancy,

$$|\delta\Phi| = \frac{3z}{8}\frac{\hbar^2}{m}(A - A_D), \tag{3.20$'$}$$

where z is the number of nearest neighbours of the vacancy.

The concentration of vacancies x is obtained by writing that the free energy $F_V = Nx\Phi - TS_V$ connected with them is minimum:

$$\frac{F_V}{N} = x\Phi + k_B T\{x \ln x + (1-x)\ln(1-x)\};$$
$$\frac{\partial}{\partial x}\left(\frac{F_V}{N}\right) = \Phi + k_B T \ln\left(\frac{x}{1-x}\right) = 0, \tag{3.21}$$

or:

$$x = [1 + \exp(\Phi/k_B T)]^{-1} \cong \exp(-\Phi/k_B T). \tag{3.22}$$

The specific heat connected with the vacancies is:

$$\frac{C_V}{N} = \frac{\partial}{\partial T}(x\Phi) \cong k_B\left(\frac{\Phi}{k_B T}\right)^2 \exp(-\Phi/k_B T). \tag{3.23}$$

There are several ways to determine the energy Φ.

(i) The thermal dilatation of solid ^{3}He is negligible at low temperature. At constant pressure the volume is then proportional to $(1+x)$ and, at constant volume, the *atomic* volume to $(1+x)^{-1} \approx 1-x$. The lattice spacing is proportional to $(1+x)^{-1/3} \approx (1-x/3)$. The temperature dependence of the lattice spacing can be measured by X-ray diffraction and then Φ is obtained from (3.22).

(ii) If the specific heat measurements are sufficiently accurate to separate the vacancies contribution (3.23) from that of the phonons, Φ can be measured.

(iii) The NMR measurements also provide an estimate of Φ to be described later.

Table 3.2 gives values for Φ measured by one of these methods together with the 'theoretical' values (3.18). These values increase with pressure and the experimental values are systematically smaller than the values (3.18). Both of these results are to be expected. The first is due to the fact that in (3.18) the first and the second term increase with pressure (T increases because δ *decreases* with pressure) whereas the cohesive energy remains small. The second result is due to the fact that (3.18) does not take into account the expansion of the wave functions of the atoms surrounding a vacancy which leads to the decrease (3.20) of their kinetic energy and overestimates the vacancy formation energy Φ by the amount (3.20′).

2. *The tunnelling frequency*

The vacancies in solid ^{3}He are mobile. It would be more accurate to say that they are not localized. Consider an atom at a site 1, nearest neighbour of a vacancy at site 2. It is quite clear that this state of the crystal is degenerate with one in which the atom is at site 2 and the vacancy at site 1. This

Table 3.2

Molar volume $V(\text{cm}^3)$	Φ(K) Theoretical	Φ(K) Experimental
b.c.c.		
23·80	26·5	6·64
22·86	28·9	7·95
21·46	34·2	10·49
20·18	40·4	13·20
19·24		17·9
h.c.p.		
19·05	47·5	
18·95		33·7
18·49		35·2
17·63		42·6
17·13	68·2	

degeneracy is lifted by a proper choice of the wave functions. The correct wave functions of the atom will be the symmetrical and antisymmetrical combinations,

$$\psi_{\pm}(r) \simeq \frac{\varphi(\mathbf{r}-\mathbf{R}_1) \pm \varphi(\mathbf{r}-\mathbf{R}_2)}{\sqrt{2}}, \tag{3.24}$$

and the tunnelling frequency $\hbar\omega_V$ is the difference in energies between the two states $\psi_{\pm}$.

This is very similar to the earlier calculation of the exchange integral (eqn (3.12)) where a state of two atoms, with spin up on site 1 and spin down on site 2, is degenerate with a state where the atoms are interchanged. However whilst the calculation of the exchange integral is formally a two-particle problem, that of the tunnelling frequency is a one-particle problem.

In this respect the calculation of the exchange integral J is similar to that of the energy difference between the symmetric (ground) state and the antisymmetric (unstable) state of the hydrogen molecule. The calculation of $\hbar\omega_V$ has the same similarity with respect to the hydrogen molecular ion H_2^+ which is also a one-particle problem.

Using the Gaussian wave functions (3.5) a calculation similar to that of J yields (Landesman, 1973):

$$\omega_V = \Omega_D \sqrt{\left(\frac{2}{3\pi}\right)\left(\frac{a}{\delta}\right)} \exp\left(-\frac{3}{8}\frac{a^2}{\delta^2}\right), \tag{3.25}$$

where we have used $A = 3/2\delta^2$ *and* $\delta = 9\hbar/4m\,\Omega_D$ (eqn (3.3)). Comparing ω_V to J given by (3.15) we find:

$$\frac{\omega_V}{|J|} = \sqrt{2}\exp\left\{\frac{3}{8}\frac{a^2+2\lambda^2}{\delta^2}\right\} \gg 1. \tag{3.26}$$

The reason for this large ratio is the fact that in the calculation of ω_V which is a one-particle problem the overlap integral $S_{12} = \int \varphi(\mathbf{r}-\mathbf{R}_1)\varphi(\mathbf{r}-\mathbf{R}_2)\,d^3\mathbf{r}$ appears in the first power whereas it appears squared in the calculation of the exchange, a two-particle problem. (This also occurs in a comparison between the hydrogen molecular ion and the hydrogen molecule.) The ratio (3.26) is further increased by the hard core repulsion constant which reduces the value of J in (3.14) and (3.15). Actually the expression (3.26) is an overestimate of $\omega_V/|J|$. It does not take into account the distortion of the atomic wave functions near the vacancy as described by (3.19). This reduces appreciably the ratio $\omega_V/|J|$ by a factor difficult to estimate (Hetherington, 1968). The tunnelling frequency ω_V, whilst much larger than $|J|$ is much smaller than Ω_D as exhibited by eqn (3.25) and since $\hbar\Omega_D$ is of the same order of magnitude as Φ (eqn (3.18')), it follows that $\omega_V \ll \Phi$. We insist once more on the fact that because of the inadequacy of the Gaussian functions (3.5), the estimates (3.25) and (3.26) for ω_V are of a qualitative nature only.

3. *Vacancy waves*

The non-local character of vacancies due to tunnelling, leads to the existence of wavelike excitations with a continuous energy spectrum.

Vacancy waves in solid ^{4}He. This is best illustrated first on the example of ^{4}He where the spin variable is absent.

We can then introduce a fictitious spin operator $\mathbf{s}^i$ for each lattice site with eigenvalues $s_z^i = -\frac{1}{2}$ if the site is occupied by an atom and $s_z^i = +\frac{1}{2}$ if it is vacant.

The energy for the formation of vacancies is then expressed by the Hamiltonian:

$$\mathcal{H}_f = \omega_f \sum_i \left(\tfrac{1}{2} + s_z^i\right) \quad \text{where } \omega_f = k_B\Phi/\hbar. \tag{3.27}$$

The tunnelling Hamiltonian which expresses the possibility of a vacancy to move from one site to another can be written as a sum over nearest neighbours of flip-flop operators of fictitious spins.

$$\mathcal{H}_t = \omega_V \sum_{i<j} \left(s_i^+ s_j^- + s_i^- s_j^+\right). \tag{3.28}$$

The Hamiltonian $\mathcal{H}_t + \mathcal{H}_f$ is formally (apart from a constant) that of a magnetic system of spins $\frac{1}{2}$ in an applied field with a Larmor frequency ω_f and an XY exchange coupling ω_V. The elementary excitations of such a system are well known. They form a band with energies (Bernier and Landesman, 1971):

$$\omega(\mathbf{k}) = \omega_f + z\omega_V\gamma(\mathbf{k}),$$

with:

$$\gamma(\mathbf{k}) = \frac{1}{z}\sum_j{}' \exp(i\mathbf{k}\,.\,\mathbf{R}_{ij}). \tag{3.29}$$

The summation is over the z nearest neighbours. The band centered at the energy Φ has a width,

$$\Delta = 2z\hbar\omega_V. \tag{3.29'}$$

Vacancy waves in solid ^{3}*He*. In ^{3}He there are two types of atoms; those with spin up and those with spin down. If the nuclear polarization is very high, say +1, we can disregard the nuclear spins down and the spectrum of the vacancies is identical to that in ^{4}He.

The general case can be treated formally by introducing for each site a fictitious spin $\mathbf{S}^i$ with $S = 1$. $S_z^i = 0$ corresponds to a vacancy and $S_z^i = \pm 1$ to an atom with its real spin $I_z^i = \pm 1/2$.

The components I_z^i and $I_\pm^i$ are easily seen to be expressed as:

$$I_z^i = \tfrac{1}{2}S_z^i; \qquad I_\pm^i = \tfrac{1}{4}(S_\pm^i)^2. \tag{3.30}$$

The operators for filling a vacancy with a spin $I_z^i = \pm 1/2$ are expressed as:

$$a_i^\dagger = \frac{S_z^i S_+^i}{\sqrt{2}}; \qquad b_i^\dagger = -\frac{S_z^i S_-^i}{\sqrt{2}}. \tag{3.31}$$

The Hamiltonians for the vacancies can be written now:

$$\mathcal{H}_f = \omega_f \sum_i (1 - (S_z^i)^2); \tag{3.32a}$$

$$\mathcal{H}_Z = \omega_0 \sum_i I_z^i = \frac{\omega_0}{2}\sum_i S_z^i; \tag{3.32b}$$

$$\mathcal{H}_t = \omega_V\Big(\sum_{i<j} a_j^\dagger a_i + b_j b_i^\dagger + \text{h.c.}\Big), \tag{3.32c}$$

where the $a_i^\dagger$, $b_i^\dagger$, a_i, b_i are given by (3.31) and Hermitian conjugates thereof. The diagonalization of (3.32) is more involved than that of (3.27) or (3.28) in ^{4}He and will not be pursued here. The band width of the spectrum of vacancy waves is still of order $\Delta = 2z\hbar\omega_V$. Since $\Delta \ll \Phi$, it is only at very low temperatures, $k_B T < \Delta \ll \Phi$ that the influence of the band structure on the concentration of vacancies becomes appreciable. Unfortunately at such low temperatures, the absolute concentration $x \approx \exp-(\Phi/k_B T)$ of the vacancies becomes too small to allow any information on the excitation spectrum. For all purposes we shall keep the relation: $x = \exp-(\Phi/k_B T)$.

(*c*) *Impurities of* ^{4}*He*

Unless very special precautions are taken solid ^{3}He contains always a small proportion, typically one hundred p.p.m. of impurities of ^{4}He.

These impurities also are mobile. The interchange of an impurity atom of ^{4}He with a neighbouring atom of ^{3}He will occur with a frequency J_{43} which can reasonably be expected to be of the same order of magnitude as the exchange integral J between two ^{3}He atoms and thus appreciably slower than the tunnelling frequency ω_V of the vacancies.

It is also possible to envisage the motion of impurities of ^{3}He in solid ^{4}He. This would lead to a tunnelling frequency J_{34} presumably comparable to but not necessarily equal to J_{43}.

Like a vacancy, an impurity atom of ^{4}He is a mobile non-magnetic point-defect in a regular lattice of ^{3}He and would presumably affect the NMR in ^{3}He in a similar manner.

The differences are threefold.

(i) As already mentioned the tunnelling frequency J_{34} is much lower than that ω_V of the vacancies.

(ii) In contrast with vacancies the concentration of impurity atoms is temperature-independent.

(iii) If the concentration of impurities is not too small, they interact with each other through the strain fields they produce in the regular lattice. This increases appreciably the heat capacity of the system of impurities and is not without consequences on the NMR of ^{3}He.

(*d*) *Phonons*

When one speaks of the temperature of a sample without further specification one means the temperature of the phonons. On the other hand in the foregoing we have listed several other heat reservoirs, to wit: Zeeman; exchange; vacancies; impurities. The question of the coupling of the phonons with these reservoirs thus arises and of the corresponding relaxation rates. For the first two, Zeeman phonons and exchange phonons the relaxation rates turn out to be very small and can be disregarded.

The calculation of the other two: vacancy phonons and impurity phonons uses a formalism very similar to that of nuclear magnetic relaxation and will be treated in section **E**(*b*)1.

C. The effects of exchange on NMR in solid ^{3}He

Introduction

The existence of exchange couplings $\mathcal{H}_e$ between nuclear spins, much stronger than their dipolar interaction $\mathcal{H}_D$ has a profound effect on the nuclear magnetic resonance of ^{3}He, a resonance which behaves far more like that of a liquid than of a solid. The liquid-like 'motion' of the spins, responsible for this, *is* the exchange. The main features of $\mathcal{H}_e$ are that it commutes with the three components of the total nuclear spin **I**, but not with

$\mathcal{H}_{\rm D}$ and has a continuous spectrum of average span $\omega_{\rm e}$, much larger than that of the dipolar energy $\mathcal{H}_{\rm D}$. As we saw in **B**(a)2, $\mathcal{H}_{\rm e}$ increases very rapidly with the molar volume. Depending on this volume and also on the magnitude of the applied field, $\omega_{\rm e}$ can be much smaller, comparable to, or even larger than the Larmor frequency ω_0.

We shall often use for $\mathcal{H}_{\rm e}$ the bilinear expression (3.11) and even give some explicit estimates for the nearest neighbours coupling J, for the sake of internal consistency of the formalism. We must remember however, as already stated in **B**(a)1 that the expression (3.11) for $\mathcal{H}_{\rm e}$ is disproved by experiments performed at very low temperatures, as will appear at the end of this chapter, and that these estimates of J should not be taken literally.

It has already been indicated in **A** that the large value of the exchange energy was related to the large zero-point amplitude of the motion of the atoms. This motion has an effect on the dipolar energy itself independently of the modulation by exchange of its spin part. The point is that in the dipolar interaction between two nuclear spins,

$$\mathcal{H}_{\rm D}^{12} = -\gamma^2\hbar r_{12}^{-3} \sum_m F_m^*(\Omega_{12})T_m(\mathbf{I}_1, \mathbf{I}_2),$$

given by the equations (1.2) to (1.8) the orbital coefficients $F_m^*(\Omega)r^{-3}$ are modulated by the zero-point motion. Their average values $\langle F_m^*(\Omega)r^{-3}\rangle$ are different from $F_m^*(\Omega_0)r_0^{-3}$ where Ω_0 and r_0 define the orientation and the length of the vector $\mathbf{r}_0 = \mathbf{R}_1 - \mathbf{R}_2$ between the lattice sites $\mathbf{R}_1$ and $\mathbf{R}_2$ of the two spins. Ratios such as $\xi^{12} = \langle F_m^*(\Omega)r^{-3}\rangle / F_m^*(\Omega_0)r_0^{-1}$ can be estimated for instance by using for the description of the zero-point motion the Gaussian wavefunctions (3.5). Ratios ξ^{12} of the order of 0.8 are thus found for nearest neighbours (Harris, 1971). These reduction factors lead to reductions, of similar magnitude and rather difficult to calculate, for the various moments of the resonance line. Since the aim of these calculations is to refine the estimate of J that can be extracted from the knowledge of the moments and since the very definition of J is somewhat doubtful we shall forego these refinements and disregard altogether the changes in the dipolar interactions introduced by the zero-point motions.

The quantities accessible to measurements by NMR are the spin-lattice relaxation time T_1 of the Zeeman energy, the transverse relaxation time or inverse line width T_2 and, a rather novel feature in the NMR of solids, the diffusion coefficient D. We shall examine in turn the contribution of exchange to these three quantities.

(a) *The theory of the spin-lattice relaxation $T_{\rm Ze}$*

Below 1 K and for reasonably pure samples the number of vacancies and of impurity atoms of ^{4}He is small and their effect on relaxation can in general be neglected.

The only reservoir of energy with which the Zeeman energy is in contact is then the exchange energy $\mathcal{H}_e$, their coupling being mediated by the dipolar Hamiltonian $\mathcal{H}_D$ which commutes with neither Z nor $\mathcal{H}_e$.

If the exchange reservoir $\mathcal{H}_e$ is tightly coupled to the thermal bath (by mechanisms to be examined later on) and can be assumed to remain at a constant temperature, the rate of change of $\langle Z \rangle$ can be taken directly from the generalized Provotorov eqns (1.98a) as the coefficient $A_{11} = A_{22}$. We can also use the equivalent equations (1.105) the first of which can be rewritten as:

$$\frac{\mathrm{d}\beta_Z}{\mathrm{d}t} = -\frac{1}{T_{Ze}}(\beta_Z - \beta_e), \tag{3.33}$$

where $1/T_{Ze} = A_{11} = a/\mathrm{Tr}\{Z^2\}$ is given by the equation (1.107) where we make $\mathcal{H}_1 = Z = \omega_0 I_z$, $\mathcal{H}_0 = Z + \mathcal{H}_e$, $V = \mathcal{H}_D$, $W = 1/T_{Ze}$ and which reads:

$$\frac{1}{T_{Ze}} = \frac{\int_0^\infty \mathrm{Tr}\{[I_z, \mathcal{H}_D] \exp(\mathrm{i}\mathcal{H}_0 t)[\mathcal{H}_D, I_z] \exp(-\mathrm{i}\mathcal{H}_0 t)\}}{\mathrm{Tr}\{I_z^2\}}. \tag{3.34}$$

We shall find it convenient to write:

$$\begin{aligned} \exp(\mathrm{i}\mathcal{H}_0 t)\mathcal{H}_D \exp(-\mathrm{i}\mathcal{H}_0 t) &= \exp(\mathrm{i}Zt) \exp(\mathrm{i}\mathcal{H}_e t)\mathcal{H}_D \exp(-\mathrm{i}\mathcal{H}_e t) \exp(-\mathrm{i}Zt) \\ &\equiv \mathcal{H}_D^*(t), \end{aligned} \tag{3.35a}$$

and:

$$\mathcal{H}_D^*(t) = \exp(\mathrm{i}Zt)\mathcal{H}_D(t) \exp(-\mathrm{i}Zt), \tag{3.35b}$$

where:

$$\mathcal{H}_D(t) = \exp(\mathrm{i}\mathcal{H}_e t)\mathcal{H}_D \exp(-\mathrm{i}\mathcal{H}_e t). \tag{3.35c}$$

We thus separate in $\mathcal{H}_D^*(t)$ the coherent precession due to the Zeeman Hamiltonian Z from the quasi-random time dependence of $\mathcal{H}_D(t)$ induced by the exchange modulation. Since I_z commute with both $Z = \omega_0 I_z$ and $\mathcal{H}_e$, (3.34) can yet be rewritten as:

$$\frac{1}{T_{Ze}} = \frac{\int_0^\infty \mathrm{Tr}\{[I_z, \mathcal{H}_D][\mathcal{H}_D^*(t), I_z]\}\,\mathrm{d}t}{\mathrm{Tr}\{I_z^2\}}. \tag{3.36}$$

To proceed further we expand $\mathcal{H}_D$ as:

$$\mathcal{H}_D = \sum_m G_m, \qquad G_m = \gamma^2 \hbar \sum_{i<j} r_{ij}^{-3} F_m^*(\mathbf{r}_{ij}) T_m^{ij}. \tag{3.37}$$

The symbols F_m and T_m have been defined in (1.3) and (1.7). They transform respectively under rotations of the orbital and spin co-ordinates like second order spherical harmonics (eqn 1.4). The letters i and j refer to the various nuclear spins.

We then use the transformation properties of the operators G_m (eqns (1.9) and (1.10)):

$$[I_z, G_m] = mG_m, \qquad \exp(\mathrm{i}\omega_0 I_z t)G_m \exp(-\mathrm{i}\omega_0 I_z t) = \exp(\mathrm{i}m\omega_0 t)G_m. \tag{3.38}$$

The expression (3.34) for $1/T_{\mathrm{Ze}}$ becomes:

$$\frac{1}{T_{\mathrm{Ze}}} = \sum_{m,m'} (m'm) \int_0^\infty Tr\{G_{-m'}G_m(t)\} \exp(\mathrm{i}m\omega_0 t)\, \mathrm{d}t / \mathrm{Tr}\{I_z^2\}, \tag{3.39}$$

where:

$$G_m(t) = \exp(\mathrm{i}\mathscr{H}_{\mathrm{e}}t)G_m \exp(-\mathrm{i}\mathscr{H}_{\mathrm{e}}t).$$

It is clear that $\mathrm{Tr}\{G_{-m'}G_m(t)\}$ vanishes unless $m = m'$ and:

$$\frac{1}{T_{\mathrm{Ze}}} = \frac{1}{2} \sum_m m^2 \int_{-\infty}^\infty \mathrm{Tr}\{(G_{-m}G_m(t)\} \exp(\mathrm{i}m\omega_0 t)\, \mathrm{d}t / \mathrm{Tr}\{I_z^2\}. \tag{3.40}$$

We have made use in (3.40) of the fact that $\mathrm{Tr}\{G_{-m}G_m(t)\}$ is an even function of t.

We define now a few symbols. The correlation functions,

$$\mathscr{G}_m(t) = \mathscr{G}_{-m}(t) = \mathrm{Tr}\{G_{-m}G_m(t)\} / \mathrm{Tr}\{I_z^2\}, \tag{3.41}$$

and the spectral densities,

$$J_m(\omega) = \int_{-\infty}^\infty \mathscr{G}_m(t) \exp(\mathrm{i}\omega t)\, \mathrm{d}t = J_{-m}(\omega). \tag{3.42}$$

From (3.40), (3.41), and (3.42) we get:

$$\frac{1}{T_{\mathrm{Ze}}} = J_1(\omega_0) + 4J_2(2\omega_0). \tag{3.43}$$

It is worth pointing out that the normalization of the $J_m(\omega)$ given in (3.42) is different from (and more rational than) that introduced (for historical reasons) in Abragam (1961) chapter VIII.

$$J_0(\omega) : J_1(\omega) : J_2(\omega) = \frac{J_0^{\mathrm{A}}(\omega)}{6} : J_1^{\mathrm{A}}(\omega) : \frac{J_2^{\mathrm{A}}(\omega)}{4}, \tag{3.44}$$

where the $J_m^{\mathrm{A}}(\omega)$ are the symbols used in Abragam (1961). From the expression (3.37) of $\mathscr{H}_{\mathrm{D}}$ we obtain for the correlation functions $\mathscr{G}_m(t)$ defined by (3.41):

$$\mathscr{G}_m(t) = \gamma^2\hbar \sum_{i<j;k<l} r_{ij}^{-3} r_{kl}^{-3} F_{-m}^*(\mathbf{r}_{ij}) F_m^*(\mathbf{r}_{kl}) \Gamma_{ijkl}(t), \tag{3.45}$$

where:

$$\Gamma_{ijkl}(t) = \mathrm{Tr}\{T_{ij}^{-m} \exp(\mathrm{i}\mathcal{H}_{\mathrm{e}}t) T_{kl}^{m} \exp(-\mathrm{i}\mathcal{H}_{\mathrm{e}}t)\}/\mathrm{Tr}\{I_z^2\}. \qquad (3.45')$$

Because of the invariance through rotation of the exchange Hamiltonian the quantities Γ_{ijkl} are independent of m. The rate $1/T_{\mathrm{Ze}}$ which, according to (3.43) is a function of the Larmor frequency ω_0, obeys the integral relation,

$$\int_0^\infty \frac{\mathrm{d}\omega}{T_{\mathrm{Ze}}(\omega)} = \int_0^\infty \mathrm{d}\omega\{J_1(\omega) + 4J_2(2\omega)\} = \int_0^\infty \mathrm{d}\omega\{J_1(\omega) + 2J_2(\omega)\}$$
$$= \pi\ \mathrm{Tr}\{G_1G_{-1} + 2G_2G_{-2}\}/\mathrm{Tr}\{I_z^2\}, \qquad (3.46)$$

which is a direct consequence of (3.40). From the expression (3.37) of $\mathcal{H}_{\mathrm{D}}$ and the normalization (1.6) of the tensors T_{ij}^m we find:

$$\int_0^\infty \frac{\mathrm{d}\omega}{T_{\mathrm{Ze}}(\omega)} = \pi\gamma^4\hbar^2 \frac{1}{2}\frac{I(I+1)}{3}\sum_j \{|F_{ij}^1|^2 + 2|F_{ij}^2|^2\} r_{ij}^{-6}$$
$$= \pi\gamma^4\hbar^2 \frac{9}{16}\sum_j r_{ij}^{-6}(1 - \cos^4\theta_{ij}), \qquad (3.47)$$

where $I = \frac{1}{2}$ and the F_{ij} are given their values (1.3). Equation (3.47) can be compared to the second moment M_2 of the resonance line:

$$M_2 = -\mathrm{Tr}\{[I_x, \mathcal{H}'_{\mathrm{D}}]^2\}/\mathrm{Tr}\{I_x^2\} = -\mathrm{Tr}\{[I_x, G_0]^2\}/\mathrm{Tr}\{I_x^2\}$$
$$= \gamma^4\hbar^2 \frac{I(I+1)}{3} \times \frac{3}{2}\sum_j |F_{ij}^0|^2 r_{ij}^{-6}$$
$$= \gamma^4\hbar^2 \frac{I(I+1)}{3}\sum_j \frac{9}{4}(3\cos^2\theta_{ij} - 1)^2 r_{ij}^{-6}$$
$$= \gamma^4\hbar^2 \frac{9}{16}\sum_j (3\cos^2\theta_{ij} - 1)^2 r_{ij}^{-6}. \qquad (3.48)$$

For a polycrystalline sample (termed a powder from now on) we have to take angular averages of the quantities $|F_{ij}^m|^2$ and since $\overline{|F_{ij}^m|^2}$ is clearly independent of m:

$$\overline{|F_{ij}^1|^2} + 2\overline{|F_{ij}^2|^2} = 3\overline{|F_{ij}^0|^2},$$

which shows by comparison of (3.47) with (3.48) that for a powder:

$$\int_0^\infty \frac{\mathrm{d}\omega}{T_{\mathrm{Ze}}(\omega)} = \pi M_2. \qquad (3.49)$$

In a powder the correlation functions $\mathcal{G}_m(t)$ and the spectral densities $J_m(\omega)$ become $\mathcal{G}(t)$ and $J(\omega)$ independent of the index m:

$$\overline{F^{-m}(\mathbf{r}_{ij})F^m(\mathbf{r}_{kl})} = \tfrac{1}{5}\sum_m F^{-m}(\mathbf{r}_{ij})F^m(\mathbf{r}_{kl})$$

$$\propto P_2\{\cos\theta_{ij,kl}\} \equiv P_2(ij, kl),$$

where $\theta_{ij,kl}$ is the angle between the vectors $\mathbf{r}_{ij}$ and $\mathbf{r}_{kl}$, and

$$\mathcal{G}(t) = \frac{\gamma^4\hbar^2}{5N}\sum_{i<j} P_2(ij, kl) r_{ij}^{-3} r_{kl}^{-3} \Gamma_{ijkl}(t). \tag{3.50}$$

A few experiments having been performed with single crystals, attempts have been made to evaluate the correlation functions $\mathcal{G}_m(t)$ and the spectral densities $J_m(\omega)$. The calculation is difficult in particular because the 4-spins correlation functions $\Gamma_{ij,kl}(t)$ which vanish for $t=0$ unless the couples (ij) and (kl) coincide (eqn (3.45′)) do not do so for $t \neq 0$. This is why the *ad hoc* assumption is sometimes made that the $J_m(\omega)$ are of the form (Harris, 1971):

$$J_m(\omega) = \mathcal{G}_m(0)\varphi(\omega), \tag{3.51}$$

where $\varphi(\omega)$ is a function independent of m. The quantities $\mathcal{G}_m(0)$ which involve only the calculation of the 4-spins correlation functions Γ_{ijij} are easier to compute and the anisotropy of $1/T_{Ze}(\omega)$ can be calculated. There is no theoretical justification for the assumption (3.51) and it is not surprising that the agreement with experiment of a theory based on this assumption is far from perfect. We shall not develop here the theory of the anisotropy of $(1/T_{Ze})$ which can be found in Deville (1976).

(*b*) *The theory of line width and transverse relaxation time*

1. *Adiabatic line width*

For an ordinary solid the absorption line shape is the Fourier transform of the FID function $G(t)$ given by (1.43′),

$$G(t) = \mathrm{Tr}\{\exp(-\mathrm{i}\mathcal{H}'_{\mathrm{D}}t)I_x \exp(\mathrm{i}\mathcal{H}'_{\mathrm{D}}t)I_x\}, \tag{3.52}$$

where $\mathcal{H}'_{\mathrm{D}} = G_0$ (eqn (3.37)) is the part of the dipolar interaction that commutes with $Z = \omega_0 I_z$. It was stated in chapter 1, section **B**(*d*)1 that, in the presence of an exchange Hamiltonian $\mathcal{H}_{\mathrm{e}}$, the latter should be added to $\mathcal{H}'_{\mathrm{D}}$ in the exponentials of eqn (3.52). This statement must be qualified as follows. We recollect first how the truncation of the dipolar Hamiltonian is justified. In the absence of exchange the evolution operator of the spin system is:

$$U(t) = \exp[-\mathrm{i}(Z + \mathcal{H}_{\mathrm{D}})t]. \tag{3.53}$$

This can be rewritten as:

$$U(t)=\exp[-\mathrm{i}Zt]T\left\{\exp\left(-\mathrm{i}\int_0^t \mathscr{H}_{\mathrm{D}}^*(t')\,\mathrm{d}t'\right)\right\}, \tag{3.53$'$}$$

where:

$$\mathscr{H}_{\mathrm{D}}^*(t)=\exp(\mathrm{i}Zt)\mathscr{H}_{\mathrm{D}}\exp(-\mathrm{i}Zt). \tag{3.53$''$}$$

Using for $\mathscr{H}_{\mathrm{D}}$ the expansion (3.37) we get:

$$\mathscr{H}_{\mathrm{D}}^*(t)=\mathscr{H}'_{\mathrm{D}}+\mathscr{H}''^*_{\mathrm{D}}(t)=G_0+\sum_{m\neq 0} G_m\exp(\mathrm{i}m\omega_0 t). \tag{3.54}$$

The second part of (3.54) has a fast time dependence which makes its contribution to the integral in (3.53′) rather ineffective and justified its neglect and the customary truncation of $\mathscr{H}_{\mathrm{D}}$. Introducing now the strong exchange coupling $\mathscr{H}_{\mathrm{e}}$, eqns (3.53) and (3.53′) become:

$$\begin{aligned} U(t)&=\exp[-\mathrm{i}(Z+\mathscr{H}_{\mathrm{e}}+\mathscr{H}_{\mathrm{D}})t]\\ &=\exp(-\mathrm{i}(Z+\mathscr{H}_{\mathrm{e}})t)T\left\{\exp\left(-\mathrm{i}\int_0^t \mathscr{H}_{\mathrm{D}}^*(t')\,\mathrm{d}t'\right)\right\}, \end{aligned} \tag{3.55}$$

where now:

$$\begin{aligned} \mathscr{H}_{\mathrm{D}}^*(t)&=\exp(\mathrm{i}\mathscr{H}_{\mathrm{e}}t)\exp(\mathrm{i}Zt)\left(\sum_m G_m\right)\exp(-\mathrm{i}Zt)\exp(-\mathrm{i}\mathscr{H}_{\mathrm{e}}t)\\ &=G_0(t)+\sum_{m\neq 0} G_m(t)\exp(\mathrm{i}m\omega_0 t). \end{aligned} \tag{3.56}$$

The question which arises now is that of the relative magnitude of the Larmor frequency ω_0 and of the average frequency ω_{e} (rather loosely defined so far) which expresses the magnitude of the exchange energy $\mathscr{H}_{\mathrm{e}}$. If $\omega_0 \gg \omega_{\mathrm{e}}$ we can argue that in (3.56) $G_0(t)$ varies much more slowly than the terms with $m\neq 0$ and that the contribution of these fast varying terms to $T\{\exp-\mathrm{i}\int_0^t \mathscr{H}_{\mathrm{D}}^*(t')\,\mathrm{d}t'\}$ can be neglected in comparison with that of the relatively slowly varying $G_0(t)$. This is called the adiabatic approximation. The FID function $G(t)$ can then be written as:

$$\begin{aligned} G(t)&=\mathrm{Tr}\{\exp(-\mathrm{i}(G_0+\mathscr{H}_{\mathrm{e}})t)I_x\exp(\mathrm{i}(G_0+\mathscr{H}_{\mathrm{e}})t)I_x\}/\mathrm{Tr}\{I_x^2\}\\ &=\frac{\langle I_x(t)\rangle_{\mathrm{adiab}}}{\mathrm{Tr}\{I_x^2\}}, \end{aligned} \tag{3.57}$$

and all the theory of line width developed in chapter 1, section **B**(d)1 can be taken over directly. In the spirit of that section we begin by calculating the second and fourth moments, M_2 (already given in (3.48)) and M_4, of the resonance line:

$$M_2 = -\frac{\mathrm{Tr}\{[G_0, I_x]^2\}}{\mathrm{Tr}\{I_x^2\}} = \gamma^4\hbar^2\tfrac{3}{2}\sum_j |F_{ij}^0|^2 r_{ij}^{-6}$$
$$= \tfrac{9}{16}\gamma^4\hbar^2\sum_j (3\cos^2\theta_{ij} - 1)^2 r_{ij}^{-6}. \tag{3.48}$$

For a powder one finds:

$$M_2 = \gamma^4\hbar^2 a^{-6} C \tag{3.58}$$

where a is the distance between nearest neighbours and $C = 12{\cdot}25$ for the b.c.c. structure and 14·45 for h.c.p.

For the fourth moment M_4 we use the expression:

$$M_4 = \mathrm{Tr}\{[\mathcal{H}_e[G_0, I_x]]^2\}/\mathrm{Tr}\{I_x^2\}, \tag{3.59}$$

where we have assumed that $\mathcal{H}_e \gg G_0$ and kept only the leading term. If we take for $\mathcal{H}_e$ the bilinear interaction (3.11) between nearest neighbours we find for a powder:

$$M_4 = C'M_2J^2, \tag{3.60}$$

with $C' = 22{\cdot}8$ for b.c.c. ^{3}He and 42·0 for h.c.p. ^{3}He (Hirschfelder, Curtiss, and Bird, 1954).

We assume that the ratio,

$$\mu = \frac{M_4}{M_2} = \frac{C'}{C}\left(\frac{J}{\gamma^2\hbar/a^3}\right)^2 \gg 1, \tag{3.61}$$

and that the memory function $K(t)$ decays much more rapidly than $G(t)$. We saw in chapter 1, section **B**(d)1 that this leads for the FID function $G(t)$ to eqn (1.125′):

$$\frac{\mathrm{d}G}{\mathrm{d}t} = -G(t)\int_0^\infty K(t')\,\mathrm{d}t', \tag{1.125$'$}$$

that is to an exponential decay for $G(t)$:

$$G(t) = \exp(-t/T_2'),$$

which corresponds to an adiabatic line width $1/T_2'$:

$$\frac{1}{T_2'} = \int_0^\infty K(t)\,\mathrm{d}t = \mathcal{K}'(0). \tag{3.62}$$

If we assume for the memory function $K(t)$ defined by the expansion (1.135) the trial Gaussian form (1.136):

$$K(t) = M_2\exp(-\tfrac{1}{2}N_2t^2), \tag{1.136}$$

eqn (3.62) yields the result (1.140) or (1.142). Since $\mu \gg 1$, we have to a good approximation:

$$\frac{1}{T_2'} = \sqrt{\left(\frac{\pi}{2}\right)}\left(\frac{M_2}{\sqrt{N_2}}\right) \cong \sqrt{\left(\frac{\pi}{2}\right)}\left(\frac{M_2}{\mu}\right)^{1/2}, \tag{3.63}$$

where μ is given by (3.61) for a powder. We shall see that there is an experimental justification for using a Gaussian memory function in h.c.p. ^{3}He.

The Gaussian shape is not the only trial choice for $K(t)$ compatible with the expansion (1.135), and as will appear shortly a better choice for $K(t)$ in the b.c.c. phase is a Lorentzian:

$$K(t) = M_2/(1+\tfrac{1}{2}N_2t^2), \tag{3.64}$$

which leads for $1/T_2'$ to the formula:

$$\frac{1}{T_2'} = M_2\int_0^{\infty}\frac{dt}{1+\frac{1}{2}N_2t^2} = \frac{\pi}{\sqrt{2}}\left(\frac{M_2}{\mu}\right)^{1/2}. \tag{3.65}$$

If the experimental conditions are such that the non-adiabatic contributions to the line width are negligible it is possible to extract from a measurement of $1/T_2'$ a value for the coefficient J in the exchange interaction (3.11).

From (3.63), (3.65), (3.58), and (3.61) we get for a powder:

$$J = \left(\frac{\gamma^2\hbar}{a^3}\right)^2 T_2'C''C^{3/2}C'^{-1/2},$$

with:

	C	C'	C''
Lorentzian, b.c.c.	12·25	22·8	$\pi/\sqrt{2}$
Gaussian, h.c.p.	14·45	42	$\sqrt{(\pi/2)}$

whence:

$$\begin{aligned} J_{\text{b.c.c.}} &= 20T_2'(\gamma^2\hbar/a^3)^2; \\ J_{\text{h.c.p.}} &= 10{\cdot}6T_2'(\gamma^2\hbar/a^3)^2. \end{aligned} \tag{3.66}$$

To sharpen the somewhat loose definition of the average exchange frequency ω_e we define it as:

$$\omega_e^2 = \int_{-\infty}^{+\infty}\omega^2\mathcal{K}'(\omega)\,d\omega\bigg/\int_{-\infty}^{+\infty}\mathcal{K}'(\omega)\,d\omega, \tag{3.67}$$

$\mathcal{K}'(\omega)$ being the cosine Fourier transform of $K(t)$ its second moment is equal to minus the second derivative of $K(t)$ at $t=0$, that is:

$$\omega_e^2 = N_2 = M_2(\mu-1) \cong M_2\mu. \tag{3.68}$$

With this notation, we find for

$$\mathcal{K}'(\omega) = \int_0^\infty K(t) \cos(\omega t)\,dt$$

and for $1/T'_2$ the following expressions:

Gaussian memory function,

$$\mathcal{K}'(\omega) = \sqrt{\left(\frac{\pi}{2}\right)}\frac{M_2}{\omega_e}\exp(-\omega^2/2\omega_e^2); \tag{3.69}$$

$$\frac{1}{T'_2} = \sqrt{\left(\frac{\pi}{2}\right)}\frac{M_2}{\omega_e}; \tag{3.70}$$

Lorentzian memory functions,

$$\mathcal{K}'(\omega) = \frac{\pi}{\sqrt{2}}\frac{M_2}{\omega_e}\exp(-\omega\sqrt{(2)}/\omega_e); \tag{3.71}$$

$$\frac{1}{T'_2} = \frac{\pi}{\sqrt{2}}\frac{M_2}{\omega_e}. \tag{3.72}$$

In the case of the so-called 'strong narrowing' when $\mathcal{H}_e \gg \mathcal{H}_D$, it is convenient to introduce besides the FID function

$$G(t) = 1 + \sum_1^\infty (-1)^n \frac{M_{2n}}{(2n)!} t^{2n},$$

and the memory function

$$K(t) = M_2\left\{1 + \sum_1^\infty (-1)^n \frac{N_{2n}}{(2n)!} t^{2n}\right\},$$

two other functions $\tilde{G}(t)$ and $\tilde{K}(t)$ defined as follows: the moments M_{2n} and N_{2n} in the expansion of $G(t)$ and $K(t)$ are replaced by M'_{2n} and N'_{2n} which are the leading terms in M_{2n} and N_{2n}, that is terms containing $\mathcal{H}'_D$ twice and $\mathcal{H}_e$ $(2n-2)$ times. In the strong narrowing case, $\tilde{G}(t)$ and $\tilde{K}(t)$ differ very little respectively from $G(t)$ and $K(t)$. Similarly their cosine Fourier transforms $\tilde{g}(\omega)$ and $\tilde{\mathcal{K}}'(\omega)$ differ very little from $g(\omega)$ and $\mathcal{K}'(\omega)$. In the following, dealing with exchange or later with vacancies, we will be confronted with correlation functions $G_i(t)$ similar to $G(t)$. We will likewise introduce besides these functions and their associated memory functions $K_i(t)$, modified functions $\tilde{G}_i(t)$ and $\tilde{K}_i(t)$ and, unless otherwise stated, drop the tilde sign and disregard the small difference between the two types of functions.

2. *Transverse relaxation: general theory*

We develop in the present section a general theory of the transverse relaxation without discarding as in **C**(*b*)1 the off-diagonal terms of $\mathcal{H}_D$.

The decay rate of the precessing component $\langle I_x \rangle$ of the nuclear magnetization is derived from the Liouville formalism, in analogy with the derivation of the longitudinal decay rate $1/T_{Ze}$, and is given by a formula analogous to (3.36). There are however significant differences. Firstly, whereas $\langle I_z \rangle$ is coupled to $\langle \mathcal{H}_e \rangle$ by the Provotorov equation:

$$\frac{d}{dt}\langle I_z \rangle = -\frac{1}{T_{Ze}}\left\{\langle I_z \rangle - \omega_e \langle \mathcal{H}_e \rangle \frac{\mathrm{Tr}\{I_z^2\}}{\mathrm{Tr}\{\mathcal{H}_e^2\}}\right\}, \tag{3.73}$$

the decay of the transverse component $\langle I_x \rangle$ is not coupled to any other expectation value:

$$\frac{d}{dt}\langle I_x \rangle = -\frac{1}{T_2}\langle I_x \rangle. \tag{3.74}$$

Secondly, since (3.74) describes the evolution of $\langle I_x \rangle$ *in the rotating frame* it is determined by the Hamiltonian in the rotating frame:

$$\mathcal{H}(t) = \mathcal{H}_e + \tilde{\mathcal{H}}_D(t), \tag{3.75}$$

with:

$$\tilde{\mathcal{H}}_D(t) = \exp(\mathrm{i}\omega_0 I_z t)\mathcal{H}_D \exp(-\mathrm{i}\omega_0 I_z t). \tag{3.76}$$

The passage to the rotating frame was not necessary for the calculation of T_{Ze}, because $\langle I_z \rangle$ is the same in the laboratory frame as in the rotating frame.

The variation of $\langle I_x \rangle$ is described by an equation of the form (1.92) with only one operator of interest $Q = I_x$, and a memory function $K(t, t')$ (1.93) where $\mathcal{H}(t)$ depends explicitly on time. We make the same approximation as for the derivation of the Provotorov equations: since $\mathcal{H}_D \ll \mathcal{H}_e$ and $[\mathcal{H}_e, I_x] = 0$, we replace the exact expression (1.93) for $K(t, t')$ by an approximation limited to the second order in $\mathcal{H}_D$. This yields:

$$K(t, t') = \frac{\langle I_x | \hat{\tilde{\mathcal{H}}}_D(t) \exp[-\mathrm{i}\hat{\mathcal{H}}_e(t-t')]\hat{\tilde{\mathcal{H}}}_D(t') | I_x \rangle}{\langle I_x | I_x \rangle}, \tag{3.77}$$

or else, in the conventional notations of the Hilbert space:

$$K(t, t') = \frac{1}{\mathrm{Tr}\{I_x^2\}} \mathrm{Tr}\{[I_x, \tilde{\mathcal{H}}_D(t)] \exp(-\mathrm{i}\mathcal{H}_e(t-t'))[\tilde{\mathcal{H}}_D(t'), I_x] \exp(\mathrm{i}\mathcal{H}_e(t-t'))\}. \tag{3.78}$$

According to (3.37) and (3.38), $\tilde{\mathcal{H}}_D(t)$ is of the form:

$$\tilde{\mathcal{H}}_D(t) = \sum_m \exp(\mathrm{i}m\omega_0 t) G_m. \tag{3.79}$$

With the help of (3.79), the properties of the trace and the definition used in **C**(a):

$$G_m(t) = \exp(\mathrm{i}\mathcal{H}_e t) G_m \exp(-\mathrm{i}\mathcal{H}_e t),$$

(3.78) can be written:

$$\begin{aligned}
K(t, t') &= \frac{1}{\mathrm{Tr}\{I_x^2\}} \mathrm{Tr}\{\exp(\mathrm{i}\mathcal{H}_e(t-t'))[I_x, \tilde{\mathcal{H}}_D(t)] \exp(-\mathrm{i}\mathcal{H}_e(t-t'))[\tilde{\mathcal{H}}_D(t'), I_x]\} \\
&= \frac{1}{\mathrm{Tr}\{I_x^2\}} \sum_{m,m'} \exp(\mathrm{i}m\omega_0 t) \exp(-\mathrm{i}m'\omega_0 t') \\
&\quad \times \mathrm{Tr}\{[I_x, G_m(t-t')][G_{-m'}(0), I_x]\} \\
&= \frac{1}{\mathrm{Tr}\{I_x^2\}} \sum_{m,m'} \exp(\mathrm{i}\omega_0(m-m')t) \exp(\mathrm{i}m'\omega_0(t-t')) \\
&\quad \times \mathrm{Tr}\{[I_x, G_m(t-t')][G_{-m'}(0), I_x]\}.
\end{aligned} \tag{3.80}$$

The rate eqns (1.92) for $\langle I_x(t)\rangle$ becomes:

$$\begin{aligned}
\frac{\mathrm{d}}{\mathrm{d}t}\langle I_x(t)\rangle = &-\frac{1}{\mathrm{Tr}\{I_x^2\}} \sum_{m,m'} \exp(\mathrm{i}\omega_0(m-m')t) \\
&\times \int_0^t \exp(\mathrm{i}m'\omega_0(t-t')) \mathrm{Tr}\{[I_x, G_m(t-t')][G_{-m'}(0), I_x]\}\langle I_x(t')\rangle \, \mathrm{d}t'.
\end{aligned} \tag{3.81}$$

We make in the integral the substitution $t - t' = \tau$ and we take advantage of the fact that the trace falls off to zero for increasing τ much faster than $1/T_2$: we replace in the integral $\langle I_x(t')\rangle = \langle I_x(t-\tau)\rangle$ by $\langle I_x(t)\rangle$ and we extend its upper limit to ∞, whence:

$$\begin{aligned}
\frac{\mathrm{d}}{\mathrm{d}t}\langle I_x(t)\rangle = &-\frac{\langle I_x(t)\rangle}{\mathrm{Tr}\{I_x^2\}} \sum_{m,m'} \exp(\mathrm{i}\omega_0(m-m')t) \\
&\times \int_0^\infty \exp(\mathrm{i}m'\omega_0\tau) \mathrm{Tr}\{[I_x, G_m(\tau)][G_{-m'}(0), I_x]\} \, \mathrm{d}\tau.
\end{aligned} \tag{3.82}$$

On the right-hand side of (3.82) the terms with $m \neq m'$ oscillate rapidly as a function of t and bring a negligible contribution to $\langle I_x(t)\rangle$. Dropping them yields:

$$\frac{\mathrm{d}}{\mathrm{d}t}\langle I_x\rangle = -\frac{1}{T_2}\langle I_x\rangle,$$

$$\frac{1}{T_2} = \frac{1}{\mathrm{Tr}\{I_x^2\}} \sum_m \int_0^\infty \exp(\mathrm{i}m\omega_0\tau) \mathrm{Tr}\{[I_x, G_m(\tau)][G_{-m}, I_x]\}. \tag{3.83}$$

The following remark can be made. The expressions (3.36) for $1/T_{Ze}$ can be rewritten in the equivalent form:

$$\frac{1}{T_{Ze}}=\frac{1}{\mathrm{Tr}\{I_z^2\}}\sum_m\int_0^\infty \exp(im\omega_0\tau)\,\mathrm{Tr}\{[I_z, G_m(\tau)][G_{-m}, I_z]\}\,\mathrm{d}\tau. \quad (3.36')$$

We see that $1/T_2$, as given by (3.83) can be written straightaway by replacing I_z by I_x in (3.36′) but *not* in the equivalent expression (3.36). This is a justification for a somewhat lengthy derivation of a result at first sight self-evident. $1/T_2$ is calculated using the commutation relations (1.9*a*) which can be rewritten:

$$[I_x, T_0]=\sqrt{(\tfrac{3}{2})}(T_1+T_{-1}); \quad (3.84a)$$

$$[I_x, T_1]=T_2+\sqrt{(\tfrac{3}{2})}T_0; \quad (3.84b)$$

$$[I_x, T_2]=T_1. \quad (3.84c)$$

Together with the definition (3.37) for G_m they yield for the transverse relaxation rate (3.83):

$$\frac{1}{T_2}=\tfrac{3}{2}J_0(0)+\tfrac{5}{2}J_1(\omega_0)+J_2(2\omega_0), \quad (3.85)$$

where the $J_m(\omega)$ have been defined in (3.42). The first term $(\frac{3}{2})\,J_0(0)$ which originates in the part $G_0(t)$ of the dipolar Hamiltonian modulated by exchange coincides with the adiabatic relaxation rate $1/T_2'=\mathcal{K}'(0)$ of eqn (3.62). More generally, $\frac{3}{2}J_0(\omega)=\mathcal{K}'(\omega)$ or to be more accurate, $\frac{3}{2}J_0(\omega)=\tilde{\mathcal{K}}'(\omega)$ where $\tilde{\mathcal{K}}'(\omega)$, practically indistinguishable from $\mathcal{K}'(\omega)$ has been defined in **C**(*b*)1. The proof which is easily obtained by taking the time derivative of eqn (3.57) is left to the reader.

$\frac{3}{2}J_0(0)$ is the only term that contributes to the line width in high field when $J_1(\omega_0)$ and $J_2(\omega_0)$ are negligible in comparison with $J_0(0)$. On the other hand in low fields, when $\omega_0\ll\omega_e$,

$$\frac{1}{T_2}\approx\tfrac{3}{2}J_0(0)+\tfrac{5}{2}J_1(0)+J_2(0).$$

In a powder where $J_m(\omega_0)$ is independent of m this is equal to $5J(0)$ which is $\frac{10}{3}$ of $1/T_2'=\frac{3}{2}J_0(0)$.

This is the so-called $\frac{10}{3}$ effect, an increase by that factor in line width as the ratio ω_0/ω_e goes from a value much greater than 1 to a value much smaller than 1.

(*c*) *Exchange and diffusion of nuclear magnetization*

The concept of spin diffusion was introduced in nuclear magnetism many years ago (Bloembergen, 1949) to explain the nuclear relaxation by paramagnetic impurities (see chapter 6). The flip-flop terms $I_+^iI_-^j$ present in the dipolar spin–spin interaction prevent the z component I_z^j of a spin j from being a good quantum number. If for some reason the nuclear magnetization at a given time is not uniform in space, if for instance there is an excess of it near an electronic impurity spin, this excess will spread out and the inhomogeneity will decay by a diffusion process mediated by successive flip-flops between neighbouring spins.

The usual practice is to assume that the diffusion on a macroscopic scale of the nuclear magnetization $\mathbf{M}(\mathbf{r}, t)$ can be described by the classical equation:

$$\frac{\partial \mathbf{M}}{\partial t} = D\nabla^2\mathbf{M} \tag{3.86}$$

A crude argument (Abragam (1961), chapter V) shows that the diffusion constant D is of the order Wa^2 where W is the probability of a flip-flop occurring between two neighbouring spins separated by a distance a. The magnitude of D for a diffusion induced by dipolar interaction is exceedingly small, of the order of 10^{-12} cm^2 s^{-1} and the propagation of magnetization exceedingly slow. Over a distance r it requires a time $\tau \approx r^2/D$, of the order of several hours for one micron.

This is to be contrasted with diffusion due to atomic or molecular motion in liquids or even in solids not too far from the melting point ($D \approx 2 \,.\, 10^{-5}$ for protons in water at 20°C). The diffusive motion of a nuclear spin in an inhomogeneous magnetic field affects the amplitude of a 180° spin echo in a way which provides a means of measuring D. It can be shown (see Abragam (1961), chapter III) that the amplitude of the echo at time 2τ is attenuated by:

$$E(2\tau) = \exp\{-\tfrac{2}{3}D\gamma^2G^2\tau^3\}, \tag{3.87}$$

where G is the gradient of the applied field over the sample. Atomic self-diffusion in many liquids has thus been studied by spin echoes in fields with known inhomogeneity.

On the other hand spin-diffusion in solids is far too slow to be studied by this method. The only exception is solid ^{3}He where the exchange coupling J, or rather $J/2\pi$, can reach 20 MHz for the largest molar volume, with diffusion coefficients D of the order of 10^{-7}.

This is not too surprising if one considers that in ^{3}He there is no real difference between spin diffusion and atomic diffusion: when atoms at neighbouring sites i and j exchange their nuclear spin orientations, the question whether this has occurred through a flip-flop of the spins, or through the interchange of the atoms themselves, has no physical meaning.

The purpose of this section is the following. In order to extract from a measurement of diffusion of nuclear magnetization a value of J (with all the reservations already made about the meaning of J itself) we want to make a theoretical estimate of the dimensionless coefficient, of order unity which relates D to Ja^2 (Redfield and Yu, 1968; 1969).

The theory is a makeshift one, not unlike that of the resonance line width of chapter 1, incorporating exact calculations of moments to more or less warranted assumptions about the form of certain correlation functions. We begin by recognizing that, as stated earlier, the z component I_z^j of a spin j is time-dependent because of the existence of exchange and that there is a correlation between I_z^j at time 0 and I_z^k at time t. This leads to the definition of a correlation function:

$$\Gamma_{jk}(t) = \mathrm{Tr}\{I_z^j I_z^k(t)\}/\mathrm{Tr}\{1\} = \mathrm{Tr}\{I_z^j \exp(\mathrm{i}\mathcal{H}_e t) I_z^k \exp(-\mathrm{i}\mathcal{H}_e t)\}, \quad (3.88)$$

with $\Gamma_{jk}(0) = \frac{1}{4}\delta_{jk}$. Equation (3.88) can be expanded in powers of t:

$$\Gamma_{jk}(t) = \sum_{n=0}^{\infty} \frac{(-t^2)^n}{(2n!)} \gamma_{jk}^{(2n)}, \quad (3.89)$$

with:

$$\sum_k \Gamma_{jk}(t) = \sum_k \mathrm{Tr}\{I_z^j \exp(\mathrm{i}\mathcal{H}_e t) I_z^k \exp(-\mathrm{i}\mathcal{H}_e t)\}/\mathrm{Tr}\{1\}$$
$$= \mathrm{Tr}\{I_z^j I_z\}/\mathrm{Tr}\{1\} = 1/4, \quad (3.90)$$

whence:

$$\sum_k \gamma_{jk}^{(2n)} = \tfrac{1}{4}\delta_{n0}. \quad (3.91)$$

Instead of $\Gamma_{jk}(t)$ we introduce through the spatial Fourier transforms,

$$I_z^{\mathbf{q}} = \frac{1}{\sqrt{N}} \sum_j I_z^j \exp(-\mathrm{i}\mathbf{q}\,.\,\mathbf{R}_j),$$

the correlation function:

$$\Gamma_{\mathbf{q}}(t) = \mathrm{Tr}\{I_z^{\mathbf{q}}(t) I_z^{-\mathbf{q}}(0)\}/\mathrm{Tr}\{1\} = \sum_k \Gamma_{jk}(t)\exp(\mathrm{i}\mathbf{q}\,.\,(\mathbf{R}_k - \mathbf{R}_j)). \quad (3.92)$$

From (3.88), (3.89), and (3.92) we get for $\Gamma_{\mathbf{q}}(t)$ the expansion:

$$\Gamma_{\mathbf{q}}(t) = \frac{1}{4}\left\{1 + \sum_{n=1}^{\infty} \frac{(-t^2)^n}{(2n)!} m_{2n}(\mathbf{q})\right\}, \quad (3.93)$$

with:

$$m_{2n}(\mathbf{q}) = 4\sum_k \gamma_{jk}^{(2n)} (\exp(\mathrm{i}\mathbf{q}\,.\,(\mathbf{R}_k - \mathbf{R}_j)) - 1). \quad (3.93')$$

These expressions are both exact and useless unless some phenomenological assumptions are made.

We concentrate on the behaviour of $\Gamma_{\mathbf{q}}(t)$ for wavelengths $1/|\mathbf{q}|$, much larger than the atomic spacing a (although much smaller than the size of the sample in order to make expansions such as (3.92) meaningful), and for times t very much larger than the only microscopic time constant of the problem, the inverse exchange frequency $1/J$.

We then make the *assumption* that the time dependence of $\Gamma_{\mathbf{q}}(t)$ is the same as that of a classical solution of the diffusion equation (3.86) with the same wave vector $\mathbf{q}$. If we try in (3.86): $M_z(\mathbf{r}, t) = M_0\Gamma_{\mathbf{q}}(t)\exp(i\mathbf{q}.\mathbf{r})$ we find:

$$\frac{d\Gamma\mathbf{q}(t)}{dt} = -Dq^2\Gamma_{\mathbf{q}}(t),$$

$$\Gamma_{\mathbf{q}}(t) = \tfrac{1}{4}\exp - \frac{|t|}{\tau_{\mathbf{q}}},$$

where

$$\frac{1}{\tau_{\mathbf{q}}} = Dq^2.$$

This expression for $\Gamma_{\mathbf{q}}(t)$ clearly contradicts the expansion (3.93) for small $t \approx 1/J$. On the other hand t comparable to $\tau_{\mathbf{q}}$ means $Dq^2t \sim 1$ and since $D \sim Ja^2$, $Ja^2q^2t \sim 1$ or $t \sim (1/J)(1/q^2a^2)$. Since we have assumed that $qa \ll 1$ we have $t \gg 1/J$ and the contradiction need not occur. To correlate $1/\tau_{\mathbf{q}} = Dq^2$ to the *exact* expansion (3.93) we resort once more to the approximation of the Gaussian memory function. We notice from (3.89) and (3.93′) that $m_2(q)$ is of order $J^2q^2a^2$ and $m_4(q)$ of order $J^4q^2a^2$. Since we have assumed that $qa \ll 1$, $\mu = m_4(q)/[m_2(q)]^2 \sim 1/q^2a^2 \gg 1$, and $1/\tau_{\mathbf{q}}$ is given by exactly the same formula as $1/T_2'$ in (3.63).

$$\frac{1}{\tau_{\mathbf{q}}} = \sqrt{\left(\frac{\pi}{2}\right)}[m_2(q)]^{3/2}[m_4(q)]^{-1/2} = Dq^2. \tag{3.94}$$

It remains to compute $m_2(q)$ and $m_4(q)$ from (3.88), (3.89), and (3.93′), and for small qa, a straightforward but cumbersome calculation. For a b.c.c. lattice it yields (Redfield and Yu 1968; 1969):

$$\begin{cases} m_2(q) = \frac{8}{3}J^2a^2q^2 \\ m_4(q) = \frac{208}{3}J^4a^2q^2, \end{cases} \tag{3.95}$$

whence from (3.94)

$$D = \frac{4}{3}\sqrt{\left(\frac{\pi}{13}\right)}Ja^2 = 4{\cdot}13\left(\frac{J}{2\pi}\right)a^2. \tag{3.96}$$

It is possible in principle to improve on this approximation by calculating $m_6(\mathbf{q})$ and incorporating its knowledge in the form selected for the memory function. In view of the uncertainty connected with the definition of J, this is hardly worth the trouble.

(*d*) *NMR and exchange: experimental results*

The criterion for the Zeeman relaxation time observed by NMR to be the parameter T_{Ze} expressed by eqns (3.34) and (3.43), is its independence of the temperature of the sample. The range of temperatures where this independence occurs has an upper limit T_{max} above which the vacancies, whose number is proportional to $\exp(-\Phi/k_BT)$, take over. There is also a lower limit T_{min} below which the bottle-neck is the coupling between the exchange reservoir and the lattice.

Since $1/T_{Ze}$ depends rather sharply on the Larmor frequency (eqn (3.43)) these temperature limits vary with the applied magnetic field, and also with the molar volume which increases the strength of exchange. The spectral densities J_1 and J_2 decrease rapidly with ω and except in low fields it is often permissible to neglect in (3.43) $J_2(2\omega)$ in comparison with $J_1(\omega)$. This spectral density exhibits a rather remarkable behaviour: over several decades it is an exponential in b.c.c. ^{3}He and a Gaussian in h.c.p. ^{3}He. Figure 3.2 shows a plot of $\ln(T_{Ze})$, against $(\omega_0/2\pi)$ in the b.c.c. phase and against $(\omega_0/2\pi)^2$ in the h.c.p. phase. Both plots are well approximated by straight lines. We had already met earlier in chapter 1, section **B**(e)3 with a spectral density which had an exponential shape over several decades, a fact for which we had no theoretical explanation. The same situation prevails in b.c.c. solid ^{3}He.

From the expressions (3.69) and (3.71) for $\mathcal{K}'(\omega)$ and the relation $\mathcal{K}'(\omega) = \frac{3}{2}J_0(\omega)$ we are led (for a powder) to the following spectral density:

$$\text{b.c.c.: } J(\omega) = \frac{\pi\sqrt{2}}{3}\frac{M_2}{\omega_e}\exp\left(-\frac{\omega\sqrt{2}}{\omega_e}\right); \qquad (3.97a)$$

$$\text{h.c.p.: } J(\omega) = \frac{\sqrt{(2\pi)}}{3}\frac{M_2}{\omega_e}\exp(-\omega^2/2\omega_e^2). \qquad (3.97b)$$

Since M_2 and M_4 are given by (3.58) and (3.60) the measurement of T_{Ze} provides a value for J. The values of J extracted from measurements of T_{Ze} are presented in Fig. 3.3 together with values obtained by other methods.

The remarkable behaviour of the correlation function $J(\omega)$ given by (3.97) breaks down for large values of the Larmor frequency ω_0. When ω_0/ω_e exceeds 7 or 8 the exponential field dependence of T_{Ze} on ω_0 is replaced in b.c.c. ^{3}He by a quadratic field dependence, which has a universal character in the sense that it depends only on the ratio (ω_0/ω_e) (Sullivan and Chapellier, 1974). Instead of T_{Ze} it is convenient to introduce a reduced relaxation time: $T'_{Ze} = T_{Ze}(21/V_M)^2$ where V_M is the molar volume. This factor removes the trivial dependence of the second moment on the molar volume. Between $\omega_0/\omega_e = 7$ and 70, T'_{Ze} is well approximated by:

$$T'_{Ze} \cong 7{\cdot}6\left(\frac{\omega_0}{\omega_e}\right)^2.$$

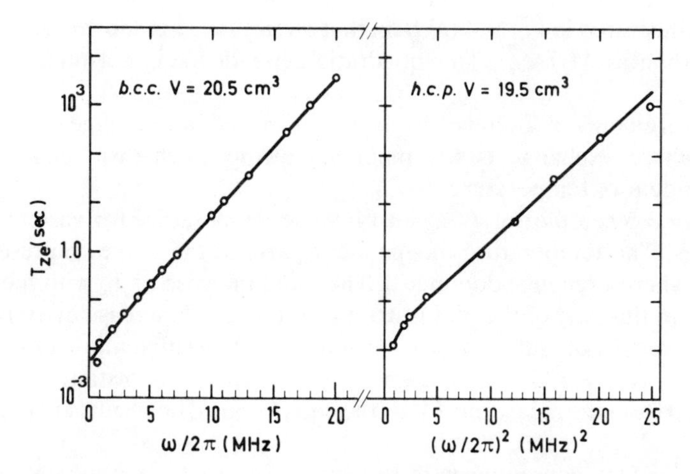

FIG. 3.2. Zeeman exchange relaxation time T_{Ze} against nuclear Larmor frequency in two samples of solid ^{3}He with crystalline structures b.c.c. and h.c.p. (After Landesman, 1973.)

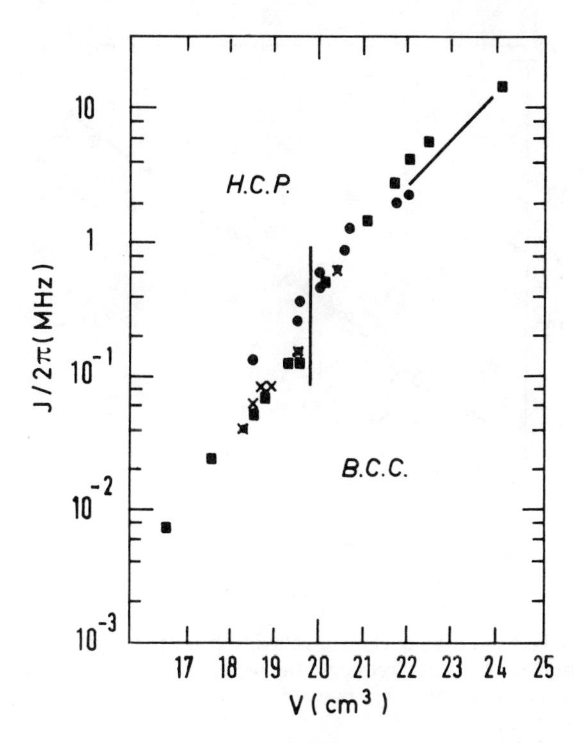

FIG. 3.3. Exchange interaction J in solid ^{3}He against molar volume in both b.c.c. and h.c.p. phases, as deduced from NMR measurements. (After Landesman, 1973.)

This is illustrated in Fig. 3.4 (*N.B.* in that reference ω_e^2 is defined as $M_4/2M_2^2$ rather than as M_4/M_2^2). This quadratic dependence is not understood at present.

Measurements of T_2 have also been performed in the range of temperatures where exchange is the main narrowing mechanism and $1/T_2$ is independent of temperature.

Figure 3.5 is a plot of T_2 against inverse temperature for various molar volumes. The temperature independent parts of the curves represent the region where exchange dominates. The rapid increase of T_2 with the molar volume in this part of the plot is witness to the rapid increase in exchanges. In the range of ratios ω_0/ω_e where the non-adiabatic contribution $(1/T_2)^{na} = 1/T_2 - 1/T_2'$ is negligible the exchange constant J can be obtained from the measured T_2' by the eqns (3.66). The results are plotted in Fig. 3.3.

The $\frac{10}{3}$ effect, consequence of the eqn (3.85), that is the increase by this factor in line width as the Larmor frequency ω_0 is reduced from a value $\omega_0 \gg \omega_e$ to $\omega_0 \ll \omega_e$, has been observed in the domain of temperatures where exchange is the dominant relaxation mechanism. In the h.c.p. phase this

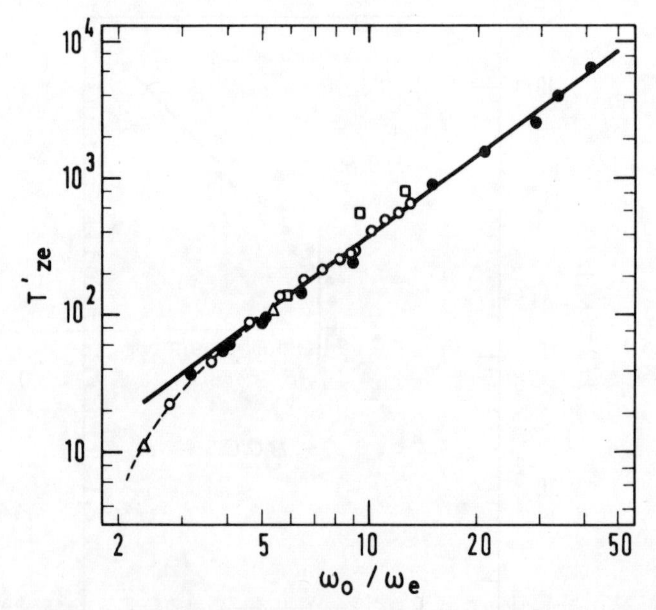

FIG. 3.4. Reduced Zeeman exchange relaxation time $T_{Ze}' = T_{Ze}(21/V_M)^2$ in high field against nuclear Larmor frequency in b.c.c. ^{3}He. Solid circles: constant field $H_0 = 2{\cdot}5T$ and molar volume V_M over the range 20·1 to 23·5 cm^3. Open circles: constant $V_M = 21$ cm^3 and $0{\cdot}5 < H_0 < 2{\cdot}5T$. Open triangles: constant $V_M = 20{\cdot}5$ cm^3 and $0{\cdot}5 < H_0 < 2{\cdot}5T$. Open squares: other measurements. (After Sullivan and Chapellier, 1974).

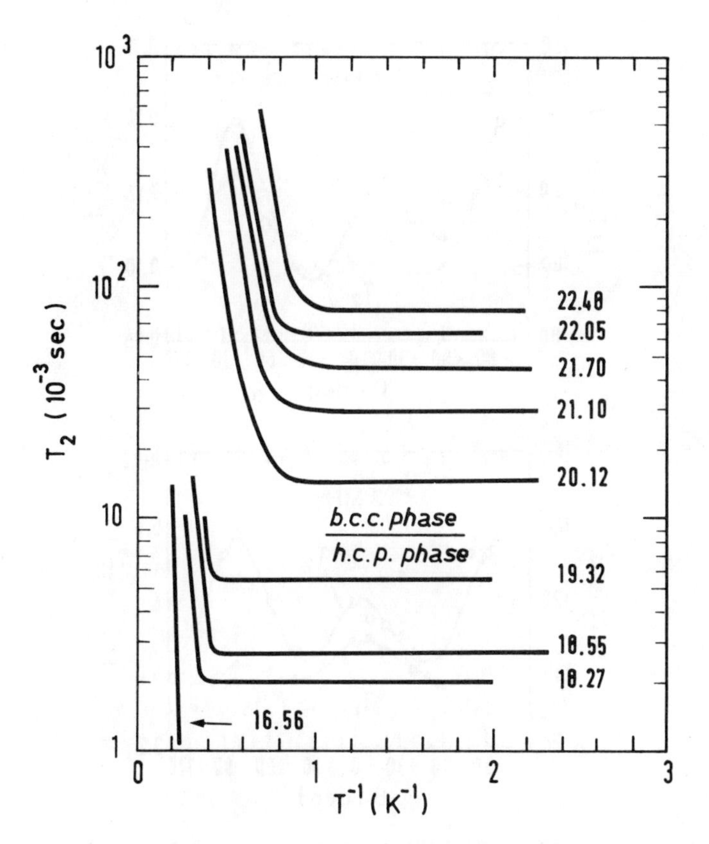

FIG. 3.5. Transverse relaxation time T_2 against temperature for various molar volumes in the b.c.c. and h.c.p. phases of solid ^{3}He. (After Landesman, 1975.)

increase occurred as $(\omega_0/\omega_e)^2$ went from approximately 30 to 0·2 and in the b.c.c. phase as (ω_0/ω_e) went from 3 to 0·15 (Richardson, Landesman, Hunt, and Meyer, 1966).

In single crystals T_1 and T_2 exhibit a certain anisotropy (Deville, 1976). The experimental conditions under which these measurements were performed do not permit a clear-cut decision as to the mechanisms responsible for these relaxations: exchange, vacancies, or most likely, a combination of the two. This renders even more uncertain the theoretical interpretation of the anisotropy, already complicated with a single mechanism, which is attempted in Deville (1976). As an illustration, Fig. 3.6 shows the kind of anisotropy observed in a hexagonal crystal. In Fig. 3.7, the angular dependence of $1/T_1$ is plotted for several values of the temperature. As the temperature goes up from 1.18 K where exchange is presumably dominant to 2·4 K, where the vacancies have taken over, $1/T_1$ increases, but its

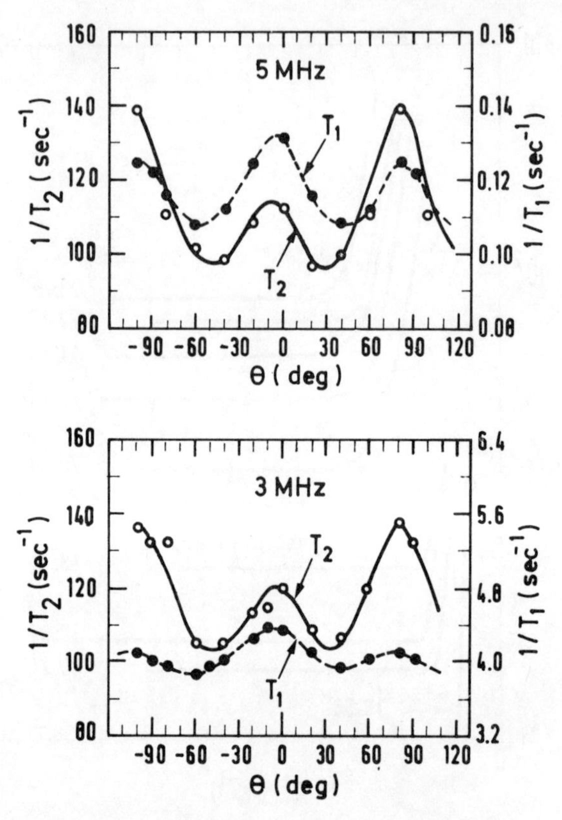

FIG. 3.6. Angular variations of the nuclear relaxation rates in a single crystal of h.c.p. ^{3}He at two different Larmor frequencies. Molar volume $V_M = 19{\cdot}46\ \mathrm{cm}^3$, temperature $T = 1{\cdot}18$ K. The angle θ is between the field direction and the c-axis. (After Deville, 1976.)

absolute anisotropy changes very little, suggesting that the relaxation caused by exchange is more anisotropic than that due to vacancies.

It must be admitted that the study of the anisotropy of relaxation, besides demonstrating its existence, has contributed little to the understanding of solid ^{3}He.

Below the temperature $T_{\min}$ mentioned at the beginning of this section, the Zeeman-exchange coupling is faster than the coupling between the exchange reservoir and the rest of the lattice. If $1/T_{eL}$ is the rate of this coupling, whatever its nature, to be considered later in this chapter, the *observed* Zeeman relaxation rate $1/T_{1Z}$ is then related to $1/T_{eL}$ by:

$$\frac{1}{T_{1Z}} = \frac{1}{T_{eL}} \frac{\langle \mathcal{H}_e \rangle}{\langle Z \rangle + \langle \mathcal{H}_e \rangle}. \tag{3.98}$$

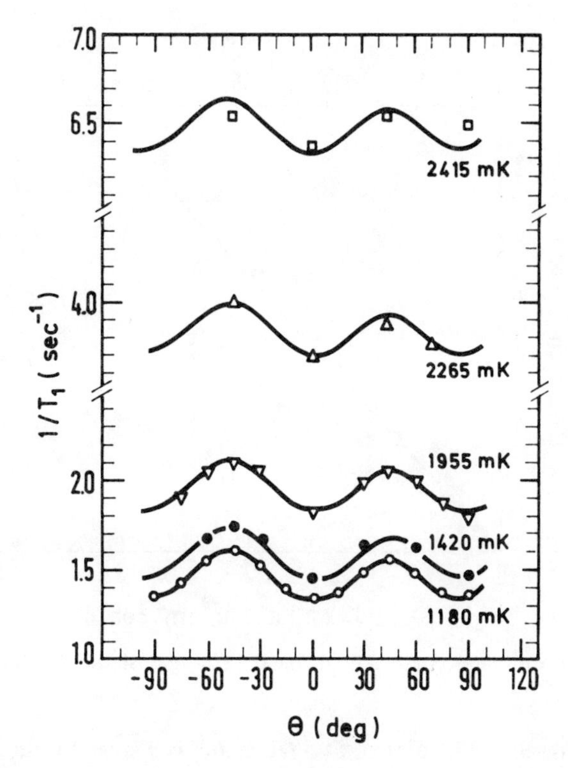

FIG. 3.7. Angular variation of the spin–lattice relaxation rate in h.c.p. ^{3}He at various temperatures. Molar volume $V_M = 19{\cdot}15$ cm^3; Larmor frequency $\omega/2\pi = 3$ MHz. (After Deville, 1976.)

At high temperatures $\langle Z\rangle$ and $\langle\mathcal{H}_e\rangle$ are given by:

$$\langle Z\rangle = -\frac{N}{4}\beta\omega_0^2; \qquad \langle\mathcal{H}_e\rangle = -\tfrac{3}{8}zN\beta J^2. \tag{3.99}$$

From the variation of the *observed* Zeeman relaxation rate with the magnetic field a value of J is obtained. This value is also plotted in Fig. 3.3 against the molar volume.

It appears on this plot that the values of J obtained by the three methods described so far, measurement of $1/T_{Ze}$, of T_2, and of $\langle\mathcal{H}_e\rangle/\langle Z\rangle$ give results in reasonable agreement with each other. All three methods give the magnitude but not the sign of J. A last determination of J is through the measurement of the diffusion coefficient D by the spin-echo method.

Figure 3.8 is a plot of D against $(Ja^2/2\pi)$ where for each molar volume for which D is measured, J is extracted from the plot of J against the molar

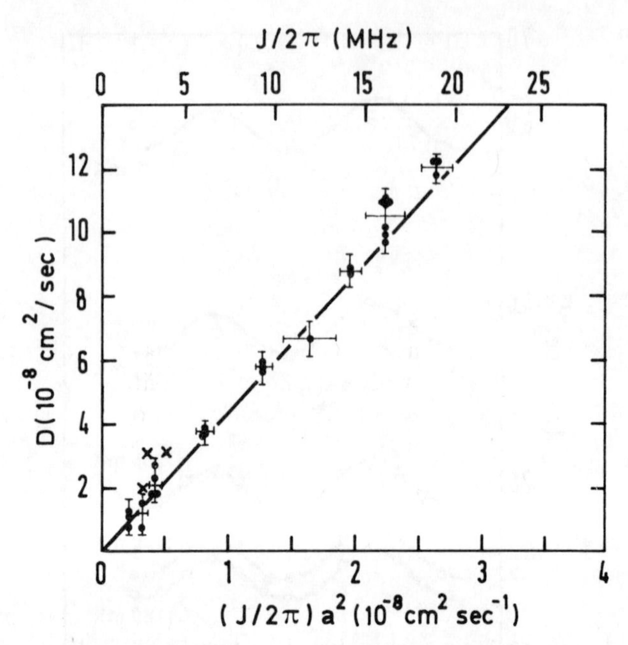

FIG. 3.8. Spin diffusion coefficient in b.c.c. ^{3}He against the product Ja^2. (After Landesman, 1973.)

volume of Fig. 3.3. The plot of D against Ja^2 is a straight line,

$$D = (4{\cdot}4 \pm 0{\cdot}4)J\frac{a^2}{2\pi}, \tag{3.100}$$

whose excellent agreement with the semi-theoretical relation (3.96) is very likely accidental.

On the whole the experimental evidence presented in this section is accounted for rather favourably by the assumption of an exchange interaction of the type (3.11).

We shall see later that other measurements, near the ordering temperature T_c or even well above, spoil this picture and require a different form of the exchange Hamiltonian.

D. The effects of vacancies on NMR in solid ^{3}He

(*a*) *Introduction*

Formally the effects of vacancies on NMR in solid ^{3}He are very similar to those of exchange.

The equivalent of the exchange Hamiltonian (3.11) is the vacancies Hamiltonian $(\mathcal{H}_f + \mathcal{H}_t)$ of eqns (3.32) where $\mathcal{H}_f$ is the Hamiltonian for the formation of the vacancies (3.32a) and $\mathcal{H}_t$ the tunnelling Hamiltonian (3.32c) responsible for their motion. The operators $\mathbf{S}_i$, a_i and b_i in (3.32) are related to the nuclear spin variables $\mathbf{I}_j$ through (3.30) and (3.31). As in the case of exchange the coupling Hamiltonian between the Zeeman reservoir and the motion of the vacancies is the dipolar spin–spin interaction.

We shall thus be able to take over most of the formalism described in section **C**. On the experimental side however there is an essential difference between the behaviour of exchange and of vacancies, namely the role of temperature.

The exchange frequencies J (in radians per second) do not exceed 10^8, which justified our neglect in section **C** of the effects of temperature on the spectral density connected with exchange. By contrast the temperature $T_v = \hbar\omega_f/k_B$ for the formation of vacancies, of the order of a few kelvins exceeds considerably the temperature at which the NMR experiments are performed in solid ^{3}He, and requires the inclusion in the calculation of the various correlation functions, of the density matrix: $\sigma_f = \exp - \beta\mathcal{H}_f$. On the other hand at temperatures high enough for the concentration of vacancies $x = \exp(-\beta\Phi)$ not to be negligibly small, the product $\beta\omega_v$, where $\omega_v \lesssim 10^{10}$ radians per second, is small and $\exp(-\beta\mathcal{H}_t)$ can be taken equal to unity. Thus the only influence of temperature on the various statistical averages that we shall be dealing with will be through the concentration,

$$x = \exp - (\Phi/k_B T).$$

This description is rather different from that encountered for ordinary solids. There, it is assumed that in its ground state the probability for a vacancy to tunnel into a neighbouring site is negligible. A jump of the vacancy or rather of a neighbouring atom into the vacancy can only occur if some extra energy is provided through thermal excitation in order to overcome a potential barrier of height Φ_m.

This leads to an exponential Arrhenius law for the motion of the vacancy with an activation energy,

$$W = \Phi + \Phi_m, \tag{3.101}$$

where Φ is the formation energy and Φ_m the height of the barrier. By identifying W with Φ we explicitly assume that the motion of the vacancies in solid ^{3}He is a quantum mechanical tunnelling. We shall consider later the experimental justification of this assumption.

We shall also consider in section **E**(b) the influence of the existence of phonons on the motion of the vacancies. Anticipating that section we recognize three ways in which this influence is perceived by the vacancies.

(i) The coupling between vacancies and phonons keeps the vacancy reservoir at the temperature of the thermal bath.

(ii) The collision between a phonon and a vacancy wave gives to the latter a finite lifetime. When this collision rate becomes larger than the tunnelling frequency ω_v, the picture of the vacancy propagating as a wave becomes invalid.

(iii) The existence of phonons spoils the periodicity of the lattice, an effect which also interferes with the motion of the vacancy considered as a wave propagation in a periodic medium. It may be worth pointing out that the effects (ii) and (iii) although connected are not the same thing. To conclude we remark that the introduction of vacancies through an increase in temperature does not suppress exchange and both should be considered simultaneously. Fortunately because of the Arrhenius law for the vacancy concentration x, the range of temperatures where both effects coexist, that is where $x\omega_v$ and J are comparable, is rather narrow and can often be disregarded.

(b) Vacancies and diffusion of nuclear magnetization

The normal treatment of diffusion by vacancies is very similar to that given for exchange in **C**(c), with some minor changes. We define the correlation function $\Gamma_{jk}(t)$ as:

$$\begin{aligned}\Gamma_{jk}(t) &= \tfrac{1}{4}\langle \exp(-it(\mathcal{H}_t+\mathcal{H}_f))S_z^j \exp(it(\mathcal{H}_t+\mathcal{H}_f))S_z^k\rangle \\ &= \tfrac{1}{4}\langle \exp(-it\mathcal{H}_t)S_z^j \exp(it\mathcal{H}_t)S_z^k\rangle. \end{aligned} \tag{3.102}$$

S_z^j is the fictitious spin defined in (3.30), which vanishes for a vacant site j and is equal to $2I_z^j$ at normal sites, and $\mathcal{H}_f$ and $\mathcal{H}_t$ are defined in (3.32). The expectation value $\langle Q\rangle$ of an operator Q is given by $\mathrm{Tr}\{\sigma Q\}$. As explained in the introduction we can take for $\langle Q\rangle$ the expression:

$$\begin{aligned}\langle Q\rangle &= \mathrm{Tr}\{\exp(-\beta(\mathcal{H}_t+\mathcal{H}_f))Q\}/\mathrm{Tr}\{\exp(-\beta(\mathcal{H}_t+\mathcal{H}_f))\} \\ &\cong \mathrm{Tr}\{\exp(-\beta\mathcal{H}_f)Q\}/\mathrm{Tr}\{\exp(-\beta\mathcal{H}_f)\}. \end{aligned} \tag{3.103}$$

We introduce $\Gamma_{\mathbf{q}}(t)$ related to $\Gamma_{jk}(t)$ by eqn (3.92) and expand $\Gamma_{\mathbf{q}}(t)$ in a power series by eqn (3.93). We then calculate $m_2(\mathbf{q})$ and $m_4(\mathbf{q})$ for a b.c.c. lattice, finding, for small qa (Landesman, 1974):

$$\begin{aligned} m_2(q) &= (\tfrac{8}{3})x\omega_v^2q^2a^2; \\ m_4(q) &= m_2(q)\times 15\omega_v^2. \end{aligned} \tag{3.104}$$

The coefficient D for the diffusion of nuclear magnetization is then obtained from eqn (3.94) by the same procedure as for exchange:

$$D = \frac{8}{6}\sqrt{\left(\frac{2\pi}{15}\right)}x\omega_v a^2 = 5{\cdot}2\times\tfrac{1}{6}x\omega_v a^2. \tag{3.105}$$

The derivation of (3.105) which has followed the same lines as that of the diffusion constant (3.96) for exchange is entirely quantum mechanical, and its physical content may not be very easy to perceive. In the usual picture of diffusion as a random walk of steps a, the atomic diffusion coefficient D is associated to a correlation time τ_a (a for atomic) by the classical relation:

$$D = a^2/6\tau_a, \tag{3.106}$$

which identified with 3.105 yields:

$$\frac{1}{\tau_a} = Ax\omega_v = 8\sqrt{\left(\frac{2\pi}{15}\right)}\, x\omega_v = 5{\cdot}2x\omega_v. \tag{3.107}$$

We can also assign a correlation time τ_v to the diffusive motion of the vacancy itself, faster by a factor x^{-1} than that of an average atom:

$$\frac{1}{\tau_v} = A\omega_v = 5{\cdot}2\omega_v. \tag{3.108}$$

One should not take too seriously the numerical value $A = 8\sqrt{(2\pi/15)}$ in (3.107) and (3.108). It originates once more in the somewhat arbitrary choice of a Gaussian shape for the memory function relative to the correlation function $\Gamma_{\mathbf{q}}(t)$.

(c) *Theory of the relaxation and narrowing by vacancies*

The formalism for Zeeman relaxation and line narrowing by vacancies is again very similar to that due to exchange.

For instance the Zeeman-vacancies relaxation rate $1/T_{Zv}$ can be treated as a coupling between the Zeeman energy $Z = \omega_0 I_z$ and the vacancy tunnelling Hamiltonian $\mathscr{H}_t$ mediated by the dipolar energy $\mathscr{H}_D$. In analogy with eqn (3.36) the relaxation rate $1/T_{Zv}$ is given by

$$\frac{1}{T_{Zv}} = \frac{\int_0^\infty \mathrm{Tr}\{\exp(-\beta\mathscr{H}_f)[Z, \mathscr{H}_D][\mathscr{H}_D^*(t), Z]\}\,dt}{\mathrm{Tr}\{Z^2\}\,\mathrm{Tr}\{\exp(-\beta\mathscr{H}_f)\}}. \tag{3.109}$$

The main difference with eqn (3.36) relative to the Zeeman-exchange relaxation resides again in the fact that $1/T_{Zv}$ is strongly temperature dependent since the formation energy $\Phi = \hbar\omega_v$ of a vacancy is much larger than the temperature at which the experiments are performed. This requires, as explained in **D**(a) the introduction in (3.109) of the statistical operator $\exp(-\beta\mathscr{H}_f)$ where $\mathscr{H}_f$ is given by eqn (3.32a), but not of $\exp(-\beta\mathscr{H}_t)$ where the tunnelling Hamiltonian $\mathscr{H}_t$ is given by (3.32c).

The operators $\mathcal{H}_t$ and $Z = \omega_0 I_z$ are expressed by means of the fictitious spin variables $\mathbf{S}_i$ as indicated in eqns (3.30) and (3.31). Finally,

$$\begin{aligned}\exp\{i(Z+\mathcal{H}_t)t\}\mathcal{H}_D \exp\{-i(Z+\mathcal{H}_t)t\} \\ = \exp(iZt)\exp(i\mathcal{H}_t t)\mathcal{H}_D \exp(-i\mathcal{H}_t t)\exp(-iZt) \\ = \exp(iZt)\mathcal{H}_D(t)\exp(-iZt),\end{aligned} \tag{3.110}$$

is similar to (3.35), the tunnelling Hamiltonian $\mathcal{H}_t$ replacing the exchange Hamiltonian $\mathcal{H}_e$.

Equation (3.109) leads to the same expression for $1/T_{Zv}$ as eqn (3.43):

$$\frac{1}{T_{Zv}} = J_1(\omega_0) + 4J_2(2\omega_0); \tag{3.109$'$}$$

$$J_m(\omega) = \int_{-\infty}^{\infty} d\omega \exp(i\omega t) \frac{\mathrm{Tr}\{\exp(-\beta\mathcal{H}_f)G_{-m}G_m(t)\}}{\mathrm{Tr}\{\exp(-\beta\mathcal{H}_f)I_z^2\}}. \tag{3.111}$$

Similarly the transverse relaxation time $1/T_2$ is given by an equation formally identical with (3.83) apart from the inclusion into the traces of (3.81) of the statistical operator $\exp(-\beta\mathcal{H}_f)$, and $1/T_2$ is again given by eqn (3.85):

$$\frac{1}{T_2} = \tfrac{3}{2}J_0(0) + \tfrac{5}{2}J_1(\omega_0) + J_2(2\omega_0). \tag{3.112}$$

The foregoing would suggest complete similarity between relaxation and motion narrowing processes induced by exchange and by vacancies apart from a trivial temperature dependence for the latter. There is however an important difference between the spectral densities $J_m(\omega)$ relative to exchange and to vacancies. The reason for this difference is by no means trivial and provides a new and interesting illustration of the formalism of the memory function, which warrants a careful discussion.

Consider for argument's sake the calculation of the adiabatic width $1/T_2'$ as it is given for exchange in **C**(b)1. We recall that it rests on the following reasoning. The FID signal $G(t)$ is related to the memory function $K(t)$ by the equation:

$$\frac{dG}{dt} = -\int_0^t K(t')G(t-t')\,dt'. \tag{1.125}$$

The power expansion of $K(t)$ is given by (1.135). It is argued that if the ratio $\mu = M_4/M_2^2$ is very much larger than unity, $K(t)$ decays very rapidly with t and the expansion (1.135) is approximated by the Gaussian

$$M_2 \exp\left(-\frac{N_2 t^2}{2}\right) \cong M_2 \exp\left(-\frac{\mu M_2 t^2}{2}\right) \tag{3.113}$$

or sometimes by the Lorentzian:

$$M_2\left(1+\frac{N_2t^2}{2}\right)^{-1}. \tag{3.113'}$$

The decay of $G(t)$, it is argued, is much slower than that of $K(t)$ and (1.125) is approximated for sufficiently large t by:

$$\frac{dG}{dt}=-G(t)\int_0^\infty K(t')\,dt'=-\frac{1}{T_2'}G(t). \tag{1.125'}$$

For $N_2t^2 \approx \mu M_2 t^2 \gg 1$, $G(t)$ behaves as an exponential with a decay constant T_2' given by (1.125'). $1/T_2' = \sqrt{(\pi/2)}(M_2/\mu)^{1/2}$ if a Gaussian shape is assumed for $K(t)$ (eqn (3.63)) and $1/T_2' = \pi/\sqrt{(2)}(M_2/\mu)^{1/2}$ for a Lorentzian shape of $K(t)$ (eqn (3.65)).

This is the procedure that was used for exchange narrowing. From measurements of $1/T_{Ze}(\omega)$ it followed, quite empirically, that the Gaussian shape of $K(t)$ was a good approximation for h.c.p. solid ^{3}He and the Lorentzian one for b.c.c. It is worth pointing out that, for the sake of internal consistency, these approximations could have been checked by calculating in the expansion (1.135) of $K(t)$ the 4th order term $N_4t^4/4!$ given by (1.134b) which depends on the sixth moment M_6 of $G(t)$. This 4th order term is different for the Gaussian and Lorentzian approximations (3.113) which allows a discrimination between them and thus also between the values (3.63) and (3.65) for $1/T_2'$. Nobody seems to have been tempted by the rather heavy calculation of M_6. It is only fair to add that it has lost some of its interest since it has been recognized that the bilinear expression (3.11) for the exchange Hamiltonian, which enters in the expressions of M_4 and M_6, has probably little validity.

Consider now the result of the same procedure for motion narrowing by vacancies. For the b.c.c. structure it is found (Redfield and Yu, 1968; 1969):

$$M_4 = 0{\cdot}13M_2^2 + \alpha_4 x\omega_v^2 M_2, \tag{3.114}$$

with:

$$\alpha_4 = 23{\cdot}5.$$

If $\alpha_4 x\omega_v^2 \gg M_2$, $\mu = M_4/M_2^2 \gg 1$ and there *is* motion narrowing, let us calculate $1/T_2'$ by, say, eqn (3.63). It yields:

$$\frac{1}{T_2'} = \frac{\pi}{\sqrt{2}}\left(\frac{M_2}{\omega_v}\right)\frac{1}{\sqrt{(x\alpha_4)}} \tag{3.115}$$

The inverse square root dependence on concentration does not look right. Indeed if instead of the picture of vacancy waves we had used a classical random walk model for vacancies, with a correlation time $\tau_v \sim \omega_v^{-1}$ a model which cannot be very wrong, the correlation time τ_a for the random walk of an atom would be longer by a factor x^{-1}, with $\tau_a \sim (\omega_v x)^{-1}$ leading to a classical motion narrowing formula:

$$\Delta\omega \sim \frac{1}{T_2'} \sim M_2 \tau_a \sim \frac{M_2}{\omega_v} \frac{1}{x}, \tag{3.116}$$

in contradiction to (3.115). Since we arrived at the conflicting result (3.115) by following the same procedure as for exchange it is essential to track down the origin of the discrepancy.

Consider first the expansion of $G(t)$. It can be rewritten:

$$G(t) \cong \tilde{G}(t) = 1 - M_2 \frac{t^2}{2} + x\alpha_4 M_2 \omega_v^2 \frac{t^4}{4!} - x\alpha_6 M_2 \omega_v^4 \frac{t^6}{6!}, \tag{3.117}$$

with $\alpha_4 = 23{\cdot}5$ and α_6 alas unknown. For each moment M_{2n} we have kept in (3.117) only the leading term, proportional to $M_2 \omega_v^{2n-2}$.

Because the tunnelling frequency ω_v is much larger than the dipolar frequency $(M_2)^{1/2}$, the ratio $\mu = M_4/M_2^2 = \alpha_4 x(\omega_v^2/M_2)$ is very large unless x is exceedingly small. Actually in order for the vacancies to be a more important narrowing and relaxation mechanism than exchange we require the far more stringent condition: $x\omega_v > J \gg M_2^{1/2}$. However because $\omega_v/|J| \gg 1$ (eqn (3.26)), $x\omega_v > J$ can also be secured with very small x.

We are used to associating a line where $\mu = M_4/(M_2)^2 \gg 1$ with a Lorentzian shape (except at the extremities) and the FID function $G(t)$ with an exponential, (except very near the origin). Why then is it wrong in the present case to attribute to it the customary width $\sim(M_2/\mu)^{1/2}$ given by (3.115)? The answer is in the analysis of the structure of the memory function $K(t)$. If we keep only the leading terms for the various moments M_{2n}, the expansion (1.135) of $K(t)$ becomes:

$$K(t) \cong \tilde{K}(t) = M_2 \left\{ 1 - \tfrac{1}{2} x\alpha_4 \omega_v^2 t^2 + \frac{1}{4!} x\alpha_6 \omega_v^4 t^4 + \cdots \right\}. \tag{3.118}$$

Since x is a small number the ratio

$$N_4/(N_2)^2 \simeq \frac{1}{x} \frac{\alpha_6}{(\alpha_4)^2}$$

is very large and it is clearly wrong to expect $K(t)$ to behave like,

$$M_2 \exp - \frac{N_2 t^2}{2} \quad \text{or} \quad M_2 \left(1 + \frac{N_2 t^2}{2}\right)^{-1},$$

which led to the incorrect concentration dependence $1/T_2' \propto x^{-1/2}$ of eqn (3.115). We can seek refuge in a search for a memory function $L(t)$ for the function $K(t)$ itself, satisfying:

$$\frac{\mathrm{d}K}{\mathrm{d}t} = -\int_0^t L(t')K(t-t')\,\mathrm{d}t'. \tag{3.119}$$

This was done in chapter 1, **B**(d)2 with a power expansion for $L(t)$:

$$L(t) = N_2\left\{1 - \frac{R_2 t^2}{2!} + \frac{R_4 t^4}{4!} + \cdots\right\},$$

where:

$$R_2 = M_2(\mu - 1)^{-1}(\nu - \mu^2) \cong M_2\frac{\nu}{\mu} = \frac{M_6}{M_4} = \frac{\alpha_6}{\alpha_4}\omega_\mathrm{v}^2. \tag{3.120}$$

It is easily checked that the leading term in R_4 is of the form $\beta\omega_\mathrm{v}^4$ and independent of x, as well as the subsequent terms. It is then at least not obviously wrong to approximate $L(t)$ as:

$$L(t) = N_2 \exp -\frac{R_2 t^2}{2}. \tag{3.121}$$

For $R_2 t^2 \gg 1$ that is for $t \gg 1/\omega_\mathrm{v}$ we rewrite (3.119) as:

$$\frac{\mathrm{d}K}{\mathrm{d}t} = -K(t)\int_0^\infty L(t')\,\mathrm{d}t' = -K(t)\sqrt{\left(\frac{\pi}{2}\right)}\frac{N_2}{\sqrt{R_2}} = -\frac{1}{\tau_\mathrm{d}}K(t). \tag{3.119$'$}$$

where:

$$\frac{1}{\tau_\mathrm{d}} = \sqrt{\left(\frac{\pi}{2}\right)}\frac{N_2}{\sqrt{R_2}} = \sqrt{\left(\frac{\pi}{2}\right)}\,x\omega_\mathrm{v}\left(\frac{\alpha_4}{\alpha_6}\right)^{1/2}, \tag{3.122}$$

is a *definition* of τ_d (d for dipolar). We see that $1/\tau_\mathrm{d}$ of the order of $x\omega_\mathrm{v}$ is much smaller than ω_v, and that it is indeed consistent to approximate (3.119) by (3.119′) and to write:

$$K(t) \cong M_2 \exp\left(-\frac{t}{\tau_\mathrm{d}}\right). \tag{3.123}$$

We then return to the eqn (1.125′) for $G(t)$ which reads:

$$\frac{\mathrm{d}G}{\mathrm{d}t} = -G(t)\int_0^\infty K(t')\,\mathrm{d}t' = -G(t)M_2\tau_\mathrm{d},$$

whence:

$$\frac{1}{T_2'} = M_2\tau_\mathrm{d} = \frac{M_2}{x\omega_\mathrm{v}}\sqrt{\left(\frac{2}{\pi}\right)}\left(\frac{\alpha_6}{\alpha_4}\right)^{1/2}, \tag{3.124}$$

which is a result physically acceptable. We see from (3.122) and (3.124) that $1/T_2' \ll 1/\tau_d$ and that $G(t)$ decays much more slowly than $K(t)$, as it should, in order for the whole procedure to be consistent.

From the relation $K(t) = M_2 \exp(-t/\tau_d)$ where τ_d is given by (3.122) we get:

$$\mathcal{K}'(\omega) = \int_0^\infty K(\tau) \cos \omega t = \frac{M_2 \tau_d}{(1+\omega^2 \tau_d^2)},$$

and

$$J_0(\omega) = \tfrac{2}{3}\mathcal{K}'(\omega) = \frac{2}{3} \frac{M_2}{1+\omega^2 \tau_d^2}. \tag{3.125}$$

For a powder we take the same expression for $J_1(\omega)$ and $J_2(\omega)$. The expression (3.125) is valid as long as (3.123) is valid for $K(t)$, that is as long as $t \gg 1/\omega_v$ or else as long as $\omega \ll \omega_v$. The cut-off frequency for the spectral densities (3.125) is $\omega_v \sim 1/x\tau_d \gg 1/\tau_d$ where $1/\tau_d$ is *defined* by (3.122).

We see that the spectral densities $J_m(\omega)$ (3.111) for relaxation by vacancies are rather different from those relative to exchange. They have a Lorentzian shape with a correlation time τ_d of order $1/x\omega_v$ and a cut-off frequency $\sim\omega_v$.

A last question is that of the numerical factor A' that relates $1/\tau_d$ to $x\omega_v$: $1/\tau_d = A'x\omega_v$. We recall that for the diffusion of nuclear magnetization by vacancies the assumption of a Gaussian memory function had led to eqn (3.107):

$$\frac{1}{\tau_a} = 8\sqrt{\left(\frac{2\pi}{15}\right)}\, x\omega_v = 5{\cdot}2x\omega_v = Ax\omega_v. \tag{3.107}$$

For relaxation of nuclear magnetization by vacancies we have:

$$\frac{1}{\tau_d} = A'x\omega_v = \sqrt{\left(\frac{\pi}{2}\right)}\left(\frac{\alpha_4}{\alpha_6}\right)^{1/2} x\omega_v = \sqrt{\left(\frac{\pi}{2}\right)}\left(\frac{23{\cdot}5}{\alpha_6}\right)^{1/2} x\omega_v, \tag{3.122}$$

but since α_6 has not been calculated we do not know what A' is.

On physical grounds if we consider both diffusion and relaxation to result from a random walk of vacancies it is not unreasonable to assume: $1/\tau_d = 2(1/\tau_a)$ and $A' = 2A$. This is because the random modulation of the dipolar interaction which is bilinear with respect to spins is expected to be twice as fast as that of the nuclear magnetization.

Once more one should not attach too much significance to the numerical values of these proportionality factors. The one result which *is* physically significant is the Lorentzian shape of the spectral densities $J_m(\omega)$ and the frequency dependence of $1/T_{Zv}$:

$$\frac{1}{T_{Zv}} = \frac{2M_2}{3}\left\{\frac{\tau_d}{1+\omega^2\tau_d^2} + \frac{4\tau_d}{1+4\omega^2\tau_d^2}\right\}. \tag{3.109''}$$

The reader may question the usefulness of lengthy calculations which lead in the end to the old BPP formula of 1948. We think that these calculations serve a useful purpose by proving or at least making very plausible the equivalence of the classical BPP approach to the purely quantum mechanical description of the vacancies by the Hamiltonian (3.32*a*) and (3.32*c*).

To conclude we recapitulate the definitions of the various correlation times related to the tunnelling frequency ω_v. τ_a is the correlation time for the diffusion of an atom, related to the diffusion D by: $D = \frac{1}{6}a^2/\tau_a$. It is given by eqn (3.107):

$$\frac{1}{\tau_a} = 8\sqrt{\left(\frac{2\pi}{15}\right)}\, x\omega_v = 5{\cdot}2x\omega_v; \tag{3.107}$$

$$\frac{1}{\tau_v} = x^{-1}\tau_a = 5{\cdot}2\omega_v, \tag{3.107'}$$

is the inverse correlation time for the diffusion of a vacancy. $1/\tau_d$ is the rate at which the dipolar interaction is modulated by the motion of the vacancies:

$$\frac{1}{\tau_d} = \sqrt{\left(\frac{\pi}{2}\right)}\, x\omega_v \left(\frac{\alpha_4}{\alpha_6}\right)^{1/2} = \sqrt{\left(\frac{\pi}{2}\right)}\, x\omega_v \left(\frac{23{\cdot}5}{\alpha_6}\right)^{1/2}, \tag{3.122}$$

where:

$$M_4 \cong x\alpha_4\omega_v^2 M_2,$$

and:

$$M_6 \cong x\alpha_6\omega_v^4 M_2.$$

Since α_6 has not been calculated $1/\tau_d$ is not known. We make the *assumption* that $2/\tau_d \simeq 1/\tau_a \cong 10x\omega_v$.

N.B. In Landesman (1975) different notations are used:

$$\tau_a \to \tau_r \qquad \tau_d \to \tau_c \qquad \tau_v \to \tau_0.$$

(*d*) *Diffusion and Zeeman relaxation caused by vacancies: experimental results*

1. *Diffusion*

Figure 3.9 is a plot of the diffusion coefficient D, measured by spin-echoes against the inverse temperature $1/T$ for various molar volumes in b.c.c. ^{3}He. The two curves with the largest molar volumes exhibit for low temperatures a temperature-independent diffusion coefficient due to exchange, of no concern to us in this section. From these curves it is possible, using (3.105) to extract the Arrhenius energy constant and the tunnelling frequency ω_v. Their values are given in Table 3.3. The Arrhenius energy is labelled W rather than Φ so as not to rule out *a priori* the existence of a potential barrier Φ_m (eqn (3.101)). The measurements are rather ancient (1963) and the

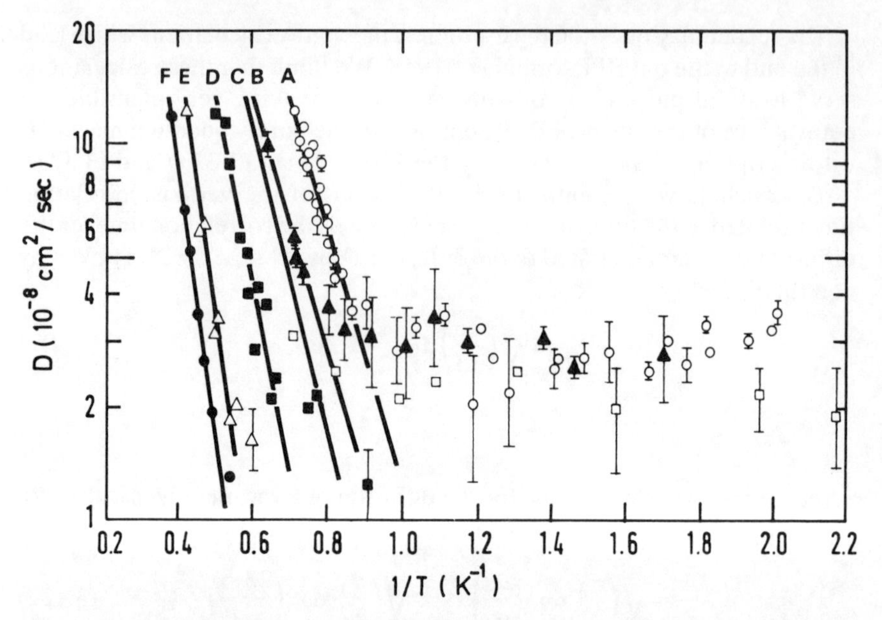

FIG. 3.9. Spin diffusion coefficient, measured by spin echoes, against inverse temperature in b.c.c. ^{3}He at various molar volumes. A: 22·48; B: 22·05; C: 21·70; D: 21·10; E: 20·12; F: 19·75 cm^3. (After Landesman, 1975.)

values of ω_v, which are extracted from the exponential factor D_0 in $D = D_0 \exp(-W/k_T)$ and are always uncertain, do not agree well with those extracted from relaxation measurements to be considered presently.

2. *Zeeman-vacancies relaxation*

Figure 3.10 is a plot of the Zeeman-vacancies relaxation time against inverse temperature for two values of the Larmor frequency. They exhibit the characteristic minimum of the BPP formula (3.109″).

Figure 3.11 represents measurements performed for several molar volumes at the much higher frequency of $\omega_0/2\pi = 80$ MHz. The variation of

Table 3.3

V(cm^3)b.c.c.	W(K)	ω_v(sec^{-1})$\times 10^{-10}$
22·48	7·82±0·72	3
22·05	7·78±0·27	1·2
21·70	9·13±0·67	2·4
21·10	11·7 ±1·2	4
20·12	13·6	4
19·75	16·5 ±1·2	7

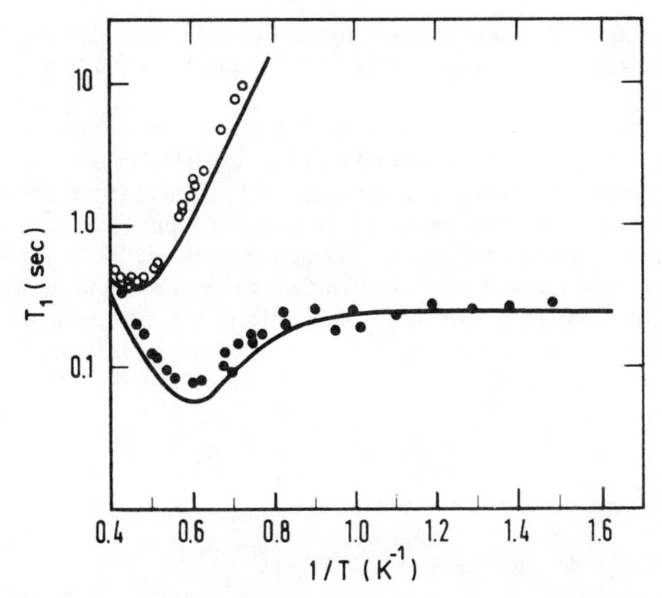

FIG. 3.10. Zeeman-vacancy relaxation time against inverse temperature in a sample of b.c.c. ^{3}He. Molar volume $V_M = 20{\cdot}12$ cm^3. The Larmor frequency is 5·224 MHz for the solid circles and 30·4 MHz for the open circles. (After Landesman, 1975.)

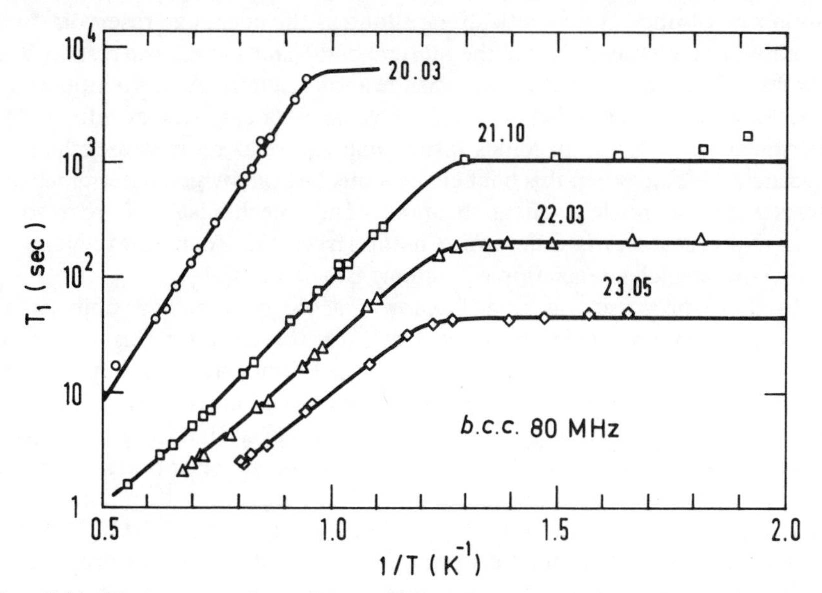

FIG. 3.11. Temperature dependence of the nuclear spin–lattice relaxation times in b.c.c. ^{3}He in high field at various molar volumes. Larmor frequency $\omega/2\pi = 80$ MHz. (After Sullivan, Deville, and Landesman, 1975.)

$1/T_{Zv}$ against $1/T$ in the temperature dependent range, is monotonic as expected for $\omega_0 > 1/\tau_d$. In the temperature independent part it is T_{Ze} that is measured.

Figure 3.12 is a plot of T_{Zv} for several molar volumes and frequencies in h.c.p. ^{3}He. (It will be remembered that this is a low molar volume (high pressure) phase where exchange is small and $1/T_{Ze}$ negligible.) From these measurements of T_{Zv}, by means of eqns (3.109″) and (3.122), values are extracted for τ_d, hence ω_v, and the Arrhenius constant W. W is plotted in Fig. 3.13 against the molar volume together with values deduced from other experiments. Whilst in the b.c.c. phase the results are compatible with $W = \Phi$, that is with quantum mechanical tunnelling of the vacancies, in the h.c.p. phase the existence of a potential barrier $\Phi_m = W - \Phi$ is not to be excluded.

E. Non-Zeeman relaxation mechanism

(*a*) *Exchange–defects coupling*

1. *Exchange–vacancies coupling*

In section **C**(*a*) we calculated the relaxation rate $1/T_{Ze}$ which couples Zeeman to exchange, and stated that it coincides with the *observed* relaxation rate of nuclear magnetization as long as the exchange reservoir itself remains at the temperature of the thermal bath. The mechanism responsible for that is the coupling between exchange and vacancies. As the temperature goes down and the number $N_v = x_v N$ of vacancies decreases, eventually the relaxation rate $1/T_{Ze}$ becomes faster than the relaxation rate exchange–vacancies $1/T_{ev}$. When this happens it is this last rate which determines the relaxation of nuclear magnetization. The mechanism of relaxation exchange–vacancies is different in nature from the Zeeman-exchange or Zeeman-vacancies relaxation of sections **C**(*a*) and **D**(*c*).

In those processes, as is well known, a flip of a nuclear spin is the accumulated result of very many weak collisions occurring at a fast rate, or to be more precise at a rate comparable with the Larmor frequency involved in a nuclear spin flip. By contrast, every time an atom jumps into a neighbouring vacant site the exchange energy changes. It is clear that for the exchange energy this is a strong collision and that the exchange relaxation rate will be of the same order of magnitude as the collision rate, that is in the present case the atomic jumping rate $1/\tau_a = Ax\omega_v$ of equation (3.107). It is the proportionality coefficient between $1/\tau_a$ and $1/T_{ev}$ that we propose to calculate in this section. The principle of the calculation, due to various authors is described in detail in Goldman (1970, p. 67), and we shall be content to sketch the derivation here.

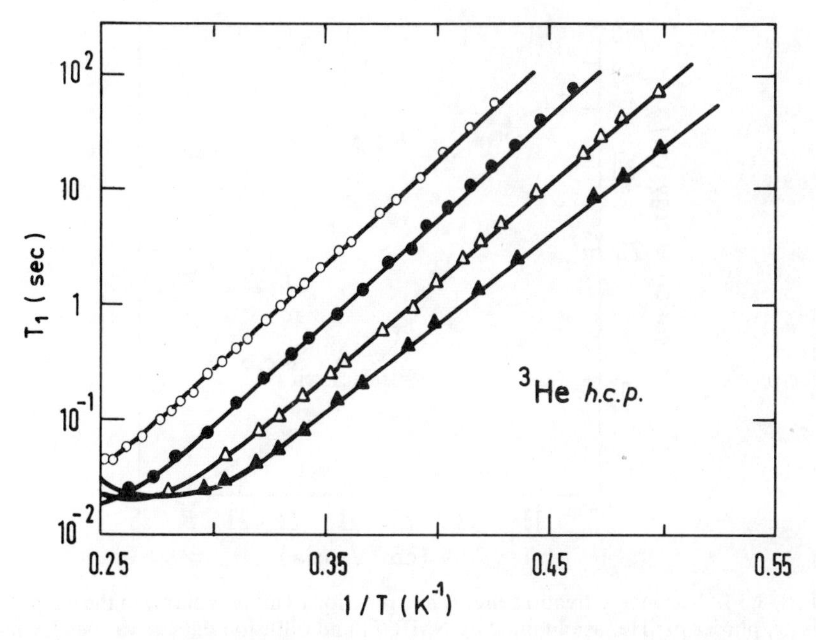

FIG. 3.12. Temperature dependence of the nuclear spin–lattice relaxation times in h.c.p. ^{3}He in low field at various molar volumes V_M. Open circles: 17·63 cm^3. Solid circles: 17·81 cm^3. Open triangles: 18·07 cm^3. Solid triangles: 18·34 cm^3. Larmor frequency: 3 MHz for the open circles and 2·125 MHz for the remainder. (After Sullivan *et al.*, 1975.)

We assume that just before an atomic jump the spin system is in a thermal equilibrium described by a density matrix $\sigma_i = 1 - \beta_i \mathcal{H}_i$ where $\mathcal{H}_i$ is the spin Hamiltonian in the presence of all the vacancies. Immediately after the jump the Hamiltonian has a slightly different form $\mathcal{H}_f$ but the density matrix which is still σ_i is no more a thermal equilibrium matrix for the new Hamiltonian $\mathcal{H}_f$. The spin system then evolves, at constant energy to a new thermal density matrix $\sigma_f = 1 - \beta_f \mathcal{H}_f$ and as shown in chapter 1, eqn (1.23).

$$\frac{\beta_f}{\beta_i} = \frac{\mathrm{Tr}\{\mathcal{H}_f \mathcal{H}_i\}}{\mathrm{Tr}\{\mathcal{H}_f^2\}},$$

or

$$\frac{\beta_i - \beta_f}{\beta_i} = -\frac{\Delta\beta}{\beta} \cong \frac{\mathrm{Tr}\{\mathcal{H}_f(\mathcal{H}_f - \mathcal{H}_i)\}}{\mathrm{Tr}\{\mathcal{H}_f^2\}}. \tag{3.126}$$

The number of vacancies in the sample is $N_v = x_v N$ and the time between two jumps of vacancies anywhere in the sample is:

$$\Delta t = \frac{\tau_v}{N_v} = \frac{\tau_v}{x_v N} = \frac{1}{N x_v A \omega_v} = \frac{\tau_a}{N}, \tag{3.127}$$

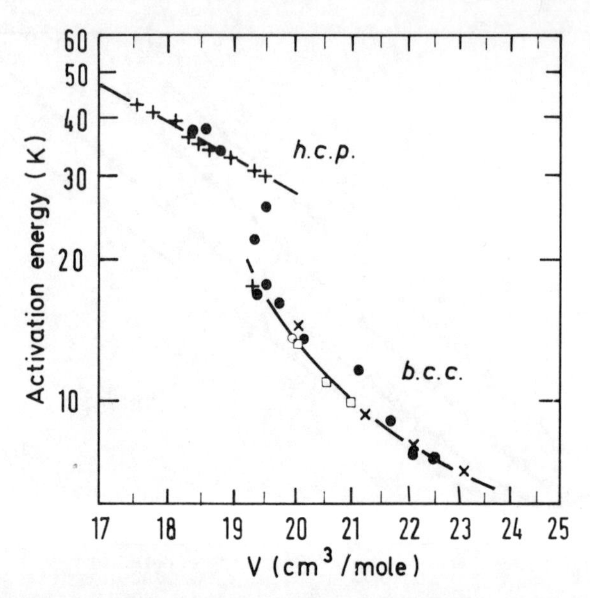

FIG. 3.13. Vacancy activation energy as a function of molar volume in the h.c.p. and b.c.c. phases of ^{3}He, as obtained by NMR T_1 and diffusion data at various Larmor frequencies (2 to 80 MHz), and by X-ray diffraction. (After Sullivan *et al.*, 1975.)

whence:

$$-\frac{\Delta\beta}{\beta}=-\frac{1}{\beta}\frac{\mathrm{d}\beta}{\mathrm{d}t}\Delta t=\frac{1}{T_{\mathrm{ev}}}\frac{\tau_{\mathrm{a}}}{N}=\frac{\mathrm{Tr}\{\mathcal{H}_{\mathrm{f}}(\mathcal{H}_{\mathrm{f}}-\mathcal{H}_i)\}}{\mathrm{Tr}\{\mathcal{H}_{\mathrm{f}}^2\}}$$

which yields:

$$\frac{1}{T_{\mathrm{ev}}}=\frac{1}{\tau_{\mathrm{a}}}\left\{N\frac{\mathrm{Tr}\{\mathcal{H}_{\mathrm{f}}(\mathcal{H}_{\mathrm{f}}-\mathcal{H}_i)\}}{Tr\{\mathcal{H}_{\mathrm{f}}^2\}}\right\}=\frac{2\xi}{\tau_{\mathrm{a}}}. \tag{3.128}$$

The dimensionless parameter 2ξ is evaluated as follows. Suppose that in the Hamiltonian $\mathcal{H}_i$ the vacancy is at site 2, and in the Hamiltonian $\mathcal{H}_{\mathrm{f}}$ it has moved to site 1. For a bilinear exchange coupling, 2ξ is given by:

$$2\xi=N\frac{\sum_{k\neq 1,2}\mathrm{Tr}\{(\mathbf{I}_1\,.\,\mathbf{I}_k)^2J_{2k}(J_{2k}-J_{1k})\}}{\mathrm{Tr}\{[\sum_{i,k}J_{ik}(\mathbf{I}_i\,.\,\mathbf{I}_k)]^2\}}. \tag{3.129}$$

For an exchange interaction (3.11), limited to nearest neighbours ξ is a purely geometrical factor:

$$\xi_{\mathrm{b.c.c.}}=\tfrac{7}{8};\qquad \xi_{\mathrm{h.c.p.}}=\tfrac{7}{11}. \tag{3.130}$$

In order for $1/T_{\mathrm{ev}}$ to be the relevant parameter in the *observed* relaxation process it is yet necessary for the vacancies reservoir to remain at a fixed temperature that is to be more closely coupled to the phonons than to exchange. This condition can be expressed as:

$$\frac{1}{T_{v\to e}} = \frac{1}{T_{ev}} \frac{\langle \mathcal{H}_e \rangle}{\langle \mathcal{H}_t \rangle} \ll \frac{1}{T_{vph}}, \tag{3.131}$$

where $1/T_{vph}$ is the vacancies–phonon relaxation-rate to be studied shortly. The ratio $\langle \mathcal{H}_e \rangle / \langle \mathcal{H}_t \rangle$ is approximately:

$$\langle \mathcal{H}_e \rangle / \langle \mathcal{H}_t \rangle \simeq \frac{NJ^2}{N_v \omega_v^2} \simeq \frac{1}{x_v} \frac{J^2}{\omega_v^2}, \tag{3.132}$$

and:

$$\frac{1}{T_{v\to e}} \simeq 2\xi x_v (A\omega_v) \frac{1}{x_v} \frac{J^2}{\omega_v^2} \approx (2\xi A) J \left(\frac{J}{\omega_v}\right). \tag{3.133}$$

In contrast to $1/T_{ev}$, $1/T_{v\to e}$ is temperature-independent. We shall see in section **E**(b) that the rate $1/T_{v\to ph}$ which is strongly temperature dependent, is in general much larger than $1/T_{v\to e}$. The significant link is then the one between exchange and vacancies and the *observed* rate for the nuclear magnetization is given by:

$$\frac{1}{T_1^{obs}} = \frac{1}{T_{ev}} \frac{\langle \mathcal{H}_e \rangle}{\langle \mathcal{H}_e \rangle + \langle Z \rangle} = \frac{1}{T_{ev}} \frac{\frac{3}{8} z J^2}{\frac{3}{8} z J^2 + \omega_0^2/4}. \tag{3.134}$$

From (3.128), (3.107), and (3.22):

$$\frac{1}{T_{ev}} = \frac{2\xi}{\tau_a} = (2\xi A)\omega_v \exp(-\Phi/k_B T). \tag{3.135}$$

Figure 3.14 is a plot of the relaxation rate $1/T_1$ observed in b.c.c. ^{3}He where $2\xi A = 2 \times \frac{7}{8} 5{\cdot}2 = 9{\cdot}1$. It has the right temperature dependence (3.135). From the results of that figure one gets for ω_v and Φ the values recorded in Table 3.3 and plotted in Fig. 3.13 in good agreement with results obtained by other methods.

2. *Exchange–impurities coupling*

In section **B**(c) we had mentioned the fact that impurities of ^{4}He in ^{3}He are expected to behave like mobile defects propagating with a tunnelling frequency J_{43} presumably of the same order of magnitude as the normal exchange frequency $J_{33} = J$. The relaxation mechanism of the exchange reservoir $\mathcal{H}_e$ by these impurities is clearly of the same nature as the one due to vacancies and described in **E**(a)1. We can straightaway write a relaxation rate $1/T_{e4}$ (4 stands for impurities of ^{4}He) by replacing in (3.135) the vacancy frequency ω_v by J_{43} and the vacancy concentration $x_v =$

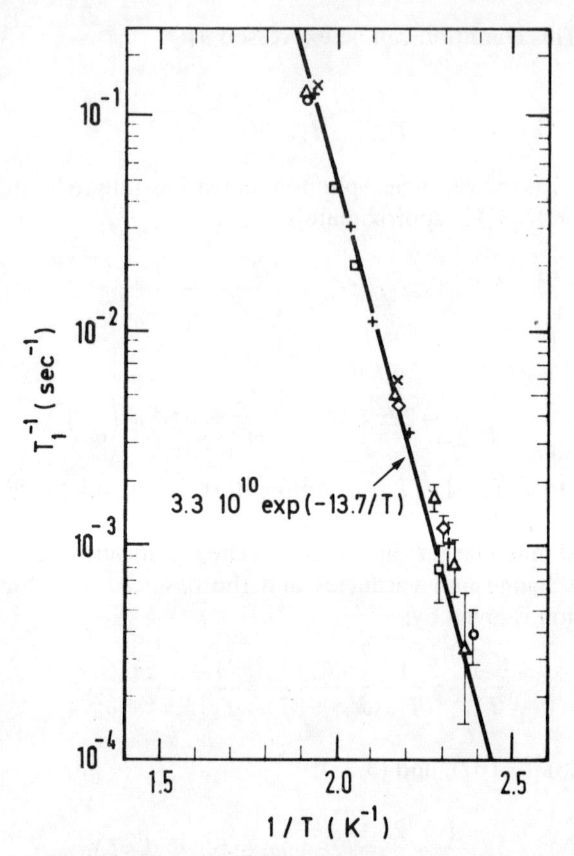

FIG. 3.14. Nuclear Zeeman relaxation rate, as determined by the exchange–vacancy coupling, against inverse temperature in b.c.c. ^{3}He at a molar volume of 20 cm^3. The concentration of ^{4}He in the various experiments ranges from 10^{-7} to $5 \,.\, 10^{-7}$. (After Landesman, 1975.)

$\exp(-\Phi/k_B T)$ by the impurities concentration x_4:

$$\frac{1}{T_{e4}} = 2\xi \times 5 \cdot 2 x_4 J_{43}. \tag{3.136}$$

In order for this mechanism to be competitive with the vacancies, we need the condition:

$$x_4 J_{43} \geqslant x_v \omega_v = \omega_v \exp(-\Phi/k_B T). \tag{3.137}$$

At very low temperatures the vacancies concentration is exceedingly small and (3.137) can be satisfied although $J_{43}/\omega_v \ll 1$, even for fairly pure samples. However, in contrast with the exchange–vacancies coupling rate $1/T_{ev}$, $1/T_{e4}$ is *not* in general the parameter observed in a relaxation process because of the very small heat capacity of the reservoir $\mathcal{H}_{4t}$ representing the tunnelling motion of the impurities. In full analogy with eqn (3.133) we find:

$$\begin{aligned}\frac{1}{T_{4\to e}} &\simeq \frac{1}{T_{e4}}\frac{\langle\mathcal{H}_e\rangle}{\langle\mathcal{H}_{4t}\rangle} \cong 2\xi x_4(AJ_{43})\frac{1}{x_4}\frac{J_{33}^2}{J_{43}^2} \\ &\cong (2\xi A)J\frac{J}{J_{43}} \cong (2\xi A)\left(\frac{J}{\omega_v}\right)\left(\frac{\omega_v}{J_{43}}\right) \\ &= \frac{1}{T_{v\to e}}\left(\frac{\omega_v}{J_{43}}\right) \gg \frac{1}{T_{v\to e}}. \end{aligned} \tag{3.138}$$

We see that $1/T_{4\to e}$ is much larger than $1/T_{v\to e}$ and, as we shall see, in general much larger than $1/T_{4ph}$.

(b) Defects–phonons coupling

1. *Relaxation*

We can use the same treatment for the relaxation of vacancies and of ^{4}He impurities by phonons, following the now familiar lines of the Provotorov-like rate equations encountered already several times in this book.

We sketch below the derivation of this relaxation rate. The unperturbed Hamiltonian $\mathcal{H}_0$ is the sum of the well-known Hamiltonian of phonons and of the tunnelling Hamiltonian $\mathcal{H}_{vt}$ or $\mathcal{H}_{4t}$ for vacancies or impurities. We introduce a fictitious spin $\frac{1}{2}$, $\mathbf{s}^i$, at each site, with $s_z^i = +\frac{1}{2}$ if the site i is occupied by a vacancy (or a ^{4}He atom) and $s_z^i = -\frac{1}{2}$ for a normal site. It will be noticed that in contrast with the descriptions of sections **B** and **D** we do *not* need here the more elaborate formalism of the fictitious spin $S = 1$ which distinguishes between sites occupied by ^{3}He atoms with spin up and spin down. The reason is that we are dealing here with the defects–phonons coupling where such a distinction is irrelevant. We can then write:

$$\mathcal{H}_{ph} = \textstyle\sum_{\mathbf{q}} \omega_{\mathbf{q}}(a_{\mathbf{q}}^{\dagger}a_{\mathbf{q}} + \frac{1}{2}); \tag{3.139a}$$

$$\mathcal{H}_t = \omega_t \sum_{i<j} (s_i^+ s_j^- + s_i^- s_j^+). \tag{3.139b}$$

where ω_t is either the tunnelling frequency ω_v of the vacancies, or the exchange constant J_{43} of the impurities as the case may be, and the $a_{\mathbf{q}}^{\dagger}$, $a_{\mathbf{q}}$ the classical phonon creation and annihilation operators. An approximate expression for the Hamiltonian V, coupling the two systems is obtained as follows. If we replace at a site 1 an atom of ^{3}He by a vacancy or an atom of ^{4}He there is a change $(\Delta M/2)\dot{u}_1^2$ in the kinetic energy relative to that site ($\Delta M/M_3 = +1$ for a ^{4}He atom and -3 for a vacancy). Naturally this is not the only change in the vibrational energy of the atoms since the forces between the atoms and thus also the potential energy of the system are affected by the defect. We take this formally into account by writing instead of ΔM, SM_3 where S is an unknown dimensionless factor of order unity. The coupling Hamiltonian V can then be written:

$$V = \sum_l \Delta_l(\tfrac{1}{2} + s_z^l), \tag{3.140}$$

where $\Delta_l = S(M_3/2)\dot{u}_l^2$ can be expressed by means of the classical expansion of the atomic displacement u_l into the phonon operators $a_{\mathbf{q}}$, $a_{\mathbf{q}}^\dagger$:

$$\Delta_l = S(4N)^{-1} \sum_{\mathbf{q},\mathbf{q}'} (\omega_{\mathbf{q}}\omega_{\mathbf{q}'})(\mathbf{e}_{\mathbf{q}} \cdot \mathbf{e}_{\mathbf{q}'})\{a_{\mathbf{q}}a_{\mathbf{q}'}^\dagger \exp[\mathrm{i}(\mathbf{q}-\mathbf{q}') \cdot \mathbf{r}_l] + \text{h.c.}\}. \tag{3.141}$$

The rate calculation involves the correlation functions:

$$\langle \Delta_l(t)\Delta_m\rangle = \langle \exp(\mathrm{i}\mathcal{H}_{\text{ph}}t)\Delta_l \exp(-\mathrm{i}\mathcal{H}_{\text{ph}}t)\Delta_m\rangle, \tag{3.142a}$$

and:

$$\langle s_z^l(t)s_z^m\rangle = \langle \exp(\mathrm{i}\mathcal{H}_{\text{t}}t)s_z^l \exp(-\mathrm{i}\mathcal{H}_{\text{t}}t)s_z^m\rangle. \tag{3.142b}$$

The calculation is made much simpler by assuming that the phonon correlation times are very short: $\tau_{\text{ph}} \sim \hbar/k_{\text{B}}T$ is much shorter than the inverse frequency ω_{t}^{-1}.

The classical formula for the relaxation rate $1/T_{\text{d,ph}}$, where d stands for defects and can be either 4 for ^{4}He or v for vacancies, is:

$$\frac{1}{T_{\text{d,ph}}} = \int_0^\infty \langle [\mathcal{H}_{\text{t}}, V][V(t), \mathcal{H}_{\text{t}}]\rangle \,\mathrm{d}t/\langle \mathcal{H}_{\text{t}}^2\rangle.$$

It contains terms such as:

$$\int_0^\infty \langle [\mathcal{H}_{\text{t}}, s_z^l][s_z^m(t), \mathcal{H}_{\text{t}}]\Delta_l(t)\Delta_m\rangle \,\mathrm{d}t, \tag{3.143}$$

which can be approximated as:

$$\int_0^\infty \langle \Delta_l(t)\Delta_m\rangle \,\mathrm{d}t \times \int_0^\infty \langle [\mathcal{H}_{\text{t}}, s_z^l][s_z^m(t), \mathcal{H}_{\text{t}}]\rangle \,\mathrm{d}t.$$

If one makes for the phonon spectrum the admittedly crude Debye approximation, one finds after a straightforward but somewhat lengthy calculation (Bernier and Landesman, 1971):

$$\frac{1}{T_{\text{d,ph}}} = 96\pi^9\left(\frac{6\pi^2 Na^3}{V}\right)^{2/3} S^2\left(\frac{T}{\theta_{\text{D}}}\right)^9 \Omega_{\text{D}}. \tag{3.144}$$

The unknown constant S can be obtained from experiment as follows. The operator Δ_l of eqn (3.141) also gives the amplitude for the scattering of a phonon from a fixed point defect. It is related to the life-time τ_{R} of a phonon of long wavelength against the so called Rayleigh scattering from this point defect by the classical Rayleigh formula:

$$\frac{1}{\tau_R}=\frac{3\pi}{2}S^2x_d\frac{\omega_q^4}{\Omega_D^3}=B_dx_d\omega_q^4, \tag{3.145}$$

where $B_d=(3\pi/2)(S^2/\Omega_D^3)$ (Bernier and Landesman, 1971). The constant B_d can be extracted from thermal conductivity measurements in solid ^{3}He containing as defects either vacancies or ^{4}He impurities, whence the value of S.

The two relaxation rates $1/T_{4,ph}$ and $1/T_{v,ph}$ differ only by the value of the constant S and are of the same order of magnitude. However as already stated and as demonstrated in eqn (3.138) the impurities of ^{4}He are much more closely coupled to the exchange reservoir than the vacancies. For the latter, because of the steep temperature dependence of $1/T_{v,ph}$, $1/T_{v,ph} \gg 1/T_{v,e}$ (except at temperatures so low that the concentration and the effect of vacancies is negligible anyway) the vacancies–phonons coupling is not observed.

By contrast, for the impurities of ^{4}He $1/T_{4,ph} \ll 1/T_{4e}$ except perhaps at temperatures so high that the effect of impurities is overshadowed by the large number of vacancies. Figure 3.15 shows diagrammatically the difference between the two relaxation schemes.

The weak link, represented by a single line, is the one which is observed. The relevant formula for vacancies has already been given in (3.134). For the relaxation of exchange by impurities the relevant formula is then:

$$\frac{1}{T_1^{obs}}=\frac{1}{T_{4,ph}}\frac{x_4zJ_{43}^2}{(\omega_0^2/4)+\frac{3}{8}zJ^2+zJ_{43}^2x_4}. \tag{3.146}$$

By combining the eqns (3.146) and (3.144) where S has been determined from (3.145), and J having been measured from one of the methods described in **C**, it is possible to obtain approximate values for J_{43}. The results show that J_{43} is approximately one half of J.

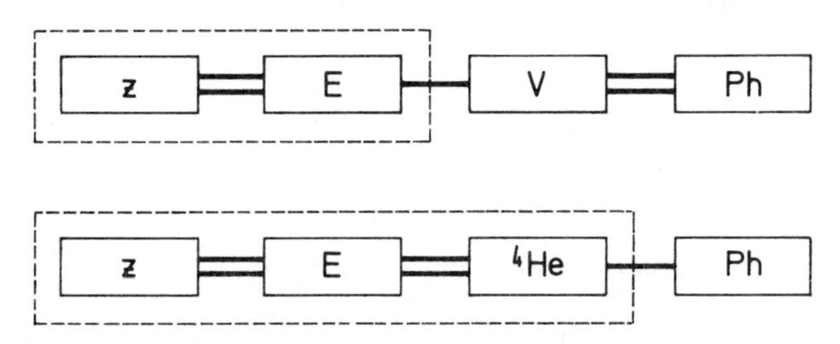

FIG. 3.15. Two relaxation schemes in solid ^{3}He. Top: The relaxation bottle-neck is the exchange–vacancy coupling. Bottom: The relaxation bottle-neck is the coupling between ^{4}He impurities and phonons.

When the concentration of ^{4}He impurity atoms increases there appears a new term in the Hamiltonian of the impurities, due to their mutual interactions. If the reservoir of these interactions is in thermal contact with the tunnelling reservoir, it brings to the heat capacity of the impurities a contribution proportional to x_4^2 observable for relatively large concentrations. The reader is referred to the literature for details (Bernier and Landesman, 1971).

2. *Vacancy waves and phonons*

In section **D**(a) we had already referred to three aspects of the coupling between vacancy waves and phonons.

The first, namely relaxation, has been dealt with in **E**(b)1 leading, for the relaxation of the tunnelling energy, to a rate given by (3.144). The second is the lifetime $\tau(\mathbf{k})$ of a vacancy wave of wave vector $\mathbf{k}$ against scattering by phonons. The scattering Hamiltonian is again V given by (3.140) and (3.141). The calculation starts from the matrix element for a transition $\mathbf{k}\to\mathbf{k}'$ and $\mathbf{q}\to\mathbf{q}'$ where $\mathbf{k}$ and $\mathbf{q}$ are the wave vectors of the two colliding waves, vacancy and phonon:

$$\begin{aligned}\langle\mathbf{q}',\mathbf{k}'|V|\mathbf{k},\mathbf{q}\rangle &= \frac{S}{4N}\sum_l \exp\{\mathrm{i}[(\mathbf{q}-\mathbf{q}')+(\mathbf{k}-\mathbf{k}')]\,.\,\mathbf{r}_l\}\omega_{\mathbf{q}}\omega_{\mathbf{q}'}(\mathbf{e}_{\mathbf{q}}\,.\,\mathbf{e}_{\mathbf{q}'})[n_{\mathbf{q}}(n_{\mathbf{q}'}+1)]^{1/2}\\ &= \frac{S}{4N}\left(\frac{2\pi}{V}\right)^3\delta(\mathbf{q}-\mathbf{q}'+\mathbf{k}-\mathbf{k}')\omega_{\mathbf{q}}\omega_{\mathbf{q}'}(\mathbf{e}_{\mathbf{q}}\,.\,\mathbf{e}_{\mathbf{q}'})[n_{\mathbf{q}}(n_{\mathbf{q}'}+1)]^{1/2}.\end{aligned} \tag{3.147}$$

The calculation of $1/\tau(\mathbf{k})$ is greatly simplified by the assumption that the vacancy frequency ω_v is much smaller on the average than the frequencies $\omega_{\mathbf{q}}$ and $\omega_{\mathbf{q}'}$ of the scattered phonon and therefore:

$$\omega_{\mathbf{q}'}\approx\omega_{\mathbf{q}}.$$

The details of the calculation can be found in the literature (Bernier and Landesman, 1971).

For $|\mathbf{k}|\to 0$, the result is:

$$\frac{1}{\tau(0)}=\frac{72\pi^7}{7}S^2\left(\frac{T}{\theta_\mathrm{D}}\right)^7\Omega_\mathrm{D}. \tag{3.148}$$

We find $1/\tau(0)=\omega_\mathrm{v}$ for:

$$\left(\frac{T}{\theta_\mathrm{D}}\right)=\frac{1}{\pi}\left(\frac{7}{72S^2}\right)^{1/7}\left(\frac{\Omega_\mathrm{D}}{\omega_\mathrm{v}}\right)^{1/7}. \tag{3.148'}$$

If we take $\theta_\mathrm{D}=25$ K and $S=1$, we find from (3.148′) that $1/\tau(0)<\omega_\mathrm{v}\sim 10^{10}$ for $T<4$ K and that below that temperature, vacancy waves can propagate without excessive scattering by phonons.

The third effect of the vacancies–phonons coupling is the lack of periodicity of the lattice introduced by the existence of phonons. It will in particular interfere with the propagation of vacancy waves by reducing the diffusion coefficient D. This can be seen as follows. D is given by eqn (3.94):

$$D = q^{-2}\sqrt{\left(\frac{\pi}{2}\right)}\,[m_2(\mathbf{q})]^{3/2}[m_4(\mathbf{q})]^{-1/2}.$$

The value (3.104) for $m_4(q)$, fourth derivative of $\Gamma_{\mathbf{q}}(t)$, resulted from the following formula, applied to b.c.c. lattice:

$$m_4(\mathbf{q}) = -\tfrac{1}{6}\sum_{k\neq j}\langle[\mathscr{H}_t[\mathscr{H}_t, S_z^j]][[S_z^k, \mathscr{H}_t], \mathscr{H}_t]\rangle q^2 R_{jk}^2, \tag{3.149}$$

(Landesman, 1975) where $\mathscr{H}_t$ is the tunnelling Hamiltonian (3.32c). In order to include the effects of phonons we must introduce the phonons–vacancies coupling V. It is not quite correct however to express it here by the Hamiltonian (3.140) which makes no distinction between ^{3}He spins up and down. We must use instead the expression:

$$V = \sum_l \Delta_l(1 - S_z^{l2}), \tag{3.150}$$

with the fictitious spin $S = 1$ as explained in section **B**(b)3. If then we replace in the expression (3.149) for $m_4(\mathbf{q})$, $\mathscr{H}_t$ by $(\mathscr{H}_t + V)$, ($m_2(\mathbf{q})$ is not affected by V), the expression (3.104) for $m_4(\mathbf{q})$ is replaced by (Landesman 1975):

$$m_4(\mathbf{q}) = m_2(\mathbf{q})\{15\omega_v^2 + \langle(\Delta_j - \Delta_k)^2\rangle\}, \tag{3.151}$$

where the difference $\langle \Delta_j - \Delta_k)^2\rangle$ is between nearest neighbours. This results in a reduction of D by a factor

$$\frac{D}{D_0} = \left\{1 + \frac{1}{15}\frac{\langle(\Delta_j - \Delta_k)^2\rangle}{\omega_v^2}\right\}^{-1/2}. \tag{3.152}$$

The appearance of the quadratic *difference* $\langle(\Delta_j - \Delta_k)^2\rangle$ in (3.152) shows clearly that it is the lack of periodicity of the lattice which is responsible for the decrease of D.

From the expression (3.141) of Δ_l it is found (Landesman, 1975):

$$\langle(\Delta_j - \Delta_k)^2\rangle = \frac{(18\pi^{16})^{1/3}}{240} S^2 \left(\frac{T}{\theta_D}\right)^4 \Omega_D^2. \tag{3.153}$$

The reduction (3.152) in the diffusion coefficient D reaches the value $\frac{1}{2}$, according to (3.152) and (3.153) for,

$$\frac{T}{\theta_D} = 1{\cdot}3\left(\frac{\omega_v}{S\Omega_D}\right)^{1/2}, \tag{3.154}$$

a condition quite different from (3.148′). For $\omega_v = 10^{10}$, $S = 1$, $\theta_D = 25$ K, one obtains $T = 1{\cdot}8$ K.

F. Multiple echoes in solid ^{3}He

The phenomenon of spin echoes has been known and studied for a long time (Abragam, 1961, p. 58): in a liquid, after an r.f. pulse the precessing transverse magnetization decays in a time comparable with the inverse of the Larmor frequency spread caused by the inhomogeneity of the applied d.c. field and usually much shorter than the transverse relaxation time T_2. A second pulse applied a time τ later modifies the magnetization orientations in the various parts of the sample in such a way that their precession in the *same* inhomogeneous field leads to a refocussing of the various transverse magnetizations at time 2τ. The theory, confirmed by experiment predicts only *one* echo after a sequence of two pulses.

In solid ^{3}He the spin tunnelling described by the exchange Hamiltonian has the same effect as the atomic motions in liquids: a washing-out of the dipolar interactions which leads to long values of T_2, of the order of a fraction of a second. This makes it possible to produce in solid ^{3}He the same spin echoes as in liquids.

An experiment initially performed with solid ^{3}He in low field ($H \sim 40$ G) and at very low temperature (1 to 20 mK) has led to a totally unexpected result: the observation after two r.f. pulses of a whole series of echoes at times $2\tau, 3\tau \ldots, n\tau$ (Bernier and Delrieu, 1977). Such a production of multiple echoes can only originate from non-linear effects in the evolution of the transverse magnetization. The origin of this non-linearity has been identified: it is the dipolar field produced at the site of each spin by the average magnetization of the neighbouring spins. This interpretation was confirmed by repeating the experiments in high field ($H \sim 9$ kG) and at high temperature ($T \sim 0{\cdot}3$ to $0{\cdot}7$ K) (Deville, Bernier, and Delrieu, 1979).

The phenomenon of multiple echoes has an interest of its own as a rather remarkable phenomenon in nuclear magnetism. Furthermore it provides an application of physical interest for solid ^{3}He: an absolute measurement of the magnetic susceptibility in an unknown quantity of solid.

In principle, the production of multiple echoes is not restricted to solid ^{3}He and could as well be observed with liquids. It is only for practical reasons as will appear later that they are much more difficult to produce in liquids than in solid ^{3}He.

The theory of multiple echoes will be developed in several steps: multiple echoes in the absence of relaxation in a sample of special shape; influence of relaxation; influence of the sample shape; influence of spin diffusion. We will then discuss the experimental results and their application.

(a) Simplified theory of multiple echoes

When in an inhomogeneous d.c. field one applies a $\pi/2$ r.f. pulse the transverse magnetizations in the various parts of the sample, initially all parallel, precess each one with its own Larmor frequency and form a spiral around the direction of the field gradient, with a pitch that varies with time. The second pulse, if different from π, changes partially the transverse magnetization into a longitudinal magnetization which is modulated in space in a fashion closely related to the space variation of the d.c. field. It is the dipolar field produced by this space-varying magnetization that is responsible for the production of multiple echoes. In this first, simplified version of the theory, we neglect relaxation and diffusion effects.

We are in high field and we use truncated dipolar interactions:

$$\mathcal{H}'_{\mathrm{D}} = \tfrac{1}{2}\sum_{i,j} A_{ij}[2I^i_z I^j_z - I^i_x I^j_x - I^i_y I^j_y].$$

The average dipolar field $\mathbf{H}^{\mathrm{D}}_i$ experienced by a spin i corresponds to a Larmor frequency $\boldsymbol{\omega}^D_i = -\gamma \mathbf{H}^{\mathrm{D}}_i$ of components,

$$\omega^{\mathrm{D}}_{iz} = 2\sum_j A_{ij}\langle I^j_z\rangle = \sum_j A_{ij} p^i_z, \tag{3.155}$$

$$\omega^{\mathrm{D}}_{ix,y} = -\sum_j A_{ij}\langle I^j_{x,y}\rangle = -\tfrac{1}{2}\sum_j A_{ij} p^i_{x,y}, \tag{3.156}$$

where the $\mathbf{p}^j$ are the polarizations of the spins $\frac{1}{2}$ of ^{3}He.

The calculation of the dipolar fields is particularly simple when the pitch of the spiral is much smaller than the dimensions of the sample. It is this case that we consider first.

We use the following space Fourier transforms:

$$p_\alpha(\mathbf{k}) = N^{-1/2}\sum_i p^i_\alpha \exp(-i\mathbf{k}\,.\,\mathbf{r}_i); \tag{3.157}$$

$$\omega^{\mathrm{D}}_\alpha(\mathbf{k}) = N^{-1/2}\sum_i \omega^{\mathrm{D}}_{i\alpha} \exp(-i\mathbf{k}\,.\,\mathbf{r}_i); \tag{3.158}$$

$$A(\mathbf{k}) = \sum_i A_{ij} \exp[-i\mathbf{k}\,.\,(\mathbf{r}_i - \mathbf{r}_j)]. \tag{3.159}$$

This last sum is independent of subscript j when $|\mathbf{k}|^{-1}$ is much smaller than the sample dimensions (Cohen and Keffer, 1955).

By carrying these expressions into (3.155) and (3.156) we obtain:

$$\omega^{\mathrm{D}}_z(\mathbf{k}) = A(\mathbf{k}) p_z(\mathbf{k}); \tag{3.160}$$

$$\omega^{\mathrm{D}}_{x,y}(\mathbf{k}) = -\tfrac{1}{2}A(\mathbf{k}) p_{x,y}(\mathbf{k}). \tag{3.161}$$

We assume (realistically) that the polarization inhomogeneities are negligible on the interatomic distance scale. Then for all values of $p(\mathbf{k})$ one has $|\mathbf{k}|a \ll 1$. Furthermore the experiments are performed in the b.c.c. phase of the ^{3}He, and for such a structure the average dipolar field at the site of each spin is zero in a sphere of homogeneous polarization.

Under these conditions it can be shown (Cohen and Keffer, 1955) that the values of $A(\mathbf{k})$ are particularly simple. They depend neither on $|\mathbf{k}| = k$ nor on the orientation of the crystal with respect to the field, and are given by:

$$A(\mathbf{k}) = \tfrac{2}{3}\pi\gamma^2\hbar n(3\cos^2\theta_{\mathbf{k}} - 1), \tag{3.162}$$

where n is the number of spins per unit volume and $\theta_{\mathbf{k}}$ the angle between $\mathbf{k}$ and $\mathbf{H}_0$.

For simplicity we take the field gradient parallel to H_0. The transverse magnetizations will spiral around this direction, and the only vectors $\mathbf{k}$ to be considered will be those parallel to $\mathbf{H}_0$, for which $\theta_{\mathbf{k}} = 0$. There will then be only one value of $A(\mathbf{k})$:

$$A(\mathbf{k}) = A = \tfrac{4}{3}\pi\gamma^2\hbar n. \tag{3.162$'$}$$

The situation is much more complicated when the pitch of the spiral is comparable with the sample dimensions, because then the right-hand side of (3.159) depends on the subscript j. In this section we choose a particular sample shape which bypasses these problems: a thin pancake perpendicular to the field gradient. It is not difficult to show that for such a shape, any distribution of polarizations that depends only on the co-ordinate z perpendicular to the pancake, corresponds to components $\mathbf{p}(\mathbf{k})$ with $\mathbf{k}$ also parallel to the axis $0z$, except for small boundary corrections that can be neglected. For almost all spins i within the sample, the transform $A(\mathbf{k})$ is then well-defined and has the value (3.162′), down to $k = 0$.

Equations (3.160) and (3.161) are then always of the form:

$$\omega_z^{\mathrm{D}}(k) = Ap_z(k); \tag{3.160$'$}$$

$$\omega_{x,y}^{\mathrm{D}}(k) = -\tfrac{1}{2}Ap_{x,y}(k). \tag{3.161$'$}$$

The inverse Fourier transform of these equations yields:

$$\omega_{iz}^{\mathrm{D}} = Ap_z^i,$$

$$\omega_{ix,y}^{\mathrm{D}} = -\tfrac{1}{2}Ap_{x,y}^i$$

or else:

$$\boldsymbol{\omega}_i^{\mathrm{D}} = -\tfrac{1}{2}A\mathbf{p}^i + \tfrac{3}{2}Ap_z^i\hat{\mathbf{z}} \tag{3.163}$$

where $\hat{\mathbf{z}}$ is the unit vector along the direction $0z$ of $\mathbf{H}_0$. The first term on the right-hand side of (3.163) corresponds to a field parallel to the spins and does not influence their motion. The second term leads to an extra precession around $\mathbf{H}_0$.

To this dipolar field one must add the internal field originating from the exchange interactions. These interactions are short-range and commute with the three components of the total spin. Insofar as the variation of the polarization takes place over distances much larger than the range of exchange, the latter does not influence the motion of the polarization, except for diffusion effects that will be considered later.

With the present geometry, the departures $\Delta(z)$ from the central Larmor frequency ω_0, as well as the polarizations p_α^i resulting from the pulses and the precession, depend only on the co-ordinate z along the direction of H_0. The wavelengths $2\pi/k$ being always much larger than the interatomic distance, we can replace the discrete distributions of polarizations and dipolar fields by continuous ones. The effective part of the dipolar field then reads:

$$\omega^{\mathrm{D}}(z)=\tfrac{3}{2}Ap_z(z). \tag{3.164}$$

The great advantage of the experimental conditions: unidirectional field gradient and appropriate sample shape, is that the problem to be solved is one-dimensional.

We analyse now what happens when we apply a sequence of two $\pi/2$ r.f. pulses separated by a time interval τ. We use a frame rotating at the central Larmor frequency ω_0, and single out a slice of the sample, of co-ordinate z. Its polarization is $\mathbf{p}$ and its frequency, purely longitudinal, is:

$$\omega_z=\omega=\Delta+\tfrac{3}{2}Ap_z. \tag{3.165}$$

We have then $\mathrm{d}p_z/\mathrm{d}t=0$ and, with the complex notation $p_+=p_x+ip_y$ the Larmor equation is:

$$\frac{\mathrm{d}}{\mathrm{d}t}p_+=\mathrm{i}\omega p_+. \tag{3.166}$$

(α) *Evolution between $t=0$ and $t=\tau$.* Before the first pulse, p_z has its thermal equilibrium value p_0. The first pulse produces a $\pi/2$ rotation around $0y$, and the polarization is aligned with $0x$:

$$p_+(0)=p_0.$$

The frequency ω reduces to Δ, because $p_z=0$ and, immediately before the second pulse ($t=\tau_-$) one has:

$$p_+(\tau_-)=p_0\exp(\mathrm{i}\Delta\tau). \tag{3.167}$$

(β) *Evolution after the second pulse.* At time τ, the second pulse rotates the spin by $\pi/2$ around $0y$, and the polarization components become:

$$p_+(\tau_+)=\mathrm{i}p_0\sin\Delta\tau; \tag{3.168a}$$

$$p_z(\tau_+)=-p_0\cos\Delta\tau. \tag{3.168b}$$

From now on, we take the origin of time after the second pulse. According to the initial condition (3.168*b*) the frequency ω is now:

$$\omega = \Delta - b \cos \Delta\tau, \tag{3.169}$$

with:

$$b = \tfrac{3}{2}Ap_0$$

that is, according to (3.162′):

$$\begin{aligned} b &= 2\pi\gamma^2\hbar n p_0 = 4\pi\gamma(\tfrac{1}{2}\gamma\hbar n p_0) \\ &= \gamma 4\pi M_0, \end{aligned} \tag{3.170}$$

where M_0 is the magnetization at thermal equilibrium.

According to (3.168*a*), (3.169), and (3.166) we have, at time t after the second pulse:

$$\begin{aligned} p_+(t) &= p_+(0) \exp[\mathrm{i}(\Delta - b \cos \Delta\tau)t] \\ &= \mathrm{i}p_0 \sin \Delta\tau \exp[\mathrm{i}(\Delta - b \cos \Delta\tau)t]. \end{aligned} \tag{3.171}$$

The term $(-\mathrm{i}b \cos \Delta\tau)$ in the exponent of the exponential is responsible for the production of multiple echoes.

Using the standard expansion:

$$\exp[-\mathrm{i}(b \cos \Delta\tau)t] = \sum_{-\infty}^{+\infty} (-\mathrm{i})^n J_n(bt) \exp(\mathrm{i}n\Delta\tau), \tag{3.172}$$

where the J_n are Bessel functions, we obtain in place of (3.171):

$$p_+(t) = \tfrac{1}{2}p_0[\exp(\mathrm{i}\Delta\tau) - \exp(-\mathrm{i}\Delta\tau)] \exp(\mathrm{i}\Delta t) \times \sum_{-\infty}^{+\infty} (-\mathrm{i})^n J_n(bt) \exp(\mathrm{i}n\Delta\tau). \tag{3.173}$$

The echoes are observed when the transverse polarizations in all parts of the sample are all in phase, that is when $p_+(t)$ is independent of Δ. According to (3.173), this occurs for:

$$\Delta\tau + \Delta t + n\Delta\tau = 0 \tag{3.174a}$$

or

$$-\Delta\tau + \Delta t + n\Delta\tau = 0. \tag{3.174b}$$

The echo of order m, for $t = m\tau$, is determined by the terms in the expansion satisfying either (3.174*a*) or (3.174*b*), that is:

$$n = -1 - m \quad \text{and} \quad n = 1 - m.$$

The amplitude of this echo is proportional to:

$$\begin{aligned} I_m &\propto (-\mathrm{i})^{(-1-m)} J_{-1-m} - (-\mathrm{i})^{(1-m)} J_{1-m} \\ &= (-\mathrm{i})^{(m+1)} J_{m+1} - (-\mathrm{i})^{m-1} J_{m-1} \\ &= (-\mathrm{i})^{(m+1)} [J_{m+1} + J_{m-1}] \end{aligned} \tag{3.175}$$

Using the following property of Bessel functions:

$$J_{m+1}(z) + J_{m-1}(z) = \frac{2m}{z} J_m(z),$$

we obtain finally:

$$I_m \propto (-\mathrm{i})^{(m+1)} \frac{2}{b\tau} J_m(mb\tau). \tag{3.176}$$

This theory shows clearly that multiple echoes result from the dipolar field associated with the longitudinal polarization produced by the second pulse. The crucial point is that the space modulation of this longitudinal polarization is strictly correlated with the space variation of the external field.

It must be noted that, contrary to the case of usual spin echoes, it is not essential that immediately after the second pulse the phases of local transverse magnetizations be variable in space. According to (3.171) and (3.172), multiple echoes would be observed even if the value of p_+ after the second pulse were uniform over the sample. The occurrence of this case will be described later on.

As a numerical example, let us consider solid ^{3}He with the following conditions:

$$V_{\mathrm{mol}} = 23\ \mathrm{cm}^3, \qquad \omega_0/2\pi = 30\ \mathrm{MHz}, \qquad T = 0{\cdot}5\ \mathrm{K}.$$

The corresponding value of b is:

$$b = 2\pi\gamma^2\hbar n p_0 \simeq 103\ \mathrm{s}^{-1},$$

which corresponds to a dipolar field:

$$b/\gamma \simeq 5\ \mathrm{mG}.$$

According to (3.176) the condition for the multiple echoes to have amplitudes comparable with that of the first one is:

$$b\tau \gtrsim 1. \tag{3.177}$$

Another limitation, discussed in the next section, is that the time interval τ between the two pulses must not be much larger than the transverse relaxation time T_2. The condition then becomes:

$$bT_2 \gtrsim 1. \tag{3.177'}$$

In the example under consideration, one has $T_2 \simeq 0{\cdot}15$ s, whence:

$$bT_2 \simeq 15.$$

One expects then the production of many echoes. This is indeed observed, as described later. One should even observe multiple echoes up to temperatures where the dipolar field is a fraction of a milligauss. The reason why such a small field can yield observable results is that the small non-linearity it produces acts continuously after the second pulse and that its accumulated coherent influence over a relatively long time can thus have sizeable effects.

As another example let us consider a liquid such as cyclohexane at room temperature in which the time T_2 for protons may be larger than 10 s. For the same frequency of 30 MHz, one finds $b \simeq 0{\cdot}8 \text{ sec}^{-1}$. One should then expect the production of multiple echoes were it not for another factor disregarded so far, the molecular diffusion which damps out non-homogeneous distributions of polarizations. It would not help to decrease the diffusion constant by increasing the viscosity of the liquid, because this would also shorten T_2. In solid ^{3}He, there is also a diffusion of polarization mediated by the exchange interactions. However the diffusion constant, although much larger than in ordinary solids, is still a thousand times less than in liquids. This unique position of solid ^{3}He, ('liquid-like' substance as far as NMR is concerned, with a long T_2, a relatively slow diffusion, and, because of its low temperature, a large polarization) makes it the only substance in which multiple echoes, in principle a general phenomenon, can easily be observed in practice. The problem of diffusion will be considered in the next section.

It must be noted that in order to make the above theory tractable, the only requirement is for the field gradient to be unidirectional. There is however no need for this gradient to be uniform. It is only at a later stage, for the calculation of diffusion, that the choice of a uniform field gradient will prove useful.

The gradient was chosen parallel to the external field. It is for this orientation that $A(\mathbf{k})$ is maximum (eqn (3.162)) and that the production of multiple echoes is the largest. This is however not essential, and any other orientation of the field gradient is acceptable.

The theory, developed for the case of $\pi/2$ pulses is easily generalized to different pulse angles. $\pi/2$ pulses are the most favourable for the production of multiple echoes because the modulated longitudinal polarization following the second pulse is then maximum.

(b) *General case*

1. *The effects of relaxation*

We assume a sample with the same shape as before, still neglect diffusion and examine how the foregoing theory is modified by relaxation. The longitudinal and transverse relaxations affect the multiple echoes in several ways.

First, the uniform relaxation of p_z towards its thermal equilibrium value p_0 produces a uniform longitudinal dipolar field which results in a global dephasing of the transverse polarizations. This process is effective during the whole sequence. Furthermore the uniform longitudinal polarization at time τ is transformed by the second pulse into a uniform transverse polarization which, according to a remark made earlier, will yield an extra contribution to the echoes.

The transverse polarization is attenuated between the two pulses by relaxation, so that the value of $p_z(\tau_+)$ produced by the second pulse is smaller than (3.168*b*). This modulated longitudinal polarization then decreases through longitudinal relaxation. The combined effect of these two factors is a decrease of efficiency for the production of multiple echoes.

Finally, the transverse relaxation after the second pulse reduces the amplitude of the successive echoes.

The inclusion of relaxation effects in the theory of multiple echoes is straightforward. The resulting expression for the amplitude of the echo of order m ($t = m\tau$) is the following:

$$|I_m| = p_0 \exp[-(m+1)\tau/T_2] J_m(\delta_m) \times \left|\frac{m}{\delta_m} + \mathrm{i}[1-\exp(-\tau/T_1)]\exp(-\tau/T_2)\right|, \qquad (3.178)$$

with:

$$\delta_m = \tfrac{3}{2} A p_0 T_1[1-\exp(-m\tau/T_1)]\exp(-\tau/T_2). \qquad (3.179)$$

2. *Sample shape and field gradient*

For an arbitrary shape of the sample, the relations (3.160′) and (3.161′) are no longer valid in the presence of a non-zero macroscopic polarization. The orientation of the local dipolar field with respect to the local polarization is then not uniform within the sample, and the motion of the polarization is complicated and cannot be treated in a simple way.

Let us first consider the dipolar field associated with the transverse polarization. The complex evolution it produces is limited to the time interval when a bulk transverse magnetization is present, and can be rendered negligible if the time T_2^* for the vanishing of the transverse signal is made short compared with the time interval τ between the pulses. This can be achieved by using a sufficiently large field gradient. As an order of magnitude, T_2^* is the time required to produce a dephasing of π between the transverse polarizations at both ends of the sample. If L is the size of the sample and (G/γ) the field gradient, one has:

$$GLT_2^* \simeq \pi,$$

and the condition is:

$$T_2^* = (\pi/GL) \ll \tau \sim T_2. \tag{3.180}$$

For a quantitative exploitation of multiple echoes, it is imperative that this condition be fulfilled.

As for the *uniform* longitudinal polarization resulting from spin-lattice relaxation, the dipolar field it creates is also inhomogeneous and produces different dephasings of the transverse polarizations in different parts of the sample. It can be shown that when the external field gradient (G/γ) is much larger than the gradient of this dipolar field the amplitudes of the successive echoes are only slightly altered. The condition is roughly the same as (3.180).

Another factor dictating the choice of the field gradient is the following. The external d.c. field is not rigorously homogeneous over the sample, and its intrinsic gradient has no reason to be unidirectional. The applied unidirectional field gradient must be sufficiently larger than this intrinsic gradient to make the total field gradient approximately unidirectional over the sample.

The tractability of the theory resulting from the use of a sufficiently large field gradient more than balances the inconvenience arising from the increase of diffusion damping with the gradient.

3. *The effects of diffusion*

We consider a uniform field gradient (G/γ) along the direction $0z$:

$$\gamma \frac{\mathrm{d}H}{\mathrm{d}z} = G.$$

Even in that simple case, the calculation of diffusion effects on multiple echoes is complicated, and we content ourselves with giving its principle.

In the frame rotating at the average Larmor frequency ω_0, the effective frequency $\omega(z)$ is:

$$\omega(z) = Gz + \tfrac{3}{2}Ap_z(z) = Gz + (b/p_0)p_z(z). \tag{3.181}$$

The evolution of the polarization is governed by the equations:

$$\frac{\mathrm{d}}{\mathrm{d}t} p_z = -\frac{1}{T_1}(p_z - p_0) + D\nabla^2 p_z; \tag{3.182a}$$

$$\frac{\mathrm{d}}{\mathrm{d}t} p_+ = \left(\mathrm{i}\omega - \frac{1}{T_2}\right) p_+ + D\nabla^2 p_+; \tag{3.182b}$$

where D is the diffusion constant.

The standard method for solving these equations is through the use of an auxiliary quantity $S(z, t)$ defined by:

$$p_+(z, t) = S(z, t)\exp(\mathrm{i}Gzt). \tag{3.183}$$

By carrying (3.183) into (3.182*b*) one obtains:

$$\frac{d}{dt}S = \left[i(b/p_0)p_z - \frac{1}{T_2}\right] S + D\left[\frac{\partial^2 S}{\partial z^2} + 2iGt\frac{\partial S}{\partial z} - G^2t^2S\right]. \quad (3.184)$$

One then proceeds by taking the space-Fourier transforms of both sides of this equation. One has

$$S_k = \int S(z)\exp(ikz)\,dz, \quad (3.185)$$

and a similar expression for p_{zk}. The complication arises from the fact that, on the right-hand side of (3.184), the Fourier transform of $p_z(z)\,S(z)$ is the convolution of the Fourier transforms of $p_z(z)$ and $S(z)$.

This causes no problem in the interval between the pulses, because p_z is uniform, with a value:

$$p_z(t) = p_0\left[1 - \exp\left(-\frac{t}{T_1}\right)\right], \quad (3.186)$$

and its only Fourier component is for $k = 0$. Equation (3.184) then becomes:

$$\frac{d}{dt}S_k = \left[i(b/p_0)p_z - \frac{1}{T_2} - D(k+Gt)^2\right]S_k. \quad (3.187)$$

Each mode k evolves independently from the others. The only mode to be considered is that present after the first pulse, i.e. $k = 0$, whose time evolution is readily obtained from (3.187) and (3.186).

After the second pulse $p_z(z)$ will contain not only the mode $k = 0$, but also the modes k_0, where $k_0 = G\tau$ is the wave vector corresponding to the helix of polarization produced by the precession in the time interval τ between the pulses. The time derivative of the mode S_k will then depend not only on S_k, but also on $S_{k\pm k_0}$.

We have then a set of coupled differential equations which must be solved with a computer. This calculation is long. In order to yield good results it must include modes $k = qk_0$ of order q much higher than the order m of the observed echo.

(*c*) *Experimental results*

The verification of the theory of multiple echoes was performed in the b.c.c. phase of solid ^{3}He, in two series of experiments.

The first series was conducted in high field ($H_0 = 9250$ G) at temperatures between 0·75 and 0·3 K. These temperatures are much higher than the magnetic transition temperature, the system is perfectly paramagnetic and its magnetization follows Curie's law. A slight difficulty arises from the fact that in high field the relaxation times T_1 and T_2 depend on the orientation of the magnetic field with respect to the crystalline axes (Deville, 1976), so that a clean verification of the theory involves the use of single crystals. The molar volumes of the sample are about 23 cm^3. The intrinsic inhomogeneity of the field over the sample is about 0·2 G cm^{-1}, and the external field gradient is comprised between 2 and 7 G cm^{-1}.

Figure 3.16 shows the echo trains observed after two $\pi/2$ pulses at $T = 320$ mK, for two different values of the interval τ between the pulses. The non-monotonic variation of the successive echo amplitude is very visible on Fig. 3.16(b).

Figure 3.17 reproduces the variation as a function of τ of the experimental and theoretical amplitudes of the first four echoes, at $T = 740$ mK. The field gradient is 3 G cm^{-1}. The value of T_1 was measured in a separate experiment (sequences of $\pi - \pi/2$ pulses). It is approximately $T_1 \simeq 500$ ms. As for the

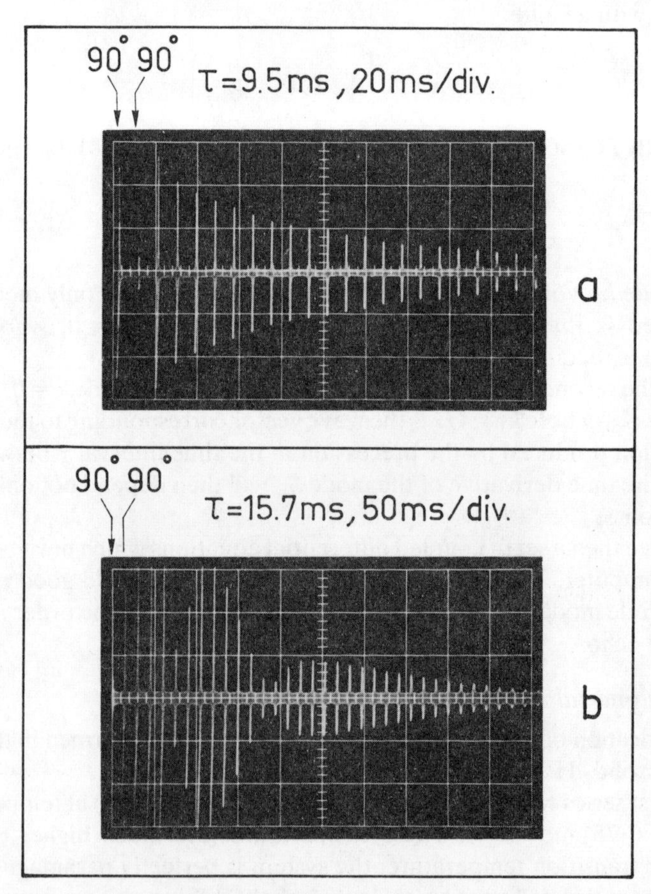

FIG. 3.16. Multiple echoes observed in solid ^{3}He after two 90° pulses, for two different time intervals between the pulses. Fields $H_0 = 9250$ G.; molar volume $V_M = 23·6$ cm^3; temperature $T = 320$ mK. (After Deville *et al.*, 1979.)

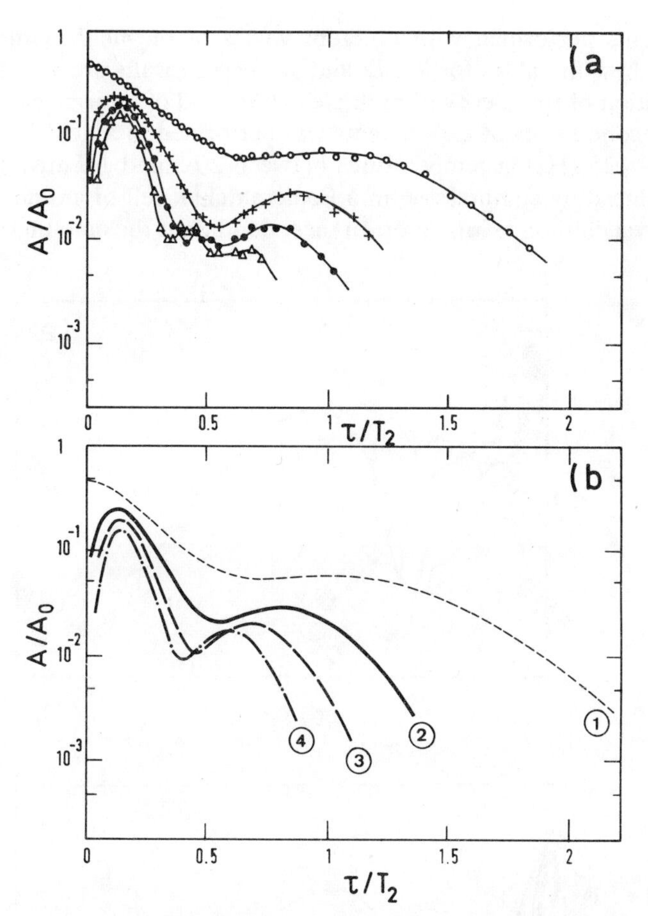

FIG. 3.17. Experimental (a) and calculated (b) amplitudes of the first four echoes following two 90° pulses, against time interval between the pulses. Field $H_0 = 9250$ G. Molar volume $V_M = 23$ cm^3. Temperature $T = 740$ mK. Field gradient $G = 3$ G cm^{-1}. In this experiment, more than 40 echoes were observed. (After Deville *et al.*, 1979.)

values of T_2, D and p_0, they are introduced as parameters in the numerical solution of eqns (3.182) and adjusted for a best fit with experiments. In the whole temperature range studied (740 to 315 mK) the agreement with theory is comparable with that exhibited in Fig. 3.17. The best-fit values of T_2 and D, temperature-independent are:

$$T_2 = 145 \text{ ms}; \qquad D = 4{\cdot}6 \times 10^{-8} \text{ cm}^2 \text{ s}^{-1}.$$

In this temperature range the best-fit values of p_0 depart by 4 per cent at most from those calculated according to Curie's law. It was also verified that experiments performed with different values of the field gradient yield identical best-fit values for T_2, D and p_0. These results are a quantitative confirmation of the theory of multiple echoes and of their physical origin.

The second series of experiments was performed in a field $H = 38{\cdot}6$ G ($\omega_0/2\pi = 125$ kHz) at temperatures between 20·5 and 1·7 mK. The solid was produced by compression in a Pomerantchuk cell of a liquid initially cooled by a dilution refrigerator. In these conditions the amount of solid at

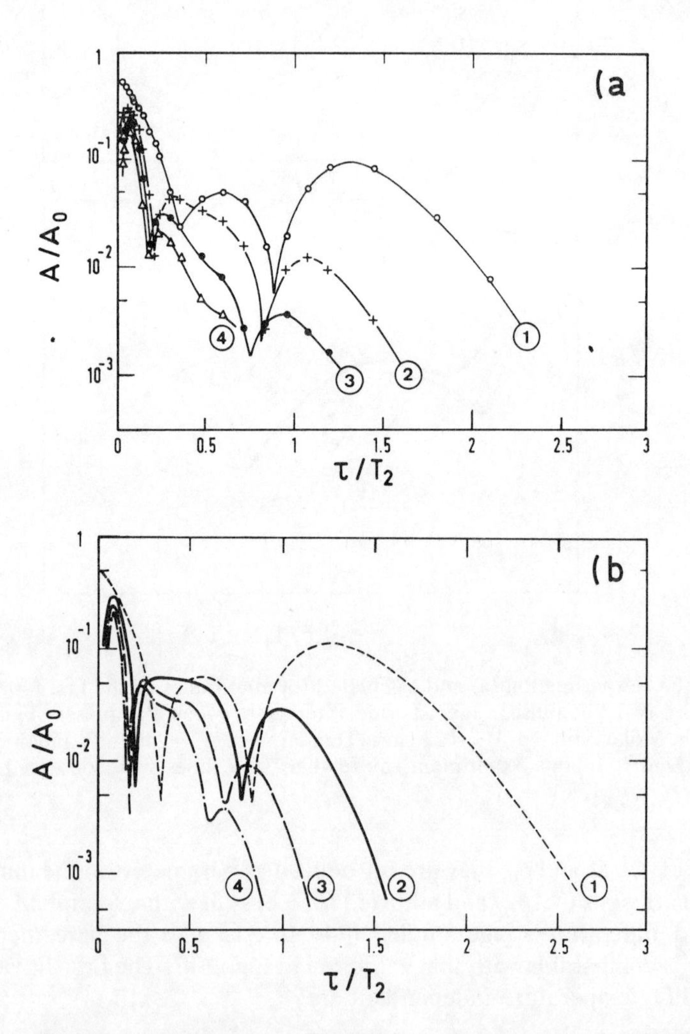

FIG. 3.18. Experimental (a) and calculated (b) amplitudes of the first four echoes following two 90° pulses, against time interval between the pulses. Field $H_0 = 38{\cdot}6$ G. Molar volume $V_M = 24{\cdot}25\ \text{cm}^3$. Temperature $T = 1{\cdot}47$ mK. (After Deville *et al.*, 1979.)

each time is unknown, and it is therefore not possible to use the intensity of its resonance signal to measure its magnetic susceptibility. This quantity can however be obtained from a study of the multiple echoes, which depend only on the magnetization of the solid and not on its bulk magnetic moment. In low fields, the values of T_1 and T_2 are isotropic and there is no need to use a single crystal.

Figure 3.18 reproduces the variation as a function of τ of the experimental and theoretical amplitudes of the first four echoes at $T = 1{\cdot}47$ mK and for $V_M = 24{\cdot}25\ \text{cm}^3$. In the temperature range investigated, T_1, T_2 and D are found to be constant:

$$T_1 = T_2 = 250\ \text{ms}; \qquad D = 15\times 10^{-8}\ \text{cm}^2\ \text{s}^{-1}.$$

The variation with temperature of the magnetic susceptibility, as deduced from these experiments is plotted in Fig. 3.19. These results provided the first evidence, later confirmed by other authors (Prewitt and Goodkind, 1977), of a departure from the Curie–Weiss law, characteristic of the inadequacy of an exchange coupling limited to a Heisenberg Hamiltonian of two-spin exchange between nearest neighbours.

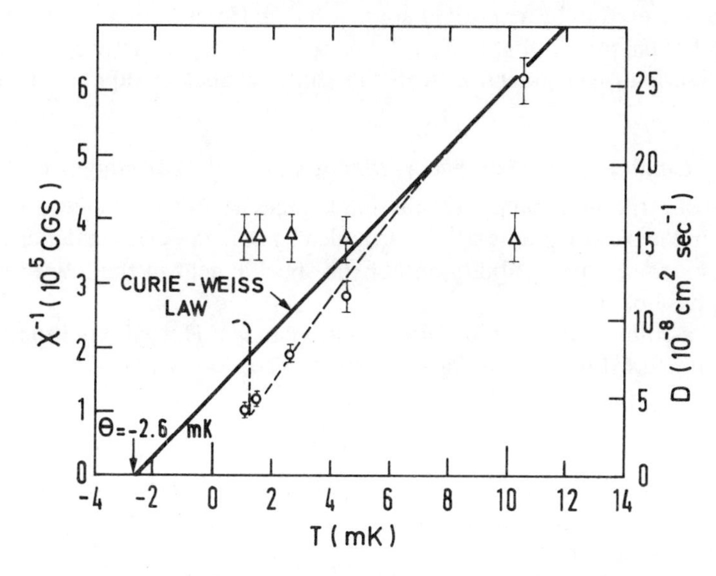

FIG. 3.19. Solid ^{3}He on the melting curve ($V_M = 24{\cdot}25\ \text{cm}^3$) in the vicinity of the spin ordering transition. Inverse susceptibility (circles) and diffusion coefficient (triangles), as measured by multiple spin echoes, as a function of temperature. Solid line: Curie–Weiss linear extrapolation from susceptibility data at higher temperature. Broken line: experimental susceptibility results of Prewitt and Goodkind (1977). (After Deville *et al.*, 1979.)

G. Low temperature properties of solid ^{3}He

The model of two-spin exchange interactions between nearest neighbours, which was extremely successful for interpreting the high-temperature properties of solid ^{3}He, breaks down completely at temperatures below a few mK.

Solid ^{3}He undergoes a phase transition to a magnetically ordered phase in the mK range. Both above and below the transition, its properties disagree violently with those predicted for a Heisenberg Hamiltonian (Landesman, 1978).

Among several attempts to account for these properties, the most promising one uses multiple-spin exchange interactions (Thouless, 1965; Hetherington and Willard, 1975; Roger, Delrieu, and Hetherington, 1980; Roger (1980) in thesis no 2297, Orsay, France).

The scope of this section is limited to two points. Firstly we describe two important properties of solid ^{3}He in the paramagnetic state which cannot possibly be accounted for by a Heisenberg Hamiltonian. Secondly, we give a brief discussion of the physical reasons why multiple-spin exchange interactions are likely to be present.

The discussion of the properties of solid ^{3}He in the ordered phases, and their interpretation by a multiple-spin exchange Hamiltonian, is deferred to section I of chapter 8, that is after the background of magnetic ordering has been described in connection with the study of nuclear dipolar magnetic ordering.

(a) *Inadequacy of the Heisenberg Hamiltonian in the paramagnetic state*

The most striking evidence for the inadequacy of the Heisenberg Hamiltonian in the paramagnetic state is provided by the temperature dependence of the magnetic susceptibility and of the specific heat in the low-pressure b.c.c. phase of ^{3}He.

If we assume that the spin–spin Hamiltonian *is* a Heisenberg interaction between nearest neighbours of the form (3.11):

$$\mathcal{H}_e = -2J \sum_{i<j} \mathbf{I}_i \,.\, \mathbf{I}_j, \tag{3.11}$$

the thermal equilibrium values of the susceptibility χ and the specific heat C can be calculated in the form of a power expansion with respect to the inverse temperature β.

For instance, to the lowest order in β, the energy in zero field is:

$$\begin{aligned}\langle \mathcal{H}_e \rangle &= -\beta \,\mathrm{Tr}\{\mathcal{H}_e^2\} \\ &= -N\beta \,.\, \tfrac{3}{8} z J^2 \\ &= -Nk_B \,.\, 3J^2/T,\end{aligned} \tag{3.188}$$

where $z = 8$ is the number of nearest neighbours of a spin in the b.c.c. lattice, and J is expressed in temperature units.

The specific heat is:

$$C = \frac{d\langle\mathcal{H}_e\rangle}{dT} = Nk_B \,.\, 3J^2/T^2,$$

which is often written in the form:

$$C = \frac{Nk_B}{4}\frac{\tilde{e}_2}{T^2} \tag{3.189}$$

with $\tilde{e}_2 = 12\,J^2$.

On the melting curve, for a molar volume $V = 24{\cdot}2$ cm^3, the experimental value of $\tilde{e}_2$ is:

$$\tilde{e}_2 = 6{\cdot}95 \text{ mK}^2,$$

whence:

$$|J| = 0{\cdot}76 \text{ mK}.$$

The susceptibility χ is obtained from the equilibrium spin polarization in an external field. The Hamiltonian is:

$$\mathcal{H} = \mathcal{H}_e + \omega I_z.$$

It is shown in chapter 5, section **D**, that to the lowest order in β and J, the average value of $\langle I_z^i \rangle$ is equal to:

$$\langle I_z^i \rangle = -\tfrac{1}{4}\beta(\omega - 2J \sum_j{}' \langle I_z^j \rangle). \tag{3.190}$$

Since all $\langle I_z^i \rangle$ are equal, $\sum_j' \langle I_z^j \rangle = 8\langle I_z^i \rangle$ and:

$$\chi \propto \langle I_z^i \rangle/\omega \propto (T - \theta)^{-1}, \tag{3.191}$$

with $\theta = 4J$.

Equation (3.191) is the well-known Curie–Weiss law. As shown in Fig. 3.19, this law is satisfied for $T \geqslant 10$ mK. The Curie–Weiss temperature θ on the melting curve is:

$$\theta = -3 \pm 0{\cdot}4 \text{ mK},$$

whence:

$$J \simeq -0{\cdot}75 \pm 0{\cdot}1 \text{ mK},$$

which is in close agreement with the value obtained from specific heat measurements. The sign of J being negative, as deduced from the susceptibility measurements, the exchange energy is minimum when neighbouring spins have opposite magnetizations, i.e. one expects an antiferromagnetic ordering at low temperature.

However, when the expansion in powers of β is pushed one step further, the agreement breaks down. The expressions for $1/\chi$ and C become:

$$1/\chi \propto T - \theta + B/T; \tag{3.192}$$

$$C = \tfrac{1}{4} N k_B \left(\frac{\tilde{e}_2}{T^2} - \frac{\tilde{e}_3}{T^3} \right). \tag{3.193}$$

The expressions for B and $\tilde{e}_3$ (Landesman, 1978) are immaterial. The important point is that $B > 0$ and $\tilde{e}_3 < 0$.

Both results are in contradiction with experiment. $B > 0$ means that when T is lowered, the decrease of $1/\chi$ should be slower than predicted by the Curie–Weiss formula (3.191), whereas it is visible in Fig. 3.19 that this decrease is *faster* than (3.191). As for the specific heat measurements below 40 mK, the value of $\tilde{e}_3$ they yield is known with poor accuracy, but its sign is definitely positive, in contradiction with the prediction based on the Heisenberg Hamiltonian. Attempts to patch up the disagreement by extending the range of the two-spin exchange interactions have been unsuccessful.

(b) Physical origin of multiple-spin exchange

The origin of exchange was discussed in section **B**(*a*): in the solid the overall wave function of the ^{3}He nuclei, which are fermions, must be antisymmetric with respect to *all* variables, according to the Pauli principle.

In section **B**(*a*), we had considered the case of two ^{3}He nuclei and showed that their eigenfunctions were either a symmetric orbital function associated with a singlet spin state, or an antisymmetric orbital function associated with a triplet spin state (eqn (3.12)). The exchange frequency was defined as the difference in energy (in frequency units) between the symmetric and the antisymmetric orbital wave functions ψ_S and ψ_A.

The exchange frequency can be described in a different way. Instead of ψ_S and ψ_A, we choose as a basis orbital wave functions where each nucleus is centred on a definite lattice site:

$$\psi_1 = \varphi(\mathbf{r}_1 - \mathbf{R}_1)\varphi(\mathbf{r}_2 - \mathbf{R}_2) g(\mathbf{r}_1 - \mathbf{r}_2)$$

$$= \frac{1}{\sqrt{2}} \{\psi_S + \psi_A\};$$

$$\psi_2 = \varphi(\mathbf{r}_1 - \mathbf{R}_2)\varphi(\mathbf{r}_2 - \mathbf{R}_1) g(\mathbf{r}_1 - \mathbf{r}_2)$$

$$= \frac{1}{\sqrt{2}} \{\psi_S - \psi_A\}.$$

In this basis, the Hamiltonian has non-diagonal matrix elements, and the exchange frequency is that of the interchange of the two ^{3}He atoms through quantum-mechanical tunnelling. This tunnelling frequency is smaller, the higher the energy barrier between the two configurations. The height of the barrier is estimated by considering the variation of the energy of the system that would result from the continuous motion in space of the two atoms all the way from their initial to their final positions. The energy variation during the interchange of the two atoms depends on the positions of the other atoms in the solid, which cannot be ignored. The calculation of exchange is therefore a many-body problem which is much more complex than described in section **B**(*a*)2.

Within this tunnelling picture, we can likewise consider more complicated tunnelling motions, corresponding to the cyclic permutations of the positions of more than two spins. Such permutations correspond to multiple-spin exchange.

The main difference between exchange in paramagnetic crystals and solid ^{3}He, is that in the former case it is the electrons which interchange positions whereas the atoms remain fixed. By contrast, in solid ^{3}He, it is the atoms themselves which exchange their positions. Helium nuclei are surrounded with electrons which experience attractive van der Waals interactions at large distances and a very strong, short range repulsive interaction with a 'hard core' of radius $r \simeq 2.14$ Å. These interactions result in a shallow attractive minimum at a distance $r_0 \simeq 2{\cdot}9$ Å. There is another limitation to the distance of approach of ^{3}He atoms, originating in the large zero-point motion of helium nuclei: any attempt to restrict too closely the positions of the nuclei would unduly increase their kinetic energy. As a consequence of the potential and kinetic barriers, the interatomic distance on the melting curve is $R_0 = 3{\cdot}65$ Å, much larger than the distance r_0 of potential minimum.

The existence of a hard core repulsion between helium atoms imposes a steric hindrance to the exchange of their positions and has therefore an important influence on the nature of the exchange. In b.c.c. solid ^{3}He, the motion of two atoms for exchanging their positions is only possible if the other nearby atoms are pushed away, which increases the tunnelling barrier for this exchange. The cyclic permutation of three, four or more atoms might cause a much smaller perturbation of the solid, resulting in a lower tunnelling barrier and thus favouring multiple-spin exchanges. On the other hand, increasing the number of atoms involved in a cyclic motion is an unfavourable feature and there is an optimum value for the number of atoms taking part in the exchange. The multiple-spin exchange interactions which are most likely to be important are the three and four spin exchange interactions.

A cyclic permutation of more than two spins can be written as a product of two-spin transpositions, and can therefore be expressed as a function of spin operators according to (3.9).

1. *Three-spin exchange*

The most compact configuration of three spins is shown in Fig. 3.20. The two cyclic permutations of three spins 1, 2, 3 are P_{123} and P_{321}. We have:

$$\begin{aligned} P_{123} &= P_{13}P_{12} \\ &= (\tfrac{1}{2}+2\mathbf{I}_1 \cdot \mathbf{I}_3)(\tfrac{1}{2}+2\mathbf{I}_1 \cdot \mathbf{I}_2). \end{aligned} \tag{3.194}$$

By using the properties of spin 1/2 operators we obtain:

$$P_{123}+P_{321} = \tfrac{1}{2}+2(\mathbf{I}_1 \cdot \mathbf{I}_2+\mathbf{I}_1 \cdot \mathbf{I}_3+\mathbf{I}_2 \cdot \mathbf{I}_3), \tag{3.195}$$

which is a sum of scalar products as for two-spin exchange. However it should be noticed that the product $\mathbf{I}_2 \cdot \mathbf{I}_3$ is between next-nearest neighbours.

2. *Four-spin exchange*

There are two types of four-spin exchanges, involving folded or planar spin cycles (Fig. 3.21). A cubic permutation P_{1234} can be written:

$$P_{1234} = P_{14}P_{123}$$

We obtain after a little algebra:

$$P_{1234}+P_{4321} = \tfrac{1}{4}+\sum_{i<j} \mathbf{I}_i \cdot \mathbf{I}_j + S_{1234}$$

with:

$$S_{1234} = \tfrac{1}{4}\{(\mathbf{I}_1 \cdot \mathbf{I}_2)(\mathbf{I}_3 \cdot \mathbf{I}_4)+(\mathbf{I}_1 \cdot \mathbf{I}_4)(\mathbf{I}_2 \cdot \mathbf{I}_3)-(\mathbf{I}_1 \cdot \mathbf{I}_3)(\mathbf{I}_2 \cdot \mathbf{I}_4)\}. \tag{3.196}$$

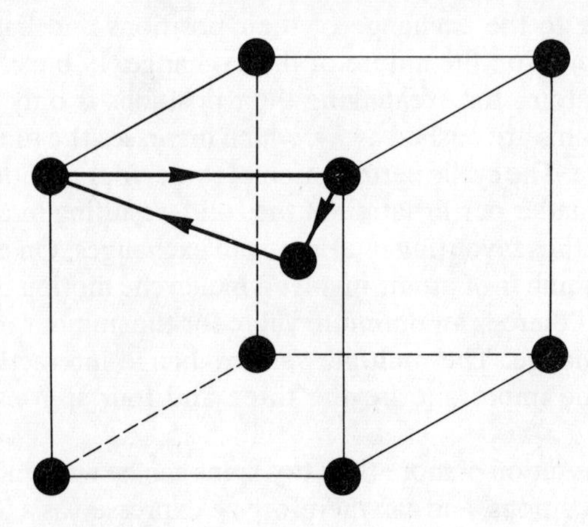

FIG. 3.20. Cycle of spins involved in 3-spin exchange in b.c.c. ^{3}He.

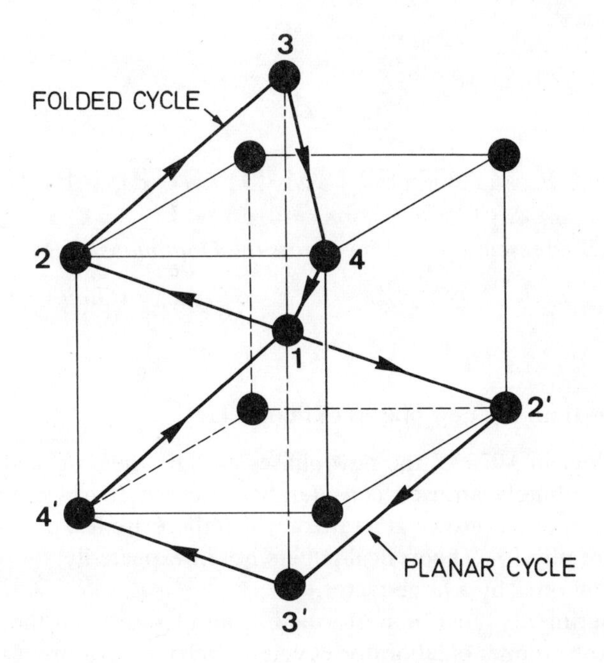

FIG. 3.21. Planar and folded cycles for 4-spin exchange in b.c.c. ^{3}He.

An approximate calculation of multiple-spin exchange from first principles suggests that 3 and 4-spin exchanges are larger than 2-spin exchange, but it cannot ascertain the relative importance of folded and planar 4-spin exchange (Roger (1980) in thesis no 2297, Orsay, France).

As stated earlier, the utilization of 3 and 4-spin exchange for the explanation of the behaviour of solid ^{3}He at very low temperatures is deferred till chapter 8.

4

NUCLEAR MAGNETISM OF SUPERFLUID ^{3}He

Let AB represent a line drawn cross the Dominions of Balnibarbi.

SWIFT (*Gulliver's Travels*)

A. Introduction: the new phases of liquid ^{3}He

The discovery in 1972 of the new phases of ^{3}He, the so-called superfluid phases immediately aroused considerable interest. It triggered a flow of publications that has grown at a rate of which there are few examples in low temperature physics. Theoretical papers not unexpectedly, outnumber the experimental ones by a large factor.

The experiments must be performed at temperatures in the millikelvin range and the number of laboratories where such temperatures are available is limited, although growing.

The large number of theoretical papers witnesses to the wealth of problems raised by the existence of the superfluid phases of ^{3}He. The subject combines on an increased level of sophistication many of the features of fields which have been very much alive in recent years: theoretical superconductivity, anisotropic fluids such as liquid crystals, quantum hydrodynamics, topology of defects, etc.

It is quite out of the question to attempt here even a superficial description of all the properties of superfluid ^{3}He. Thus, somewhat paradoxically, nothing will be said about the superfluidity of ^{3}He nor about any other of its orbital properties. Why then bring up the subject at all?

The answer is this. Everywhere in this book, as well as in our former books we have shown that the Zeeman interaction of nuclear spins with applied external fields and their mutual dipolar interactions are essentially what nuclear magnetism is about.

It turns out that in the new phases of ^{3}He both of these interactions have a behaviour and are responsible for phenomena which have no analogue in other media. No serious student of modern nuclear magnetism should ignore this behaviour and these phenomena. This chapter is an attempt at their description, as self-contained as possible, made by non-specialists in the field of superfluids for such students.

In this endeavour we lean heavily on the work of the masters: Leggett who has done so much for the theory and written most extensively about it, but also Anderson and Brinkman, Balian and Werthamer, and others. Unpublished lectures by Delrieu have been a great help. The formalism used in the theory is elementary throughout and does not go beyond the use of fermion creation and annihilation operators. On the other hand, intermediate steps of calculations which are not obvious are indicated. We shall not describe the dramatic discovery of the new phases of ^{3}He, as they appeared, like accidents on the solid–liquid equilibrium curve of pressure versus time, observed for the first time during an adiabatic compression of the solid–liquid mixture by Osheroff, Richardson, and Lee (1972), at temperatures in the 2 mK range. Instead, we shall give here the phase diagram of the various phases, which results from the contributions of many workers and contains most of the background that we shall require (Figs. 4.1 and 4.2).

At the melting pressure of 33 bars, as the temperature is lowered, there is a second order transition near 2·7 mK from the normal phase to a superfluid phase, which has been identified with the phase known theoretically as ABM, first mentioned here in section **F**(d). As the temperature is lowered

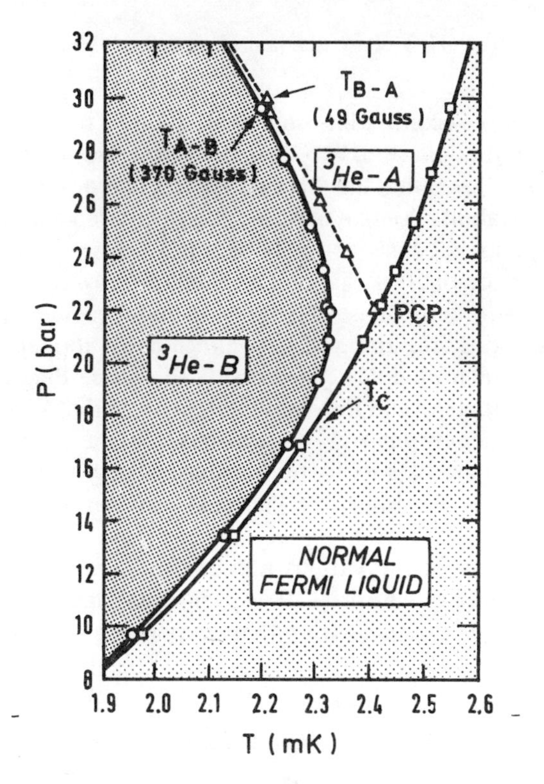

FIG. 4.1. Pressure–temperature phase diagram of superfluid ^{3}He for two values of the magnetic field.

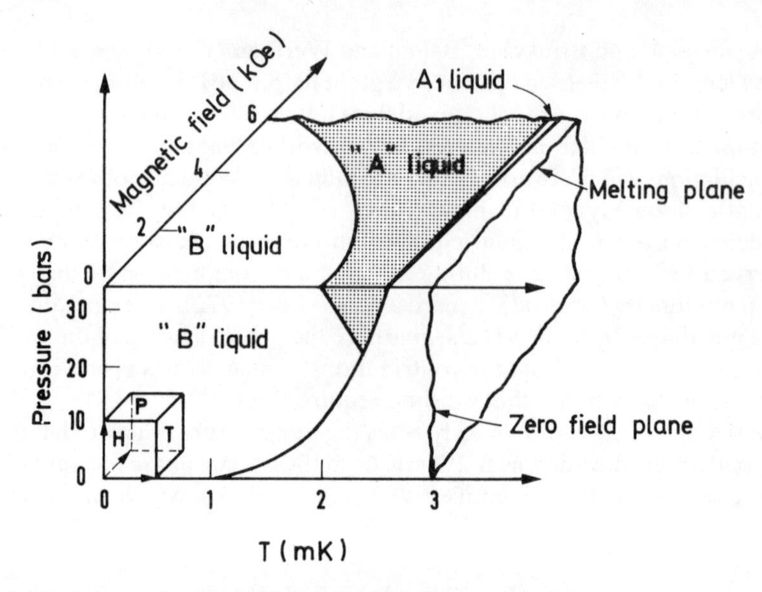

FIG. 4.2. Schematic P–H–T phase diagram of superfluid ^{3}He.

further there is a second transition near 2·1 mK to a phase called B, identified with the phase BW of $\mathbf{F}(d)$. There are reasons to believe (hysteresis, latent heat) that the AB transition is first order. As the pressure is lowered, the stability domain of phase A is decreased and reduces to zero at a pressure near 22 bars and a temperature near 2·45 mK. The corresponding point on the P–T diagram is called the polycritical point or PCP. Below this pressure only phase B is stable in zero-magnetic field.

These transitions are dramatically affected by the application of a magnetic field which modifies considerably the phase diagram. The polycritical point disappears and the stability domain of the A phase broadens. At the same time a new phase A_1 appears between the normal fluid and the phase A. The domain of stability of this phase is very narrow, being at the melting pressure of the order of 60 microkelvins per Tesla. (It cannot be seen on the scale of Fig. 4.2.)

The identification of the various phases has been achieved by several physical means among which NMR experiments play a leading part. As stated earlier the theory underlying these experiments is the object of this chapter.

B. Normal ^{3}He and the Landau interactions

We summarize first the basic facts about the Landau theory of Fermi fluids as it applies to liquid ^{3}He in its normal state.

Atoms of ^{3}He are fermions with zero electronic spin and a nuclear spin $\frac{1}{2}$. Let us first assume, an unrealistic assumption to be sure, that we can neglect all the interactions and consider liquid ^{3}He as a gas of free fermions.

The Hamiltonian of the system can be written in the well-known operator form of second quantization,

$$\mathscr{H} = \sum_{\mathbf{k},\sigma} \varepsilon^0_{k\sigma} a^\dagger_{\mathbf{k}\sigma} a_{\mathbf{k}\sigma}, \tag{4.1}$$

where $\mathbf{k}$ is the wave vector of a free fermion, σ its spin index, up or down, $\varepsilon^0_{k\sigma}$ its kinetic energy counted from the Fermi energy μ, $a^\dagger_{\mathbf{k}\sigma}$ and $a_{\mathbf{k}\sigma}$ the well-known creation and annihilation operators. The Fermi energy μ is related to the Fermi momentum $p_{\mathrm{F}} = \hbar k_{\mathrm{F}}$ by $\mu = \hbar^2 k_{\mathrm{F}}^2/2m$ where m is the mass of the individual fermion.

In 1956 Landau had shown that much of this description could be retained for liquids if the concept of quasi-particles as elementary excitations of the fluid was introduced. In this description it is assumed that a one to one correspondence can be established between the states of a free Fermi gas and those of the interacting fluid, starting from the former and switching the interaction adiabatically. In the same way as the state of the gas is defined by the occupation numbers $n_{\mathbf{k},\sigma}$ for the various particles, the corresponding state of the interacting fluid is defined by the same occupation numbers but referring to the quasi-particles rather than to particles.

If the occupation numbers $n_{\mathbf{k}\sigma}$ describing the ground state of the fluid are slightly varied by $\delta n_{\mathbf{k}\sigma}$, the energy per unit volume of the fluid will undergo a change:

$$\delta E = V^{-1} \sum_{\mathbf{k},\sigma} \varepsilon_{\mathbf{k}\sigma}\, \delta n_{\mathbf{k}\sigma}. \tag{4.2}$$

But, in the theory of Landau dealing with quasi-particles, $\varepsilon_{\mathbf{k}\sigma}$ in (4.2) depends itself on all increments $\delta n_{\mathbf{k}'\sigma'}$ of the occupation numbers.

If the $\delta n_{\mathbf{k}\sigma}$ are small one can use a linear approximation and write:

$$\varepsilon_{\mathbf{k}\sigma} = \varepsilon^0_{\mathbf{k}\sigma} + \frac{1}{V} \sum_{\mathbf{k}'\sigma'} f_{\mathbf{k}\sigma,\mathbf{k}'\sigma'}\, \delta n_{\mathbf{k}'\sigma'}, \tag{4.3}$$

where the functions $f_{\mathbf{k}\sigma,\mathbf{k}'\sigma'}$ are characteristic of the fluid. The theory of Landau is only valid near the Fermi surface so that the length of all wave vectors $\mathbf{k}$ in (4.2) or (4.3) is very nearly k_{F}. Since the occupation number of fermions can only be zero or one it may seem strange to speak of small increments $\delta n_{\mathbf{k}\sigma}$. The way out is to define $n_{\mathbf{k}\sigma}$ and $\delta n_{\mathbf{k}\sigma}$ as continuous variables, averages of the discrete occupation numbers over small regions in k space. Another important parameter in the description of a Fermi fluid is the density of states (for both spins) per unit volume at the Fermi level $(\mathrm{d}n/\mathrm{d}\varepsilon)_0$ which for free fermions is given by $(\mathrm{d}n/\mathrm{d}\varepsilon)_0 = (3N/2\mu)$ where N is the number of particles per unit volume.

In nuclear magnetism it is convenient to describe the behaviour of a spin $\frac{1}{2}$ by 2×2 density matrices. The two increments $\delta n_{\mathbf{k}\uparrow}$ and $\delta n_{\mathbf{k}\downarrow}$ of the occupation numbers correspond to the special case of a diagonal matrix $\delta\hat{n}_{\mathbf{k}}$. (We use the superscript ^ to represent a 2×2 matrix. We shall use the superscript $\hat{\hat{}}$ for a 4×4 matrix later on.)

$$\delta\hat{n}(\mathbf{k}) = \tfrac{1}{2}\,\delta\tilde{n}(\mathbf{k})\hat{1} + \hat{\boldsymbol{\sigma}}\,.\,\mathbf{s}(\mathbf{k}). \tag{4.4}$$

$\hat{\boldsymbol{\sigma}}$ is the vector operator whose 3 components are the Pauli matrices $\hat{\sigma}_1, \hat{\sigma}_2, \hat{\sigma}_3$; $\delta\tilde{n}(\mathbf{k})$ and the 3 components $s_1(\mathbf{k}), s_2(\mathbf{k}), s_3(\mathbf{k})$ of the vector $\mathbf{s}(\mathbf{k})$ are c-numbers which contain all the statistical information about the changes $\delta n_{\mathbf{k}\sigma}$ of the occupation numbers.

From (4.4) we get:

$$\delta\tilde{n}(\mathbf{k}) = \mathrm{Tr}\{\delta\hat{n}(\mathbf{k})\}; \qquad \mathbf{s}(\mathbf{k}) = \mathrm{Tr}\{\tfrac{1}{2}\hat{\boldsymbol{\sigma}}\,.\,\delta\hat{n}(\mathbf{k})\}. \tag{4.5}$$

(a) Molecular fields

We can represent the quasi-particle energies $\varepsilon_{\mathbf{k}\sigma}$ of (4.3) in matrix notation also, as we did for $\delta n_{\mathbf{k},\sigma}$ by writing:

$$\hat{\varepsilon}(\mathbf{k}) - \hat{\varepsilon}_0(\mathbf{k}) = \tilde{\varepsilon}(\mathbf{k})\hat{1} + \mathbf{m}(\mathbf{k})\,.\,\hat{\boldsymbol{\sigma}}. \tag{4.6}$$

The relationship (4.3) between $\varepsilon_{\mathbf{k}\sigma}$ and $\delta n_{\mathbf{k}'\sigma'}$ can then also be rewritten compactly in matrix form as a relation between the $\delta\hat{n}(\mathbf{k})$ matrix (4.4) and the $\hat{\varepsilon}(\mathbf{k})$ matrix (4.6). Because of the rotational invariance in spin space, the coupling operator whose matrix elements are the $f_{\mathbf{k}\sigma,\mathbf{k}'\sigma'}$ is necessarily of the form:

$$\hat{\hat{F}}(\mathbf{k}, \mathbf{k}'; \hat{\boldsymbol{\sigma}}, \hat{\boldsymbol{\sigma}}') = f(\mathbf{k}, \mathbf{k}')\hat{\hat{1}} + \hat{\boldsymbol{\sigma}}\,.\,\hat{\boldsymbol{\sigma}}' g(\mathbf{k}, \mathbf{k}'), \tag{4.7}$$

where $f(\mathbf{k}, \mathbf{k}')$ and $g(\mathbf{k}, \mathbf{k}')$ are spin-independent, and (4.3) becomes:

$$\hat{\varepsilon}(\mathbf{k}) = \varepsilon^0(\mathbf{k})\hat{1} + \sum_{\mathbf{k}'} \mathrm{Tr}_{\sigma'}\{\hat{\hat{F}}(\mathbf{k}, \mathbf{k}'; \hat{\boldsymbol{\sigma}}, \hat{\boldsymbol{\sigma}}')\,\delta\hat{n}(\mathbf{k}', \hat{\boldsymbol{\sigma}}')\}. \tag{4.8}$$

The parameters $\tilde{\varepsilon}(\mathbf{k})$ and $\mathbf{m}(\mathbf{k})$ in eqn (4.6) are then related to $\delta\tilde{n}(\mathbf{k})$ and $\mathbf{s}(\mathbf{k})$ of eqn (4.4) by the following equations:

$$\begin{cases} \tilde{\varepsilon}(\mathbf{k}) = \sum_{\mathbf{k}'} f(\mathbf{k}, \mathbf{k}')\,\delta\tilde{n}(\mathbf{k}'); \\ \mathbf{m}(\mathbf{k}) = 2\sum_{\mathbf{k}'} g(\mathbf{k}, \mathbf{k}')\mathbf{s}(\mathbf{k}'). \end{cases} \tag{4.9}$$

Near the Fermi surface the functions $f(\mathbf{k}, \mathbf{k}')$ and $g(\mathbf{k}, \mathbf{k}')$ depend only on the angle θ between the vectors $\mathbf{k}$ and $\mathbf{k}'$ and can be expanded in a series of Legendre polynomials.

Since we are not concerned with orbital phenomena, only the expansion of $g(\mathbf{k}, \mathbf{k}')$ is of interest to us:

$$g(\mathbf{k}, \mathbf{k}') = (\mathrm{d}n/\mathrm{d}\varepsilon)_0^{-1} \sum_l F_l^a P_l(\cos\theta). \tag{4.10}$$

We introduced $(\mathrm{d}_n/\mathrm{d}\varepsilon)_0^{-1}$ on the right-hand side of (4.10) in order for the coefficients F_l^a to be dimensionless. If we apply to the fluid a uniform magnetic field H_0, it is polarized by that field and its total spin $\mathbf{S}$ acquires a non-vanishing expectation value:

$$\langle \mathbf{S} \rangle = \tfrac{1}{2} \sum_{\mathbf{k}} \mathrm{Tr}\{\hat{\boldsymbol{\sigma}} \,.\, \delta \hat{n}_{\mathbf{k}}\} = \sum_{\mathbf{k}} \mathbf{s}(\mathbf{k}). \tag{4.11}$$

This polarization is clearly uniform so that all the $\mathbf{s}(\mathbf{k})$ are independent of $\mathbf{k}$ and only the coefficient F_0^a of expansion (4.10) comes in. From eqns (4.9) and (4.10) we get:

$$\mathbf{m}(\mathbf{k}) = 2(\mathrm{d}n/\mathrm{d}\varepsilon)_0^{-1} F_0^a \langle \mathbf{S} \rangle, \tag{4.12}$$

and from (4.6):

$$\hat{\varepsilon}(\mathbf{k}) - \hat{\varepsilon}_0(\mathbf{k}) = \tilde{\varepsilon}(\mathbf{k})\hat{1} + 4\left(\frac{\mathrm{d}n}{\mathrm{d}\varepsilon}\right)_0^{-1} F_0^a \langle \mathbf{S} \rangle \,.\, \frac{\hat{\boldsymbol{\sigma}}}{2}. \tag{4.13}$$

We define a molecular field $\mathbf{H}_{\mathrm{mol}}$ resulting from Landau interactions by rewriting the second term of (4.13) as $-\frac{1}{2}\gamma\hbar\mathbf{H}_{\mathrm{mol}} \,.\, \hat{\boldsymbol{\sigma}}$ where $\mathbf{H}_{\mathrm{mol}}$ is then:

$$\mathbf{H}_{\mathrm{mol}} = -4(\gamma\hbar)^{-1}\left(\frac{\mathrm{d}n}{\mathrm{d}\varepsilon}\right)_0^{-1} F_0^a \langle \mathbf{S} \rangle = -\frac{F_0^a \mathbf{M}}{\frac{1}{4}(\gamma\hbar)^2(\mathrm{d}n/\mathrm{d}\varepsilon)_0}, \tag{4.14}$$

and $\mathbf{M} = \gamma\hbar\langle \mathbf{S} \rangle$ is the magnetization of the fluid.

(*b*) *Free energy of a Landau fluid*

For a system of fermions in thermal equilibrium, describable by a Hamiltonian of independent particles of the form,

$$\mathcal{H} = \sum_{\mathbf{k},\sigma} \varepsilon_{\mathbf{k}\sigma} a^\dagger_{\mathbf{k}\sigma} a_{\mathbf{k}\sigma}, \tag{4.15}$$

the following results are well known.

The free energy F is given by:

$$F = -(1/\beta) \sum_{\mathbf{k},\sigma} \ln\{1 + \exp(-\beta\varepsilon_{\mathbf{k}\sigma})\}, \tag{4.16}$$

where $\beta = (k_{\mathrm{B}}T)^{-1}$.

The occupation numbers are:

$$n_{\mathbf{k}\sigma} = \langle a^\dagger_{\mathbf{k}\sigma} a_{\mathbf{k}\sigma} \rangle = \frac{\partial F}{\partial \varepsilon_{\mathbf{k}\sigma}} = f(\varepsilon_{\mathbf{k}\sigma}) = [\exp(\beta\varepsilon_{\mathbf{k}\sigma}) + 1]^{-1}. \tag{4.17}$$

It is worth pointing out that eqns (4.16) and (4.17) are valid even if the $\varepsilon_{\mathbf{k}\sigma}$ are temperature-dependent as can be the case for quasi-particles, a by no means obvious result.

(*c*) *Static magnetic susceptibility*

The magnetic susceptibility is given by the relation

$$\chi = -\frac{1}{V}\left(\frac{\partial^2 F}{\partial H^2}\right)_{H=0}$$

where H is an applied field:

$$-\frac{\partial F}{\partial H} = -\sum_{\mathbf{k},\sigma} \frac{\partial F}{\partial \varepsilon_{\mathbf{k}\sigma}} \frac{\partial \varepsilon_{\mathbf{k}\sigma}}{\partial H},$$

where:

$$\frac{\partial \varepsilon_{\mathbf{k}\sigma}}{\partial H} = -\gamma\hbar\frac{\sigma}{2};$$

$$-\frac{\partial F}{\partial H} = \frac{\gamma\hbar}{2}\sum_{\mathbf{k},\sigma} f(\varepsilon_{\mathbf{k}\sigma})\sigma; \tag{4.18}$$

$$-\frac{\partial^2 F}{\partial H^2} = -\left(\frac{\gamma\hbar}{2}\right)^2 \sum_{\mathbf{k},\sigma} \frac{\partial f(\varepsilon_{\mathbf{k}\sigma})}{\partial \varepsilon_{\mathbf{k}\sigma}}.$$

For $H=0$, $\varepsilon_{\mathbf{k}\sigma} = \varepsilon_k$ and the summation $\sum_{\mathbf{k},\sigma}$ can be replaced by $2\sum_{\mathbf{k}}$:

$$\chi = -\frac{1}{V}\left(\frac{\partial^2 F}{\partial H^2}\right)_{H=0} = -\left(\frac{\gamma\hbar}{2}\right)^2 \frac{2}{V}\sum_{\mathbf{k}} \frac{\partial f(\varepsilon_{\mathbf{k}})}{\partial \varepsilon_{\mathbf{k}}}$$

$$= -\left(\frac{\gamma\hbar}{2}\right)^2 \int_{-\infty}^{\infty} (\mathrm{d}n/\mathrm{d}\varepsilon)\frac{\partial f}{\partial \varepsilon}\,\mathrm{d}\varepsilon. \tag{4.19}$$

At low temperatures where f is very nearly a step function and $\mathrm{d}f/\mathrm{d}\varepsilon$ a δ function this is approximately:

$$\chi = -\left(\frac{\gamma\hbar}{2}\right)^2\left(\frac{\mathrm{d}n}{\mathrm{d}\varepsilon}\right)_0 \int_{-\infty}^{\infty} \frac{\partial f}{\partial \varepsilon}\,\mathrm{d}\varepsilon = \left(\frac{\gamma\hbar}{2}\right)^2\left(\frac{\mathrm{d}n}{\mathrm{d}\varepsilon}\right)_0 \{f(-\infty) - f(\infty)\};$$

or, since $f(\infty) = 0$, $f(-\infty) = 1$,

$$\chi = \chi_{n0} = \left(\frac{\gamma\hbar}{2}\right)^2\left(\frac{\mathrm{d}n}{\mathrm{d}\varepsilon}\right)_0. \tag{4.20}$$

This is the well known formula for the temperature-independent Pauli susceptibility. The molecular field resulting from Landau interactions and given by eqn (4.14) can be rewritten using (4.20):

$$\mathbf{H}_{\mathrm{mol}} = -F_0^a \frac{\mathbf{M}}{\chi_{n0}}. \tag{4.21}$$

Using eqn (4.21) it is easy to take into account the changes in the susceptibility due to the Landau interactions. We can write for the nuclear magnetization M: $\mathbf{M}=\chi_{n0}\mathbf{H}_{\text{tot}}$ where $\mathbf{H}_{\text{tot}}=\mathbf{H}_0+\mathbf{H}_{\text{mol}}$ is the total field, applied plus molecular, 'seen' by the nuclear spins. On the other hand $\mathbf{M}=\chi\mathbf{H}_0$, where χ is the true susceptibility of the fluid. From:

$$M=\chi H_0=\chi_{n0}\left\{H_0-F_0^a\frac{M}{\chi_{n0}}\right\}=\chi_{n0}\left\{H_0-F_0^a\frac{\chi H_0}{\chi_{n0}}\right\}, \tag{4.22}$$

we find

$$\chi=\frac{\chi_{no}}{1+F_0^a}. \tag{4.23}$$

One often finds in the literature the notation: $Z_0=4F_0^a$.

(d) Some numerical data on liquid ^{3}He

The Larmor frequency of ^{3}He is $\gamma/2\pi=3{\cdot}2435$ KHz per gauss.

The density of states $(dn/d\varepsilon)_0$, necessary for the calculation of the susceptibility can be written as:

$$(dn/d\varepsilon)_0=3\left(\frac{N_0}{V}\right)\frac{1}{2k_BT_F},$$

where N_0 is the Avogadro number, V the molar volume and T_F the Fermi temperature. The latter can be obtained from the specific heat per mole which for very low temperatures is linear in temperature and given by: $C/RT=\frac{1}{2}\pi^2T_F^{-1}$. Finally a 'magnetic' Fermi temperature T^* is defined by $T^*=\frac{2}{3}T_F(1+F_0^a)=\frac{2}{3}T_F(1+Z_0/4)$ so that according to (4.23):

$$\chi=\tfrac{2}{3}\chi_{n0}\frac{T_F}{T^*}$$

T^* is sometimes found in the tables.

As an example we give the values of all these parameters for two values of the pressure (Wheatley, 1975).

V cm^3	P bar	T^*	T_F	$\chi_{n0}\times10^8$ *emu*	Z_0	$\chi\times10^8$ *emu*
37	0	0·36	1·64	1·25	−2·69	3·81
27	24	0·20	1·13	2·47	−2·95	9·42

C. The Cooper pairs and the BCS theory applied to ^{3}He

In a famous paper (Cooper, 1956) Leon Cooper has shown that two fermions on the Fermi surface interacting with each other, but not (for the simplicity of the argument) with the other fermions of the Fermi sea, will necessarily form a bound state if their mutual interaction, however weak, is attractive. Cooper showed that the energy of the bound state as a function of the attractive potential V has a mathematical singularity for $V = 0$ and thus could not be calculated by a perturbation method. A crude calculation shows that the critical temperature T_c at which the whole Fermi fluid undergoes a phase transition, has a similar dependence with respect to V, with as a consequence the fact that a relatively small error in V, which is not well known, may lead to an error on T_c by several orders of magnitude. This explains why, although between 1959 and 1972 many theorists had pointed out that a new phase of liquid ^{3}He, which is a Fermi fluid, should exist at sufficiently low temperatures, they were unable to predict the critical temperature even in order of magnitude. This is why the historical discovery of the new phases of liquid ^{3}He by Osheroff, Richardson, and Lee was at first mistaken for a phase transition of solid ^{3}He.

The BCS theory (Bardeen, Cooper, Schrieffer) has given to the qualitative concept of Cooper pairs the mathematical basis of the second quantization formalism which takes automatically into account the antisymmetrical character of the fermion wave function. Its success in describing superconductivity as a transition toward a superfluid state of the conduction electrons of a metal, considered as Fermi fluid, is well known.

What are the main differences between this 'fluid' and that made of the atoms of ^{3}He? First and foremost naturally the absence of charge for helium atoms and thus also of the two most spectacular aspects of superconductivity: the Meissner effect, that is the expulsion of magnetic flux from the superconductor and the vanishing of its electrical resistance. Second, the nature of the attraction between the quasi-particles of the two Fermi fluids under comparison.

For conduction electrons it can be pictured as follows: an electron e attracts towards itself the surrounding positive ions. This concentration of positive charges around e attracts a second electron e'. Since the motion of the electrons is much faster than that of the ions this concentration of positive charge will not be able to 'follow' the electron e which concentrated it, and the effective interaction between the electrons e and e' will be a retarded interaction whose Fourier transform depends on both the momentum $\mathbf{q}$ and the energy ω exchanged by the two electrons.

In ^{3}He there is the intrinsic van der Waals attraction between the electronic shells of the atoms. On this attraction is superimposed a hard core repulsion which prevents atoms from coming too close together and thus favours pair wave functions which vanish at the origin, that is functions with an orbital momentum $l > 0$. This contrasts with Cooper pairs of conduction electrons where the stable state is an isotropic s-state with $l = 0$. The spin state of a Cooper pair in ^{3}He is thus expected to be either a triplet (spin-symmetrical) if l is odd or a singlet if l is even.

To the intrinsic interaction adds an indirect, spin-dependent interaction which takes its origin in the Landau interactions. We saw in the last section that to a uniform polarization $\langle \mathbf{S} \rangle$ of the fluid one could associate a molecular magnetic field given by eqn (4.14). This result can in principle be generalized as follows. To a spin $\mathbf{s}$ localized in space-time at $(\mathbf{r}, t)$ corresponds a local magnetization $\mathbf{M}(\mathbf{r}, t)$ and through the Landau interactions a molecular field $\mathbf{H}_{\text{mol}}(\mathbf{r}, t)$. The latter induces through a non-local susceptibility $\chi(\mathbf{r}-\mathbf{r}', t-t')$ at another point $(\mathbf{r}', t')$ a magnetization $\mathbf{M}(\mathbf{r}', t')$ to which corresponds another molecular field $\mathbf{H}_{\text{mol}}(\mathbf{r}', t')$. This field in turn interacts with a spin $\mathbf{s}'$ in $(\mathbf{r}', t')$ and the whole procedure results in an indirect interaction bilinear in $\mathbf{s}$ and $\mathbf{s}'$ which depends on $\mathbf{r}-\mathbf{r}'$ and $t-t'$ and turns out to be attractive for parallel spins.

A last point must be stressed in the comparison between superconducting electrons and ^{3}He. In the latter the polarizable medium responsible for the indirect interactions is the fluid itself rather than, as in the case of electrons, a foreign medium, the positive ions. It follows that the contribution of the indirect interactions to the total energy will depend very strongly on the state assumed for the fluid: the comparison of the expectation values of the interaction energy for two different states with different polarizabilities is complicated by the fact that not only the wave functions but also the interactions themselves will be different for these two states.

This situation is sometimes referred to as strong coupling. Its effects may be to reverse the relative order of energies calculated in the weak coupling approximation, that is using the *same* potential for the two states under comparison. This is actually the case for two ordered phases A and B of ^{3}He.

D. The order parameter

In the original BCS theory the superconducting singlet state is described by the famous BCS wave function:

$$|\psi\rangle = \prod_{\mathbf{k}} (u_{\mathbf{k}} + v_{\mathbf{k}} a^{\dagger}_{\mathbf{k}\uparrow} a^{\dagger}_{-\mathbf{k}\downarrow})|0\rangle. \tag{4.24}$$

In this equation $|0\rangle$ is the vacuum state and the coefficients $u_{\mathbf{k}}$ and $v_{\mathbf{k}}$ satisfy the normalization condition $|u_{\mathbf{k}}|^2 + |v_{\mathbf{k}}|^2 = 1$ and $u_{\mathbf{k}} = u_{-\mathbf{k}}$, $v_{\mathbf{k}} = v_{-\mathbf{k}}$.

It is clear that the state $|\psi\rangle$ does not correspond to a definite number of particles since the expansion of the infinite product (4.24) yields states with $2, 4, 6 \ldots, N$ particles. This is a procedure familiar in statistical mechanics which leads to no practical difficulties. Although (4.24) is not an eigenstate of the operator number of particles, $\mathcal{N} = \sum_{\mathbf{k},\sigma} a^{\dagger}_{\mathbf{k}\sigma} a_{\mathbf{k}\sigma}$ and thus, strictly speaking, cannot describe a system with a fixed number of particles, it can be shown that the relative fluctuation of its expectation value $\langle \mathcal{N} \rangle = \langle \psi | \mathcal{N} | \psi \rangle$ is very small when the number of particles is large. Equation (4.24) is used by

BCS as a trial function which minimizes the energy if one looks for the ground state at zero temperature, or the free energy for a description of the system at a finite temperature. In the latter case the coefficients $u_{\mathbf{k}}$ and $v_{\mathbf{k}}$ are temperature-dependent.

In an important review article Leggett (1975) has generalized the BCS wave function (4.24) for the case when the spin state of the pair is a triplet and has consistently based the description of superfluid ^{3}He on this wave function. This treatment is probably an excellent shortcut for someone already familiar with the theory of superconductivity. Since we do not assume this familiarity for the reader, we use the more compact formalism of Balian and Werthamer where the description of a spin triplet state is tackled from the start rather than as a generalization of the spin singlet.

This formalism can be introduced as follows. Let us go back for a moment to the state $|\psi\rangle$ defined by eqn (4.24) and consider the expectation value:

$$\langle a_{\mathbf{k}\uparrow}a_{-\mathbf{k}\downarrow}\rangle = \langle\psi|a_{\mathbf{k}\uparrow}a_{-\mathbf{k}\downarrow}|\psi\rangle.$$

It is easy to verify that this expectation value is not zero. This contrasts with a normal state of a Fermi fluid where expectation values of 'anomalous' products of two annihilation (or creation) operators vanish. We shall characterize more generally a state formed of Cooper pairs by the condition that 'anomalous' averages, such as

$$\hat{x}(\mathbf{k})_{\sigma\sigma'} = \langle a_{\mathbf{k}\sigma}a_{-\mathbf{k}\sigma'}\rangle \tag{4.25}$$

are different from zero. The average in (4.25) will be understood not only as a quantum mechanical expectation value but as a statistical (and thus temperature dependent) average as well.

We have defined through (4.25) a 2×2 matrix $\hat{x}(\mathbf{k})$. From its definition and the anticommutation rule of the operators $a_{\mathbf{k}\sigma}$ it is clear that:

$$\langle a^{\dagger}_{-\mathbf{k}\sigma'}a^{\dagger}_{\mathbf{k}\sigma}\rangle = \hat{x}^{\dagger}(\mathbf{k})_{\sigma'\sigma}; \tag{4.26a}$$

$$\tilde{\hat{x}}(\mathbf{k}) = -\hat{x}(-\mathbf{k}); \tag{4.26b}$$

$$\hat{x}^{\dagger}(\mathbf{k}) = -\hat{x}^{*}(-\mathbf{k}); \tag{4.26c}$$

where $\tilde{\hat{x}}$ is the matrix transposed of $\hat{x}$ and $\hat{x}^{\dagger} = \tilde{\hat{x}}^{*}$ its Hermitian conjugate. Since the direction along which are quantized the spins of the fermions is arbitrary, it is essential to know how the matrices $\hat{x}(\mathbf{k})$ transform under rotation of the spin co-ordinate axes.

Let us define two kinds of 2-component spinor operators:
column operators,

$$a(\mathbf{k}) = \begin{pmatrix} a_{\mathbf{k}\uparrow} \\ a_{\mathbf{k}\downarrow} \end{pmatrix}, \qquad a^{\dagger}(\mathbf{k}) = \begin{pmatrix} a^{\dagger}_{\mathbf{k}\uparrow} \\ a^{\dagger}_{\mathbf{k}\downarrow} \end{pmatrix}, \tag{4.27}$$

and line operators,

$$\begin{aligned}\tilde{a}(\mathbf{k}) &= (a_{\mathbf{k}\uparrow}a_{\mathbf{k}\downarrow}),\\ \tilde{a}^{\dagger}(\mathbf{k}) &= (a^{\dagger}_{\mathbf{k}\uparrow}a^{\dagger}_{\mathbf{k}\downarrow}).\end{aligned} \tag{4.27'}$$

A product such as $\tilde{a}_{\mathbf{k}}a_{\mathbf{k}}$ is a scalar:

$$\tilde{a}_{\mathbf{k}}a_{\mathbf{k}} = a_{\mathbf{k}\uparrow}a_{\mathbf{k}\uparrow} + a_{\mathbf{k}\downarrow}a_{\mathbf{k}\downarrow}$$

whereas a product such as $a_{\mathbf{k}}\tilde{a}_{\mathbf{k}}$ is a matrix:

$$a_{\mathbf{k}}\tilde{a}_{\mathbf{k}} = \begin{pmatrix} a_{\mathbf{k}\uparrow}a_{\mathbf{k}\uparrow} & a_{\mathbf{k}\uparrow}a_{\mathbf{k}\downarrow} \\ a_{\mathbf{k}\downarrow}a_{\mathbf{k}\uparrow} & a_{\mathbf{k}\downarrow}a_{\mathbf{k}\downarrow} \end{pmatrix}.$$

Under a rotation of the spin co-ordinate axes, a column spinor is transformed into:

$$a'(\mathbf{k}) = \hat{C}a(\mathbf{k}) \tag{4.28}$$

and a line spinor is transformed into:

$$\tilde{a}'(\mathbf{k}) = \tilde{a}(\mathbf{k})\tilde{\hat{C}}, \tag{4.28'}$$

where $\tilde{\hat{C}}$ is the transpose of the rotation matrix $\hat{C}$.

Using these notations, eqns (4.25) and (4.26a) can be written:

$$\hat{x}(\mathbf{k}) = \langle a(\mathbf{k})\tilde{a}(-\mathbf{k})\rangle; \tag{4.25'}$$

$$\hat{x}^{\dagger}(\mathbf{k}) = \langle a^{\dagger}(-\mathbf{k})\tilde{a}^{\dagger}(\mathbf{k})\rangle. \tag{4.26'}$$

According to (4.25′), (4.28), and (4.28′) a rotation of spin co-ordinate axes transforms the matrix $\hat{x}$ into:

$$\hat{x}' = \hat{C}\hat{x}\tilde{\hat{C}}. \tag{4.28''}$$

For a rotation of angle ω around a unit vector $\mathbf{n}$, the matrix $\hat{C}$ can be written in operator form:

$$\hat{C} = \exp\left[\mathrm{i}\frac{\omega}{2}(\mathbf{n}\,.\,\hat{\boldsymbol{\sigma}})\right] = \cos\frac{\omega}{2}\hat{1} + \mathrm{i}(\mathbf{n}\,.\,\hat{\boldsymbol{\sigma}})\sin\frac{\omega}{2}. \tag{4.29}$$

If instead of $\tilde{\hat{C}}$ we had in (4.28″) $\hat{C}^{\dagger} = \hat{C}^{-1}$ we could make use of the relations between spinors and vectors:

$$\hat{C}\hat{\boldsymbol{\sigma}}\hat{C}^{-1} = R^{-1}\hat{\boldsymbol{\sigma}} \tag{4.30}$$

where R is the 3×3 real orthogonal matrix describing the rotation $(\omega, \mathbf{n})$. We can make $\hat{C}^{-1}$ appear in (4.28″) by using well-known properties of the Pauli matrices $\hat{\sigma}_1, \hat{\sigma}_2, \hat{\sigma}_3$ and the definition (4.29):

$$\begin{cases} \hat{\sigma}_2\tilde{\hat{\sigma}}_i\hat{\sigma}_2 = -\hat{\sigma}_i \\ \hat{\sigma}_2\{\mathrm{i}\widetilde{(\mathbf{n}\,.\,\hat{\boldsymbol{\sigma}})}\}\hat{\sigma}_2 = -\mathrm{i}(\mathbf{n}\,.\,\hat{\boldsymbol{\sigma}}) = [\mathrm{i}(\mathbf{n}\,.\,\hat{\boldsymbol{\sigma}})]^{\dagger} \\ \hat{\sigma}_2\tilde{\hat{C}}\hat{\sigma}_2 = \hat{C}^{\dagger} = \hat{C}^{-1}. \end{cases} \tag{4.31}$$

Going back to $\hat{x}$, it is clear that any 2×2 complex matrix $\hat{x}$ can always be written in the form:

$$\hat{x} = \{x_0\hat{1} + \mathbf{x}\,.\,\hat{\boldsymbol{\sigma}}\}\hat{\sigma}_2 \tag{4.32}$$

where nothing can as yet be stated about the transformation law of the 4 complex constants x_0, x_1, x_2, x_3. A symbol such as $\mathbf{x}\,.\,\hat{\boldsymbol{\sigma}}$ is so far simply a shorthand notation for $x_1\hat{\sigma}_1 + x_2\hat{\sigma}_2 + x_3\hat{\sigma}_3$. Let x_0' and $\mathbf{x}'$ be the values in the frame of reference derived from the previous one by the rotation $(\omega, \mathbf{n})$. From (4.28″) and (4.31) we get:

$$\begin{aligned}\hat{x}' &= \{x_0'\hat{1} + \mathbf{x}'\,.\,\hat{\boldsymbol{\sigma}}\}\hat{\sigma}_2 = \hat{C}\{x_0\hat{1} + (\mathbf{x}\,.\,\hat{\boldsymbol{\sigma}})\}\hat{\sigma}_2\tilde{\hat{C}} \\ &= \hat{C}\{x_0\hat{1} + (\mathbf{x}\,.\,\hat{\boldsymbol{\sigma}})\}\hat{\sigma}_2\tilde{\hat{C}}\hat{\sigma}_2\hat{\sigma}_2 = \hat{C}\{x_0\hat{1} + (\mathbf{x}\,.\,\hat{\boldsymbol{\sigma}})\}\hat{C}^{-1}\hat{\sigma}_2 \\ &= \{x_0\hat{1} + \hat{C}(\mathbf{x}\,.\,\hat{\boldsymbol{\sigma}})\hat{C}^{-1}\}\hat{\sigma}_2.\end{aligned} \tag{4.33}$$

It follows from (4.33) that $x_0 = x_0'$ is invariant by rotation. From (4.30) and (4.33) we get:

$$(\mathbf{x}'\,.\,\hat{\boldsymbol{\sigma}}) = \hat{C}(\mathbf{x}\,.\,\hat{\boldsymbol{\sigma}})\hat{C}^{-1} = \mathbf{x}\,.\,(\hat{C}\hat{\boldsymbol{\sigma}}\hat{C}^{-1}) = \mathbf{x}\,.\,(R^{-1}\boldsymbol{\sigma}) = (R\mathbf{x}\,.\,\hat{\boldsymbol{\sigma}}). \tag{4.34}$$

From the relation $\mathbf{x}' = R\mathbf{x}$ we see that $\mathbf{x}$ behaves indeed as a vector under rotation of the spin-co-ordinate axes. Consider first the contribution to $\hat{\mathbf{x}}$ of:

$$x_0\hat{\sigma}_2 = \begin{pmatrix} 0 & -\mathrm{i}x_0 \\ \mathrm{i}x_0 & 0 \end{pmatrix}.$$

The products $a_{\mathbf{k}\uparrow}a_{-\mathbf{k}\downarrow}$ and $a_{\mathbf{k}\downarrow}a_{-\mathbf{k}\uparrow}$ have equal and opposite expectation values. This is characteristic of a singlet state whose wave function $\alpha(1)\beta(2) - \alpha(2)\beta(1)$ is antisymmetric with respect to the spins of the two fermions.

As will appear later on (in section **G**) there is unmistakable experimental evidence that the pairs in superfluid ^{3}He are in a triplet state and we shall drop the term $x_0\hat{\sigma}_2$ from now on. The remaining term,

$$(\mathbf{x}\,.\,\hat{\boldsymbol{\sigma}})\hat{\sigma}_2 = x_2\hat{1} - \mathrm{i}x_3\hat{\sigma}_1 + \mathrm{i}x_1\hat{\sigma}_3 = \begin{pmatrix} x_2 + \mathrm{i}x_1 & -\mathrm{i}x_3 \\ -\mathrm{i}x_3 & x_2 - \mathrm{i}x_1 \end{pmatrix}, \tag{4.35}$$

is symmetrical as befits a spin triplet.

We can define a new order parameter, a normalized vector $\mathbf{d}(\mathbf{n})$ which depends on the direction but not on the magnitude of the wave vector $\mathbf{k} = |\mathbf{k}|\mathbf{n} = k\mathbf{n}$, by summing $\mathbf{x}(\mathbf{k})$ over the magnitude k of the vector $\mathbf{k}$: the summation $\sum_k$ being normalized by the definition $\sum_{\mathbf{k}} = \int (\mathrm{d}\Omega/4\pi)\sum_k$:

$$\sum_k \mathbf{x}(\mathbf{k}) = \psi\mathbf{d}(\mathbf{n}); \qquad \int |\mathbf{d}(\mathbf{n})|^2 \frac{\mathrm{d}\Omega}{4\pi} = 1, \tag{4.36}$$

where the constant ψ depends on the temperature but is independent of the orientation of the vector $\mathbf{n}$ on the Fermi surface and can always be made real by including its phase in $\mathbf{d}(\mathbf{n})$. The wave function of a pair described by the order parameter $\mathbf{d}(\mathbf{n})$ can be written, replacing in (4.35) $\mathbf{x}$ by $\mathbf{d}$:

$$\varphi = \frac{1}{\sqrt{2}}\{(d_2 + \mathrm{i}d_1)|+,+\rangle + (d_2 - \mathrm{i}d_1)|-,-\rangle - \mathrm{i}d_3[|+,-\rangle + |-,+\rangle]\}$$

$$= \frac{1}{\sqrt{2}}\{(d_2 + \mathrm{i}d_1)|1\rangle + (d_2 - \mathrm{i}d_1)|-1\rangle - \mathrm{i}d_3|0\rangle\}, \tag{4.37}$$

where $|1\rangle, |-1\rangle, |0\rangle$ are eigenstates of the component S_z of the spin of the pair. Equation (4.37) can yet be rewritten:

$$\varphi = d_1\left\{\frac{|-1\rangle - |1\rangle}{\mathrm{i}\sqrt{2}}\right\} + d_2\left\{\frac{|1\rangle + |-1\rangle}{\sqrt{2}}\right\} - \mathrm{i}d_3|0\rangle$$

$$= d_1|X_1\rangle + d_2|X_2\rangle + d_3|X_3\rangle, \tag{4.38}$$

where the wave functions $|X_1\rangle, |X_2\rangle, |X_3\rangle$ defined by (4.38) transform under rotation of the spin axes like the three components of a vector in real space. The following relation is well known (and easily verified from (4.38)).

$$\langle X_p|S_r|X_q\rangle = -\mathrm{i}\varepsilon_{pqr} \tag{4.39}$$

where S_r is the r component of the spin S of the pair. It follows from (4.39) that the expectation value of a component S_r in the state given by (4.38) is proportional to:

$$\langle\varphi|S_r|\varphi\rangle = -\mathrm{i}\varepsilon_{pqr}d_p^*d_q,$$

whence:

$$\langle\varphi|\mathbf{S}|\varphi\rangle = -\mathrm{i}(\mathbf{d}^* \wedge \mathbf{d}). \tag{4.40}$$

If the vector $\mathbf{d}$ is real in the sense that $\mathbf{d}^* \wedge \mathbf{d} = 0$, that is if its three components are real within a common phase factor, the expectation value of any component of $\mathbf{S}$ vanishes. States with a real vector $\mathbf{d}$ are termed in the literature, somewhat unfelicitously, unitary.

If $\mathbf{d}$ is real it is easy to see that φ is eigenstate, with the eigenvalue zero of the component of $\mathbf{S}$ along $\mathbf{d}$. To see it, it is sufficient to make $d_1 = d_2 = 0$ in (4.37). This can only be done if $\mathbf{d}$ is real. Otherwise the vectors $\mathbf{d}' = \mathrm{Re}\,\mathbf{d}$ and $\mathbf{d}'' = \mathrm{Im}\,\mathbf{d}$ are not parallel and cannot be made simultaneously parallel to $0z$.

The assumption in (4.25) of the existence of an order parameter is fairly far-reaching and is sufficient to show that the nuclear dipolar Hamiltonian does not vanish in first order in superfluid ^{3}He, in contrast to the normal phase. This is a key feature of the NMR properties of the superfluid and the formalism of the vector order parameter $\mathbf{d}(\mathbf{n})$ is sufficient to obtain an analytical expression for the dipolar Hamiltonian and even to write down the celebrated Leggett equations of motions. We shall however postpone this study until we have established the nature of the energy spectrum of superfluid ^{3}He and the existence of the energy gap and calculated the magnetic susceptibility.

E. Elementary excitations and energy gap in superfluid ^{3}He

(a) Linearization of the interaction Hamiltonian

The Hamiltonian of ^{3}He contains a single particle part and an interaction part. We keep in the latter only the so-called pair-scattering part represented by the diagram of Fig. 4.3. The total Hamiltonian can be written:

$$\mathcal{H} = \sum_{\mathbf{k},\sigma} \varepsilon_{\mathbf{k}\sigma} a^{\dagger}_{\mathbf{k}\sigma} a_{\mathbf{k}\sigma} + \tfrac{1}{2} \sum_{\mathbf{k},\sigma,\mathbf{k}',\sigma'} V(\mathbf{k},\mathbf{k}')(a^{\dagger}_{\mathbf{k}\sigma} a_{\mathbf{k}'\sigma})(a^{\dagger}_{-\mathbf{k}\sigma'} a_{-\mathbf{k}'\sigma'})$$

$$= \sum_{\mathbf{k},\sigma} \varepsilon_{\mathbf{k}\sigma} a^{\dagger}_{\mathbf{k}\sigma} a_{\mathbf{k}\sigma} + \tfrac{1}{2} \sum_{\mathbf{k},\sigma,\mathbf{k}',\sigma'} V(\mathbf{k},\mathbf{k}') a^{\dagger}_{\mathbf{k}\sigma} a^{\dagger}_{-\mathbf{k}\sigma'} a_{-\mathbf{k}'\sigma'} a_{\mathbf{k}'\sigma}. \tag{4.41}$$

The form chosen in (4.41) for the interaction energy does not imply that the pair scattering term is the only one or even the main one, in the interaction between quasi-particles. Rather, it is the only one expected to give very different results for the normal state, where it vanishes, and the superfluid state, and since all one can hope to estimate is the difference in energy between the two, this is probably a legitimate procedure.

To find the elementary excitations we must diagonalize the Hamiltonian (4.41). The main difficulty comes from the interaction term which is a sum of products of four fermion operators. It can be linearized, that is reduced to a quadratic form by making the assumption that a product such as $a_{\mathbf{k}\sigma} a_{-\mathbf{k}\sigma'}$ differs little from its mean value $\hat{x}(\mathbf{k})_{\sigma\sigma'}$ (eqn (4.25)). By writing:

$$\begin{aligned} a_{\mathbf{k}\sigma} a_{-\mathbf{k}\sigma'} &= \hat{x}(\mathbf{k})_{\sigma\sigma'} + \{a_{\mathbf{k}\sigma} a_{-\mathbf{k}\sigma'} - \hat{x}(\mathbf{k})_{\sigma\sigma'}\}; \\ a^{\dagger}_{-\mathbf{k}\sigma'} a^{\dagger}_{\mathbf{k}\sigma} &= \hat{x}^{\dagger}(\mathbf{k})_{\sigma'\sigma} + \{a^{\dagger}_{-\mathbf{k}\sigma'} a^{\dagger}_{\mathbf{k}\sigma} - \hat{x}^{\dagger}(\mathbf{k})_{\sigma'\sigma}\}; \end{aligned} \tag{4.42}$$

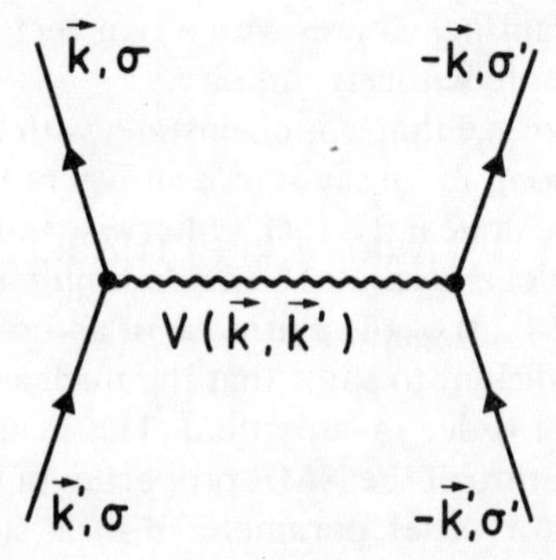

FIG. 4.3. Pair-scattering terms of the liquid ^{3}He Hamiltonian responsible for superfluidity.

considering the terms in curly brackets as small and neglecting their products, one finds:

$$V=-\frac{1}{2}\sum_{\mathbf{k},\mathbf{k}'} V(\mathbf{k},\mathbf{k}')[\{a^{\dagger}_{\mathbf{k}\sigma}\hat{x}(\mathbf{k}')_{\sigma\sigma'}a^{\dagger}_{-\mathbf{k}\sigma'}+a_{-\mathbf{k}\sigma'}\hat{x}^{\dagger}(\mathbf{k}')_{\sigma'\sigma}a_{\mathbf{k}\sigma}\}$$
$$+\operatorname{Tr}\{\hat{x}^{\dagger}(\mathbf{k})\,.\,\hat{x}(\mathbf{k}')\}]. \tag{4.43}$$

To write (4.43) we have made use of the anticommutation relations of creation and annihilation operators. Equation (4.43) can be written in a more compact form if we introduce the 2-components spinors $a(\mathbf{k})$ and $a^{\dagger}(\mathbf{k})$ defined by (4.27) and if we define a 2×2 matrix $\hat{\Delta}(\mathbf{k})$, called, for reasons to appear shortly, the gap:

$$\hat{\Delta}(\mathbf{k})=-\sum_{\mathbf{k}'} V(\mathbf{k},\mathbf{k}')\hat{x}(\mathbf{k}'). \tag{4.44}$$

The linearized Hamiltonian (4.43) then takes the more compact form:

$$V=\frac{1}{2}\sum_{\mathbf{k}}\tilde{a}^{\dagger}(\mathbf{k})\,.\,\hat{\Delta}(\mathbf{k})\,.\,a^{\dagger}(-\mathbf{k})+\frac{1}{2}\sum_{\mathbf{k}}\tilde{a}(-\mathbf{k})\,.\,\hat{\Delta}^{\dagger}(\mathbf{k})\,.\,a(\mathbf{k})$$
$$+\frac{1}{2}\sum_{\mathbf{k}}\operatorname{Tr}\{\hat{\Delta}^{\dagger}(\mathbf{k})\,.\,\hat{x}(\mathbf{k})\}. \tag{4.45}$$

The single particle part of the Hamiltonian (4.41) can be rewritten as:

$$\sum_{\mathbf{k}}\tilde{a}^{\dagger}(\mathbf{k})\,.\,\hat{\varepsilon}(\mathbf{k})\,.\,a(\mathbf{k})$$

where $\hat{\varepsilon}(\mathbf{k})$ is the 2×2 matrix:

$$\hat{\varepsilon}(\mathbf{k})=\begin{pmatrix}\varepsilon_{\mathbf{k}\uparrow} & 0\\ 0 & \varepsilon_{\mathbf{k}\downarrow}\end{pmatrix}.$$

In the absence of a magnetic field $\varepsilon_{\mathbf{k}\uparrow}=\varepsilon_{\mathbf{k}\downarrow}$ and $\hat{\varepsilon}(\mathbf{k})=\varepsilon_{\mathbf{k}}\hat{1}$ is a multiple of the unit matrix. The Hamiltonian can then be written:

$$\mathcal{H}=\sum_{\mathbf{k}}\tilde{a}^{\dagger}(\mathbf{k})\,.\,\hat{\varepsilon}(\mathbf{k})\,.\,a(\mathbf{k})+\frac{1}{2}\tilde{a}^{\dagger}(\mathbf{k})\,.\,\hat{\Delta}(\mathbf{k})\,.\,a^{\dagger}(-\mathbf{k})$$
$$+\frac{1}{2}\tilde{a}(-\mathbf{k})\,.\,\hat{\Delta}^{\dagger}(\mathbf{k})\,.\,a(\mathbf{k})+\frac{1}{2}\operatorname{Tr}\{\hat{\Delta}^{\dagger}(\mathbf{k})\,.\,\hat{x}(\mathbf{k})\}. \tag{4.46}$$

Making use of the anticommutation relations and of the symmetry between $\mathbf{k}$ and $-\mathbf{k}$ the first term of (4.46) can be rewritten:

$$\sum_{\mathbf{k}}\tilde{a}^{\dagger}(\mathbf{k})\,.\,\hat{\varepsilon}(\mathbf{k})\,.\,a(\mathbf{k})=\frac{1}{2}\sum_{\mathbf{k}}\tilde{a}^{\dagger}(\mathbf{k})\,.\,\hat{\varepsilon}(\mathbf{k})\,.\,a(\mathbf{k})$$
$$-\frac{1}{2}\sum_{\mathbf{k}}\tilde{a}(-\mathbf{k})\,.\,\hat{\varepsilon}(\mathbf{k})\,.\,a^{\dagger}(-\mathbf{k})+\frac{1}{2}\sum_{\mathbf{k}}\operatorname{Tr}\{\hat{\varepsilon}(\mathbf{k})\}. \tag{4.47}$$

Following Balian and Werthamer we introduce 4-components column spinors $\mathscr{A}(\mathbf{k})$ and $\mathscr{A}^+(\mathbf{k})$

$$\mathscr{A}(\mathbf{k})=\begin{pmatrix} a(\mathbf{k}) \\ a^{\dagger}(-\mathbf{k}) \end{pmatrix}=\begin{pmatrix} a_{\mathbf{k}\uparrow} \\ a_{\mathbf{k}\downarrow} \\ a^{\dagger}_{-\mathbf{k}\uparrow} \\ a^{\dagger}_{-\mathbf{k}\downarrow} \end{pmatrix}; \qquad \mathscr{A}^+(\mathbf{k})=\begin{pmatrix} a^{\dagger}(\mathbf{k}) \\ a(-\mathbf{k}) \end{pmatrix}=\begin{pmatrix} a^{\dagger}_{\mathbf{k}\uparrow} \\ a^{\dagger}_{\mathbf{k}\downarrow} \\ a_{-\mathbf{k}\uparrow} \\ a_{-\mathbf{k}\downarrow} \end{pmatrix}, \tag{4.48}$$

and the line spinors $\tilde{\mathscr{A}}(\mathbf{k})$ and $\tilde{\mathscr{A}}^{\dagger}(\mathbf{k})$. Using (4.47) the linearized Hamiltonian (4.46) can be rewritten in terms of the 4-components spinors (4.48):

$$\mathscr{H}=\tfrac{1}{2}\sum_{\mathbf{k}}\tilde{\mathscr{A}}^{\dagger}(\mathbf{k})\,.\,\hat{\hat{\mathscr{E}}}(\mathbf{k})\,.\,\mathscr{A}(\mathbf{k})+\tfrac{1}{2}\sum_{\mathbf{k}}\mathrm{Tr}\{\hat{\varepsilon}(\mathbf{k})+\hat{\Delta}^{\dagger}(\mathbf{k})\,.\,\hat{\chi}(\mathbf{k})\}, \tag{4.49}$$

where the 4×4 matrix $\hat{\hat{\mathscr{E}}}(\mathbf{k})$ is defined as:

$$\hat{\hat{\mathscr{E}}}(\mathbf{k})=\begin{pmatrix} \hat{\varepsilon}(\mathbf{k}) & \hat{\Delta}(\mathbf{k}) \\ \hat{\Delta}^{\dagger}(\mathbf{k}) & -\hat{\varepsilon}(\mathbf{k}) \end{pmatrix}. \tag{4.50}$$

From the relations $\hat{\varepsilon}(\mathbf{k})=\hat{\varepsilon}(-\mathbf{k})=\hat{\varepsilon}(-\mathbf{k})^*$ and $\hat{\Delta}^{\dagger}(\mathbf{k})=-\hat{\Delta}^*(-\mathbf{k})$ (consequence of (4.26) and of (4.44)) we find that an interchange between the first two and the last two lines and also between the first two and the last two columns turns $\hat{\hat{\mathscr{E}}}(\mathbf{k})$ into $-\hat{\hat{\mathscr{E}}}^*(-\mathbf{k})$. This operation can be expressed as:

$$\hat{\hat{\theta}}\hat{\hat{\mathscr{E}}}(\mathbf{k})\hat{\hat{\theta}}^{-1}=-\hat{\hat{\mathscr{E}}}^*(-\mathbf{k}), \tag{4.51}$$

where

$$\hat{\hat{\theta}}=\begin{pmatrix} 0 & \hat{1} \\ \hat{1} & 0 \end{pmatrix}.$$

(*b*) *Diagonalization of the energy matrix*

We want to transform the Hamiltonian (4.49) into a diagonal form by a substitution on the 4-components spinors $\mathscr{A}(\mathbf{k})$ and $\mathscr{A}^+(\mathbf{k})$:

$$\mathscr{A}(\mathbf{k})=\hat{\hat{U}}(\mathbf{k})A(\mathbf{k}), \qquad \tilde{\mathscr{A}}^{\dagger}(\mathbf{k})=\tilde{A}^{\dagger}(\mathbf{k})\hat{\hat{U}}^{\dagger}(\mathbf{k}); \tag{4.52}$$

$$A(\mathbf{k})=\begin{pmatrix} \alpha(\mathbf{k}) \\ \alpha^{\dagger}(-\mathbf{k}) \end{pmatrix}; \qquad A^{\dagger}(\mathbf{k})=\begin{pmatrix} \alpha^{\dagger}(\mathbf{k}) \\ \alpha(-\mathbf{k}) \end{pmatrix}. \tag{4.52$'$}$$

$A(\mathbf{k})$ are new 4-components spinors, $\alpha(\mathbf{k})$ new 2-components spinors. $\hat{\hat{U}}(\mathbf{k})$ is a unitary 4×4 matrix which from the relations (4.48), (4.51), and (4.52) is necessarily of the form:

$$\hat{\hat{U}}(\mathbf{k})=\begin{pmatrix} \hat{u}(\mathbf{k}) & \hat{v}(\mathbf{k}) \\ \hat{v}^*(-\mathbf{k}) & \hat{u}^*(-\mathbf{k}) \end{pmatrix}, \tag{4.53}$$

where $\hat{u}(\mathbf{k})$ and $\hat{v}(\mathbf{k})$ are unitary 2×2 matrices. $\hat{\hat{U}}(\mathbf{k})$ transforms the energy matrix $\hat{\hat{\mathscr{E}}}(\mathbf{k})$ into a real diagonal matrix $\hat{\hat{E}}(\mathbf{k})$ by the relation:

$$\hat{\hat{E}}(\mathbf{k})=\hat{\hat{U}}^{-1}(\mathbf{k})\hat{\hat{\mathscr{E}}}(\mathbf{k})\hat{\hat{U}}(\mathbf{k}). \tag{4.54}$$

From (4.51) and (4.53) we obtain:

$$\hat{\hat{\theta}}\hat{\hat{E}}(\mathbf{k})\hat{\hat{\theta}}^{-1}=-\hat{\hat{E}}^*(-\mathbf{k})=-\hat{\hat{E}}(-\mathbf{k}), \tag{4.55}$$

and $\hat{\hat{E}}(\mathbf{k})$ has thus the following form:

$$\hat{\hat{E}}(\mathbf{k})=\begin{pmatrix} E'_{\mathbf{k}} & & & \\ & E''_{\mathbf{k}} & & \\ & & -E'_{-\mathbf{k}} & \\ & & & -E''_{-\mathbf{k}} \end{pmatrix}=\begin{pmatrix}\hat{E}(\mathbf{k}) & \\ & -\hat{E}(-\mathbf{k})\end{pmatrix}, \tag{4.56}$$

where $\hat{E}(\mathbf{k})$ is the 2×2 matrix:

$$\hat{E}(\mathbf{k})=\begin{pmatrix}E'_{\mathbf{k}} & \\ & E''_{\mathbf{k}}\end{pmatrix}. \tag{4.56'}$$

The first part of the Hamiltonian (4.49) can now be written as:

$$\begin{aligned}\tfrac{1}{2}\sum_{\mathbf{k}}\tilde{A}^{\dagger}(\mathbf{k})\,.\,\hat{\hat{E}}(\mathbf{k})\,.\,A(\mathbf{k})&=\tfrac{1}{2}\sum_{\mathbf{k}}\tilde{\alpha}^{\dagger}(\mathbf{k})\,.\,\hat{E}(\mathbf{k})\,.\,\alpha(\mathbf{k})\\
&\quad-\tfrac{1}{2}\sum_{\mathbf{k}}\tilde{\alpha}(-\mathbf{k})\,.\,\hat{E}(-\mathbf{k})\,.\,\alpha^{\dagger}(-\mathbf{k})\\
&=\tfrac{1}{2}\sum_{\mathbf{k}}\{\tilde{\alpha}^{\dagger}(\mathbf{k})\hat{E}(\mathbf{k})\alpha(\mathbf{k})\\
&\quad+\tilde{\alpha}^{\dagger}(-\mathbf{k})\hat{E}(-\mathbf{k})\alpha(-\mathbf{k})-\mathrm{Tr}[\hat{E}(-\mathbf{k})]\}.\end{aligned} \tag{4.57}$$

We have used anticommutation relations for the last step of (4.57). The Hamiltonian $\mathscr{H}$ can be rewritten as:

$$\begin{aligned}\mathscr{H}&=\sum_{\mathbf{k}}\tilde{\alpha}^{\dagger}(\mathbf{k})\,.\,\hat{E}(\mathbf{k})\,.\,\alpha(\mathbf{k})+\tfrac{1}{2}\,\mathrm{Tr}\{\hat{\varepsilon}(\mathbf{k})-\hat{E}(\mathbf{k})+\hat{\Delta}^{\dagger}(\mathbf{k})\,.\,\hat{x}(\mathbf{k})\}\\
&=\sum_{\mathbf{k}}E'_{\mathbf{k}}\alpha'^{\dagger}_{\mathbf{k}}\alpha'_{\mathbf{k}}+E''_{\mathbf{k}}\alpha''^{\dagger}_{\mathbf{k}}\alpha''_{\mathbf{k}}+\tfrac{1}{2}\,\mathrm{Tr}\{\hat{\varepsilon}(\mathbf{k})-\hat{E}(\mathbf{k})+\hat{\Delta}^{\dagger}(\mathbf{k})\,.\,\hat{x}(\mathbf{k})\}.\end{aligned} \tag{4.58}$$

It is now the Hamiltonian of an assembly of independent fermions. It is convenient to introduce an index $\sigma=\pm1$ to label the operators α' and α'' and the energies E' and E'' and to rewrite (4.58) as:

$$\mathscr{H}=\sum_{\mathbf{k},\sigma}.\,E_{\mathbf{k}\sigma}\alpha^{\dagger}_{\mathbf{k}\sigma}\alpha_{\mathbf{k}\sigma}+\tfrac{1}{2}\,\mathrm{Tr}\{\hat{\varepsilon}(\mathbf{k})-\hat{E}(\mathbf{k})+\Delta^{\dagger}(\mathbf{k})\hat{x}(\mathbf{k})\}. \tag{4.58'}$$

The indices $\sigma=\pm1$ of the $\alpha_{\mathbf{k}\sigma}$ are not to be identified with the original indices $\uparrow$ and $\downarrow$ of the $a_{\mathbf{k}\sigma}$.

The calculation of the eigenvalues $E_{\mathbf{k}\sigma}$ which involves the diagonalization of the 4×4 matrix $\hat{\hat{\mathscr{E}}}(\mathbf{k})$ (4.50) is much simpler in the absence of a magnetic field. Then $\varepsilon_{\mathbf{k}\uparrow}=\varepsilon_{\mathbf{k}\downarrow}$ and the matrix $\hat{\varepsilon}(\mathbf{k})=\varepsilon_{\mathbf{k}}\hat{1}$ is a multiple of the unit matrix.

The square of the 4×4 matrix $\hat{\hat{\mathscr{E}}}(\mathbf{k})$ is:

$$\hat{\hat{\mathscr{E}}}(\mathbf{k})^2=\begin{pmatrix}\hat{\varepsilon}(\mathbf{k})^2+\hat{\Delta}(\mathbf{k})\hat{\Delta}^\dagger(\mathbf{k}) & [\hat{\varepsilon}(\mathbf{k}),\hat{\Delta}(\mathbf{k})]\\ [\hat{\Delta}^\dagger(\mathbf{k}),\hat{\varepsilon}(\mathbf{k})] & \hat{\varepsilon}(\mathbf{k})^2+\hat{\Delta}^\dagger(\mathbf{k})\hat{\Delta}(\mathbf{k})\end{pmatrix}. \tag{4.59}$$

If $\hat{\varepsilon}(\mathbf{k})$ is a multiple of the unit matrix, the off-diagonal elements in (4.59) vanish and $\hat{\hat{\mathscr{E}}}(\mathbf{k})$ breaks into two 2×2 matrices. The energies of the elementary excitations are then obtained from the eigenvalues of the 2×2 matrices:

$$\hat{\mathscr{E}}'(\mathbf{k})^2=\varepsilon_{\mathbf{k}}^2\hat{1}+\hat{\Delta}(\mathbf{k})\hat{\Delta}^\dagger(\mathbf{k});\qquad \hat{\mathscr{E}}''(\mathbf{k})^2=\varepsilon_{\mathbf{k}}^2\hat{1}+\hat{\Delta}^\dagger(\mathbf{k})\hat{\Delta}(\mathbf{k}). \tag{4.60}$$

We shall make a further simplification by assuming (in agreement with experiment) that with one exception, to be mentioned later in this chapter, the only states observed in ^{3}He are such that the matrix $\hat{\Delta}^\dagger(\mathbf{k})\hat{\Delta}(\mathbf{k})$ is a c-number $|\Delta(\mathbf{k})|^2\hat{1}$ (also equal naturally to $\hat{\Delta}(\mathbf{k})\hat{\Delta}^\dagger(\mathbf{k})$). We shall see later that these are the unitary states, already mentioned in **D**, for which the expectation value $\langle\mathbf{S}\rangle$ of the spin of the Cooper pairs vanishes. For these states, according to (4.60) the spectrum of the elementary excitations is given by:

$$E_{\mathbf{k}}^2=\varepsilon_{\mathbf{k}}^2+|\Delta(\mathbf{k})|^2. \tag{4.60$'$}$$

We choose in the following $E_{\mathbf{k}}$ to have the sign of $\varepsilon_{\mathbf{k}}$. In contrast to the normal fluid, the energy of the elementary excitations never falls below a minimum value $|\Delta(\mathbf{k})|$, whence the term of gap or energy gap used for Δ.

(*c*) *The self-consistent equation*

The Hamiltonian (4.58) is temperature-dependent since both its eigenvalues and eigenvectors are derived from the diagonalization of the energy matrix $\hat{\hat{\mathscr{E}}}(\mathbf{k})$ itself a function of the matrices $\hat{x}(\mathbf{k})$ which are temperature-dependent statistical averages of products of fermion operators $a_{\mathbf{k}\sigma}$ (eqn (4.25)). Nonetheless as pointed out in connection with eqn (4.17), the form of the Hamiltonian (4.58), as that of an assembly of free fermions is sufficient to predict the values of statistical average of products of operators: $\alpha'_{\mathbf{k}},\alpha''_{\mathbf{k}},\alpha'^\dagger_{\mathbf{k}},\alpha''^\dagger_{\mathbf{k}}$ (or as in (4.58$'$) $\alpha_{\mathbf{k}\sigma},\alpha^\dagger_{\mathbf{k}\sigma}$).

These are:

$$\begin{cases}\langle\alpha'^\dagger_{\mathbf{k}}\alpha'_{\mathbf{k}}\rangle=1-\langle\alpha'_{\mathbf{k}}\alpha'^\dagger_{\mathbf{k}}\rangle=[1+\exp(\beta E'_{\mathbf{k}})]^{-1}=f(E'_{\mathbf{k}})\\ \langle\alpha''^\dagger_{\mathbf{k}}\alpha''_{\mathbf{k}}\rangle=1-\langle\alpha''_{\mathbf{k}}\alpha''^\dagger_{\mathbf{k}}\rangle=[1+\exp(\beta E''_{\mathbf{k}})]^{-1}=f(E''_{\mathbf{k}}),\end{cases} \tag{4.61}$$

with all the other products of any two fermion operators being zero. The relations (4.61) can be transcribed in a more compact form using the 4-components spinors $\mathscr{A}(\mathbf{k})$ and $A(\mathbf{k})$. The eqn (4.25) had been written:

$$\langle a(\mathbf{k})\tilde{\mathbf{a}}(-\mathbf{k})\rangle=\hat{x}(\mathbf{k}). \tag{4.25$'$}$$

We build similarly 4×4 matrices from the 4-dimensional spinors $\mathscr{A}(\mathbf{k})$ of eqn (4.48):

$$\hat{\hat{X}}(\mathbf{k})=\langle\mathscr{A}(\mathbf{k})\tilde{\mathscr{A}}^{\dagger}(\mathbf{k})\rangle=\begin{pmatrix}\langle a(\mathbf{k})\tilde{a}^{\dagger}(\mathbf{k})\rangle & \langle a(\mathbf{k})\tilde{a}(-\mathbf{k})\rangle\\ \langle a^{\dagger}(-\mathbf{k})\tilde{a}^{\dagger}(\mathbf{k})\rangle & \langle a^{\dagger}(-\mathbf{k})\tilde{a}(-\mathbf{k})\rangle\end{pmatrix}. \tag{4.62}$$

We transcribe, similarly to (4.25′), (4.61) as:

$$\begin{aligned}\langle\alpha(\mathbf{k})\tilde{\alpha}^{\dagger}(\mathbf{k})\rangle&=\begin{pmatrix}\langle\alpha'(\mathbf{k})\alpha'^{\dagger}(\mathbf{k})\rangle & \\ & \langle\alpha''(\mathbf{k})\alpha''^{\dagger}(\mathbf{k})\rangle\end{pmatrix}=\begin{pmatrix}1-f(E'_{\mathbf{k}}) & \\ & 1-f(E''_{\mathbf{k}})\end{pmatrix},\\ &=\hat{1}-f(\hat{E}_{(\mathbf{k})})\end{aligned} \tag{4.63}$$

where $\hat{E}(\mathbf{k})$ is defined in (4.56), and,

$$\begin{aligned}\langle A(\mathbf{k})\tilde{A}^{\dagger}(\mathbf{k})\rangle&=\begin{pmatrix}\langle\alpha(\mathbf{k})\tilde{\alpha}^{\dagger}(\mathbf{k})\rangle & \\ & \langle\alpha^{\dagger}(-\mathbf{k})\tilde{\alpha}(-\mathbf{k})\rangle\end{pmatrix}=\begin{pmatrix}\hat{1}-f(\hat{E}(\mathbf{k})) & \\ & f(\hat{E}(-\mathbf{k}))\end{pmatrix}\\ &=\begin{pmatrix}\hat{1}-f(\hat{E}(\mathbf{k})) & \\ & \hat{1}-f(-\hat{E}(-\mathbf{k}))\end{pmatrix}=\hat{\hat{1}}-f(\hat{\hat{E}}(\mathbf{k})).\end{aligned} \tag{4.64}$$

From the form (4.61) of the Fermi function f, eqn (4.64) can be rewritten

$$\langle A(\mathbf{k})\tilde{A}^{\dagger}(\mathbf{k})\rangle=\frac{1}{2}\left\{\hat{\hat{1}}+\tanh\left(\frac{\beta\hat{\hat{E}}(\mathbf{k})}{2}\right)\right\}. \tag{4.64′}$$

Reversing the transforms (4.52) and (4.54) we get:

$$\langle\mathscr{A}(\mathbf{k})\tilde{\mathscr{A}}^{\dagger}(\mathbf{k})\rangle=\hat{\hat{1}}-\hat{\hat{f}}(\hat{\hat{\mathscr{E}}}(\mathbf{k}))=\frac{1}{2}\left\{\hat{\hat{1}}+\tanh\left(\frac{\beta\hat{\hat{\mathscr{E}}}(\mathbf{k})}{2}\right)\right\}. \tag{4.65}$$

The left-hand side of (4.65) is a 4×4 matrix written out in (4.62) and has as matrix elements 2×2 matrices which are themselves average statistical values of bilinear products of fermion operators. The right-hand side is by (4.50) a function of $\hat{\Delta}(\mathbf{k})$ itself related by (4.44) to the mean value matrix $\hat{x}(\mathbf{k})$. Equation (4.65) is thus a self-consistency equation from which the $\hat{x}(\mathbf{k})$ and the $\hat{\Delta}(\mathbf{k})$ must be obtained.

As for the calculation of the energy spectrum there is considerable simplification in the absence of a magnetic field. Not only $\hat{\hat{\mathscr{E}}}(\mathbf{k})^2$ as in (4.59) but any even matrix function of $\hat{\hat{\mathscr{E}}}(k)$ breaks into two 2×2 matrices.

In particular:

$$F(\hat{\hat{\mathscr{E}}}(\mathbf{k})^2)=\frac{\tanh\left(\dfrac{\beta}{2}\hat{\hat{\mathscr{E}}}(\mathbf{k})\right)}{\hat{\hat{\mathscr{E}}}(\mathbf{k})}=\begin{pmatrix}F(\hat{\mathscr{E}}'(\mathbf{k})^2) & \\ & F(\hat{\mathscr{E}}''(\mathbf{k})^2)\end{pmatrix}, \tag{4.66}$$

where $\hat{\mathscr{E}}'(\mathbf{k})^2$ and $\hat{\mathscr{E}}''(\mathbf{k})^2$ are given by (4.60). It is convenient to rewrite eqn (4.65) as:

$$2\langle \mathscr{A}(\mathbf{k})\tilde{\mathscr{A}}^{\dagger}(\mathbf{k})\rangle - \hat{\hat{1}} = \hat{\hat{\mathscr{E}}}(\mathbf{k})\frac{\tanh\left(\frac{\beta}{2}\hat{\hat{\mathscr{E}}}(\mathbf{k})\right)}{\hat{\hat{\mathscr{E}}}(\mathbf{k})} = \hat{\hat{\mathscr{E}}}(\mathbf{k})F(\hat{\hat{\mathscr{E}}}(\mathbf{k})^2). \tag{4.67}$$

Equation (4.67) is an equality between two 4×4 matrices built from elements which are 2×2 matrices. We replace in (4.67) the left-hand side by (4.62), $\hat{\hat{\mathscr{E}}}(\mathbf{k})$ by (4.50), and $F(\hat{\hat{\mathscr{E}}}(\mathbf{k})^2)$ by (4.66). Equating on both sides the (1, 2) matrix element (a 2×2 matrix itself) we find:

$$\langle a(\mathbf{k})\tilde{a}(-\mathbf{k})\rangle = \hat{x}(\mathbf{k}) = \hat{\Delta}(\mathbf{k})\frac{\tanh\left(\frac{\beta}{2}\hat{\mathscr{E}}''(\mathbf{k})\right)}{2\hat{\mathscr{E}}''(\mathbf{k})}. \tag{4.68}$$

If we multiply both sides of (4.68) by $V(\mathbf{k}, \mathbf{k}')$ and sum over $\mathbf{k}$ (and then interchange $\mathbf{k}$ and $\mathbf{k}'$) we obtain

$$\hat{\Delta}(\mathbf{k}) = -\sum_{\mathbf{k}'} V(\mathbf{k}, \mathbf{k}')\frac{\tanh\left(\frac{\beta}{2}\hat{\mathscr{E}}''(\mathbf{k}')\right)}{2\hat{\mathscr{E}}''(\mathbf{k}')}\hat{\Delta}(\mathbf{k}'). \tag{4.69}$$

This is the generalized BCS equation. For unitary states where $\hat{\Delta}\hat{\Delta}^{\dagger}$ is a c-number it simplifies further.

$$\left\{\begin{aligned} &\hat{\Delta}(\mathbf{k}) = -\sum_{\mathbf{k}'} V(\mathbf{k}, \mathbf{k}')\frac{\tanh\left(\frac{\beta}{2}E_{\mathbf{k}'}\right)}{2E_{\mathbf{k}'}}\hat{\Delta}(\mathbf{k}') \\ &E_{\mathbf{k}}^2 = |\Delta(\mathbf{k})|^2 + \varepsilon_{\mathbf{k}}^2 \\ &\hat{x}(\mathbf{k}) = \hat{\Delta}(\mathbf{k})\frac{\tanh\left(\frac{\beta}{2}E_{\mathbf{k}}\right)}{2E_{\mathbf{k}}}. \end{aligned}\right. \tag{4.70}$$

Equation (4.70) is the BCS equation for unitary states. It must be stressed that eqns (4.68), (4.69), and (4.70) are valid *only* in zero magnetic field.

F. Solutions of the BCS equations and the Ginzburg–Landau approximation

(*a*) *The critical temperature*

We shall deal in this section with unitary states only, described by the eqn (4.70). The critical temperature $T_c = (k_B\beta_c)^{-1}$ is that for which the order parameter $\hat{x}(\mathbf{k})$ and the gap $\Delta(\mathbf{k})$ vanish so that the first eqn (4.70) becomes:

$$\hat{\Delta}(\mathbf{k}) = -\sum_{\mathbf{k}'} V(\mathbf{k}, \mathbf{k}') \frac{\tanh(\frac{1}{2}\beta_c \varepsilon_{\mathbf{k}'})}{2\varepsilon_{\mathbf{k}'}} \hat{\Delta}(\mathbf{k}'). \tag{4.71}$$

This is a linear equation for $\hat{\Delta}(\mathbf{k})$ and it is convenient to expand into spherical harmonics the potential $V(k, k')$ which depends only on the angle between the vectors $\mathbf{k} = k\mathbf{n}$ and $\mathbf{k}' = k'\mathbf{n}'$:

$$V(\mathbf{k}, \mathbf{k}') = \sum_{l} V_l(k, k') \sum_{m} Y_l^m(\mathbf{n}) Y_l^{-m}(\mathbf{n}'). \tag{4.72}$$

Let $Y_{l'}(\mathbf{n}) = \sum_m a_m Y_{l'}^m(\mathbf{n})$ be any combination of harmonics of order l'. It is clear that:

$$\int \mathrm{d}\Omega' \sum_{m} Y_l^m(\mathbf{n}) Y_l^{-m}(\mathbf{n}') Y_{l'}(\mathbf{n}') = Y_l(\mathbf{n}) \delta_{ll'}.$$

If we look for a solution of (4.71) of the form:

$$\hat{\Delta}(\mathbf{k}) = \hat{\Delta}_l(k) Y_l(\mathbf{n}), \tag{4.73}$$

$\hat{\Delta}_l(k)$ is a solution of the equation:

$$\hat{\Delta}_l(k) = -\sum_{k'} V_l(k, k') \frac{\tanh(\frac{1}{2}\beta_c \varepsilon_{k'})}{2\varepsilon_{k'}} \hat{\Delta}_l(k'). \tag{4.74}$$

The linear character of eqn (4.71) makes it possible to treat separately the various spherical harmonics V_l in the expansion (4.72) of the potential. It is important to notice that for a given l all the solutions (4.73) have the same critical temperature, whatever the angular part $Y_l(\mathbf{n})$ of (4.73). Naturally components $V_l(k, k')$ of the interaction energy with different values of l yield different values for the critical temperature.

The calculation of β_c can be performed very simply if one chooses for $V_l(k, k')$ the so-called BCS model:

$$V_l(k, k') = -V_l \ (V_l > 0) \quad \text{for } |\varepsilon_k|, |\varepsilon_{k'}| < \varepsilon_0,$$

where ε_0 is a certain cut-off energy and $V_l(k, k') = 0$ for other values of ε_k, $\varepsilon_{k'}$. It is assumed that ε_0 is smaller than the Fermi energy ε_F and much larger than the critical energy $k_B T_c$. It follows from (4.74) that $\hat{\Delta}_l(k)$ has the same dependence on energy as V_l (a constant value $\hat{\Delta}_l$ for $|\varepsilon_k| < \varepsilon_0$ and zero elsewhere).

The eqn (4.74) reduces to:

$$1 = V_l \sum_{k'}{}' \frac{\tanh(\frac{1}{2}\beta_c \varepsilon_{k'})}{2\varepsilon_{k'}}, \tag{4.75}$$

where $\sum'$ means a sum restricted to $|\varepsilon_{k'}| < \varepsilon_0$.

$$1 = V_l \frac{1}{2}\left(\frac{dn}{d\varepsilon}\right)_0 \int_{-\varepsilon_0}^{\varepsilon_0} \frac{\tanh(\frac{1}{2}\beta_c\varepsilon)}{2\varepsilon}\, d\varepsilon$$

$$= V_l \frac{1}{2}\left(\frac{dn}{d\varepsilon}\right)_0 \int_0^{\beta_c\varepsilon_0/2} \frac{\tanh x}{x}\, dx, \qquad (4.76)$$

where $\beta_c\varepsilon_0 \gg 1$.

The integral in (4.76) reduces to:

$$\approx \ln\left(\frac{\beta_c\varepsilon_0}{2}\right) - \int_0^{\beta_c\varepsilon_0/2} \ln x \,\mathrm{sech}^2 x\, dx$$

$$\approx \ln\left(\frac{\beta_c\varepsilon_0}{2}\right) - \int_0^{\infty} \ln x \,\mathrm{sech}^2 x\, dx = \ln(1{\cdot}14\beta_c\varepsilon_0),$$

whence the famous BCS relation:

$$1 = V_l \frac{1}{2}\left(\frac{dn}{d\varepsilon}\right)_0 \ln(1{\cdot}14\beta_c\varepsilon_0),$$

or:

$$k_B T_c = 1{\cdot}14\varepsilon_0 \exp\left\{\frac{-1}{V_l \frac{1}{2}(dn/d\varepsilon)_0}\right\}. \qquad (4.77)$$

Equation (4.77) confirms the statement made in section **C** about the singular dependence of the critical temperature on the strength of the attractive potential. This very steep dependence of β_c on V_l makes it possible to retain in the expansion (4.72) of the potential the term with the largest V_l and the highest critical temperature disregarding all the others.

(b) Solutions below the critical temperature

Below the critical temperature the equation for $\hat{\Delta}(\mathbf{k})$ is not linear any more. The relevant equation is now (4.70) where on the right-hand side $E_{\mathbf{k}'}$ depends explicitly on $|\Delta(\mathbf{k}')|^2$. Although not linear, it is still separable: if we assume that the potential energy reduces to a single spherical harmonic in the expansion (4.72) it is easy to verify that the various solutions of the BCS equations contain only spherical harmonics of order l. However they are not degenerate any more, and to discriminate among the various $\hat{\Delta}_l(\mathbf{k})$, obtained for a given $V_l(k, k')$, with respect to their relative stability, it will be necessary to compute their free energy.

Once the gap $\hat{\Delta}_l(\mathbf{k})$ is calculated from the first eqn (4.70), the last eqn (4.70) gives the order parameter. Below the critical temperature but very near to it $\hat{\Delta}_l(\mathbf{k})$ is very small, $E_{\mathbf{k}}$ is very nearly $\varepsilon_{\mathbf{k}}$ and $\hat{x}(\mathbf{k})$ is proportional to $\hat{\Delta}_l(\mathbf{k})$. Like the latter it will have no admixture from harmonics $l' \neq l$.

However as one goes down in temperature $\tanh((\beta/2)E_{\mathbf{k}})/2E_{\mathbf{k}}$ becomes more and more dependent on $|\Delta_l(\mathbf{k})|^2$ with an angular dependence which admixes into $\hat{x}(\mathbf{k})$ harmonics with $l' \neq l$. Still, sufficiently near to T_c this admixture can be disregarded and we shall treat in the following both $\hat{\Delta}(\mathbf{k})$ and $\hat{x}(\mathbf{k})$ as purely l-functions.

To sum up:

(a) The critical temperature depends sharply on the magnitude V_l of the attractive potential and we retain in the expansion (4.72) only the term with the highest $V_l > 0$.

(b) For a given V_l all the states $\hat{\Delta}_l(\mathbf{k})$ solutions of the BCS equation have the same critical temperature and their angular variation is undetermined.

(c) For a potential V_l with a single value l of the orbital momentum below the critical temperature, the various solutions are still made of l-harmonics only. However their degeneracy is lifted by the non-linear character of the BCS equation and they have different free energies.

(d) The order parameter is strictly proportional to the gap at the critical temperature and thus also strictly an l-function. Below the critical temperature T_c it contains admixtures from other harmonics l' which are small if $(T_c - T)/T_c$ is small.

We recall that we had defined in (4.33) a vector $\mathbf{x}(\mathbf{k})$ by the relation:

$$\hat{x}(\mathbf{k}) = (\mathbf{x}(\mathbf{k}) \cdot \hat{\boldsymbol{\sigma}})\hat{\sigma}_2, \tag{4.78}$$

a relation which can be reversed:

$$\mathbf{x}(\mathbf{k}) = \tfrac{1}{2}\,\mathrm{Tr}\{\hat{x}(\mathbf{k})\hat{\sigma}_2\hat{\boldsymbol{\sigma}}\} \tag{4.78'}$$

to give:

$$\sum_k \mathbf{x}(\mathbf{k}) = \psi\mathbf{d}(\mathbf{n}),$$

where $\mathbf{d}(\mathbf{n})$ is normalized:

$$\int |\mathbf{d}(\mathbf{n})|^2 \frac{\mathrm{d}\Omega}{4\pi} = 1.$$

We can similarly define:

$$\boldsymbol{\Delta}(\mathbf{k}) = \tfrac{1}{2}\,\mathrm{Tr}\{\hat{\Delta}(\mathbf{k})\hat{\sigma}_2\hat{\boldsymbol{\sigma}}\}. \tag{4.78''}$$

In the BCS model where the gap does not depend on the magnitude of $\mathbf{k}$ we can write:

$$\boldsymbol{\Delta}(\mathbf{k}) = \Delta\mathbf{d}'(\mathbf{n}) \tag{4.79}$$

where $\mathbf{d}'(\mathbf{n})$ is normalized.

If we are sufficiently near the critical temperature to be able to replace in the last eqn (4.70) $E_{\mathbf{k}}$ by ε_k we obtain:

$$\mathbf{x}(\mathbf{k}) = \boldsymbol{\Delta}(\mathbf{k})\frac{\tanh(\frac{1}{2}\beta_c\varepsilon_k)}{2\varepsilon_k} = \Delta\mathbf{d}'(\mathbf{n})\frac{\tanh(\frac{1}{2}\beta_c\varepsilon_k)}{2\varepsilon_k}. \tag{4.80}$$

Summing (4.80) over k we get from (4.36):

$$\psi\mathbf{d}(\mathbf{n}) \simeq \Delta\mathbf{d}'(n)\sum_k{}' \frac{\tanh(\tfrac{1}{2}\beta_c\varepsilon_k)}{2\varepsilon_k}, \tag{4.81}$$

which shows that the normalized vectors: $\mathbf{d}(\mathbf{n})$ for the order parameter (eqn (4.36)) and $\mathbf{d}'(\mathbf{n})$ for the gap are equal. This is an exact relation at the critical temperature and an approximate one not too much below. It follows then from (4.75) to (4.77) and (4.81) that:

$$\psi = (V_l)^{-1}\Delta = \frac{\Delta}{2}\left(\frac{\mathrm{d}n}{\mathrm{d}\varepsilon}\right)_0 \ln(1{\cdot}14\beta_c\varepsilon_0). \tag{4.82}$$

The gap matrix $\hat{\Delta}(\mathbf{k})$ can be written in the BCS model,

$$\hat{\Delta}(\mathbf{k}) = \Delta\{\mathbf{d}(\mathbf{n})\,.\,\hat{\boldsymbol{\sigma}}\}\hat{\sigma}_2. \tag{4.83}$$

Both ψ defined in (4.82) and Δ defined in (4.79) are functions of the temperature, $\psi(T)$ and $\Delta(T)$, which vanish at the critical temperature. Their temperature dependence is obtained from the Ginzburg–Landau equations that we consider next.

Remark In the singlet state which is not observed in superfluid ^{3}He but exists in superconductors, it was stated in section **D** that the order parameter $\hat{x}(\mathbf{k})$ is an antisymmetric matrix $x_0(\mathbf{k})\hat{\sigma}_2$. In analogy with (4.83) the gap matrix for the singlet is:

$$\hat{\Delta}(\mathbf{k}) = \Delta d_0(\mathbf{n})\hat{\sigma}_2, \tag{4.83'}$$

where $d_0(\mathbf{n})$ is invariant through rotation of the spin axes.

(c) *The free energy and the Ginzburg–Landau approximation*

The expression for the free energy can be written straightaway using the free fermions Hamiltonian (4.58) and the formula (4.16).

$$F = \sum_{\mathbf{k}} -\frac{1}{\beta}\{\ln[1+\exp(-\beta E'_{\mathbf{k}})] + \ln[1+\exp(-\beta E''_{\mathbf{k}})]\}$$
$$+\tfrac{1}{2}\,\mathrm{Tr}\{\hat{\varepsilon}(\mathbf{k}) - \hat{E}(\mathbf{k}) + \hat{\Delta}^{\dagger}(\mathbf{k})\hat{x}(\mathbf{k})\}, \tag{4.84}$$

where:

$$\hat{E}(\mathbf{k}) = \begin{pmatrix} E'(\mathbf{k}) & \\ & E''(\mathbf{k}) \end{pmatrix}.$$

In the absence of a magnetic field and for unitary states,

$$E'(\mathbf{k}) = E''(\mathbf{k}) = E(\mathbf{k}) = \pm[\varepsilon_{\mathbf{k}}^2 + |\Delta(\mathbf{k})|^2]^{1/2}$$

(eqn (4.70)) and F can be rewritten:

$$F = -\frac{2}{\beta}\sum_{\mathbf{k}}\{\ln[1+\exp-(\beta E_{\mathbf{k}})]\} + \sum_{\mathbf{k}}\tfrac{1}{2}\mathrm{Tr}\{\hat{\varepsilon}(\mathbf{k}) - \hat{E}(\mathbf{k}) + \hat{\Delta}^{\dagger}(\mathbf{k})\hat{x}(\mathbf{k})\}$$

$$= -\frac{2}{\beta}\sum_{\mathbf{k}}[\ln\cosh(\tfrac{1}{2}\beta E_{\mathbf{k}}) + \ln 2] + \sum_{\mathbf{k}}(\varepsilon(\mathbf{k}) + \tfrac{1}{2}\mathrm{Tr}(\hat{\Delta}^{\dagger}(\mathbf{k})\,.\,\hat{x}(\mathbf{k})). \qquad (4.85)$$

It can be shown that F considered as a functional of $\hat{\Delta}(\mathbf{k})$ (to which $\hat{x}(\mathbf{k})$ is related by (4.44)), is stationary for small variations of $\hat{\Delta}(\mathbf{k})$ when the latter obeys the BCS eqn (4.70). We shall not reproduce this proof here.

In the neighbourhood of the critical temperature T_c the difference $(F - F_0)$ between the free energy of the superfluid and that of the normal state, which is a functional of $\hat{\Delta}(\mathbf{k})$ can be expanded in powers of the latter. For brevity we shall write F for $F - F_0$. After some algebra that we omit the following result is obtained, for unitary states (and per unit volume):

$$\frac{F}{V} = \frac{1}{2}\left(\frac{dn}{d\varepsilon}\right)_0\left\{\frac{T-T_c}{T_c}\Delta^2 + \frac{1}{2}\kappa\frac{1}{(3{\cdot}06)^2}\frac{1}{(k_B T_c)^2}\Delta^4\right\}. \qquad (4.86)$$

The Δ in (4.86) is that defined in eqn (4.79) by:

$$\mathbf{\Delta}(\mathbf{k}) = \Delta\mathbf{d}'(\mathbf{n}) \cong \Delta\mathbf{d}(\mathbf{n})$$

where:

$$\int |\mathbf{d}(\mathbf{n})|^2\frac{d\Omega}{4\pi} = 1,$$

and $\mathbf{d}(\mathbf{n})$ is defined in eqn (4.36).

and

$$\begin{cases} \kappa = \displaystyle\int |\mathbf{d}(\mathbf{n})|^4\frac{d\Omega}{4\pi}, \\ \displaystyle\frac{1}{(3{\cdot}06)^2} = \frac{1}{9{\cdot}3} = -\frac{1}{8}\int_0^{\infty}\frac{d}{dz}\left(\frac{\tanh z}{z}\right)\frac{dz}{z}. \end{cases} \qquad (4.87)$$

(The last integral in (4.87) originates from an expansion of the integral $\int_0^{\infty}[\tanh(\tfrac{1}{2}\beta E_{\mathbf{k}})\,d\varepsilon/E_{\mathbf{k}}]$ in powers of Δ^2.)

One should not take eqn (4.86) too seriously. It is a consequence of the weak-coupling approximation, which as we shall see shortly is not valid in superfluid ^{3}He. Minimizing (4.86) with respect to Δ^2 it is easy to find:

$$\Delta = \left(1 - \frac{T}{T_c}\right)^{1/2} k_B T_c (3{\cdot}06)\kappa^{-1/2}. \qquad (4.88)$$

The corresponding minimum of the free energy is given by:

$$\frac{F - F_0}{V} = -\frac{(3{\cdot}06)^2}{4}\left(\frac{dn}{d\varepsilon}\right)_0 (k_B T_c)^2\left(1 - \frac{T}{T_c}\right)^2\frac{1}{\kappa}. \qquad (4.89)$$

If we remember that $(dn/d\varepsilon)_0 = (3N/2k_BT_F)$, we see that the free energy per quasi-particle is of the order of:

$$k_BT_c\left(\frac{T_c}{T_F}\right)\left(1-\frac{T}{T_c}\right)^2.$$

(d) The $l = 1$ solutions of the BCS equation: the ABM and BW phases

It is well established now, in particular from susceptibility measurements, that the Cooper pairs in superfluid ^{3}He are in a triplet state. The lowest value of l compatible with this fact is $l = 1$: the pairs are in a p-state. This assignment is in good agreement with experiment and will be made henceforth. It can be expressed in a simple form by stating that the normalized order parameter $\mathbf{d}(\mathbf{n})$ is a linear function of $\mathbf{n}$ the unit vector of the wave-vector $\mathbf{k}$:

$$d_i = \sum_\alpha A_{i\alpha} n_\alpha, \tag{4.90}$$

where $A_{i\alpha}$ is a complex matrix.

The problem of finding the solutions of the BCS equation for p-waves is greatly simplified by the assumption that the states are unitary. It can be shown then that there are 4 types of solutions each of which yields a different value for the free energy (4.85) or (4.86). In the latter expression the differences between the values of the free energy for the different solutions are contained in the anisotropy constant $\kappa = \int |\mathbf{d}(\mathbf{n})|^4 \, d\Omega/4\pi$. A rotation in spin space and/or in co-ordinate space represented respectively by the substitutions: $d_i = R_{ij}d'_j$ and $n_\alpha = Q_{\alpha\beta}n'_\beta$ where R and Q are orthogonal matrices, transforms a solution into a solution of the same type. In the absence of external anisotropy such as the presence of a magnetic field or a wall, and also of a spin–orbit coupling energy such as, as we shall see later, the dipolar magnetic interaction, all the solutions of the same type have the same free energy. Even so, unless the spin and space rotations R and Q are the same, they have different properties. As stated in the previous section the 4 types of solution have the same critical temperature but different domains of stability below T_c. We now list the analytical expressions (4.90) for the 4 solutions. The fact that they are solutions of the BCS equation can be verified by direct inspection.

The Balian–Werthamer state (BW)

A particularly simple solution of this type is:

$$\mathbf{d}(\mathbf{n}) = \mathbf{n},$$

and the more general one:

$$\mathbf{d}(\mathbf{n}) = R\mathbf{n}, \tag{4.91}$$

where R is an orthogonal matrix. It is clear that $|\mathbf{d}(\mathbf{n})|^2 = 1$ as well as $\langle|\mathbf{d}(\mathbf{n})|^2\rangle = \int |\mathbf{d}|^2 \, \mathrm{d}\Omega/4\pi$ and $\kappa = \int |\mathbf{d}(\mathbf{n})|^4 \, \mathrm{d}\Omega/4\pi$.

The Anderson–Brinkman–Morel state (ABM)
The simplest form is:

$$\mathbf{d}(\mathbf{n}) = \begin{cases} 0 \\ 0 \\ \sqrt{(\tfrac{3}{2})}(n_x + \mathrm{i}n_y); \end{cases} \tag{4.92}$$

$$|\mathbf{d}(\mathbf{n})|^2 = \tfrac{3}{2}(n_x^2 + n_y^2); \qquad \langle|\mathbf{d}(\mathbf{n})|^2\rangle = 1;$$

$$\langle|\mathbf{d}(\mathbf{n})|^4\rangle = \kappa = \tfrac{6}{5}.$$

The more general form of the ABM solution is:

$$\mathbf{d}(\mathbf{n}) = (\tfrac{3}{2})^{1/2} \, \mathbf{d}(\boldsymbol{\alpha}_1 \, . \, \mathbf{n} + \mathrm{i}\boldsymbol{\alpha}_2 \, . \, \mathbf{n}). \tag{4.93}$$

$\boldsymbol{\alpha}_1$, $\boldsymbol{\alpha}_2$, $\mathbf{d}$ are three constant unitary vectors with $(\boldsymbol{\alpha}_1 \, . \, \boldsymbol{\alpha}_2) = 0$.

For each pair d($\mathbf{n}$) one can define a vector $\mathbf{l}$ which has the form of an orbital momentum:

$$\mathbf{l} = \frac{1}{i} \int \frac{\mathrm{d}\Omega}{4\pi} \sum_i d_i^*(\mathbf{n}) \left(\mathbf{n} \wedge \frac{\partial}{\partial \mathbf{n}} \right) d_i(\mathbf{n}). \tag{4.94}$$

It follows from (4.93) that the vector $\mathbf{l}$ of the ABM pairs is

$$\mathbf{l} = \boldsymbol{\alpha}_1 \wedge \boldsymbol{\alpha}_2, \tag{4.95}$$

and that $\mathbf{l}$ defines in $\mathbf{n}$ space the direction along which the ABM order parameter, and the gap vanish.

The planar solution

$$d_x = (\tfrac{3}{2})^{1/2} n_x, \qquad d_y = (\tfrac{3}{2})^{1/2} n_y, \qquad d_z = 0;$$

$$\kappa = \tfrac{6}{5}.$$

The polar solution

$$d_z = \sqrt{(3)} n_z, \qquad d_x = d_y = 0;$$

$$\kappa = \tfrac{9}{5}.$$

From the values of κ listed above and the free energy formula (4.89) one would expect that below the critical temperature the stablest solution would be BW, followed by ABM and planar, and then by polar. This is not what happens. We saw in the introduction that just below T_c the phase A identified with ABM is more stable than the phase B identified with BW.

The $\mathbf{A}_1$ *Phase*
This phase in a magnetic field is, near the transition temperature, more stable than the phase ABM. It is assumed that its orbital wave function has

the same form (4.93) as the phase ABM and that it contains superfluid pairs with a single value of S_z along the field. It is not known at present whether it is $S_z = \pm 1$. Let us assume for argument's sake that it is $S_z = +1$.

Its order parameter can be obtained from eqn (4.37):

$$d_2 + \mathrm{i}d_1 = \sqrt{2}; \qquad d_2 - \mathrm{i}d_1 = 0; \qquad d_3 = 0$$

or:

$$d_1 = -\frac{\mathrm{i}}{\sqrt{2}}; \qquad d_2 = \frac{1}{\sqrt{2}}; \qquad d_3 = 0.$$

More generally for an arbitrary orientation of the field, **d** in (4.93) is a complex vector with:

$$\mathbf{d} = \mathbf{d}' + \mathrm{i}\mathbf{d}''; \qquad |\mathbf{d}'|^2 = |\mathbf{d}''|^2 = \tfrac{1}{2}; \qquad (\mathbf{d}' \cdot \mathbf{d}'') = 0. \tag{4.96}$$

A_1 is clearly a non unitary state. From eqn (4.83) we get:

$$\begin{aligned}[\hat{\Delta}, \hat{\Delta}^\dagger] &= [\Delta(\mathbf{d} \cdot \hat{\boldsymbol{\sigma}})\hat{\sigma}_2, \Delta\hat{\sigma}_2(\mathbf{d}^* \cdot \boldsymbol{\sigma})] \\ &= 2\mathrm{i}\Delta^2 \hat{\boldsymbol{\sigma}} \cdot (\mathbf{d} \wedge \mathbf{d}^*) = 4\Delta^2 \hat{\boldsymbol{\sigma}} \cdot (\mathbf{d}' \wedge \mathbf{d}''),\end{aligned} \tag{4.97}$$

$\hat{\Delta} \cdot \hat{\Delta}^\dagger$ is *not* a c-number.

G. Magnetic susceptibility

(a) Susceptibility of a single pair **d(n)**

In the presence of a small field **H**, the magnetization is a linear function of the field:

$$\mathbf{M} = \chi \mathbf{H} \tag{4.98}$$

whence:

$$\chi = \left(\frac{\mathrm{d}\mathbf{M}}{\mathrm{d}\mathbf{H}}\right)_{\mathbf{H}=0}. \tag{4.99}$$

In this section we select Cooper pairs with a given orientation of **n** and thus of **d(n)** and sum only over the length k of the wave vectors. As a shorthand notation the total susceptibility of all pairs with given **n** is called the susceptibility of the pair **d(n)**. In the next section we will average the susceptibilities over all orientations of **n** and **d(n)** in the superfluid phase. Since the susceptibility is a tensor it will be sufficient to consider $\mathbf{d}(\mathbf{n}) \| \mathbf{H}$ and $\mathbf{d}(\mathbf{n}) \perp \mathbf{H}$. We choose the z-axis along the applied field.

The magnetization of the set of fermions that we consider is given by:

$$M_z = \mu \sum_k \{\langle a^\dagger_{\mathbf{k}\uparrow} a_{\mathbf{k}\uparrow}\rangle - \langle a^\dagger_{\mathbf{k}\downarrow} a_{\mathbf{k}\downarrow}\rangle\} = \mu \sum_k \langle \tilde{a}^\dagger_{\mathbf{k}} \hat{\sigma}_z a_{\mathbf{k}}\rangle, \tag{4.100}$$

where $\mu = \gamma\hbar/2$.

An obvious method for calculating (4.100) would be to express it as a function of the elementary excitation occupation numbers $\langle\alpha^{\dagger}_{\mathbf{k}\sigma}\alpha_{\mathbf{k}\sigma}\rangle$ and $\langle\alpha^{\dagger}_{-\mathbf{k}\sigma}\alpha_{-\mathbf{k}\sigma}\rangle$ and to use for the latter the thermal averages (4.61). We will see below how this complicated procedure can be avoided. The existence of a non-vanishing magnetization is a consequence of the change induced by the magnetic field on the 4×4 energy matrix (4.50). In the low-field limit where we look for a magnetization M_z proportional to H, it is sufficient to retain only the terms linear in field in the modified energy matrix.

The effect of the field is to replace in (4.50) $\hat{\varepsilon}$ by $\hat{\varepsilon}-\mu H\hat{\sigma}_3$, and also to induce a change in the gap $\hat{\Delta}=\Delta(\mathbf{d}\,.\,\hat{\boldsymbol{\sigma}})\hat{\sigma}_2$. It will be shown in the appendix at the end of this section that the latter change is negligible in small fields, so that we keep for $\hat{\Delta}_{\mathbf{k}}$ and $\hat{\Delta}^{\dagger}_{\mathbf{k}}$ the same form as in zero field. We use then an energy matrix of the form:

$$\hat{\hat{\mathscr{E}}}(\mathbf{k},H)=\hat{\hat{\mathscr{E}}}(\mathbf{k},0)-\mu H\begin{pmatrix}\hat{\sigma}_3 & \\ & -\hat{\sigma}_3\end{pmatrix}. \tag{4.101}$$

The problem of diagonalizing (4.101) for the purpose of calculating M_z can be short-circuited by noting that, according to (4.48) and to the form (4.101) of the field-dependence of $\hat{\hat{\mathscr{E}}}$ on H, eqn (4.100) can be rewritten as:

$$M_z=-\tfrac{1}{2}\sum_k\left\langle\tilde{\mathscr{A}}^{\dagger}(\mathbf{k})\,.\,\frac{\partial}{\partial H}\,.\,\hat{\hat{\mathscr{E}}}(\mathbf{k},H)\,.\,\mathscr{A}(\mathbf{k})\right\rangle,$$

(using the anticommutation relations of the operators $a^{\dagger}_{\mathbf{k}}$ and $a_{\mathbf{k}}$), or else, through the unitary transformation that diagonalizes $\hat{\hat{\mathscr{E}}}(\mathbf{k})$:

$$\begin{aligned}M_z&=-\tfrac{1}{2}\sum_k\left\langle\tilde{A}^{\dagger}(\mathbf{k})\,.\,\frac{\partial}{\partial H}\hat{\hat{E}}(\mathbf{k},H)\,.\,A(\mathbf{k})\right\rangle\\&=\sum_{k,\sigma}\langle\alpha^{\dagger}_{\mathbf{k}\sigma}\alpha_{\mathbf{k}\sigma}\rangle\frac{\partial}{\partial H}E(\mathbf{k},\sigma)\\&=\sum_{k,\sigma}f(E_{\mathbf{k}\sigma})\frac{\partial}{\partial H}E(\mathbf{k},\sigma).\end{aligned} \tag{4.102}$$

$\sigma=\pm1$ is the label index introduced in (4.58′).

(α) **d** *parallel to* **H**, $d_1=d_2=0$

This is the case when the pair is in a state $S_z=0$. We have:

$$\hat{\Delta}=\Delta d_3\hat{\sigma}_3\hat{\sigma}_2=\mathrm{i}\Delta d_3\hat{\sigma}_1=\mathrm{i}\Delta_3\hat{\sigma}_1,$$

and the 4×4 energy matrix $\hat{\hat{\mathscr{E}}}$ can be written:

$$\hat{\hat{\mathscr{E}}}(\mathbf{k},H)=\begin{pmatrix}\varepsilon_k\hat{1}-\mu H\hat{\sigma}_3 & \mathrm{i}\Delta_3\hat{\sigma}_1\\ -\mathrm{i}\Delta_3^{*}\hat{\sigma}_1 & -(\varepsilon_k\hat{1}-\mu H\hat{\sigma}_3)\end{pmatrix}. \tag{4.103}$$

Writing out explicitly this 4×4 matrix it is easy to verify that it factors into two 2×2 matrices and to find its eigenvalues:

$$\begin{cases} E_{\mathbf{k}\sigma} = E_{\mathbf{k}} - \mu H\sigma \\ E_{\mathbf{k}} = \pm[\varepsilon_k^2 + |\Delta_3|^2]^{1/2}, \end{cases} \tag{4.104}$$

where $E_{\mathbf{k}}$ has the same sign as ε_k, and $\sigma = \pm 1$.

According to (4.102) and (4.104) we have:

$$M_z = \mu \sum_{k,\sigma} \sigma f(E_{\mathbf{k}\sigma}) \tag{4.104'}$$

and:

$$\chi = \left.\frac{\mathrm{d}M_z}{\mathrm{d}H}\right)_{H=0} = \mu \sum_{k,\sigma} \sigma \left.\frac{\partial f(E_{\mathbf{k}\sigma})}{\partial E_{\mathbf{k}\sigma}} \frac{\partial E_{\mathbf{k}\sigma}}{\partial H}\right)_{H=0},$$

$$= -\mu^2 \sum_{k,\sigma} \left.\frac{\partial f(E_{\mathbf{k}\sigma})}{\partial E_{\mathbf{k}\sigma}}\right)_{H=0},$$

or else, since $H = 0$:

$$\chi = -2\mu^2 \sum_k \frac{\partial f(E_{\mathbf{k}})}{\partial E_{\mathbf{k}}}$$

$$= -2\mu^2 \frac{1}{2}\left(\frac{\mathrm{d}n}{\mathrm{d}\varepsilon}\right)_0 \times \int_{-\infty}^{\infty} \frac{\partial f}{\partial E_{\mathbf{k}}}\,\mathrm{d}\varepsilon. \tag{4.105}$$

From:

$$f(E_{\mathbf{k}}) = [\exp(\beta E_{\mathbf{k}}) + 1]^{-1},$$

we get:

$$\frac{\partial f}{\partial E_{\mathbf{k}}} = -\frac{\beta}{4}\operatorname{sech}^2\left(\frac{\beta E_{\mathbf{k}}}{2}\right),$$

an even function of $E_{\mathbf{k}}$ and of ε, whence:

$$\chi = \frac{\gamma^2\hbar^2}{4}\left(\frac{\mathrm{d}n}{\mathrm{d}\varepsilon}\right)_0 \int_0^{\infty} \frac{\beta}{2}\operatorname{sech}^2\left(\frac{\beta E_{\mathbf{k}}}{2}\right)\mathrm{d}\varepsilon$$

$$= \chi_{n_0} Y(\mathbf{n}, T). \tag{4.106}$$

$Y(\mathbf{n}, T)$ is the so-called Yosida function:

$$Y(\mathbf{n}, T_c) = \int_0^{\infty} \frac{\beta_c}{2}\operatorname{sech}^2\left(\frac{\beta_c \varepsilon_k}{2}\right)\mathrm{d}\varepsilon = 1. \tag{4.107}$$

In the neighbourhood of T_c an expansion of (4.106) in powers of $|\Delta(\mathbf{n})|^2$ yields:

$$1-Y(\mathbf{n},T)=\frac{2}{9\cdot 3}\,\frac{\Delta^2|\mathbf{d}(\mathbf{n})|^2}{(k_B T_c)^2}. \tag{4.108}$$

If Δ^2 is given its BCS value (4.88),

$$1-Y(\mathbf{n},T)\simeq\frac{2}{\kappa}\left(1-\frac{T}{T_c}\right)|\mathbf{d}(\mathbf{n})|^2. \tag{4.109}$$

As temperature goes to zero, $Y(\mathbf{n},T)$ goes to zero very rapidly (as $\exp(-T/T_c)$) a feature well known from superconductivity where Cooper pairs are in a singlet state. It is not surprising that for triplet pairs with $S_z=0$ along the field, the susceptibility should be the same as for the singlet pairs where $S_z=0$ along any direction.

This result is easy to prove formally: The matrix energy for the singlet is according to (4.83′) of the form:

$$\begin{pmatrix} \varepsilon_k\hat{1}-\mu H\hat{\sigma}_3 & \Delta d_0(\mathbf{n})\hat{\sigma}_2 \\ \Delta d_0^*(\mathbf{n})\hat{\sigma}_2 & -(\varepsilon_k\hat{1}-\mu H\hat{\sigma}_3) \end{pmatrix}. \tag{4.110}$$

It has a structure similar to (4.103), its eigenvalues are given by the same eqn (4.104) and the singlet susceptibility is given by the same formula (4.106).

(β) **d** *perpendicular to* **H**, $d_3=0$

This is the case of pairs which are superpositions with equal weights of states $S_z=\pm 1$ (eqn (4.37)).

Without restricting the generality we can take the axis $0y$ along the vector **d**, making $d_1=0$. The gap matrix becomes:

$$\hat{\Delta}=\Delta d_2\hat{1}=\Delta_2\hat{1}.$$

The energy matrix $\hat{\mathscr{E}}(\mathbf{k})$ is:

$$\hat{\mathscr{E}}(\mathbf{k})=\begin{pmatrix} \varepsilon_k\hat{1}-\mu H\hat{\sigma}_3 & \Delta_2\hat{1} \\ \Delta_2^*\hat{1} & -(\varepsilon_k\hat{1}-\mu H\hat{\sigma}_3) \end{pmatrix}. \tag{4.111}$$

The eigenvalues of (4.111) are easily found to be:

$$\begin{aligned} E_{\mathbf{k}\sigma} &= \pm[(\varepsilon_k-\mu H\sigma)^2+|\Delta_2|^2]^{1/2} \\ &= \pm[\varepsilon_{\mathbf{k}\sigma}^2+|\Delta_2|^2]^{1/2}, \end{aligned} \tag{4.112}$$

whence:

$$\frac{\partial E_{\mathbf{k}\sigma}}{\partial H}=-\mu\sigma\frac{\partial E_{\mathbf{k},\sigma}}{\partial \varepsilon_{\mathbf{k}\sigma}}. \tag{4.112′}$$

Equations (4.102) and (4.112′) yield:

$$M_z = \mu \sum_{k,\sigma} \sigma f(E_{\mathbf{k}\sigma}) \frac{\partial E_{\mathbf{k}\sigma}}{\partial \varepsilon_{\mathbf{k}\sigma}}; \tag{4.113a}$$

$$\begin{aligned} \chi &= -\mu^2 \sum_{k,\sigma} \frac{\partial}{\partial \varepsilon_{\mathbf{k}\sigma}} \left\{ f(E_{\mathbf{k}\sigma}) \frac{\partial E_{\mathbf{k}\sigma}}{\partial \varepsilon_{\mathbf{k}\sigma}} \right\} \bigg) _{H=0} \\ &= -2\mu^2 \sum_k \frac{\partial}{\partial \varepsilon_k} \left\{ f(E_{\mathbf{k}}) \frac{\partial E_{\mathbf{k}}}{\partial \varepsilon_k} \right\} \\ &= -2\mu^2 \frac{1}{2} \left(\frac{\mathrm{d}n}{\mathrm{d}\varepsilon}\right)_0 \int_{-\infty}^{+\infty} \frac{\partial}{\partial \varepsilon} \left\{ f(E_k) \frac{\partial E_k}{\partial \varepsilon} \right\} \mathrm{d}\varepsilon. \end{aligned} \tag{4.113b}$$

$f(+\infty) = 0, f(-\infty) = 1, \partial E_{\mathbf{k}}/\partial \varepsilon\,(-\infty) = 1$ (since we have chosen in (4.60′) for $E_{\mathbf{k}}$ the same sign as for $\varepsilon_{\mathbf{k}}$) and the integral (in 4.113b) is equal to -1. We have then:

$$\chi = \mu^2 \left(\frac{\mathrm{d}n}{\mathrm{d}\varepsilon}\right)_0 = \chi_{n0}. \tag{4.114}$$

Pairs for which $S_z = \pm 1$ along the applied field have the same susceptibility as the normal phase.

The fact that the susceptibility of the superfluid is reduced with respect to the normal phase for pairs with $S_z = 0$ and unchanged for pairs with $S_z = \pm 1$ can be understood intuitively as follows (Leggett, 1975): for a pair with $S_z = 0$, be it triplet or singlet, the application of a magnetic field displaces the energies of the up and down spins with respect to each other. If the spins kept the same populations as in the normal state this would decrease the number of Cooper pairs with opposite spin and momentum which could be formed in the superfluid and thus would reduce the energy of condensation into the superfluid phase. In weak fields at least, it is energetically more advantageous for the system to lose some magnetic energy and thus some polarization in order to regain some condensation energy, whence a reduced susceptibility. For a pair with $S_z = \pm 1$ along the applied field, the spins of the two fermions of the pair are parallel and this problem does not arise.

Appendix A1

We prove the statement that Δ is not affected by the field. More precisely, since we are interested in the linear response to the field, all we need to show is that Δ contains no term linear in H. In the following, all calculations are therefore limited to the first order in H.

The proof goes as follows:

1. Assuming that $\hat{\Delta}(\mathbf{k})$ has the same value as in zero field, we calculate $\hat{x}(\mathbf{k})$ through eqns (4.67) and (4.62).

2. We insert this value into the definition (4.44) of $\hat{\Delta}(\mathbf{k})$:

$$\hat{\Delta}(\mathbf{k}) = -\sum_{\mathbf{k}'} V(\mathbf{k}, \mathbf{k}')\hat{x}(\mathbf{k}'),$$

and show that, within the BCS approximation for $V(\mathbf{k}, \mathbf{k}')$, it is consistent with the independence of $\hat{\Delta}$ on H. It is not possible to use the simpler eqn (4.68) which is not valid in the presence of a field, as shown below. $\hat{\Delta}(\mathbf{k})$ depends on the $\hat{x}(\mathbf{k}')$ corresponding to *all* directions of $\mathbf{k}'$. Since in an actual superfluid the direction of $\mathbf{d}(\mathbf{n})$ may depend on $\mathbf{n}$, we cannot limit ourselves to a particular orientation of $\mathbf{d}$.

In the presence of the field, the 4×4 matrix $\hat{\hat{\mathscr{E}}}(\mathbf{k})^2$ is no longer diagonal. It is, to the first order in H, of the form:

$$\hat{\hat{\mathscr{E}}}(\mathbf{k})^2 = \begin{pmatrix} E_{\mathbf{k}}^2 - 2\mu H\varepsilon_k\hat{\sigma}_3 & -\mu H[\hat{\sigma}_3, \hat{\Delta}] \\ -\mu H[\hat{\Delta}^\dagger, \hat{\sigma}_3] & E_{\mathbf{k}}^2 - 2\mu H\varepsilon_k\hat{\sigma}_3 \end{pmatrix}$$

$$= E_k^2\hat{\hat{1}} - \mu H\begin{pmatrix} 2\varepsilon_k\hat{\sigma}_3 & [\hat{\sigma}_3, \hat{\Delta}] \\ [\hat{\Delta}^\dagger, \hat{\sigma}_3] & 2\varepsilon_k\hat{\sigma}_3 \end{pmatrix}; \qquad (4.115)$$

(remember the provisional assumption: $\hat{\Delta}$ independent of H) with $E_{\mathbf{k}}^2 = \varepsilon_k^2 + |\Delta(\mathbf{k})|^2$.

The matrix:

$$\frac{\tanh\left[\dfrac{\beta}{2}\hat{\hat{\mathscr{E}}}(\mathbf{k})\right]}{2\hat{\hat{\mathscr{E}}}(\mathbf{k})} = \sum \lambda_n \hat{\hat{\mathscr{E}}}(\mathbf{k})^{2n},$$

is to the first order in H, of the form:

$$\frac{\tanh\left[\dfrac{\beta}{2}\hat{\hat{\mathscr{E}}}(\mathbf{k})\right]}{2\hat{\hat{\mathscr{E}}}(\mathbf{k})} = A_{\mathbf{k}}\hat{\hat{1}} - \mu H B_{\mathbf{k}}\begin{pmatrix} 2\varepsilon_k\hat{\sigma}_3 & [\hat{\sigma}_3, \hat{\Delta}] \\ [\Delta^\dagger, \hat{\sigma}_3] & 2\varepsilon_k\hat{\sigma}_3 \end{pmatrix}, \qquad (4.116)$$

where:

$$A_{\mathbf{k}} = \sum \lambda_n E_{\mathbf{k}}^{2n};$$
$$B_{\mathbf{k}} = \sum n\lambda_n E_{\mathbf{k}}^{2n-2};$$

are even functions of $E_{\mathbf{k}}$, that is of ε_k. Equation (4.116) is *not* of the form (4.66) and this is why (4.68) is not valid in the present case.

According to (4.116), eqn (4.67) yields:

$$\hat{x}(\mathbf{k}) = A_{\mathbf{k}}\hat{\Delta}(\mathbf{k}) - \mu H B_{\mathbf{k}}\varepsilon_{\mathbf{k}}\{\hat{\sigma}_3, \hat{\Delta}(\mathbf{k})\}, \qquad (4.117)$$

whence, according to (4.44) and (4.72):

$$\hat{\Delta}(\mathbf{k}) = -V_l \sum_{m,\mathbf{k}'} Y_l^m(\mathbf{n})Y_l^{-m}(\mathbf{n}')[A_{\mathbf{k}'}\hat{\Delta}(\mathbf{k}') - \mu H B_{\mathbf{k}'}\varepsilon_{\mathbf{k}'}\{\hat{\sigma}_3, \hat{\Delta}(\mathbf{k}')\}]. \qquad (4.118)$$

The summation on the right-hand side of (4.118) can be performed first on k' at constant $\mathbf{n}'$, and then on $\mathbf{n}'$. We show that in the first summation the term in H vanishes. Since $\Delta(\mathbf{k}')$ depends only on $\mathbf{n}'$, we have indeed:

$$\sum_{k'} B_{\mathbf{k}'}\varepsilon_k\{\hat{\sigma}_3, \hat{\Delta}(\mathbf{k}')\} = \{\hat{\sigma}_3, \hat{\Delta}(\mathbf{n}')\} \sum_{k'} B_{\mathbf{k}'}\varepsilon_{k'}$$

$$\propto \{\hat{\sigma}_3, \hat{\Delta}(\mathbf{n}')\}\left(\frac{\mathrm{d}n}{\mathrm{d}\varepsilon}\right)_0 \int_{-\infty}^{+\infty} B(\varepsilon)\varepsilon \,\mathrm{d}\varepsilon.$$

The integral vanishes because its integrand is odd in ε, which proves the statement.

(b) Susceptibility of a superfluid phase

We saw in the previous section that the magnetization of a superfluid pair was $\chi_{n0}H$ for $\mathbf{d}(\mathbf{n}) \perp \mathbf{H}$ and $\chi_{n0}Y(\mathbf{n}, T)H$ for $\mathbf{d}(\mathbf{n}) \parallel \mathbf{H}$.

When the angle between $\mathbf{H}_0$ and $\mathbf{d}$ is arbitrary the magnetization, a linear function of the field, is:

$$\mathbf{M} = \chi_\parallel \mathbf{H}_\parallel + \chi_\perp \mathbf{H}_\perp$$

where $\mathbf{H}_\parallel$ and $\mathbf{H}_\perp$ are the components of $\mathbf{H}$ respectively parallel and perpendicular to $\mathbf{d}$. Let $\mathbf{u}$ be the unit vector along $\mathbf{d}$. Then:

$$\mathbf{H}_\parallel = (\mathbf{u}\,.\,\mathbf{H})\mathbf{u}; \qquad \mathbf{H}_\perp = \mathbf{H} - (\mathbf{u}\,.\,\mathbf{H})\mathbf{u},$$

whence:

$$\mathbf{M} = \chi_\perp \mathbf{H} + (\chi_\parallel - \chi_\perp)(\mathbf{u}\,.\,\mathbf{H})\mathbf{u}. \tag{4.119}$$

We can write:

$$\mathbf{M} = \boldsymbol{\chi}\,.\,\mathbf{H}$$

where $\boldsymbol{\chi}$ is a tensor. According to (4.119), its components are:

$$\boldsymbol{\chi}_{ij} = [\chi_\perp \delta_{ij} + (\chi_\parallel - \chi_\perp)u_i u_j]. \tag{4.120}$$

According to (4.91), (4.93), (4.106), and (4.114) this yields, for both the BW and the ABM phases:

$$\chi_{ij}\{\mathbf{d}(\mathbf{n})\} = \chi_{n0}\left\{\delta_{ij} - (1 - Y(\mathbf{n}, T))\frac{d_i(\mathbf{n})d_j^*(\mathbf{n})}{|\mathbf{d}(\mathbf{n})|^2}\right\}. \tag{4.121}$$

In order to calculate the susceptibility for a given phase, rather than a given pair, we must average (4.121) over the orientations of all the pairs.

1. *BW phase*

$\mathbf{d}(\mathbf{n})$ is given by $\mathbf{d}(\mathbf{n}) = R\mathbf{n}$ (eqn (4.91)). Then $|\mathbf{d}(\mathbf{n})|^2 = 1$ independently of $\mathbf{n}$. It follows that $\Delta\Delta^\dagger$ and $E_\mathbf{k}$ are independent of $\mathbf{n}$, as well as the Yosida function:

$$Y(\mathbf{n}, T) = Y(T).$$

It is clear that the average:

$$\left\langle \frac{d_i^*(\mathbf{n})d_j(\mathbf{n})}{|\mathbf{d}(\mathbf{n})|^2} \right\rangle = \int \frac{\mathrm{d}\Omega}{4\pi} d_i^*(\mathbf{n})d_j(\mathbf{n}) = \tfrac{1}{3}\delta_{ij},$$

whence an isotropic susceptibility that we write:

$$\left(\chi^0_{\mathrm{BW}}\right)_{ij} = \chi_{n0}\left\{1 - \frac{(1-Y(T))}{3}\right\}\delta_{ij} = \chi_{n0}\left(\frac{2+Y(T)}{3}\right)\delta_{ij}. \quad (4.122)$$

In the neighbourhood of T_c, according to (4.109):

$$1 - Y(T) \simeq \frac{2}{\kappa}\left(1 - \frac{T}{T_c}\right) = 2\left(1 - \frac{T}{T_c}\right). \quad (4.123)$$

2. *ABM phase*

With the order parameter defined by eqn (4.93):

$$\mathbf{d}(\mathbf{n}) = \left(\tfrac{3}{2}\right)^{1/2}\mathbf{d}(\boldsymbol{\alpha}_1 \,.\, \mathbf{n} + \mathrm{i}\boldsymbol{\alpha}_2 \,.\, \mathbf{n}),$$

where $\mathbf{d}$ is a real unit vector, the orientation of $\mathbf{d}(\mathbf{n})$ is independent of $\mathbf{n}$, but not its modulus, and the Yosida function $Y(\mathbf{n}, T)$ depends on $\mathbf{n}$. It is easy to verify that:

$$(\chi^0_{\mathrm{ABM}})_{ij} = \chi_{n0}\{\delta_{ij} - (1 - Y(T))d_i d_j\}, \quad (4.124)$$

where:

$$Y(T) = \int \frac{\mathrm{d}\Omega}{4\pi} Y(\mathbf{n}, T).$$

(*c*) *Landau corrections to the susceptibility*

The susceptibilities (4.122) and (4.124) do not take into account the effect of the molecular field (4.14), and this is why we wrote them with the superscript zero. Using a tensor notation we can write:

$$\boldsymbol{\chi}^0 = \chi_{n0}\boldsymbol{\theta}^0 \quad (4.125)$$

where:

$$\chi_{n0} = \frac{\gamma^2\hbar^2}{4}\left(\frac{\mathrm{d}n}{\mathrm{d}\varepsilon}\right)_0,$$

and $\boldsymbol{\theta}^0$ is a tensor defined by (4.122) and (4.124) for the phases BW and ABM. We include the molecular field by writing the nuclear magnetization

M as:

$$\mathbf{M}=\boldsymbol{\chi}H_0=\boldsymbol{\chi}^0(\mathbf{H}_0+\mathbf{H}_{\text{mol}})=\chi_{n0}\boldsymbol{\theta}^0(\mathbf{H}_0+\mathbf{H}_{\text{mol}})$$

$$=\chi_{n0}\boldsymbol{\theta}^0\left(\mathbf{H}_0-\frac{F_0^a\mathbf{M}}{\chi_{n0}}\right)=\chi_{n0}\boldsymbol{\theta}^0\left(\mathbf{H}_0-\frac{F_0^a\boldsymbol{\chi}\mathbf{H}_0}{\chi_{n0}}\right), \tag{4.126}$$

whence:

$$\boldsymbol{\chi}=\chi_{n0}\frac{\boldsymbol{\theta}^0}{1+F_0^a\boldsymbol{\theta}^0}=\frac{\chi_{n0}\boldsymbol{\theta}^0}{1+\dfrac{Z_0}{4}\boldsymbol{\theta}^0}. \tag{4.127}$$

It is convenient to introduce $\chi_n=\chi_{n0}[1+(Z_0/4)]^{-1}$ which is the susceptibility of the normal state corrected for the Landau interactions.

This gives:

$$\boldsymbol{\chi}=\left[\chi_n\left(1+\frac{Z_0}{4}\right)\boldsymbol{\theta}^0\right]\bigg/\left(1+\frac{Z_0}{4}\boldsymbol{\theta}^0\right). \tag{4.128}$$

This gives the following results for the BW and ABM phases:

$$(\chi^S_{\text{BW}})_{ij}=\left\{\left[\chi_n\left(1+\frac{Z_0}{4}\right)\frac{2+Y}{3}\right]\bigg/\left(1+\frac{Z_0}{4}\frac{2+Y}{3}\right)\right\}\delta_{ij}. \tag{4.129}$$

For the ABM phase it is found from (4.124) and (4.128)

$$(\chi^S_{\text{ABM}})_{ij}=\chi_n\left\{\delta_{ij}-\frac{d_id_j(1-Y)}{1+(Z_0/4)Y}\right\}. \tag{4.130}$$

To verify that (4.130) is equivalent to (4.128) for the ABM phase, it is sufficient to compare the principal values of both tensors.

(*d*) *Experimental results*

1. A *phase*

The equilibrium susceptibility of the A phase is equal to the susceptibility χ_n of the normal phase and is independent of temperature. This fact demonstrates unambiguously that in the A phase the pairs are in a triplet spin state: In a singlet state the uncorrected susceptibility is, according to eqn (4.106), given by $\chi_{n0}Y(\mathbf{n}, T)$ and after the Landau correction by:

$$\frac{\chi_n(1+Z_0/4)Y(\mathbf{n},T)}{1+(Z_0/4)Y(\mathbf{n},T)}.$$

Both these expressions decrease with temperature and vanish with it, which rules out a singlet state for the A phase. We shall see in section **I** eqn (4.166) that the minimum of the magnetic energy $-\frac{1}{2}\mathbf{H}\,.\,\boldsymbol{\chi}\,.\,\mathbf{H}$ is obtained in the phase ABM when the vector **d** is perpendicular to **H**. Then, according to (4.130), where we take **H** along the z axis:

$$(\chi^{S}_{\text{ABM}})_{33}=\chi_n\left\{1-\frac{d_3^2(1-Y)}{1+(Z_0/4)Y}\right\},$$

and if **d** is perpendicular to $\mathbf{H}_0$ and therefore $d_3=0$, $\chi^{S}_{\text{ABM}}=\chi_n$ as in the A phase. This justifies the identification of the phase A with ABM.

2. B *phase*

In the phase B the susceptibility decreases with temperature and for very low temperatures falls to approximately one third of the normal value χ_n (Fig. 4.4). This is in good agreement with formula (4.129) for the BW phase where for $T=0$ and thus $Y=0$, $\chi_S(0)$ is $\chi_n/3$ if we make $Z_0=-3$ which is approximately its value between 15 and 30 bars.

This justifies the identification of the B phase with the phase BW.

(*e*) *The adiabatic susceptibility*

This is a concept introduced by Leggett and Takagi (1977) for their phenomenological theory of nuclear relaxation in the superfluid to be considered later in this chapter. When an external field H_0 is applied to

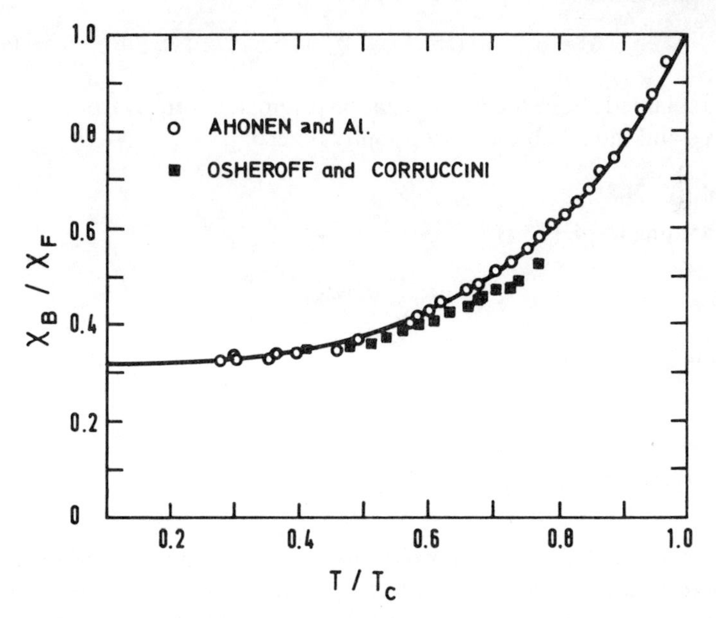

FIG. 4.4. Reduced magnetic susceptibility against reduced temperature in the superfluid phase B of ^{3}He. (After Ahonen, Krusius, and Paalanen, 1975.)

superfluid ^{3}He, the latter acquires a magnetization the origin of which is twofold.

(i) In the independent fermions' Hamiltonian which describes the behaviour of superfluid ^{3}He after linearization and diagonalization, the application of H_0 changes the energies of the elementary excitations from original values $E^0_{\mathbf{k}}$ to new values $E_{\mathbf{k}\sigma}$ given by (4.104) or (4.112). If there is a relaxation mechanism present which can induce transitions, the occupation numbers of the elementary excitations, originally equal to $f(E^0_{\mathbf{k}})$ will in thermal equilibrium become $f(E_{\mathbf{k}\sigma}) \neq f(E^0_{\mathbf{k}})$. The contribution to the magnetization proportional to the changes in the occupation numbers: $f(E_{\mathbf{k}\sigma}) - f(E^0_{\mathbf{k}})$ corresponds to the non-adiabatic part of the susceptibility.

(ii) The application of the magnetic field changes the unitary transform $\hat{U}$ of (4.53) which leads from the original creation and annihilation operators $a_{\mathbf{k}}, a_{-\mathbf{k}}, a^\dagger_{\mathbf{k}}, a^\dagger_{-\mathbf{k}}$, to the new operators $\alpha_{\mathbf{k}\sigma}, \alpha^\dagger_{\mathbf{k}\sigma}$ in (4.58′). Even if the collisions are absent and the new elementary excitations keep the same occupation numbers as the old ones, that is if the application of the field occurs adiabatically, there may still be a magnetization. The corresponding contribution to the susceptibility is called the adiabatic susceptibility. We now proceed to apply this concept to the cases examined earlier.

1. $\mathbf{d}(\mathbf{n}) \| \mathbf{H}_0$

The magnetization is given by eqn (4.104′)

$$\mathbf{M} = \mu \sum_{\mathbf{k},\sigma} \sigma f(E_{\mathbf{k}\sigma}) = \mu \sum_{\mathbf{k},\sigma} \sigma \{ f(E_{\mathbf{k}\sigma}) - f(E^0_{\mathbf{k}}) \}. \tag{4.131}$$

As explained above the magnetization originates entirely from population changes and the adiabatic susceptibility is zero.

2. $\mathbf{d}(\mathbf{n}) \perp \mathbf{H}_0$

According to (4.113a):

$$\mathbf{M} = \mu \sum_{\mathbf{k},\sigma} \sigma f(E_{\mathbf{k}\sigma}) \frac{\partial E_{\mathbf{k}\sigma}}{\partial \varepsilon_{\mathbf{k}\sigma}}.$$

In the limit of small fields:

$$\mathbf{M} = \mathbf{M}_{na} + \mathbf{M}_a = \mu \sum_{\mathbf{k},\sigma} \sigma [f(E_{\mathbf{k}\sigma}) - f(E^0_{\mathbf{k}})] \frac{\partial E^0_{\mathbf{k}}}{\partial \varepsilon^0_{\mathbf{k}}}$$

$$+ \sigma f(E^0_{\mathbf{k}}) \left[\frac{\partial E_{\mathbf{k}\sigma}}{\partial \varepsilon_{\mathbf{k}\sigma}} - \frac{\partial E^0_{\mathbf{k}}}{\partial \varepsilon^0_{\mathbf{k}}} \right]. \tag{4.132}$$

The first part of (4.132) corresponds to the change in the occupation numbers induced by collisions and is thus interpreted as non-adiabatic. The second part of (4.132) where occupation numbers are unchanged is the adiabatic magnetization.

Going to the susceptibilities, we see that for a pair with $\mathbf{d}(\mathbf{n}) \| \mathbf{H}_0$ the adiabatic susceptibility vanishes; for a pair with $\mathbf{d}(\mathbf{n}) \perp \mathbf{H}_0$ it is (the notation χ_{p0} is that of Leggett):

$$\chi_{p0} = \left(\frac{\partial M_a}{\partial H}\right)_{H=0} = -2\mu^2 \sum_{\mathbf{k}} f(E^0_{\mathbf{k}}) \frac{\partial^2 E_{\mathbf{k}}}{\partial \varepsilon^2_{\mathbf{k}}};$$

$$\chi_{p0} = -2\mu^2 \frac{1}{2}\left(\frac{\mathrm{d}n}{\mathrm{d}\varepsilon}\right)_0 \int_{-\infty}^{\infty} f(E^0_{\mathbf{k}}) \frac{\partial^2 E^0_{\mathbf{k}}}{\partial \varepsilon^2} \mathrm{d}\varepsilon$$

$$= -\chi_{n0} \int_{-\infty}^{\infty} \left\{\frac{\partial}{\partial \varepsilon}\left[f(E)\frac{\partial E}{\partial \varepsilon}\right] - \frac{\partial f}{\partial E}\left(\frac{\partial E}{\partial \varepsilon}\right)^2\right\} \mathrm{d}\varepsilon$$

$$= \chi_{n0}\left\{1 + 2\int_0^{\infty} \frac{\partial f}{\partial E}\left(\frac{\partial E}{\partial \varepsilon}\right)^2 \mathrm{d}\varepsilon\right\}; \tag{4.133}$$

$$\chi_{p0} = \chi_{n0}\left\{1 - \int_0^{\infty}\left(\frac{\varepsilon}{E}\right)^2 \frac{\beta}{2}\operatorname{sech}^2\left(\frac{\beta E}{2}\right) \mathrm{d}\varepsilon\right\}$$

$$= \chi_{n0}\{1 - Z(\mathbf{n}, T)\}, \tag{4.134}$$

where:

$$Z(\mathbf{n}, T_c) = \int_0^{\infty} \frac{\beta}{2}\operatorname{sech}^2\left(\frac{\beta \varepsilon}{2}\right) \mathrm{d}\varepsilon = 1. \tag{4.135}$$

In the integrals (4.133) and (4.134), $E^0_{\mathbf{k}} = E(\varepsilon, \mathbf{n})$ and upon integration over ε a function of $\mathbf{n}$ remains.

We now proceed to calculate the adiabatic susceptibility χ_{p0} for the phases BW and ABM. For the phase ABM the orientation of $\mathbf{d}(\mathbf{n})$, defined by the vector $\mathbf{d}$ is independent of $\mathbf{n}$ and we define

$$Z(T) = \int \frac{\mathrm{d}\Omega}{4\pi} Z(\mathbf{n}, T). \tag{4.136}$$

The adiabatic susceptibility is 0 if $\mathbf{d} \| \mathbf{H}$ and $\chi_{n0}[1 - Z(T)]$ if $\mathbf{d} \perp \mathbf{H}$. The tensor $\boldsymbol{\chi}_{p0}$ is then:

$$(\chi_{p0})_{ij} = \chi_{n0}\{[1 - Z(T)](\delta_{ij} - d_i d_j)\}. \tag{4.137}$$

For the phase BW, $E^0_{\mathbf{k}}$ does not depend on $\mathbf{n}$, nor does Z:

$$Z(\mathbf{n}, T) = Z(T).$$

The susceptibility χ_{p0} is clearly isotropic and given by:

$$\chi_{p0} = \chi_{n0}\tfrac{2}{3}[1 - Z(T)]. \tag{4.138}$$

We see from (4.135), (4.136), (4.137), and (4.138) that χ_{p0} vanishes at the critical temperature for both superfluid phases as expected.

We shall defer the discussion of the Landau effects on the adiabatic susceptibilities until the section on relaxation.

H. The magnetic dipole–dipole interactions in superfluid ^{3}He

(*a*) *The local field*

It is a well known fact that in isotropic liquids, because of the fast relative motion of nuclear spins, the average value of the dipolar interaction vanishes. This is evidenced for instance by the extreme narrowness of NMR lines in liquids of low viscosity, contrasting with the broad NMR lines in solids where such motion is absent. More precisely, what the narrowness of the NMR lines in liquids demonstrates is not only the vanishing of the average value of the dipolar energy but the extreme efficiency of the averaging which reduces considerably the fluctuations around the vanishing mean value. In anisotropic liquids such as liquid crystals the average value of the dipolar energy is not zero anymore. The NMR lines are still narrow however because the fluctuations of the dipolar energy around its mean value, now different from zero, are still negligible.

In all studies of NMR a very useful concept is that of local field, which is the field 'seen' by a spin and produced by the dipolar coupling with its neighbours. Under usual NMR conditions of very weak nuclear polarizations, in isotropic liquids the local field seen by any given spin has a finite *instantaneous* value H_L of the order of a few gauss at most. It vanishes if averaged over a correlation time τ_c short compared to $(\gamma H_L)^{-1}$. There is no displacement of the line in first order and the width is of the order of $(\gamma H_L)^2\tau_c$ in frequency units.

In anisotropic liquids (liquid crystals) the local field also averages over a time τ_c but to a finite value rather than to zero, resulting in splittings of the NMR line of the order of γH_L.

In solids the local fields do not average out in a time short compared to $(\gamma H_L)^{-1}$ and give rise to lines of width γH_L.

For high nuclear polarizations an overall displacement of the NMR line dependent on the shape of the sample has to be added to all the previous effects.

However none of the broadenings, splittings or displacements considered so far can exceed the value (in field units) of a few gauss that correspond to the local field.

The situation is radically different in superfluid ^{3}He as will appear shortly. This is due to the coherence that exists in the superfluid phase between the spin states of the various Cooper pairs, a coherence expressed by the existence of a common order parameter $\mathbf{d}(\mathbf{n})$ for these pairs.

(b) The dipolar interactions in superfluid ^{3}*He*

The expression of the dipolar energy is:

$$D = \frac{\gamma^2\hbar^2}{8} \sum_{m,n} (r^{mn})^{-3}\{\hat{\boldsymbol{\sigma}}^{(m)} \cdot \hat{\boldsymbol{\sigma}}^{(n)} - 3(\hat{\boldsymbol{\sigma}}^{(m)} \cdot \mathbf{u}^{mn})(\hat{\boldsymbol{\sigma}}^{(n)} \cdot \mathbf{u}^{mn})\}, \tag{4.139}$$

where $\mathbf{r}^{(m)}$ and $\hat{\boldsymbol{\sigma}}^{(m)}$ are relative to an atom at site (m). $\mathbf{u}^{mn}$ is the unit vector:

$$\mathbf{u}^{mn} = (\mathbf{r}^{(m)} - \mathbf{r}^{(n)})/|\mathbf{r}^{(m)} - \mathbf{r}^{(n)}| = \mathbf{r}^{mn}/|\mathbf{r}^{mn}|.$$

Equation (4.139) can be rewritten:

$$D = \tfrac{1}{2} \sum_{m,n} \sum_{i,j} g_{ij}(\mathbf{r}^{(m)} - \mathbf{r}^{(n)})\hat{\sigma}_i^{(m)}\hat{\sigma}_j^{(n)}, \tag{4.140}$$

where:

$$g_{ij}(\mathbf{r}) = \frac{\gamma^2\hbar^2}{4} r^{-3}\{\delta_{ij} - 3u_iu_j\}; \tag{4.141}$$

i and j are indices for spatial co-ordinates (1.2.3.).

We want to calculate the mean value of (4.140) taken over the state of the superfluid ^{3}He, defined by the order parameter $\mathbf{d}(\mathbf{n})$.

Going over into Fourier space and second quantization, (4.140) becomes:

$$D = \tfrac{1}{2} \sum_{\mathbf{k},\mathbf{k}'} \sum_{i,j} G_{ij}(\mathbf{k} - \mathbf{k}')(a^\dagger_{\mathbf{k}}\hat{\sigma}_i\tilde{a}_{\mathbf{k}'})(a^\dagger_{-\mathbf{k}}\hat{\sigma}_j\tilde{a}_{-\mathbf{k}'}), \tag{4.142}$$

where $G_{ij}(\mathbf{k})$ is the Fourier transform of $g_{ij}(\mathbf{r})$.

As we did in the expression (4.41) for the potential energy of the superfluid we do not keep in (4.142) the whole of dipolar energy in operator form but only that part of it which will give different results for the normal and the superfluid states. Equation (4.142) can be rewritten (summation over repeated indices):

$$D = \tfrac{1}{2} \sum_{\mathbf{k},\mathbf{k}'} G_{ij}(\mathbf{k} - \mathbf{k}')a^\dagger_{\mathbf{k}\alpha}(\hat{\sigma}_i)_{\alpha\gamma}a_{\mathbf{k}'\gamma}a^\dagger_{-\mathbf{k}\beta}(\hat{\sigma}_j)_{\beta\delta}a_{-\mathbf{k}'\delta}; \tag{4.143}$$

$$\langle D\rangle \cong \tfrac{1}{2} \sum_{\mathbf{k},\mathbf{k}'} G_{ij}(\mathbf{k} - \mathbf{k}')\langle a^\dagger_{-\mathbf{k}\beta}a^\dagger_{\mathbf{k}\alpha}\rangle\langle a_{\mathbf{k}'\gamma}a_{-\mathbf{k}'\delta}\rangle \times (\hat{\sigma}_i)_{\alpha\gamma}(\hat{\sigma}_j)_{\beta\delta}; \tag{4.143'}$$

or, using (4.25) and (4.26):

$$\langle D\rangle = \tfrac{1}{2} \sum_{\mathbf{k},\mathbf{k}'} (\hat{x}^\dagger(\mathbf{k}))_{\beta\alpha}(\hat{\sigma}_i)_{\alpha\gamma}(\hat{x}(\mathbf{k}'))_{\gamma\delta}(\tilde{\hat{\sigma}}_j)_{\delta\beta}G_{ij}(\mathbf{k} - \mathbf{k}'), \tag{4.144}$$

(note the transpose on the last matrix $\hat{\sigma}_j$),

$$\langle D\rangle = \tfrac{1}{2} \sum_{\mathbf{k},\mathbf{k}'} \mathrm{Tr}\{\hat{x}(\mathbf{k}')\tilde{\hat{\sigma}}_j\hat{x}^\dagger(\mathbf{k})\hat{\sigma}_i\}G_{ij}(\mathbf{k} - \mathbf{k}'). \tag{4.145}$$

Using (4.32) and its Hermitian conjugate (triplet part only) we introduce the vectors $\mathbf{x}(\mathbf{k}')$ and $\mathbf{x}^*(\mathbf{k})$ by:

$$\hat{x}(\mathbf{k}') = (\mathbf{x}(\mathbf{k}') \,.\, \hat{\sigma})\hat{\sigma}_2; \qquad \hat{x}^\dagger(\mathbf{k}) = \hat{\sigma}_2(\mathbf{x}^*(\mathbf{k}) \,.\, \hat{\boldsymbol{\sigma}});$$

$$\langle D\rangle = \tfrac{1}{2} \sum_{\mathbf{k},\mathbf{k}'} \mathrm{Tr}\{(\mathbf{x}(\mathbf{k}') \,.\, \hat{\boldsymbol{\sigma}})\hat{\sigma}_2\tilde{\hat{\sigma}}_j\hat{\sigma}_2(\mathbf{x}^*(\mathbf{k}) \,.\, \hat{\boldsymbol{\sigma}})\hat{\sigma}_i\}G_{ij}(\mathbf{k}-\mathbf{k}'). \tag{4.146}$$

Or remembering that $\hat{\sigma}_2\tilde{\hat{\sigma}}_j\hat{\sigma}_2 = -\hat{\sigma}_j$:

$$\langle D\rangle = \tfrac{1}{2} \sum_{\mathbf{k},\mathbf{k}'} G_{ij}(\mathbf{k}-\mathbf{k}')\mathbf{x}(\mathbf{k}')_p\mathbf{x}^*(\mathbf{k})_q\Lambda_{pjqi}, \tag{4.147}$$

where:

$$\Lambda_{pjqi} = -\mathrm{Tr}\{\hat{\sigma}_p\hat{\sigma}_j\hat{\sigma}_q\hat{\sigma}_i\} = 2\{\delta_{pq}\delta_{ij} - \delta_{pj}\delta_{qi} - \delta_{pi}\delta_{qj}\}, \tag{4.148}$$

whence:

$$\langle D\rangle = \sum_{\mathbf{k},\mathbf{k}'} G_{ij}(\mathbf{k}-\mathbf{k}')\{\mathbf{x}(\mathbf{k}') \,.\, \mathbf{x}^*(\mathbf{k})\delta_{ij} - (\mathbf{x}(\mathbf{k}')_i\mathbf{x}^*(\mathbf{k})_j - \mathbf{x}(\mathbf{k}')_j\mathbf{x}^*(\mathbf{k})_i)\}. \tag{4.149}$$

It remains to calculate the Fourier transform:

$$G_{ij}(\mathbf{q}) = \int g_{ij}(\mathbf{r}) \exp(\mathrm{i}\mathbf{q} \,.\, \mathbf{r})\,\mathrm{d}^3r = \frac{\gamma^2\hbar^2}{4}\int \exp(\mathrm{i}\mathbf{q} \,.\, \mathbf{r})(\delta_{ij} - 3u_iu_j)r^{-3}\,\mathrm{d}^3r. \tag{4.150}$$

$G_{ij}(\mathbf{q})$ is a symmetrical tensor with a vanishing trace. Taking the z-axis along $\mathbf{q}$ we get from (4.150) after a little algebra:

$$G_{zz} = -2G_{xx} = -2G_{yy} = \frac{8\pi}{3}\frac{\gamma^2\hbar^2}{4} \tag{4.151}$$

$G_{ij}(\mathbf{q})$ does not depend on the magnitude of $\mathbf{q}$ but only on the orientation of the unit vector:

$$\mathbf{v} = \frac{\mathbf{q}}{|\mathbf{q}|} = \frac{\mathbf{k}-\mathbf{k}'}{|\mathbf{k}-\mathbf{k}'|} = \frac{\mathbf{n}-\mathbf{n}'}{|\mathbf{n}-\mathbf{n}'|}. \tag{4.152}$$

For a general choice of axes:

$$G_{ij}(\mathbf{v}) = -\frac{4\pi}{3}(\delta_{ij} - 3v_iv_j)\frac{\gamma^2\hbar^2}{4}. \tag{4.153}$$

From eqn (4.36): $\sum_k \mathbf{x}(\mathbf{k}) = \psi\mathbf{d}(\mathbf{n})$, (4.149) becomes:

$$\langle D\rangle = \psi^2\int\frac{\mathrm{d}\Omega}{4\pi}\int\frac{\mathrm{d}\Omega'}{4\pi}G_{ij}(\mathbf{v})\{\mathbf{d}^*(\mathbf{n}) \,.\, \mathbf{d}(\mathbf{n}')\delta_{ij} - 2\,\mathrm{Re}\,d_i^*(\mathbf{n})d_j(\mathbf{n}')\}. \tag{4.154}$$

It can be shown (Appendix 2) that for $l=1$ this expression reduces to:

$$\langle D\rangle = \frac{\pi}{2}\psi^2\gamma^2\hbar^2\int\frac{d\Omega}{4\pi}\{3|(\mathbf{n}\,.\,\mathbf{d}(\mathbf{n}))|^2 - |\mathbf{d}(\mathbf{n})|^2\}; \tag{4.155}$$

$$\langle D\rangle = g_D\langle 3|(\mathbf{n}\,.\,\mathbf{d}(\mathbf{n}))|^2 - |\mathbf{d}(\mathbf{n})|^2\rangle;$$

$$g_D = \frac{\pi}{2}\gamma^2\hbar^2\psi^2. \tag{4.155'}$$

(*c*) *Estimate of the constant* g_D

As stated in section **D** only the assumption of the existence of an order parameter is sufficient to derive straightaway for $\langle D\rangle$ the expression (4.154). However an estimate of the constant ψ does require the developments of sections **E** and **F**.

From eqn (4.82) we get:

$$g_D = \frac{\pi}{2}\gamma^2\hbar^2[\ln(1.14\beta_c\varepsilon_0)]^2\Delta^2\left[\frac{1}{2}\left(\frac{dn}{d\varepsilon}\right)_0\right]^2, \tag{4.156}$$

and from the value (4.88) of Δ:

$$g_D = \frac{\pi}{2}\gamma^2\hbar^2\frac{9\cdot 3}{\kappa}[\ln(1\cdot 14\beta_c\varepsilon_0)]^2\left(1-\frac{T}{T_c}\right)\left[\frac{1}{2}\left(\frac{dn}{d\varepsilon}\right)_0\right]^2(k_BT_c)^2. \tag{4.157}$$

To get a feeling for the magnitude of g_D we write: $(dn/d\varepsilon)_0 = 3N_0/2k_BT_F$ where T_F is the Fermi temperature and N_0 the number of atoms per unit volume. We also define $M_0=(\gamma\hbar/2)N_0$ where M_0 is the nuclear magnetization of a unit volume of ^{3}He with all its spins parallel. g_D becomes:

$$g_D = \alpha M_0^2\left(\frac{T_c}{T_F}\right)^2\left(1-\frac{T}{T_c}\right)\frac{1}{\kappa}, \tag{4.158}$$

where α is a numerical constant,

$$\alpha \simeq 30\,[\ln(1\cdot 14\beta_c\varepsilon_0)]^2, \tag{4.159}$$

and $M_0 \approx 0\cdot 2$ gauss.

There is considerable arbitrariness in the choice of the cut-off energy ε_0, very strongly attenuated, however by the logarithmic dependence of g_D on $\beta_c\varepsilon_0$. One should not take in any case too seriously the formula (4.159) because of the arbitrariness of the BCS cut-off model and of the shortcomings of the weak coupling theory. In practice it is best to take T_F from the specific heat as explained in section **B** and to adjust the numerical constant α to the experimental values of g_D, whose measurement will be described in the next section. These values are of the order of $10^{-3}(1-T/T_c)$.

Simple minded estimate of g_D

The dipolar energy per Cooper pair is of the order of $\gamma^2\hbar^2/a^3 \sim \gamma^2\hbar^2 N_0$ where a is the distance between the two atoms of the pairs. The number of Cooper pairs $\simeq$ number N_0 of atoms, times the square of the probability for each atom to belong to a Cooper pair that is, $(\Delta/\varepsilon_F)^2 \simeq (T_c/T_F)^2[1-(T/T_c)]$, according to (4.98). Whence:

$$g_D \sim \gamma^2\hbar^2 N_0 \times N_0 \left(\frac{T_c}{T_F}\right)^2 \left(1-\frac{T}{T_c}\right)$$

$$\sim M_0^2 \left(\frac{T_c}{T_F}\right)^2 \left(1-\frac{T}{T_c}\right), \tag{4.160}$$

which is the same order of magnitude as (4.158).

Appendix A2: proof of the equivalence of eqns (4.154) and (4.155).

According to (4.153), the right-hand side of eqn (4.154) is equal to:

$$\frac{2\pi}{3}\psi^2\gamma^2\hbar^2 P$$

where:

$$P = \sum_{i,j} P_{ij} = \sum_{i,j} \langle d_i^*(\mathbf{n}) d_j(\mathbf{n}')(\delta_{ij} - 3v_i v_j)\rangle$$

$$= \langle \mathbf{d}^*(\mathbf{n}) \,.\, \mathbf{d}(\mathbf{n}') - 3(\mathbf{v} \,.\, \mathbf{d}^*(\mathbf{n}))(\mathbf{v} \,.\, \mathbf{d}(\mathbf{n}'))\rangle, \tag{4.161}$$

whereas the right-hand side of eqn (4.155) is equal to:

$$-\frac{\pi}{2}\psi^2\gamma^2\hbar^2 Q,$$

where:

$$Q = \langle |\mathbf{d}(\mathbf{n})|^2 - 3|(\mathbf{n} \,.\, \mathbf{d}(\mathbf{n}))|^2\rangle$$

$$= \sum_{i,j} Q_{ij} = \sum_{i,j} \langle d_i^*(\mathbf{n}) d_j(\mathbf{n})(\delta_{ij} - 3n_i n_j)\rangle. \tag{4.162}$$

We show that if the components $d_i(\mathbf{n})$ are spherical harmonics of order l, that is of the form:

$$d_i(\mathbf{n}) = \sum_m A_{im} Y_l^m(\mathbf{n}),$$

the ratio P/Q is a numerical constant which depends on l but not on the actual form of the $d_i(\mathbf{n})$.

Let us compare P_{ij} and Q_{ij}. The terms $(\delta_{ij} - 3v_i v_j)$ and $(\delta_{ij} - 3n_i n_j)$ transform under rotation of co-ordinate axes as components of a traceless tensor of order 2:

$$(\delta_{ij} - 3n_i n_j) = \sum_q K_{ij}^q Y_2^q(\mathbf{n}); \tag{4.163}$$

$$(\delta_{ij} - 3v_i v_j) = \sum_q K_{ij}^q Y_2^q(\mathbf{v});$$

where $\mathbf{v} = \mathbf{n} - \mathbf{n}'/|\mathbf{n} - \mathbf{n}'|$.

In order for P_{ij} to be different from zero, the term,

$$d_i^*(\mathbf{n}) d_j(\mathbf{n}') = \sum_{m,m'} A_{im}^* A_{jm'} Y_l^{m^*}(\mathbf{n}) Y_l^{m'}(\mathbf{n}'),$$

must have a non-vanishing part transforming under rotation of the co-ordinate axes like D_2. This part is:

$$B_{ij} = \sum_q C_q F_2^q(\mathbf{n}, \mathbf{n}') \tag{4.164}$$

with:

$$C_q = \sum_{m,m'} \delta_{m'-m,q} (-1)^m A_{im}^* A_{jm'} \langle -ml, m'l | q, 2 \rangle,$$

where the $\langle \cdots | \cdots \rangle$ are Clebsch–Gordan coefficients and $F_2^q(\mathbf{n}, \mathbf{n}')$ is a function of $\mathbf{n}$ and $\mathbf{n}'$ which transforms under rotation of the co-ordinate axes like Y_2^q. Likewise, the part of $d_i^*(\mathbf{n}) d_j(\mathbf{n})$ transforming like D_2 is obtained from (4.164) by the replacement of $F_2^q(\mathbf{n}, \mathbf{n}')$ by $F_2^q(\mathbf{n}, \mathbf{n})$.

According to the properties of irreducible tensors the only non-vanishing averages are of the form,

$$\langle F_2^q Y_2^{-q} \rangle,$$

and these averages are independent of q. We have then:

$$\frac{P_{ij}}{Q_{ij}} = \frac{\xi \langle F_2^0(\mathbf{n}, \mathbf{n}') Y_2^0(\mathbf{v}) \rangle}{\xi \langle F_2^0(\mathbf{n}, \mathbf{n}) Y_2^0(\mathbf{n}) \rangle} = \frac{\langle F_2^0(\mathbf{n}, \mathbf{n}') Y_2^0(\mathbf{v}) \rangle}{\langle F_2^0(\mathbf{n}, \mathbf{n}) Y_2^0(\mathbf{n}) \rangle}, \tag{4.165}$$

where ξ is of the form:

$$\xi = \sum_{m,m'} (-1)^m A_{im}^* A_{jm'} \langle -ml, m'l | m' - m, 2 \rangle K_{ij}^{(m-m')}.$$

The ratio,

$$\frac{P}{Q} = \frac{\sum_{ij} P_{ij}}{\sum_{ij} Q_{ij}} = \frac{P_{ij}}{Q_{ij}},$$

is therefore independent of the coefficients $A^*_{im}A_{jm'}$, that is of the form of the $d_i(\mathbf{n})$ as long as they are l harmonics. This ratio can be computed by taking any special form for $\mathbf{d}(\mathbf{n})$.

For $l=1$ we choose $\mathbf{d}(\mathbf{n})=\mathbf{n}$, whence:

$$P=\langle \mathbf{n}\,.\,\mathbf{n}'-3(\mathbf{v}\,.\,\mathbf{n})(\mathbf{v}\,.\,\mathbf{n}')\rangle$$

$$=3\left\langle \sin^2\left(\frac{\widehat{\mathbf{n}\,.\,\mathbf{n}'}}{2}\right)\right\rangle=\tfrac{3}{2};$$

$$Q=\langle n^2-3n^2\rangle=-2;$$

$$P/Q=-3/4;$$

which ends the proof of the equality of (4.154) and (4.155).

I. Magnetic resonance: the various energies, the commutation relations, the equations of motion

(*a*) *Magnetic and dipolar energy: equilibrium states*

It was stated in section **F** that two superfluid phases of ^{3}He were known: ABM and BW. We saw that either phase was considerably degenerate. Thus all the phases ABM deduced from that defined by eqn (4.92) by a rotation R_s of the spin–axes and a rotation R_O of the orbital axes, and represented by the more general equations (4.93) have the same free energy. In particular this free energy is independent of the angle between the vectors $\mathbf{d}$ and $\boldsymbol{\alpha}_1\wedge\boldsymbol{\alpha}_2=l$ of (4.93).

Similarly all the BW states defined by $\mathbf{d}(\mathbf{n})=R\,\mathbf{n}$ have the same free energy whatever the rotation R. Things change when the Zeeman energy of coupling with an external field H and the dipolar energy given by (4.155) are taken into account.

Equilibrium position in phase ABM

From eqn (4.130) we find that the anisotropic part of the susceptibility tensor provides an orientation energy given by:

$$\Delta F^m=-\tfrac{1}{2}\delta\chi_{ij}H_iH_j=\tfrac{1}{2}\chi_n\frac{[1-Y(T)](\mathbf{d}\,.\,\mathbf{H})^2}{1+(Z_0Y/4)}, \qquad (4.166)$$

where $\mathbf{d}$ is the vector which defines the orientations of $\mathbf{d}(\mathbf{n})$ (eqn (4.93)).

$$\mathbf{d}(\mathbf{n})=\sqrt{(\tfrac{3}{2})}\,\mathbf{d}((\boldsymbol{\alpha}_1\,.\,\mathbf{n})+\mathrm{i}(\boldsymbol{\alpha}_2\,.\,\mathbf{n}))$$

ΔF^m is minimum for $\mathbf{d}\perp\mathbf{H}$.

The dipolar energy can be written:

$$\langle D\rangle = g_D\langle 3|(\mathbf{n}\,.\,\mathbf{d}(\mathbf{n}))|^2 - |\mathbf{d}(\mathbf{n})|^2\rangle$$
$$= g_D\{\langle 3|(\mathbf{n}\,.\,\mathbf{d}(\mathbf{n}))|^2 - 1\rangle\},$$

or using (4.93):

$$\begin{aligned}3\langle|(\mathbf{n}\,.\,\mathbf{d}(\mathbf{n}))|^2\rangle &= \tfrac{9}{2}\langle(\mathbf{n}\,.\,\mathbf{d})^2[(\boldsymbol{\alpha}_1\,.\,\mathbf{n})^2+(\boldsymbol{\alpha}_2\,.\,\mathbf{n})^2]\rangle \\ &= \tfrac{9}{2}\langle(\mathbf{n}\,.\,\mathbf{d})^2[1-(\mathbf{l}\,.\,\mathbf{n})^2]\rangle = \tfrac{3}{2}-\tfrac{9}{2}\langle(\mathbf{n}\,.\,\mathbf{d})^2(\mathbf{n}\,.\,\mathbf{l})^2\rangle \\ &\quad -\tfrac{9}{2}\langle\mathbf{n}\,.\,\mathbf{d})^2(\mathbf{n}\,.\,\mathbf{l})^2\rangle = -\tfrac{9}{2}d_\alpha d_\beta l_\gamma l_\delta\langle n_\alpha n_\beta n_\gamma n_\delta\rangle \qquad (4.167)\\ &= -\tfrac{9}{2}d_\alpha d_\beta l_\gamma l_\delta \tfrac{1}{15}\{\delta_{\alpha\beta}\delta_{\gamma\delta}+\delta_{\alpha\gamma}\delta_{\beta\delta}+\delta_{\alpha\delta}\delta_{\beta\gamma}\} \\ &= -\tfrac{9}{2}\tfrac{1}{15}\{d^2l^2+2(\mathbf{d}\,.\,\mathbf{l})^2\} = -\tfrac{3}{5}(\mathbf{d}\,.\,\mathbf{l})^2+cst;\end{aligned}$$

$$\langle D\rangle = -\tfrac{3}{5}g_D(\mathbf{d}\,.\,\mathbf{l})^2+cst = -\tfrac{3}{5}g_D\cos^2\theta+cst, \qquad (4.168)$$

where θ is the angle between **d** and **l**. From (4.166) and (4.168) we see that we minimize the total energy if $\mathbf{d}\perp\mathbf{H}$ and $\mathbf{d}\|\mathbf{l}$ and therefore $\mathbf{l}\perp\mathbf{H}$. This still leaves undetermined the position of $\mathbf{d}\|\mathbf{l}$ in the plane perpendicular to **H**. This position may be fixed by small external perturbations such as some restriction on the geometry of the sample.

Equilibrium position in phase BW

The dipolar energy lifts to some extent the degeneracy of the BW phase:

$$\langle D\rangle = 3g_D\langle|(\mathbf{n}\,.\,\mathbf{d}(\mathbf{n}))|^2\rangle+cst$$
$$= 3g_D\langle|R_{\alpha\beta}n_\alpha n_\beta|^2\rangle = 3g_D R_{\alpha\beta}R_{\gamma\delta}\langle n_\alpha n_\beta n_\gamma n_\delta\rangle+cst,$$

or using (4.167):

$$\langle D\rangle = \frac{g_D}{5}\{R_{\alpha\alpha}R_{\gamma\gamma}+R_{\alpha\beta}R_{\alpha\beta}+R_{\alpha\beta}R_{\beta\alpha}\}+cst. \qquad (4.169)$$

Since the matrix R is orthogonal $R_{\alpha\beta}R_{\alpha\beta}=3$ and:

$$\langle D\rangle = \frac{g_D}{5}\{(\mathrm{Tr}(R))^2+\mathrm{Tr}(R^2)\}+cst. \qquad (4.170)$$

If θ is the angle of rotation $\mathrm{Tr}(R)=1+2\cos\theta$; $\mathrm{Tr}(R^2)=1+2\cos 2\theta$ and:

$$\langle D\rangle = \frac{g_D}{5}\{(1+2\cos\theta)^2+1+2\cos 2\theta\}+cst = \tfrac{4}{5}g_D\{2\cos^2\theta+\cos\theta\}+cst. \qquad (4.171)$$

$\langle D\rangle$ is minimum for $\partial\langle D\rangle/\partial\theta = -(4g_D/5)\sin\theta(1+4\cos\theta) = 0$ that is for:

$$\cos\theta_0 = -\tfrac{1}{4}; \qquad \theta = \theta_0 = 104°30'; \tag{4.172}$$

$$\left(\frac{\partial^2\langle D\rangle}{\partial\theta^2}\right)_{\theta=\theta_0} = 3g_D > 0.$$

We see that the minimization of the dipolar energy restricts the rotations R in (4.91) to those with the angle θ_0. On the other hand the axis of rotation is undetermined, and since the susceptibility of the phase BW given by (4.129) is isotropic it cannot lift this degeneracy.

There is however in the presence of a magnetic field a small advantage in orientating the axis of rotation parallel to that field. If $\mathbf{u}$ is the unit vector along the axis of rotation there is a magnetic energy of the order of $-g_D(\mu H/\Delta)^2\,(\mathbf{u}\,.\,\mathbf{H}/H)^2$, minimized when $\mathbf{u}\parallel\mathbf{H}$.

Qualitatively this energy arises as follows (Leggett, 1975): It can be shown that the effect of a magnetic field parallel to $0z$ is to reduce the proportion of pairs with $S_z = 0$, that is those with $\mathbf{d}\parallel\mathbf{H}$ with respect to the pairs with $S_z = \pm 1$, that is those with $\mathbf{d}\perp\mathbf{H}$. This reduction, of order $(\mu H/\Delta)^2$, does not appear in the approximation of vanishingly small fields used in section **G**. The component d_z is thus reduced with respect to d_x and d_y. In the rotation of angle θ_0 which minimizes the dipolar energy, it is more advantageous to have the correct orientation, that is an angle θ_0 between $\mathbf{n}$ and $\mathbf{d}(\mathbf{n})$ for the larger components of $\mathbf{d}$, that is for d_x and d_y rather than for d_z and thus to take the axis of rotation parallel to $\mathbf{H}$. This orientation energy is of the order of $4\,.\,10^{-2}$ per Gauss^2. It can be calculated (Engelsberg, Brinkman, and Anderson, 1974) by adding to the $G-L$ free energy the dipolar energy and the magnetic energy and minimizing the sum. We omit this calculation because we are not concerned here with the actual value of the magnetic anisotropy energy but only with the fact that in the absence of other causes it is capable of selecting from all the BW phases the one with minimum energy.

(*b*) *The adiabatic approximation*

We have just seen that superfluid phases which differ from each other by an overall rotation of their parameters differ in energy. These anisotropy energy differences, magnetic and dipolar in origin, are very much smaller than the differences in energy between a superfluid and a normal phase or between two superfluid phases of different nature such as ABM and BW, which are of the same order as the gap Δ. In nuclear magnetic resonance one investigates the dynamics of these rotations. Saying that the energy differences betwen the various configurations are small is tantamount to saying that the corresponding motions are slow, with periods that can be much longer than the microscopic time constants required for each configuration to come to thermal equilibrium. This is the adiabatic approximation where the total spin S of the superfluid as well as the collection of order parameter vectors $\mathbf{d}(\mathbf{n})$ rotate as a whole. On the other hand a relative change of orientation of a vector $\mathbf{d}(\mathbf{n})$ with respect to another vector $\mathbf{d}(\mathbf{n}')$ would change the microscopic nature of the superfluid phase and require too much energy to be able to occur. Such at any rate are the basic assumptions of the phenomenological theory of NMR in superfluids, due to Leggett, and well verified by experiment, that we outline now.

The only dynamical variables of the problem are the orientation of the total spin $\mathbf{S}$ and of the configuration $\mathbf{d}(\mathbf{n})$. To describe their motions we need two things: their commutation relations and their Hamiltonian.

Since the theory is dealing with adiabatic departures from equilibrium, what is needed is a phenomenological Hamiltonian that accounts for the change of energy resulting from such adiabatic departures from equilibrium, and which therefore is *not* the true Hamiltonian. It must satisfy the condition that in the equilibrium configuration the energy is minimum.

The relevant terms in this Hamiltonian are the following.

(i) The dipolar energy $E_{\mathrm{d}} = \langle D \rangle$ given in general by (4.155) and more specifically, for the ABM phase by (4.168) and for the BW phase by (4.171).

(ii) The Zeeman energy $Z = -\gamma\hbar\mathbf{S}\,.\,\mathbf{H}$.

(iii) A phenomenological internal magnetic energy, due to the existence of a finite nuclear magnetization, quadratic in the components of the total spin S. Its form can be determined as follows. The total magnetic energy E_m can be written:

$$E_m = -\gamma\hbar\mathbf{S}\,.\,\mathbf{H} + \mathbf{S}\,.\,\mathbf{A}\,.\,\mathbf{S} \tag{4.173}$$

where $\mathbf{A}$ is a second rank tensor since the superfluid phase may be anisotropic. The tensor $\mathbf{A}$ can be determined by writing that at equilibrium the magnetic energy (4.173) is minimum:

$$\frac{\partial E_m}{\partial S_i} = 2A_{ij}S_j - \gamma\hbar H_i = 0; \qquad \mathbf{AS} = \frac{\gamma\hbar}{2}\mathbf{H};$$

or:

$$\mathbf{S} = \frac{\gamma\hbar}{2}\mathbf{A}^{-1}\,.\,\mathbf{H}. \tag{4.174}$$

On the other hand: at equilibrium $\gamma\hbar\mathbf{S} = \mathbf{M} = \boldsymbol{\chi}\mathbf{H}$, whence: $\mathbf{A} = (\gamma^2\hbar^2/2)\boldsymbol{\chi}^{-1}$ and:

$$E_{\mathrm{S}} = \frac{\gamma^2\hbar^2}{2}\mathbf{S}\,.\,\boldsymbol{\chi}^{-1}\,.\,\mathbf{S}. \tag{4.175}$$

We shall come back to this last term when we write down Leggett's equations of motion.

(c) The commutation relations

In order to write these equations we must find the commutation relations of the dynamic variables of the motion, namely the total spin S and the ensemble of the order parameter vectors $\{\mathbf{d}(\mathbf{n})\}$. This can be done in two ways:

(i) We start from the expressions of $\mathbf{S}$ and $\mathbf{d}(\mathbf{n})$ in the formalism of second quantization:

$$\mathbf{S}=\tfrac{1}{2}\sum_{\mathbf{k}}\mathrm{Tr}\{\tilde{a}_{\mathbf{k}}^{\dagger}\,.\,\hat{\boldsymbol{\sigma}}\,.\,a_{\mathbf{k}}\},$$

that is:

$$S_i=\tfrac{1}{2}\sum_{\mathbf{k},\alpha,\beta} a_{\mathbf{k}\alpha}^{\dagger}(\hat{\sigma}_i)_{\alpha\beta}a_{\mathbf{k}\beta}. \tag{4.176}$$

We can also define $S_i(\mathbf{n})$

$$S_i(\mathbf{n})=\tfrac{1}{2}\sum_{|\mathbf{k}|,\alpha,\beta} a_{\mathbf{k}\alpha}^{\dagger}(\hat{\sigma}_i)_{\alpha\beta}a_{\mathbf{k}\beta}. \tag{4.176$'$}$$

$\mathbf{d}(\mathbf{n})$ has been defined as:

$$\mathbf{d}(\mathbf{n})=\psi^{-1}\sum_{|\mathbf{k}|}\mathbf{x}(\mathbf{k}),$$

where:

$$\mathbf{x}(\mathbf{k})=\tfrac{1}{2}\,\mathrm{Tr}\{(\hat{x}(\mathbf{k})\,.\,\hat{\sigma}_2\hat{\boldsymbol{\sigma}})\};$$

$$x_i(\mathbf{k})=\tfrac{1}{2}\sum_{\alpha,\beta,\gamma}\langle a_{\mathbf{k}\alpha}a_{-\mathbf{k}\beta}\rangle(\hat{\sigma}_2)_{\beta\gamma}(\hat{\sigma}_i)_{\gamma\alpha}. \tag{4.177}$$

$x_i(\mathbf{k})$ as defined by (4.177) is a c-number, in contrast with S_i defined by (4.176) as an operator. We can however drop the averaging sign $\langle\;\;\rangle$ in the expressions for $\mathbf{x}(\mathbf{k})$ and $\mathbf{d}(\mathbf{n})$: the fluctuations of an operator around its mean value are negligible for systems with a large number of particles.

We can thus write:

$$d_i(\mathbf{n})=\frac{\psi^{-1}}{2}\sum_{|\mathbf{k}|}a_{\mathbf{k}\alpha}a_{-\mathbf{k}\beta}(\hat{\sigma}_2)_{\beta\gamma}(\hat{\sigma}_i)_{\gamma\alpha}. \tag{4.178}$$

From the explicit expressions of S_i and $d_i(\mathbf{n})$ as functions of creation and annihilation of fermion operators and the known commutation relations of the latter, the commutators:

$$[S_i, S_j]; \qquad [S_i, d_j(\mathbf{n})]; \qquad [d_i(\mathbf{n}), d_j(\mathbf{n}')];$$

can be computed with a little algebra.

$$[S_i, S_j] = i\varepsilon_{ijl} S_l;$$
$$[S_i, d_j] = i\varepsilon_{ijl} d_l; \tag{4.179}$$
$$[d_i(\mathbf{n}), d_j(\mathbf{n}')] = \frac{i\varepsilon_{ijl}}{2\psi^2}\{S_l(\mathbf{n}) + S_l(-\mathbf{n})\}\{\delta(\mathbf{n}-\mathbf{n}') + \delta(\mathbf{n}+\mathbf{n}')\};$$

where ε_{ijl} is the totally antisymmetric symbol with $\varepsilon_{123} = 1$.

The first two commutators in (4.179) can be obtained directly from the fact that the total spin **S** represents an infinitesimal rotation for all the spin variables.

Let **A** be any vector (say **S** itself or **d**(**n**)) relative to spin variables, and let $\delta\boldsymbol{\omega}$ be a vector defining an infinitesimal rotation in spin space.

The corresponding infinitesimal rotation operator is:

$$\exp(\mathrm{i}(\delta\boldsymbol{\omega} \,.\, \mathbf{S}) \approx 1 + \mathrm{i}(\delta\boldsymbol{\omega} \,.\, \mathbf{S}).$$

The increment $\delta\mathbf{A}$ of the vector **A**, due to the rotation is:

$$\mathbf{A} + \delta\mathbf{A} = \exp(\mathrm{i}\delta\boldsymbol{\omega} \,.\, \mathbf{S})\mathbf{A}\exp(-\mathrm{i}(\delta\boldsymbol{\omega} \,.\, \mathbf{S})) \cong \mathbf{A} + \mathrm{i}[\delta\boldsymbol{\omega} \,.\, \mathbf{S}, \mathbf{A}]. \tag{4.180}$$

On the other hand, from classical kinematics $\delta\mathbf{A} = \delta\boldsymbol{\omega} \wedge \mathbf{A}$ whence:

$$\delta A_i = \mathrm{i}\delta\omega_j[S_j, A_i] = \varepsilon_{ijl}\delta\omega_j A_l. \tag{4.181}$$

Equating the values (4.180) and (4.181) of δA_i we obtain:

$$[S_j, A_i] = -\mathrm{i}\varepsilon_{ijl}A_l = \mathrm{i}\varepsilon_{jil}A_l. \tag{4.182}$$

Replacing **A** by either **S** or **d**(**n**) in (4.182) we obtain the first two commutators in (4.179).

We are now almost ready to write out the equations of motion. With $\mathscr{H} = Z + E_d + E_S$ we have:

$$\dot{\mathbf{S}} = -\frac{\mathrm{i}}{\hbar}[\mathbf{S}, \mathscr{H}]; \qquad \dot{\mathbf{d}}(\mathbf{n}) = -\frac{i}{\hbar}[\mathbf{d}(\mathbf{n}), \mathscr{H}]. \tag{4.183}$$

Before we compute these commutators a question must be settled: in the adiabatic motion of the spin variables what should one assume about the motion of orbital parameters such as $\mathbf{l} = \boldsymbol{\alpha}_1 \wedge \boldsymbol{\alpha}_2$ in the ABM phase? At the beginning of an NMR experiment **l** has in general reached an equilibrium orientation determined by some, possibly competing effects such as the magnetic energy, the dipolar energy, as well as some influence of the geometry of the sample. Leggett assumes that the inertia associated with the orbital motion of **l** is too large for **l** to move from its equilibrium position, where it remains through the NMR experiment. We shall henceforth make this assumption which seems in agreement with experiment.

(d) The equations of motion

The first Leggett equation:

$$\dot{\mathbf{S}} = -\frac{\mathrm{i}}{\hbar}[\mathbf{S}, \mathcal{H}],$$

can be rewritten:

$$\dot{\mathbf{S}} = -\frac{\mathrm{i}}{\hbar}[\mathbf{S}, -\gamma\hbar\mathbf{S}\,.\,\mathbf{H} + E_s + E_d], \tag{4.184}$$

that is, according to (4.175):

$$\dot{\mathbf{S}} = \gamma\mathbf{S} \wedge (\mathbf{H} - \gamma\hbar\boldsymbol{\chi}^{-1}\,.\,\mathbf{S}) + \mathbf{R}_d, \tag{4.185}$$

where $\mathbf{R}_d$ is defined as:

$$\mathbf{R}_d = -\frac{\mathrm{i}}{\hbar}[\mathbf{S}, E_d],$$

and:

$$E_d = \int \frac{\mathrm{d}\Omega}{4\pi} \mathcal{H}_D\{\mathbf{d}(\mathbf{n})\},$$

is given by (4.155). The computation of $\mathbf{R}_d$ is made as follows. If the spin system undergoes a small rotation $\delta\boldsymbol{\omega}$ the change in E_d is:

$$\delta E_d = \mathrm{i}[(\delta\boldsymbol{\omega}\,.\,\mathbf{S}), E_d] = \mathrm{i}\delta\boldsymbol{\omega}\,.\,[\mathbf{S}, E_d] = -\hbar\delta\boldsymbol{\omega}\,.\,\mathbf{R}_d. \tag{4.186}$$

On the other hand:

$$\begin{aligned} \delta E_d &= \int \frac{\mathrm{d}\Omega}{4\pi} \frac{\partial \mathcal{H}_D\{\mathbf{d}(\mathbf{n})\}}{\partial \mathbf{d}(\mathbf{n})}\,.\,\delta\mathbf{d}(\mathbf{n}) \\ &= \int \frac{\mathrm{d}\Omega}{4\pi} \frac{\partial \mathcal{H}_D\{\mathbf{d}(\mathbf{n})\}}{\partial \mathbf{d}(\mathbf{n})}\,.\,[\delta\boldsymbol{\omega} \wedge \mathbf{d}(\mathbf{n})] \\ &= \delta\boldsymbol{\omega}\,.\int \frac{\mathrm{d}\Omega}{4\pi}\left\{\mathbf{d}(\mathbf{n}) \wedge \frac{\partial \mathcal{H}_D\{\mathbf{d}(\mathbf{n})\}}{\partial \mathbf{d}(\mathbf{n})}\right\}, \end{aligned}$$

whence:

$$\mathbf{R}_d = -\frac{1}{\hbar}\int \frac{\mathrm{d}\Omega}{4\pi}\left[\mathbf{d}(\mathbf{n}) \wedge \frac{\partial \mathcal{H}_D\{\mathbf{d}(\mathbf{n})\}}{\partial \mathbf{d}(\mathbf{n})}\right]. \tag{4.187}$$

The expressions (4.186) and (4.187) for $\mathbf{R}_d$ are general: we shall derive in the next section a form specialized to the ABM and BW phase.

The second Leggett equation for $\dot{\mathbf{d}}(\mathbf{n})$ (or rather the set of equations for the various $\dot{\mathbf{d}}(\mathbf{n})$) can be written:

$$\dot{\mathbf{d}}(\mathbf{n}) = -\frac{\mathrm{i}}{\hbar}[\mathbf{d}(\mathbf{n}), -\gamma\hbar\mathbf{S}\,.\,\mathbf{H} + E_s + E_d]. \tag{4.188}$$

The first two terms yield:

$$\dot{\mathbf{d}}(\mathbf{n}) = \gamma \mathbf{d}(\mathbf{n}) \wedge (\mathbf{H} - \gamma\hbar \boldsymbol{\chi}^{-1} . \mathbf{S}). \tag{4.189}$$

The last term $-\mathrm{i}/\hbar[\mathbf{d}(\mathbf{n}), E_{\mathrm{d}}]$ can be neglected for two reasons. Using the last commutation relation (4.179) we see that this term is of order $g_{\mathrm{D}}Sd/\hbar\psi^2$ to be compared with the second term in (4.189) of the order of $\gamma^2\hbar Sd/\chi$. Their ratio is of the order of $g_{\mathrm{D}}\chi/(\gamma^2\hbar^2\psi^2)$. With $g_{\mathrm{D}}/\psi^2 \sim \gamma^2\hbar^2$ (eqn (4.155′)), this ratio is of the order of $\chi \sim 10^{-7}$ and utterly negligible.

One can also argue that the contributions to $\mathbf{d}(\mathbf{n})$ arising from the commutator $[\mathbf{d}(\mathbf{n}), E_{\mathrm{d}}]$ correspond to a motion where relative positions of vectors $\mathbf{d}(\mathbf{n})$ and $\mathbf{d}(\mathbf{n}')$ are altered and lead to changes in the microscopic structure of the superfluid phase which requires too much energy to occur.

In the BW phase, the susceptibility is isotropic, and the tensor $\boldsymbol{\chi}$ in (4.185) and (4.189) reduces to a scalar χ whose value is given by (4.129). We show that in the ABM phase, this tensor $\boldsymbol{\chi}$ can be replaced by the equilibrium susceptibility χ, a scalar given by $\chi_n = \chi_{n0}\,[1 + (Z_0/4)]^{-1}$.

In the ABM phase, all vectors $\mathbf{d}(\mathbf{n})$ are parallel to the same vector $\mathbf{d}$. We use (4.185) and (4.189) to calculate the time derivative of the scalar product $\mathbf{d} . \mathbf{S}$. Anticipating the next section where we prove that in the ABM phase $\mathbf{d} . \mathbf{R}_{\mathrm{d}} = 0$, we obtain:

$$\begin{aligned} \mathbf{d} . \dot{\mathbf{S}} + \dot{\mathbf{d}} . \mathbf{S} &= \gamma \mathbf{d} . \mathbf{S} \wedge [\mathbf{H} - \gamma\hbar\boldsymbol{\chi}^{-1} . \mathbf{S}] + \gamma \mathbf{S} . \mathbf{d} \wedge [\mathbf{H} - \gamma\hbar\boldsymbol{\chi}^{-1} . \mathbf{S}] \\ &= \gamma[\mathbf{d} \wedge \mathbf{S} + \mathbf{S} \wedge \mathbf{d}] . [\mathbf{H} - \gamma\hbar\boldsymbol{\chi}^{-1} . \mathbf{S}] \\ &= 0, \end{aligned}$$

so that if initially $\mathbf{d} \perp \mathbf{S}$ as at equilibrium, these vectors will remain orthogonal afterwards. The susceptibility tensor $\boldsymbol{\chi}$ being axially symmetric around $\mathbf{d}$, with principal values $\chi_{\parallel}$ and $\chi_{\perp}$, we have then at any time:

$$\boldsymbol{\chi}^{-1} . \mathbf{S} = \chi_{\perp}^{-1}\mathbf{S} = \chi^{-1}\mathbf{S},$$

where $\chi = \chi_{\perp}$ is the equilibrium susceptibility.

The Leggett equations become finally, for both the BW and ABM phases:

$$\dot{\mathbf{S}} = \gamma \mathbf{S} \wedge \mathbf{H} + \mathbf{R}_{\mathrm{d}}; \tag{4.185′}$$

$$\dot{\mathbf{d}}(\mathbf{n}) = \gamma \mathbf{d}(\mathbf{n}) \wedge \left(\mathbf{H} - \frac{\gamma\hbar}{\chi}\mathbf{S}\right). \tag{4.189′}$$

We can introduce the vectors:

$$\mathbf{S}' = \mathbf{S} - \frac{\chi}{\gamma\hbar}\mathbf{H}, \tag{4.190}$$

and:

$$\mathbf{K} = \gamma\left[\mathbf{H} - \frac{\gamma\hbar}{\chi}\mathbf{S}\right] = -\frac{\gamma^2\hbar}{\chi}\mathbf{S}'. \tag{4.190′}$$

The second Leggett equation can be written:

$$\dot{\mathbf{d}}(\mathbf{n}) = -\frac{\gamma^2\hbar}{\chi}\mathbf{d}(\mathbf{n}) \wedge \mathbf{S}' = \mathbf{d}(\mathbf{n}) \wedge \mathbf{K}. \tag{4.190$''$}$$

These forms are convenient for many applications.

J. Magnetic resonance in the ABM and BW phases

(a) Equations of motion

1. *The ABM phase*: $\mathbf{d}(\mathbf{n}) = \sqrt{(\frac{3}{2})}\mathbf{d}[(\boldsymbol{\alpha}_1 \,.\, \mathbf{n}) + \mathrm{i}(\boldsymbol{\alpha}_2 \,.\, \mathbf{n})]$

The dipolar energy in the ABM phase is given by:

$$E_\mathrm{d} = -\tfrac{3}{5}g_\mathrm{D}(\mathbf{d} \,.\, \mathbf{l})^2 = -\frac{\lambda}{2}(\mathbf{d} \,.\, \mathbf{l})^2 = -\frac{\lambda}{2}\cos^2\theta. \tag{4.168}$$

The change of E_d in a small rotation $\delta\boldsymbol{\omega}$ is:

$$\delta E_\mathrm{d} = -\lambda(\mathbf{d} \,.\, \mathbf{l})(\mathbf{l} \,.\, \delta\mathbf{d}) = -\lambda(\mathbf{d} \,.\, \mathbf{l})[\mathbf{l} \,.\, (\delta\boldsymbol{\omega} \wedge \mathbf{d})] \tag{4.191}$$

$$= -\lambda(\mathbf{d} \,.\, \mathbf{l})(\delta\boldsymbol{\omega} \,.\, (\mathbf{d} \wedge \mathbf{l})) = -\hbar(\delta\boldsymbol{\omega} \,.\, \mathbf{R}_\mathrm{d}); \tag{4.186}$$

$$\mathbf{R}_\mathrm{d} = \frac{\lambda}{\hbar}(\mathbf{d} \wedge \mathbf{l})(\mathbf{d} \,.\, \mathbf{l}); \tag{4.192}$$

$(\mathbf{d} \,.\, \mathbf{R}_\mathrm{d}) = 0$ as anticipated in $\mathbf{I}(d)$.

The Leggett eqns (4.185′) and (4.189′) become:

$$\dot{\mathbf{S}} = \gamma\mathbf{S} \wedge \mathbf{H} + \frac{\lambda}{\hbar}(\mathbf{d} \wedge \mathbf{l})(\mathbf{d} \,.\, \mathbf{l}); \tag{4.193a}$$

$$\dot{\mathbf{d}} = \mathbf{d} \wedge \mathbf{K} = \gamma\mathbf{d} \wedge \left(\mathbf{H} - \frac{\gamma\hbar\mathbf{S}}{\chi}\right) = -\frac{\gamma^2\hbar}{\chi}\mathbf{d} \wedge \mathbf{S}'; \tag{4.193b}$$

where:

$$\mathbf{S}' = \mathbf{S} - \frac{\chi}{\gamma\hbar}\mathbf{H} = \mathbf{S} - \mathbf{S}_0, \qquad \lambda = \tfrac{6}{5}g_\mathrm{D}.$$

2. *The BW phase*: $\mathbf{d}(\mathbf{n}) = R\mathbf{n}$

Let the rotation R which connects each unit vector $\mathbf{n} = \mathbf{k}/k$ to $\mathbf{d}(\mathbf{n})$ be defined by the unit vector $\mathbf{u}$ of its axis and its angle θ. From an inspection of Fig. 4.5 we get:

$$\mathbf{d}(\mathbf{n}) = \mathbf{n}\cos\theta + \mathbf{u}(\mathbf{n} \,.\, \mathbf{u})(1 - \cos\theta) + \sin\theta(\mathbf{u} \wedge \mathbf{n}). \tag{4.194}$$

From the *infinite set* of equations $\dot{\mathbf{d}}(\mathbf{n}) = \mathbf{d}(\mathbf{n}) \wedge \mathbf{K}$, we want to derive *two* equations for the rates of change $\mathbf{u}$ and θ.

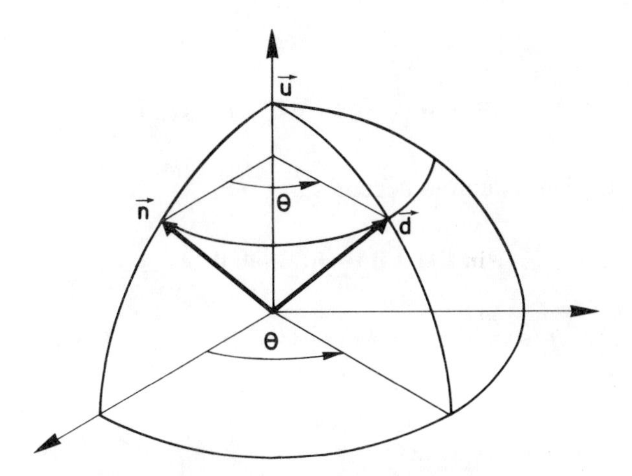

FIG. 4.5. Superfluid phase B of ^{3}He. The vector **d** is connected to **n** by a rotation of angle θ around the vector **u**.

This is done as follows.

(i) We take the time derivative of (4.194);

$$\begin{aligned}\dot{\mathbf{d}}(\mathbf{n}) &= \dot{\theta}\{-\mathbf{n}\sin\theta+\mathbf{u}(\mathbf{n}\,.\,\mathbf{u})\sin\theta+\cos\theta(\mathbf{u}\wedge\mathbf{n})\}\\ &\quad+\dot{\mathbf{u}}(\mathbf{n}\,.\,\mathbf{u})(1-\cos\theta)+\mathbf{u}(\mathbf{n}\,.\,\dot{\mathbf{u}})(1-\cos\theta)+(\dot{\mathbf{u}}\wedge\mathbf{n})\sin\theta.\end{aligned}\tag{4.195}$$

(ii) We take the angular averages of $P_s=\langle\mathbf{n}\,.\,\dot{\mathbf{d}}(\mathbf{n})\rangle$ and $\mathbf{P}_v=\langle\mathbf{n}\wedge\dot{\mathbf{d}}(\mathbf{n})\rangle$. In P_s and P_v we replace $\dot{\mathbf{d}}(\mathbf{n})$ either by (4.195) or by $\dot{\mathbf{d}}(\mathbf{n})=(\mathbf{d}(\mathbf{n})\wedge\mathbf{K})$ where $\mathbf{d}(\mathbf{n})$ is given by (4.194).

(iii) In the course of the averaging one should remember that for any two vectors **A** and **B** independent of **n**:

$$\langle(\mathbf{n}\,.\,\mathbf{A})(\mathbf{n}\,.\,\mathbf{B})\rangle=\tfrac{1}{3}(\mathbf{A}\,.\,\mathbf{B});$$

$$\langle(\mathbf{n}\,.\,\mathbf{A})(\mathbf{n}\wedge\mathbf{B})\rangle=\tfrac{1}{3}(\mathbf{A}\wedge\mathbf{B}).$$

With a little algebra we then find:

$$\dot{\theta}=-\mathbf{u}\,.\,\mathbf{K}=-\gamma\mathbf{u}\,.\left(\mathbf{H}-\frac{\gamma\hbar\mathbf{S}}{\chi}\right)=\frac{\gamma^2\hbar}{\chi}(\mathbf{u}\,.\,\mathbf{S}');\tag{4.196}$$

$$\begin{cases}\dot{\mathbf{u}}=\tfrac{1}{2}\mathbf{u}\wedge\mathbf{K}+\tfrac{1}{2}\operatorname{cotan}\left(\dfrac{\theta}{2}\right)\mathbf{u}\wedge(\mathbf{u}\wedge\mathbf{K})\text{ or:}\\[2ex] \dot{\mathbf{u}}=-\dfrac{\gamma^2\hbar}{2\chi}(\mathbf{u}\wedge\mathbf{S}')-\dfrac{\gamma^2\hbar}{2\chi}\operatorname{cotan}\left(\dfrac{\theta}{2}\right)[\mathbf{u}\wedge(\mathbf{u}\wedge\mathbf{S}')].\end{cases}\tag{4.197}$$

From eqn (4.171) $E_d=\frac{4}{5}g_D(2\cos^2\theta+\cos\theta)$ we find that its change δE_d for an infinitesimal rotation: $\delta\boldsymbol{\omega}=\mathbf{u}\delta\theta$ is:

$$\delta E_d=-\hbar(\mathbf{R}_d\,.\,\delta\boldsymbol{\omega})=-\hbar(\mathbf{R}_d\,.\,\mathbf{u})\delta\theta=-\tfrac{4}{5}g_D\sin\theta(1+4\cos\theta)\delta\theta,$$

whence:

$$\mathbf{R}_{\mathrm{d}} = \mathbf{u}\frac{4}{5\hbar}g_{\mathrm{D}}\sin\theta(1+4\cos\theta). \tag{4.198}$$

The first Leggett equation becomes:

$$\dot{\mathbf{S}} = \gamma \boldsymbol{S}\wedge \boldsymbol{H} + \boldsymbol{u}\frac{4}{5\hbar}g_{\mathrm{D}}\sin\theta(1+4\cos\theta). \tag{4.199}$$

$\mathbf{R}_{\mathrm{d}}$ vanishes for θ_0 such that $1+4\cos\theta_0=0$ and for $\theta-\theta_0$ small:

$$\mathbf{R}_{\mathrm{d}} \cong -\frac{3}{\hbar}g_{\mathrm{D}}(\theta-\theta_0)\mathbf{u},$$

and:

$$\dot{\mathbf{S}} \approx \gamma\mathbf{S}\wedge\mathbf{H} - \frac{3}{\hbar}g_{\mathrm{D}}(\theta-\theta_0)\mathbf{u}. \tag{4.199'}$$

In the phase BW the motion of the spins in the superfluid phase is described by the three equations: (4.196), (4.197), and (4.199).

(b) Motions of small amplitude

1. *The ABM phase*

The dipolar energy E_{d} is minimum and the vector $\mathbf{R}_{\mathrm{d}}$ given by (4.192) vanishes when $\mathbf{d}\|\mathbf{l}$ which is a position of stable equilibrium. We take the direction of $\mathbf{l}$ as the x-axis and the magnetic field which at equilibrium is perpendicular to $\mathbf{l}$ along the z-axis. If we assume that the amplitudes of the motions away from equilibrium are small we can take $d_x=(d_x)_0 \simeq 1$ and d_y, d_z, S_x, S_y, S_z' small. We rewrite the equations (4.193a) as:

$$\dot{\mathbf{S}}' = \omega_0(\mathbf{z}\wedge\mathbf{S}') + \frac{\lambda}{\hbar}(\mathbf{d}\wedge\mathbf{x})(\mathbf{d}\,.\,\mathbf{x})$$

$$\simeq \omega_0(\mathbf{z}\wedge\mathbf{S}') + \frac{\lambda}{\hbar}(\mathbf{d}\wedge\mathbf{x}); \qquad (\omega_0=-\gamma H_0), \tag{4.200}$$

where $\mathbf{z}$ and $\mathbf{x}$ are unit vectors along $0z$ and $0x$.

Its time derivative is according to (4.193):

$$\ddot{\mathbf{S}}' \simeq \omega_0(\mathbf{z}\wedge\dot{\mathbf{S}}') + \frac{\lambda}{\hbar}(\dot{\mathbf{d}}\wedge\mathbf{x}) = \omega_0(\mathbf{z}\wedge\dot{\mathbf{S}}') + \frac{\lambda}{\hbar}[(\mathbf{d}\wedge\mathbf{K})\wedge\mathbf{x}]$$

$$\simeq \omega_0(\mathbf{z}\wedge\dot{\mathbf{S}}') + \frac{\lambda}{\hbar}[(\mathbf{x}\wedge\mathbf{K})\wedge\mathbf{x}];$$

$$\ddot{\mathbf{S}}' \simeq \omega_0(\mathbf{z} \wedge \dot{\mathbf{S}}') + \frac{\lambda}{\hbar}[\mathbf{y}\mathrm{K}_y + \mathbf{z}\mathrm{K}_z]$$

$$= \omega_0(\mathbf{z} \wedge \dot{\mathbf{S}}') - \frac{\lambda\gamma^2}{\chi}[\mathbf{y}S'_y + \mathbf{z}S'_z]; \tag{4.201}$$

$$\ddot{S}'_z = -\frac{\lambda\gamma^2}{\chi}S'_z; \qquad \ddot{S}'_y = -\omega_0\dot{S}'_x - \frac{\lambda\gamma^2}{\chi}S'_y; \qquad \ddot{S}'_x = \omega_0\dot{S}'_y. \tag{4.202}$$

The first eqn (4.202) shows that S'_z, the component along the field, oscillates with a frequency $\Omega^{\mathrm{L}}_{\mathrm{ABM}}$ given by:

$$(\Omega^{\mathrm{L}}_{\mathrm{ABM}})^2 = \lambda\gamma^2/\chi. \tag{4.203}$$

The motion of the two other components is clearly elliptical with a frequency Ω given by:

$$\Omega^2 = (\Omega^{\mathrm{L}}_{\mathrm{ABM}})^2 + \omega_0^2. \tag{4.204}$$

2. *The BW phase*

We stated in section **I** that in the presence of a magnetic field there was a rather weak interaction of the order of $-g_{\mathrm{D}}(\mu H/\Delta)^2(\mathbf{u}\,.\,\mathbf{H}/H)^2$ that oriented the rotation axis $\mathbf{u}$ along the field. In our study of the small oscillations in phase BW we shall assume that $\mathbf{u}$ remains along the field $\mathbf{H}$ through the course of the motion. We take then the z-axis along $\mathbf{H}$ and $\mathbf{u}$ and use the eqns (4.196) and (4.199′) which yield:

$$\dot{S}'_x = -\omega_0 S'_y; \qquad \dot{S}'_y = \omega_0 S'_x; \qquad \dot{S}'_z = -\frac{3g_{\mathrm{D}}}{\hbar}(\theta - \theta_0); \tag{4.205a}$$

$$\dot{\theta} = \frac{\gamma^2\hbar}{\chi}S'_z. \tag{4.205b}$$

The motion of S'_x and S'_y is the classical Larmor precession whereas for S'_z we get:

$$\ddot{S}'_z = -\frac{3g_{\mathrm{D}}}{\hbar}\dot{\theta} = -\frac{3g_{\mathrm{D}}\gamma^2}{\chi}S'_z. \tag{4.206}$$

The longitudinal component of S oscillates with a frequency $\Omega^{\mathrm{L}}_{\mathrm{BW}}$:

$$(\Omega^{\mathrm{L}}_{\mathrm{BW}})^2 = \frac{3g_{\mathrm{D}}\gamma^2}{\chi}. \tag{4.207}$$

In the absence of a magnetic field if $\mathbf{u}$ is fixed by some other cause and we take its direction as the z-axis we still have a linear oscillation along that axis.

To conclude at this point the study of small oscillations in the two phases, the results are as follows:

(i) Both phases present a resonance in the absence of a magnetic field, with frequencies given by:

$$(\Omega^{\rm L}_{\rm ABM})^2 = \tfrac{6}{5} g_{\rm D} \frac{\gamma^2}{\chi}; \qquad (\Omega^{\rm L}_{\rm BW})^2 = 3 g_{\rm D} \frac{\gamma^2}{\chi};$$

$$[(\Omega^{\rm L}_{\rm BW})/(\Omega^{\rm L}_{\rm ABM})]^2 = \tfrac{5}{2}\left(\frac{\chi_{\rm ABM}}{\chi_{\rm BW}}\right) \tag{4.208}$$

$$\simeq \tfrac{5}{2} \text{ very near } T_{\rm c}.$$

(We have assumed that $(g_{\rm D})_{\rm BW}/(g_{\rm D})_{\rm ABM} = 1$, a result in agreement with the weak coupling theory and also with experiment.)

(ii) In the presence of a magnetic field the component of **S** parallel to the field has the same frequency as in the absence of the field.

(iii) The frequency of the transverse motion is unshifted in phase BW and given in phase ABM by: $\Omega^2 = \omega_0^2 + \Omega_{\rm L}^2$.

Remark: on the motion of the axis of rotation in phase BW.

One may ask to what extent the assumption that the orientation of **u** remains fixed in space is compatible with the eqn (4.197). This last equation is not complete since we have not included the stabilizing influence of the small energy $-g_{\rm D}(\mu H/\Delta)^2(\mathbf{u}\,.\,\mathbf{H}/H)^2$. Even in the absence of this term we can see from (4.197) that $\dot{\mathbf{u}}$ is proportional to the transverse components S'_x and S'_y of S'. The latter oscillate with the Larmor frequency and if the amplitudes of S'_x and S'_y are small, the contributions to u are small and periodical and it is a good approximation to assume that **u** is time-independent. This would not be true in general with large transverse components of **S**.

Leggett has given a general theory of small motions in superfluid ^{3}He of which the results obtained here for the phases ABM and BW are special cases (Leggett, 1974).

(*c*) *Experimental results*

Shortly after the discovery of the strange behaviour of the melting curve of ^{3}He, referred to in section **A** (Osheroff *et al.* 1972*b*) a large shift of the NMR frequency from the Larmor frequency ω_0 was observed (Osheroff, Gully, Richardson, and Lee, 1972) in phase A. It was interpreted by Leggett (1972) as conclusive evidence for the existence of a new phase of *liquid* ^{3}He. Such a shift is predicted by eqn (4.204) and confirms the identification, already made in section **G** of the A phase with theoretical phase ABM.

The magnitude of $\Omega_{\rm A}^2$ according to (4.203) is given by $\Omega_{\rm A}^2 = \lambda\gamma^2/\chi = \frac{6}{5} g_{\rm D}(\gamma^2/\chi)$ where $g_{\rm D}$, given by (4.157) or (4.158), is proportional to $(1 - T/T_{\rm c})$. We saw in section **H** that $g_{\rm D}$ is of the order of 10^{-3} $(1 - T/T_{\rm c})$ and

$f_A = \Omega_A/2\pi$ is thus of the order of $f_A \simeq 200\ (1-T/T_c)^{1/2}$ kHz. Shortly afterwards Leggett (1973) suggested that a longitudinal resonance, observable with an r.f. field parallel to the d.c. field H_0, or even in the absence of H_0 should be observable in the A phase, with the *same* frequency Ω_A as that which corresponds to the quadratic shift (4.204) of the transverse resonance.

Figure 4.6 shows a plot of f_A against $(1-T/T_c)$ obtained at or near the melting pressure. There is no observable difference between the longitudinal frequencies and those extracted from the shifts of the transverse resonances. Figure 4.7 exhibits the dependence of f_A^2 on the pressure. We shall not discuss it in any detail. It is seen that the lower the pressure, the lower the frequency. This is consistent with the proportionality of g_D and thus of Ω_A^2 to T_c^2 and with the decrease, apparent on the phase diagram (Fig. 4.1), of T_c with the pressure.

In phase B a longitudinal resonance is observed as predicted by eqn (4.207). Figure 4.8 is a plot of f_B^2 against T/T_c for various pressures. The experimental ratio Ω_B^2/Ω_A^2 is in good agreement with the theoretical expression (4.208). Unless special circumstances such as the presence of walls (see section **K**) force the rotation vector **u** away from the magnetic field, no shift is observed for the transverse resonance in the B phase, in agreement with eqn (4.205*a*).

A last word about the absolute magnitude of f_A and f_B is in order. A frequency shift of the order of 100 kHz, would correspond to a field of the order of thirty gauss, and no arrangement of nuclear spins can produce a dipolar field of this magnitude. The field δH producing such shifts is

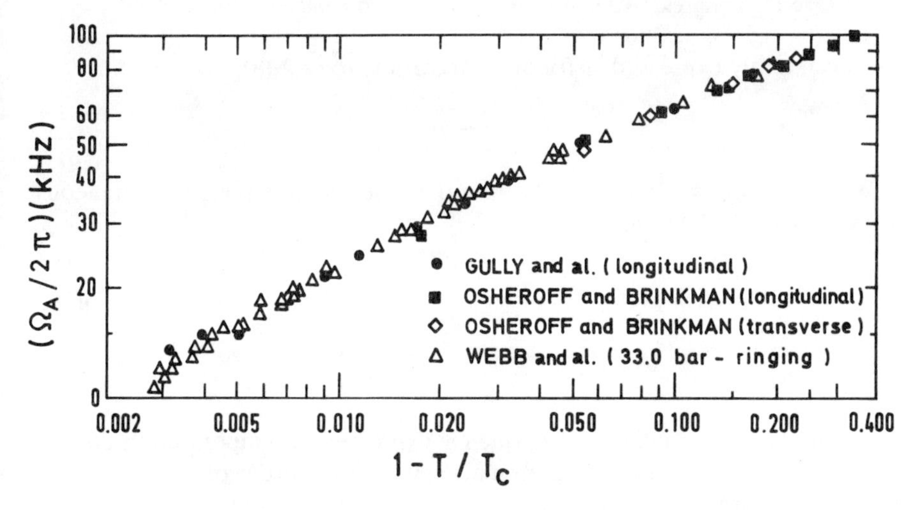

FIG. 4.6. Superfluid phase A of ^{3}He. Variation of the frequency f_A against $(1-T/T_c)$ near the melting pressure, as obtained by longitudinal or transverse resonance, or ringing experiments. (After Gully, Gould, Richardson, and Lee, 1976.)

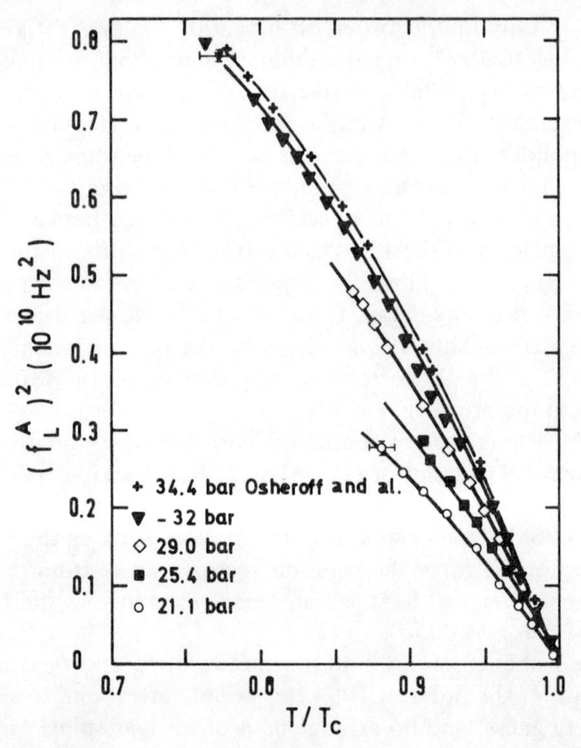

FIG. 4.7. Superfluid phase A of ^{3}He. Variation of f_A^2 against reduced temperature for several pressures. (After Ahonen, Krusius, and Paalanen, 1975.)

conveniently expressed as follows. According to (4.208),

$$\delta H = \frac{\Omega}{\gamma} \sim (g_D/\chi)^{1/2}.$$

g_D is given by (4.158), where M_0 is of the order of the dipolar instantaneous local field H_L:

$$M_0 = \xi H_L,$$

with the numerical factor ξ of order unity. As for the susceptibility, it is of the order of:

$$\chi \sim \frac{\mu M_0}{k_B T_F} = \xi \frac{\mu H_L}{k_B T_F} \sim \frac{\theta}{T_F},$$

where θ is the dipolar local frequency expressed in temperature units. As discussed in chapter 8, it is extremely small, of the order of a μK.

We have then:

$$\delta H \sim H_L \left(\frac{\alpha}{\kappa}\right)^{1/2} \left(1 - \frac{T}{T_c}\right)^{1/2} \left(\frac{T_c}{T_F}\right) \left(\frac{T_F}{\theta}\right)^{1/2}.$$

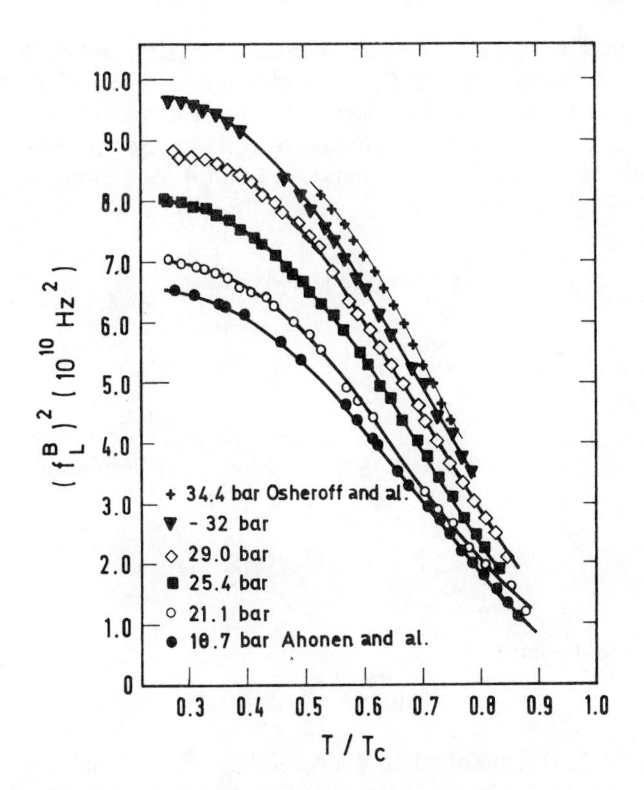

FIG. 4.8. Superfluid phase B of ^{3}He. Variation of f_B^2 against reduced temperature for several pressures. (After Ahonen, Krusius, and Paalanen, 1975.)

This is much larger than H_L, because of the extreme smallness of $\theta \ll T_c$. It is worth pointing out that the reason why the shifts are not even much larger than they are is purely accidental and follows from the smallness of (T_c/T_F). The existence of such large frequency shifts is a unique feature in nuclear magnetism.

In conclusion it can be said that the interpretation or the prediction of the above features of NMR in superfluid ^{3}He, and the detailed agreement between theory and experiment are among the most beautiful happenings in nuclear magnetism since the discovery of NMR in late 1945.

(*d*) *Motions of large amplitude*

1. *Longitudinal ringing in ABM phase*

What is meant thereby is an experiment where, after the system has come to equilibrium in a finite field H_0, the latter is modified suddenly by an increment ΔH.

Since, prior to this change, $\gamma\hbar S_z$ was equal to χH_0 and S'_z to zero, this increment of H_0 is equivalent to a sudden increase of S'_z by $-\chi\Delta H/\gamma\hbar$. According to the eqn (4.193b) this starts **d** into a rotation around **S′**, that is in the present case around $\mathbf{H}_0$. This in turn by eqn (4.193*a*) induces a change in S_z (and S'_z). Let θ be the angle between **l** and **d**. Equations (4.193*a*) and (4.193*b*) can be written:

$$\dot{S}'_z = -\left(\frac{\lambda}{\hbar}\right)\sin\theta\cos\theta; \tag{4.209a}$$

$$\dot{\theta} = \frac{\gamma^2\hbar}{\chi} S'_z, \tag{4.209b}$$

with the initial condition $\theta_0 = 0$, $(S'_z)_0 = -\chi\Delta H/\gamma\hbar$. Eliminating S'_z between the two equations (4.209) we obtain for $\varphi = 2\theta$:

$$\ddot{\varphi} = -\frac{\gamma^2\lambda}{\chi}\sin\varphi, \tag{4.210}$$

with the initial condition:

$$(\dot{\varphi})_0 = -2\gamma\Delta H.$$

Equation (4.210) is simply that of a pendulum. For φ small, $\sin\varphi \approx \varphi$ and the frequency of oscillation is $\Omega = (\gamma^2\lambda/\chi)^{1/2}$ already given in eqn (4.203). This will happen if $|(\dot{\varphi})_0| = |2\gamma\Delta H| \ll \Omega$.

If on the other hand we make $|(\dot{\varphi})_0| = |2\gamma\Delta H| > 2\Omega$ the oscillatory motion of the pendulum becomes rotary and for $2|\gamma\Delta H| \gg 2\Omega$ it is very nearly uniform with an angular velocity $2|\gamma\Delta H|$. This is then also the frequency of the oscillation of S_z. One sees that as $|\gamma\Delta H|$ goes from a value much smaller than Ω to a value much larger than Ω, the frequency of oscillation of the longitudinal magnetization should go from $\Omega_L = (\lambda\gamma^2/\chi)^{1/2}$ to $2\gamma\Delta H$. The intermediate steps are somewhat complicated. The experimental results are in qualitative agreement with the theory.

The attempts to observe longitudinal ringing in the BW phase have led to the observation of a complicated behaviour which cannot be understood without considering the influence of the walls. This will be done in section **K**.

2. *Free precession in the ABM phase* (Brinkman and Smith, 1975*a*).

In this experiment, performed in a magnetic field H_0 after the system has come to equilibrium, a pulse of angle φ is applied by means of a rotating field $H_1 \ll H_0$. At equilibrium, the vector **d** is perpendicular to $\mathbf{H}_0$ and parallel to **l**, whereas **S** is parallel to $\mathbf{H}_0$ and therefore $(\mathbf{S}\,.\,\mathbf{d})(0) = 0$. It follows from the equations of motion (4.193) that:

$$\mathbf{d}\,.\,\dot{\mathbf{S}}+\mathbf{S}\,.\,\dot{\mathbf{d}}=\frac{\mathrm{d}}{\mathrm{d}t}(\mathbf{S}\,.\,\mathbf{d})=0,$$

and therefore $\mathbf{S}\,.\,\mathbf{d}=0$ at all times.

We assume that the applied field is large enough for the term proportional to λ in eqn (4.193a) to be treated as a perturbation. We disregard this term for the time being, and we solve the approximate system:

$$\begin{aligned}\dot{\mathbf{S}}&=\gamma\mathbf{S}\wedge\mathbf{H};\\ \dot{\mathbf{d}}&=\gamma\mathbf{d}\wedge\left(\mathbf{H}-\frac{\gamma\hbar}{\chi}\mathbf{S}\right).\end{aligned} \tag{4.211}$$

The field $\mathbf{H}$ consists of the d.c. field $\mathbf{H}_0$ and, during the pulse $(0\leqslant t\leqslant\tau)$ of a field $\mathbf{H}_1$ perpendicular to $\mathbf{H}_0$, rotating at the frequency $\boldsymbol{\omega}_0=-\gamma\mathbf{H}_0$ and of magnitude:

$$H_1=\varphi/\gamma\tau.$$

The system (4.211) is solved by performing two successive changes of reference frame. We first consider a frame 1 rotating with frequency $\boldsymbol{\omega}_0$ with respect to the laboratory frame, and then a frame 2 which, during the pulse, rotates with respect to the latter with frequency $\omega_1=\varphi/\tau$ around $-\mathbf{H}_1$, and ceases rotating at the end of the pulse. In this last frame, the system (4.211) becomes:

$$\begin{aligned}\dot{\mathbf{S}}&=0;\\ \dot{\mathbf{d}}&=-\frac{\gamma^2\hbar}{\chi}\mathbf{d}\wedge\mathbf{S}.\end{aligned} \tag{4.211$'$}$$

These equations are valid during the pulse as well as afterwards. The vector $\mathbf{S}$ is constant and the vector $\mathbf{d}$ precesses around $\mathbf{S}$ with the frequency $\gamma^2\hbar|\mathbf{S}|/\chi$. The initial condition being that of equilibrium $\gamma\hbar\mathbf{S}=\chi\mathbf{H}_0$, this frequency is equal to $-\omega_0$. The solution of (4.211′) is straightforward and yields the components of $\mathbf{S}$ and $\mathbf{d}$ in a basis of unit vectors of frame 2. It is a simple matter to express the unit vectors of frame 2 as a function of basic unit vectors in the laboratory frame. We give only the result. Let $\mathbf{x}$, $\mathbf{y}$, $\mathbf{z}$ be a basis of unit vectors in the laboratory frame. We have initially $\mathbf{d}(0)=\mathbf{l}=\mathbf{y}$. If α is the angle between $-\mathbf{H}_1(0)$ and $\mathbf{y}$ one obtains after the pulse, that is for $t>\tau$:

$$\begin{aligned}S_x&=S_0\sin\varphi\,\cos(\omega_0t+\alpha);\\ S_y&=S_0\sin\varphi\,\sin(\omega_0t+\alpha);\\ S_z&=S_0\cos\varphi;\end{aligned} \tag{4.212}$$

where $S_0=(\chi/\gamma\hbar)H_0$:

$$d_x=(\cos\varphi-1)\sin(\omega_0 t+\alpha)\cos(\omega_0 t+\alpha);$$
$$d_y=\cos^2(\omega_0 t+\alpha)+\cos\varphi\,\sin^2(\omega_0 t+\alpha); \tag{4.213}$$
$$d_z=-\sin\varphi\,\sin(\omega_0 t+\alpha).$$

We choose as a new initial time the shortest time after the end of the pulse for which $\omega_0 t+\alpha=2\pi n$. Starting from this 'initial' condition, we solve eqn (4.193*a*) by introducing the perturbation in λ. We use the reduced variable $\mathbf{s}=\mathbf{S}/S_0$ and, according to (4.203), the equations (4.193) become:

$$\dot{\mathbf{s}}=-\omega_0\mathbf{s}\wedge\mathbf{z}-\frac{\Omega_A^2}{\omega_0}(\mathbf{d}\wedge\mathbf{l})(\mathbf{d}\,.\,\mathbf{l}); \tag{4.214a}$$

$$\dot{\mathbf{d}}=\omega_0\mathbf{d}\wedge(\mathbf{s}-\mathbf{z}). \tag{4.214b}$$

We seek a steady-state solution of these equations corresponding to a periodic motion of s and d with a frequency $\Omega=\omega_0+\delta$. In a frame rotating around $\mathbf{z}$ with frequency Ω, eqns (4.214*a*) become:

$$\dot{s}_x=\delta s_y+\frac{\Omega_A^2}{\omega_0}d_z\cos\Omega t[d_x\sin\Omega t+d_y\cos\Omega t];$$

$$\dot{s}_y=-\delta s_x-\frac{\Omega_A^2}{\omega_0}d_z\sin\Omega t[d_x\sin\Omega t+d_y\cos\Omega t]; \tag{4.215}$$

$$\dot{s}_z=-\frac{\Omega_A^2}{\omega_0}[d_x\cos\Omega t-d_y\sin\Omega t][d_x\sin\Omega t+d_y\cos\Omega t].$$

For the components of the vector $\mathbf{d}$, which enter (4.215) only as a perturbation, we keep the same amplitudes as in the unperturbed solution (4.213), replacing $(\omega_0 t+\alpha)$ by Ωt. In the rotating frame, their expression is:

$$d_x\simeq\cos\varphi\,\sin\Omega t;$$
$$d_y\simeq\cos\Omega t; \tag{4.216}$$
$$d_z\simeq-\sin\varphi\,\sin\Omega t.$$

As for the vector $\mathbf{s}$ we seek to make its components time independent in the rotating frame by a suitable choice of the frequency Ω. In accordance with the initial conditions (4.212) they should be:

$$s_x=\sin\varphi$$
$$s_y=0 \tag{4.212$'$}$$
$$s_z=\cos\varphi.$$

We carry the values (4.216) into (4.215) and we obtain, after a little algebra:

$$\dot{s}_x = \delta s_y - \frac{\Omega_A^2}{8\omega_0} \sin\varphi[2(1+\cos\varphi)\sin 2\Omega t + (1-\cos\varphi)\sin 4\Omega t];$$

$$\dot{s}_y = -\delta s_x + \frac{\Omega_A^2}{8\omega_0} \sin\varphi[(1+3\cos\varphi) - 4\cos\varphi\cos 2\Omega t - (1-\cos\varphi)\cos 4\Omega t];$$

$$(4.217)$$

$$\dot{s}_z = \frac{\Omega_A^2}{8\omega_0}(1-\cos\varphi)[2(1+\cos\varphi)\sin 2\Omega t + (1-\cos\varphi)\sin 4\Omega t].$$

The frequency $\Omega \sim \omega_0$ being much larger than the frequencies Ω_A^2/ω_0 and δ involved in the evolution of **s**, the oscillating terms on the right-hand sides of (4.217) are ineffective and can be discarded (the argument is the same as for the truncation of $\mathscr{H}_D$ in the rotating frame). The remaining terms are made to vanish by the choice (4.212′):

$$s_x \simeq \sin\varphi; \qquad s_y = 0; \qquad s_z \simeq \cos\varphi;$$

consistent with the initial condition, and:

$$\delta = \frac{\Omega_A^2}{2\omega_0}\frac{1+3\cos\varphi}{4}. \tag{4.218}$$

δ is the frequency shift of the free precession of **s** around **z**. This frequency shift depends on the pulse angle φ.

The reader can check that the assumption that **d** moves at a frequency Ω can be made consistent with eqn (4.214*b*).

3. *Free precession in BW phase*

Whereas in phase A the frequency of the free precession is shifted by an amount $(\Omega_A^2/2\omega_0)(1+3\cos\varphi)/4$ where φ is the angle of the pulse, the behaviour of phase BW is very different. There is no shift as long as φ is smaller than the magic angle $\theta_0 = 104°30'$ such that $1+4\cos\theta_0 = 0$, and a shift,

$$\Delta\omega \simeq -\frac{4}{15}\frac{\Omega_B^2}{\omega_0}(1+4\cos\varphi), \tag{4.219}$$

for $\varphi > \theta_0$.

We shall describe a simplified theory of this phenomenon (Brinkman and Smith, 1975*b*). We start from the Leggett equations (4.196), (4.197), and (4.199) for phase BW, where we introduce explicitly the frequency $\Omega_B^2 = 3g_D\gamma^2/\chi$, and we add a relaxation term for θ towards its equilibrium value

$\theta_0 = \cos^{-1}(-\frac{1}{4})$.

$$\dot{\mathbf{S}} = \gamma \mathbf{S} \wedge \mathbf{H} + \mathbf{u} \frac{4}{15} \frac{\Omega_B^2 \chi}{\hbar \gamma^2} \sin\theta (1 + 4\cos\theta),$$

$$\dot{\theta} = -\mathbf{u} \,.\, \mathbf{K} - \frac{1}{\tau}(\theta - \theta_0), \tag{4.220}$$

$$\dot{\mathbf{u}} = \tfrac{1}{2}\mathbf{u} \wedge \mathbf{K} + \tfrac{1}{2}\,\text{cotan}\left(\frac{\theta}{2}\right)\{\mathbf{u}(\mathbf{u} \,.\, \mathbf{K}) - \mathbf{K}\}.$$

We look for a steady-state, after the pulse of angle φ, that corresponds to a precession of $\mathbf{S}$ of frequency $(\omega_0 + \delta)$ with $\delta \ll \omega_0$, around the axis $\mathbf{z}$ (the direction of $\mathbf{H}$). We introduce the same reduced variable as in the preceding section:

$$\mathbf{s} = \mathbf{S}/S_0.$$

In the frame rotating at the frequency $\omega_0 + \delta$, the system of equations (4.220) becomes (remember $\omega_0 = -\gamma H_0$):

$$\dot{\mathbf{s}} = \delta \mathbf{s} \wedge \mathbf{z} - \mathbf{u} \frac{4}{15} \frac{\Omega_B^2}{\omega_0} \sin\theta (1 + 4\cos\theta); \tag{4.221a}$$

$$\dot{\theta} = -\omega_0[\mathbf{u} \,.\, (\mathbf{s} - \mathbf{z})] - \frac{1}{\tau}(\theta - \theta_0); \tag{4.221b}$$

$$\dot{\mathbf{u}} = \delta \mathbf{u} \wedge \mathbf{z} + \frac{\omega_0}{2} \mathbf{u} \wedge (\mathbf{s} + \mathbf{z}) + \frac{\omega_0}{2}\,\text{cotan}\left(\frac{\theta}{2}\right)\{\mathbf{u}[\mathbf{u} \,.\, (\mathbf{s} - \mathbf{z})] - (\mathbf{s} - \mathbf{z})\}. \tag{4.221c}$$

Since $\delta \ll \omega_0$, as will appear later, we neglect the term $\delta \mathbf{u} \wedge \mathbf{z}$ on the right-hand side of (4.221c).

We define two unitary vectors:

$$\mathbf{a}_+ = (\mathbf{s} + \mathbf{z}) \Big/ \left(2\cos\left(\frac{\varphi}{2}\right)\right);$$
$$\mathbf{a}_- = (\mathbf{s} - \mathbf{z}) \Big/ \left(2\sin\left(\frac{\varphi}{2}\right)\right); \tag{4.222}$$

and we have:

$$\mathbf{a}_+ \wedge \mathbf{a}_- = \mathbf{y}$$
$$\mathbf{y} \wedge \mathbf{a}_+ = \mathbf{a}_-$$
$$\mathbf{a}_- \wedge \mathbf{y} = \mathbf{a}_+.$$

We expand the vector $\mathbf{u}$ as:

$$\mathbf{u} = u_+ \mathbf{a}_+ + u_- \mathbf{a}_- + u_y \mathbf{a}_y, \tag{4.223}$$

and we look for a stationary solution of (4.221) with:

$$s_x = \sin\varphi; \qquad s_y = 0; \qquad s_z = \cos\varphi.$$

The eqns (4.221*b*) and (4.221*c*) become:

$$\dot{\theta} = -2\omega_0 \sin\left(\frac{\varphi}{2}\right) u_- - \frac{1}{\tau}(\theta - \theta_0); \tag{4.224a}$$

$$\dot{u}_+ = \omega_0 \sin\left(\frac{\varphi}{2}\right) \operatorname{cotan}\left(\frac{\theta}{2}\right) u_+ u_-; \tag{4.224b}$$

$$\dot{u}_- = \omega_0 \cos\left(\frac{\varphi}{2}\right) u_y - \omega_0 \sin\left(\frac{\varphi}{2}\right) \operatorname{cotan}\left(\frac{\theta}{2}\right)(1 - u_-^2); \tag{4.224c}$$

$$\dot{u}_y = -\omega_0 u_- \left\{\cos\left(\frac{\varphi}{2}\right) - \sin\left(\frac{\varphi}{2}\right) \operatorname{cotan}\left(\frac{\theta}{2}\right) u_y\right\}. \tag{4.224d}$$

For a given pulse angle φ the steady state solution of eqns (4.224) will yield the value of the frequency shift δ and of the angle θ which describes the BW phase.

When $\dot{\theta} = 0$, eqn (4.224*a*) yields:

$$u_- = \frac{\theta_0 - \theta}{2\omega_0\tau \sin(\varphi/2)}, \tag{4.225}$$

and, if $\omega_0\tau \gg 1$, u_- is very small. By inserting (4.225) into (4.224*b*, *c*, *d*) and neglecting terms in u_-^2, we obtain:

$$0 = \dot{u}_+ = \frac{\theta_0 - \theta}{2\tau} \operatorname{cotan}\left(\frac{\theta}{2}\right) u_+; \tag{4.226a}$$

$$0 = \dot{u}_- = \omega_0 \cos\left(\frac{\varphi}{2}\right) u_y - \omega_0 \sin\left(\frac{\varphi}{2}\right) \operatorname{cotan}\left(\frac{\theta}{2}\right); \tag{4.226b}$$

$$0 = \dot{u}_y = \frac{\theta_0 - \theta}{2\tau} \left\{\operatorname{cotan}\left(\frac{\theta}{2}\right) u_y - \operatorname{cotan}\left(\frac{\varphi}{2}\right)\right\}. \tag{4.226c}$$

The condition (4.226*b*) yields:

$$u_y = \frac{\tan(\varphi/2)}{\tan(\theta/2)}, \tag{4.227}$$

and since $|\mathbf{u}| = 1$, one must have $\theta \geqslant \varphi$. If $\theta > \varphi$, the curly bracket in (4.226*c*) does not vanish and we have then:

$$\theta = \theta_0.$$

This case therefore occurs when the pulse angle φ is smaller than the magic angle θ_0. Equation (4.226*a*) is satisfied and, since $(1 + 4\cos\theta_0) = 0$ the condition $\dot{\mathbf{s}} = 0$ (eqn (4.221*a*)) yields $\delta = 0$: the precession frequency of the magnetization is unshifted.

If $\varphi > \theta_0$ one has necessarily $\theta = \varphi$. Equation (4.226*c*) yields:

$$u_y = 1, \qquad u_+ = 0.$$

Equation (4.221*a*) is then of the form:

$$\dot{s}_y = -\delta \sin\varphi - \frac{4}{15}\frac{\Omega_B^2}{\omega_0}\sin\varphi(1+4\cos\varphi),$$

and the frequency shift, corresponding to $\dot{s}_y = 0$, is:

$$\delta = -\frac{4}{15}\frac{\Omega_B^2}{\omega_0}(1+4\cos\varphi). \tag{4.228}$$

(*e*) *Experimental results*

1. *Longitudinal ringing in phase* A

Figure 4.9 shows the ringing frequency after a sudden change of the applied field by an amount ΔH. For small ΔH the ringing frequency is as expected Ω_A, given by (4.203). For $|\gamma\Delta H| = \Omega_A$ which corresponds to the transition from oscillatory to rotary motion of the pendulum whose equa-

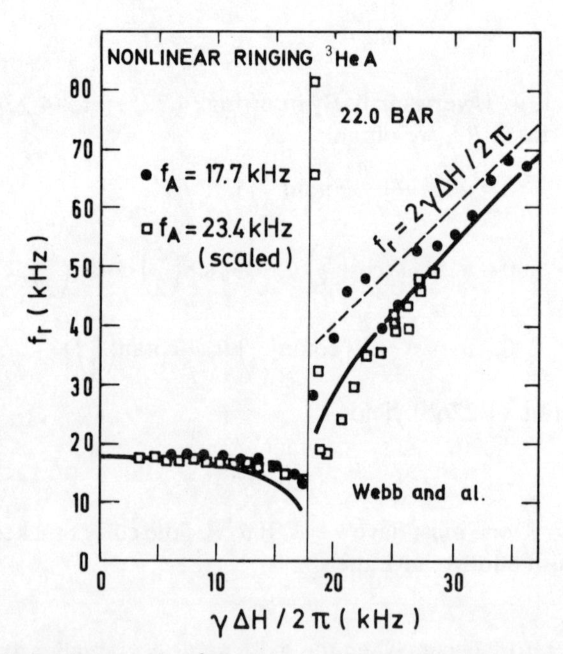

FIG. 4.9. Superfluid phase A of ^{3}He. Ringing frequency after a sudden field jump, against the corresponding Larmor frequency jump. The results for $f_A = 23{\cdot}4$ kHz are scaled down to those for $f_A = 17{\cdot}7$ kHz. (After Webb, Kleinberg, and Wheatley, 1974.)

tions of motion are identical to eqn (4.210), the ringing frequency is expected to reduce to zero. It does not do so in fact. On the other hand the trend for the ringing frequency to become equal to $|2\gamma\Delta H|$ for $|2\gamma\Delta H| \gg \Omega_A$ is qualitatively correct. There is on the whole qualitative agreement with the simple theory of $\mathbf{J}(d)$1 but not the remarkable quantitative agreement that exists for small amplitudes.

2. *Free precession in the* A *phase*

Figure 4.10 exhibits the beautiful agreement between the theoretical frequency shift (4.218) and experiment.

3. *Free precession in the* B *phase*

The agreement with theory is not so good as in the A phase. Whereas for a pulse angle φ smaller than $\theta_0 = 104°30'$ the precession frequency is the Larmor frequency ω_0 as predicted, for $\varphi > \theta_0$ the agreement of the simple theory of $\mathbf{J}(d)$3 and of the frequency shift (4.228) with experiment is only qualitative, as appears in Fig. 4.11.

K. Walls

(*a*) *General*

We have assumed implicitly so far that the superfluid medium was infinite. In the BW phase the presence of a magnetic field was then sufficient for a

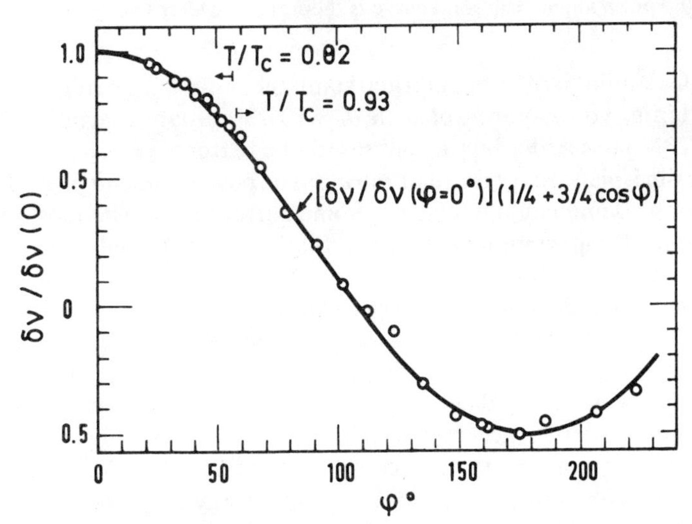

FIG. 4.10. Superfluid phase A of ^{3}He. Variation of the reduced free precession frequency against the pulse angle. The solid curve is theoretical. (After Osheroff and Corruccini, 1975.)

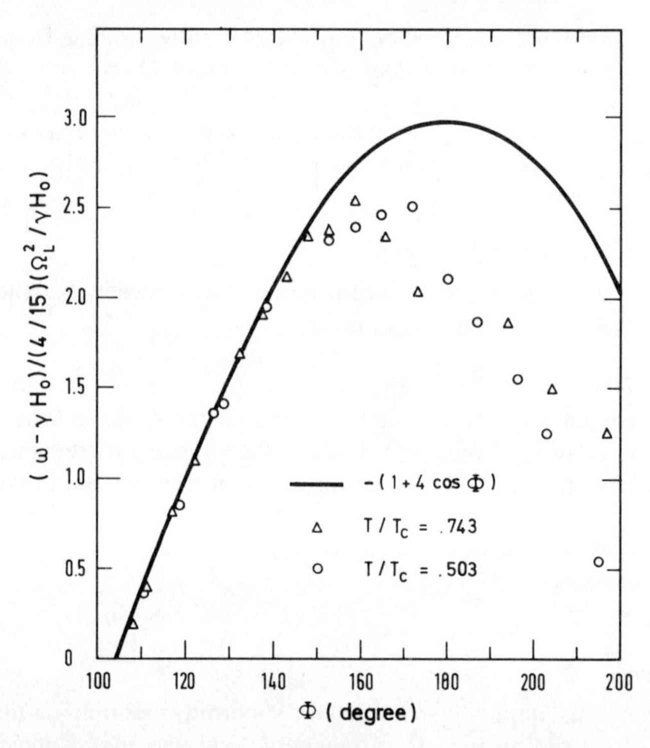

FIG. 4.11. Superfluid phase B of ^{3}He. Variation of the free precession frequency against the pulse angle. The solid curve is theoretical. (After Osheroff, 1975.)

complete definition of its equilibrium state; the order parameter vector $\mathbf{d}(\mathbf{n})$ resulted from $\mathbf{n}$ by a rotation of angle $\theta_0 = 104°30'$ around magnetic field $\mathbf{H}_0$. In the ABM phase a degeneracy subsisted: the vector $\mathbf{d}$ was parallel to $\mathbf{l}$ and $\mathbf{l}$ was lying in a plane perpendicular to $\mathbf{H}_0$ but its position in that plane was left undetermined and could be fixed by some perturbation. The most obvious and the most important of these perturbations and the only one we shall consider in any detail is the presence of walls.

The influence of a wall on a Cooper pair $\mathbf{d}(\mathbf{n})$ can be formulated approximately as follows (Ambegaokar, DeGennes, and Rainer, 1974): consider a pair undergoing a specular reflection on a wall. It will change the sign of the component of $\mathbf{n}$ perpendicular to the wall and will not affect the other components. Interference between incoming and outgoing waves tends therefore to decrease $\mathbf{n}_\perp$ in the vicinity of the wall. In the ABM phase, the way to avoid a loss in superfluid condensation energy is to choose the vector $\mathbf{l}$ perpendicular to the wall, so that the order parameter vector $\mathbf{d}(\mathbf{n})$, which depends on the components of $\mathbf{n}$ perpendicular to $\mathbf{l}$, will not be affected by the reflection.

The case of the BW phase is more complex. Brinkman, Smith, Osheroff, and Blount (1974) use the tentative prescription:

$$\mathbf{d}(\mathbf{n}) = \sqrt{(\tfrac{3}{2})}R\,\mathbf{n}_{\|} = \sqrt{(\tfrac{3}{2})}R\{\mathbf{n} - (\mathbf{n}\,.\,\mathbf{N})\mathbf{N}\}, \tag{4.229}$$

where $\mathbf{n}_{\|}$ is the component of $\mathbf{n}$ parallel to the wall and $\mathbf{N}$ is the normal to the wall. The factor $\sqrt{\frac{3}{2}}$ ensures that $\langle|\mathbf{d}(\mathbf{n})|^2\rangle$ is still unity. Equation (4.229) can also be written:

$$\begin{aligned} d_i(\mathbf{n}) &= \sqrt{(\tfrac{3}{2})}R_{i\alpha}(\mathbf{n}_{\|})_\alpha \\ &= \sqrt{(\tfrac{3}{2})}R_{i\beta}\{\delta_{\beta\alpha} - N_\beta N_\alpha\}n_\alpha. \end{aligned} \tag{4.229'}$$

If we assume now that both a magnetic field and a wall are present, two types of situation can be considered.

(i) No conflict: this is when the field and the wall lead to the same constraints. Thus in phase ABM a field parallel to the wall is a case in point: a vector $\mathbf{l}$ perpendicular to the wall and thus to the field will satisfy the wall boundary condition while minimizing the dipolar plus magnetic energy, and a uniform structure can be expected for the superfluid.

(ii) In general there is conflict between the field and wall requirements. To take up again the ABM phase, but now with the field perpendicular to the wall, near the wall $\mathbf{l}$ will be perpendicular to it, but far away from it, $\mathbf{l}$ will be perpendicular to $\mathbf{H}_0$, in order to minimize the dipolar plus magnetic energy, and thus parallel to the wall. It is clear that it will have to twist somehow in between. The general description of the macroscopic inhomogeneous structures introduced in the superfluid phase by these conflicting requirements as well as by others such as heat flow, superfluid currents, etc., which have been given the general name of textures is one of the most vigorous and promising subfields in the study of superfluid ^{3}He. Sophisticated theories, using the resources of topology have been applied to the prediction and classification of these structures. As explained earlier, these studies are outside the scope of this book and we shall be content to describe a few situations where the observed motion of nuclear magnetization can be explained in simple terms as a consequence of the boundary conditions due to the presence of the walls.

A useful concept is that of healing distance. The superfluid phase does not go abruptly from the structure which satisfies the boundary condition near the wall, to the structure which far from the wall minimizes the magnetic and the dipolar energies, because an abrupt change would cost too much energy. It is a common feature of ordered phases that the free energy density expression contains besides terms quadratic (and quartic) with respect to the order parameter (which is now a function of the space co-ordinate $\mathbf{r}$), also a term proportional to the square of its gradient, which makes rapid spatial changes of this parameter energetically disadvantageous. (Actually in an anisotropic superfluid where the order parameter is not a scalar the gradient term is to be replaced by a more complicated quadratic form constructed from space derivatives of the order parameter (Leggett, 1975).)

What is then meant by healing distance is the distance from the wall where its influence is more or less forgotten and the orientation of the superfluid phase is the same as in the bulk.

Let $\gamma|\nabla\Delta(\mathbf{r})|^2$ be the density of free energy corresponding to the gradient term and δF_b be the bulk orientation energy per unit volume. More precisely δF_b is the difference between the orientation energy calculated at the wall and its minimum value in the bulk.

The healing length R can be estimated in order of magnitude by a crude argument. The contribution of the gradient term to the free energy in a cylinder perpendicular to the wall, of height R and cross section unity is of the order of $\gamma R|\nabla\Delta(r)|^2 \sim \gamma R(\Delta^2/R^2) \sim \gamma\Delta^2/R$. The contribution of the bulk orientation energy is $\sim R\delta F_b$. Their sum is minimum when they are equal that is for:

$$R \sim \left(\frac{\gamma\Delta^2}{\delta F_b}\right)^{1/2}. \tag{4.230}$$

The main point of eqn (4.230) is that the weaker the orientation energy δF_b the greater the healing length. This energy is of the order of χH^2 in the ABM phase, that is of the order of $10^{-7}H^2$ (eqn (4.166)) and only of the order of $g_D(\mu H/\Delta)^2$ in the BW phase, that is of the order of $4.10^{-12}H^2$. According to (4.230) the healing distance is between two and three orders of magnitude greater in the BW phase than in the ABM phase and turns out to be actually of macroscopic size. This is why conspicuous wall effects are much easier to observe in the BW phase than in the ABM phase.

(*b*) *Nuclear magnetism of the BW phase in restricted geometry*

The simplest geometry which exhibits clearly the influence of the walls is a stack of parallel plates with a separation a between adjacent plates.

There are three lengths involved in the problem: the plate separation a, the healing length R, and the so-called coherence length ξ well known in the theory of superconductivity.

Near the wall ξ is the distance over which the relationship $\mathbf{d}(\mathbf{n}) = R\mathbf{n}$ characteristic of the BW phase is to be replaced by (4.229). ξ is a microscopic quantity of the order of a few hundred Angströms, always much smaller than a and R.

Because of (4.229), the dipolar energy $E_d = 3g_D\langle|(\mathbf{n}\,.\,\mathbf{d}(\mathbf{n})|^2\rangle$ defined in (4.155) which in the BW phase is given apart from a constant term, by eqn (4.171):

$$E_d = \tfrac{4}{5}g_D(2\cos^2\theta + \cos\theta), \tag{4.231}$$

will have a different form near the wall. Replacing in (4.155) $\mathbf{d}(\mathbf{n})$ by (4.229) and using (4.194), one finds after a little algebra,

$$E_{\mathrm{d}} = \tfrac{3}{5}g_{\mathrm{D}}\{4\cos^2\theta + (1-\cos\theta)(1+4\cos\theta)\sin^2\lambda + (1-\cos\theta)^2\sin^4\lambda\}, \tag{4.232}$$

where θ is the rotation angle in R and λ the angle of the rotation axis with the normal to the wall. Thus, in contrast to what happens in the bulk, near the wall the dipolar energy depends explicitly on the orientation of the rotation axis with respect to the wall.

The coherence length ξ over which (4.232) is valid, being much smaller than the plate spacing a, over most of the volume it is more advantageous to take $\theta = \theta_0$ which minimizes the dipolar energy (4.231).

Because it would cost far too much gradient energy to have in the thin shell of thickness ξ near the wall a different value of θ, that would minimize (4.232), θ will be equal to θ_0 at the wall also. Equation (4.232) becomes then:

$$E_{\mathrm{d}}(\theta_0) = \tfrac{3}{20}g_{\mathrm{D}}\{1 + \tfrac{25}{4}\sin^4\lambda\}. \tag{4.233}$$

Equation (4.233) is minimum for $\lambda = 0$ that is with the rotation axis $\mathbf{u}$ perpendicular to the wall. And since the bulk dipolar energy (4.231) is indifferent to the orientation of $\mathbf{u}$, it will save gradient energy to have $\mathbf{u}$ perpendicular to the wall through the whole sample.

The situation just described prevails in the absence of the magnetic field. We consider now what happens in the presence of a strong magnetic field parallel to the wall. What is meant by strong is such that the Larmor frequency ω_0 is much larger than the oscillation frequency Ω_{L}, a condition where the dipolar energy is negligible in comparison with the Zeeman energy. Near the wall the susceptibility of the superfluid is anisotropic.

Let us define a frame x, y, z, where z is the normal to the wall and x the direction of the field, and three other vectors:

$$\mathbf{X} = R\mathbf{x}; \qquad \mathbf{Y} = R\mathbf{y}; \qquad \mathbf{Z} = R\mathbf{z}.$$

For every pair $\mathbf{d}(\mathbf{n})$ defined by (4.229′) the susceptibility tensor is axially symmetric around $\mathbf{d}$ with principal values $\chi_\parallel$ and $\chi_\perp$ and according to (4.106) and (4.114):

$$\frac{\chi_\parallel}{\chi_\perp} < 1. \tag{4.234}$$

The magnetic energy for this pair, is:

$$\begin{aligned} E_{\mathrm{m}} &= -\tfrac{1}{2}\{\chi_\parallel(\mathbf{H}\cdot\hat{\mathbf{d}})^2 + \chi_\perp[H^2 - (\mathbf{H}\cdot\hat{\mathbf{d}})^2]\} \\ &= -\tfrac{1}{2}\{\chi_\perp H^2 + (\chi_\parallel - \chi_\perp)(\mathbf{H}\cdot\hat{\mathbf{d}})^2\}, \end{aligned} \tag{4.235}$$

where $\hat{\mathbf{d}}$ is the unit vector $\mathbf{d}/|\mathbf{d}|$.

According to (4.234) the energy is minimum when $(\mathbf{H}\,.\,\hat{\mathbf{d}})^2$ is minimum. We have according to (4.229′):

$$\mathbf{d}(\mathbf{n}) = \surd(\tfrac{3}{2})(n_x\mathbf{X} + n_y\mathbf{Y}),$$

and:

$$\langle(\mathbf{H}\,.\,\hat{\mathbf{d}})^2\rangle = \left\langle\frac{(n_x\mathbf{H}\,.\,\mathbf{X} + n_y\mathbf{H}\,.\,\mathbf{Y})^2}{n_x^2 + n_y^2}\right\rangle$$
$$= \tfrac{1}{2}[(\mathbf{H}\,.\,\mathbf{X})^2 + (\mathbf{H}\,.\,\mathbf{Y})^2]$$
$$= \tfrac{1}{2}[H^2 - (\mathbf{H}\,.\,\mathbf{Z})^2],$$

and the total magnetic energy is therefore minimum when $(\mathbf{H}\,.\,\mathbf{Z})^2 = (\mathbf{H}\,.\,R\mathbf{z})^2$ is maximum, that is if the axis of rotation is such as to bring the normal z to the plane, parallel to the magnetic field. From (4.194) where we replace $\mathbf{n}$ by $\mathbf{z}$ and $\mathbf{d}(\mathbf{n})$ by $\mathbf{h} = \mathbf{H}/|\mathbf{H}|$, remembering that $(\mathbf{h}\,.\,\mathbf{z}) = 0$ (field parallel to the wall) we find for the rotation axis $\mathbf{u}$:

$$(\mathbf{u}\,.\,\mathbf{z})^2 = \cos^2\alpha = (\mathbf{u}\,.\,\mathbf{h})^2 = \frac{-\cos\theta_0}{1-\cos\theta_0} = \tfrac{1}{5}. \tag{4.236}$$

Near the wall the rotation axis must make with the field an angle α given by (4.236). This conflicts with the bulk requirement that $\mathbf{u}$ be parallel to the field in order to minimize the small orientation energy $\delta F_b = -(\mathbf{h}\,.\,\mathbf{u})^2(\mu H/\Delta)^2$. If the spacing a between plates is much smaller than the healing length (4.230) the orientation of the rotation axes $\mathbf{u}$ with respect to the wall, namely an angle α with the field (and also with the normal to the wall) such that $\cos^2\alpha = 1/5$, will prevail through the whole sample. If the magnetic energy is much higher than the dipolar energy this requirement will override the requirement $\mathbf{u}\|\mathbf{z}$ which minimizes the dipolar energy (4.233) near the wall.

The value of the angle α can be obtained from the resonance frequency of the superfluid. We take new axes: z along $\mathbf{H}$ and $\mathbf{u}$ in the xz plane. The equation of motion (4.196) and (4.199′) yield:

$$\dot{\theta} = \frac{\gamma^2\hbar}{\chi}(S_z' \cos\alpha + S_x \sin\alpha);$$

$$\dot{\mathbf{S}} = \gamma\mathbf{S}\wedge\mathbf{H} - \frac{3g_D}{\hbar}(\theta - \theta_0)\mathbf{u}; \tag{4.237}$$

where $S_z' = S_z - (\gamma\hbar)^{-1}\chi H$:

$$\dot{S}_z = -\frac{3g_D}{\hbar}(\theta - \theta_0)\cos\alpha;$$

$$\dot{S}_x = \omega S_y - \frac{3g_D}{\hbar}(\theta - \theta_0)\sin\alpha; \tag{4.238}$$

$$\dot{S}_y = -\omega S_x;$$

$$\ddot{S}_z = \ddot{S}'_z = -\frac{3g_D}{\hbar}\dot{\theta}\cos\alpha = -\Omega_L^2\cos\alpha\,(S'_z\cos\alpha + S_x\sin\alpha);$$

$$\ddot{S}_x = -\omega^2 S_x - \Omega_L^2\sin\alpha\,(S'_z\cos\alpha + S_x\sin\alpha); \tag{4.239}$$

where Ω_L^2 is given by (4.207).

The eigenfrequencies Ω of (4.239) are given by:

$$(\Omega^2 - \Omega_L^2\cos^2\alpha)(\Omega^2 - \omega^2 - \Omega_L^2\sin^2\alpha) - \Omega_L^4\sin^2\alpha\cos^2\alpha = 0;$$

$$\Omega^2 = \tfrac{1}{2}\{(\omega^2 + \Omega_L^2) \pm [(\omega^2 + \Omega_L^2)^2 - 4\omega^2\Omega_L^2\cos^2\alpha]^{1/2}\}. \tag{4.240}$$

If $\omega^2 \gg \Omega_L^2$ one of the frequencies is given by:

$$\Omega^2 \cong \omega^2 + \Omega_L^2\sin^2\alpha = \omega^2 + \tfrac{4}{5}\Omega_L^2. \tag{4.241}$$

The relation (4.241), has been verified quite precisely in slab geometries where the healing distance was much larger than the distance a between plates using the value $\frac{1}{5}$ for $\cos^2\alpha$.

(c) *The wall-pinned mode*

This is another instance of the influence of the wall in the superfluid BW phase (Brinkman, 1974). In this example, the superfluid comes to equilibrium in a field $\mathbf{H} = H\mathbf{x}$, parallel to the wall but so low that $|\omega_0| = |-\gamma H_0| \ll \Omega_L$. Since the dipolar energy is then the dominant one, according to (4.233) the equilibrium orientation $\mathbf{u}_0$ of the rotation axis $\mathbf{u}$ is parallel to the normal z to the wall. After $\mathbf{S}$ has reached its equilibrium value: $\mathbf{S} = \chi\mathbf{H}/\gamma\hbar$, $S = -\chi\omega_0/\gamma^2\hbar$, the field is cut off and the motion of the nuclear magnetization is observed. It is convenient to introduce the unit vector $\mathbf{s} = \mathbf{S}/S = -(\gamma^2\hbar/\chi\omega_0)\mathbf{S}$. The equations (4.196), (4.197), and (4.199) yield:

$$\dot{\theta} = \frac{\gamma^2\hbar}{\chi}(\mathbf{u}\,.\,\mathbf{S}) = -\omega_0(\mathbf{s}\,.\,\mathbf{u});$$

$$\dot{\mathbf{u}} = \frac{\omega_0}{2}(\mathbf{u}\wedge\mathbf{s}) - \omega_0\frac{\sin\theta}{2(1-\cos\theta)}\{\mathbf{s} - \mathbf{u}(\mathbf{u}\,.\,\mathbf{s})\};$$

$$\dot{\mathbf{s}} = -\frac{4}{15}\frac{\Omega_L^2}{\omega_0}\sin\theta(1 + 4\cos\theta)\mathbf{u}. \tag{4.242}$$

At time $t = 0$ when the field is suppressed, $\mathbf{s}$ is parallel to $\mathbf{H}$ and $\mathbf{u}$ is perpendicular to the wall and thus to $\mathbf{H}$ and to $\mathbf{s}$. At time $t = 0$ $(\mathbf{s}\,.\,\mathbf{u})_0 = 0$. It is possible to find a stationary solution of (4.242) such that $(\mathbf{s}\,.\,\mathbf{u}) = 0$ at all times. This is the wall-pinned mode of Brinkman.

The condition $\mathrm{d}/\mathrm{d}t\,(\mathbf{s}\,.\,\mathbf{u}) = (\mathbf{s}\,.\,\dot{\mathbf{u}}) + (\mathbf{u}\,.\,\dot{\mathbf{s}}) = 0$ requires:

$$\frac{\omega_0\sin\theta}{2(1-\cos\theta)} = -\frac{4}{15}\frac{\Omega_L^2}{\omega_0}\sin\theta(1 + 4\cos\theta), \tag{4.243}$$

or

$$1+4\cos\theta=-\frac{15}{8}\frac{\omega_0^2}{\Omega_L^2}\frac{1}{1-\cos\theta}. \tag{4.243'}$$

Since $\omega_0 \ll \Omega_L$ this condition is approximately compatible with: $1+4\cos\theta=0$ which minimizes the dipolar energy in the bulk. The equations of motion of **u** and **s** become:

$$\dot{\mathbf{s}}=\operatorname{cotan}\left(\frac{\theta}{2}\right)\frac{\omega_0}{2}\mathbf{u}$$

$$\dot{\mathbf{u}}=\frac{\omega_0}{2}(\mathbf{u}\wedge\mathbf{s})-\operatorname{cotan}\left(\frac{\theta}{2}\right)\frac{\omega_0}{2}\mathbf{s}. \tag{4.244}$$

The combined motion of **u** and **s** can be described as a rotation around a fixed vector $\mathbf{\Omega}$:

$$\dot{\mathbf{s}}=\mathbf{\Omega}\wedge\mathbf{s};\qquad \dot{\mathbf{u}}=\mathbf{\Omega}\wedge\mathbf{u}. \tag{4.245}$$

To show (4.245) we introduce $\mathbf{n}=\mathbf{u}\wedge\mathbf{s}$. Equations (4.244) become:

$$\dot{\mathbf{s}}=\frac{\omega_0}{2}\operatorname{cotan}\left(\frac{\theta}{2}\right)\mathbf{u};$$

$$\dot{\mathbf{u}}=\frac{\omega_0}{2}\mathbf{n}-\frac{\omega_0}{2}\operatorname{cotan}\left(\frac{\theta}{2}\right)\mathbf{s}; \tag{4.246}$$

$$\dot{\mathbf{n}}=-\frac{\omega_0}{2}\mathbf{u}.$$

Expanding $\mathbf{\Omega}=p\mathbf{s}+q\mathbf{n}+r\mathbf{u}$ and carrying $\mathbf{\Omega}$ into (4.245), we get from (4.246):

$$p=-\frac{\omega_0}{2};\qquad q=-\frac{\omega_0}{2}\operatorname{cotan}\left(\frac{\theta}{2}\right);\qquad r=0.$$

The derivative $\dot{\mathbf{\Omega}}=-(\omega_0/2)\dot{\mathbf{s}}-(\omega_0/2)\operatorname{cotan}(\theta/2)\dot{\mathbf{n}}$ vanishes according to (4.246) and $\mathbf{\Omega}$ is indeed a fixed vector:

$$\mathbf{\Omega}=\mathbf{\Omega}_0=-\frac{\omega_0}{2}\mathbf{x}-\frac{\omega_0}{2}\operatorname{cotan}\left(\frac{\theta}{2}\right)\mathbf{y}. \tag{4.247}$$

The frequency of the motion is given by:

$$\Omega^2=\frac{\omega_0^2}{4}\left\{1+\operatorname{cotan}^2\left(\frac{\theta}{2}\right)\right\}=\frac{\omega_0^2}{4}\left\{1+\left(\frac{\sin\theta}{1-\cos\theta}\right)^2\right\}\simeq\frac{\omega_0^2}{4}\left(1+\frac{15}{25}\right)=2\frac{\omega_0^2}{5};$$

$$\Omega=\omega_0\sqrt{\tfrac{2}{5}}. \tag{4.248}$$

We calculate the energy stored in the wall-pinned mode: It is the sum of the magnetic energy,

$$E_{\rm m}=\frac{\gamma^2\hbar^2S^2}{2\chi}=\tfrac{1}{2}\omega_0^2\frac{\chi}{\gamma^2}=\frac{5}{4}\frac{\Omega^2\chi}{\gamma^2}, \tag{4.249}$$

and of the dipolar energy $E_{\rm d}(\theta)$. In view of the fact that θ is very near θ_0, $E_{\rm d}$ can be approximated as:

$$E_{\rm d}=\frac{3g_{\rm D}}{2}(\theta-\theta_0)^2.$$

From (4.243) we get:

$$\theta-\theta_0\cong\frac{1}{\sin\theta_0(1-\cos\theta_0)}\frac{15}{32}\frac{\omega_0^2}{\Omega_{\rm L}^2}=\frac{3}{2\sqrt{15}}\frac{\omega_0^2}{\Omega_{\rm L}^2};$$

$$E_{\rm d}=\frac{3g_{\rm D}}{2}\frac{3}{20}\left(\frac{\omega_0^2}{\Omega_{\rm L}^2}\right)^2; \tag{4.250}$$

or, using:

$$3g_{\rm D}=\frac{\Omega_{\rm L}^2\chi}{\gamma^2}; \tag{4.107}$$

$$E_{\rm d}=\frac{3}{40}\frac{\omega_0^2\chi}{\gamma^2}\left(\frac{\omega_0^2}{\Omega_{\rm L}^2}\right). \tag{4.251}$$

From a comparison of (4.249) and (4.251) we see that $E_{\rm d}\ll E_{\rm m}$ and $E\simeq E_{\rm m}$.

In the next section we shall calculate the relaxation rate of the energy and compare it with experimental results.

L. Relaxation

In this section we shall describe briefly the phenomenological theory of relaxation due to Leggett and Takagi (1977). Besides its relative simplicity it has to its credit a satisfactory description of some experimental results such as the relaxation of the wall-pinned mode, and the c.w. resonance line width. It must be recognized however that the results of some pulse experiments do not agree with its predictions and that a general theory of spin relaxation in superfluid ^{3}He is still lacking.

(a) *The theory*

1. *Superfluid and normal magnetizations*

It is well known that in normal ^{3}He, as in all normal fluids the component of the total spin along a static magnetic field cannot change in the absence of relaxation induced by collisions between quasi-particles.

Things are different in the superfluid, as explained in **G**(e): the application of a magnetic field changes the form of the creation and annihilation operators $\alpha^\dagger_{\mathbf{k}}$, $\alpha_{\mathbf{k}}$, for the elementary excitations as well as their energies $E_{\mathbf{k}}$. If the field can be applied adiabatically, that is in the absence of collisions, the elementary excitations keep the same populations as in zero field but, as a consequence of the change of their relationships with the $a^\dagger_{\mathbf{k}\sigma}$, $a_{\mathbf{k}\sigma}$, the superfluid exhibits a net magnetization, the so-called adiabatic magnetization.

In **G**(e) we derived the adiabatic susceptibility χ_{p0} in the absence of Fermi liquid effects. These are as usual taken into account by introducing the molecular field:

$$\mathbf{H}_{\mathrm{mol}} = -\frac{Z_0}{4}\frac{\gamma\hbar}{\chi_{n0}}\mathbf{S},$$

and writing:

$$\begin{aligned}\mathbf{M}_p = \gamma\hbar\mathbf{S}_p = \chi_p\mathbf{H} &= \chi_{p0}(\mathbf{H}+\mathbf{H}_{\mathrm{mol}}) \\ &= \chi_{p0}\left(\mathbf{H}-\frac{Z_0}{4}\frac{1}{\chi_{n0}}\chi_p H\right),\end{aligned} \tag{4.252}$$

where $\mathbf{S}_p$ is the spin in the absence of collisions whence:

$$\chi_p = \chi_{p0}\left(1+\frac{Z_0}{4}\frac{1}{\chi_{n0}}\chi_{p0}\right)^{-1},$$

or:

$$\chi_p^{-1} = \chi_{p0}^{-1} + \frac{Z_0}{4\chi_{n0}}. \tag{4.253}$$

On the other hand, at thermal equilibrium, that is when the populations of the elementary excitations are equal to the thermal equilibrium values corresponding to their new energies, the magnetization differs from the preceding one and is equal to:

$$\begin{aligned}\mathbf{M} = \chi\mathbf{H} &= \chi_0(\mathbf{H}+\mathbf{H}_{\mathrm{mol}}) \\ &= \chi_0\left(\mathbf{H}-\frac{Z_0}{4}\frac{1}{\chi_{n0}}\chi\mathbf{H}\right),\end{aligned} \tag{4.254}$$

whence:

$$\chi^{-1} = \chi_0^{-1} + \frac{Z_0}{4\chi_{n0}}. \tag{4.255}$$

Through a microscopic analysis of the superfluid wavefunction, Leggett and Takagi are led to use a two-fluid model somewhat analogous to that used for superfluid ^{4}He: in the superfluid phase, liquid ^{3}He consists of a superfluid and a normal component. The magnetization of the fluid is conceptually separated into two parts: the magnetization $\mathbf{M}_p$ associated with the change of the form of the elementary excitations is identified with the magnetization of the superfluid component, and the magnetization $\mathbf{M}_q = \mathbf{M} - \mathbf{M}_p$ originating from the change of populations is identified with the magnetization of the normal component. We consider the corresponding spin moments:

$$\mathbf{S}_p = \mathbf{M}_p/\gamma\hbar \quad \text{and} \quad \mathbf{S}_q = \mathbf{M}_q/\gamma\hbar.$$

The ratio of their thermal equilibrium values is derived as follows. At equilibrium we have:

$$\mathbf{S}_p = \frac{1}{\gamma\hbar}\chi_{p0}(\mathbf{H}+\mathbf{H}_{\text{mol}}); \tag{4.256}$$

$$\mathbf{S} = \frac{1}{\gamma\hbar}\chi_0(\mathbf{H}+\mathbf{H}_{\text{mol}}); \tag{4.256$'$}$$

whence:

$$\mathbf{S}_q = \mathbf{S} - \mathbf{S}_p = \frac{1}{\gamma\hbar}\chi_{q0}(\mathbf{H}+\mathbf{H}_{\text{mol}}), \tag{4.256$''$}$$

with:

$$\chi_{q0} = \chi_0 - \chi_{p0}. \tag{4.257}$$

The tensorial nature of the susceptibilities no longer appears in these equilibrium equations: in the BW phase the susceptibilities are isotropic, and in the ABM phase they correspond to the equilibrium configuration $\mathbf{d} \perp \mathbf{H}$.

We have, according to (4.256) and (4.256$'$):

$$(\mathbf{S}_p/\mathbf{S})_{\text{eq.}} = \chi_{p0}/\chi_0 = \lambda, \tag{4.258}$$

or else:

$$(\mathbf{S}_p/\mathbf{S}_q)_{\text{eq.}} = \lambda/(1-\lambda). \tag{4.258$'$}$$

The ratio λ is that of the *bare* susceptibilities *without* the corrections for Fermi-liquid effects. The reason is that the spins respond to the total field they experience, external plus molecular, and that both superfluid and normal magnetizations contribute to the latter. For instance, in eqn (4.256):

$$\mathbf{S}_p = \frac{1}{\gamma\hbar}\chi_{p0}\left[\mathbf{H} - \frac{Z_0\gamma\hbar}{4\chi_{n0}}(\mathbf{S}_p+\mathbf{S}_q)\right], \tag{4.259}$$

that is, according to (4.253):

$$\mathbf{S}_p = \frac{1}{\gamma\hbar}\chi_p\left[\mathbf{H} - \frac{Z_0\gamma\hbar}{4\chi_{n0}}\mathbf{S}_q\right]. \tag{4.259$'$}$$

This is different from $(1/\gamma\hbar)\chi_p\mathbf{H}$, which is the value of $\mathbf{S}_p$ if the collisions are absent.

2. *The equations of motion*

We consider as in section **I**(b) motions where the various vectors $\mathbf{d}(\mathbf{n})$ characterizing the superfluid state keep fixed relative orientations with respect to one another. In sections **I**(b) to **I**(d), we described the macroscopic behaviour of the superfluid through a macroscopic motion of the vectors $\mathbf{d}(\mathbf{n})$ and $\mathbf{S}$, that is we tacitly assumed that throughout the evolution of the system $\mathbf{S}_p$ and $\mathbf{S}_q$ remained parallel and kept their equilibrium ratio (4.258′). This last restriction is lifted in the present section, and the vectors $\mathbf{S}_p$ and $\mathbf{S}_q$ are treated as independent variables. The phenomenological theory of **I**(c) and **I**(d) is generalized in two respects. Firstly, we characterize the state of the system by the set of vectors $\mathbf{d}(\mathbf{n})$ plus two spin vectors, $\mathbf{S}_p$ and $\mathbf{S}_q$, and secondly we supplement the equations of motion with relaxation terms in addition to the adiabatic terms considered in **I**(d).

For the adiabatic terms, Leggett and Takagi assume that, since the vectors $\mathbf{d}(\mathbf{n})$ are purely superfluid in nature, they are coupled only to the superfluid magnetization. Furthermore they assume that we may replace the tensorial susceptibilities everywhere by their scalar equilibrium values. This is justified in the BW phase where these susceptibilities are isotropic. It can be justified in the ABM phase by arguments similar to those of section **I**(d), which will not be developed here.

We write then, in place of eqn (4.185):

$$\dot{\mathbf{S}}_{p\text{ adiab.}} = \gamma\mathbf{S}_p \wedge (\mathbf{H}+\mathbf{H}_{\text{mol}}) + \mathbf{R}_d; \tag{4.260a}$$

$$\dot{\mathbf{S}}_{q\text{ adiab.}} = \gamma\mathbf{S}_q \wedge (\mathbf{H}+\mathbf{H}_{\text{mol}}). \tag{4.260b}$$

During the evolution, $\mathbf{S}_p$ and $\mathbf{S}_q$ will not necessarily remain parallel to each other, nor will the molecular field $\mathbf{H}_{\text{mol}}$ be parallel to either one and it will not cancel out of the equations, as in (4.185′).

As for the torque acting on the vector $\mathbf{d}(\mathbf{n})$, it is determined by the total field, $\mathbf{H}+\mathbf{H}_{\text{mol}}$ plus a phenomenological field, linear in $\mathbf{S}_p$ and adjusted so as to annul the torque when $\mathbf{S}_p$ has its equilibrium value (4.259′). We use then, in place of (4.189):

$$\dot{\mathbf{d}}(\mathbf{n})_{\text{adiab.}} = \gamma\mathbf{d}(\mathbf{n}) \wedge \left(\mathbf{H}+\mathbf{H}_{\text{mol}} - \frac{\gamma\hbar}{\chi_{p0}}\mathbf{S}_p\right)$$
$$= \gamma\mathbf{d}(\mathbf{n}) \wedge \left(\mathbf{H} - \frac{\gamma\hbar}{\chi_p}\mathbf{S}_p - \frac{Z_0\gamma\hbar}{4\chi_{n0}}\mathbf{S}_q\right). \tag{4.260c}$$

The procedure used for deriving (4.260c) differs slightly from that of **I**(b) to **I**(d). There, we have used in the Hamiltonian a phenomenological magnetic energy adjusted so as to minimize the total energy when $\mathbf{S}$ had its equilibrium value, and derived the equation of motion of $\mathbf{d}(\mathbf{n})$ through its commutator with this Hamiltonian. Here we adjust the torque so as to annul $\dot{\mathbf{d}}(\mathbf{n})$ at equilibrium. The two approaches are equivalent.

The presence of a term in $\mathbf{S}_q$ on the right-hand side of (4.260*c*) is not contradictory with the fact that $\mathbf{d}(\mathbf{n})$ is coupled only to the superfluid spin $\mathbf{S}_p$. It simply describes the influence of the molecular field produced by $\mathbf{S}_q$ on the equilibrium value of $\mathbf{S}_p$.

In the case of very fast collisions $\mathbf{S}_p$ and $\mathbf{S}_q$ are constantly related through the relative equilibrium condition (4.258′) and the eqns (4.260) reduce to (4.185′) and (4.189′). On the other hand, in the absence of collisions only the superfluid magnetization is present, ($\mathbf{S}_q = 0$) and these eqns (4.260) take again a form analogous to (4.185′) and (4.189′) but with $\mathbf{S}$ replaced by $\mathbf{S}_p$ and χ replaced by χ_p.

As for the dissipative processes giving rise to relaxation, Leggett and Takagi assume that the dominant ones are those producing an evolution of $\mathbf{S}_p$ and $\mathbf{S}_q$ towards relative equilibrium, defined by eqn (4.258′), at constant total magnetization. In accordance with this assumption introducing the vector,

$$\boldsymbol{\eta} = (1-\lambda)\mathbf{S}_p - \lambda\,\mathbf{S}_q = \mathbf{S}_p - \lambda\,\mathbf{S}, \tag{4.261}$$

which describes the departure of $\mathbf{S}_p$ and $\mathbf{S}_q$ from relative equilibrium, we write:

$$(\dot{S}_p + \dot{S}_q)_{\text{rel.}} = \dot{S}_{\text{rel.}} = 0; \tag{4.262}$$

$$\dot{\boldsymbol{\eta}}_{\text{rel.}} = -\frac{1}{\tau}\,\boldsymbol{\eta}. \tag{4.263}$$

The time τ is expected to be comparable with the characteristic quasi-particle lifetime and to have no singularity in the limit $T \to T_c$. The eqns (4.260), (4.262), and (4.263) are finally combined and expressed as a function of the vectors $\mathbf{d}(\mathbf{n})$, $\mathbf{S}$ and $\boldsymbol{\eta}$, to yield the Leggett–Takagi equations of motion:

$$\dot{\mathbf{S}} = \gamma\mathbf{S} \wedge \mathbf{H} + \mathbf{R}_{\mathrm{d}}; \tag{4.264a}$$

$$\dot{\mathbf{d}}(\mathbf{n}) = \mathbf{d}(\mathbf{n}) \wedge \mathbf{K}; \tag{4.264b}$$

$$\dot{\boldsymbol{\eta}} = \gamma\boldsymbol{\eta} \wedge \left[\mathbf{H} - \frac{Z_0\gamma\hbar}{4\chi_{n0}}\mathbf{S}\right] + (1-\lambda)\mathbf{R}_{\mathrm{d}} - \frac{\boldsymbol{\eta}}{\tau}; \tag{4.264c}$$

with:

$$\mathbf{K} = \gamma\left[\mathbf{H} - \frac{\gamma\hbar}{\chi}\mathbf{S} - \frac{\gamma\hbar}{\chi_{p0}}\boldsymbol{\eta}\right], \tag{4.265}$$

as deduced from eqns (4.253), (4.255), (4.260*c*), (4.262), and (4.263).

3. *Relaxation of energy*

We calculate the rate of damping of the energy in the L–T theory. We first assume that the relaxation rate τ^{-1} is much faster than the evolution rates of both E and S. We can then replace $\boldsymbol{\eta}$ by its steady-state value. Since according to this hypothesis $\omega\tau \ll 1$, we have approximately from eqn (4.264*c*):

$$\boldsymbol{\eta} \simeq (1-\lambda)\tau \mathbf{R}_{\mathrm{d}}. \tag{4.266}$$

This vector being small, the system is not far from equilibrium and its energy is approximately:

$$E = -\gamma\hbar(\mathbf{H}\,.\,\mathbf{S}) + \frac{\gamma^2\hbar^2}{2\chi}S^2 + E_{\mathrm{d}}, \tag{4.267}$$

and:

$$\dot{E} = -\hbar\dot{\mathbf{S}}\,.\left\{\gamma\mathbf{H} - \frac{\gamma^2\hbar}{\chi}\mathbf{S}\right\} + \dot{E}_{\mathrm{d}}. \tag{4.267$'$}$$

We recall the definition of $\mathbf{R}_{\mathrm{d}}$:

$$\mathrm{d}E_{\mathrm{d}} = -\hbar(\boldsymbol{\omega}\,.\,\mathbf{R}_{\mathrm{d}})\,\mathrm{d}t, \tag{4.268}$$

where $\boldsymbol{\omega}$ is the instantaneous rotation of the superfluid phase, equal to $-\mathbf{K}$ according to (4.264*b*). By using (4.264*a*), (4.266), and (4.268), eqn (4.267′) becomes:

$$\dot{E} = -\hbar\mathbf{R}_{\mathrm{d}}\,.\left\{\mathbf{K} + \frac{\gamma^2\hbar}{\chi_{p0}}\boldsymbol{\eta}\right\} + \hbar(\mathbf{R}_{\mathrm{d}}\,.\,\mathbf{K})$$

$$= -\frac{\gamma^2\hbar^2}{\lambda\chi_0}(\boldsymbol{\eta}\,.\,\mathbf{R}_{\mathrm{d}});$$

$$\frac{\mathrm{d}E}{\mathrm{d}t} = -\frac{(1-\lambda)}{\lambda\chi_0}\tau\gamma^2\hbar^2 R_{\mathrm{d}}^2. \tag{4.269}$$

(*b*) *Relaxation of the wall-pinned mode*

Since according to (4.243′), θ is near θ_0 we can use for $\hbar R_{\mathrm{d}}$ the approximate expression $-3g_{\mathrm{D}}(\theta-\theta_0)$ (eqn (4.199′)) whence, using (4.207), (4.250), and (4.248):

$$\hbar^2 R_{\mathrm{d}}^2 = 9g_{\mathrm{D}}^2(\theta-\theta_0)^2 = \frac{3}{20}\frac{\chi^2}{\gamma^4}\omega_0^4 = \frac{15}{16}\frac{\chi^2}{\gamma^4}\Omega^4. \tag{4.270}$$

Since in the wall-pinned mode $E_{\mathrm{m}} \gg E_{\mathrm{d}}$ we can take $E = E_{\mathrm{m}} + E_{\mathrm{d}} \cong E_{\mathrm{m}} = (5/4)\Omega^2\chi/\gamma^2$ (eqn (4.249)) and carrying (4.249) and (4.270) into (4.269) we obtain:

$$\frac{d\Omega^2}{dt} = -\frac{3}{4}\frac{1-\lambda}{\lambda}\tau\left(\frac{\chi}{\chi_0}\right)\Omega^4,$$

which integrates to:

$$\Omega^{-2}(t) = \Omega^{-2}(0) + \frac{3}{4}\frac{\chi}{\chi_0}\frac{1-\lambda}{\lambda}\tau t. \tag{4.271}$$

The square of the period $T = 2\pi/\Omega$ of the wall-pinned mode increases during the damping linearly with time, like αt, where:

$$\alpha = \frac{3}{4}\left(\frac{\chi}{\chi_{p0}} - \frac{\chi}{\chi_0}\right)(2\pi)^2\tau. \tag{4.272}$$

The remarkable proportionality of the square of the period of the wall-pinned mode to the time elapsed after the cut-off of the d.c. field predicted by eqn (4.271), has been well verified experimentally. Figure 4.12 exhibits this dependence for several values of ΔH and several values of temperature. It will be noticed that the square of the period for $t = 0$ depends only on the value of ΔH but *not* on the temperature in agreement with the theory of the wall-pinned mode. There is a 10 per cent discrepancy between the experimental value of $f_R(0)/(\gamma\Delta H/2\pi) = 0{\cdot}688$ and its theoretical value which according to eqn (4.248) is $\frac{2}{5} = 0{\cdot}632$.

Another interesting verification of the L–T theory of relaxation is the singularity of α in eqn (4.272) at the critical temperature. The adiabatic susceptibility χ_{p0} given in phase B by eqn (4.138) vanishes at $T = T_c$. It is easy to see from (4.134) that near T_c it behaves like $(T_c - T)^{1/2}$. The relaxation coefficient α behaves near T_c as $(T_c - T)^{-1/2}$ and a plot of $\alpha^{-2}(T)$ as a function of $1 - T/T_c$ should be a straight line going through the origin. That this is indeed so can be seen on Fig. 4.13.

The value of τ the collision time, cannot be extracted unambiguously from eqn (4.272) and the measured values of α, since experimentally there appears a dependence of α on ΔH absent from (4.272). An approximate value $\tau \simeq 7{\cdot}5 \times 10^{-8}$ is obtained for the collision time at $T = T_c$.

(c) *Relaxation of oscillations of small amplitude*

Consider first the ABM phase. We start from eqn (4.201) where in order to take relaxation into account $\mathbf{K} = (-\gamma^2\hbar/\chi)\mathbf{S}'$ has to be replaced by its expression (4.265) with $\boldsymbol{\eta}$ being given the value (4.266),

$$\boldsymbol{\eta} = (1-\lambda)\tau\mathbf{R}_d = (1-\lambda)\tau\{\dot{\mathbf{S}}' - \omega_0(\mathbf{z} \wedge \mathbf{S}')\}. \tag{4.273}$$

The modified eqn (4.201) becomes using (4.203):

$$\ddot{\mathbf{S}}' = \omega_0(\mathbf{z} \wedge \dot{\mathbf{S}}') - \Omega_A^2(\mathbf{y}S'_y + \mathbf{z}S'_z);$$

$$-\frac{\chi\tau}{\lambda\chi_0}(1-\lambda)\Omega_A^2\{\dot{S}'_y\mathbf{y} + \dot{S}'_z\mathbf{z} - \omega_0(\mathbf{z} \wedge \mathbf{S}')_y\mathbf{y}\}; \tag{4.274}$$

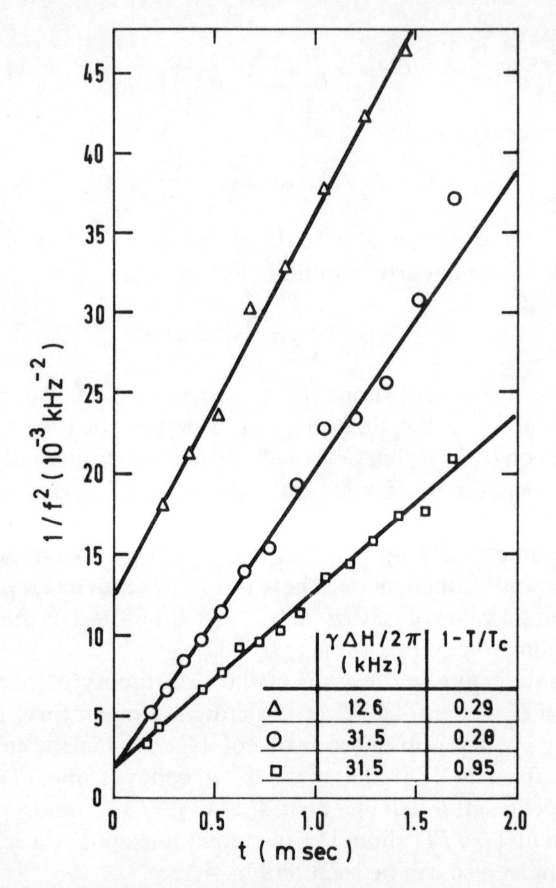

FIG. 4.12. Superfluid phase B of ^{3}He. Variation of the square of the period of the wall-pinned mode against time in the course of relaxation, for various field jumps and temperatures. (After Webb, Sager, and Wheatley, 1975.)

that is:

$$\ddot{S}'_z = -\Omega_A^2 S'_z - \Omega_A^2 \left(\frac{\chi}{\chi_0}\right)\left(\frac{1-\lambda}{\lambda}\right)\tau \dot{S}'_z; \qquad (4.275a)$$

$$\ddot{S}'_x = -\omega_0 \dot{S}'_y; \qquad (4.275b)$$

$$\ddot{S}'_y = \omega_0 \dot{S}'_x - \Omega_A^2 S'_y - \Omega_A^2 \left(\frac{\chi}{\chi_0}\right)\frac{1-\lambda}{\lambda}\tau[\dot{S}'_y - \omega_0 S'_x]. \qquad (4.275c)$$

We look for solutions of (4.275) with a complex time dependence $\exp-(\mathrm{i}\Omega t) = \exp-\mathrm{i}(\Omega_0 - (\mathrm{i}\Gamma/2))t$ with $\Gamma \ll \Omega_0$ (we write $\Gamma/2$ to follow Leggett's notation).

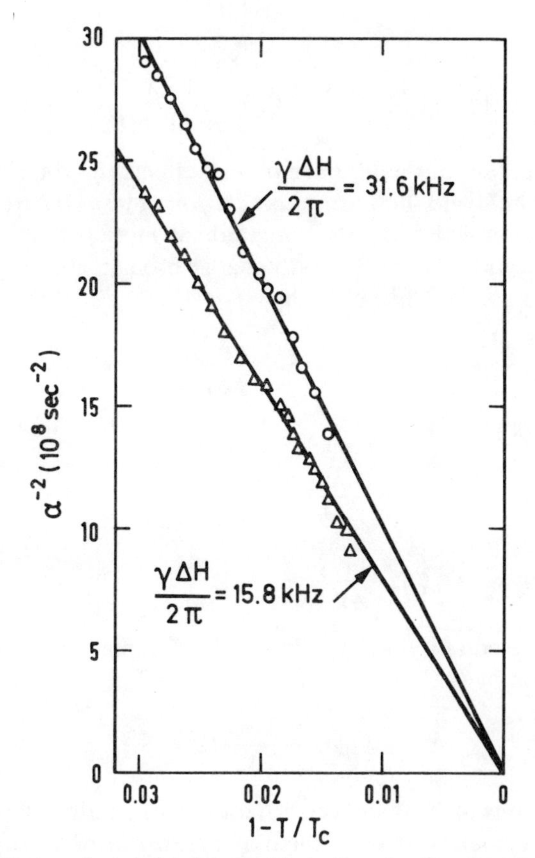

FIG. 4.13. Wall-pinned mode in phase B. Variation of α^{-2} against $(1-T/T_c)$ for two different field jumps, where α is the time-derivative of the square of the period. (After Webb, Sager, and Wheatley, 1975.)

For S_z we find from (4.275*a*):

$$\Omega^2-\Omega_A^2+i\frac{\chi}{\chi_0}\left(\frac{1-\lambda}{\lambda}\right)\Omega\tau\Omega_A^2=0,$$

whence:

$$\Omega_\parallel^2=\Omega_A^2,\qquad \Gamma_\parallel=\frac{\chi}{\chi_0}\left(\frac{1-\lambda}{\lambda}\right)\Omega_A^2\tau. \tag{4.276}$$

Since we assumed $\Omega_A\tau \ll 1$ the assumption $\Gamma \ll \Omega_0$ is justified. The symbols $\Omega_\parallel$ and $\Gamma_\parallel$ refer to components along the field. From (4.275*b*) and (4.275*c*) we get similarly the secular equation:

$$\Omega^2(\Omega^2-\Omega_A^2-\omega_0^2)+i\frac{\chi}{\chi_0}\frac{1-\lambda}{\lambda}\Omega_A^2\tau(\Omega^2-\omega_0^2)\Omega=0, \tag{4.277}$$

which yields:

$$\Omega_{\perp}^2 = \Omega_{\mathrm{A}}^2 + \omega_0^2; \qquad \Gamma_{\perp} = \frac{\chi}{\chi_0}\frac{1-\lambda}{\lambda}\frac{\Omega_{\mathrm{A}}^4 \tau}{\Omega_{\mathrm{A}}^2 + \omega_0^2}. \tag{4.278}$$

In the BW phase, in the absence of wall effects the rotation vector **u** is parallel to the field and the Larmor precession of transverse components is completely uncoupled from the longitudinal motion along the field, as appears from eqns (4.205*a*). The only change brought about by relaxation is the replacement in (4.205*b*) of S_z' by

$$S_z' + \frac{\chi}{\lambda\chi_0}\eta_z = S_z' + \frac{\chi}{\lambda\chi_0}(1-\lambda)\tau\dot{S}_z'.$$

The equation of motion of S_z is:

$$\ddot{S}_z = -\frac{3g_{\mathrm{D}}}{\hbar}\dot{\theta} = -\frac{3g_{\mathrm{D}}}{\hbar}\frac{\gamma^2\hbar}{\chi}\left\{S_z' + \frac{\chi}{\lambda\chi_0}(1-\lambda)\tau\dot{S}_z'\right\};$$

$$\ddot{S}_z = -\Omega_{\mathrm{B}}^2\left\{S_z' + \frac{\chi}{\chi_0}\frac{1-\lambda}{\lambda}\tau\dot{S}_z'\right\}. \tag{4.279}$$

This is the same equation as (4.275*a*) with Ω_{B}^2 replacing Ω_{A}^2 and it is clear that $\Gamma_{\|\mathrm{B}}$ is given by (4.276):

$$\Gamma_{\|\mathrm{B}} = \frac{\chi}{\chi_0}\frac{1-\lambda}{\lambda}\Omega_{\mathrm{B}}^2\tau. \tag{4.280}$$

The expressions (4.276) and (4.280) for the longitudinal line width are in qualitative agreement with experiment if a value of τ of the order of 5.10^{-8} is used, that is of the same order of magnitude as that extracted from the relaxation of the wall-pinned mode.

For the transversal line width $\Gamma_{\perp}$ the relation,

$$\Gamma_{\perp} = \Gamma_{\|}\frac{\Omega_{\mathrm{A}}^2}{\Omega_{\mathrm{A}}^2 + \omega_0^2}, \tag{4.281}$$

a consequence of (4.276) and (4.278) is reasonably well verified by experiment.

We shall not describe the relaxation of the magnetization after *large* pulses which has a very peculiar behaviour at variance with the L–T theory (Brinkman and Smith, 1975*b*).

5

SPIN SYSTEMS AT LOW TEMPERATURE

The spin who came in from the cold.

In the first three chapters, we were concerned as a rule with spin systems whose temperature (or temperatures) was 'high' in the sense that one could use an expansion of the density matrix limited to the first order in inverse spin temperatures. Exceptions to this high temperature case are found at the end of chapter 3, for the description of multiple echoes in solid ^{3}He, and of the approach to the magnetic transition. Chapter 4 is somewhat special: in superfluid ^{3}He, the orbital and spin properties are closely intermingled, which requires a theoretical treatment departing markedly from that used in the rest of nuclear magnetism.

Beginning with this chapter, we consider spin systems whose temperatures are so low that it is no longer permissible to use a linear expansion of the density matrix. These low spin temperatures are produced through the use of dynamic polarization, which is studied in chapter 6.

The present chapter serves mainly as an introduction to the following ones, and is to a large extent devoted to forging several theoretical tools of nuclear magnetic resonance adapted to the low temperature domain, which will be useful in the following chapters.

We consider a spin system in high field, subjected to Zeeman and truncated spin–spin interactions:

$$\mathcal{H} = Z + \mathcal{H}'_{\mathrm{D}}. \tag{5.1}$$

The system may contain more than one spin species.

In section **A**, we derive the general expression for the absorption signal of the spin species I at low temperature.

In section **B** we specialize the analysis of the absorption signal to the so-called pure Zeeman case, that is to the case when the initial density matrix depends only on the Zeeman interaction, and not on the spin–spin interactions. We calculate the first few moments of the resonance line and describe their use for the absolute calibration of the nuclear polarizations.

In section **C** we derive several rigorous results of nuclear magnetic resonance, valid at all temperatures and even in the presence of spin ordering, and we describe some of their applications.

Finally, section **D** describes *several thermodynamic* results in the paramagnetic state, as derived from an expansion of the density matrix to an order higher than the first in inverse temperatures.

A. Formal expression of the resonance signal

The state of the system at the time $t=0$ is described by a density matrix σ_0. We suppose that it is invariant with respect to a rotation of the spins around $0z$, the direction of the d.c. field H_0, i.e. that $[\sigma_0, Z]=0$. We suppose further that in the absence of external perturbation the state of the system is stable, i.e. that for every operator Q one has:

$$\begin{aligned}\langle Q\rangle_0(0) &= \mathrm{Tr}\{\sigma_0 Q\}\\ &= \mathrm{Tr}\{\sigma(t)Q\} = \langle Q\rangle_0(t).\end{aligned} \tag{5.2}$$

This is obviously the case when σ_0 has the canonical form:

$$\sigma_0 = \exp\left(-\sum_\mu \beta_{Z_\mu} Z_\mu - \beta_D \mathscr{H}'_D\right)\Big/ \mathrm{Tr}(\cdots), \tag{5.3}$$

where Z_μ is the Zeeman Hamiltonian of a spin species μ, for we have then:

$$\sigma(t) = \exp(-\mathrm{i}\mathscr{H}t)\sigma_0 \exp(\mathrm{i}\mathscr{H}t) = \sigma_0. \tag{5.4}$$

However, we will consider in chapter 8 systems whose density matrix is *not* of the form (5.3). This is the case for instance for antiferromagnetic phases which break the symmetry of the Hamiltonian. These phases are nevertheless stable: the properties of an antiferromagnet isolated from the lattice do not depend on time, i.e. eqn (5.2) is valid.

The calculation of the resonance signal follows the same lines as in chapter 1, section **A**(*c*). (In section **C**(*f*) we give a different derivation.)

We use the rotating frame picture, i.e., the frame is rotating with respect to each spin species at its respective Larmor frequency. The effective Hamiltonian reduces to $\mathscr{H}'_D$. Following an infinitely sharp pulse of infinitely small angle θ around $0x$, the density matrix σ_0 is transformed into:

$$\sigma(0) = \sigma_0 - i\theta[I_x, \sigma_0], \tag{5.5}$$

and then varies according to:

$$\sigma(t) = \exp(-\mathrm{i}\mathscr{H}'_D t)\sigma(0)\exp(\mathrm{i}\mathscr{H}'_D t).$$

The response to this excitation is:

$$\begin{aligned}R_0(t) &= \langle I_+\rangle(t)\\ &= -\mathrm{i}\theta\, \mathrm{Tr}\{\exp(-\mathrm{i}\mathscr{H}'_D t)[I_x, \sigma_0]\exp(\mathrm{i}\mathscr{H}'_D t) I_+\}\\ &= -\mathrm{i}\theta\, \mathrm{Tr}\{[I_x, \sigma_0]\tilde{I}_+(t)\}\end{aligned} \tag{5.6}$$

where $\tilde{I}_+(t)$ is defined as:

$$\tilde{I}_+ \equiv \tilde{I}_+(t) = \exp(\mathrm{i}\mathcal{H}'_D t) I_+ \exp(-\mathrm{i}\mathcal{H}'_D t). \tag{5.7}$$

Equation (5.6) can be written:

$$\begin{aligned} R_0(t) &= \mathrm{i}\theta \operatorname{Tr}\{\sigma_0[I_x, \tilde{I}_+(t)]\} \\ &= \mathrm{i}\theta \langle [I_x, \tilde{I}_+(t)] \rangle_0 \\ &= \mathrm{i}\frac{\theta}{2} \langle [I_-, \tilde{I}_+(t)] \rangle_0, \end{aligned} \tag{5.8}$$

where $\langle Q \rangle_0$ is defined by (5.2). In (5.8), since $[\sigma_0, I_z] = 0$, the term $\langle [I_+, \tilde{I}_+] \rangle$ vanishes. According to the property (5.2), we have:

$$\begin{aligned} \langle [I_-, \tilde{I}_+(t)] \rangle_0 &= \langle [\tilde{I}_-(-t), I_+] \rangle_0 \\ &= \langle [I_-, \tilde{I}_+(-t)] \rangle_0^*. \end{aligned} \tag{5.9}$$

The linear response to a field H_1 rotating with frequency $\omega = \omega_0 - \Delta$ is, according to eqns (1.49) and (1.49′):

$$R(t) = \langle I_+ \rangle(\Delta) \exp(-\mathrm{i}\Delta t);$$

$$\langle I_+ \rangle(\Delta) = \frac{\omega_1}{\theta} \int_0^\infty R_0(t') \exp(\mathrm{i}\Delta t')\, \mathrm{d}t';$$

or else, according to (5.8):

$$\langle I_+ \rangle(\Delta) = \frac{\mathrm{i}}{2} \omega_1 \int_0^\infty \langle [I_-, \tilde{I}_+(t)] \rangle_0 \exp(\mathrm{i}\Delta t)\, \mathrm{d}t, \tag{5.10}$$

whence the absorption signal:

$$\langle I_y \rangle(\Delta) = \frac{\omega_1}{2} \operatorname{Re}\left\{ \int_0^\infty \langle [I_-, \tilde{I}_+(t)] \rangle_0 \exp(\mathrm{i}\Delta t)\, \mathrm{d}t \right\}.$$

If we make use of (5.9), we obtain finally:

$$\langle I_y \rangle(\Delta) = \frac{\omega_1}{4} \int_{-\infty}^{+\infty} \langle [I_-, \tilde{I}_+(t)] \rangle_0 \exp(\mathrm{i}\Delta t)\, \mathrm{d}t, \tag{5.11}$$

which is the general form of the absorption signal under the conditions that the state of the system be stable and invariant by rotation around $0z$.

We can perform an integration of both sides of eqn (5.11) with respect to Δ. Using the well-known result,

$$\int_{-\infty}^{+\infty} \exp(\mathrm{i}\Delta t)\, \mathrm{d}\Delta = 2\pi\, \delta(t), \tag{5.12}$$

we obtain:

$$\int_{-\infty}^{+\infty} \langle I_y \rangle(\Delta)\, \mathrm{d}\Delta = \frac{\pi}{2}\omega_1 \langle [I_-, I_+] \rangle_0$$

$$= -\pi\omega_1 \langle I_z \rangle_0. \tag{5.13}$$

The area of the absorption signal is proportional to the nuclear polarization:

$$p = \langle I_z \rangle_0 / I N_I$$

of the spins I.

B. The Zeeman resonance signal

As a result of dynamic polarization, a spin system can be brought to a low temperature in the laboratory frame. Its Zeeman and dipolar temperatures are equal but, since the field is high, the spin–spin order is very much smaller than the Zeeman order, and its influence on the resonance signal is negligible. It will therefore make no difference if we assume an infinite spin–spin temperature.

If the system contains only one spin species I, the density matrix σ_0 is:

$$\sigma_0 = \exp(-\beta\omega_0 I_z)/\mathrm{Tr}\{\exp(-\beta\omega_0 I_z)\}, \tag{5.14}$$

which is a product of individual spin matrices:

$$\sigma_0 = \prod_i \sigma_{0i} \tag{5.15}$$

with:

$$\sigma_{0i} = \exp(-\beta\omega_0 I_z^i)/\mathrm{Tr}_i\{\exp(-\beta\omega_0 I_z^i)\}. \tag{5.15$'$}$$

When the system contains several spin species with different Larmor frequencies and possibly different Zeeman temperatures, the density matrix σ_0 is a product of matrices of the form (5.14), i.e. σ_0 is of the form (5.3) with $\beta_{\mathrm{D}} = 0$.

The absorption signal given by eqn (5.11), will first be cast into two slightly different forms. One of them is convenient for a qualitative description of the signal shape, and the second is well adapted to the calculation of its moments. These forms are based on a property derived below.

For σ_0 given by (5.3) we can rewrite,

$$\langle I_- \tilde{I}_+ \rangle_0 = \mathrm{Tr}\{\sigma_0 I_- \tilde{I}_+\},$$

as:

$$\langle I_- \tilde{I}_+ \rangle_0 = \mathrm{Tr}\{\sigma_0 I_- \exp(\beta\omega_0 I_z)\exp(-\beta\omega_0 I_z)\tilde{I}_+\}$$

$$= \mathrm{Tr}\{\exp(-\beta\omega_0 I_z) I_- \exp(\beta\omega_0 I_z)\sigma_0 \tilde{I}_+\}. \tag{5.16}$$

Using the relation:

$$\exp(-\beta\omega_0 I_z) I_- \exp(\beta\omega_0 I_z) = \exp(\beta\omega_0) I_-, \tag{5.17}$$

and the fact that the trace of a product of operators is invariant by cyclic permutation, eqn (5.16) yields:

$$\langle I_- \tilde{I}_+(t)\rangle_0 = \exp(\beta\omega_0)\langle \tilde{I}_+(t) I_-\rangle_0, \tag{5.18}$$

whence we obtain the following relations:

$$\langle [I_-, \tilde{I}_+(t)]\rangle_0 = (\exp(\beta\omega_0) - 1)\langle \tilde{I}_+(t) I_-\rangle_0; \tag{5.19}$$

$$\langle \{I_-, \tilde{I}_+(t)\}\rangle_0 = (\exp(\beta\omega_0) + 1)\langle \tilde{I}_+(t) I_-\rangle_0, \tag{5.20}$$

where $\{A, B\}$ is the anticommutator,

$$\{A, B\} = AB + BA.$$

By combining (5.19) and (5.20) we have further:

$$\langle [I_-, \tilde{I}_+(t)]\rangle_0 = \tanh\left(\frac{\beta\omega_0}{2}\right)\langle \{I_-, \tilde{I}_+(t)\}\rangle_0. \tag{5.21}$$

According to (5.19) and (5.21), eqn (5.11) can be written under the following forms:

$$\langle I_y\rangle(\Delta) = \frac{\omega_1}{4}(\exp(\beta\omega_0) - 1)\int_{-\infty}^{+\infty} \langle \tilde{I}_+(t) I_-\rangle_0 \exp(\mathrm{i}\Delta t)\,\mathrm{d}t, \tag{5.22}$$

and:

$$\langle I_y\rangle(\Delta) = \frac{\omega_1}{4}\tanh\left(\frac{\beta\omega_0}{2}\right)\int_{-\infty}^{+\infty} \langle \{I_-, \tilde{I}_+(t)\}\rangle_0 \exp(\mathrm{i}\Delta t)\,\mathrm{d}t. \tag{5.23}$$

From eqn (5.23) it is possible to get a feeling for the absorption shape and to see how it is modified when the sign of the temperature is changed.

By analogy with (5.9) we have:

$$\langle \{I_-, \tilde{I}_+(-t)\}\rangle_0 = \langle \{I_-, \tilde{I}_+(t)\}\rangle_0^*. \tag{5.24}$$

On the other hand the value of the trace:

$$\langle \{I_-, \tilde{I}_+(t)\}\rangle_0 = \mathrm{Tr}[\sigma_0\{I_-, \tilde{I}_+(t)\}],$$

does not change if we perform a rotation R of angle π around $0x$. Under this rotation:

$$\begin{aligned} RI_+R^\dagger &= I_-; \\ RI_-R^\dagger &= I_+; \\ RI_zR^\dagger &= -I_z; \\ R\mathcal{H}'_{\mathrm{D}}R^\dagger &= \mathcal{H}'_{\mathrm{D}}. \end{aligned} \tag{5.25}$$

The last result is due to the form of $\mathscr{H}'_D$, bilinear in spin operators and invariant by rotation around $0z$.

We have, according to (5.25):

$$\begin{aligned}\langle\{I_-, \tilde{I}_+(t)\}\rangle_0 &= \mathrm{Tr}[R\sigma_0\{I_-, \tilde{I}_+(t)\}R^\dagger]\\ &= \mathrm{Tr}[R\sigma_0 R^\dagger\{I_+, \tilde{I}_-(t)\}]\\ &= \mathrm{Tr}[R\sigma_0 R^\dagger\{\tilde{I}_+(-t), I_-\}]\\ &= \mathrm{Tr}[R\sigma_0 R^\dagger\{I_-, \tilde{I}_+(t)\}^*],\end{aligned} \tag{5.26}$$

where in the last line, we use (5.24) and the fact that $\{A, B\} = \{B, A\}$.

According to the third relation (5.25), $R\sigma_0 R^\dagger$ differs from σ_0 by the replacement in (5.14) of β by $-\beta$.

We write:

$$\langle\{I_-, \tilde{I}_+(t)\}\rangle_0 = H(\beta, t) - \mathrm{i}F(\beta, t), \tag{5.27}$$

where H and F are real.

We have, according to (5.24):

$$H(\beta, -t) + \mathrm{i}F(\beta, -t) = H(\beta, t) - \mathrm{i}F(\beta, t)$$

and, according to (5.26) and the form of $R\sigma_0 R^\dagger$:

$$H(\beta, t) + \mathrm{i}F(\beta, t) = H(-\beta, t) - \mathrm{i}F(-\beta, t),$$

that is, $H(\beta, t)$ is even in β and t, and $F(\beta, t)$ is odd in β and t.

Equation (5.23) becomes:

$$\langle I_y\rangle(\Delta) = \frac{\omega_1}{4}\tanh\left(\frac{\beta\omega_0}{2}\right)\int_{-\infty}^{+\infty}[H(\beta, t)\cos\Delta t + F(\beta, t)\sin\Delta t]\,\mathrm{d}t. \tag{5.28}$$

The absorption signal is the sum of a term even in Δ, originating from $\tanh(\beta\omega_0/2)H(\beta, t)$ and a term odd in Δ, originating from $\tanh(\beta\omega_0/2)F(\beta, t)$. The latter vanishes at high temperature: since $F(\beta, t)$ is odd in β, $F(0, t) = 0$ and, for β very small, it is very small.

The first qualitative result of the theory is that when the temperature is lowered, the absorption signal ceases to be symmetrical with respect to the Larmor frequency.

This is physically plausible: at high temperature the spin-polarization is very small. Each spin has on the average as many neighbours up as down, and local fields of opposite signs have the same statistical weight, resulting in a symmetrical resonance curve. This is not true any more when the polarization is not vanishingly small.

Secondly, as a consequence of the parity with respect to β of $H(\beta, t)$, $F(\beta, t)$, and $\tanh(\beta\omega_0/2)$, one has:

$$\langle I_y\rangle(\Delta, \beta) = -\langle I_y\rangle(-\Delta, -\beta). \tag{5.29}$$

When the sign of the temperature is changed, the new signal is the centrosymmetric of the old one with respect to the Larmor frequency.

(a) The moments of the absorption signal

The moments of the absorption signal are defined with respect to the Larmor frequency ω_0, through the use of the imaginary part of the susceptibility:

$$\chi''(\omega) = \chi''(\omega_0 - \Delta) = \frac{1}{\omega_1}\langle I_y\rangle(\Delta). \tag{5.30}$$

Their formal expression is:

$$M_n = \int_{-\infty}^{+\infty} (\omega - \omega_0)^n \chi''(\omega)\, d\omega \bigg/ \int_{-\infty}^{+\infty} \chi''(\omega)\, d\omega \tag{5.31}$$

or else, according to (5.30):

$$M_n = (-1)^n \int_{-\infty}^{+\infty} \Delta^n \langle I_y\rangle(\Delta)\, d\Delta \bigg/ \int_{-\infty}^{+\infty} \langle I_y\rangle(\Delta)\, d\Delta. \tag{5.32}$$

We introduce the Fourier transform $G(t)$ of $\langle I_y\rangle(\Delta)$ normalized to $G(0)=1$:

$$G(t) = \int_{-\infty}^{+\infty} \langle I_y\rangle(\Delta) \exp(-i\Delta t)\, d\Delta \bigg/ \int_{-\infty}^{+\infty} \langle I_y\rangle(\Delta)\, d\Delta. \tag{5.33}$$

According to (5.22), this function is equal to:

$$G(t) = \langle \tilde{I}_+(t) I_-\rangle_0 / \langle I_+ I_-\rangle_0. \tag{5.34}$$

By taking the successive derivatives of both sides of (5.33) we have:

$$\begin{aligned} \frac{d^n}{dt^n} G(t)\bigg|_{t=0} &= (-i)^n \int_{-\infty}^{+\infty} \Delta^n \langle I_y\rangle(\Delta)\, d\Delta \bigg/ \int_{-\infty}^{+\infty} \langle I_y\rangle(\Delta)\, d\Delta \\ &= (i)^n M_n. \end{aligned} \tag{5.35}$$

On the other hand, we have according to (5.7):

$$\frac{d^n}{dt^n} \langle \tilde{I}_+(t) I_-\rangle_0 \bigg|_{t=0} = (i)^n \operatorname{Tr}\{\sigma_0 [\mathcal{H}_D'^{(n)}, I_+] I_-\}, \tag{5.36}$$

where we use the notation:

$$[\mathcal{H}_D'^{(n)}, A] = \underbrace{[\mathcal{H}_D', [\mathcal{H}_D', [\cdots [\mathcal{H}_D', A] \cdots]]}_{n \text{ times}}. \tag{5.37}$$

Since $[\sigma_0, \mathcal{H}_D'] = 0$, we have for every operator A:

$$\sigma_0 [\mathcal{H}_D'^{(n)}, A] = [\mathcal{H}_D'^{(n)}, \sigma_0 A]. \tag{5.38}$$

With the help of (5.38), and of the property:

$$\operatorname{Tr}\{A[B, C]\} = \operatorname{Tr}\{[A, B] C\}, \tag{5.39}$$

we obtain:

$$\langle[\mathcal{H}_D'^{(n)}, I_+]I_-\rangle_0 = \langle[\mathcal{H}_D'^{(q)}, I_+][I_-, \mathcal{H}_D'^{(n-q)}]\rangle_0. \tag{5.40}$$

Equations (5.34), (5.35), (5.36), and (5.40) yield finally an expression for M_n adapted to calculation:

$$M_n = \langle[\mathcal{H}_D'^{(q)}, I_+][I_-, \mathcal{H}_D'^{(n-q)}]\rangle_0/\langle I_+I_-\rangle_0. \tag{5.41}$$

As a consequence of σ_0 being a product of matrices of individual spins, (eqn (5.15)), the expectation value of a product of operators of different spins is equal to the product of the expectation values of the individual spin operators. For two spins, for instance:

$$\langle I_\alpha^i I_\beta^j\rangle_0 = \langle I_\alpha^i\rangle_0\langle I_\beta^j\rangle_0; \qquad i \neq j.$$

With σ_{0i} of the form (5.15′) the only non-vanishing expectation value is $\langle I_z^i\rangle_0$.

In particular:

$$\langle I_+^i I_-^j\rangle_0 = \delta_{ij}\langle I_+^i I_-^i\rangle_0,$$

and the denominator of (5.41) is of the form:

$$\begin{aligned}\langle I_+I_-\rangle_0 &= \sum_{i,j}\langle I_+^i I_-^j\rangle_0 = \sum_i \langle I_+^i I_-^i\rangle_0\\ &= N_I\langle I_+^i I_-^i\rangle_0\end{aligned} \tag{5.42}$$

where N_I is the number of spins I.

In the following we consider the case when the spin–spin interactions are purely dipolar.

The secular dipolar interaction between like spins I is, according to (1.27), (1.3a) and (1.7a) of the form:

$$\begin{aligned}\mathcal{H}_D'^{II} &= \tfrac{1}{2}\sum_{i,j} A_{ij}[3I_z^i I_z^j - \mathbf{I}^i \cdot \mathbf{I}^j]\\ &= \tfrac{1}{2}\sum_{i,j} A_{ij}[2I_z^i I_z^j - I_x^i I_x^j - I_y^i I_y^j],\end{aligned} \tag{5.43}$$

with:

$$A_{ij} = \tfrac{1}{2}\gamma_I^2\hbar(1-3\cos^2\theta_{ij})r_{ij}^{-3}. \tag{5.44}$$

The secular dipolar interaction between unlike spins I and S is, according to (1.27′), of the form:

$$\mathcal{H}_D'^{IS} = \sum_{i,\mu} B_{i\mu} 2I_z^i S_z^\mu, \tag{5.45}$$

with:

$$B_{i\mu} = \tfrac{1}{2}\gamma_I\gamma_S\hbar(1-3\cos^2\theta_{i\mu})r_{i\mu}^{-3}. \tag{5.46}$$

(b) The first moment

The first moment of the absorption signal of spins I is, according to (5.41):

$$M_1 = \langle [\mathscr{H}'_D, I_+] I_- \rangle_0 / \langle I_+ I_- \rangle_0. \tag{5.47}$$

1. *One spin species*

We consider first the case when the system contains one spin species only. The Hamiltonian $\mathscr{H}'_D$ is then given by (5.43). Since $I_+ = \sum_i I^i_+$ commutes with $\mathbf{I}^i \cdot \mathbf{I}^j$, we have:

$$[\mathscr{H}'_D, I_+] = 3 \sum_{i,j} A_{ij} I^i_+ I^j_z, \tag{5.48}$$

which, when inserted in (5.47), yields:

$$M_1 = 3 \sum_{i,j,k} A_{ij} \langle I^i_+ I^j_z I^k_- \rangle_0 / N_I \langle I^l_+ I^l_- \rangle_0. \tag{5.49}$$

The only non-vanishing terms of the numerator are those for which $i = k \neq j$ (remember that $A_{ii} = 0$).

We have:

$$\langle I^i_+ I^j_z I^i_- \rangle_0 = \langle I^j_z \rangle_0 \langle I^i_+ I^i_- \rangle_0. \tag{5.50}$$

The first term on the right-hand side of (5.50) is equal to:

$$\langle I^j_z \rangle_0 = Ip, \tag{5.51}$$

where p is the polarization of the spins I. The second term cancels out with the denominator, and we get:

$$M_1 = 3Ip \left(\frac{1}{N_I} \sum_{i,j} A_{ij} \right). \tag{5.52}$$

This result can be given a simple physical interpretation in the Weiss theory of magnetism (see e.g. Smart, 1966). The average field seen by a spin I_i as a result of its coupling with the other spins is, according to (5.43), equal to:

$$\mathbf{H}_i = -\left(\frac{1}{\gamma_I} \right) \sum_j A_{ij} [2 \langle I^j_z \rangle \mathbf{n}_z - \langle I^j_x \rangle \mathbf{n}_x - \langle I^j_y \rangle \mathbf{n}_y], \tag{5.53}$$

where $\mathbf{n}_x$, $\mathbf{n}_y$, $\mathbf{n}_z$ are unit vectors along the axes $0x$, $0y$, and $0z$.

The shift of the resonance frequency of the spin I_i produced by this average, or Weiss field is not, as one might naïvely expect, given by $-\gamma_I H_{iz}$, but rather by:

$$-\tfrac{3}{2} \gamma_I H_{iz} = 3 \sum_j A_{ij} \langle I^j_z \rangle_0 = 3Ip \sum_j A_{ij}.$$

This '3/2 effect' originates from the transverse Weiss fields which during the resonance precess at the same frequency as the transverse component of the spin $\mathbf{I}_i$ and act upon it (Kittel, 1948).

We define the Weiss-field factor:

$$a_i = \sum_j A_{ij}, \tag{5.54}$$

whose average value over the spins is:

$$\bar{a} = \frac{1}{N_I} \sum_i a_i. \tag{5.54'}$$

The first moment M_1 (eqn (5.52)) then takes the form:

$$M_1 = 3Ip\bar{a}, \tag{5.55}$$

which is precisely equal to the shift predicted by the Weiss-field approximation. It is worth pointing out that the result (5.55) is *rigorous* and independent of the shape of the absorption signal.

The local Weiss fields are uniform in the following special case (Landau and Lifschitz, 1960):

All spins are crystallographically equivalent, and the sample has the shape of an ellipsoid.

We have in that case:

$$a_i = \bar{a} = a.$$

a_i is computed as follows. The sample is separated into two regions: a sphere centred on the spin I_i, of radius much larger than the interatomic spacing, called 'internal region', and the remainder of the sample, called 'external region' (Fig. 5.1). Let a_i(int.) and a_i(ext.) be the contributions of the two regions to the sum (5.54). a_i(int.) must be computed by a discrete summation over the spins. As for a_i(ext.), the distances r_{ij} are much larger than the interatomic spacing, and the summation can be replaced by an integral which does not depend on the crystalline structure, but only on the concentration of the spins I and the shape of the sample.

For an ellipsoid, $a_i(\text{ext.}) = a(\text{ext.})$ does not depend on the position of the spin I_i. We list below its value for various shapes:

Sphere

$$\bar{a}(\text{ext.}) = 0. \tag{5.56}$$

Infinitely elongated ellipsoid

Field H_0 parallel to the long axis:

$$\bar{a}(\text{ext.}) = -\frac{2\pi}{3}\gamma_I^2 \hbar n_I, \tag{5.57}$$

where n_I is the number of spins I per unit volume.

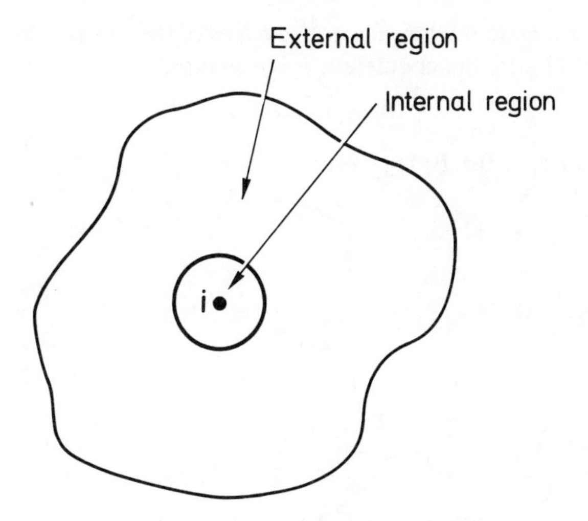

FIG. 5.1. Internal and external regions contributing to the average dipolar field experienced by a given spin I_i.

Field H_0 perpendicular to the axis:

$$\bar{a}(\text{ext.}) = \frac{\pi}{3}\gamma_I^2 \hbar n_I. \tag{5.58}$$

Infinite flat ellipsoid
Field H_0 parallel to the short axis:

$$\bar{a}(\text{ext.}) = \frac{4\pi}{3}\gamma_I^2 \hbar n_I. \tag{5.59}$$

Field H_0 perpendicular to the short axis:

$$\bar{a}(\text{ext.}) = -\frac{2\pi}{3}\gamma_I^2 \hbar n_I. \tag{5.60}$$

For an arbitrary sample shape, if we write:

$$\bar{a}(\text{ext.}) = \xi\frac{2\pi}{3}\gamma_I^2 \hbar n_I, \tag{5.61}$$

it can be shown that:

$$-1 \leqslant \xi \leqslant 2.$$

Besides the case of an ellipsoid, the coefficient ξ can also be expressed in a closed form in the simple case when the sample is a parallelepiped and the field H_0 is parallel to one of its edges (Roinel and Bouffard, 1975). The calculation is trivial but cumbersome and we give only its result.

Let l_x, l_y, and l_z be the lengths of the edges of the sample. With: $a = l_x/l_z$, $b = l_y/l_z$, and $H_0 \| 0z$, the coefficient ξ is equal to:

$$\xi = f(a, b) + f(b, a) - 1, \tag{5.62}$$

where the function $f(a, b)$ is given by:

$$\begin{aligned} f(a, b) = &\left(\frac{a}{\pi b}\right)\{a + 3[(1+a^2+b^2)^{1/2} - (1+a^2)^{1/2} - (a^2+b^2)^{1/2}]\} \\ &+ \left(\frac{1}{\pi ab}\right)[(a^2+b^2)^{3/2} + (1+a^2)^{3/2} \\ &+ (1+b^2)^{3/2} - (1+a^2+b^2)^{3/2} - 1] \\ &+ \left(\frac{3}{\pi a}\right) \ln \frac{b + (1+a^2+b^2)^{1/2}}{(1+a^2)^{1/2}[b + (1+b^2)^{1/2}]} \\ &- \left(\frac{3a}{\pi}\right) \ln \frac{a[b + (1+a^2+b^2)^{1/2}]}{(1+a^2)^{1/2}[b + (a^2+b^2)^{1/2}]} \\ &+ \left(\frac{3}{2\pi}\right) \cos^{-1}\left[1 - \frac{2a^2b^2}{(1+a^2)(1+b^2)}\right]. \end{aligned} \tag{5.63}$$

An experimental check of eqn (5.55) was performed on a sample of calcium fluoride (Roinel and Bouffard, 1975), with:

$$(l_x; l_y; l_z) = (1{\cdot}9; 4{\cdot}7; 0{\cdot}7) \pm 0{\cdot}05 \text{ mm}.$$

Equation (5.62) yields the value,

$$\xi = 0{\cdot}934 \pm 0{\cdot}005,$$

and the simple cubic structure of the ^{19}F spins in CaF_2 corresponds to $a_i\,(\text{int.}) = 0$.

According to (5.61), the theoretical value (5.55) of the first moment, expressed in Gauss, is:

$$m_{1F} = -\xi \times 4{\cdot}09 p_F + C_F \tag{5.64}$$

where C_F is a constant depending on the origin chosen for the field.

Figure 5.2 is a plot of the experimental values of m_{1F} as a function of $4{\cdot}09 p_F$, for values of p_F between 0 and 0·5. The fluorine polarization was calibrated by comparison with the thermal-equilibrium absorption signal at 29 kG and 4·17 K. The slope of the straight line, equal to $-0{\cdot}940$ is very close to the computed one.

Conversely, one could use the linear variation of m_{1F} as a function of the absorption signal area, for an absolute calibration of the polarization. This calibration differs by 7 per cent from that obtained from the thermal-equilibrium signal.

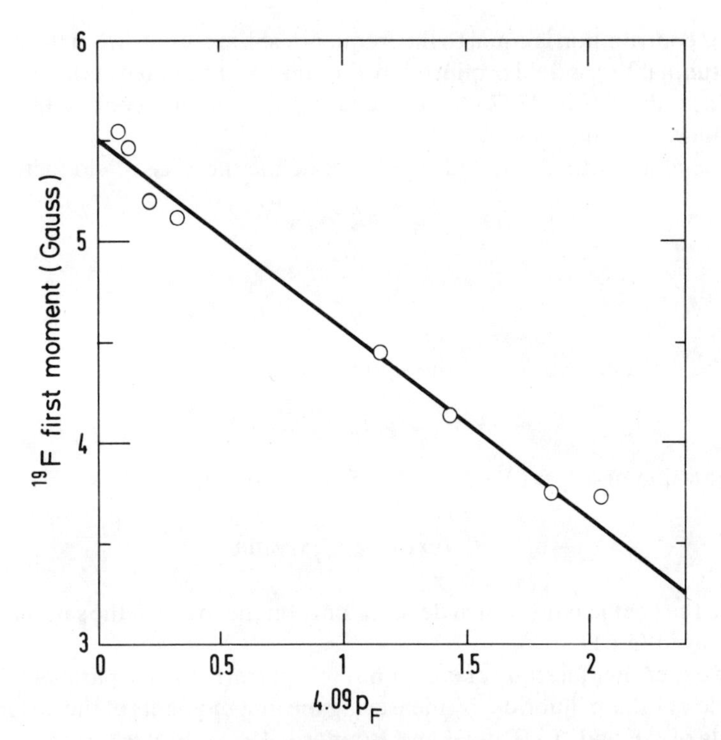

FIG. 5.2. First moment of the ^{19}F absorption signal versus ^{19}F polarization in a parallelepiped of CaF_2. (After Roinel and Bouffard, 1975.)

2. *Several spin species*

We consider the case when the system contains two spin species I and S. The extension to more than two species is straightforward.

We observe the resonance signal of the spins I. The contribution of $\mathcal{H}_D'^{II}$ (eqn (5.43)) to M_1^I is the same as for the one-species case (eqn (5.52)). As for the coupling $\mathcal{H}_D'^{IS}$ (eqn (5.45)), its commutator with I_+ is:

$$[\mathcal{H}_D'^{IS}, I_+] = 2\sum_{i,\mu} B_{i\mu} S_z^\mu I_+^i. \tag{5.65}$$

When inserted into (5.47), it yields the contribution:

$$\begin{aligned} M_1^{IS} &= 2\sum_{i\mu} B_{i\mu} \langle S_z^\mu I_+^i I_-^i \rangle_0 / N_I \langle I_+^l I_-^l \rangle_0 \\ &= \frac{2}{N_I} \sum_{i\mu} B_{i\mu} \langle S_z^\mu \rangle_0 \\ &= 2Sp_S\left(\frac{1}{N_I}\sum_{i,\mu} B_{i\mu}\right). \end{aligned} \tag{5.66}$$

This contribution is equal to the frequency shift resulting from the average longitudinal Weiss field produced by the spins S at the sites of the spins I. In this case, there is no '3/2 effect' because $\mathscr{H}'^{IS}_{\mathrm{D}}$ does not contain transverse couplings for unlike spins.

By analogy with (5.54) and (5.54') we define the Weiss-field factors:

$$a_i^{IS} = \sum_{\mu} B_{i\mu}; \tag{5.67}$$

$$\bar{a}^{IS} = \frac{1}{N_I} \sum_i a_i^{IS}; \tag{5.67'}$$

and the first moment M_1^I of the signal of the spins I is:

$$M_1^I = 3Ip_I\bar{a}^{II} + 2Sp_S\bar{a}^{IS}. \tag{5.68}$$

The value of $\bar{a}^{II}$ (ext.) is given by (5.61), and that of $\bar{a}^{IS}$ (ext.) is:

$$\bar{a}^{IS}(\text{ext.}) = \xi \frac{2\pi}{3} \gamma_I \gamma_S \hbar n_S \tag{5.69}$$

where the coefficient ξ, which depends only on the shape of the sample, is the *same* as in (5.61).

An experimental study of eqn (5.68) was performed on a parallelepipedic sample of lithium fluoride, by measuring the first moments of the absorption signals of ^{19}F and ^{7}Li (Roinel and Bouffard, 1975). In this system:

$$\bar{a}^{II}(\text{int.}) = \bar{a}^{IS}(\text{int.}) = 0.$$

The sample dimensions: $(l_x, l_y, l_z) = (0{\cdot}295; 2{\cdot}32; 3{\cdot}74) \pm 0{\cdot}005$ mm yield the value:

$$\xi = -0{\cdot}775 \pm 0{\cdot}005.$$

The theoretical first moments of the ^{19}F and ^{7}Li resonance signals, are, in Gauss units:

$$\begin{aligned} m_{1\mathrm{F}} &= 0{\cdot}775(5p_{\mathrm{F}} + 3{\cdot}9p_{\mathrm{Li}}) + C_{\mathrm{F}}; \\ m_{1\mathrm{Li}} &= 0{\cdot}775(3{\cdot}35p_{\mathrm{F}} + 5{\cdot}85p_{\mathrm{Li}}) + C_{\mathrm{Li}}; \end{aligned} \tag{5.70}$$

where C_{F} and C_{Li} are experimental constants.

These equations are used for an absolute calibration of the polarizations as follows. For a series of different couples of polarizations $(p_{\mathrm{F}}, p_{\mathrm{Li}})$, one measures the first moments and the areas of the absorption signals, proportional to the polarizations:

$$p_{\mathrm{F}} = k_{\mathrm{F}} A_{\mathrm{F}}; \qquad p_{\mathrm{Li}} = k_{\mathrm{Li}} A_{\mathrm{Li}};$$

where k_{F} and k_{Li} are the calibration constants to be determined.

The eqns (5.70) are then of the form:

$$m_1 = k_1 a_1 A_1 + k_2 b_1 A_2 + C_1;$$
$$m_2 = k_1 a_2 A_1 + k_2 b_2 A_2 + C_2; \tag{5.71}$$

where the indices 1 and 2 stand for ^{19}F and ^{7}Li.

The coefficients k_1, k_2, C_1, C_2 are unknown whereas a_i and b_i have known theoretical values (eqns (5.70)), and the moments m_i and the areas A_i are measured. Calibrated values of k_F and k_{Li} can be extracted from (5.71) if a sufficient number of measurements of A_1, A_2, m_1, m_2 are performed. In the present experiment nine such sets of measurements were made with values of p_F ranging from $-0{\cdot}49$ to $+0{\cdot}74$ and values of p_{Li} from $-0{\cdot}33$ to $+0{\cdot}51$. Figure 5.3 reproduces the variation of the experimental first moment as a function of those calculated from (5.71) with the best-fit values of k_F and k_{Li}.

By comparing this calibration with that obtained from the high temperature thermal equilibrium signals, one finds:

$$p_F(\text{cal.})/p_F(\text{H.T.}) = 1{\cdot}03;$$
$$p_{Li}(\text{cal.})/p_{Li}(\text{H.T.}) = 0{\cdot}99;$$

which is a very satisfactory agreement.

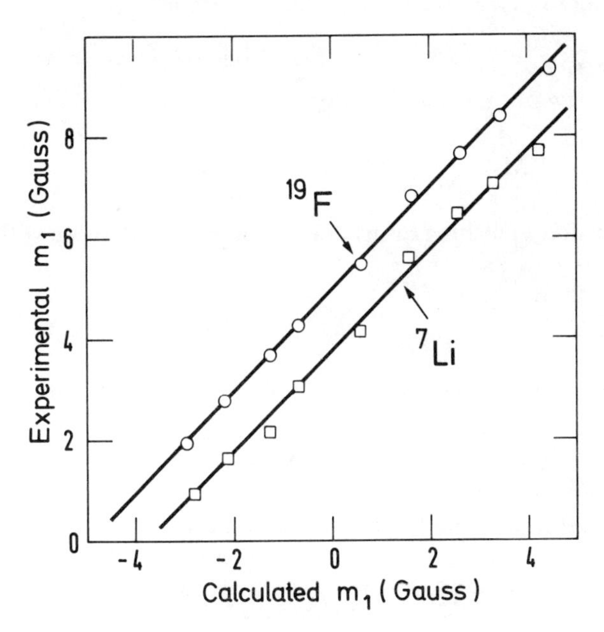

FIG. 5.3. Experimental first moments of ^{7}Li and ^{19}F (to within an arbitrary constant) against their best-fit values in a parallelepiped of LiF. (After Roinel and Bouffard, 1975.)

It must be noted that besides the spins ^{19}F and ^{7}Li, the system contains two more spin species: the ^{6}Li spins, representing 7 per cent of the lithium concentration, and paramagnetic F centres used for the dynamic polarization (chapter 6). Each spin species yields a contribution to the first moments of ^{19}F and ^{7}Li of the form (5.66). The contribution of ^{6}Li is negligible, because of its low abundance and of its small magnetic moment. As for the paramagnetic centres, their polarization in the conditions of the experiment is equal to unity, and their constant contribution to the first moments is included in C_1 and C_2.

The exploitation of the paramagnetic contribution to the nuclear first moments of the NMR lines will be described in chapter 6.

(c) *The second moment*

The calculation of the second moment will be limited to the case of a simple cubic system of like spins 1/2 in a spherical sample. This second moment will then be used for the calibration of the polarization of the ^{19}F spins in CaF_2. In that case the first moment is zero and cannot be used for a calibration of the polarization. The generalization of the calculation of M_2 when $I > 1/2$, and when $M_1 \neq 0$ is given in Abragam, Chapellier, Jacquinot, and Goldman (1973).

According to (5.41), the second moment is equal to:

$$M_2 = \langle[\mathcal{H}'_D, I_+][I_-, \mathcal{H}'_D]\rangle_0 / \langle I_+ I_-\rangle_0, \tag{5.72}$$

or else, in view of (5.42) and (5.48):

$$M_2 = \frac{9}{N_I} \sum_{i,j,k,l} A_{ij} A_{kl} \langle I_z^i I_+^j I_-^k I_z^l\rangle_0 / \langle I_+^m I_-^m\rangle_0. \tag{5.73}$$

The only non-vanishing terms of the numerator are those with $j = k$ and, since $A_{ii} = 0$, $j \neq i$ and $j \neq l$.

We have then:

$$\langle I_z^i I_+^j I_-^j I_z^l\rangle_0 = \langle I_z^i I_z^l\rangle_0 \langle I_+^j I_-^j\rangle_0.$$

The last term cancels out with the denominator of (5.73), and we obtain:

$$M_2 = \frac{9}{N_I} \sum_{i,j,l} A_{ij} A_{jl} \langle I_z^i I_z^l\rangle_0, \tag{5.74}$$

or else, by separating the terms $i = l$ and $i \neq l$:

$$\begin{aligned} M_2 &= \frac{9}{N_I}\left\{\sum_{i,j} A_{ij}^2 \langle (I_z^i)^2\rangle_0 + \sum_{i \neq j \neq l} A_{ij} A_{jl} \langle I_z^i\rangle_0 \langle I_z^l\rangle_0\right\} \\ &= \frac{9}{N_I}\left\{\langle (I_z^i)^2\rangle_0 \sum_{i,j} A_{ij}^2 + (\langle I_z^i\rangle_0)^2 \sum_{i \neq j \neq l} A_{ij} A_{jl}\right\}. \end{aligned} \tag{5.75}$$

The last sum can be written:

$$\sum_{i\neq j\neq l} A_{ij}A_{jl} = \sum_{i,j,l} A_{ij}A_{jl} - \sum_{i,j} A_{ij}^2$$
$$= \Big(\sum_{i,j} A_{ij}\Big)^2 - \sum_{i,j} A_{ij}^2. \tag{5.76}$$

In the system under consideration: simple cubic lattice and spherical sample:

$$\sum_{i,j} A_{ij} = 0. \tag{5.77}$$

By inserting (5.77) into (5.76) we get:

$$M_2 = \frac{9}{N_I}\sum_{i,j} A_{ij}^2\{\langle I_z^{i2}\rangle_0 - (\langle I_z^i\rangle_0)^2\}. \tag{5.78}$$

Since all spins are crystallographically equivalent:

$$\sum_{i,j} A_{ij}^2 = N_I \sum_j A_{ij}^2$$

and, for spins $\frac{1}{2}$:

$$\langle (I_z^i)^2\rangle = \tfrac{1}{4}; \qquad \langle I_z^i\rangle_0 = \tfrac{1}{2}p.$$

We obtain finally:

$$M_2(p) = \tfrac{9}{4}\sum_j A_{ij}^2(1-p^2)$$
$$= M_2(0)(1-p^2), \tag{5.79}$$

where $M_2(0)$ is the second moment at high temperature ($p \ll 1$).

What is expected is therefore a linear variation of M_2 as a function of the square of the absorption signal area (for $A \propto p$), whence we can obtain an absolute calibration of the polarization.

This experiment was performed with two different spherical samples of CaF_2, with $H_0 \| [100]$. Each sample contains a small, and different, amount of magnetic centres Tm^{2+} and Tm^{3+}. The results are shown in Fig. 5.4. A linear variation of M_2 as a function of A^2 is actually observed, but the values of M_2 extrapolated to zero area are different in each sample: $15{\cdot}35\,G^2$ and $13{\cdot}35\,G^2$ in samples A and B, respectively. Furthermore these values differ from the theoretical value:

$$M_2(0)_{\text{th.}} = 12{\cdot}95\,G^2.$$

The excess of second moment is attributed to the influence of the paramagnetic impurities. The contribution of these impurities to M_2 must be analysed in detail if M_2 is to be used for calibration purposes. Such an analysis is given in Abragam, Chapellier, Jacquinot, and Goldman (1973). It

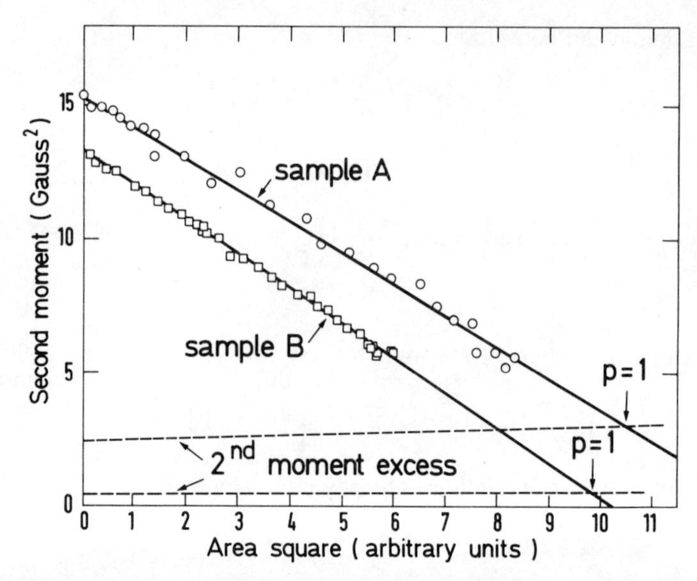

FIG. 5.4. Second moment of the ^{19}F absorption signal against its area square, for two spherical samples of CaF_2 with $H_0 \| [100]$. (After Abragam, Chapellier, Jacquinot, and Goldman, 1973.)

will not be repeated here, and we content ourselves here with a few qualitative indications.

It can be shown that a small concentration of randomly distributed paramagnetic impurities produces at the nuclear sites a Lorentzian distribution of magnetic fields. The observed absorption signal is a convolution of this Lorentzian shape with the signal that would be observed in the absence of impurities. It has Lorentzian wings which yield an important contribution to the experimental second moment. In practice the computation of M_2 is limited to a 'window' of width $2\Delta_0$ (in frequency units) centred on the signal, outside of which the signal is indistinguishable from the noise. If δ is the half-width in frequency units of the Lorentzian distribution of fields, one obtains the following relationship between the experimental second moment M_2' and the theoretical second moment M_2:

$$\begin{aligned} M_2' &\simeq M_2 + \frac{2}{\pi}\delta\Delta_0\left(1 - \frac{2M_2}{\Delta_0^2}\right) \\ &= M_2(0)(1-p^2) + \frac{2}{\pi}\delta\Delta_0\left[1 - \frac{2M_2(0)}{\Delta_0^2}(1-p^2)\right]. \end{aligned} \tag{5.80}$$

This correction is valid in the limit:

$$(\delta/\Delta_0);\ (M_2/\Delta_0^2) \ll 1.$$

The correction term is adjusted to the excess second moment at zero polarization. It is apparent on (5.80) that M_2' varies linearly with p^2, but that this dependence involves the correction due to the impurities. This correction is shown in Fig. 5.4. When the latter is taken into account, one obtains the curves shown in Fig. 5.5 for the variation of M_2' as a function of the absolute polarization.

This calibration method yields the absolute value of the polarization with an accuracy between 3 and 5 per cent.

(*d*) *Higher moments*

We give without proof the forms of M_3 and M_4 in a spherical sample of a simple cubic system of spins $\frac{1}{2}$ (Abragam, Chapellier, Jacquinot, and Goldman, 1973).

$$M_3 \simeq -0{\cdot}39[M_2(0)]^{3/2}p(1-p^2); \tag{5.81}$$

$$M_4 \simeq 2{\cdot}18[M_2(0)]^2(1-p^2)(1-0{\cdot}42p^2). \tag{5.82}$$

The non-vanishing value of M_3 is characteristic of the signal asymmetry when the polarization increases. Its measurement should in principle allow an independent absolute calibration of the polarization: M_3, proportional to $p(1-p^2)$, goes through a maximum for $p = 1/\sqrt{3}$. In practice the experimental value of M_3 is very sensitive to the admixture of a small dispersion component, and no reliable calibration could be obtained by this method.

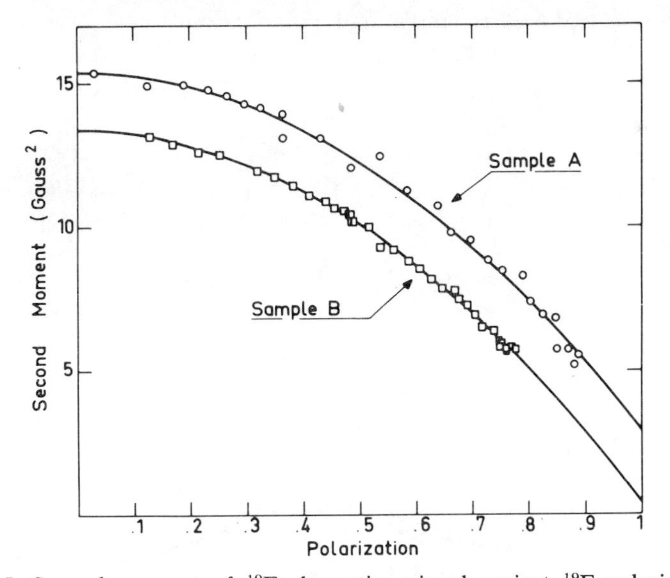

FIG. 5.5. Second moment of ^{19}F absorption signal against ^{19}F polarization as deduced from the results of Fig. 5.4 for the two samples of CaF_2. (After Abragam, Chapellier, Jacquinot, and Goldman, 1973.)

As for the fourth moment M_4, its variation with p at large polarizations shows that:

$$M_4/(M_2)^2 \sim (1-p^2)^{-1} \gg 1.$$

The increase of M_4/M_2^2 when $(1-p^2)$ goes to zero suggests, in accordance with chapter 1, section **B**(d)2, a progressive change in the lineshape from roughly Gaussian to Lorentzian. In this model the linewidth falls off as $(1-p)$, that is faster than $\sqrt{M_2}$ which goes as $(1-p)^{1/2}$. This qualitative discussion neglects the complication due to the non-vanishing third moment.

A Lorentzian shape for polarizations very nearly unity is not unexpected. The very few spins which point in the wrong direction can be considered as dilute impurities in a sea of rigid uniform magnetization which in a sphere produce neither shift nor broadening. It is well known (see Abragam, 1961, chapter IV) that under great dilution the lineshape tends to be Lorentzian.

To conclude this discussion on the NMR of highly polarized spins one can make the following admittedly crude statements. The effect of a large polarization is to shift and to narrow the NMR lines, the shift being proportional to p and the width to $(1-p)$. The calculation and the measurement of the first and the second moments is the correct way to measure and to describe this shift and this narrowing.

On Fig. 5.6 are shown the signal shapes observed in CaF_2 with $H_0 \| [100]$ at various polarizations. They exhibit an asymmetry at intermediate polarizations, and a narrowing at high polarization.

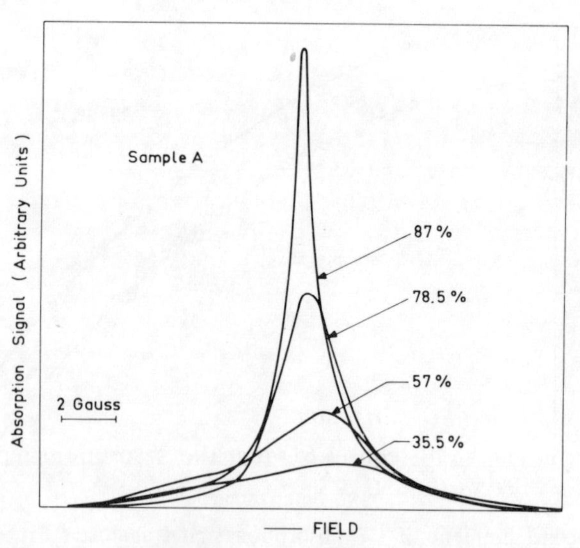

FIG. 5.6. Shape of ^{19}F absorption signal in CaF_2 with $H_0 \| [100]$ at various polarizations. (After Abragam, Chapellier, Jacquinot, and Goldman, 1973.)

C. Rigorous NMR results

In this section we derive several NMR results which are valid rigorously in the whole temperature range and even in the presence of spin ordering. Under ideal conditions they allow the unambiguous determination of well-defined physical properties of the spin system. Actual experimental conditions are often not too far from ideal, and rigorous results are then still valuable tools yielding information with reasonable accuracy.

Two rigorous results have already been established (eqns (1.61) and (5.13)). They are recalled in this section for completeness.

(*a*) *Area of the absorption signal*

The area of the absorption signal $I_y(\Delta)$ is related to the nuclear polarization by eqn (5.13):

$$\int_{-\infty}^{+\infty} \langle I_y \rangle(\Delta)\, \mathrm{d}\Delta = -\pi\omega_1 \langle I_z \rangle_0. \tag{5.13}$$

What is experimentally recorded is a voltage proportional to $\langle I_y \rangle$:

$$v(\Delta) = \xi \langle I_y(\Delta) \rangle,$$

where ξ will be called the gain of the receiver. The area of the observed signal is then:

$$\int_{-\infty}^{+\infty} v(\Delta)\, \mathrm{d}\Delta = -\xi\pi\omega_1 \langle I_z \rangle_0. \tag{5.83}$$

The observation of a pure Zeeman signal, when the polarization is known through one of the absolute calibration methods described in section **B**, yields according to (5.83) a calibration of the receiver gain, or more precisely of the product $\xi\omega_1$. As will be shown later, when the state of the system is modified, by complete or partial adiabatic demagnetization for instance, the absorption signal observed with the same receiver and the same r.f. field is still proportional to $\xi\omega_1$. The knowledge of this factor makes it possible to use the absorption signal for deriving calibrated values of various physical quantities.

If necessary, the r.f. amplitude can itself be calibrated by a method described in section **C**(*c*)1.

(*b*) *Rate of change of the polarization*

It was shown in chapter 1, eqn (1.61) that the absorption signal is quite generally proportional to the time-derivative of the longitudinal polarization:

$$\frac{\mathrm{d}}{\mathrm{d}t}\langle I_z \rangle = \omega_1 \langle I_y \rangle. \tag{1.61}$$

In the course of an irradiation which produces a variation of the longitudinal polarization, the latter can be monitored by observing the absorption signal on the run. We have according to (1.61):

$$\delta\langle I_z\rangle = \omega_1 \int \langle I_y\rangle \, \mathrm{d}t. \tag{5.84}$$

This result is used in the next section. Another example will be described in chapter 8.

(c) Slightly saturating linear passages

We consider a linear sweep of the external field through the resonance line of spins I in the presence of r.f. irradiation. It would be equivalent to perform a linear sweep of the r.f. frequency. This sweep produces a saturation, both of the polarization of the spins I and of the dipolar energy. We suppose that ω_1 is small enough and the sweep fast enough for this saturation to be small. The duration of the passage through the line is however much longer than T_2. In other words if,

$$\mathrm{d}\Delta/\mathrm{d}t = \gamma \, \mathrm{d}H/\mathrm{d}t = \dot{\Delta},$$

is the constant sweep rate, the frequency interval $\dot{\Delta}T_2$ swept in a time T_2 is much less than the absorption linewidth T_2^{-1}:

$$\dot{\Delta}T_2^2 \ll 1. \tag{5.85}$$

We want to calculate under these conditions the saturation produced by a passage. This problem was analysed in the high-temperature domain in Goldman, 1970, chapter 5. Its extension to low temperatures is described in Appendix C of Abragam, Chapellier, Jacquinot, and Goldman (1973).

1. *Saturation of the polarization*

The variation of $\langle I_z\rangle$ caused by a linear passage is calculated through (5.84), which is a consequence of (1.61). The relation (1.61) is general and rigorous, but the value of $\langle I_y\rangle$ on its right-hand side depends on the experimental conditions. When the field is fixed, that is when Δ is constant, it requires a time of the order of T_2 after the application of the r.f. field for $\langle I_y\rangle$ to reach its steady-state value (5.11). When the field sweep is so slow as to satisfy the condition (5.85), $\langle I_y\rangle$ will at all times be very nearly equal to the steady-state value (5.11) corresponding to the instantaneous value of Δ. The time-dependence of $\langle I_y\rangle$ being then entirely determined by that of Δ, (5.84) can be written:

$$\delta\langle I_z\rangle = \frac{\omega_1}{\dot{\Delta}} \int_{-\infty}^{+\infty} \langle I_y\rangle(\Delta) \, \mathrm{d}\Delta. \tag{5.86}$$

Although the total sweep is finite, the integral in (5.86) can be extended from $-\infty$ to $+\infty$ if the sweep starts well below the resonance line and ends up well above it.

According to (5.13), eqn (5.86) yields:

$$\frac{\delta\langle I_z\rangle}{\langle I_z\rangle_0}=-\pi\frac{\omega_1^2}{\dot{\Delta}}=-\pi\frac{\gamma H_1^2}{\mathrm{d}H/\mathrm{d}t}=-\varepsilon. \quad (5.87)$$

The relative decrease of polarization due to a passage depends only on ω_1 and $\dot{\Delta}$. It depends neither on the polarization, nor on the absorption line shape.

This result is valid only insofar as:

$$|\delta\langle I_z\rangle/\langle I_z\rangle|=\varepsilon\ll 1.$$

Equation (5.87) allows an absolute calibration of H_1. The experimental procedure consists in measuring the loss of polarization caused by a succession of n saturating passages. One has:

$$\frac{\langle I_z\rangle(n)}{\langle I_z\rangle(0)}=(1-\varepsilon)^n\simeq\exp(-n\varepsilon). \quad (5.88)$$

Although ε is very small, if the number n of passages is large enough the decrease of polarization is important and can be measured accurately, whence an accurate determination of ε and ω_1 (or H_1).

An experimental verification of eqn (5.88) has been performed on the ^{19}F resonance in CaF_2 (Abragam, Chapellier, Jacquinot, and Goldman, 1973). On Fig. 5.7 is shown the experimental variation of the polarization as a function of the number of passages. The variation is exponential from $p\simeq 0{\cdot}90$ to $p\simeq 0{\cdot}03$, which shows that eqn (5.87) is indeed independent of the polarization.

2. *Saturation of the dipolar energy*

We consider first the rate of change of the dipolar energy when the system is irradiated with an r.f. field of constant frequency $\omega=\omega_0-\Delta$. In the frame rotating with frequency ω, the effective Hamiltonian is given by (1.34):

$$\mathcal{H}=\Delta I_z+\mathcal{H}'_{\mathrm{D}}+\omega_1 I_x. \quad (1.34)$$

By analogy with (1.60) the rate of change of $\langle\mathcal{H}'_{\mathrm{D}}\rangle$ is:

$$\begin{aligned}\frac{\mathrm{d}}{\mathrm{d}t}\langle\mathcal{H}'_{\mathrm{D}}\rangle&=\mathrm{Tr}\left\{\mathcal{H}'_{\mathrm{D}}\frac{\mathrm{d}\sigma}{\mathrm{d}t}\right\}=\mathrm{Tr}\{\mathcal{H}'_{\mathrm{D}}(-\mathrm{i}[\mathcal{H},\sigma])\}\\&=\mathrm{Tr}\{-\mathrm{i}[\mathcal{H}'_{\mathrm{D}},\mathcal{H}]\sigma\}=\langle-\mathrm{i}[\mathcal{H}'_{\mathrm{D}},\mathcal{H}]\rangle,\end{aligned} \quad (5.89)$$

that is, according to (1.34):

$$\frac{\mathrm{d}}{\mathrm{d}t}\langle\mathcal{H}'_{\mathrm{D}}\rangle=-\mathrm{i}\omega_1\langle[\mathcal{H}'_{\mathrm{D}},I_x]\rangle. \quad (5.89')$$

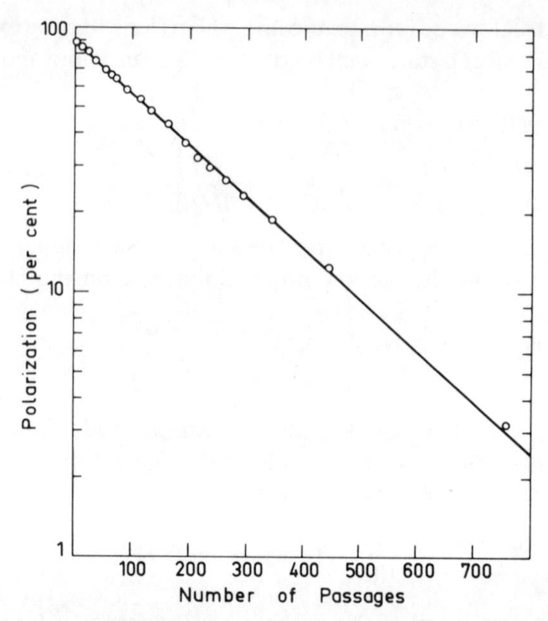

FIG. 5.7. ^{19}F nuclear polarization in CaF_2 as a function of number of slightly saturating passages. (After Abragam, Chapellier, Jacquinot, and Goldman, 1973.)

The right-hand side of (5.89′) can be related to $\langle I_y \rangle$ as follows. As long as Δ is unchanged, the energy associated with the effective Hamiltonian (1.34) is constant. Since ω_1 is small, we neglect the energy $\omega_1\langle I_x \rangle$ and we have:

$$\frac{\mathrm{d}}{\mathrm{d}t}(\langle \Delta I_z + \mathcal{H}'_{\mathrm{D}} \rangle) = \Delta \frac{\mathrm{d}}{\mathrm{d}t}\langle I_z \rangle + \frac{\mathrm{d}}{\mathrm{d}t}\langle \mathcal{H}'_{\mathrm{D}} \rangle = 0, \tag{5.90}$$

that is:

$$\frac{\mathrm{d}}{\mathrm{d}t}\langle \mathcal{H}'_{\mathrm{D}} \rangle = -\Delta \frac{\mathrm{d}}{\mathrm{d}t}\langle I_z \rangle, \tag{5.90′}$$

whence, according to (1.61) and (5.89′):

$$\langle [\mathcal{H}'_{\mathrm{D}}, I_x] \rangle = \Delta \langle I_y \rangle. \tag{5.91}$$

The gist of (5.91) is that its validity is independent of the value given to Δ. Therefore it will remain valid if Δ is swept at the very slow rate satisfying the condition (5.85), and we will have:

$$\frac{\mathrm{d}}{\mathrm{d}t}\langle \mathcal{H}'_{\mathrm{D}} \rangle = -\Delta \omega_1 \langle I_y \rangle, \tag{5.92}$$

where $\langle I_y \rangle$, having its steady-state value (5.11), depends on time only through the variation of Δ.

By analogy with (5.86), the integral of (5.92) over a linear passage is of the form:

$$\delta\langle \mathcal{H}'_{\mathrm{D}}\rangle = -\frac{\omega_1}{\dot{\Delta}}\int_{-\infty}^{+\infty} \Delta\langle I_y\rangle(\Delta)\,\mathrm{d}\Delta. \tag{5.93}$$

This result is independent of whether the system contains several spin species or not.

The integral on the right-hand side of (5.93) is a non-normalized first moment, in contrast with the moments used in section **B**(a) (eqn (5.32)). It will be noted by a capital script $\mathcal{M}_1$:

$$\mathcal{M}_1 = \int_{-\infty}^{+\infty} \Delta\langle I_y\rangle(\Delta)\,\mathrm{d}\Delta. \tag{5.94}$$

Since $\langle I_y\rangle$ is proportional to ω_1, so is $\mathcal{M}_1$. The actual value of the moment $\mathcal{M}_1$ will be calculated in the next section.

Equation (5.93) can be used for a calibration of the energy as follows (Goldman, Roinel, and Bouffard, to be published).

Two types of linear passages are performed: observation passages with an r.f. field $H_1^a = -\omega_1^a/\gamma$ so small that it produces a negligible saturation, and saturating passages with an r.f. field $H_1^b = -\omega_1^b/\gamma$, each one producing a small but non-negligible saturation.

These two types of passages are first performed on the system in a state of pure Zeeman order, for calibration purposes. The observation of the signal with the r.f. field H_1^a when the polarization is known, yields according to (5.83) a calibrated value of $\xi\omega_1^a$. The monitoring of the saturation of polarization due to passages performed with the r.f. field H_1^b yields, according to (5.88), a calibration of ε and ω_1^b.

When the system is in a demagnetized state, let us consider the following sequence.

(i) An observation passage with the r.f. field H_1^a, from which we compute the integral:

$$\mathcal{M}'_1 = \int \Delta v(\Delta)\,\mathrm{d}\Delta = \xi\mathcal{M}_1(\omega_1^a). \tag{5.95}$$

(ii) A saturating passage with the r.f. field H_1^b, which produces a variation of the dipolar energy:

$$\delta\langle \mathcal{H}'_{\mathrm{D}}\rangle = -\frac{\omega_1^b}{\dot{\Delta}}\mathcal{M}_1(\omega_1^b). \tag{5.96}$$

The signal $\langle I_y\rangle_b(\Delta)$ is *not* measured during this passage. However, ω_1^b is small enough for $\langle I_y\rangle_b(\Delta)$ to be linear in ω_1, whence:

$$\mathcal{M}_1(\omega_1^b)/\mathcal{M}_1(\omega_1^a) = \omega_1^b/\omega_1^a. \tag{5.97}$$

Equation (5.96) can then be written:

$$\delta\langle\mathcal{H}'_{\mathrm{D}}\rangle = -\frac{(\omega_1^b)^2}{\dot{\Delta}}\frac{1}{\omega_1^a}\mathcal{M}_1(\omega_1^a),$$

or else, according to (5.87) and (5.95):

$$\delta\langle\mathcal{H}'_{\mathrm{D}}\rangle = -\frac{\varepsilon}{\pi}\frac{\mathcal{M}'_1}{\xi\omega_1^a}. \qquad (5.98)$$

The factors $\xi\omega_1^a$ and ε having been calibrated as explained above, the measurement of $\mathcal{M}'_1$ yields a *calibrated* value for the energy change due to a passage.

The practical procedure consists in performing an observation passage after each sequence of n slightly saturating passages, until complete saturation of the energy, as witnessed by the vanishing of the signal, is achieved. By interpolation, one determines the energy change due to each passage. Starting from a very large number N of passages, which leads to $\langle\mathcal{H}'_{\mathrm{D}}\rangle = 0$, and performing the calculation backwards with respect to the passages, one obtains a calibration of the energy as a function of the first moment of the absorption signal of the spins I. This procedure is schematically pictured in Fig. 5.8.

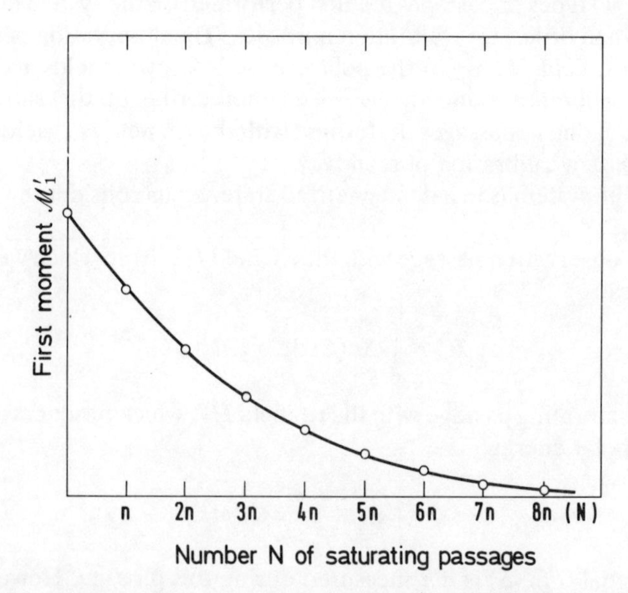

FIG. 5.8. Schematic variation of the first moment $\mathcal{M}'_1$ of the dipolar resonance signal with the number of slightly saturating passages. (After Goldman, Roinel, and Bouffard, to be published.)

It must be stressed that, even when several spin species are present, linear passages through the resonance line of *one* of these species is sufficient to calibrate the *total* dipole–dipole energy. When the system contains only one spin species, the dipolar energy can be measured much more simply, as shown in the next section.

(*d*) *First moment of the absorption signal*

The first moment $\mathcal{M}_1$ of the absorption signal is defined by (5.94). It is calculated through the use of (5.11). We have:

$$\begin{aligned}\Delta\langle I_y\rangle(\Delta) &= \frac{\omega_1}{4}\int_{-\infty}^{+\infty}\langle\!\langle[I_-,\tilde{I}_+(t)]\rangle\!\rangle_0\Delta\exp(\mathrm{i}\Delta t)\,\mathrm{d}t\\ &= -\mathrm{i}\frac{\omega_1}{4}\int_{-\infty}^{+\infty}\langle\!\langle[I_-,\tilde{I}_+(t)]\rangle\!\rangle_0\frac{\mathrm{d}}{\mathrm{d}t}\exp(\mathrm{i}\Delta t)\,\mathrm{d}t.\end{aligned} \tag{5.99}$$

We perform an integration by parts. Taking into account that:

$$\langle\!\langle[I_-,\tilde{I}_+(t)]\rangle\!\rangle_0\to 0\quad \text{for } t\to\pm\infty,$$

we obtain:

$$\Delta\langle I_y\rangle(\Delta) = \mathrm{i}\frac{\omega_1}{4}\int_{-\infty}^{+\infty}\frac{\mathrm{d}}{\mathrm{d}t}(\langle\!\langle I_-,\tilde{I}_+(t)]\rangle\!\rangle_0)\exp(\mathrm{i}\Delta t)\,\mathrm{d}t,$$

or else, since according to (5.7):

$$\begin{aligned}&\frac{\mathrm{d}}{\mathrm{d}t}\tilde{I}_+(t) = \mathrm{i}[\mathcal{H}'_{\mathrm{D}},\tilde{I}_+(t)] = -\mathrm{i}[\tilde{I}_+(t),\mathcal{H}'_{\mathrm{D}}],\\ &\Delta\langle I_y\rangle(\Delta) = \frac{\omega_1}{4}\int_{-\infty}^{+\infty}\langle\!\langle[I_-,[\tilde{I}_+(t),\mathcal{H}'_{\mathrm{D}}]]\rangle\!\rangle_0\exp(\mathrm{i}\Delta t)\,\mathrm{d}t.\end{aligned} \tag{5.100}$$

This form is inserted into (5.94) and, according to (5.12), we obtain (Goldman, 1975):

$$\mathcal{M}_1 = \frac{\pi}{2}\omega_1\langle\!\langle[I_-,[I_+,\mathcal{H}'_{\mathrm{D}}]]\rangle\!\rangle_0. \tag{5.101}$$

One should pay attention to the difference between (5.101) and the somewhat similar eqn (5.47). There is first a difference in normalization since M_1 in (5.47) is normalized to unit area whereas $\mathcal{M}_1$ is not normalized. More important, the special form of a purely Zeeman density matrix makes it possible to write (5.11) in the form (5.22), and as a result the numerator of (5.47) involves a single commutator whereas (5.101), which does *not* assume a purely Zeeman density matrix, involves a double commutator.

The simplest case is that when the system contains only one spin species.

The Hamiltonian $\mathscr{H}'_D$, of the form (1.27), is a sum of second rank tensorial operators, and we have:

$$[I_-, [I_+, \mathscr{H}'_D]] = 6\mathscr{H}'_D, \tag{5.102}$$

whence:

$$\mathscr{M}_1 = 3\pi\omega_1 \langle \mathscr{H}'_D \rangle_0. \tag{5.103}$$

What is actually measured is:

$$\mathscr{M}'_1 = \xi \mathscr{M}_1 = 3\pi\xi\omega_1 \langle \mathscr{H}'_D \rangle_0,$$

and when $\xi\omega_1$ is calibrated as in section **C**(a), the measurement of $\mathscr{M}'_1$ yields the dipolar energy $\langle \mathscr{H}'_D \rangle_0$. The more complex procedure of section **C**(c)2 is not necessary in that case.

When the system contains two spin species I and S, the dipolar Hamiltonian is:

$$\mathscr{H}'_D = \mathscr{H}'_{II} + \mathscr{H}'_{IS} + \mathscr{H}'_{SS},$$

where $\mathscr{H}'_{IS}$ is according to (1.27′) a first-rank tensor with respect to the spins I, so that:

$$[I_-, [I_+, \mathscr{H}'_{IS}]] = 2\mathscr{H}'_{IS}. \tag{5.104}$$

Since $\mathscr{H}'_{SS}$ commutes with all spin I operators,

$$\langle [I_-, [I_+, \mathscr{H}'_D]] \rangle_0 = 6\langle \mathscr{H}'_{II} \rangle_0 + 2\langle \mathscr{H}'_{IS} \rangle_0.$$

The first moment of the absorption signal of the spins I is then:

$$\mathscr{M}^I_1 = \pi\omega^I_1 (3\langle \mathscr{H}'_{II} \rangle_0 + \langle \mathscr{H}'_{IS} \rangle_0). \tag{5.105}$$

Similarly, the first moment of the absorption signal of the spin S is:

$$\mathscr{M}^S_1 = \pi\omega^S_1 (3\langle \mathscr{H}'_{SS} \rangle_0 + \langle \mathscr{H}'_{IS} \rangle_0). \tag{5.106}$$

These moments yield only linear combinations of contributions to the energy.

Their measurement, combined with the saturation method of section **C**(c)2, yield the separate values of $\langle \mathscr{H}'_{II} \rangle_0$, $\langle \mathscr{H}'_{IS} \rangle_0$ and $\langle \mathscr{H}'_{SS} \rangle_0$.

When the system contains more than two spin species, (5.105) is generalized to:

$$\mathscr{M}^I_1 = \pi\omega^I_1 (3\langle \mathscr{H}'_{II} \rangle_0 + \sum_\alpha \langle \mathscr{H}'_{IS\alpha} \rangle_0). \tag{5.107}$$

The measurement of all first moments, complemented with the saturation method, is then unable to yield a complete separation of the various contributions to the energy.

Going back to the case when there is one spin species only, eqn (5.103) inserted into (5.93) yields, for the saturation of dipolar energy by a linear passage:

$$\delta\langle\mathcal{H}'_{\mathrm{D}}\rangle = -3\pi\frac{\omega_1^2}{\dot{\Delta}}\langle\mathcal{H}'_{\mathrm{D}}\rangle_0. \tag{5.108}$$

The relative saturation of dipolar energy by a linear passage is independent of the energy. According to (5.87), it is 3 times larger than the saturation of polarization under the same conditions. These results are no longer true when there are more than one spin species.

(e) Qualitative shape of the absorption signal

Consider for simplicity a system with one spin species I, prepared by a partial ADRF stopped at the distance Δ_0 from resonance. In the frame rotating with frequency ω, the Hamiltonian is:

$$\mathcal{H}_0 = \Delta_0 I_z + \mathcal{H}'_{\mathrm{D}}. \tag{5.109}$$

We assume that the density matrix σ_0 has the canonical form:

$$\sigma_0 = \exp(-\beta\mathcal{H}_0)/\mathrm{Tr}\{\exp(-\beta\mathcal{H}_0)\}. \tag{5.110}$$

We show that the absorption signal observed with a small r.f. field vanishes at the distance Δ_0 from resonance, and that it changes sign with $\Delta' = \Delta - \Delta_0$.

According to the relation:

$$\exp(\mathrm{i}\Delta I_z t) I_+ \exp(-\mathrm{i}\Delta I_z t) = I_+ \exp(\mathrm{i}\Delta t), \tag{5.111}$$

and to the definition (5.7), $\langle I_y\rangle(\Delta)$ as given by (5.11) can be written:

$$\langle I_y\rangle(\Delta) = \langle I_y\rangle(\Delta_0 + \Delta')$$

$$= \frac{\omega_1}{4}\int_{-\infty}^{+\infty} \mathrm{Tr}\{[\sigma_0, I_-]\exp(\mathrm{i}\mathcal{H}_0 t) I_+ \exp(-\mathrm{i}\mathcal{H}_0 t)\}\exp(\mathrm{i}\Delta' t)\,\mathrm{d}t. \tag{5.112}$$

Let us choose a representation $|E, M\rangle$ where $\mathcal{H}'_{\mathrm{D}}$ and I_z are separately diagonal. Equation (5.112) can be rewritten as:

$$\langle I_y(\Delta_0 + \Delta')\rangle = \frac{\omega_1}{4}\int_{-\infty}^{+\infty} \sum_{E,E',M} \langle E, M|[\sigma_0, I_-]|E', M+1\rangle$$

$$\times\langle E', M+1|I_+|E, M\rangle \exp(\mathrm{i}(E' - E + \Delta')t)\,\mathrm{d}t,$$

or, since:

$$\int_{-\infty}^{+\infty} \exp(\mathrm{i}(E'-E+\Delta')t)\,\mathrm{d}t = 2\pi\,\delta(E'-E+\Delta');$$

$$\langle I_y(\Delta_0+\Delta')\rangle \propto \sum_{E,M} \langle E, M|[\sigma_0, I_-]|E-\Delta', M+1\rangle \times \langle E-\Delta', M+1|I_+|E, M\rangle$$

$$\propto \sum_{E,M} \exp(-\beta E)(1-\exp(\beta\Delta')) \times |\langle E, M|I_-|E-\Delta', M+1\rangle|^2. \tag{5.113}$$

Equation (5.113) shows that $\langle I_y(\Delta_0+\Delta')\rangle$ vanishes for $\Delta'=0$ and has opposite signs for opposite signs of Δ', a result well known for high temperatures (Goldman, 1970, chapter 4).

This result remains valid when other spin species are present. The calculation differs little from that given above and will not be developed here. The same result is also believed to be true in the presence of a spin ordering which breaks the symmetry of the Hamiltonian (chapter 8).

The importance of the result that $\langle I_y\rangle(\Delta)$ vanishes only for $\Delta=\Delta_0$ is that the absorption signal determines the effective frequency Δ_0 of the Hamiltonian (5.109) which enters into the form (5.110) of the density matrix.

In the particular case when $\Delta_0=0$, the absorption signal is antisymmetric. This is shown as follows. Since $\langle I_y\rangle$ is real, eqn (5.11) can be rewritten:

$$\langle I_y\rangle(\Delta) = \frac{\omega_1}{4}\int_{-\infty}^{+\infty} \langle[I_-, \tilde{I}_+(t)]\rangle_0^* \exp(-\mathrm{i}\Delta t)\,\mathrm{d}t. \tag{5.11'}$$

The trace $\langle[I_-, \tilde{I}_+(t)]\rangle_0$ (see eqn 5.11′) is invariant by rotation of π around $0x$. The density matrix depends only on $\mathscr{H}'_{\mathrm{D}}$, which is invariant through such a rotation. According to (5.25) and (5.9) we have then:

$$\begin{aligned}\langle[I_-, \tilde{I}_+(t)]\rangle_0^* &= \langle[I_+, \tilde{I}_-(t)]\rangle_0^* \\ &= -\langle[I_-, \tilde{I}_+(t)]\rangle_0.\end{aligned}$$

When inserted into (5.11′), this yields:

$$\langle I_y\rangle(\Delta) = -\langle I_y\rangle(-\Delta).$$

(*f*) *Linear response and transverse spin susceptibility*

The conceptual definition of the transverse spin susceptibility can be formulated through two seemingly different approaches, directly related to two different observation methods. The purpose of this section is to prove that these two approaches yield the same value for the susceptibility.

This property merely results from the application to the case of the transverse susceptibility of a general result of linear-response theory (Kubo, 1957) that we derive below.

We consider a system with Hamiltonian $\mathcal{H}_0$ in a state of thermal equilibrium, that is with a density matrix of the form:

$$\sigma_0 = \exp(-\beta\mathcal{H}_0)/\mathrm{Tr}\{\exp(-\beta\mathcal{H}_0)\} \tag{5.114}$$

and an operator Q such that:

$$\langle Q\rangle_0 = \mathrm{Tr}\{\sigma_0 Q\} = 0.$$

In the presence of a perturbation εV, where $\langle V\rangle_0 = 0$ and $\varepsilon \ll 1$ is a very small dimensionless parameter, $\langle Q\rangle$ no longer vanishes, but its calculation can be limited to the first order in ε (linear response), that is:

$$\langle Q\rangle = \varepsilon\chi_{V\to Q} = \varepsilon\chi, \tag{5.115}$$

where the susceptibility χ depends on the temperature of the system. Two approaches can be used for calculating $\langle Q\rangle$.

In the first, dynamical approach, we calculate to the first order in ε the variation $\delta\sigma$ of the density matrix produced by the perturbation εV, and we obtain:

$$\langle Q\rangle = \mathrm{Tr}\{\delta\sigma Q\}. \tag{5.116}$$

The calculation of $\delta\sigma$ is standard (Abragam, 1961, chapter IV; Goldman 1970, chapter 4): In the interaction representation defined by the unitary operator:

$$U(t) = \exp(\mathrm{i}\mathcal{H}_0 t),$$

the evolution of the density matrix is given by:

$$\frac{\mathrm{d}}{\mathrm{d}t}\tilde{\sigma}(t) = -\mathrm{i}\varepsilon[\tilde{V}(t), \tilde{\sigma}(t)], \tag{5.117}$$

where for every operator Q:

$$\tilde{Q} = U(t)QU^{\dagger}(t).$$

To the first order in ε, (5.117) gives:

$$\delta\tilde{\sigma}(t) = \tilde{\sigma}(t) - \sigma_0 = -\mathrm{i}\varepsilon\int_0^t [\tilde{V}(t'), \sigma_0]\,\mathrm{d}t',$$

or else, in the initial representation:

$$\begin{aligned}\delta\sigma(t) &= U^{\dagger}(t)\,\delta\tilde{\sigma}(t)U(t)\\ &= -\mathrm{i}\varepsilon U^{\dagger}(t)\int_0^t [\tilde{V}(t'), \sigma_0]\,\mathrm{d}t' U(t)\\ &= \mathrm{i}\varepsilon\int_0^t [\sigma_0, \tilde{V}(t'-t)]\,\mathrm{d}t',\end{aligned} \tag{5.118}$$

whence, according to (5.116):

$$\langle Q\rangle = \mathrm{i}\varepsilon \int_0^t \mathrm{Tr}\{[\sigma_0, \tilde{V}(t'-t)]Q\}\,\mathrm{d}t'.$$

If t is much larger than the correlation time of the trace under the integral we can extend the integral to infinity, and we obtain:

$$\langle Q\rangle = \mathrm{i}\varepsilon \int_0^\infty \langle[\exp(-\mathrm{i}\mathcal{H}_0\tau)V\exp(\mathrm{i}\mathcal{H}_0\tau), Q]\rangle_0\,\mathrm{d}\tau. \tag{5.119}$$

This method is an alternative to that used in section **A** for the calculation of the absorption and dispersions signals.

The second approach is thermodynamic. We assume that the density matrix is:

$$\sigma = \exp[-\beta(\mathcal{H}_0+\varepsilon V)]/\mathrm{Tr}\{\exp[\cdots]\}, \tag{5.120}$$

and we expand it to the first order in ε. We have the relation:

$$\exp[-\beta(\mathcal{H}_0+\varepsilon V)] = \exp(-\beta\mathcal{H}_0)\left\{1-\varepsilon\int_0^\beta \exp(\lambda\mathcal{H}_0)V\exp(-\lambda\mathcal{H}_0)\,\mathrm{d}\lambda\right\}. \tag{5.121}$$

The proof of (5.121) is entirely similar to that of the well-known first order perturbation theory formula:

$$\exp[-\mathrm{i}(\mathcal{H}_0+\mathcal{H}_1)t] = \exp(-\mathrm{i}\mathcal{H}_0 t)\left\{1-\mathrm{i}\int_0^t \tilde{\mathcal{H}}_1(t')\,\mathrm{d}t\right\}.$$

According to (5.121), we have:

$$\begin{aligned}\mathrm{Tr}\{\exp[-\beta(\mathcal{H}_0+\varepsilon V)]\} &= \mathrm{Tr}\{\exp(-\beta\mathcal{H}_0)\}\{1-\beta\varepsilon\langle V\rangle_0\}\\ &= \mathrm{Tr}\{\exp(-\beta\mathcal{H}_0)\},\end{aligned}$$

and we obtain:

$$\begin{aligned}\langle Q\rangle &= \mathrm{Tr}\{\sigma Q\}\\ &= -\varepsilon\int_0^\beta \langle\exp(\lambda\mathcal{H}_0)V\exp(-\lambda\mathcal{H}_0)Q\rangle_0\,\mathrm{d}\lambda.\end{aligned} \tag{5.122}$$

The right-hand sides of (5.119) and (5.122) are equal: if $|k\rangle$, $|l\rangle$ is a basis of eigenkets of $\mathcal{H}_0$ with $\langle k|\sigma_0|k\rangle = \sigma_k$ and $\langle k|\mathcal{H}_0|k\rangle = E_k$, each of these expressions is easily found to be equal to:

$$\varepsilon\sum_{k,l}\frac{\sigma_k-\sigma_l}{E_k-E_l}\langle k|V|l\rangle\langle l|Q|k\rangle.$$

The case of the transverse susceptibility corresponds to an unperturbed Hamiltonian:

$$\mathcal{H}_0 = \Delta_0 I_z + \mathcal{H}'_D,$$

and to a perturbation:

$$\varepsilon V = \omega_1 I_x = \frac{\omega_1}{D} D I_x;$$

i.e.:

$$V = DI_x \quad \text{and} \quad \varepsilon = \omega_1/D \ll 1,$$

where D is the local frequency.

We define here the transverse susceptibility $\chi_\perp$ as:

$$\chi_\perp = p_x/\omega_1 = p_x/(D\varepsilon), \tag{5.123}$$

where p_x is the transverse polarization per spin I:

$$p_x = \langle I_x \rangle / IN_I,$$

i.e. $Q = I_x/(IN_I D)$ in (5.119) and (5.122). Equation (5.123) departs in several points from the conventional definition of susceptibility.

(i) It refers to polarization rather than to magnetization.

(ii) It is *not* dimensionless.

The identity between the two formal definitions of the susceptibility is believed to remain true when the spin system is in an ordered state that breaks the symmetry of the Hamiltonian.

The thermodynamic approach is well suited to the description of the experiment where one observes the dispersion signal $\langle I_x \rangle$ in the course of an ADRF. The r.f. field is large enough for the system to be at all times at internal thermal equilibrium, that is for σ to be of the form (5.120). It is however small enough for $\langle I_x \rangle$ to be linear in ω_1.

The dynamical approach is well suited to the description of the following experiment: when an ADRF is stopped at the distance Δ_0 from resonance and the saturating r.f. field is removed, the density matrix reaches after several times T_2 the equilibrium form (5.114). A small non-saturating r.f. field is then applied at Δ_0 and the steady-state value of $\langle I_x \rangle$ is measured.

In practice, it is not necessary to actually observe the dispersion signal, whose value at Δ_0 can be obtained from the absorption signal $\langle I_y(\Delta) \rangle$. According to (5.10), the dispersion signal $\langle I_x \rangle$ is the Kramers–Kronig transform of the absorption signal $\langle I_y \rangle$ (see e.g. Abragam, 1961, p. 93):

$$\langle I_x \rangle(\Delta_0) = -\frac{1}{\pi} \mathcal{P} \int_{-\infty}^{+\infty} \frac{\langle I_y \rangle(\Delta)\, d\Delta}{\Delta - \Delta_0}. \tag{5.124}$$

The advantage of using this transform is twofold. First, and foremost, it is in the absorption mode that the product $\xi\omega_1$ is calibrated, and the transform of the experimental absorption signal:

$$-\frac{1}{\pi}\mathscr{P}\int_{-\infty}^{+\infty}\frac{v(\Delta)\,\mathrm{d}\Delta}{\Delta-\Delta_0}\,\mathrm{d}\Delta=\xi\omega_1 IN_I\chi_\perp, \tag{5.124'}$$

yields a calibrated value of $\chi_\perp$. According to section **C**(e), the origin Δ_0 is the frequency at which the absorption signal vanishes.

Secondly the observation of only one signal, the absorption signal, is sufficient to obtain two pieces of information: information on the energy from the first moment $\mathcal{M}_1$, and transverse susceptibility.

(g) *Cotanh transform of the absorption signal. Measurement of the dipolar temperature*

This cryptic name refers to the fluctuation–dissipation theorem (see e.g. Brout, 1965a, p. 110), as applied to the absorption signal (Goldman, Roinel, and Bouffard, to be published).

We consider a system whose density matrix has the canonical form (5.3), where the Zeeman coupling Z_μ is of the form:

$$Z_\mu=\omega_\mu S_z^\mu.$$

Since $\mathscr{H}'_{\mathrm{D}}$ commutes with all Zeeman interactions, we can use a reference frame rotating with respect to each spin species with a frequency $\omega_\mu-\Delta_\mu$ chosen so as to have:

$$\beta_{Z\mu}\omega_\mu=\beta\Delta_\mu, \tag{5.125}$$

where β is the dipolar inverse temperature. The effective Hamiltonian in this frame is:

$$\mathscr{H}_0=\mathscr{H}'_{\mathrm{D}}+\Delta_0 I_z+\sum_\mu\Delta_\mu S_z^\mu, \tag{5.126}$$

where we single out the spin species I whose absorption signal is observed. The density matrix (5.3) is equal to:

$$\sigma_0=\exp(-\beta\mathscr{H}_0)/\mathrm{Tr}\{\exp(-\beta\mathscr{H}_0)\}. \tag{5.127}$$

The absorption signal $\langle I_y\rangle$ is given by (5.11), or alternatively by (5.112):

$$\langle I_y\rangle(\Delta)=\frac{\omega_1}{4}\int_{-\infty}^{+\infty}\mathrm{Tr}\{[\sigma_0,I_-]\exp(\mathrm{i}\mathscr{H}_0t)I_+\exp(-\mathrm{i}\mathscr{H}_0t)\}\exp(\mathrm{i}\Delta' t)\,\mathrm{d}t, \tag{5.112}$$

with $\Delta'=\Delta-\Delta_0$.

Let $|k\rangle$, $|l\rangle\cdots$ be a basis of eigenkets of the Hamiltonian $\mathscr{H}_0$. We use as before the notations:

$$\langle k|\sigma_0|k\rangle=\sigma_k;\qquad \langle k|\mathscr{H}_0|k\rangle=E_k.$$

Equation (5.112) can be explicitly written in the form:

$$\langle I_y\rangle(\Delta)=\frac{\omega_1}{4}\int_{-\infty}^{+\infty} \mathrm{d}t\, \exp(\mathrm{i}\Delta' t)\times\sum_{k,l}(\sigma_k-\sigma_l)\exp(-\mathrm{i}(E_k-E_l)t)\langle k|I_-|l\rangle\langle l|I_+|k\rangle, \tag{5.128}$$

or else, according to (5.12):

$$\langle I_y\rangle(\Delta)=\frac{\pi}{2}\omega_1\sum_{k,l}(\sigma_k-\sigma_l)\langle k|I_-|l\rangle\langle l|I_+|k\rangle\,\delta(\Delta'-E_k+E_l). \tag{5.129}$$

The expression (5.129) states that the rate of Zeeman energy absorption, induced by an r.f. field rotating with frequency Δ', through transitions between states $|k\rangle$ and $|l\rangle$ of energies E_k and E_l differing by Δ' is proportional to the square of the matrix element $|\langle k|I_-|l\rangle|$, times the population difference $(\sigma_k-\sigma_l)$. It is sometimes used as a *starting point* for the derivation of (5.112), *from* which it has been deduced here. The density matrix being of the form (5.127), we have:

$$\begin{aligned}\sigma_l/\sigma_k&=\exp[-\beta(E_l-E_k)];\\ \sigma_k/\sigma_l&=\exp[-\beta(E_k-E_l)].\end{aligned} \tag{5.130}$$

The presence of the delta function on the right-hand side of (5.129) makes it possible to replace (E_k-E_l) by Δ'. Equation (5.129) can then be written under the two following forms:

$$\langle I_y\rangle(\Delta)=\frac{\pi}{2}\omega_1\sum_{k,l}\sigma_l(\exp(-\beta\Delta')-1)\langle k|I_-|l\rangle\langle l|I_+|k\rangle\,\delta(\Delta'-E_k+E_l), \tag{5.131}$$

and:

$$\langle I_y\rangle(\Delta)=\frac{\pi}{2}\omega_1\sum_{k,l}\sigma_k(1-\exp(\beta\Delta'))\langle k|I_-|l\rangle\langle l|I_+|k\rangle\,\delta(\Delta'-E_k+E_l). \tag{5.131$'$}$$

In (5.131) we divide both sides by $[\exp(-\beta\Delta')-1]$ and integrate over Δ, which yields:

$$\begin{aligned}K_1&=\int_{-\infty}^{+\infty}\frac{\langle I_y\rangle(\Delta)\,\mathrm{d}\Delta}{\exp(-\beta(\Delta-\Delta_0))-1}\\ &=\frac{\pi}{2}\omega_1\sum_{k,l}\sigma_l\langle k|I_-|l\rangle\langle l|I_+|k\rangle\\ &=\frac{\pi}{2}\omega_1\sum_{l}\sigma_l\langle l|I_+I_-|l\rangle\\ &=\frac{\pi}{2}\omega_1\langle I_+I_-\rangle_0.\end{aligned} \tag{5.132}$$

Similarly, we divide both sides of (5.131′) by $[1-\exp(\beta\Delta')]$ and integrate over Δ:

$$K_2 = \int_{-\infty}^{+\infty} \frac{\langle I_y\rangle(\Delta)\,\mathrm{d}\Delta}{1-\exp(\beta(\Delta-\Delta_0))}$$

$$= \frac{\pi}{2}\omega_1 \sum_{k,l} \sigma_k \langle k|I_-|l\rangle\langle l|I_+|k\rangle$$

$$= \frac{\pi}{2}\omega_1 \sum_{k} \sigma_k \langle k|I_-I_+|k\rangle$$

$$= \frac{\pi}{2}\omega_1 \langle I_-I_+\rangle_0. \tag{5.132'}$$

The difference $(K_2 - K_1)$ is equal to:

$$K_2 - K_1 = \int_{-\infty}^{+\infty} \langle I_y\rangle(\Delta)\,\mathrm{d}\Delta$$

$$= \frac{\pi}{2}\omega_1 \langle I_-I_+ - I_+I_-\rangle_0$$

$$= -\pi\omega_1 \langle I_z\rangle_0, \tag{5.133}$$

which is nothing but the relation (5.13) between the area of the absorption signal and the polarization.

The combination $-(K_1+K_2) \equiv \mathscr{C}(\beta)$ is equal to:

$$\mathscr{C}(\beta) \equiv -(K_1+K_2) = \int_{-\infty}^{+\infty} \langle I_y\rangle(\Delta)\,\mathrm{cotanh}\left[\frac{\beta}{2}(\Delta-\Delta_0)\right] d\Delta$$

$$= -\frac{\pi}{2}\omega_1 \langle I_-I_+ + I_+I_-\rangle_0,$$

or else:

$$\mathscr{C}(\beta) \equiv \int_{-\infty}^{+\infty} \langle I_y\rangle(\Delta)\,\mathrm{cotanh}\left[\frac{\beta}{2}(\Delta-\Delta_0)\right] \mathrm{d}\Delta = -\pi\omega_1 \langle I_x^2 + I_y^2\rangle_0. \tag{5.134}$$

The integral $\mathscr{C}(\beta)$ is called the cotanh transform of the absorption signal. The origin Δ_0 is the frequency at which the absorption signal vanishes. As before we assume that (5.134) remains valid in the presence of nuclear ordering.

In (5.134) $\langle I_x^2\rangle_0$ is of the form:

$$\langle I_x^2\rangle_0 = \sum_i \langle I_x^{i2}\rangle_0 + \sum_{i\neq j} \langle I_x^i I_x^j\rangle_0. \tag{5.135}$$

A case of particular interest is that when, besides abundant spins, the system contains a spin species of low abundance. This is the case of CaF_2 where besides the ^{19}F spins, 0·13 per cent of the calcium consists of the isotope ^{43}Ca, whereas the abundant ^{40}Ca nuclei are spinless. This is also the case in LiH, where besides the abundant ^1H and ^7Li species the sample contains a few per cent of the isotope ^6Li. In both cases, the rare spins have magnetic moments much smaller than that of the abundant spins.

In that case, the dipolar coupling between the rare spins can be neglected. Since the average distance between rare spins is much larger than that between abundant spins, we have approximately:

$$\langle I_x^i I_x^j \rangle_0 \# \langle I_x^i \rangle_0 \langle I_x^j \rangle_0,$$

and since there is no transverse coupling between the rare spins and the (unlike) abundant spins, each $\langle I_x^i \rangle_0$ vanishes. It is then a good approximation to neglect the cross terms in (5.135) and in the similar expression for $\langle I_y^2 \rangle_0$. This yields:

$$\langle I_x^2 + I_y^2 \rangle_0 \# \sum_i \langle (I_x^i)^2 + (I_y^i)^2 \rangle_0$$

$$= \sum_i [I(I+1) - \langle (I_z^i)^2 \rangle_0],$$

or else, if all spins I are equivalent:

$$\langle I_x^2 + I_y^2 \rangle_0 = N_I [I(I+1) - \langle (I_z^i)^2 \rangle_0],$$

whence:

$$\mathscr{C}(\beta) = -N_I \pi \omega_1 [I(I+1) - \langle (I_z^i)^2 \rangle_0]. \tag{5.136}$$

When $I = \frac{1}{2}$, $\langle (I_z^i)^2 \rangle_0 = \frac{1}{4}$. When $I > \frac{1}{2}$, $\langle (I_z^i)^2 \rangle_0$ depends on β even when $\langle I_z^i \rangle_0 = 0$: it will be shown in chapter 8, section **D**(c)1, that even in that case as a result of their coupling with the abundant spins, the rare spins acquire at thermal equilibrium a non-vanishing quadrupole alignment. It will be shown however, on the example of ^{43}Ca in CaF_2, that as long as $(\beta \Delta \omega_I)^2 \ll 1$, where $\Delta \omega_I$ is the coupling energy of a rare spin I with its abundant neighbours, this alignment is negligibly small, i.e.:

$$\langle (I_z^i)^2 \rangle_0 \# \mathrm{Tr}\{(I_z^i)^2\} = \tfrac{1}{3} I(I+1). \tag{5.137}$$

When $\langle I_z^i \rangle_0 \neq 0$, as long as $|p_I| = \langle I_z^i \rangle_0 / I$ is small, the departure of $\langle (I_z^i)^2 \rangle_0$ from the value (5.137) is small, being quadratic in p_I.

For $|p_I|$ either zero or small, we then have approximately:

$$\mathscr{C}(\beta) \# -N\pi\omega_1 \times \tfrac{2}{3} I(I+1). \tag{5.136'}$$

Equation (5.136′) can be used for an absolute measurement of the dipolar temperature, as follows. Given an experimental signal, we compute the function,

$$\mathscr{C}(\lambda) = \int_{-\infty}^{+\infty} \langle I_y \rangle(\Delta) \operatorname{cotanh}\left(\lambda \frac{\Delta}{2}\right) \mathrm{d}\lambda,$$

for several values of λ and determine by interpolation the value λ_0 for which $\mathscr{C}(\lambda_0)$ is equal to the right-hand side of (5.136′). This value λ_0 is equal to the inverse dipolar temperature β. The procedure is schematically shown in Fig. 5.9.

As an example of application of this method, Fig. 5.10 is a plot of the dipolar temperature, as deduced from the cotanh transform of the ^{6}Li signal, as a function of the first moment $\mathscr{M}_1$ of the ^{7}Li signal. The field H_0 is parallel to a [001] axis. The system was prepared by an ADRF at negative temperature of both the ^{1}H and ^{7}Li spin species, and then allowed to warm up through spin–lattice relaxation. The system is initially antiferromagnetic. The transition to paramagnetism, as observed by neutron diffraction, is noted by the arrow (see chapter 8).

This method of measurement of the dipolar temperature will be compared with an alternative, more direct, method described in Jacquinot, Wenckebach, Goldman, and Abragam (1974). In that method, once the abundant spins I are demagnetized by ADRF, one applies an r.f. field at a distance Δ

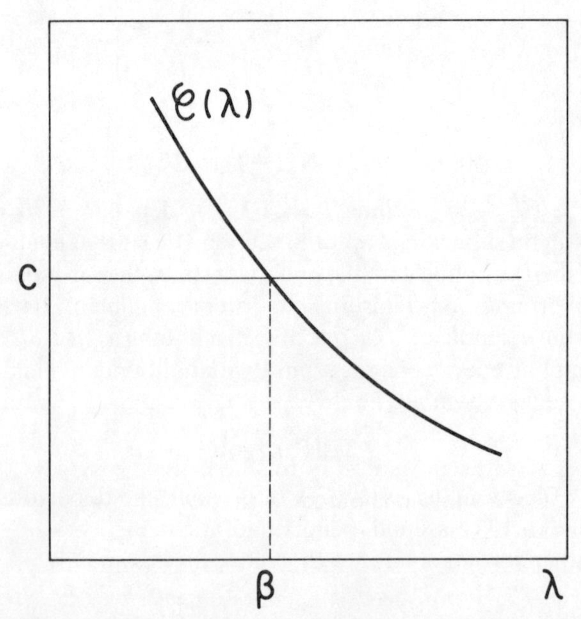

FIG. 5.9. Schematic variation of the cotanh transform $\mathscr{C}(\lambda)$ of the absorption signal of the rare spins, and determination of the dipolar inverse temperature. (After Goldman, Roinel, and Bouffard, to be published.)

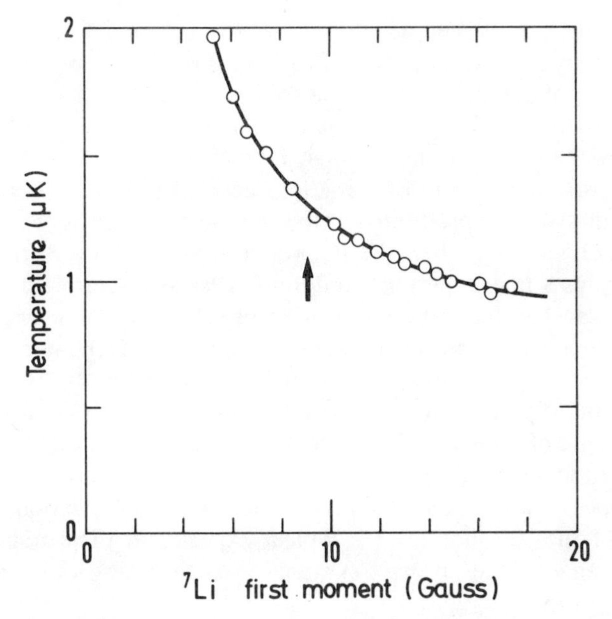

FIG. 5.10. Experimental variation of the dipolar temperature T with the first moment $\mathcal{M}_1$ of the ^{7}Li resonance signal in a sample of LiH. (After Goldman, Roinel, and Bouffard, to be published.)

from the resonance of the rare spins S. The Hamiltonian in the rotating frame is:

$$\mathcal{H} = \mathcal{H}'_D + \Delta S_z + \omega_1 S_x.$$

Through the thermal mixing between ΔS_z and $\mathcal{H}'_D$ mediated by the perturbation $\omega_1 S_x$, the spins S acquire a polarization P_S related to the dipolar inverse temperature β by the well-known Brillouin function:

$$P_S = B_S(\beta\Delta). \tag{5.138}$$

The measurement of P_S then yields the value of β.

This process of polarization of rare spins is identical with that described in chapter 1, section **B**(e).

The spins S must be rare in order for the heat capacity of their Zeeman reservoir ΔS_z to be small compared to that of $\mathcal{H}'_D$ the thermal reservoir whose temperature is being measured.

This method suffers from several drawbacks. First of all, eqn (5.138) is an approximation which requires that the external effective field $h = -\Delta/\gamma_S$ be much larger than the spread of dipolar fields produced at the sites of the spins S by the abundant spins I. It is then necessary to apply the r.f. field at a relatively large distance Δ from resonance, a condition under which the rate of thermal mixing between ΔS_z and $\mathscr{H}'_D$ is slow (chapter 1, section **B**(*e*)). As an example, in CaF_2 with the field $H_0 \| [111]$, it requires 3 min. to achieve an equilibrium polarization of the ^{43}Ca spins when an r.f. field $H_1 = 0{\cdot}7$ G is applied at 30 G from the ^{43}Ca resonance. This method can therefore be used only in systems where the dipolar relaxation time T_{1D} is very long. Furthermore the necessity of using a large H_1 produces a heating of the sample which may be prohibitive if the sample has to be kept at low lattice temperature. Finally, when the abundant spins are in an ordered state, different spins *S* may experience different average dipolar fields. This is the case for ^{43}Ca when the state of the ^{19}F spins is ferromagnetic with domains, and for ^{6}Li in ferro- or antiferromagnetic states of LiH (chapter 8). The field that must be used in (5.138) is the total field experienced by the spins *S*, external plus dipolar. The latter, determined from the splitting of the resonance line of the spins *S*, is difficult to estimate with accuracy.

This method has been used in ferromagnetic CaF_2 on the resonance of the ^{43}Ca isotope. Figure 5.11 shows the variation of the polarization of ^{43}Ca as a function of the distance Δ in a typical experiment. The polarization is calibrated against the absorption signal area by fitting the experimental

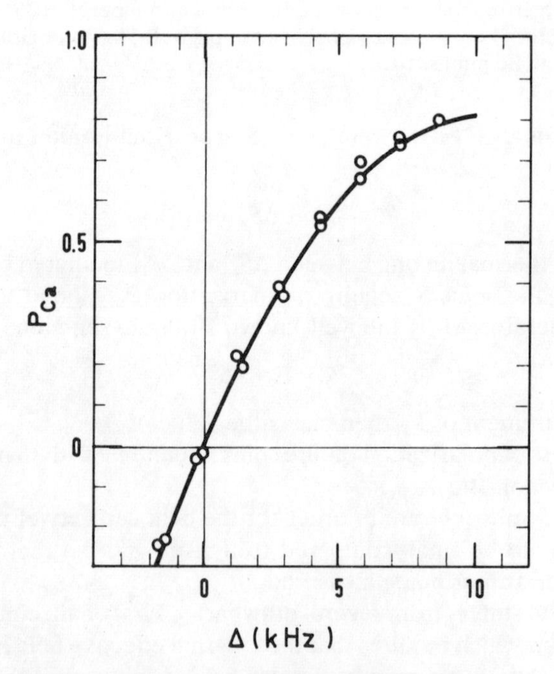

FIG. 5.11. ^{43}Ca polarization, produced by thermal mixing in the rotating frame, against ^{43}Ca effective frequency Δ. The ^{43}Ca polarization is calibrated by a best fit with a Brillouin function (solid curve). (After Jacquinot *et al.*, 1974.)

curve to the theoretical Brillouin function for spins $\frac{7}{2}$. The best-fit temperature derived from Fig. 5.11 is:

$$T_D = -0{\cdot}32\ \mu\text{K}.$$

D. Non-linear effects in spin temperature

The purpose of this section is to analyse the thermodynamic properties of paramagnetic systems in the non-linear domain, that is when it is no longer possible to expand, as in chapter 1, the density matrix to the first order with respect to the inverse temperature (or temperatures).

We limit ourselves to the simplest case, that of a system containing only one spin species, of spin $\frac{1}{2}$, in a high magnetic field and subjected to dipole–dipole interactions. As recalled in chapter 1, it is a good approximation to truncate the dipolar interaction and to use the approximate Hamiltonian:

$$\mathcal{H} \simeq \omega_0 I_z + \mathcal{H}'_D. \tag{1.26}$$

Since $[I_z, \mathcal{H}'_D] = 0$, the Zeeman and secular dipolar terms are separately constants of the motion.

The hypothesis of spin temperature consists in assuming that in the steady state, the properties of the system are well described by a density matrix whose off-diagonal matrix elements vanish (in a basis of eigenkets common to I_z and $\mathcal{H}'_D$), and whose diagonal matrix elements are such as to maximize the entropy:

$$S = -k_B \operatorname{Tr}\{\sigma \ln \sigma\}, \tag{1.13}$$

in a way consistent with the conservation of the two constants of the motion.

The standard mathematical procedure is to look for the unrestricted maximum of:

$$\frac{S}{k_B} - \beta_Z \omega_0 \langle I_z \rangle - \beta \langle \mathcal{H}'_D \rangle = -\operatorname{Tr}\{\sigma[\ln \sigma - \beta_Z \omega_0 I_z - \beta \mathcal{H}'_D]\}, \tag{5.139}$$

where β_Z and β are Lagrange multipliers. Associated with the condition $\operatorname{Tr}(\sigma) = 1$, the maximum of (5.139) corresponds to the density matrix:

$$\sigma = \exp(-\beta_Z \omega_0 I_z - \beta \mathcal{H}'_D)/\operatorname{Tr}\{\exp(\cdots)\}. \tag{1.28}$$

The Lagrange multipliers β_Z and β are the inverse Zeeman and dipolar temperatures, respectively.

If we introduce the partition function:

$$\mathcal{Z}(\beta_Z, \beta) = \operatorname{Tr}\{\exp(-\beta_Z \omega_0 I_z - \beta \mathcal{H}'_D)\}, \tag{5.140}$$

we have:

$$\omega_0\langle I_z\rangle = \omega_0 \operatorname{Tr}\{\sigma I_z\} = -\frac{\partial}{\partial \beta_Z} \ln \mathscr{Z}; \tag{5.141}$$

$$\langle \mathscr{H}'_D\rangle = \operatorname{Tr}\{\sigma \mathscr{H}'_D\} = -\frac{\partial}{\partial \beta} \ln \mathscr{Z}. \tag{5.142}$$

It can be surmised from (5.141) and (5.142) (and confirmed by detailed calculations later) that both $\langle I_z\rangle$ and $\langle \mathscr{H}'_D\rangle$ depend on *both* β_Z and β. It is worthwhile insisting on this important point: the very existence of separate and well-defined Zeeman and dipolar temperatures is a consequence of the fact that Zeeman and dipolar energies are separately constants of the motion, yet these energies depend on both temperatures.

The physical origin of this is that I_z and $\mathscr{H}'_D$ depend on the *same* spin variables and that, except in the high-temperature limit, their expectation values are not independent of each other. The most striking example is that when the polarization is unity (say $\langle I_z^i\rangle = \frac{1}{2}$), which corresponds to $|\beta_Z| = \infty$ and β arbitrary but different from $\pm\infty$. $\mathscr{H}'_D$ being of the form (5.43), its expectation value is:

$$\langle \mathscr{H}'_D\rangle = \tfrac{1}{4}\sum_{i,j} A_{ij}.$$

It has a well-defined value entirely determined by the Zeeman order and independent of β. In this limiting case, the concept of dipolar temperature loses its meaning.

All thermodynamic quantities can be obtained from the partition function $\mathscr{Z}(\beta_Z, \beta)$: the energies are derived from (5.141) and (5.142). As for the entropy, it is, according to (1.13) of the form:

$$\frac{S}{k_B} = \beta_Z\omega_0\langle I_z\rangle + \beta\langle \mathscr{H}'_D\rangle - \ln \mathscr{Z}(\beta_Z, \beta). \tag{5.143}$$

From now on, we drop the Boltzmann constant k_B, that is we express S in units of k_B.

Insofar as we are looking for a departure from the high-temperature domain, we can attempt to use an expansion of $\mathscr{Z}(\beta_Z, \beta)$ in powers of β_Z and β. We are immediately confronted with a difficulty: it can be shown that a term such as:

$$\beta^{2n} \operatorname{Tr}(\mathscr{H}'_D)^{2n} \sim (\beta D)^{2n} N^n, \tag{5.144}$$

where N is the number of spins in the sample. N is very large (in the thermodynamic limit it tends to infinity) so that even when $\beta D < 1$, (5.144) is very large, and the expansion of (5.140) does not converge. However, thermodynamic quantities depend not on $\mathscr{Z}$ itself, but on $\ln \mathscr{Z}$. In thermodynamics, $\ln \mathscr{Z}$ is assumed to be an extensive quantity, i.e. proportional to N, which is verified in all practical cases where it was computed. It is therefore directly for $\ln \mathscr{Z}$ that one must perform a power expansion.

A diagrammatic method was devised by Horwitz and Callen (1961) for the calculation of $\ln \mathscr{Z}$ in the particular case of a system of spins $\frac{1}{2}$ with bilinear spin–spin interactions that commute with I_z.

However, as noted by Englert (1963), it turns out to be somewhat simpler to compute the expectation value of an operator Q. By choosing $Q = I_z$ or $Q = \mathscr{H}'_D$, the form of $\ln \mathscr{Z}$ can then be obtained from (5.141) and (5.142). (Section **D**(b)3.)

We begin by describing the diagrammatic method for calculating the expectation value of a spin operator, then we compute several properties of the spin system, and finally we present several experimental illustrations of the theory.

(a) *The diagrammatic method*

The diagrammatic method to be used here was developed by Stinchcombe, Horwitz, Englert, and Brout (1963). It is reviewed by Brout (1965a, b) and is a particular case of the general diagrammatic methodology developed for many-body physics.

The use of diagrams is not widespread in nuclear magnetism, and the analysis of the method is presented here for the convenience of the reader.

Let us then consider an operator Q which is a product of spin operators. Its expectation value is:

$$\langle Q \rangle = \frac{\mathrm{Tr}\{\exp(-\beta_Z \omega_0 I_z - \beta \mathscr{H}'_D) Q\}}{\mathrm{Tr}\{\exp(-\beta_Z \omega_0 I_z - \beta \mathscr{H}'_D)\}}. \tag{5.145}$$

All the difficulty arises from the existence of the term $\mathscr{H}'_D$. If we had $\beta = 0$, the computation of:

$$\langle Q \rangle_0 = \frac{\mathrm{Tr}\{\exp(-\beta_Z \omega_0 I_z) Q\}}{\mathrm{Tr}\{\exp(-\beta_Z \omega_0 I_z)\}}, \tag{5.146}$$

would be straightforward. According to (5.15), a purely Zeeman density matrix factorizes into a product of matrices of individual spins and $\langle Q \rangle_0$ is computed according to section **B**(a):

$$\begin{aligned} &\langle I_z^i I_z^j \rangle_0 = \langle I_z^i \rangle_0 \langle I_z^j \rangle_0; \\ &\langle I_z^i \rangle_0 = \tfrac{1}{2} \tanh(-\tfrac{1}{2}\beta_Z \omega_0) = \tfrac{1}{2} t; \\ &\langle I_x^i \rangle_0 = \langle I_y^i \rangle_0 = 0. \end{aligned} \tag{5.147}$$

The aim of the method is to calculate (5.145) in the form of a power expansion with respect to the dipolar inverse temperature β, but valid to all orders with respect to the Zeeman inverse temperature β_Z:

$$\langle Q\rangle=\frac{\langle\exp(-\beta\mathcal{H}'_D)Q\rangle_0}{\langle\exp(-\beta\mathcal{H}'_D)\rangle_0}=\frac{\sum_0^\infty(-\beta)^n\langle\mathcal{H}'^n_D Q\rangle_0/n!}{\sum_0^\infty(-\beta)^n\langle\mathcal{H}'^n_D\rangle_0/n!}. \tag{5.148}$$

As noted earlier, neither the expansion of the numerator nor that of the denominator are convergent. The achievement of the diagrammatic method is to enable one to write the numerator in the form of a sum:

$$\langle\exp(-\beta\mathcal{H}'_D)Q\rangle_0=\sum_n\beta^n f_n(\beta_Z)\langle\exp(-\beta\mathcal{H}'_D)\rangle_0. \tag{5.148'}$$

$\langle\exp(-\beta\mathcal{H}'_D)\rangle_0$ cancels out with the denominator, and we obtain:

$$\langle Q\rangle=\sum_n\beta^n f_n(\beta_Z).$$

We consider first the case when the spin–spin interaction consists only of longitudinal terms of the form $I^i_z I^j_z$, and then treat the general case when flip-flop terms $I^i_+ I^j_-$ are also present.

1. *The longitudinal case*

We consider a spin–spin interaction of the form:

$$\mathcal{H}'=\tfrac{1}{2}\sum_{i,j}v_{ij}\sigma^i_z\sigma^j_z, \tag{5.149}$$

where $\sigma^i_z=2I^i_z$ is a Pauli operator, and $v_{ii}=0$.

A given term of the numerator of (5.148) is of the form:

$$\frac{1}{n!}(\tfrac{1}{2})^n(-\beta)^n v_{ij}v_{kl}\cdots v_{mn}\langle\sigma^i_z\sigma^j_z\sigma^k_z\cdots Q\rangle_0. \tag{5.150}$$

This term is non-zero if Q contains operators σ_z only. We choose Q to be an operator σ^i_z or a product of such operators. As stated earlier the outstanding problem is to cast the summation of (5.150) over all the indices i, j, k, l, etc. into a form which leads to the factorization described in eqn (5.148′). This essential task is by no means simple and involves a certain number of intermediate steps that will be outlined now. There are two categories of readers who will be well advised to skip the description of those steps: those already familiar with the diagrammatic method, and those who are only interested in the results obtained by this method in the field of nuclear magnetism, to be given in later sections.

Consider in (5.150) the expectation value of a term:

$$\langle\sigma^i_z\sigma^j_z\sigma^k_z\cdots\sigma^1_z\rangle_0, \tag{5.151}$$

where σ^1_z is our choice of the operator Q whose expectation value we seek. Some of the indices in (5.151), say i and k or j and l etc. may coincide. When all the indices which are the same have been gathered together (5.151) takes the form:

$$\langle(\sigma_z^{\alpha_1})^{p_1}\rangle_0\langle(\sigma_z^{\alpha_2})^{p_2}\rangle_0\cdots\langle(\sigma_z^{\alpha_q})^{p_q}\rangle_0, \tag{5.151'}$$

where the α_1, α_2, etc, are all different. Equation (5.151′) is then just:

$$\langle(\sigma_z)^{p_1}\rangle_0\langle(\sigma_z)^{p_2}\rangle_0\cdots\langle(\sigma_z)^{p_q}\rangle_0. \tag{5.152}$$

An expression such as $\langle(\sigma_z)^p\rangle_0$ is very simple:

$$\begin{aligned}\langle(\sigma_z)^p\rangle_0 &= 1 \quad \text{for } p \text{ even,}\\ &= t \quad \text{for } p \text{ odd,}\end{aligned} \tag{5.152'}$$

and summing (5.150) using (5.152) and (5.152′) would appear as a very dull but straightforward task. This will not however factor out the denominator in (5.148) and a different approach is needed. This requires a few definitions.

(α) *The semi-invariants.* We expand $\langle(\sigma_z)^n\rangle_0$ into a polynomial of quantities M_i^0 called semi-invariants according to the equation:

$$\langle(\sigma_z)^n\rangle_0 = \sum_{i_1+i_2+\cdots+i_p=n} M_{i_1}^0 M_{i_2}^0 \cdots M_{i_p}^0. \tag{5.153}$$

The sum in (5.153) is over *all* partitions $(i_1, i_2 \cdots i_p)$ of n *different* objects into sets containing respectively $i_1, i_2 \cdots i_p$ objects.

For instance, for $n = 3$ there is one partition into a triplet, three partitions into a doublet and a singlet and one partition into three singlets. Thus:

$$\langle(\sigma_z)^3\rangle_0 = M_3^0 + 3M_1^0 M_0^2 + (M_1^0)^3, \tag{5.154}$$

and also:

$$\begin{aligned}\langle\sigma_z\rangle_0 &= M_1^0;\\ \langle(\sigma_z)^2\rangle_0 &= M_2^0 + (M_1^0)^2;\\ \langle(\sigma_z)^4\rangle_0 &= M_4^0 + 4M_1^0M_3^0 + 3(M_2^0)^2 + 6(M_1^0)^2M_2^0 + (M_1^0)^4.\end{aligned} \tag{5.154'}$$

The M_i^0 can thus be calculated one after the other, M_n^0 appearing first in the expansion of $\langle(\sigma_z)^n\rangle_0$.

The semi-invariants defined in (5.153) have an interesting mathematical property: they are the cumulants of the function $\langle\exp(x\sigma_z)\rangle_0$. We recall the general definition of the cumulants of a function $F(x)$ expandable in a power series:

$$F(x) = 1 + \sum_1^\infty a_n x^n/n!. \tag{5.155}$$

The cumulants of $F(x)$ are the successive derivatives for $x = 0$ of the function:

$$G(x) = \ln[F(x)] = \sum_1^\infty M_n^0 x^n/n!. \tag{5.155'}$$

It can be shown (we omit the proof here) that:

$$a_n = \sum_{i_1+i_2+\cdots+i_p=n} M^0_{i_1} M^0_{i_2} \cdots M^0_{i_p}, \tag{5.156}$$

the summation being defined in the same way as in (5.153).

The semi-invariants M^0_n in (5.153) can thus be computed explicitly as:

$$M^0_n = \frac{\mathrm{d}^n}{\mathrm{d}x^n} (\ln\langle\exp(x\sigma_z)\rangle_0)_{x=0}. \tag{5.157}$$

From the form of the purely Zeeman density matrix σ_0:

$$\sigma_0 = \prod_i \sigma_{i0} = \prod_i (\exp(-\beta_Z \omega_0 I^i_z))/\mathrm{Tr}(\cdots), \tag{5.158}$$

and introducing $\alpha = -\frac{1}{2}\beta_Z\omega_0$, M^0_n can be rewritten:

$$M^0_n = \frac{\mathrm{d}^n}{\mathrm{d}x^n} (\ln \mathrm{Tr}\{\exp((\alpha+x)\sigma_z)\} - \ln \mathrm{Tr}\{\exp(\alpha\sigma_z)\})_{x=0}$$

$$= \frac{\mathrm{d}^n}{\mathrm{d}\alpha^n} (\ln \mathrm{Tr}\{\exp(\alpha\sigma_z)\}) = \frac{\mathrm{d}^n}{\mathrm{d}\alpha^n} (\ln \cosh \alpha), \tag{5.157$'$}$$

whence, with $\tanh \alpha = t$:

$$M^0_1 = t; \qquad M^0_2 = 1 - t^2; \qquad M^0_3 = -2t(1-t^2) \cdots. \tag{5.157$''$}$$

(β) *Contractions and diagrams.* We illustrate the expansion procedure on a definite example where we select, as the operator Q, σ^1_z and take in the numerator of (5.148) the fourth order term which according to (5.150) is equal to $(1/4!)(-\beta/2)^4$ times the sum over all indices of:

$$\langle \underbrace{v_{ij}\sigma^i_z\sigma^j_z}_{①} \times \underbrace{v_{kl}\sigma^k_z\sigma^l_z}_{②} \times \underbrace{v_{mn}\sigma^m_z\sigma^n_z}_{③} \times \underbrace{v_{pq}\sigma^p_z\sigma^q_z}_{④} \times \sigma^1_z\rangle_0, \tag{5.159}$$

where the interaction terms of $\mathcal{H}'$ are labelled ① to ④. According to eqns (5.151) to (5.153), (5.159) is a polynomial with respect to the semi-invariants M^0_i. A single product of semi-invariants can be extracted from this polynomial by a set of operations called 'contractions' defined by a set of contracting symbols shown in Fig. 5.12(b): a symbol such as ⌐ corresponds to the replacement in (5.159) of the two spins it links by the semi-invariant M^0_2, and similarly a symbol such as ⫪ corresponds to a semi-invariant M^0_3.

A set of contractions linking all the spins is shown in Fig. 5.12(b). Since semi-invariants originate in the expansion of a product of spin operators belonging to the *same* spin (eqn (5.153)) the contractions of Fig. 5.12(b) lead to the following conditions for the indices of (5.159):

$$i = l; \qquad j = k; \qquad m = p; \qquad n = q = 1;$$

but there is nothing to prevent i or j from being equal to either m or 1.

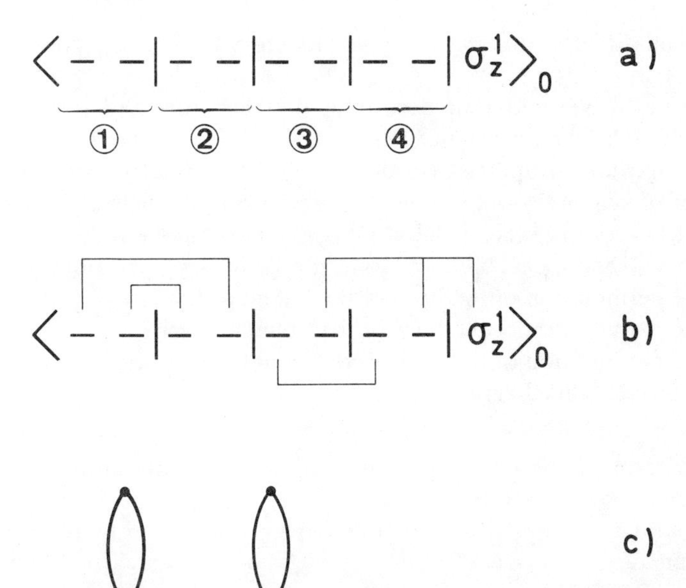

FIG. 5.12. (a) Schematic representation of the term (5.159), in the expansion of the numerator of (5.148) (see text). (b) A given set of contractions of this term. (c) The corresponding diagram.

The contribution of Fig. 5.12(b) to the summation of (5.159) is thus:

$$(M_2^0)^3 M_3^0 \sum_{i,j,m} v_{ij}^2 v_{ml}^2, \tag{5.160}$$

where the sum over i, j, m is *unrestricted*: each index runs over all nuclear sites independently of the others. As will appear later, this point is of major importance. The various semi-invariants are independent of the spin indices and have accordingly been extracted from the summation in (5.160).

The total contribution in β^4 to the numerator of (5.148) is obtained by performing all possible contractions on the term (5.159), summing their contributions, analogous to (5.160), and multiplying by $1/4!(-\beta/2)^4$.

Each contraction is represented by a diagram. The contraction of Fig. 5.12(b) is represented by the diagram of Fig. 5.12(c), which illustrates the following rules:

each vertex represented by a point corresponds to a spin index i;

special vertices, represented by a circle, correspond to spins belonging to the operator Q;

each line corresponds to an interaction v_{ij};

to each point vertex with ν lines attached to it, there corresponds a cumulant M_ν^0;

to each circle vertex with ν lines attached to it, there corresponds a cumulant $M^0_{\nu+1}$.

It is easily verified that according to these rules, (5.160) is indeed represented by the diagram of Fig. 5.12(c).

An important property of a diagram is its symmetry order g, i.e. the number of symmetry operations that transform the diagram into itself. In Fig. 5.12(c) for instance, which comprises a part linked to the operator σ^1_z, that is to Q, and a part unlinked to it, the symmetry operations are:

the permutation of the lines of the linked part;

the permutation of the lines of the unlinked part;

the permutation of the vertices of the unlinked part.

We have for this diagram:

$$g = (2!)^3 = 8.$$

Each term of the cumulant expansion of the numerator of (5.148) is represented by a well-defined diagram. However, a given diagram may correspond to several terms of this expansion, whose number must be determined.

We shall show presently that when its weight is taken into account, the contribution of a given diagram to the numerator of (5.148) is of the form:

$$\frac{1}{g}(-\beta)^n \prod_{\nu_i} M^0_{\nu_i} \times \sum_{i,j,k} v_{ij} v_{kl} \cdots, \tag{5.161}$$

where the summation is over the indices attached to the point vertices, and excludes those attached to the circle vertices.

As exemplified by the diagram of Fig. 5.12(c), the number of symmetry operations g of a diagram is the product of that of the linked part by that of the unlinked part. According to (5.161), this leads to the following very important theorem.

Theorem. The contribution of a diagram to the numerator of (5.148) is the product of the contributions of the linked part and of the unlinked part of this diagram.

The consequence of this theorem is the following. Let us consider the set of diagrams consisting of the *same* linked part and all possible unlinked parts. The contribution of this set to the numerator of (5.148) is the product of the linked part by the sum of the contributions of all possible unlinked parts. It is clear that the latter is precisely equal to the denominator of (5.148) and cancels out with it in the calculation of $\langle Q \rangle$.

We obtain finally the result we were looking for: $\langle Q \rangle$ is equal to the sum of the contributions of all linked diagrams. The term in β^n of $\langle Q \rangle$ is obtained by summing the contributions of all diagrams with n lines.

Remark. It is worth coming back to the reason for using the semi-invariant expansion which at first sight might appear as an unnecessary complication.

It would seem far more natural to assume that in a term such as (5.150) a contraction linking, say, two spin indices i and l means the replacement of $\sigma_z^i\sigma_z^l$ by $\langle(\sigma_z^i)^2\rangle_0$.

The diagram of Fig. 5.12(c) would then represent the term:

$$\frac{1}{4!}\left(-\frac{\beta}{2}\right)^4 \sum_{i,j,m} (v_{ij})^2(v_{m1})^2\langle(\sigma_z^i)^2\rangle_0\langle(\sigma_z^j)^2\rangle_0\times\langle(\sigma_z^m)^2\rangle_0\langle(\sigma_z^1)^3\rangle_0,$$

the weight of this diagram would be the same, but the sum would have to be restricted to $i \neq j \neq m \neq 1$. As a consequence, the contribution of the diagram would *not* be the product of the contributions of the linked and unlinked parts, and it would not be possible to factor out $\langle\exp(-\beta\mathcal{H}'_D)\rangle_0$ in the numerator of (5.148).

The only way out is to find an artifice allowing an unrestricted summation over the spin indices. This is precisely what is achieved by the semi-invariant expansion (5.153).

We now come back to the proof of (5.161) by listing all the terms which correspond to a given diagram.

Will be represented by the same diagram:

1. All contractions differing from a given contraction by a permutation of the interactions. For instance, the contraction of Fig. 5.13(a) differs from that of Fig. 5.12(b) by a permutation of the interactions ① and ③. In a term of order n, there are $n!$ permutations.

2. All contractions differing from a given contraction by a permutation of the spin indices in a given interaction. For instance the contraction of Fig. 5.13(b) differs from that of Fig. 5.12(b) by a permutation of the spin indices in the interaction ③, i.e. the replacement of $v_{mn}\sigma_z^m\sigma_z^n$ by $v_{nm}\sigma_z^n\sigma_z^m$. In a term of order n, there are 2^n such permutations.

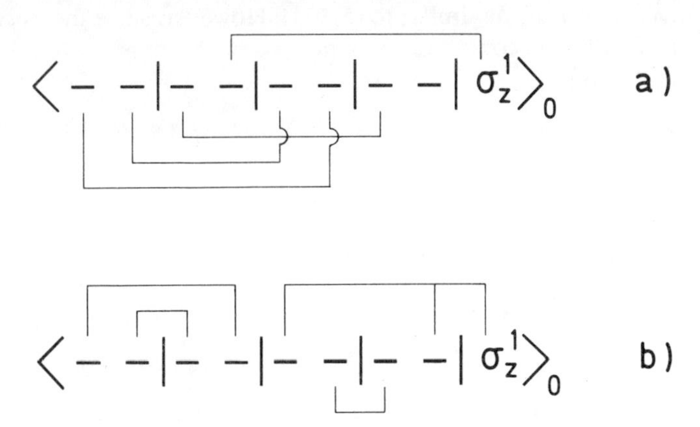

FIG. 5.13. Two sets of contractions of the term of Fig. 5.12(a) represented by the same diagram (Fig. 5.12(c)) as the set of contraction of Fig. 5.12(b).

The combinatorial product $2^n n!$ is an overestimate of the weight of a diagram, because of the terms that are counted several times by the operations 1 and 2. For instance, in Fig. 5.12(b), the permutations of ① and ②, or of ③ and ④ do not correspond to new terms, because the same spin indices are present in ① and ②, and in ③ and ④. The number of these extra terms is precisely equal to the symmetry order g of the diagram. The proper weight of the diagram is therefore $2^n n!/g$, whence the equation (5.161).

2. *The general case*

We consider a spin–spin interaction of the form:

$$\mathcal{H}' = \tfrac{1}{2} \sum_{i,j} v_{ij} \sigma^i_z \sigma^j_z + u_{ij} \sigma^i_+ \sigma^j_-. \tag{5.162}$$

If $\mathcal{H}'$ is the dipolar interaction (5.43) we have:

$$\begin{aligned} v_{ij} &= \tfrac{1}{2} A_{ij}; \\ u_{ij} &= -\tfrac{1}{4} A_{ij}. \end{aligned} \tag{5.163}$$

We can try to use the same method as in the longitudinal case. A general term of the numerator of (5.148) is of the form:

$$\frac{1}{n!}\left(-\frac{\beta}{2}\right)^n v_{ij} u_{kl} \cdots \langle \sigma^i_z \sigma^j_z \sigma^k_+ \sigma^l_- \cdots Q \rangle_0. \tag{5.164}$$

By grouping together the spin operators of the same individual spins, the expectation value in (5.164) is a product of terms of the form:

$$\langle \sigma_z \cdots \sigma_z \sigma_+ \sigma_z \cdots \sigma_z \cdots \sigma_- \cdots \rangle_0,$$

replacing $\langle (\sigma_z)^n \rangle_0$ of the longitudinal case, which can be expanded in a sum of semi-invariants in a way similar to (5.151). However, since the operators σ_z, σ_+ and σ_- do not commute, it is necessary to define ordered semi-invariants as follows:

$$\langle \sigma_z \cdots \sigma_z \sigma_+ \sigma_z \cdots \sigma_- \cdots \rangle_0 = \sum_{i_1 + i_2 + \cdots + i_p = n} M_{i_1} M_{i_2} \cdots M_{i_p}. \tag{5.165}$$

The sum is over all the partitions of n distinct *ordered* objects into *ordered* sets of $i_1, i_2 \cdots i_p$ objects.

For instance:

$$\begin{aligned} \langle \sigma_z \sigma_+ \sigma_- \sigma_z \rangle_0 = {} & M_4(\sigma_z \sigma_+ \sigma_- \sigma_z) + M_3(\sigma_z \sigma_+ \sigma_-) M_1(\sigma_z) + M_1(\sigma_z) M_3(\sigma_+ \sigma_- \sigma_z) \\ & + M_2(\sigma_z^2) M_2(\sigma_+ \sigma_-) + M_1(\sigma_z) M_2(\sigma_+ \sigma_-) M_1(\sigma_z), \end{aligned} \tag{5.166}$$

where $M_n(\sigma_z^n)$ is equal to the cumulant M_n^0 defined in the preceding section. The expectation value (5.165) is non-zero only if there are as many operators σ_- as operators σ_+. It is easily found that the same is true for the ordered semi-invariants. Vanishing semi-invariants are omitted in (5.166). There is no closed form expression analogous to (5.157′) for computing the ordered semi-invariants. They must be computed step by step. For instance:

$$\langle \sigma_+\sigma_-\rangle_0 = 2(1+t) = M_2(\sigma_+\sigma_-) \tag{5.167}$$

$$\langle \sigma_-\sigma_+\rangle_0 = 2(1-t) = M_2(\sigma_-\sigma_+) \tag{5.167′}$$

$$\begin{aligned}\langle \sigma_z\sigma_+\sigma_-\rangle_0 &= 2(1+t)\\ &= M_3(\sigma_z\sigma_+\sigma_-) + M_1^0 M_2(\sigma_+\sigma_-),\end{aligned} \tag{5.167″}$$

whence, according to (5.167):

$$M_3(\sigma_z\sigma_+\sigma_-) = 2(1-t^2). \tag{5.167‴}$$

A short list of semi-invariants is given in Table 5.1.

Table 5.1

List of semi-invariants

M_1^0	$=t$	$M_4(\sigma_-\sigma_+\sigma_z^2)$	
M_2^0	$=1-t^2$	$M_4(\sigma_z^2\sigma_-\sigma_+)$	$\Big\}=4t(1-t^2)$
M_0^3	$=-2t(1-t^2)$	$M_4(\sigma_z\sigma_-\sigma_+\sigma_z)$	
M_4^0	$=-2(1-t^2)(1-3t^2)$	$M_4(\sigma_+\sigma_z\sigma_-\sigma_z)$	$\Big\}=-4(1+t)(1-t^2)$
$M_2(\sigma_+\sigma_-)$	$=2(1+t)$	$M_4(\sigma_z\sigma_+\sigma_z\sigma_-)$	
$M_2(\sigma_-\sigma_+)$	$=2(1-t)$	$M_4(\sigma_-\sigma_z\sigma_+\sigma_z)$	$\Big\}=-4(1-t)(1-t^2)$
$M_3(\sigma_z\sigma_+\sigma_-)$	$\Big\}=2(1-t^2)$	$M_4(\sigma_z\sigma_-\sigma_z\sigma_+)$	
$M_3(\sigma_+\sigma_-\sigma_z)$		$M_4(\sigma_+\sigma_z^2\sigma_-)$	$=4t(1+t)^2$
$M_3(\sigma_z\sigma_-\sigma_+)$	$\Big\}=-2(1-t^2)$	$M_4(\sigma_-\sigma_z^2\sigma_+)$	$=-4t(1-t)^2$
$M_3(\sigma_-\sigma_+\sigma_z)$		$M_4(\sigma_+^2\sigma_-^2)$	$=-8(1+t)^2$
$M_3(\sigma_+\sigma_z\sigma_-)$	$=-2(1+t)^2$	$M_4(\sigma_-^2\sigma_+^2)$	$=-8(1-t)^2$
$M_3(\sigma_-\sigma_z\sigma_+)$	$=2(1-t)^2$	$M_4(\sigma_+\sigma_-\sigma_+\sigma_-)$	$=-8t(1+t)$
$M_4(\sigma_+\sigma_-\sigma_z^2)$		$M_4(\sigma_-\sigma_+\sigma_-\sigma_+)$	$=8t(1-t)$
$M_4(\sigma_z^2\sigma_+\sigma_-)$	$\Big\}=-4t(1-t^2)$	$M_4(\sigma_-\sigma_+^2\sigma_-)$	$\Big\}=-8(1-t^2)$
$M_4(\sigma_z\sigma_+\sigma_-\sigma_z)$		$M_4(\sigma_+\sigma_-^2\sigma_+)$	

Equation (5.164) is a sum of products of semi-invariants and each term of this sum can be represented by a diagram, as in the longitudinal case. There are however three differences with that case. Firstly there are two types of lines: longitudinal lines corresponding to $v_{ij}\sigma_z^i\sigma_z^j$, and transverse lines corresponding to $u_{ij}\sigma_+^i\sigma_-^j$. The latter are noted by an arrow going from σ_+ to σ_-. Secondly, the order in which these interactions appear in (5.164) must be specified, because the ordered semi-invariants depend on this order. Thirdly, the nature of the operators in Q must also be specified. The following notations are used: ○ for σ_z, ⊕ for σ_+ and ⊖ for σ_-.

We consider as an example the diagrams of Fig. 5.14, with $Q=\sigma_+^1\sigma_-^2$. All three diagrams correspond to:

$$\frac{1}{6!}\left(-\frac{\beta}{2}\right)^6 v_{1i}u_{1i}u_{2i}u_{jk}^2v_{jk}, \tag{5.168}$$

multiplied by a product of semi-invariants. In diagram (a) this product is:

$$M_3(\sigma_z\sigma_-\sigma_+)M_2(\sigma_+\sigma_-)M_3(\sigma_z\sigma_+\sigma_-)M_3(\sigma_+\sigma_z\sigma_-)M_3(\sigma_-\sigma_z\sigma_+), \tag{5.169}$$

where the successive semi-invariants correspond respectively to the vertices 1, 2, i, j, k.

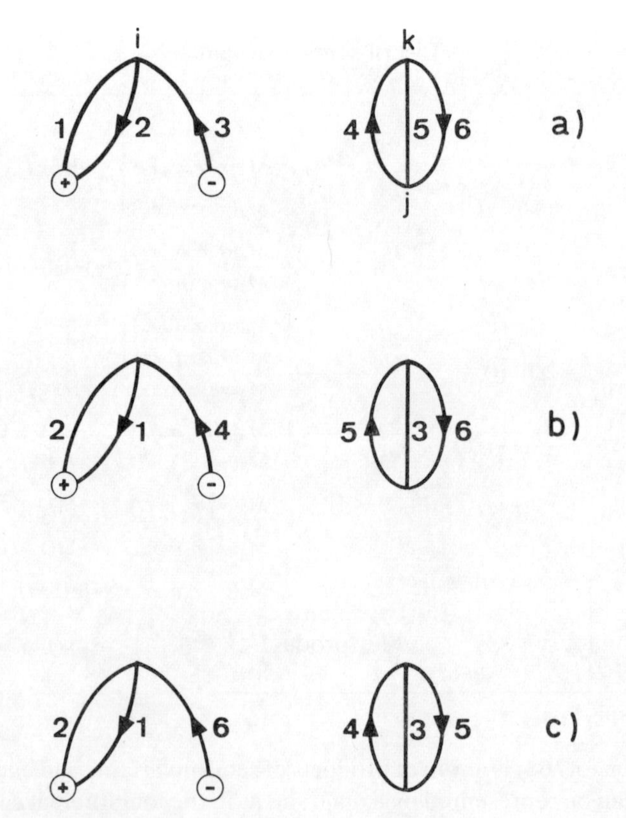

FIG. 5.14. Three 'time'-ordered diagrams differing only by the order of the 'times' associated with the various interactions.

In diagram (b) this product is:

$$M_3(\sigma_-\sigma_z\sigma_+)M_2(\sigma_+\sigma_-)M_3(\sigma_+\sigma_z\sigma_-)M_3(\sigma_z\sigma_+\sigma_-)M_3(\sigma_z\sigma_-\sigma_+). \quad (5.169')$$

It is easily found that the diagram (c) yields the same contribution (5.169′) as the diagram (b).

As in the longitudinal case, our aim is that the contribution of a diagram to the numerator of (5.148) be the product of the contributions of the linked part (that is connected to the operators of Q) and of the unlinked part. When this is achieved, the contribution of the set of all diagrams with the same linked part is equal to the contribution of this linked part multiplied by the denominator of (5.148) and, as a consequence the expectation value of $\langle Q\rangle$ is obtained by summing the contributions of all linked diagrams.

To achieve this aim is more complicated than in the longitudinal case, and is done as follows. We redefine a single diagram as the set of all diagrams with the same structure but differing by the order of the interactions. The most convenient way of defining this set is to use the formalism of time-ordered operators, a practice current when dealing with non-commuting operators. We write:

$$\exp(\alpha I_z - \beta\mathcal{H}') = \exp(\alpha I_z)T\left\{\exp\left(-\int_0^\beta \mathcal{H}'(\beta')\,\mathrm{d}\beta'\right)\right\}, \quad (5.170)$$

where T is a 'time-ordering' operator which, in any term of the expansion of the last exponential, orders the 'times' in decreasing order from left to right.

By comparison with (1.73), the coefficient β in (5.170) plays the role of an imaginary time.

A typical term of the numerator of (5.148) is then:

$$(-\tfrac{1}{2})^n v_{ij}u_{kl}\cdots v_{pq}\times T\int_0^\beta \mathrm{d}\beta_1\int_0^\beta \mathrm{d}\beta_2\cdots\int_0^\beta \mathrm{d}\beta_n;$$

$$\langle\sigma_z^i(\beta_1)\sigma_z^j(\beta_1)\sigma_+^i(\beta_2)\sigma_-^i(\beta_2)\cdots\sigma_z^p(\beta_n)\sigma_z^q(\beta_n)Q\rangle_0. \quad (5.171)$$

The expectation value is expanded in a polynomial of products of semi-invariants built from 'time' dependent spin operators. Each term of this sum is represented by a diagram.

It can be shown that after Feynman integration over the 'times' β_i, the contribution of a diagram is the product of that of the linked and the unlinked part.

We omit the derivation of this result here and we give without proof the rules for calculating the contribution to $\langle Q\rangle$ of a given linked diagram of order m.

1. To each longitudinal line is assigned the factor v_{ij} and a time β_p.
2. To each transverse line is assigned the factor $\frac{1}{2}u_{ij}$ and a 'time' β_q.
3. To each vertex is assigned a semi-invariant depending on several

'times': the time of a spin operator entering the definition of the semi-invariant is that of the corresponding interaction line. The operators of Q are assigned the time 0.

4. Divide by g, the number of symmetry operations of the diagram, and multiply by $(-1)^n$.

5. Perform a Feynman integration over the various 'times'.

6. Perform an unrestricted summation over the indices of the vertices other than those attached to Q.

These rules are best understood by considering a definite example: the contribution of the diagram of Fig. 5.15 is:

$$\begin{aligned}
&\frac{(-1)^2}{2^2}\sum_i u_{1i}u_{2i}\times T\int_0^{\beta}\mathrm{d}\beta_1\int_0^{\beta}\mathrm{d}\beta_2\, M_2[\sigma_-(\beta_1)\sigma_+(0)]\times M_2[\sigma_+(\beta_1)\sigma_-(\beta_2)]\\
&\qquad\times M_2[\sigma_+(\beta_2)\sigma_-(0)]\\
&\quad=\frac{\beta^2}{8}\sum_i u_{1i}u_{2i}M_2(\sigma_-\sigma_+)M_2(\sigma_+\sigma_-)\times\{M_2(\sigma_+\sigma_-)+M_2(\sigma_-\sigma_+)\}\\
&\quad=\frac{\beta^2}{8}\sum_i u_{1i}u_{2i}\times 2(1-t)\times 2(1+t)\times 4\\
&\quad=2\beta^2(1-t^2)\sum_i u_{1i}u_{2i},
\end{aligned}\tag{5.172}$$

where we have used (5.167) and (5.167′).

A last point must be examined in this diagrammatic method. Consider a vertex with only one line leaving it. Since $M_1(\sigma_+)=M_1(\sigma_-)=0$, the semi-invariant associated with this vertex must be $M_1(\sigma_z)=M_1^0$, i.e. the line is longitudinal. To this line corresponds the factor v_{ij}, where j is the vertex at the other end of the line. The summation over i yields the factor:

$$\sum_i v_{ij}=q,$$

i.e. the Weiss-field factor of the spin j in the sample.

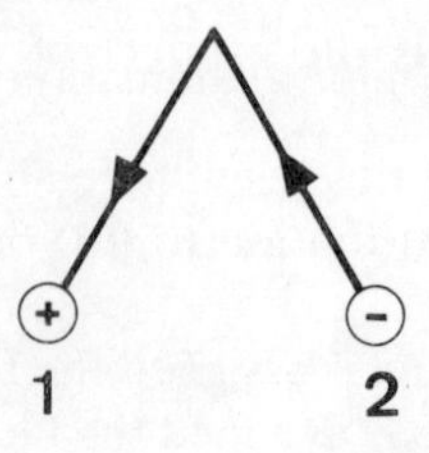

FIG. 5.15. Example of a diagram yielding a contribution in β^2 to $\langle\sigma_+^1\sigma_-^2\rangle$.

The simplest case is that when $q = 0$ (i.e. in a spherical sample of CaF_2). In that case, only diagrams with no free end must be considered, since all others yield vanishing contributions. When $q \neq 0$, it can be shown that it is possible to obtain contributions to $\langle Q \rangle$ valid to all orders with respect to the Weiss-field by using only diagrams with no free end and replacing in the expressions for the semi-invariants $t = \tanh(-\frac{1}{2}\beta_Z\omega_0)$ by:

$$t' = \tanh(-\tfrac{1}{2}\beta_Z\omega_0 - \tfrac{1}{2}\beta q t'). \tag{5.173}$$

The procedure used for deriving this result, known as 'vertex decoration by Cayley trees', is described in Stinchcombe *et al.* (1963) and will not be repeated here.

We have outlined in this section a set of rules which can be used to calculate by the diagrammatic method an expansion in powers of β of the expectation value $\langle Q(\beta_Z, \beta) \rangle$ of operators Q of physical interest. For the benefit of the reader we have also sketched the derivation of these rules. It is clear however that it is only by applying these rules to definite calculations that their use can be mastered.

(b) Thermodynamic properties of the spin system

The diagrammatic method described in the preceding section is used to obtain expansions in powers of β of various thermodynamic quantities, in a system of identical spins $\frac{1}{2}$ with Hamiltonian (1.26) and density matrix (1.28) (Goldman, Jacquinot, Chapellier, and Vu Hoang Chau, 1975).

We suppose that the nuclei are located on a Bravais lattice with one spin per unit cell. For practical purposes, we introduce the space-Fourier transforms of the coefficients A_{ij} of the dipolar Hamiltonian (5.43):

$$A(\mathbf{k}) = \sum_j A_{ij} \exp(i\mathbf{k} \cdot \mathbf{r}_{ij}) \tag{5.174}$$

where $\mathbf{k}$ is one of the N vectors of the reciprocal lattice belonging to the first Brillouin zone. Equation (5.174) is meaningful only insofar as the summation is independent of the index i. This is the case for all values of $\mathbf{k}$ except when $|\mathbf{k}|^{-1}$ is comparable with the linear size of the sample (Cohen and Keffer, 1955). The 'pathological' values of $\mathbf{k}$ are restricted to a very small part of the Brillouin zone around the origin, and very little error results if they are ignored when performing a summation over $\mathbf{k}$ of a function of $A(\mathbf{k})$. The component $A(0)$ plays a special role, and (5.174) is made meaningful for $\mathbf{k} = 0$ by choosing the sample shape to be an ellipsoid.

The calculations are, as a rule restricted to the case when:

$$\sum_j A_{ij} = A(0) = 0.$$

This is the case for instance for a cubic array of spins in a spherical sample.

1. *Dipole–dipole energy*

The dipole–dipole energy per spin is of the form:

$$\frac{1}{N}\langle \mathscr{H}'_{\mathrm{D}}\rangle = \sum_n \beta^n G_n(t), \tag{5.175}$$

where t is given by (5.147). The first three terms were calculated for the case $A(0)=0$. They are equal to:

$$G_1 = -\tfrac{1}{4}D^2(1-t^2)(1-\tfrac{2}{3}t^2); \tag{5.176}$$

$$G_2 = \frac{(1-t^2)}{64}[3(K_3-I_3)+t^2(11I_3-8K_3)+4t^4(K_3-2I_3)]; \tag{5.177}$$

$$\begin{aligned} G_3 = {}& \frac{(1-t^2)}{256}\times\{[-7K_4-10I_4+\tfrac{5}{3}K_2I_2+\tfrac{1}{6}D^4] \\ &+\tfrac{1}{3}t^2[71K_4+148I_4-214K_2I_2-\tfrac{13}{6}D^4] \\ &+t^4[-24K_4-\tfrac{260}{3}I_4+64K_2I_2+\tfrac{8}{9}D^4]+t^6[8K_4+48I_4+\tfrac{1}{3}D^4]\}; \end{aligned} \tag{5.178}$$

where the various coefficients are:

$$D^2 = \tfrac{3}{4}\sum_j A_{ij}^2 = \frac{3}{4N}\sum_{\mathbf{k}}[A(\mathbf{k})]^2;$$

$$K_3 = \sum_{j,l} A_{ij}A_{jl}A_{li} = \frac{1}{N}\sum_{\mathbf{k}}[A(\mathbf{k})]^3;$$

$$I_3 = \sum_j (A_{ij})^3;$$

$$K_4 = \sum_{j,l,m} A_{ij}A_{jl}A_{lm}A_{mi} = \frac{1}{N}\sum_{\mathbf{k}}[A(\mathbf{k})]^4;$$

$$I_4 = \sum_j (A_{ij})^4$$

$$\begin{aligned} K_2I_2 &= \sum_{j,l}(A_{ij})^2A_{jl}A_{li} \\ &= \frac{1}{N}\sum_{j,\mathbf{k}}(A_{ij})^2[A(\mathbf{k})^2\exp(\mathrm{i}\mathbf{k}\,.\,\mathbf{r}_{ij})]. \end{aligned}$$

As an example, when in a simple cubic system the external field $\mathbf{H}_0$ is along a direction [100], the values of the coefficients are the following:

$$D = 3{\cdot}16(\gamma^2\hbar/2a^3);$$

$$K_3 = 0{\cdot}737D^3;$$

$$I_3 = -0{\cdot}350D^3;$$

$$K_4 = 4{\cdot}500 D^4;$$

$$I_4 = 0{\cdot}367 D^4;$$

$$K_2 I_2 = 0{\cdot}373 D^4;$$

where a is the lattice parameter.

According to these figures, the first few terms G_n in (5.175) are, apart from their dependence on t, of the order of:

$$G_n \sim D^{n+1}.$$

It is to be expected that the same remains true for higher order terms, which substantiates the choice of βD as the characteristic parameter of dipolar order, and justifies the criterion (1.17) for the validity of the linear expansion of the density matrix.

In the case when $A(0) \neq 0$, the dipolar energy per spin is:

$$\frac{1}{N}\langle \mathcal{H}'_{\mathrm{D}} \rangle = \tfrac{1}{4} A(0) t'^2 + \sum_n \beta^n G_n(t'), \tag{5.179}$$

where t' is defined by (5.173) with $q = A(0)$.

In the rest of this section, we limit ourselves to $A(0) = 0$.

2. *Spin polarization*

According to (5.141) and (5.142) we have:

$$\begin{aligned} \omega_0 \frac{\partial}{\partial \beta} \langle I_z \rangle &= \frac{\partial}{\partial \beta_{\mathrm{Z}}} \langle \mathcal{H}'_{\mathrm{D}} \rangle \\ &= \frac{\partial}{\partial t} \langle \mathcal{H}'_{\mathrm{D}} \rangle \times \frac{\mathrm{d}t}{\mathrm{d}\beta_{\mathrm{Z}}} \\ &= -\tfrac{1}{2}\omega_0 (1 - t^2) \frac{\partial}{\partial t} \langle \mathcal{H}'_{\mathrm{D}} \rangle, \end{aligned} \tag{5.180}$$

and, since:

$$\begin{aligned} \langle I_z \rangle &= \frac{N}{2} p; \\ \frac{\partial p}{\partial \beta} &= -(1 - t^2) \frac{1}{N} \frac{\partial}{\partial t} \langle \mathcal{H}'_{\mathrm{D}} \rangle. \end{aligned} \tag{5.181}$$

This yields upon integration:

$$\begin{aligned} p(\beta_{\mathrm{Z}}, \beta) &= p(\beta_{\mathrm{Z}}, 0) + \int_0^\beta \frac{\partial}{\partial \lambda} p(\beta_{\mathrm{Z}}, \lambda)\, \mathrm{d}\lambda \\ &= p(\beta_{\mathrm{Z}}, 0) - \frac{(1 - t^2)}{N} \int_0^\beta \frac{\partial}{\partial t} \langle \mathcal{H}'_{\mathrm{D}} \rangle (t, \lambda)\, \mathrm{d}\lambda. \end{aligned} \tag{5.182}$$

We have:

$$p(\beta_Z, 0) = t,$$

whence, according to (5.182) and (5.175):

$$p = t - (1-t^2)\sum_n \frac{\beta^{n+1}}{n+1}\frac{\mathrm{d}G_n}{\mathrm{d}t}. \tag{5.183}$$

We treat as an example the case when the system has been demagnetized in high field by an ADRF down to a very small effective frequency Δ. The density matrix is:

$$\sigma = \exp[-\beta(\Delta I_z + \mathscr{H}'_{\mathrm{D}})]/\mathrm{Tr}\{\cdots\},$$

so that:

$$\beta_Z\omega_0 = \beta\Delta,$$

and $t = \tanh(-\frac{1}{2}\beta_Z\omega_0) = \tanh(-\frac{1}{2}\beta\Delta)$.

In the limit of small Δ, the polarization is proportional to Δ:

$$p = \chi_\parallel\Delta, \tag{5.184}$$

which defines the longitudinal susceptibility $\chi_\parallel$:

$$\begin{aligned}\chi_\parallel &= \left.\frac{\partial p}{\partial\Delta}\right|_{\Delta=0} = \left[\frac{\mathrm{d}t}{\mathrm{d}\Delta}\frac{\partial p}{\partial t}\right]_{t=0} \\ &= -\tfrac{1}{2}\beta\left[(1-t^2)\frac{\partial p}{\partial t}\right]_{t=0},\end{aligned} \tag{5.185}$$

that is, according to (5.183)

$$\chi_\parallel = -\tfrac{1}{2}\beta\left\{1 - \sum_n \frac{\beta^{n+1}}{n+1}\left.\frac{\mathrm{d}^2G_n}{\mathrm{d}t^2}\right|_{t=0}\right\}. \tag{5.186}$$

Up to β^3, this yields:

$$\chi_\parallel = -\tfrac{1}{2}\beta(1 - \tfrac{5}{12}\beta^2D^2). \tag{5.187}$$

3. *Free energy*

We begin by calculating the logarithm of the partition function (5.140). According to (5.142) we have:

$$\ln\mathscr{Z}(\beta_Z, \beta) = \ln\mathscr{Z}(\beta_Z, 0) - \int_0^\beta \langle\mathscr{H}'_{\mathrm{D}}\rangle(\beta_Z, \lambda)\,\mathrm{d}\lambda. \tag{5.188}$$

The straightforward result:

$$\begin{aligned}\ln\mathscr{Z}(\beta_Z, 0) &= \ln\{\mathrm{Tr}[\exp(-\beta_Z\omega_0 I_z)]\} \\ &= N\{\ln 2 - \tfrac{1}{2}\ln(1-t^2)\},\end{aligned} \tag{5.189}$$

associated with (5.175), yields:

$$\frac{1}{N}\ln \mathscr{Z}(\beta_{\mathrm{Z}},\beta)=\ln 2-\tfrac{1}{2}\ln(1-t^2)-\sum_n \frac{\beta^{n+1}}{n+1}G_n(t). \tag{5.190}$$

The free energy F is defined for a system with a single temperature β^{-1} through:

$$F=-\frac{1}{\beta}\ln \mathscr{Z}. \tag{5.191}$$

To define it in the present case, it is necessary to use a rotating frame in which the effective frequency Δ is such that:

$$\beta_{\mathrm{Z}}\omega_0=\beta\Delta.$$

The effective Hamiltonian is then:

$$\mathscr{H}=\Delta I_z+\mathscr{H}'_{\mathrm{D}},$$

and the density matrix (1.28):

$$\sigma=\exp(-\beta\mathscr{H})/\mathrm{Tr}\{\exp(-\beta\mathscr{H})\},$$

is characterized by a single temperature. Since $\ln \mathscr{Z}$ (eqn (5.190)) depends on β_{Z} only through,

$$t=\tanh(-\tfrac{1}{2}\beta_{\mathrm{Z}}\omega_0)=\tanh(-\tfrac{1}{2}\beta\Delta),$$

it is not modified by this change of reference frame.

The free energy per spin in this frame is:

$$F/N=-\frac{1}{\beta}\{\ln 2-\tfrac{1}{2}\ln(1-t^2)\}+\sum_n \frac{\beta^n}{n+1}G_n(t). \tag{5.192}$$

4. *Entropy*

The entropy S is given by (5.143). According to (5.147) we write $\beta_{\mathrm{Z}}\omega_0$ in the form:

$$\begin{aligned}\beta_{\mathrm{Z}}\omega_0&=-2\tanh^{-1}(t)\\&=-\ln\left(\frac{1+t}{1-t}\right).\end{aligned} \tag{5.193}$$

By inserting (5.175), (5.183), (5.190), and (5.193) into (5.143) we obtain:

$$\begin{aligned}\frac{S}{Nk_{\mathrm{B}}}&=\{\ln 2-\tfrac{1}{2}[(1+t)\ln(1+t)+(1-t)\ln(1-t)]\}\\&\quad+\sum_n \frac{\beta^{n+1}}{n+1}\left[nG_n-\tfrac{1}{2}(1-t^2)\ln\left(\frac{1+t}{1-t}\right)\frac{\mathrm{d}G_n}{\mathrm{d}t}\right].\end{aligned} \tag{5.194}$$

The curly bracket in (5.194) is the entropy per spin of a system of non-interacting spins $\frac{1}{2}$ of polarization t, and is therefore equal to $S(\beta_Z, 0)/Nk_B$. Since the entropy depends on β_Z only through t, it is not modified in the rotating frame picture.

5. *Transverse susceptibility*

The transverse susceptibility is defined by (5.123): it is proportional to the transverse spin polarization in the presence of a small transverse field. This transverse spin polarization is most conveniently calculated according to (5.122), with $Q = 2I_x/N_D$, $\varepsilon V = \omega_1 I_x$. We limit ourselves to the case when the system has been demagnetized to zero field, i.e. when the density matrix is:

$$\sigma = \exp(-\beta\mathcal{H}'_D)/\mathrm{Tr}\{\exp(-\beta\mathcal{H}'_D)\}.$$

For spins $\frac{1}{2}$, the transverse susceptibility is then:

$$\chi_\perp = -\frac{2}{N}\left\langle \int_0^\beta \exp(\lambda\mathcal{H}'_D)I_x \exp(-\lambda\mathcal{H}'_D)I_x \, d\lambda \right\rangle. \qquad (5.195)$$

We expand the integral on the right-hand side in powers of β:

$$\int_0^\beta \exp(\lambda\mathcal{H}'_D)I_x \exp(-\lambda\mathcal{H}'_D)I_x \, d\lambda$$
$$\simeq \beta I_x^2 + \tfrac{1}{2}\beta^2[\mathcal{H}'_D, I_x]I_x + \tfrac{1}{6}\beta^3[\mathcal{H}'_D, [\mathcal{H}'_D, I_x]]I_x + \cdots. \qquad (5.196)$$

In order to have a consistent approximation, if for instance we limit the expansion of $\chi_\perp$ to the term in β^3, $\langle I_x^2\rangle$ has to be computed up to terms in β^2, $\langle[\mathcal{H}'_D, I_x]I_x\rangle$ up to terms in β and $\langle[\mathcal{H}'_D, [\mathcal{H}'_D, I_x]]I_x\rangle 0$ to zeroth order. To this approximation, the result is:

$$\chi_\perp = -\tfrac{1}{2}\beta(1 - \tfrac{1}{6}\beta^2 D^2). \qquad (5.197)$$

The transverse susceptibility is *not* equal to $\chi_\parallel$ (eqn (5.187)), which means that if the r.f. irradiation is applied slightly off resonance, the spin polarization will *not* be aligned with the effective field in the rotating frame, a typical non-linear effect.

(c) *Experimental illustration*

We describe several experimental illustrations of non-linear effects in spin temperature, performed on the ^{19}F spin system in spherical samples of CaF_2, with the external field $\mathbf{H}_0$ parallel to a fourfold crystalline axis [001]: variation of the dipolar energy with polarization, in high effective field, and susceptibilities in zero effective field.

1. *Variation of dipolar energy with polarization in high effective field*

Following an irradiation at the distance Δ_0 from resonance, applied either suddenly at Δ_0 or through an ADRF started far away from resonance and stopped at Δ_0, the density matrix is of the form:

$$\sigma = \exp[-\beta(\Delta_0 I_z + \mathcal{H}'_D)]/\mathrm{Tr}\{\cdots\}. \tag{5.198}$$

At low entropy (5.198) is valid only insofar as Δ_0 is sufficiently larger than the effective frequency D, otherwise the system undergoes a transition to an ordered state (chapter 8). In the experiment, $\Delta_0/\gamma \simeq 7$ G, whereas $D/\gamma \simeq 2$ G, and the system remains paramagnetic up to the highest initial polarizations.

The aim of the experiment is to measure the dipolar energy $\langle \mathcal{H}'_D \rangle$ as a function of polarization, and to compare its variation with that predicted by (5.175) expanded up to β^3, and (5.183) expanded up to β^4.

Ideally, p and $\langle \mathcal{H}'_D \rangle$ could be measured when the equilibrium state described by (5.198) is achieved, by removing the strong r.f. field, recording the absorption signal with a small non-saturating r.f. field and measuring its area and first moment, as explained in sections **C**(a) and **C**(d). This is not practical in the present case, because when Δ_0 is large, the Zeeman reservoir is much more ordered than the dipolar reservoir, the absorption signal differs but slightly from a purely Zeeman signal and the measurement of its first moment is very inaccurate.

The experimental procedure takes advantage of the fact that in the samples used and under the experimental conditions of field and lattice temperature ($H_0 \simeq 2{\cdot}7T$; $T_L \simeq 0{\cdot}3$ K), whereas the Zeeman relaxation time is unobservably long, the dipolar relaxation time is relatively short: $T_{1D} \simeq 3$ min.

The experiment goes in several steps.

1. The system having only Zeeman order, its polarization is measured from the area of the absorption signal.

2. An r.f. irradiation is applied at the distance Δ_0 from resonance, during a time much shorter than T_{1D}, but long enough for the density matrix σ to become of the form (5.198).

3. The r.f. field is turned off, and the dipolar energy is allowed to relax to zero during a time interval of several times T_{1D}.

4. The new polarization is measured.

The sequence is then started again at step 2.

The decrease of polarization due to the irradiation yields directly the value of the dipolar energy: in the rotating frame, the Hamiltonian is time-independent and the evolution during the r.f. irradiation takes place at constant effective energy. If ω_1 is small, we have approximately:

$$\begin{aligned}\langle \mathcal{H} \rangle &= \Delta_0 \langle I_z \rangle + \langle \mathcal{H}'_D \rangle \\ &= \frac{N}{2}\Delta_0 p + \langle \mathcal{H}'_D \rangle.\end{aligned} \tag{5.199}$$

Before the irradiation of step 2, we have $p = p_i$ and $\langle \mathcal{H}'_D \rangle = 0$. After irradiation for a time t, with $T_2 \ll t \ll T_{1D}$, we have $p = p_f$, and $\langle \mathcal{H}'_D \rangle \neq 0$ is related to p_f through (5.175) and (5.183). By equating the initial and final effective energies:

$$\frac{N}{2}\Delta_0 p_i = \frac{N}{2}\Delta_0 p_f + \langle \mathcal{H}'_D \rangle,$$

we obtain:

$$\langle \mathcal{H}'_D \rangle = \frac{N}{2}\Delta_0 (p_i - p_f). \tag{5.200}$$

In the high temperature limit, the ratio of dipolar to effective Zeeman energies is constant. According to (1.30) and (1.31) this ratio is:

$$\frac{\langle \mathcal{H}'_D \rangle}{\langle Z \rangle} = \frac{D^2}{\Delta_0^2},$$

or else:

$$\langle \mathcal{H}'_D \rangle = \frac{N}{2}\frac{D^2}{\Delta_0} p. \tag{5.201}$$

This is no longer true when the temperature is lowered. When $\Delta_0 \gg D$, $\beta D < 1$ up to high values of p and the trend of the variation of $\langle \mathcal{H}'_D \rangle$ is already apparent when limiting the expansions of (5.175) and (5.183) to the term linear in β, that is:

$$\begin{aligned} p &\simeq t; \\ \langle \mathcal{H}'_D \rangle &\simeq -\frac{N}{4}\beta D^2 (1 - t^2)(1 - \tfrac{2}{3}t^2) \\ &\simeq -\frac{N}{4}\beta D^2 (1 - p^2)(1 - \tfrac{2}{3}p^2) \\ &= \frac{N}{4}\frac{D^2}{\Delta_0} \ln\left(\frac{1+p}{1-p}\right) \times (1 - p^2)(1 - \tfrac{2}{3}p^2). \end{aligned} \tag{5.202}$$

As a function of p, $\langle \mathcal{H}'_D \rangle$ first increases linearly, goes through a maximum for $p \sim 0{\cdot}5$ and then decreases to zero when $p = 1$.

The variation (5.202) is represented by curve 1 in Fig. 5.16. A very slightly different curve is obtained when using an expansion to β^3 in (5.175) and to β^4 in (5.183) (curve 2 of Fig. 5.16), with $\Delta_0 = 7{\cdot}12$ G, $T > 0$.

The experimental results, shown on the same figure, differ significantly from these predictions. The departure is attributed to the presence of paramagnetic impurities which produce a distribution of fields at the sites of the nuclear spins.

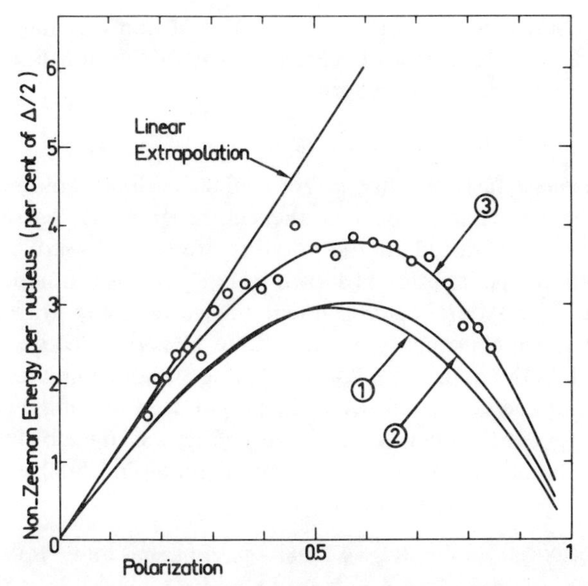

FIG. 5.16. Non-Zeeman energy per fluorine spin as a function of nuclear polarization, in a spherical sample of CaF_2 at thermal equilibrium in the rotating frame in an effective field of 7.12 G. $H_0\|[100]$ and $T>0$. Curve 1: 1st order approximation with dipolar interactions only. Curve 2: 3rd order approximation with dipolar interactions only. Curve 3: best-fit addition to curve 2 of a contribution from impurities. (After Goldman *et al.*, 1975.)

In that case, the dipolar interaction must be supplemented with the distribution of Zeeman interactions to yield a non-Zeeman term:

$$\mathcal{H}_{nZ}=\mathcal{H}'_D+\sum_i \delta_i I_{iz}.$$

A similar non-Zeeman reservoir, associated with the paramagnetic impurities, is considered at length in chapter 6. It is $\langle\mathcal{H}_{nZ}\rangle$ which is measured in the experiment, rather than $\langle\mathcal{H}'_D\rangle$. Insofar as:

$$d^2=\frac{1}{N}\sum_i \delta_i^2 \ll D^2,$$

it can be shown that $\langle\mathcal{H}'_D\rangle$ and p are not affected by this extra term, and that a first order expansion in β can be used for the expectation value of the latter:

$$\sum_i \delta_i\langle I_{iz}\rangle \simeq -\frac{N}{2}\beta d^2(1-t^2).$$

Using a best-fit value for d^2, we obtain the curve 3 of Fig. 5.16, which is in satisfactory agreement with experiment. The departure from the theoretical curve 2 increases when using a sample with a larger concentration of impurities, which substantiates the above interpretation of the results.

The departure from constancy of the ratio of non-Zeeman to Zeeman energies at high polarizations is characteristic of the non-linear effect of Zeeman order on the dipolar energy.

2. *Susceptibilities in zero effective field*

The transverse susceptibility in zero effective field was measured by observing the dispersion signal at the centre of a fast passage, i.e. an adiabatic demagnetization in the rotating frame followed by adiabatic remagnetization. Its variation is shown in Fig. 5.17 as a function of initial polarization. The latter is a non-linear measure of the entropy in high effective field, which keeps the same value in zero effective field if the passage is adiabatic. In practice, the fast passage is not completely adiabatic, as evidenced by the fact that it results in a slight decrease of the polarization ($|\delta p/p| \sim 13$ per cent). A correction for this effect was made by replacing the actual initial polarization by the average of initial and final polarizations. The entropy was computed from:

$$S/Nk_B = \ln 2 - \tfrac{1}{2}\{(1+p_i)\ln(1+p_i) + (1-p_i)\ln(1-p_i)\}.$$

The value of β in zero effective field, where $p = 0$, was deduced from the expansion (5.194) up to β^4, and this value was used in (5.197). The resulting theoretical curve is shown in Fig. 5.17.

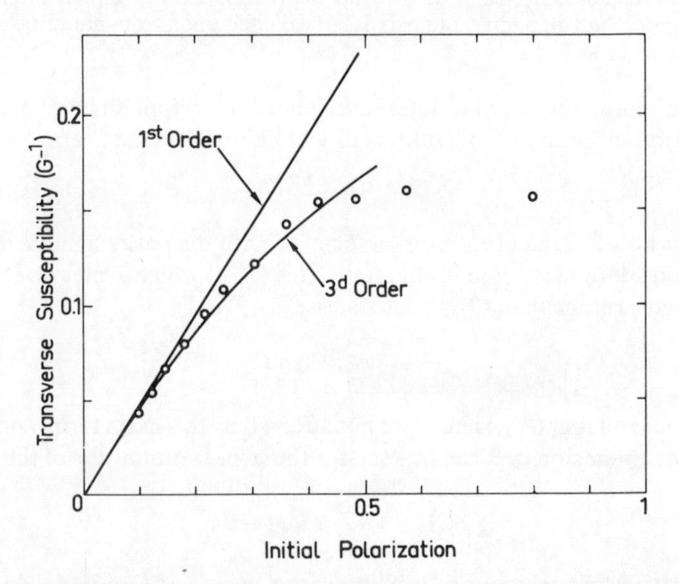

FIG. 5.17. Transverse susceptibility of ^{19}F, in a spherical sample of CaF_2, after ADRF to zero effective field, against initial polarization, together with 1st and 3rd order expansions in inverse temperature. $H_0 \| [100]$, $T < 0$. (After Goldman *et al.*, 1975.)

At high temperature, we have:

$$S/Nk_B \simeq \ln 2 - \tfrac{1}{2}p_i^2;$$

$$p_i \simeq -\tfrac{1}{2}\beta D;$$

$$\chi_\perp \simeq -\tfrac{1}{2}\beta;$$

so that $\chi_\perp$ varies linearly with p_i. The departure from linearity as p_i increases is due to the non-linear dependence on β of both $\chi_\perp$ and $(\ln 2 - S)^{1/2}$.

As a function of p_i, $\chi_\perp$ is observed to increase and then to reach a plateau, which originates in the onset of antiferromagnetism (chapter 8). The value of $\chi_\perp$ was not calibrated, but adjusted to fit the Weiss field prediction for the antiferromagnetic state in the plateau. Within this adjustment, there is a reasonable agreement between theory and experiment in the paramagnetic domain, that is for $p_i < 0{\cdot}4$.

The longitudinal susceptibility could be obtained in principle by demagnetizing the system down to a small effective frequency Δ and measuring its polarization from the area of the absorption signal. This method would be very difficult in the present case because of the shortness of T_{1D}, and a different approach was used (Jacquinot, Chapellier, and Goldman, 1974).

After demagnetization, and with the r.f. field still present, the external field was subjected to a square-wave modulation around resonance, with a small amplitude (1 G) and a period of several seconds, sufficiently long for the system to reach a new equilibrium between successive jumps of the field.

Following the sudden change of the effective frequency from, say, $-\Delta_0$ to Δ_0, the longitudinal polarization varies from $-\Delta_0\chi_\parallel$ to $\Delta_0\chi_\parallel$. The polarization variation $2\Delta_0\chi_\parallel$ is monitored by integrating the absorption signal following the field jump, in accordance with eqn (5.84), when the value of $\chi_\parallel$ is deduced. $\chi_\perp$ is simultaneously measured by observing the dispersion signal.

In the course of the experiment, the system warms up, and its energy decreases, both because of spin–lattice relaxation and of the sudden, non-adiabatic field jumps. The evolution of the dipolar energy can be determined: its relaxation decay is measured in a separate experiment, by measuring the first moment of the absorption signal as a function of time. As for variations due to the sudden field jumps, they are related to the variation of the polarization. The analysis of the evolution of the dipolar energy in the course of the experiment is described in Goldman (1977) and will not be repeated here.

What is finally obtained from this experiment is the variation of the susceptibilities $\chi_\perp$ and $\chi_\parallel$ as a function of dipolar energy. The experimental results are shown in Figs 5.18 and 5.19, respectively. $\chi_\perp$ is adjusted to the Weiss-field value for the antiferromagnetic state, in the high energy plateau, and $\chi_\parallel$ is calibrated against $\chi_\perp$ in the low energy limit when the two

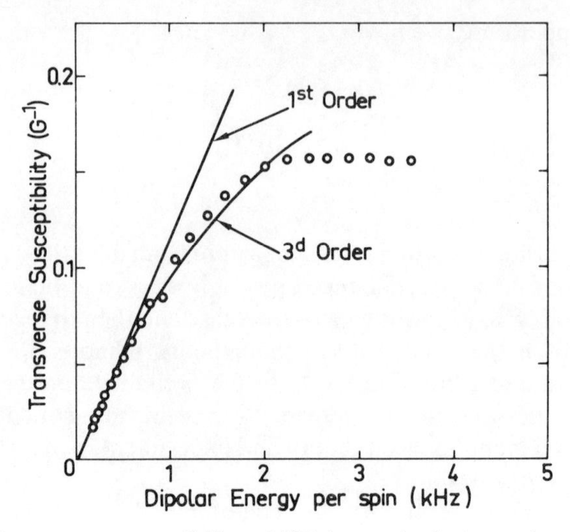

FIG. 5.18. Transverse susceptibility of ^{19}F, in a spherical sample of CaF_2 in zero effective field, against dipolar energy, together with 1st and 3rd order approximations. $H_0 \| [100]$, $T < 0$. (After Goldman *et al.*, 1975.)

susceptibilities are equal. The theoretical curves correspond to the high temperature approximation (1st order) and to the expansions to β^3 of eqns (5.175), (5.187), and (5.197). The non-linearities observed in the paramagnetic state are in reasonable agreement with theory.

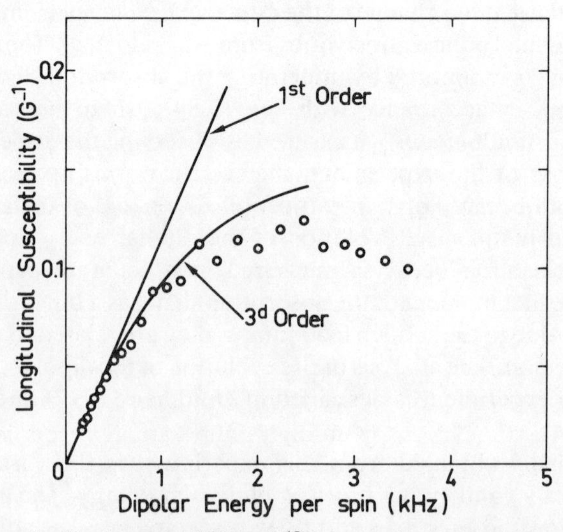

FIG. 5.19. Longitudinal susceptibility of ^{19}F, in a spherical sample of CaF_2 in zero effective field, against dipolar energy, together with 1st and 3rd order approximations. $H_0 \| [100]$, $T < 0$. (After Goldman *et al.*, 1975.)

6

THERMAL CONTACT BETWEEN NUCLEAR SPINS AND PARAMAGNETIC IMPURITIES: NUCLEAR RELAXATION AND DYNAMIC POLARIZATION

Big Brother is watching you.

GEORGE ORWELL (*Nineteen eighty four*)

Introduction

THE formalism of nuclear spin temperature introduced in chapter 1, was extended in chapter 5 to the so-called low temperature case, that is beyond the lowest order expansion of the Boltzmann exponentials into power series of inverse temperature β. The reason for this extension was naturally the existence of new experimental situations where the low order expansion was inadequate. The methods whereby such situations are created are known as dynamic nuclear polarization abbreviated as DNP. These methods make use of the interaction between nuclear spins and electronic spins in low concentration, under conditions of field and lattice temperature, such that the electronic spins are almost fully polarized. Because of the vast difference between electronic and nuclear gyromagnetic ratios, under the same conditions nuclear spins have very small polarizations.

The detailed understanding of the ways in which the large electronic polarizations are transferred to the nuclear spins requires a knowledge of the spin–lattice relaxation of the electronic spins, of their interactions with each other and with nuclear spins. The present chapter is devoted to the description of these interactions, of their manifold aspects, and of their use for DNP. In contrast to the high temperature case, where, once the spin temperature assumption is made, most of its consequences can be calculated, we shall sometimes have to rely for experimental predictions on questionable extrapolations from high temperatures or on crude models, whose reasonable agreement with experiment is the best justification. Before starting on these studies we outline the very simple principle of DNP by the so-called 'solid-effect' which started it all (Abragam and Proctor, 1958).

Consider an assembly of nuclear spins $I = \frac{1}{2}$ of Larmor frequency ω_I embedded in a diamagnetic solid that contains a few paramagnetic impurities with spins $S = \frac{1}{2}$ of Larmor frequency ω_S, coupled to neighbouring nuclear spins by dipolar interactions.

Assume for simplicity (but quite realistically) values of H and T such that, to a good approximation, the electronic spins S are completely polarized, say all 'up', and the nuclear spins I are completely unpolarized, as many 'up' as 'down'. The dipolar interaction permits simultaneous reversals of S and I in opposite directions, or flip-flops, and also reversals in the same direction, which we shall call flip-flips. However, in such reversals the total energy of the spin system changes by an amount $\hbar(\omega_S \pm \omega_I)$ and the process will not occur unless the missing energy is supplied by the crystalline lattice, usually in the form of one or several phonons. It is precisely, as we shall see later, these simultaneous reversals that are responsible for the classical mechanism of nuclear relaxation by paramagnetic impurities. The rate of these processes can be very small at low temperature ($1/T_n \simeq 10^{-3}\ \mathrm{s}^{-1}$ is typical for polarized target materials). On the other hand, the reversal of an electronic spin *alone*, caused by its coupling to the lattice, occurs at a much faster rate $1/T_e (1/T_e \simeq 10^3\ \mathrm{s}^{-1}$ is a typical value). Suppose now that an external source of microwave energy at a frequency $\Omega = \omega_S \pm \omega_I$ is capable of inducing either flip-flops ($\Omega = \omega_S - \omega_I$) or flip-flips ($\Omega = \omega_S + \omega_I$). Assume also that the electronic linewidth $\Delta\omega_S$ is much smaller than the nuclear frequency ω_I so that when the flip-flops occur ($\Omega = \omega_S - \omega_I$), flip-flips are impossible because they are off-resonance with the driving frequency Ω, and vice versa.

In practice, a microwave magnetic field is used. A flip-flop is then a forbidden transition to a first approximation, but, as for the nuclear relaxation process, it becomes allowed through the dipolar interaction which scrambles the electronic and nuclear states. Assume then that we drive, say, forced flip-flops and that the strength of the source is such that the rate at which they occur is much greater than the nuclear relaxation rate $1/T_n$. We shall show that it is possible in that way to force 'up' all the spins I.

Consider first a spin I that is up. All the spins S being up, the spin I could only do a flip-flip, which is forbidden as it is off-resonance. On the other hand, a spin I that is down may do a flip-flop with a spin S that is up, ending in a situation where I is up and S is down. This spin S, which is now down, is a danger for all the I spins that are up, since it could bring one of them down through a forced flip-flop. Fortunately, before any harm is done, its powerful relaxation mechanism will have brought this spin S to its 'up' position of thermal equilibrium and the cycle can start again until all the I spins are up.

It is easy to see that if the source frequency $\Omega = \omega_S + \omega_I$ drives flip-flips rather than flip-flops, the I spins will also go 'down' with a polarization opposite to that of the S spins. It is also easy to see that for an incomplete electronic polarization $|P_e| < 1$, these processes lead to a nuclear polarization $P_n = \pm P_e$. For small electronic concentrations, each electronic spin S must 'service' a large number N_I/N_S of nuclear spins. In order to be effective it must be able after each forced flip-flop (flip-flip) to flip back into its thermal equilibrium position before *any* of the N_I/N_S nuclear spins in its sphere of influence have flipped through a nuclear relaxation mechanism. The condition for this is clearly:

$$f = \left(\frac{N_I}{T_n}\right)\left(\frac{N_S}{T_e}\right)^{-1} \ll 1. \tag{6.1}$$

It will be shown in section **B**(b) that this condition is always true if the nuclear relaxation of the spins I has no other origin than their couplings with the spins S. If, however, other nuclear relaxation mechanisms, sometimes called leakage relaxation, are present, caused either by couplings with another species of electronic spins S' with a Larmor frequency $\omega'_S \neq \omega_S$, or by a purely nuclear mechanism, the condition may be violated and the nuclear polarization P_n could be much smaller than P_e.

This model will be expanded and put on a more quantitative basis in section **C**. This is for two reasons. Firstly, there are experimental situations where it provides a reasonably accurate description of the physical reality. Secondly, even when more complicated mechanisms come into play it still contains the essential features of the phenomenon of DNP, namely forced flips of electron spins of Larmor frequency ω_e, induced by means of photons of frequency $\omega \neq \omega_e$. The necessary conservation of energy involves flips of nuclear spins with an upward to downward ratio different from that for flips induced by thermal relaxation. Whatever the complexity of the intermediate processes it is in these *forced* nuclear flips that the gist of DNP resides.

A. Structure and relaxation of paramagnetic centres

This problem is treated in detail in a book (Abragam and Bleaney, 1970). As elsewhere in the present book, in order to make it reasonably self-contained, we gather here the information relevant to the purpose outlined in the introduction.

There are essentially two types of paramagnetic centres, those with an odd number of unpaired electrons outside closed shells, called Kramers centres and those with an even number of such electrons, called non-Kramers. A theorem due to Kramers states that all the electronic states of the Kramers centres are at least twofold degenerate in the absence of magnetic fields. Their Larmor frequencies are little affected by any irregularities caused by imperfections in the electric crystal field and as a consequence they exhibit relatively narrow resonance lines.

They are the ones which are introduced purposely into the sample in controlled concentrations as agents of DNP. The non-Kramers centres are often present in small uncontrolled amounts. They are difficult to detect by electron paramagnetic resonance (EPR) and their presence is often felt indirectly through the leakage relaxation of the nuclear spins.

(a) Spin Hamiltonian and hyperfine structure

The simplest of the Kramers centres is the F-centre which is a single electron trapped in the vacancy of a negative ion. These centres are conveniently produced by irradiation of ionic crystals by fast electrons or γ-rays. They have been widely used for DNP in LiH and LiF, and also in KCl and ^{6}LiD. The ground state of an F-centre has a spin $\frac{1}{2}$ and in the presence of a magnetic field H is described by a Zeeman Hamiltonian,

$$\hbar Z_S = g\mu_B \mathbf{H} \,.\, \mathbf{s} = \hbar\omega_S s_z, \tag{6.2}$$

where the isotropic gyromagnetic factor g is very nearly that of the free electron. Kramers ions of transition elements have an odd number of unpaired d-electrons (iron group) or f-electrons (rare earth group). The ground state is a doublet described in the presence of a field H by a Zeeman Hamiltonian:

$$\hbar Z_S = \sum_{i,\mu} \mu_B \frac{g_{i\mu}}{2} H_i \sigma_\mu, \tag{6.3}$$

where $\sigma_1, \sigma_2, \sigma_3$ are the 3 Pauli matrices and the nine coefficients $g_{i\mu}$ depend on the electronic structure of the centre in its ground state. It can be shown that under fairly general conditions (Abragam and Bleaney, 1970, p. 650) (6.3) can be rewritten as:

$$\hbar Z_S = \mu_B \mathbf{H} \,.\, \mathbf{g} \,.\, \mathbf{s}, \tag{6.4}$$

where $\mathbf{g}$ is a symmetrical tensor and the operator $\mathbf{s} = \frac{1}{2}\boldsymbol{\sigma}$, called the fictitious spin, is a vector in spin space. If the environment of the centre has cubic symmetry, Z_S has the same isotropic form as in (6.2). However the gyromagnetic factor g of the fictitious spin can be very different from that of the free electron. For axial symmetry (6.4) can be written:

$$\hbar \mathcal{H} = \mu_B [g_\parallel H_Z s_Z + g_\perp (H_X s_X + H_Y s_Y)]. \tag{6.5}$$

In an environment of sufficiently low symmetry, the ground state of a non-Kramers centre is always a singlet. For somewhat higher symmetry the ground state can be degenerate or nearly degenerate with p closely lying sublevels. It is then convenient to introduce a fictitious spin S whose $(2S+1)$ substates span the ground manifold. A simple case is $S = 1$, where the ground manifold can be described by a spin Hamiltonian:

$$\hbar \mathcal{H} = \mu_B \mathbf{H} \,.\, \mathbf{g} \,.\, \mathbf{S} + D(S_Z^2 - \tfrac{1}{3}S(S+1)) + E(S_X^2 - S_Y^2). \tag{6.6}$$

E vanishes if $\mathcal{H}$ has axial symmetry. The ground manifold then breaks into a doublet $S_Z = \pm 1$ and a singlet $S_Z = 0$, separated by D. For $S > 2$, quartic terms appear in the spin Hamiltonian. We shall not write them out, for, with one exception, we shall be concerned with Kramers centres only.

If the nucleus of the paramagnetic centre (atom or ion) has a non-zero spin I, the multiplicity of the ground manifold is increased by a factor $(2I+1)$ corresponding to the possible orientations of the nuclear spin interacting with the electronic cloud. Under fairly general conditions this interaction, called the hyperfine interaction can be written:

$$\hbar\mathcal{H}_{\text{hfs}} = \mathbf{S} \cdot \mathcal{A} \cdot \mathbf{I} \tag{6.7}$$

where $\mathcal{A}$ is a symmetrical tensor. For axial or cubic symmetry it can be written respectively as:

$$\hbar\mathcal{H}_{\text{hfs}} = \mathcal{A}_{\|} S_Z I_Z + \mathcal{A}_{\perp}(S_X I_X + S_Y I_Y); \tag{6.7'}$$

$$\hbar\mathcal{H}_{\text{hfs}} = A\mathbf{S} \cdot \mathbf{I}. \tag{6.7''}$$

Last, there is the Zeeman Hamiltonian of the nucleus:

$$\hbar Z_n = -\gamma_n \hbar(\mathbf{I} \cdot \mathbf{H}).$$

It is not the only term in the spin Hamiltonian bilinear in $\mathbf{I}$ and $\mathbf{H}$. There is another term, the pseudo-nuclear Zeeman coupling which has the same origin as the chemical shift described in chapter 2. The only difference is in the magnitude of this interaction which instead of being a small fraction of the nuclear Zeeman energy as for the chemical shift can be a good deal larger.

$$\hbar Z'_n = -\hbar\mathbf{I} \cdot \boldsymbol{\gamma}'_n \cdot \mathbf{H}. \tag{6.8}$$

We can describe the pseudo-nuclear Zeeman coupling as follows: the nuclear magnetic moment polarizes the electronic cloud: the extra electronic magnetization that obtains, proportional to $\mathbf{I}$, interacts with the external magnetic field. It is because paramagnetic centres have a much higher magnetic polarizability than diamagnetic atoms or molecules, that the interaction (6.8) is so much larger than the chemical shift. Finally for $\mathbf{I} > \frac{1}{2}$, there is the well-known interaction of the nuclear quadrupole moment with the electric field gradient but it need not concern us in this chapter.

Under most experimental conditions:

$$Z_S \gg \mathcal{H}_{\text{hfs}} \gg Z_n, Z'_n. \tag{6.9}$$

As an illustration Fig. 6.1 shows the energy levels of an electronic spin $\frac{1}{2}$, real or fictitious, interacting with a nuclear spin $\frac{1}{2}$, where we have assumed an isotropic gyromagnetic factor g and an isotropic hyperfine constant A (assumed negative in the figure). In first order perturbation the hfs energy

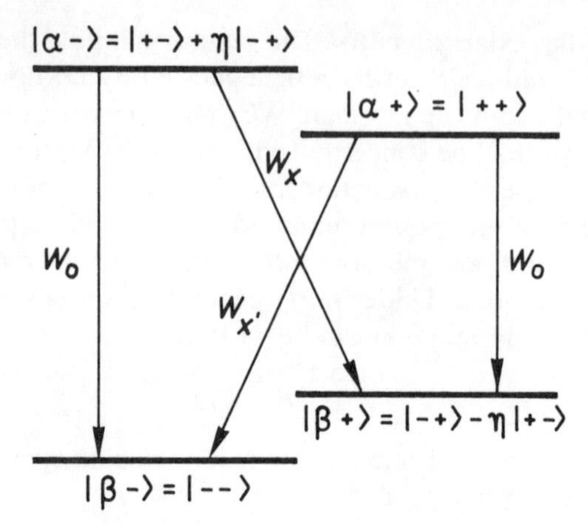

FIG. 6.1. Energy spectrum of a system $S=\frac{1}{2}$, $I=\frac{1}{2}$ in high field coupled by a scalar interaction.

levels are given by the term AI_zS_z which commutes with the electronic Zeeman energy: $\hbar\omega_S s_z = \mu_B gHs_z$, but the off-diagonal part $(A/2)(s_+I_- + s_-I_+)$ scrambles the two states $|+,-\rangle$ and $|-,+\rangle$. A general tensor hfs coupling $\mathbf{s}\,.\,\mathcal{A}\,.\,\mathbf{I}$ would scramble all four states.

The EPR spectrum is made of two lines of frequency $\omega_s \pm A/2$, $|\alpha_+\rangle \rightleftarrows |\beta_+\rangle$ and $|\alpha_-\rangle \rightleftarrows |\beta_-\rangle$. There are also forbidden low frequency transitions, the so-called Endor transitions, between the states $|\alpha_+\rangle$ and $|\alpha_-\rangle$ and between $|\beta_+\rangle$ and $|\beta_-\rangle$, of frequencies $A/2 \pm \omega_n$. Finally there are the high frequency forbidden skew transitions: $|\alpha_+\rangle \rightleftarrows |\beta_-\rangle$ and $|\alpha_-\rangle \rightleftarrows |\beta_+\rangle$.

The transition probabilities of the forbidden lines are reduced with respect to the allowed EPR transitions by a factor of the order of $\eta^2 = (A/2g\beta H)^2$ with the exception of the line $|\alpha_+\rangle \rightleftarrows |\beta_-\rangle$ which is strictly forbidden for an isotropic hfs coupling.

(b) *Electronic spin–lattice relaxation: spins and phonons*

1. *Pure electronic relaxation*

At the temperatures and fields used for DNP (below 1 K, above 1 T) the electronic relaxation is caused by the direct process, i.e. absorption or emission of a single phonon of energy $\hbar\omega_S$ equal to the Larmor energy of the spin, together with a spin flip. For Kramers centres, this process is forbidden in the absence of a magnetic field and its probability with respect to a non-Kramers centre is reduced by a factor of the order of $(\hbar\omega_S/\Delta)^2$ where Δ is the distance to the first excited state of the centre. The relaxation rate is also proportional to $(2X+1)$ where X is the number of phonons of energy $\hbar\omega_S$ per mode. At thermal equilibrium $(2\bar{X}+1) = \coth(\hbar\omega_S/2kT)$. There is in the expression for $1/T_{1e}$ an extra factor ω^3 due to the shape of the phonon spectrum. Altogether the Kramers relaxation rate can be written:

$$\frac{1}{T_{1e}}=\xi\left(\frac{\omega}{v}\right)^5 \coth\left(\frac{\hbar\omega_S}{2k_BT}\right)\frac{\hbar}{\rho}, \tag{6.10}$$

where v is the velocity of sound in the crystal, ρ is its density and ξ is a dimensionless coefficient that depends on the structure of the impurity and for paramagnetic ions is between 1 and 10^2 (it can be very much smaller when as in F-centres the first excited state of the centre is far away). For typical values of the parameters, $\omega=3\times10^{11}$ Hz, $v=3\times10^5$ m s^{-1}, $\rho=$ 2 gm cm^{-3}, $1/T_{1e}\simeq 500\xi$. In (6.10) $1/T_{1e}$ is the sum of the upward transition probability, proportional to $\bar{X}=[\exp(\hbar\omega_S/k_BT)-1]^{-1}$ and of the downward one proportional to $(\bar{X}+1)$. Their ratio,

$$r=\bar{X}/(\bar{X}+1)=\exp-(\hbar\omega_S/k_BT), \tag{6.11}$$

becomes very small for $\hbar\omega_S \gg k_BT$, that is when the equilibrium electron polarization is nearly unity. Then the only contribution to $1/T_{1e}$ is the spontaneous lifetime of the higher electronic state against the emission of a phonon, independent of lattice temperature.

Equation (6.10) describes correctly the rate of change of the electron magnetization if the number of phonons per mode, of energy $\hbar\omega_S$ (on 'speaking terms' with the spins), is equal to its equilibrium value $\bar{X}=[\exp(\hbar\omega_S/k_BT)-1]^{-1}$. This requires that the rate of flow of magnetic energy E_S from the spins to the phonons, namely $N_S\hbar\omega_S P/T_{1e}$, be small compared to the rate of flow of energy from the phonons, 'on speaking terms' with the spins, to the thermal bath, whatever its nature, that gives the phonons their temperature, namely (E_{ph}/τ_{ph}). In the Debye approximation:

$$E^0_{ph}\approx 9N_I\left(\frac{\omega_S}{\Omega_D}\right)^3\left(\frac{\Delta\omega_e}{\omega_S}\right)\hbar\omega_S(2\bar{X}+1)=9N_I\left(\frac{\omega_S}{\Omega_D}\right)^3\left(\frac{\Delta\omega_e}{\omega_S}\right)\frac{\hbar\omega_S}{P_0}. \tag{6.12}$$

In this formula N_I is the number of atoms per unit volume in the crystal, identified here with the number of nuclei I, Ω_D is the Debye frequency, $\Delta\omega_e$ is the electronic linewidth, τ_{ph} is the phonon lifetime and P_0 the equilibrium electron polarization.

We define the ratio:

$$\sigma=\frac{(E^0_S/T_{1e})}{(E^0_{ph}/\tau_{ph})}=\frac{1}{9}\frac{N_S}{N_I}\left(\frac{\Omega_D}{\omega_S}\right)^3\frac{\omega_S}{\Delta\omega_e}\frac{\tau_{ph}}{T_{1e}}P_0^2, \tag{6.13}$$

as the phonon bottleneck coefficient. From (6.10) and (6.12) we see that it is proportional to $C\omega_S^3P_0$ (C is the concentration of impurities). If σ is much larger than unity, the spin relaxation process can be described as follows. In a first step, in a very short time, of the order of τ_{ph}/σ, spins and phonons reach a common temperature. In a second step, much slower, this whole system relaxes towards the temperature of the bath at a rate:

$$\frac{1}{T_1^*}=\left(\frac{1}{\tau_{\text{ph}}}\right)E_{\text{ph}}(E_{\text{ph}}+E_S)^{-1}\simeq\frac{1}{\sigma}\left(\frac{1}{T_{1\text{e}}}\right). \tag{6.14}$$

Near equilibrium when $(P-P_0)$ is small the relaxation of the electronic polarization is then described by the equation:

$$\frac{\mathrm{d}P}{\mathrm{d}t}=-\frac{1}{\sigma T_{1\text{e}}}(P-P_0). \tag{6.15}$$

A more elaborate calculation (Faughnan and Strandberg, 1961) shows that this relaxation is actually non-linear and is described by:

$$\frac{\mathrm{d}P}{\mathrm{d}t}=-\frac{1}{T_{1\text{e}}}\frac{(P-P_0)}{1+\sigma(P/P_0)}, \tag{6.16}$$

which reduces to (6.15) when $P/P_0\to 1$. As an example, a direct measurement of σ in CaF_2 (Abragam, Bouffard, and Roinel, 1976) doped with Tm^{2+} at a frequency of 140 GHz, and a concentration $C\sim 2\times 10^{-5}$ with a relaxation time of $T_{1\text{e}}\simeq 10^{-3}$ s yielded $\sigma\simeq 10$ in qualitative agreement with formula (6.13) if reasonable values $\tau_{\text{ph}}\simeq 10^{-6}$ K, $\theta_{\text{D}}=(\hbar\Omega_{\text{D}}/k_{\text{B}})\sim 500$ K are chosen. Much larger values of σ are possible for larger concentrations and shorter relaxation times.

We shall see in section **C** the effect of the phonon bottleneck on dynamic polarization.

2. *Nuclear hyperfine relaxation*

We shall describe in chapter 7 an experimental method, the so-called pseudomagnetic neutron precession, which sometimes permits a direct measurement of the polarization of the nucleus of a paramagnetic ion. This polarization reaches its equilibrium value through relaxation transitions different from those of the electron polarization, outlined in **A**(b)1. Taking for argument's sake the case $S=\frac{1}{2}$, $I=\frac{1}{2}$ of Fig. 6.1 we see that the relevant relaxation transitions are those represented by the skew arrows, with probabilities W_x and $W_{x'}$ as well as the reverse probabilities rW_x, $rW_{x'}$ with $r=\exp-(\hbar\omega_S/k_{\text{B}}T)$. It can be shown that the 'skew' transition probabilities W_x and $W_{x'}$ (Fig. 6.1) are of the order of $\eta^2 W_0$ (Abragam, Jacquinot, Chapellier, and Goldman, 1972). (*N.B.* the probability $W_{x'}$ for the skew transition: $|\alpha_+\rangle\to|\beta_-\rangle$ is zero only for a uniform microwave field but not for a relaxation transition induced by phonons.) To these must be added the low frequency Endor relaxation transitions $|\beta_+\rangle\rightleftarrows|\beta_-\rangle$ and $|\alpha_+\rangle\rightleftarrows|\alpha_-\rangle$. For $(\hbar\omega_S/k_{\text{B}}T)\gg 1$ the states $|\alpha_+\rangle$ and $|\alpha_-\rangle$ are kept practically empty by electron relaxation and the transitions $|\alpha_+\rangle\rightleftarrows|\alpha_-\rangle$ can then be disregarded. A direct relaxation transition $|\beta_+\rangle \rightleftarrows |\beta_-\rangle$ which involves the emission or absorption of a phonon of low frequency: $\omega_{\text{endor}} \sim A/2 - \omega_n$ will have a probability: W_{endor} smaller than W_x (or $W_{x'}$) in the ratio:

$$W_{\text{endor}}/W_x \approx (\omega_{\text{endor}}/\omega_S)^3 \approx \eta^3. \tag{6.17}$$

The two transitions have comparable spin matrix elements and the ratio (6.17) originates in the phonon spectrum as explained in the derivation of eqn (6.10). The probability $W_{\text{endor}} \approx \eta^5 W_0$ is very small and usually plays a negligible part in the relaxation of the nuclear spin **I**. As an example for Tm^{2+} in CaF_2: $A \approx 1{\cdot}1$ GHz; $\omega_S \approx 140$ GHz; $\eta = A/2\omega_S \simeq 4 \,.\, 10^{-3}$; $W_0 \approx 10^3\ \text{s}^{-1}$:

$$W_{\text{endor}} \approx \eta^5 W_0 \approx 10^{-9}\ \text{s}^{-1}.$$

The relaxation of the nuclear spin then passes through the higher electronic states $|\alpha_+\rangle$ and $|\alpha_-\rangle$. It is easily seen that for a sufficiently fast electronic relaxation rate W_0 and for $P_0 \approx 1$, its rate is given by:

$$\frac{1}{T_{1I}} = r(W_x + W_{x'}) = \exp - (\hbar\omega_S/k_B T)(W_x + W_{x'}), \tag{6.18}$$

with its characteristic temperature dependence. The direct rate W_{endor} becomes comparable to (6.18) for $r \approx \eta^3$ that is $r \simeq 0{\cdot}6 \times 10^{-7}$, but by that time relaxation becomes so slow that some other processes probably take over.

When a paramagnetic centre has a hyperfine structure and one of its $(2I+1)$ EPR transitions is used for DNP by the solid effect the 'skew' relaxation may interfere with the process.

The importance for DNP of the existence of a 'skew' relaxation stems from the fact that when r is very small and only the lower electronic levels are populated, even a very slight saturation of one of the EPR lines, say $|\alpha_-\rangle \rightleftarrows |\beta_-\rangle$ leads to an almost complete transfer of all the spins into the other line. This saturation transfers some spins from $|\beta_-\rangle$ into the almost empty state $|\alpha_-\rangle$. Most of these will fall back into $|\beta_-\rangle$ by straight electron relaxation at a rate W_0 but a small proportion will fall into $|\beta_+\rangle$ at a rate W_x. Once there, they are trapped, for upward transitions occur at a very slow rate proportional to r. As soon as r is smaller than the saturation coefficient s of the line which is being irradiated, all the spins will eventually spill into the other line. This depopulation can be compensated by driving by a microwave field the incompletely forbidden transition $|\alpha_-\rangle \rightleftarrows |\beta_+\rangle$ with $\Delta(M+m) = 0$. This will bring the spins back from $|\beta_+\rangle$ into $|\alpha_-\rangle$ whence they will fall back into $|\beta_-\rangle$ by straight electron relaxation. The consequences for DNP will appear in section **C**.

As a matter of terminology: a relaxation process for a low frequency transition ω_0, which involves a passage through an intermediate state $|\delta\rangle$, with emission and absorption of two high energy phonons ω' and ω'' such that $|\omega'-\omega''|=\omega_0$, is called a Raman process if energy is not conserved in the intermediate step and an Orbach process if it is conserved. It is clear that the indirect relaxation process described by eqn (6.18) is a nuclear Orbach process.

In spite of what has been said earlier about the weakness of direct relaxation between adjacent hyperfine levels of a paramagnetic centre (what we call the nuclear direct process), at least one instance of such relaxation has been observed in holmium ethylsulphate $Ho(C_2H_5SO_4)_3.9H_2O$ where both g and A are very large (Abragam, Bacchella, Glättli, Meriel, Piesvaux, and Pinot, 1976).

The spin Hamiltonian of the Ho^{3+} ion is that described by the eqns (6.6) and (6.7′) with:

$$g_{\|}=7{\cdot}71;\quad g_{\perp}=4;\quad \frac{A_{\|}}{h}=5\text{ GHz};\quad \left|\frac{A_{\perp}}{h}\right|\approx 2{\cdot}7\text{ GHz};$$

$$\frac{D}{h}=-165\text{ GHz};\quad E=0. \tag{6.19}$$

The nuclear spin I of ^{165}Ho is $\frac{7}{2}$. The nuclear relaxation of ^{165}Ho is measured by the pseudomagnetic precession of the neutron spins of a beam crossing a single crystal of holmium ethylsulphate in a field H of 2·5 T, a method to be described in chapter 7. If the field H_0 is along the c-axis, which is the Z-axis of eqn (6.6), the distance between the ground state $S_z=-1$ and the first excited state $S_z=0$ is:

$$\frac{\Delta_{\|}}{h}=-\frac{D}{h}+\frac{g_{\|}\mu_B H}{h}=450\text{ GHz}. \tag{6.20}$$

The admixture of the excited state $S_z=0$, $I_z=\frac{5}{2}$ into the ground state $S_z=-1$, $I_z=\frac{7}{2}$ by the off diagonal part $(A_{\perp}/2)(S_+I_-+S_-I_+)$ of the hyperfine Hamiltonian, is:

$$\eta_{\|}=-(\tfrac{7}{2})^{1/2}\frac{A_{\perp}}{\Delta_{\|}}\simeq -1{\cdot}1\times 10^{-2}. \tag{6.21}$$

The criterion for the nuclear direct process to be faster than the nuclear Orbach process is:

$$\eta_{\|}^{3}\geqslant r=\exp-(\Delta_{\|}/k_B T)$$

or

$$1{\cdot}33\times 10^{-6}\times\exp\left(\frac{21}{T}\right)\geqslant 1.$$

For $T = 1$ K this expression is already of the order of 10^3 and below this temperature, the indirect process is utterly negligible in comparison with the direct process.

Both the absolute value of the observed nuclear relaxation time of ^{165}Ho, of the order of $T_1(^{165}\text{Ho}) \sim 10^3$ s for $T = 0{\cdot}25$ K, and, more important, its near constancy for $k_B T \ll A_\parallel$ that is for $T \ll 0{\cdot}25$ K, are consistent with a direct nuclear process.

B. Nuclear relaxation by isolated paramagnetic impurities

(*a*) *An isolated electron–nucleus pair*

1. *The random field approach*

The interaction between a nuclear spin I and a neighbouring electronic spin S can be written most generally as $\mathbf{S} \,.\, \mathbf{a} \,.\, \mathbf{I}$ where $\mathbf{a}$ is a tensor. If the wavefunction of the electron vanishes at the site of the nucleus, $\mathbf{a}$ is the usual dipolar interaction which in the simplest case can be written:

$$\mathscr{D}_{ij} = \gamma_I \gamma_S \hbar (\delta_{ij} - 3\hat{\mathbf{r}}_i \,.\, \hat{\mathbf{r}}_j) r^{-3}, \tag{6.22}$$

where r is the distance between the two spins and $\hat{\mathbf{r}} = (\mathbf{r}/r)$. If the gyromagnetic factor of the electron is an anisotropic tensor $\mathbf{g}$, the vector $\mathbf{S}$ in the interaction $\mathscr{H}_1 = \mathbf{S} \,.\, \mathscr{D} \,.\, \mathbf{I}$ must be replaced by $\frac{1}{2}\mathbf{g} \,.\, \mathbf{S}$. On the other hand, if the wavefunction of the electron does spill over the nuclear site, the tensor $\mathbf{a}$ is more complicated and will contain a scalar component $a\mathbf{I} \,.\, \mathbf{S}$ sometimes larger than the dipolar coupling (6.22). In the following, unless otherwise stated, we shall retain for $\mathbf{a}$ the simple expression (6.22). The interaction $\mathbf{S} \,.\, \mathscr{D} \,.\, \mathbf{I}$ can be written as $-\gamma_I \hbar \mathbf{I} \,.\, \mathbf{H}_r$ where $\mathbf{H}_r$ (or at least a part of it) can be viewed as a random fluctuating local field produced by the electron and 'seen' by the nucleus. If the power spectrum of this field, given by the tensor:

$$J_{ij}(\omega) = \int_{-\infty}^{\infty} \langle H_r^i(t) H_r^j(t-\tau) \rangle \exp(-i\omega\tau) \, d\tau,$$

(where the symbol $\langle \quad \rangle$ means thermal average), has non-vanishing values at the Larmor frequency ω_I of the nucleus, nuclear relaxation transitions will occur, leading to the establishment of an equilibrium nuclear polarization. By equating the resonance frequency of the nucleus to its Larmor frequency $\omega_I = -\gamma_I H_0$ we make the assumption that the time independent part $\langle H_r^z \rangle$ of the electronic field 'seen' by the nucleus can be neglected. We shall come back to this point in section $\mathbf{B}(b)$.

The fluctuations of the local field $\mathbf{H}_r(t) = -\gamma_S \hbar \mathbf{a} \,.\, \mathbf{S}$ can originate either in the time dependence of the tensor $\mathbf{a}$ or in that of the electronic spin $\mathbf{S}$. The two situations are sometimes called in the literature, relaxation of the first and the second type. A time dependence for the tensor $\mathbf{a}$ implies that there is relative motion, rotation and/or translation of the interacting electron and nucleus, as for instance, in the relaxation mechanism of nuclei by conduction electrons in metals or by paramagnetic ions in solutions. On the other hand, a time dependence of **S** means a fast relaxation of the electronic spin itself. When both types of time dependence are present, the relevant one for the nuclear relaxation is the faster of the two. At low temperatures (1 K or less) relative motions of unpaired electrons and nuclei are frozen out, except for conduction electrons in metals (or some quantum-mechanical tunnelling motions) and we are dealing with the problem of relaxation of nuclei by fixed paramagnetic impurities. In that case the correlation time of the random local field $\mathbf{H}_r$ is the relaxation time T_{1e} of the electron. Using the expression (6.22) for the tensor $\mathcal{D}$ it is easily found using the classical formulae of nuclear relaxation:

$$T_{1n}^{-1}(\mathbf{r}) = \tfrac{9}{2}\gamma_S^2\gamma_I^2\hbar^2 r^{-6}\sin^2\theta\cos^2\theta\int_{-\infty}^{\infty}\langle S_z(0)S_z(t)\rangle\exp(-\mathrm{i}\omega_I t)\,\mathrm{d}t$$

$$= 3\gamma_S^2\gamma_I^2\hbar^2 r^{-6}\sin^2\theta\cos^2\theta\, S(S+1)T_{1e}(1+\omega_I^2T_{1e}^2)^{-1}. \quad (6.23)$$

In equation (6.23) θ is the angle between the applied field and the vector $\mathbf{r}$. There are other contributions to T_{1n}^{-1}, resulting from the interaction $\mathbf{S}\,.\,\mathcal{D}\,.\,\mathbf{I}$ but they involve frequencies $(\omega_S \pm \omega_I)$ in (6.23) instead of ω_I and are therefore smaller than (6.23) by a factor of the order of $(\gamma_S/\gamma_I)^2 \approx 10^6$. In most practical cases $\omega_I T_{1e} \gg 1$ and (6.23) can be rewritten as:

$$T_{1n}^{-1}(\mathbf{r}) = 3\gamma_S^2\hbar^2 r^{-6}\sin^2\theta\cos^2\theta S(S+1)H_0^{-2}T_{1e}^{-1}$$

$$\simeq \left(\frac{H_e}{H_0}\right)^2\left(\frac{1}{T_{1e}}\right), \quad (6.24)$$

where: $H_e = \gamma_S\hbar S r^{-3}$ gives the order of magnitude of the local electronic dipolar field 'seen' by the nucleus. There is a major flaw in eqn (6.24). At low temperature and in high fields the electronic relaxation rate T_{1e}^{-1} is due to the direct process and becomes temperature-independent. So, according to (6.24), should be T_{1n}^{-1} in sharp contradiction with experiment. The error goes back to formula (6.23) where we took incorrectly:

$$\langle S_z(0)S_z(t)\rangle = \tfrac{1}{3}S(S+1)\exp(-t/T_{1e}). \quad (6.25)$$

Actually this formula should apply to the *fluctuating* part of S_z, namely $(S_z - S_0)$ where $S_0 = \langle S_z\rangle_L$, the thermal equilibrium value of $\langle S_z\rangle$:

$$\langle (S_z(0)-S_0)(S_z(t)-S_0)\rangle = \langle S_z - S_0\rangle^2\exp(-t/T_{1e})$$

$$= (\langle S_z^2\rangle - S_0^2)\exp(-t/T_{1e}). \quad (6.26)$$

For a spin $S = \tfrac{1}{2}$, the only case to be considered from now on, this leads to the replacement in (6.25) of $\tfrac{1}{3}S(S+1) = \tfrac{1}{4}$ by $\tfrac{1}{4}(1-P_0^2)$ where $P_0 =$

$\tanh(\gamma_S \hbar H/2k_B T)$ is the electronic thermal equilibrium polarization, and eqn (6.24) should be replaced by:

$$\frac{1}{T_{1n}}(\mathbf{r}) = 3\frac{\gamma_S^2\hbar^2}{r^6}\frac{\sin^2\theta\cos^2\theta S(S+1)}{H_0^2}\frac{1}{T_{1e}}(1-P_0^2)$$

$$\approx\left(\frac{H_e}{H_0}\right)\left(\frac{1}{T_{1e}}\right)(1-P_0^2). \tag{6.27}$$

When the electronic polarization tends toward unity the nuclear relaxation rate vanishes. This is physically reasonable as the electron spins, being all frozen in the lowest energy level, cannot relax the nuclei any more. This feature of nuclear relaxation by paramagnetic impurities has been confirmed experimentally for the proton relaxation in lanthanum magnesium double nitrate (LMN) doped with neodymium (Gunter and Jeffries, 1966, 1967) and dysprosium (Odehnal, 1967).

2. *The scrambled states approach*

There is another approach to the theory of nuclear relaxation by paramagnetic impurities that we present now because it leads by a straightforward generalization to the theory of the 'solid effect' (Abragam, 1961, chapter IX).

Consider the energy levels of an electron–nucleus pair of spins $S=\frac{1}{2}, I=\frac{1}{2}$ in a strong magnetic field H. In the absence of dipolar interaction between S and I the four states are pure: $|a_0\rangle=|++\rangle$; $|b_0\rangle=|+-\rangle$; $|c_0\rangle=|-+\rangle$; $|d_0\rangle=|--\rangle$. The dipolar interaction,

$$\gamma_I\gamma_S\hbar r^{-3}[\mathbf{I}\,.\,\mathbf{S}-3(\mathbf{I}\,.\,\mathbf{n})(\mathbf{S}\,.\,\mathbf{n})],$$

scrambles these four states, making them into the perturbed states $|a\rangle$, $|b\rangle$, $|c\rangle$, $|d\rangle$ represented in Fig. 6.2. The coefficients p and q calculated by first-order perturbation theory are given by:

$$q=\frac{3}{2}\frac{\gamma_I\gamma_S\hbar}{r^3}\frac{1}{\omega_I}\sin\theta\cos\theta\exp(i\phi)\ll 1;$$
$$p=(1-qq^*)^{1/2}\simeq 1. \tag{6.28}$$

It is worth pointing out the differences between Fig. 6.2 and Fig. 6.1 which resembles it.

(i) In Fig. 6.1 the nucleus is that of the paramagnetic centre itself and the two frequencies $|\alpha_-\rangle\rightleftarrows|\beta_-\rangle$ and $|\alpha_+\rangle\rightleftarrows|\beta_+\rangle$ differ by the hyperfine constant A which is usually much larger than the nuclear Larmor frequency ω_n. The two low-frequency transitions $|\alpha_+\rangle\rightleftarrows|\alpha_-\rangle$ and $|\beta_+\rangle\rightleftarrows|\beta_-\rangle$ also have different frequencies, $(A/2\pm\omega_n)$.

In Fig. 6.2 the nucleus is sufficiently removed from the paramagnetic centre for the two high frequency transitions $|a\rangle\rightleftarrows|c\rangle$ and $|b\rangle\rightleftarrows|d\rangle$ to have

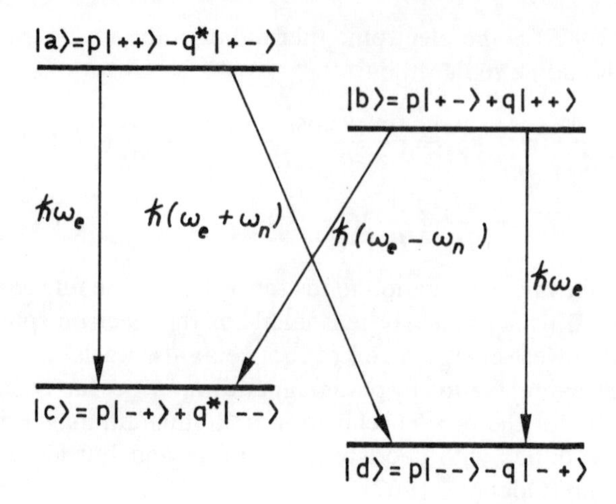

FIG. 6.2. Eigenvalues of a nucleus–electron pair of spins $\frac{1}{2}$ in high field coupled by a dipole–dipole interaction.

practically the same frequency ω_e. Similarly the two low frequency transitions $|a\rangle \rightleftarrows |b\rangle$ and $|c\rangle \rightleftarrows |d\rangle$ have practically the same frequency ω_n.

(ii) The scalar coupling $A\mathbf{I}\,.\,\mathbf{s}$ scrambles only two of the four states $|\alpha_\pm\rangle$, $|\beta_\pm\rangle$ of Fig. 6.1. The dipolar coupling $\mathbf{s}\,.\,\mathscr{D}\,.\,\mathbf{I}$, where $\mathscr{D}$ is given by (6.22), scrambles all four states of Fig. 6.2. What is more, it scrambles them far more efficiently than a scalar interaction since it mixes together pairs of states $|a\rangle$, $|b\rangle$ and $|c\rangle$, $|d\rangle$ which are separated by a small interval ω_n rather than by a large interval $(\omega_e \pm A/2)$ as in Fig. 6.1. We are then justified in neglecting, for each of the four states, admixtures from other states, removed through ω_e or $(\omega_e \pm \omega_n)$. Among these levels the spin–phonon coupling induces the following thermal relaxation transitions:

$$\begin{aligned}
&|a\rangle \to |c\rangle \text{ and } |b\rangle \to |d\rangle \text{ with probability } W_L;\\
&|c\rangle \to |a\rangle \text{ and } |d\rangle \to |b\rangle \text{ with probability } rW_L;\\
&|a\rangle \to |d\rangle \text{ and } |b\rangle \to |c\rangle \text{ with probability } U_L;\\
&|d\rangle \to |a\rangle \text{ and } |c\rangle \to |b\rangle \text{ with probability } rU_L;
\end{aligned} \tag{6.29}$$

$$\begin{aligned}
r &= \exp(-\hbar\omega_e/k_B T) \simeq \exp[-\hbar(\omega_e \pm \omega_n)/k_B T];\\
U_L &= 4|q|^2 W_L \ll W_L.
\end{aligned} \tag{6.30}$$

Let a, b, c, d be the populations of the four levels with $a+b+c+d=1$. They can be related to the electronic and nuclear polarizations P_e and P_n through:

$$P_e = (a+b)-(c+d); \qquad P_n = (a+c)-(b+d), \tag{6.31}$$

or:

$$c=\tfrac{1}{4}(1-P_e)(1+P_n); \qquad d=\tfrac{1}{4}(1-P_e)(1-P_n);$$
$$a=\tfrac{1}{4}(1+P_e)(1+P_n); \qquad b=\tfrac{1}{4}(1+P_e)(1-P_n). \tag{6.32}$$

To establish (6.32) we had to assume that at all times $a/c=b/d$. This equality imposed by the equality of relaxation probabilities $|a\rangle \rightleftarrows |c\rangle$ and $|b\rangle \rightleftarrows |d\rangle$ *cannot* be upset by an r.f. field, which must necessarily affect the two transitions $|a\rangle \rightleftarrows |c\rangle$ and $|b\rangle \rightleftarrows |d\rangle$ in the same way. The existence of transitions (6.29) results in rate equations for the populations a, b, c, d. Using equations (6.29) and (6.32) we obtain, with a little algebra, rate equations for the electronic and nuclear polarizations P_e and P_n:

$$\frac{dP_e}{dt}=-(1+r)(W_L+U_L)(P_e-P_0);$$
$$\frac{dP_n}{dt}=-(1+r)U_L P_n(1-P_eP_0); \tag{6.33}$$

where:

$$P_0=-\tanh(\tfrac{1}{2}\beta_L\omega_e)=-(1-r)(1+r)^{-1}, \tag{6.34}$$

is the electronic equilibrium polarization. The nuclear equilibrium polarization is taken to be equal to zero in the approximation where as in eqn (6.30) we neglect ω_n in comparison with ω_e. We notice that as soon as the product P_eP_0 is not very small, the rate eqns (6.33), (6.34) are not linear any more. They can be rewritten in the customary form:

$$\frac{dP_e}{dt}=-\frac{1}{T_{1e}}(P_e-P_0);$$
$$\frac{dP_n}{dt}=-\frac{1}{T_{1n}}P_n;$$
$$\frac{1}{T_{1e}}=(1+r)(W_L+U_L)\simeq(1+r)W_L;$$
$$\frac{1}{T_{1n}}=(1+r)U_L(1-P_eP_0)\approx 4|q|^2(1-P_eP_0)\frac{1}{T_{1e}}. \tag{6.35}$$

If in the last equation (6.35) we assume that the electron polarization P_e has its equilibrium value P_0 we find that T_{1n}^{-1} as given by (6.35) with q given by (6.28), is identical with the expression (6.27) derived by a seemingly different method.

(b) The relaxation of an assembly of nuclear spins by non-interacting paramagnetic centres

Formulae (6.23), (6.24), and (6.27) have been derived for an isolated electron–nucleus pair. In practice, electronic spins will flip with each other at a rate $1/T_{2e}$ which will be slower than the electronic relaxation rate $1/T_{1e}$ for very low electronic concentrations only, an assumption that we shall make in this section. On the other hand, each nuclear spin will flip-flop with a neighbour at a rate $1/T_{2n}$ very much faster than $1/T_{1n}$ as given by (6.23) or (6.24).

A direct check on these formulae is still possible under special circumstances. Nuclei in the immediate neighbourhood of an electronic impurity have their resonance frequency strongly shifted by the static part $\gamma_S \hbar S_z(1-3\cos^2\theta)r^{-3}$ of the electronic field $\mathbf{H}_e$. Thus they are not on 'speaking terms' with the 'normal' or bulk nuclei further removed from the impurity, which have negligible frequency shifts (smaller than the nuclear linewidth ΔH_n), and do not flip-flop with them. The large frequency shift of the 'anomalous' nuclei makes it possible to observe their resonance and measure their relaxation independently of those of the bulk nuclei. From the value of this shift the values of r and θ in (6.23) or (6.24) could be obtained, and after measuring $(1/T_{1e})$ independently by ESR, an absolute quantitative check of these formulae was made in yttrium ethyl sulphate doped with ytterbium (King, Wolfe, and Nallard, 1972). It should be pointed out that this is a difficult experiment because the 'anomalous' nuclei are few and their NMR signal is weak.

On the other hand, it is clear that equations (6.23), (6.24), and (6.27) are not directly applicable to the bulk nuclei. To handle those a simple-minded model valid for high fields and low temperatures is used.

In its simplest version, it is assumed that rapid flip-flops between 'normal' nuclei maintain at all times a uniform polarization and spin temperature among them; that nuclei start to be 'normal' at a distance b from the impurity called the diffusion barrier, such that their frequency shift is equal to the nuclear width; that the unique relaxation rate $1/T_{1n}$ of the normal nuclei is the mean of all the rates $1/T_{1n}(\mathbf{r})$ given by (6.24) or (6.27) in a volume comprised between a sphere of radius b and a sphere of volume V such that $VN_S = 1$ where N_S is the number of impurities per unit volume:

$$\frac{1}{T_{1n}} = \frac{1}{V}\int \frac{1}{T_{1n}}(\mathbf{r})\,\mathrm{d}^3\mathbf{r} \simeq \frac{8\pi}{5}\frac{N_S}{b^3}\frac{\gamma_S^2\hbar^2}{H_0^2}\frac{S(S+1)}{3}\frac{1}{T_{1e}}(1-P_0^2). \tag{6.36}$$

The nuclear linewidth ΔH_n is related to the nuclear density N_I in order of magnitude by $\Delta H_n \sim \gamma_I \hbar N_I$. From the definition of the diffusion barrier b: $\gamma_S \hbar/b^3 \sim \Delta H_n$. This enables (6.36) to be rewritten as:

$$\frac{1}{T_{1n}} \simeq \frac{8\pi}{5}\frac{S(S+1)}{3}C\left(\frac{\gamma_S}{\gamma_I}\right)\left(\frac{\Delta H_n}{H_0}\right)^2\left(\frac{1}{T_{1e}}\right)(1-P_0^2), \tag{6.37}$$

where $C = N_S/N_I$ is the relative concentration of impurities. It is apparent from (6.37) that $(T_{1e}/T_{1n}) \ll C$, which means that not only is the relaxation rate of a nuclear spin much slower than that of an electronic spin but also that in the sample as a whole there are far fewer nuclear flips per unit time than electronic ones, in spite of the much smaller concentration of the latter.

Actually the above model underestimates the possibilities of nuclear spin diffusion: the condition $\gamma_S \hbar/b^3 \sim \Delta H_n$ must be replaced by:

$$a\frac{\partial}{\partial r}(\gamma_S \hbar/r^3)_{r=b} \simeq \Delta H_n, \tag{6.38}$$

where a is the lattice spacing of the order of $(N_I)^{-1/3}$. Equation (6.38) states that spin diffusion can still occur between two neighbouring nuclear spins which 'see' local electronic fields differing by no more than the nuclear linewidth. This reduces the diffusion barrier b and causes T_{1n}^{-1} as given by (6.37) to increase by a factor of the order of $(\gamma_S/\gamma_I)^{1/4}$. It introduces into the frequencies of the 'anomalous' nuclei, on 'speaking terms' through diffusion with the normal nuclei, a spread of the order of $\Delta \simeq \gamma_I\, \Delta H_n(\gamma_S/\gamma_I)^{1/4}$. A careful experimental study of this problem has been performed by Cox, Read and Wenkebach (1977).

C. The well-resolved solid effect

(a) The rate equations

In this section we consider in more detail the ideal situation outlined in the introduction where the electronic linewidth $\Delta\omega_e$ is much smaller than the nuclear Larmor frequency ω_n. It is assumed that the transitions of frequencies $\omega_e - \omega_n$, $\omega_e + \omega_n$ and ω_e, termed respectively electron–nucleus flip-flop, electron–nucleus flip-flip and single-electron flip, do not overlap and that a microwave field tuned to any one of them cannot drive either of the other two (an assumption we shall give up in later sections). Consider first an isolated electron–nucleus pair with its four levels represented in Fig. 6.2 and suppose for argument's sake that we drive the transitions $|b\rangle \rightleftarrows |c\rangle$ (flip-flop) with a transition probability v by means of a microwave field H_1. The rate equations (6.33) and (6.34) have to be augmented by a contribution $-v(P_e - P_n)$ to dP_e/dt and $-v(P_n - P_e)$ to dP_n/dt and read:

$$\begin{aligned}\frac{dP_e}{dt} &= -v(P_e - P_n) - (1+r)(W_L + U_L)(P_e - P_0);\\ \frac{dP_n}{dt} &= -v(P_n - P_e) - (1+r)U_L P_n(1 - P_e P_0).\end{aligned} \tag{6.39}$$

We have now to take into account (as we already did for relaxation in **B**(b)) the fact that there are far more nuclei than electrons and that each electron has to 'service' a number $n = 1/C = N_n/N_e$ nuclei. We keep the assumption of $\mathbf{B}(b)$ that nuclear spin–spin interactions maintain a uniform polarization P_n for all nuclear spins. The rate eqns (6.39) must be modified as follows:

$$\begin{aligned}\frac{dP_e}{dt} &= -(1+r)\Big(W_L + \sum_i U_L^i\Big)(P_e - P_0) - \sum_i v_i(P_e - P_n);\\ \frac{dP_n}{dt} &= -\frac{1}{n}\sum_i v_i(P_n - P_e) - \frac{(1+r)}{n}\sum_i U_L^i P_n(1 - P_e P_0) - \frac{P_n}{T'_{1n}}.\end{aligned} \tag{6.40}$$

The coefficients U_L^i and v_i are relative to the coupling of the nucleus i with the electron. If V is the transition probability for the microwave field H_1 to induce the allowed transition, we have:

$$v_i = 4|q_i|^2 V \qquad U_L^i = 4|q_i|^2 W_L,$$

where q_i is given by eqn (6.28):

$$q_i = \frac{3}{2}\frac{\gamma_I \gamma_S \hbar}{r_i^3}\frac{1}{\omega_I}\sin\theta_i \cos\theta_i \exp(i\phi_i).$$

The last term added to eqn (6.40) $-P_n/T'_{1n}$, corresponds to so-called 'leakage' nuclear relaxation originating in processes other than the coupling with the electrons of Larmor frequency ω_e.

The summation of the coefficients $n^{-1}\sum_i v_i$ and $n^{-1}(1+r)\sum_i U_L^i$ is performed along the same lines as in equations (6.36) and (6.37) and leads to the result:

$$\begin{aligned}&\frac{1}{n}\sum_i v_i = C\alpha V = V_-;\\ &\frac{1}{n}(1+r)\sum_i U_L^i = C\alpha/T_{1e};\\ &1/T_{1e} = (1+r)W_L;\\ &\alpha \cong \frac{8\pi}{5}\frac{1}{4}\left(\frac{\gamma_S}{\gamma_I}\right)\left(\frac{\Delta H_n}{H_0}\right)^2 \ll 1.\end{aligned} \tag{6.41}$$

Before we rewrite the eqns (6.40) we define the following quantities: $1/T_n^0 = C\alpha/T_{1e}$, which is the nuclear relaxation rate (6.37) due to the coupling with electrons but divided by $(1-P_0^2)$; $s = VT_{1e}$, which is the saturation that would be induced by the microwave field H_1 if it were driving the allowed transition ω_e instead of the forbidden transition $\omega_e - \omega_n$; $f = T_n^0/T'_{1n}$, which is the so-called leakage coefficient. With these definitions the rate eqns (6.40) can be rewritten as:

$$\frac{\mathrm{d}P_e}{\mathrm{d}t} = -\frac{1}{T_{1e}}[\alpha s(P_e - P_n) + (P_e - P_0)];$$
$$\frac{\mathrm{d}P_n}{\mathrm{d}t} = -\frac{1}{T_n^0}[s(P_n - P_e) + P_n(f + 1 - P_e P_0)]. \tag{6.42}$$

It is easy to see that if we were to drive the *other* forbidden transition, of frequency $(\omega_e + \omega_n)$, we would have the same equations as (6.42) with transition probability $V_+ = V_- = C\alpha V = C\alpha s/T_{1e}$ and the only change in eqns (6.42) would be the replacement of P_n by $-P_n$.

With respect to the order of magnitude of the dimensionless parameters α, s, and f, the following comments can be made. α as given by (6.41) is small, of the order of 10^{-3}–10^{-4} and the saturation parameter s can be quite large. It is, however, unlikely to become as large as $1/\alpha$, such large microwave fields leading to an unacceptable heating of the sample. The leakage coefficient:

$$f = T_n^0/T'_{1n} = (T_{1e}/T'_{1n})(1/\alpha C),$$

can have any value, large or small. In particular, if f is so large that,

$$f\alpha = (1/T'_{1n})(C/T_{1e})^{-1} = (N_n/T'_{1n})(N_e/T_{1e})^{-1} \geqslant 1,$$

the *total* number of nuclear relaxation flips occurring per second is larger than the corresponding number of electronic relaxation flips and, as suggested in the introduction, we shall see that the limiting nuclear dynamic polarization is reduced.

We now proceed to solve eqns (6.42). Since the rate of change of P_e in the first equations (6.42) is much faster than in the second we can assume that $P_e(t)$ has at all times the quasi-equilibrium value

$$P_e = \frac{\alpha s P_n + P_0}{1 + \alpha s}. \tag{6.43}$$

If, as will be the case except for very large microwave power, $\alpha s \ll 1$, $P_e \approx P_0$ then the second eqn (6.42) can be written:

$$\frac{\mathrm{d}P_n}{\mathrm{d}t} = -\frac{1}{T_n^0}[(s + f + 1 - P_0^2)P_n - sP_0]$$
$$= -\left(\frac{s}{T_n^0} + \frac{1}{T'_{1n}} + \frac{1 - P_0^2}{T_n^0}\right)(P_n - P_{eq}), \tag{6.44}$$

with:

$$P_{eq} = \frac{P_0 s}{s + f + 1 - P_0^2} = P_0\left(1 + \frac{f}{s} + \frac{1 - P_0^2}{s}\right)^{-1}.$$

If $f \ll 1/\alpha$ we can make $s \gg 1$ and $f/s \ll 1$ without violating the conditions $s\alpha \ll 1$ and $P_e = P_0$ under which (6.44) was derived. The polarization time τ will be given by:

$$\frac{1}{\tau} = \frac{1}{T_n^0}(s + f + 1 - P_0^2) \approx \frac{s}{T_n^0} \quad \text{for } s \gg f,\ 1, \tag{6.45}$$

and $P_{eq} \approx P_0$.

To sum up this case, if $f \ll 1/\alpha$, which because of the smallness of α, defined by the last eqn (6.41), does not preclude $f \gg 1$, that is strong leakage, it is possible at least in principle to choose a microwave power level such that simultaneously $s\alpha \ll 1$, $s \gg 1$, and $s/f \gg 1$. Then the first eqn (6.42) yields $P_e \approx P_0$ (what we call the linear approximation) whilst by the last equation:

$$P_n = P_{eq} \simeq P_0.$$

On the other hand if the leakage is so strong that f becomes comparable with $1/\alpha$ we shall see that the limiting nuclear polarization P_{eq} is smaller than P_0 even in the limit of infinite microwave power. Carrying (6.43) into the second eqn (6.42) where we neglect $1 - P_e P_0$ in comparison with f, we get:

$$\frac{dP_n}{dt} \simeq -\frac{1}{T_n^0}\left[\left(\frac{s}{1+\alpha s} + f\right)P_n - P_0\frac{s}{1+\alpha s}\right]$$
$$= -\frac{1}{\tau}(P_n - P_{eq}), \tag{6.46}$$

with:

$$\frac{1}{\tau} = \frac{1}{T_n^0}\left(f + \frac{s}{1+\alpha s}\right);$$
$$P_{eq} = P_0\left[1 + f\left(\alpha + \frac{1}{s}\right)\right]^{-1}. \tag{6.47}$$

The nuclear polarization is reduced from P_0 by at least:

$$\frac{1}{1+f\alpha} = [1 + (N_n/T'_{1n})(N_e/T_{1e})^{-1}]^{-1},$$

even in the limit of infinite power, in agreement with the conclusion of the introduction.

If a strong phonon bottleneck is present the first eqn (6.42) in accordance with (6.16) has to be replaced by:

$$\frac{dP_e}{dt} = -\frac{1}{T_{1e}}\left[\alpha s(P_e - P_n) + \frac{P_e - P_0}{1 + \sigma(P_e/P_0)}\right], \tag{6.48}$$

the second eqn (6.42) remaining unchanged. This implies the assumption of a strong phonon bottleneck only for the allowed relaxation transition but not for the forbidden flip-flop and flip-flip transitions. This case has been discussed by several authors and in particular by Abragam and Goldman (1978). Their conclusions are the following. The relevant parameter against which the leakage coefficient f is to be measured is $1/\alpha\sigma$ rather than $1/\alpha$. If $f \ll 1/\alpha\sigma$ the limiting nuclear polarization is still P_0. On the other hand if $f > 1/\alpha\sigma$, the limiting nuclear polarization is less than P_0.

(*b*) *Experimental results*

(i) A classical example of the well-separated solid effect is provided by the observation of DNP in LMN doped with Nd^{3+}. Figure 6.3 exhibits plots of dynamic proton polarizations against the distance $(H - H_0)$ from the ESR resonance of Nd^{3+}, observed at various microwave powers in a field $H_0 \simeq$ 20 kOe (Schmugge and Jeffries, 1965). The two peaks do coincide with the theoretical values $H - H_0 = \pm H_0\gamma_n/\gamma_e$. (The difference in heights between the two peaks is probably accidental.)

(ii) A very clean example of a well-separated solid effect is provided by proton DNP in $Ca(OH)_2$ doped with paramagnetic (O_2^-) centres created by fast electron irradiation of the sample (Marks, Wenckebach, and Poulis, to be published). Figure 6.4 shows the proton polarization after 1 hr of microwave irradiation. The positions of the two peaks are in exact

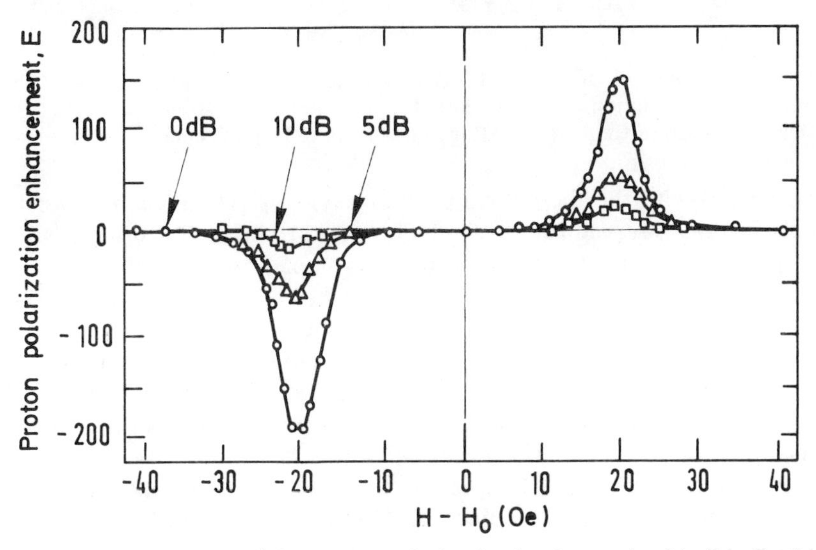

FIG. 6.3. Enhancement of the proton polarization by the resolved 'solid effect' in LMN: Nd subjected to 4 mm microwave irradiation as a function of applied d.c. field for various irradiation levels. (After Schmugge and Jeffries, 1965.)

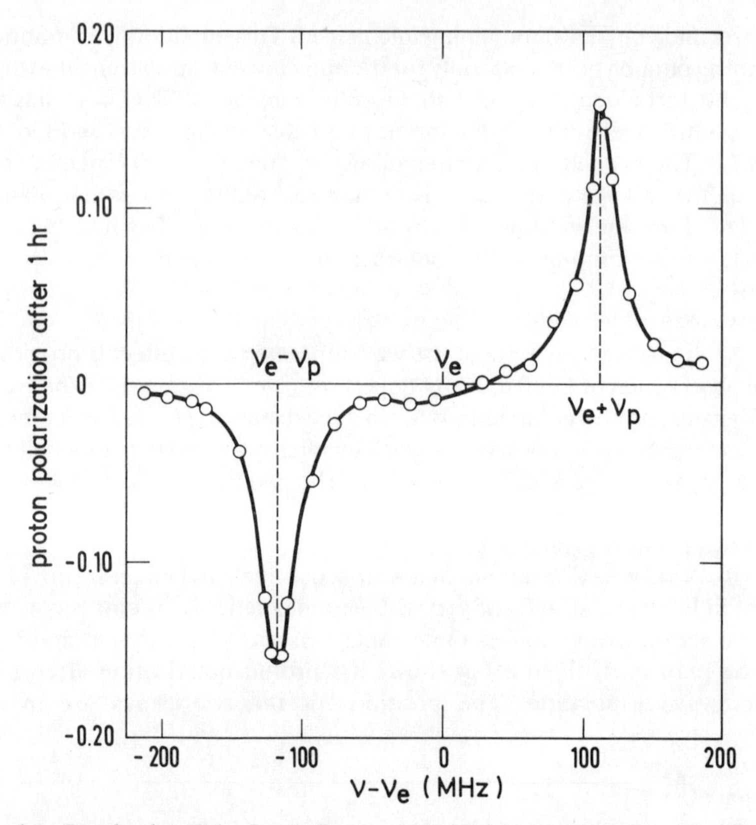

FIG. 6.4. Resolved 'solid effect'. Proton polarization in $Ca(OH)_2$ after 1 h of microwave irradiation at 4 mm plotted against microwave frequency offset from electronic Larmor frequency. (After Marks *et al.* to be published.)

agreement with the theoretical positions: $\gamma_e \pm \gamma_P$. The maximum nuclear polarization observed for $\nu_e \sim 75$ HGz, $\gamma_P \sim 116$ MHz, $T \approx 0{\cdot}65$ K is $P_n =$ 94 per cent. We shall come back to this example in section **F**(e).

(iii) A third example of interest is the DNP of ^{19}F in CaF_2 doped with U^{3+}, which exhibits qualitatively but unambiguously the effects of a phonon bottleneck on DNP (Abragam, Chapellier, Goldman, Jacquinot, and Vu Hoang Chau, 1971). The electronic half-width is $\Delta\omega_e \simeq 50$ MHz whilst in the field $H_0 \simeq 27$ kOe the nuclear frequency $\omega_n \simeq 107$ MHz results in a well-separated solid effect. The observation of the ESR of U^{3+} shows the possibility of 'burning' a hole in the line, demonstrating the slowness of spectral diffusion across the line (Vu Hoang Chau, Chapellier, and Goldman, 1969).

At 1 K, DNP yields an optimum limiting value of $P_n(^{19}F) \simeq 25$ per cent, reached in a time of the order of one hour. The relative smallness of this value can be attributed either to the 'inhomogeneous' character of the line, resulting in the saturation of the spin packets which participate in the DNP process or to a phonon bottleneck. The discrimination between the two causes is achieved by the following procedure. During the DNP a simultaneous modulation of the field *and* of the microwave frequency of amplitude Δ is applied, adjusted in phase and amplitude so as to satisfy at all times the condition $\omega(t) = -(\gamma_e - \gamma_n)H(t)$.

In this way, it is always the *same* spin packets which participate in the DNP process and they should be as much saturated as without modulation. However, the phonons emitted or absorbed by the relaxation of these spin packets cover now a width Δ greater than in the absence of modulation, thus diminishing the phonon bottleneck coefficient σ. The fact that in the presence of modulation the nuclear polarization rises to 45 per cent shows unambiguously that the polarization was limited by the phonon bottleneck rather than by the lack of spectral diffusion.

(iv) A fourth example is the DNP of ^{19}F in CaF_2 doped with Tm^{2+} in a field of 27 kOe at a frequency $\omega_e \simeq 130$ GHz ($g = 3{\cdot}45$ for Tm^{2+} in a cubic environment) and a temperature of 0·7 K. The nuclear frequency ω_n is 107 MHz and the electronic linewidths of the order of 50 MHz so that we are again dealing with a well-separated solid effect.

Experimentally it is observed that the DNP obtained by a microwave irradiation at a distance $|\omega_n|$ from one of the two hyperfine lines of the ESR spectrum of Tm^{2+} is poor. This is due to the fact that even the very weak saturation of the chosen hyperfine line, produced by DNP, that is a small departure from the linear approximation $P_e = P_0$, is sufficient to depopulate that line almost completely, as explained in **A**(*b*)2 and to render the DNP process inefficient by reducing considerably the number of electron spins participating in it.

The remedy, described in **A**(*b*)2 namely the repopulation of the line through microwave driving of the forbidden transition $\Delta(M+m) = 0$, is put into practice as follows.

The frequency of the microwave for the DNP is square-modulated: during 96 per cent of the time it has the value $\omega_e - \frac{1}{2}A - \omega_n$ corresponding to the solid-effect polarization from the line $|\alpha_-\rangle \rightleftarrows |\beta_-\rangle$ (Fig. 6.1) but every millisecond and for 40 μs (4 per cent of the time) it is brought to the value ω_e and drives the transition $\Delta(M+m) = 0$ which repopulates this line. Under these conditions a polarization of 90 per cent can be obtained for ^{19}F in three hours.

(v) The saturation of one of the hyperfine lines accompanied by the spilling of the spins into the other line corresponds to a 100 per cent polarization of the nuclear spin of the paramagnetic centre. This suggests that if by some artifice its Endor frequency can be made equal to the Larmor frequency of the 'normal' nuclei to be polarized (that is those beyond the diffusion barrier) these nuclei will be polarized by nuclear spin diffusion across the diffusion barrier (Abragam, 1965).

Thus for instance in solid HD, where hydrogen atoms are produced by γ irradiation, the Endor transition $(-\frac{1}{2}, -1) \rightarrow (-\frac{1}{2}, 0)$ of a deuterium atom can be made to coincide with the Larmor frequency of the proton in a field of the order of 30·5 kOe. The main interest of this method is a considerable reduction in microwave power with respect to that required by solid effect. The proposal relative to HD was never carried out, because the leakage relaxation of the protons (due to a small proportion of ortho H_2 molecules) prevents any type of DNP in HD (and also to be truthful, because for an unknown reason, only ^{1}H atoms are stable in irradiated solid HD).

A variant of this method was tried in CaF_2 (Jacquinot, Lounasmaa, and Urbina, 1978). In a sample doped with Tm^{2+}, the nearest neighbours of a paramagnetic centre could be given Endor frequencies equal to that of normal nuclei, for selected orientations of the magnetic field with respect to the crystalline axes.

It was indeed found that for moderate microwave power and for these preferred orientations of the field, the rate of growth of the nuclear polarization could exceed that of the normal solid effect by as much as a factor of five.

(vi) A phenomenon sometimes observed in DNP by the well separated solid effect is the appearance of secondary polarization peaks at frequencies distant from the ESR line by $2|\omega_n|$ rather than $|\omega_n|$ (Read, 1962*). They correspond to simultaneous flips of *two* nuclear spins, rather than one, accompanying an electron flip. These peaks correspond to transitions more highly forbidden than those of the normal solid effect and are accordingly smaller. Even for very large microwave power their intensity would be smaller than that of the main peaks: the maximum inverse nuclear spin temperature is $\beta_L|\omega_e/2\omega_n|$ rather than $\beta_L|\omega_e/\omega_n|$ for the normal peaks.

D. Electronic spin–spin interactions and electronic spin–spin temperature

A necessary condition for the validity of the theory of nuclear relaxation culminating in eqn (6.37) and of the theory of DNP by the solid effect, expressed in the rate equation (6.42), is a very low concentration of paramagnetic centres. This is understandable since nowhere in the sections **B** and **C** did we take into account the interactions between electronic spins. We do so now for the relaxation problem.

If the electronic concentration $C = N_S/N_I$ is not very low, electronic flip-flops will occur at a rate T_{2e}^{-1} much faster than T_{1e}^{-1}. It is tempting to argue that T_{2e} rather than T_{1e} is the correlation time of the local electronic field 'seen' by the nuclei and to replace $1/T_{1e}$ by $1/T_{2e}$ in (6.37).

* S. F. J. Read, University of Oxford, thesis, 1962.

For $1/T_{2e}$ a crude estimate is the Larmor frequency of an electronic spin in the local field produced by a neighbouring electronic spin:

$$\frac{1}{T_{2e}} \sim \gamma_S \, \Delta H_e \sim \gamma_S(\gamma_S \hbar N_S). \tag{6.49}$$

Introducing,

$$\Delta H_n = \gamma_I \hbar N_I,$$

we get:

$$\frac{1}{T_{2e}} \sim C(\gamma_S/\gamma_I)\gamma_S \, \Delta H_n, \tag{6.49'}$$

and feeding (6.49′) into (6.37) one would obtain a relaxation rate that, for reasons to appear shortly, we shall call $1/T_{I,SS}$:

$$\begin{aligned}\frac{1}{T_{I,SS}} &\approx \frac{2\pi}{5} C\left(\frac{\gamma_S}{\gamma_I}\right)\left(\frac{\Delta H_n}{H_0}\right)^2 (1-P_0^2)\frac{1}{T_{2e}} \\ &\approx \frac{2\pi}{5} C^2\left(\frac{\gamma_S}{\gamma_I}\right)^2\left(\frac{\Delta H_n}{H_0}\right)^2 (1-P_0^2)(\gamma_S \, \Delta H_n).\end{aligned} \tag{6.50}$$

It is clear that the rate given by (6.50) has nothing to do with the coupling of the nuclear spins to the lattice since no parameters relative to the lattice appear in that formula, nor to the electronic Zeeman energy, since an electronic flip-flop leaves it unchanged. What it expresses is the existence of a coupling between the electronic spin–spin energy $\mathcal{H}_{SS} = \sum_{i,j} S_i \cdot \mathbf{B}_{ij} \cdot \mathbf{S}_j$ (or rather its secular part $\mathcal{H}'_{SS}$ that commutes with the Zeeman electronic energy Z_S) and the nuclear Zeeman energy $-\sum_i \gamma_I \hbar \mathbf{H}_0 \cdot \mathbf{I}_i = Z_I$.

We must therefore take a closer look at the behaviour of this electronic spin–spin energy. Just as we did throughout this book for the nuclear spins, we make the assumption of an electronic spin temperature. The quantitative formulation of this assumption is the following expression for the electronic density matrix:

$$\sigma_S = \exp[-(\alpha_Z \omega_S S_Z + \beta \mathcal{H}'_{SS})]/\mathrm{Tr}\{\exp[-(\alpha_Z \omega_S S_Z + \beta \mathcal{H}'_{SS})]\}. \tag{6.51}$$

The fact that when $\alpha_Z \omega_S$ is not small, $\langle \mathcal{H}'_{SS} \rangle = \mathrm{Tr}\{\sigma_S \mathcal{H}'_{SS}\}$ is a function of *both* α_Z and β, was discussed in chapter 5 and we need not dwell on it here.

On the other hand the fact that the spins S instead of forming a regular lattice are dilute and distributed at random on a few of the sites of a regular lattice is a novel feature. The validity of (6.51) is thus a *new* assumption which requires an experimental check.

Another novel feature with respect to nuclear spins systems is the existence of the so-called inhomogeneous EPR broadening. There are essentially two types of inhomogeneous broadening. The simpler of the two is a spread of the Zeeman frequencies of an assembly of electronic spins, around a central value, due to a distribution of their g factors.

The electronic Hamiltonian $\sum_i \omega_e^i S_z^i$ can be rewritten as $\omega_e S_Z - \sum_i \Delta_i S_z^i$ where $\omega_e^i = \omega_e - \Delta_i$ and $\sum_i \Delta_i = 0$.

The second type of broadening is due to unresolved hyperfine structure from couplings between an electron spin $\mathbf{S}^i$ and the nuclear spins $\mathbf{I}_\mu^i$ which surround it. These 'anomalous' spins $\mathbf{I}_\mu^i$ have Endor frequencies sufficiently different from those of the normal nuclear spins to be unable to communicate with them by nuclear spin diffusion. The secular part of this coupling can be written: $\sum_{i,\mu} S_z^i A_\mu^i I_{\mu z}^i$. In the absence of the spin–spin interaction $\mathcal{H}_{SS}$ the EPR line is made of spin packets which do not communicate with each other. The Larmor frequency of a spin packet is $(\omega_e - \Delta_i)$ for a broadening of the first type and $\omega_e + \sum_\mu A_\mu^i m_\mu^i$, where m_μ^i are nuclear quantum numbers, for an inhomogeneous broadening of the second kind. If we assume that the nuclear environment is the same for all the electronic spins $\mathbf{S}^i$ we can write A_μ instead of A_μ^i.

The Hamiltonian of the electronic spins can be written:

$$\mathcal{H}_e = \omega_e S_z + \mathcal{H}'_{SS} - \sum_i \Delta_i S_z^i + \sum_{i,\mu} A_\mu m_\mu^i S_z^i, \tag{6.52}$$

where we have assumed that both types of inhomogeneous broadening are present. The first term of (6.52) is not on speaking terms with any of the others, the Zeeman quanta ω_e being several orders of magnitude larger than Δ_i or A_μ. On the other hand, thanks to the flip-flops initiated by $\mathcal{H}'_{SS}$, energy can flow from one spin packet to another by a phenomenon known as cross-relaxation (see Goldman, 1970, chapter 6). We lump the inhomogeneous broadening terms of (6.52), together with $\mathcal{H}'_{SS}$ into a single Hamiltonian $\mathcal{H}'^*_{SS}$, that we call the non-Zeeman Hamiltonian, and rewrite (6.52) as:

$$\mathcal{H}_e = \omega_e S_z + \mathcal{H}'_{SS} + \mathcal{H}'_{IB} = \omega_e S_z + \mathcal{H}'^*_{SS} = Z_S + \mathcal{H}'^*_{SS}. \tag{6.53}$$

In (6.53) $\mathcal{H}'_{IB}$ (IB stands for inhomogeneous broadening) represents the last two terms of (6.52).

We then make the far reaching, and by no means obvious assumption that the expression (6.51) for the density matrix is still valid if $\mathcal{H}'_{SS}$ is replaced by $\mathcal{H}'^*_{SS}$. The problem is one of time-scale. If energy is brought to a fraction of the electronic spins, for instance, by saturating a few spin-packets in the EPR line and if this energy can spread out to all spin-packets in a time short compared to the electronic spin–lattice relaxation time, the expression (6.51) is likely to be correct since it represents the most probable distribution of populations in the system. A discussion of this problem for the high temperature case using the standard rate equations for cross-relaxation is given by Rodak (1973).

On the experimental side the validity of the assumption of a temperature for a non-Zeeman electronic Hamiltonian, distinct from the Zeeman temperature was carefully checked by Atsarkin (1970) on the EPR of Ce^{3+} in a crystal of $CaWO_4$. The choice of this matrix is dictated by the absence of nuclear spins. The inhomogeneous broadening is then exclusively of the first type and can be changed at will within certain limits by changing the orientation of the magnetic field with respect to the crystal axes. One of the checks rests on the shape of the EPR signal observed with a non-saturating microwave field immediately after a strong saturation of the spin system at a distance $-\Delta_0$ from resonance.

Consider all pairs of levels $|a\rangle$, $|b\rangle$ whose energies E_a and E_b differ by $E_b - E_a = \hbar(\omega_S - \Delta)$. Their Zeeman energies cannot differ by anything but $\hbar\omega_S$, and therefore their non-Zeeman energies differ by $-\hbar\Delta$. According to (6.51) the ratio of their populations is:

$$p_a/p_b = \exp(\alpha_Z \omega_S - \beta\Delta). \tag{6.54}$$

Immediately after the saturation this ratio is unity for those pairs of levels for which $\Delta = \Delta_0$ whence:

$$\alpha_Z \omega_S = \beta\Delta_0. \tag{6.54'}$$

For all the other pairs this ratio is then:

$$\frac{p_a}{p_b} = \exp[\beta(\Delta_0 - \Delta)]. \tag{6.55}$$

The EPR signal observed immediately after the saturation, proportional to $(p_a - p_b)$, vanishes for $\Delta = \Delta_0$, has the normal absorption form for $\Delta < 0$ but is a reversed emission signal for $\Delta > \Delta_0$. (This proof bears a close analogy to that of eqn (5.113).) This behaviour, characteristic of the existence of a temperature for the spin–spin reservoir, very different from the Zeeman temperature, was indeed observed.

E. Electronic spin–spin reservoir and nuclear relaxation

(a) *The coupling between the nuclear Zeeman and the electronic non-Zeeman reservoirs*

To the assumptions on the electronic spin–spin (or non-Zeeman) temperature made in section **D**, we add the following: the electronic non-Zeeman reservoir $\mathcal{H}'^*_{SS}$ and the nuclear Zeeman reservoir Z_I are on 'speaking terms' and reach rapidly a common temperature, rapidly meaning here, on a time scale shorter than any other processes affecting $\langle Z_I \rangle$ and $\langle \mathcal{H}'^*_{SS} \rangle$.

The semi-classical calculation of the time-constant for this coupling leading to eqn (6.50) has been put on a quantum mechanical basis by

Melikiya (1968) and Goldman *et al.* (1974*b*). The principle of their derivation is sketched below.

The thermal coupling between the two Hamiltonians Z_I and $\mathcal{H}'^*_{SS} = \mathcal{H}'_{SS} + \mathcal{H}'_{IB}$ (eqns (6.52), (6.53)) is handled using essentially the formalism of the generalized Provotorov equations presented in chapter 1. The high temperature approximation is made for the nuclear Zeeman temperature α_I and for the electronic non-Zeeman temperature β, but not for the electronic Zeeman temperature α_{ZS}. The coupling Hamiltonian V between Z_I and $\mathcal{H}'^*_{SS}$ can be written:

$$V = \sum_{i,\mu} \mathcal{F}^i_\mu S^i_z I^\mu_+ + h.c. \tag{6.56}$$

$\hbar(\mathcal{F}^i_\mu/\gamma_I)S^i_z \sim \gamma_S r^{-3}_{i\mu} S^i_z$ is the component of the magnetic field produced by the electron spin $\mathbf{S}^i$ (assumed $\frac{1}{2}$ for simplicity) and 'seen' by the nuclear spin $\mathbf{I}^\mu$ in accordance with the random field approach of eqn (6.23). The formalism of chapter 1, section **A**(c)3, modified to take into account the high polarization of the electronic spins leads in a straightforward manner to a value for the coupling coefficient $W = 1/T_{I,SS}$:

$$W = 1/T_{I,SS} = \tfrac{1}{2}C(1-P_e^2)\mathcal{F}^2 f(\omega_I). \tag{6.57}$$

In (6.57) C is the relative impurity concentration: $C = (N_S/N_I)$. $\mathcal{F}^2 = \sum_\mu \mathcal{F}^2_{i\mu}$, independent of i if all spins $\mathbf{S}_i$ have the same environment and $f(\omega_I)$ is the Fourier transform of:

$$F(\tau) = 4\left\langle\left(S^i_z(\tau) - \frac{P_e}{2}\right)\left(S^i_z - \frac{P_e}{2}\right)\right\rangle \Big/ (1-P_e^2), \tag{6.58}$$

the correlation function of the fluctuating part $(S^i_z(\tau) - P_e/2)$ of the z component of the spin $\mathbf{S}_i$, normalized to unity for $\tau = 0$:

$$S^i_z(\tau) = \exp -(\mathrm{i}\mathcal{H}'^*_{SS}\tau)S^i_z \exp(\mathrm{i}\mathcal{H}'^*_{SS}\tau).$$

To proceed with an estimate of (6.57) Melikiya (1968) and Goldman *et al.* (1974*b*) calculate the second and fourth moments of the shape function $f(\omega)$ in (6.57), that is the second and fourth derivative of $F(\tau)$ in (6.58). There is an important point to realize in this connection: $F(\tau)$ is the correlation function of a z component of an electronic spin $\mathbf{S}^i$, rather than that of an x component as for a magnetic resonance signal. As a consequence S^i_z commutes with the Hamiltonian $\mathcal{H}'_{IB}$ of eqn (6.52), which therefore does not bring any contribution to the second moment M_2 of $f(\omega)$, determined solely by the true spin–spin coupling:

$$\mathcal{H}'_{SS} = \tfrac{1}{2}\sum_{i,j} B_{ij}\{3S^i_z S^j_z - \mathbf{S}^i \cdot \mathbf{S}^j\}. \tag{6.59}$$

An elementary calculation gives:

$$M_2 = -\left(\frac{\partial^2 F}{\partial t^2}\right)_{t=0} = \tfrac{1}{2}c' B^2 \quad \text{with} \quad B^2 = \sum_j B_{ij}^2, \tag{6.60}$$

the sum being taken over *all* the sites available to the impurities; c' is the ratio of the sites occupied by the impurities to those available to them. For instance in CaF_2 $c' = C/2$. For the fourth moment, it is found:

$$M_4 = c'\{\tfrac{1}{2}B'^4 + B^2 A^2\},$$

with:

$$B'^4 = \sum_j B_{ij}^4; \qquad A^2 = \frac{1}{N_S}\sum_i \Delta_i^2 + \frac{I(I+1)}{3}\sum_\mu A_\mu^2. \tag{6.61}$$

A^2 is the second moment of the inhomogeneously broadened EPR line. The usual ratio $M_4/(M_2)^2$ is:

$$\frac{M^4}{(M_2)^2} = \frac{1}{c'}\frac{(B'^4/2) + B^2 A^2}{B^4}. \tag{6.62}$$

$M_4/(M_2)^2$ is much larger than unity because of the smallness of c' due to the great dilution of the impurities. The shape function $f(\omega)$ of (6.57) can then be tentatively represented by a truncated Lorentzian with a width of the order of $\sqrt{(M_2)}(M_2^2/M_4)^{1/2} \ll \sqrt{M_2}$ and a high frequency cut-off $\alpha \sim (M_4/M_2)^{1/2}$.

In an order of magnitude calculation we can take $B' \approx B$. Then:

$$\delta \approx c'B/\sqrt{(1 + (B^2/A^2))}; \qquad \alpha \approx B\sqrt{(1 + (B^2/A^2))}. \tag{6.63}$$

We see that the cut-off α of the shape function $f(\omega)$ occurs at a frequency of (at least) order B, that is of the order of an EPR line width where *all* the sites available to impurities are occupied. This result, surprising at first sight is only possible because however dilute the impurities, a pair of them can always sit on neighbouring sites, providing a flip-flop frequency of order B.

It is well to clarify once more a possible confusion. The shape function $f(\omega)$ in (6.57) is *not*, repeat *not*, the shape function $f_{\text{EPR}}(\omega)$ of the EPR resonance. The contributions of $\mathcal{H}'_{SS}$, the true spin–spin interaction to the second moments of $f(\omega)$ and $f_{\text{EPR}}(\omega)$ are comparable and of the order of $c'B^2$. However the contribution of $\mathcal{H}'_{\text{IB}}$ to the second moment of $f(\omega)$ vanishes, whereas it is $A^2 \gg c'B^2$ for $f_{\text{EPR}}(\omega)$.

The fourth moment M_4 is of order $c'B^2(B^2 + A^2) \gg M_2^2 = c'^2B^4$ for $f(\omega)$, but it is of order $A^4 \approx M_2^2$ for $f_{\text{EPR}}(\omega)$. This is why $f(\omega)$ is Lorentzian with a width of the order of $\delta \sim c'B/\sqrt{(1 + (B^2/A^2))}$ and with long wings, cut off at $\sqrt{(B^2 + A^2)}$ whereas $f_{\text{EPR}}(\omega)$ is roughly Gaussian with a width A. The line width δ of $f(\omega)$ is usually much smaller than $\omega_I = -\gamma_I H_0$.

The transition probability $W = 1/T_{I,SS}$ is thus by (6.57) of the order of:

$$W \approx C(1-P^2)\mathcal{F}^2 \frac{\delta}{\pi} \frac{1}{\omega_I^2+\delta^2} \sim C(1-P^2)\mathcal{F}^2 \frac{\delta}{\omega_I^2}. \tag{6.64}$$

In the expression $\mathcal{F}^2 = \sum_\mu \mathcal{F}_{i\mu}^2$ the summation begins with nuclei *outside* the diffusion barrier b defined by:

$$\hbar \frac{\gamma_S}{b^3} \sim \Delta H_n, \tag{6.65}$$

exactly as in the semi-classical derivation of (6.37) and (6.50). (Nuclei *inside* the diffusion barrier belong to $\mathcal{H}'_{IB}$.) The summation yields:

$$\mathcal{F}^2 \sim \gamma_S \gamma_I (\Delta H_n)^2, \tag{6.66}$$

where use has been made of the $r_{i\mu}^{-6}$ dependence of $\mathcal{F}_{i\mu}^2$ and of the definition (6.65) of b. Replacing in (6.64) δ by:

$$\delta \sim CB \sim C\gamma_S N_I \sim C\frac{\gamma_S}{\gamma_I}\Delta H_n, \tag{6.67}$$

we find within numerical factors of order unity the semi-classical formula (6.50).

One may question the usefulness of a lengthy discussion which itself is an oversimplification of the original papers and whose final result was already known. The answer is that, through the essential distinction between $f(\omega)$, spectral density of $S_z^i(\tau)$ and of $f_{EPR}(\omega)$, EPR lineshape, this discussion leads to a better understanding of the respective roles of $\mathcal{H}'_{SS}$, essential for the $Z_I \rightleftarrows \mathcal{H}'^*_{SS}$ coupling, but negligible for the EPR broadening and that of $\mathcal{H}_{IB}$, essential for the EPR broadening but unessential, if not negligible, for the $Z_I \rightleftarrows \mathcal{H}'^*_{SS}$ coupling.

It might be added that the assimilation of $f(\omega)$ to a Lorentzian on the strength of two moments only, is to say the least, somewhat questionable. Its best justification is a not unreasonable agreement with experiment (Goldman *et al.* 1974*b*).

(*b*) *Nuclear Zeeman relaxation*

The tight coupling between $\langle \mathcal{H}'^*_{SS} \rangle$ and $\langle Z_I \rangle$ leads to a new mechanism for nuclear spin lattice relaxation, based on the following physical picture. Assume first that inhomogeneous broadening is absent from the EPR line so that $\langle \mathcal{H}'^*_{SS} \rangle = \langle \mathcal{H}'_{SS} \rangle$. $\langle \mathcal{H}'_{SS} \rangle$ has a heat capacity much smaller than that of $\langle Z_I \rangle$ for two reasons: (i) $\langle \mathcal{H}'_{SS} \rangle$ is proportional to C^2 where C, the concentration of the paramagnetic centres, is small; (ii) $\langle \mathcal{H}'_{SS} \rangle$ must be understood as the part of the expectation value of $\mathcal{H}'_{SS}$, due to the existence of short range dipolar order. If α_Z and β are the inverse temperatures of the electronic Zeeman energy Z_S and of the dipolar energy $\mathcal{H}'_{SS}$, what is meant by $\langle \mathcal{H}'_{SS} \rangle$ is:

$$\langle \mathcal{H}'_{SS} \rangle = \frac{\mathrm{Tr}\{\mathcal{H}'_{SS} \exp[-(\alpha_Z Z_S + \beta \mathcal{H}'_{SS})]\}}{\mathrm{Tr}\,\{\exp[-(\alpha_Z Z_S + \beta \mathcal{H}'_{SS})]\}} - \frac{\mathrm{Tr}\{\mathcal{H}'_{SS} \exp(-\alpha_Z Z_S)\}}{\mathrm{Tr}\,\{\exp(-\alpha_Z Z_S)\}}. \quad (6.68)$$

When the electronic polarization becomes unity ($\alpha_Z \to \infty$) only the lowest energy state becomes available to the electron spins and $\langle \mathcal{H}'_{SS} \rangle$ as defined by (6.68) goes to zero as $(1 - P_0^2)$. For large electron polarizations which are those considered in this chapter $\langle \mathcal{H}'_{SS} \rangle$ is much smaller than $\langle Z_I \rangle$. As a consequence $\langle \mathcal{H}'_{SS} \rangle$ is very strongly coupled to $\langle Z_I \rangle$, much more strongly than to the lattice. $\langle Z_I \rangle$ and $\langle \mathcal{H}'_{SS} \rangle$ can thus be considered at all times as a single reservoir with a common temperature. $\langle \mathcal{H}'_{SS} \rangle$ is also coupled to the lattice with a time constant of the same order of magnitude as the electronic Zeeman relaxation time T_{1e}: it is clear that each time a single electronic spin flips because of T_{1e} a change occurs in the spin–spin energy $\langle \mathcal{H}'_{SS} \rangle$. The nuclear Zeeman energy $\langle Z_I \rangle$ acts as a 'load' on $\langle \mathcal{H}'_{SS} \rangle$ and their combined relaxation rate is:

$$\frac{1}{T_{1n}} \approx \frac{1}{T_{1,SS}} \frac{\langle \mathcal{H}'_{SS} \rangle}{\langle \mathcal{H}'_{SS} \rangle + \langle Z_I \rangle} \approx \frac{1}{T_{1,SS}} \frac{\langle \mathcal{H}'_{SS} \rangle}{\langle Z_I \rangle}. \quad (6.69)$$

This is analogous to the electron relaxation with a strong phonon bottleneck. The phonons on speaking terms with the electronic spins immediately take the temperature of the spins and it is the combined spin–phonon system that relaxes to the bath at a rate $(1/\tau_{\mathrm{ph}})E_{\mathrm{ph}}/E_S$ (eqn 6.14). $1/T_{1n}$ has thus the same proportionality to $(1 - P_0^2)$ as $\langle \mathcal{H}'_{SS} \rangle$.

If an inhomogeneous broadening $\mathcal{H}'_{\mathrm{IB}}$ is present, we replace in (6.69) $\langle \mathcal{H}'_{SS} \rangle$ by $\langle \mathcal{H}'^{*}_{SS} \rangle$.

There is a considerable amount of experimental evidence demonstrating that in a sample doped with paramagnetic impurities the spin–spin energy and the nuclear Zeeman energy have the same temperature. We present here a few experimental results among the most convincing. Some more will be discussed in the next section in connection with DNP.

(i) We have shown that strong saturation of an EPR line at a frequency $\omega_S - \Delta_0$ resulted in an inverse dipolar temperature β much higher than the Zeeman inverse temperature α_Z (eqn 6.54′). The EPR line then becomes asymmetrical (Atsarkin, 1970). It was shown in chapter 1 (eqn 1.51b) that in the high temperature approximation the antisymmetrical part of the line is proportional to β. In LMN doped with Nd^{3+} it has been observed that the decay time of this antisymmetrical part of the ESR line coincides within experimental error with the nuclear relaxation time of the protons, thus demonstrating that electronic non-Zeeman and nuclear Zeeman energies are indeed in thermal equilibrium all through the spin–lattice decay (Wenckebach, van den Heuvel, Hoogstraate, Swanenburg, and Poulis, 1969).

(ii) In the experiment (i) the time constant $T_{I,SS}$, defined in equation (6.50), for the thermal mixing between these two energies was too short to be observed. A direct measurement of this time constant could be made in lithium fluoride (LiF) (Cox, Bouffard, and Goldman, 1973), where the paramagnetic impurities responsible for the DNP were F centres created by electron bombardment of the crystal. Their equilibrium polarization P_0 was sufficiently near unity to expect for $T_{I,SS}$, proportional to $(1-P_0^2)^{-1}$, a value much longer than in the previous low-field experiments. The experiment ran as follows:

(a) Nuclei of ^{7}Li and ^{19}F are polarized dynamically and the polarizing microwave is then cut off.

(b) The magnetization of one of the nuclear species, say ^{7}Li, is saturated by a burst of r.f. power.

(c) The magnetization of both species are monitored by their NMR signals.

It was found that $M(^7\text{Li})$ increased and $M(^{19}\text{F})$ decreased, with a time constant of the order of one minute, toward limiting values describable by the *same* spin temperature for both species, a temperature much lower than that of the lattice. From there on, the decay of both magnetizations could be described as a decay of this common temperature, towards the lattice temperature at a rate about a hundred times slower than that of the first phase. The results of a typical experiment are depicted in Fig. 6.5.

The first phase of the experiment is interpreted as a thermal mixing of the Zeeman energy of each nuclear species with the electron non-Zeeman energy, and thus with each other. The $(1-P_0^2)$ dependence is well verified

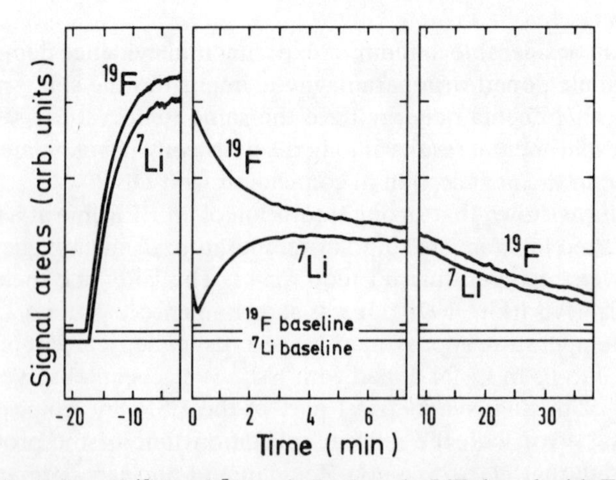

FIG. 6.5. Evolution of ^{19}F and ^{7}Li polarizations in LiF doped with F centers. (a) Dynamic polarization. (b) After saturation of ^{7}Li the two polarizations vary toward equal temperature values as a result of indirect thermal mixing through the non-Zeeman reservoir. (c) Spin–lattice relaxation. (After Cox *et al.* 1973.)

experimentally and the absolute value of the time constant $T_{I,SS}$ agrees in order of magnitude with the theoretical estimate (Goldman, Cox, and Bouffard, 1974).

The second phase is the spin–lattice relaxation of the combined spin system.

(iii) An interesting example of nuclear relaxation by thermal contact with a spin–spin reservoir is provided by the study of proton relaxation in holmium ethylsulphate at very low lattice temperatures down to 30 mK (Abragam, Bacchella, Glättli, Meriel, Piesvaux, and Pinot, 1976).

The electronic spin Hamiltonian of Ho^{3+} is given by (6.6) and (6.7′) where **I** is the spin ^{165}Ho and the values of the various constants are given by (6.19). For each holmium nucleus there are 33 protons. Their relaxation rate, measured, as that of ^{165}Ho described in **A**(b)2 by pseudomagnetic neutron precession (see chapter 7), shows a remarkable temperature dependence, when the applied field is perpendicular to the crystalline c-axis. Between 100 mK and 30 mK $T_1(^1\mathrm{H})$ is represented reasonably well by:

$$T_1(^1\mathrm{H}) \approx 100 \exp(-0{\cdot}09/T)\ \mathrm{s}. \tag{6.70}$$

It is clear that an energy splitting Δ of the order of 0.09 k_B is involved in this relaxation and only a hyperfine splitting can have so small a value. The energy splitting between the lowest state of the Ho^{3+} ion and its first excited state is indeed a hyperfine splitting $\mathscr{A}_\perp$, obtained by diagonalization of (6.6) and (6.7′) for H perpendicular to c.

Its value is: $\mathscr{A}_\perp \simeq 0{\cdot}85\, A_\perp \approx 0{\cdot}11\, k_\mathrm{B}$ which within experimental errors can be identified with the splitting Δ responsible for the proton relaxation.

We are thus led to assume that the *nuclear* spins of ^{165}Ho play, with respect to the protons, the role of paramagnetic impurities, and that their spin–spin interaction provides the reservoir through which the protons relax to the lattice. An order of magnitude calculation shows that the factor in front of the exponential in (6.70) is not unreasonable. To make this plausible one should remember that: (a) the concentration of ^{165}Ho nuclei with respect to protons is quite high (1/33); (b) for H perpendicular to the c-axis the nuclear states of ^{165}Ho are strongly scrambled by electronic admixtures and their nuclear flip-flops have appreciable electronic contributions; (c) the ion Ho^{3+} being non-Kramers is rather strongly coupled to the lattice.

(c) *Nuclear dipolar relaxation*

There is another reservoir whose relaxation has not been considered yet, namely $\mathscr{H}'_{II}$, the secular dipolar energy of the *nuclear* spins. Although very much smaller than Z_I, $\mathscr{H}'_{II}$ is responsible for the existence of ordered nuclear states at temperatures of $\mathscr{H}'_{II}$ which are several orders of magnitude lower than those of Z_I, $\mathscr{H}'_{IS}$ and $\mathscr{H}'_{SS}$. The 'spin–lattice' relaxation time of $\mathscr{H}'_{II}$ can be identified as the time constant for its coupling with these spin reservoirs much 'hotter' and 'heavier'.

As thus defined it is a time constant $T_{1D} = T_{II,SS}$ analogous to that given in (6.50) for the coupling between Z_I and $\mathcal{H}'_{SS}$. There is however a difference: the formulae (6.37) for T_{1n} or (6.50) for $T_{I,SS}$ had been obtained by assuming that the nuclei were 'seeing' a random electronic field with a correlation time T_{1e} or T_{2e} and thus with a power spectrum:

$$J(\omega) = \frac{T_{1e}}{1+\omega^2 T_{1e}^2} \quad \text{or} \quad \frac{T_{2e}}{1+\omega^2 T_{2e}^2}. \tag{6.71}$$

To calculate the coupling of the Zeeman nuclear energy Z_I to the lattice in the first case or to the spin–spin energy $\mathcal{H}'_{SS}$ in the second case we took for ω in the power spectrum the nuclear Larmor frequency $\omega_n = -\gamma_n H_0$. In the case of the nuclear dipolar energy, the relevant frequency is practically zero and in the expression for $T_{II,SS}$ we use $J(0) = T_{2e}$ rather than:

$$J(\omega_n) \simeq (\omega_n^2 T_{2e})^{-1}. \tag{6.71'}$$

It follows that $1/T_{1D} = 1/T_{II,SS}$ is deduced from $1/T_{I,SS}$ as given by eqn (6.50) by multiplying it by:

$$(\omega_n T_{2e})^2 = (\gamma_n H_0 T_{2e})^2.$$

With a little algebra we get (using for T_{2e}^{-1} the expression (6.49)):

$$\frac{1}{T_{1D}} = \frac{1}{T_{II,SS}} \simeq \frac{2\pi}{5} \frac{\gamma_I}{\gamma_S} (\gamma_I \, \Delta H_n)(1 - P_0^2), \tag{6.72}$$

or, writing $1/T_{2n}$ for $\gamma_I \, \Delta H_n$:

$$\frac{1}{T_{1D}} \simeq \frac{1}{T_{2n}} \frac{\gamma_I}{\gamma_S} (1 - P_0^2), \tag{6.73}$$

which exhibits a remarkable independence of the concentration of paramagnetic impurities.

This feature, surprising at first sight, has a simple physical explanation. The rate $1/T_{II,SS}$ is proportional to the product of the spectral density $J(0) = T_{2e}$ of the fluctuating electronic field due to the electronic flip-flops, by the concentration C of the impurities:

$$1/T_{II,SS} \propto C T_{2e}.$$

However, the frequency T_{2e}^{-1} of the electronic flip-flops is also proportional to the concentration C, whence the independence of $T_{II,SS}$ from this concentration.

In order for the coupling constant (6.73) between $\mathcal{H}'_{II}$ and $\mathcal{H}'_{SS}$ to be considered as the dipolar relaxation rate, i.e. as the rate of decay of the nuclear dipolar energy one must show either that the heat capacity $\langle \mathcal{H}'_{SS} \rangle$ as defined in eqn (6.68) is much larger than $\langle \mathcal{H}'_{II} \rangle$ or that $\langle \mathcal{H}'_{SS} \rangle$ is itself coupled sufficiently strongly to another large spin reservoir. For electron polarizations P_0 very near unity it is possible for $\langle \mathcal{H}'_{SS} \rangle$, proportional to $(1 - P_0)$, to be smaller than $\langle \mathcal{H}'_{II} \rangle$. However, in that case we may recollect that Z_I is coupled to $\mathcal{H}'_{SS}$ with a time constant $T_{I,SS}$ given by eqn (6.50). The rate of flow of energy *from* $\mathcal{H}'_{SS}$ *to* Z_I is of the order of:

$$\frac{\langle \mathcal{H}'_{SS} \rangle}{T_{SS,I}} \simeq \frac{\langle Z_I \rangle}{\langle \mathcal{H}'_{SS} \rangle} \frac{\langle \mathcal{H}'_{SS} \rangle}{T_{I,SS}} = \frac{\langle Z_I \rangle}{T_{I,SS}}. \tag{6.74}$$

This has to be larger than the rate of flow of energy from $\mathcal{H}'_{II}$ to $\mathcal{H}'_{SS}$, namely $\langle \mathcal{H}'_{II} \rangle / T_{II,SS}$. We recollect that we derived $1/T_{II,SS}$ in eqn (6.72) by noticing that its ratio to $1/T_{I,SS}$ was $(\omega_n T_{2e})^2$. On the other hand, the ratio $\langle Z_I \rangle / \langle \mathcal{H}'_{II} \rangle$ is approximately $(\omega_n T_{2n})^2$.

The condition:

$$\langle \mathcal{H}'_{II} \rangle / T_{II,SS} \ll \langle Z_I \rangle / T_{I,SS},$$

becomes:

$$\left(\frac{T_{2e}}{T_{2n}}\right)^2 \ll 1 \quad \text{or} \quad [C(\gamma_S/\gamma_I)^2]^2 \gg 1. \tag{6.75}$$

For $C \geqslant 10^{-5}$ eqn (6.75) is verified and $1/T_{II,SS}$ in eqn (6.73) is indeed the operational 'relaxation rate' of the dipolar energy.

F. Electronic spin–spin reservoir and nuclear dynamic polarization

(*a*) *Historical background*

In section **C** we considered the ideal case of the well-resolved solid effect with $\omega_n \gg \Delta\omega_e$. Actually, in contrast with the few examples discussed in that section this condition is violated in most substances where DNP is observed, even in fairly high fields ($H_0 \approx 25$ to 50 kOe) and for nuclei with high magnetic moments such as protons and fluorine.

The causes of these large electron widths are manifold. For samples that are not single crystals such as most target materials, there is a magnetic width due to a small anisotropy in the *g* factors of the electron spins that we have termed inhomogeneous broadening of the first type.

A relative anisotropy of the order of $\gamma_I/\gamma_S \ll 1$ will be sufficient to violate the condition $\omega_n \gg \Delta\omega_e$ at *all* fields. Secondly, there are the hyperfine couplings of the electron spins with the neighbouring nuclei, particularly large when the electronic wavefunction spills over the neighbouring nuclei (inhomogeneous broadening of the second type). Among the paramagnetic impurities widely used in DNP, this is the case for free radicals in organic materials (polarized targets) and for F centres in ionic crystals (LiF, LiH) but not in general for rare earth ions.

Finally, there are the magnetic spin–spin interactions between electrons. Although their contribution to the linewidth is, in most cases, much smaller than those of g anisotropy and hyperfine couplings, the role of these interactions is essential. As explained in detail in sections **D** and **E** they are responsible for the existence of a single temperature for all the broadening interactions of the electron resonance lumped together into the electronic non-Zeeman reservoir, and also for the coupling of this thermal reservoir to the nuclear Zeeman reservoir.

The present picture of the process of DNP has evolved through several steps; it is instructive to retrace these steps briefly for a better perspective, even though some of them rest on unrealistic assumptions.

In the earliest picture the ESR line was an assembly of independent narrow spin packets with an envelope $h(\omega)$ describing the amplitudes of the individual packets. With microwave power applied at a frequency ω (Fig. 6.6) only the two spin packets of frequencies $\omega+\omega_n$ and $\omega-\omega_n$ would give to the nuclear polarization contributions proportional to $h(\omega+\omega_n)$ and $-h(\omega-\omega_n)$ with a net result of relative value:

$$h(\omega+\omega_n)-h(\omega-\omega_n)\approx 2\omega_n(dh/d\omega) \quad \text{if} \quad \Delta\omega_e \gg \omega_n. \tag{6.76}$$

Although fairly unrealistic, this picture, sometimes called the differential solid effect, held some favour for a while because two of its predictions were

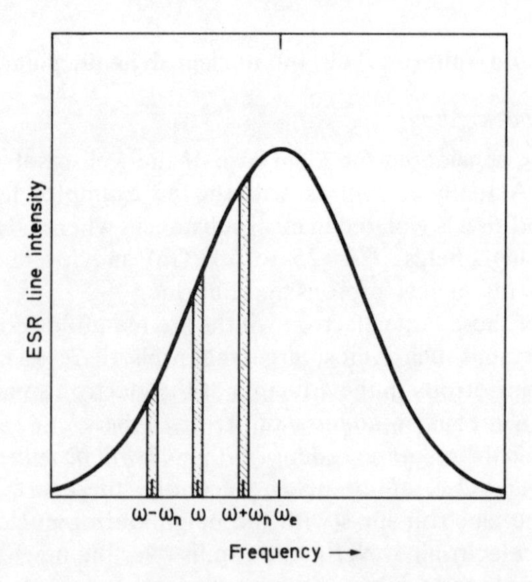

FIG. 6.6. Example of spin packets involved in a differential solid effect, or cross effects, when the ESR line is inhomogeneously broadened.

in agreement with experiment. Firstly, the enhancement of the nuclear polarization measured as a function of the irradiation frequency was an antisymmetrical curve not unlike the derivative of the envelope $h(\omega)$ in agreement with (6.76). Secondly, the enhancements of the nuclear polarizations for two different nuclear species such as, say, ^{7}Li and ^{19}F in LiF turned out to be nearly equal in spite of the large difference of the gyromagnetic factor γ of the two species. This too is in agreement with (6.76). According to this equation the dynamically increased polarization is proportional to ω_n, i.e. to $\gamma_n H_0$. Since the thermal equilibrium polarization is also proportional to $\gamma_n H_0$ (at high temperatures) the enhancement, which is the ratio of the two, is independent of γ_n in agreement with experiment.

Another model, the so-called 'differential cross effect' (Hwang and Hill, 1967), rests on the remark that two electronic spin packets whose frequencies differ by ω_n can perform a flip-flop if a nuclear spin flips with them to conserve energy. If one of the packets is saturated the nuclear spins flip preferentially in one direction inducing a nuclear polarization. It is easy to see in Fig. 6.6 that the net result of the two spin packets of frequencies $(\omega+\omega_n)$ and $(\omega-\omega_n)$ flipping with the saturated packet of frequency ω is again an enhanced polarization proportional to:

$$h(\omega+\omega_n)-h(\omega-\omega_n)\approx 2\omega_n(\mathrm{d}h/\mathrm{d}\omega) \quad \text{if } \Delta\omega_e \gg \omega_n.$$

The weakness of these models is two-fold: the assumption of independent spin packets is unrealistic and contradicted by experiment and these models do not permit a reasonable calculation of the absolute value of the enhanced polarization.

An important step towards a deeper understanding of DNP was the link suggested by Solomon (1963) between DNP and the existence of a spin temperature in the rotating frame, a concept due to Redfield (1955). This concept can be utilized even for the well-resolved solid effect where it is simply the statement in a different language of results already known. In a frame rotating with respect to the spins S at the frequency of the microwave field the effective Hamiltonian is:

$$\mathcal{H}^* = \Delta S_Z + \omega_n I_Z + \omega_1 S_x + \mathcal{H}'^*_{SS}, \tag{6.77}$$

where $\Delta=\omega_S-\omega$, $\omega_1=-\gamma_S H_1$ and $\mathcal{H}'^*_{SS}$ is the non-Zeeman Hamiltonian. The well-resolved solid effect corresponds to $\omega_n \gg \Delta\omega_e$. In that case in the rotating frame, the electronic spins S and the nuclear spins I appear to have Larmor frequencies Δ and ω_n that can be made equal or opposite. Energy conserving flip-flops or flip-flips between them convey to the nuclei a polarization equal or opposite to that of the electrons. These flip-flops are induced by the interplay of the microwave Hamiltonian $\omega_1 S_x$ with the elements $S_Z I_\pm$ of the dipolar electron–nuclear coupling. Their transition probabilities can be seen to be identical with the forbidden transition probabilities $V_{\pm}=C\alpha V$ defined in section **C**, eqn (6.41). When ω_n is not large compared to $\Delta\omega_e$ this picture breaks down. In his thermodynamical approach Solomon, following Redfield, states that the whole of the Hamiltonian $\mathscr{H}^*$ *including* the nuclear Zeeman energy reaches an inverse temperature β_S related to the inverse lattice temperature β_L by a relation of the form:

$$\frac{\beta_S}{\beta_L}=\frac{\omega_S\Delta}{\Delta^2+2\omega_L^2}, \tag{6.78}$$

where $\omega_L^2=-\beta_S^{-1}\langle\mathscr{H}_{nZ}\rangle$ and the factor of two in front of it is the ratio assumed by Solomon for the relaxation rates of $\mathscr{H}'^*_{SS}$ and Z_e. In this picture the microwave field serves a double role:

(i) It 'cools' the electronic spin system in the rotating frame.

(ii) It produces a thermal contact between this 'cold' system and the nuclear spins in the laboratory frame. Solomon's idea was an important breakthrough in the understanding of DNP. Equation (6.78) explained in a natural manner the fact that the nuclear polarization, which for high temperatures ($\beta\omega_n \ll 1$) is proportional to the inverse temperature, measured against the irradiation frequency was an antisymmetrical curve of width comparable to the electronic linewidth. It also explained why two different nuclear spin species present in the sample, being at the same temperature, would exhibit the same polarization enhancement (in the high-temperature limit).

Solomon's theory considered only the so-called limiting Redfield case of very strong r.f. fields. Provotorov (1961) had shown how to deal with r.f. fields of moderate strength where the inverse temperatures α and β of the Zeeman and spin–spin energy were different. Borghini extended this theory to DNP by introducing a third inverse temperature γ for the nuclear Zeeman energy. With α and β being coupled by the allowed transition probability V, and α and γ by the forbidden transition probabilities $V_{\pm}$ he was able to write and solve three rate equations for α, β, γ and to get the dependence of the nuclear polarization on the applied power. These equations and their solutions can be found in Abragam and Borghini (1964). Although basically sound, these equations lack the most important element whose role in nuclear relaxation was stressed in section **E**: the existence of a strong coupling between the electronic spin–spin energy (or more generally non-Zeeman energy) on the one hand and the nuclear Zeeman energy on the other, *and this even in the absence of irradiation*, a point whose significance for dynamic polarization was first realized by Kozhushner and Provotorov (1964) and by Buishvili (1965).

In a model for DNP which takes into account this last feature, the role of the microwave field, which saturates the electron resonance off-centre is to cool the electronic spin–spin (non-Zeeman) reservoir. The latter, through its coupling with the nuclear Zeeman energy, cools and thus polarizes the nuclei.

(b) High temperature case

The inverse temperature β of the electronic non-Zeeman reservoir $\mathcal{H}'^{*}_{SS}$ which is also that of Z_I because of its strong coupling with $\mathcal{H}'^{*}_{SS}$ can be obtained from the Provotorov eqns (1.66), suitably modified to take into account this coupling. We introduce the electronic Zeeman inverse temperature α in the frame rotating at the irradiation frequency, $\omega = \omega_S - \Delta$ as $\alpha = \alpha_Z \omega_S/\Delta$.

The Provotorov equations then read:

$$\frac{d\alpha}{dt} = -W(\alpha - \beta) - \frac{1}{T_{1e}}\left(\alpha - \frac{\omega_S}{\Delta}\beta_L\right),$$

$$\frac{d\beta}{dt} = W\frac{\Delta^2}{D'^2}(\alpha - \beta) - \frac{1}{T_{1n}}\beta. \tag{6.79}$$

In (6.79)

$$D'^2 = [\mathrm{Tr}\{\mathcal{H}'^{*2}_{SS}\} + \omega_I^2\,\mathrm{Tr}\{I_z^2\}]/\mathrm{Tr}\{S_z^2\}$$

$$= D^2 + \omega_I^2 \frac{N_I}{N_S}\frac{I(I+1)}{S(S+1)}, \tag{6.80}$$

where the first term is proportional to $\langle\mathcal{H}'^{*}_{SS}\rangle$ and the second, much larger, to $\langle Z_I\rangle$.

$1/T_{1n}$ the nuclear relaxation rate is given by eqn (6.69):

$$\frac{1}{T_{1n}} \approx \frac{1}{T^*_{1,SS}}\frac{\langle\mathcal{H}'^{*}_{SS}\rangle}{\langle\mathcal{H}'^{*}_{SS}\rangle + \langle Z_I\rangle} + \frac{1}{T'_{1n}} = \frac{1}{T^*_{1,SS}}\frac{D^2}{D'^2}(1+f), \tag{6.81}$$

where we have introduced the leakage coefficient f to take into account nuclear relaxation mechanisms other than the coupling of Z_I with $\mathcal{H}'^{*}_{SS}$.

In the first equation (6.79) β_L is the inverse lattice temperature.

We have introduced in (6.81) the relaxation rate $1/T^*_{1,SS}$ of $\langle\mathcal{H}'^{*}_{SS}\rangle$ which can be defined by the relation:

$$\frac{\langle\mathcal{H}'^{*}_{SS}\rangle}{T^*_{1,SS}} = \frac{\langle\mathcal{H}'_{SS}\rangle}{T_{1,SS}} + \frac{\langle\mathcal{H}'_{IB}\rangle}{T_{1,IB}}. \tag{6.82}$$

If the inhomogeneous broadening $\mathcal{H}'_{IB}$ is of the first type $\mathcal{H}'_{IB} = \sum_i \Delta_i S^i_z$ and $1/T_{1,IB} = 1/T_{1e}$.

The relaxation rate $1/T_{1,SS}$ of the true dipolar energy $\mathcal{H}'_{SS}$ is equal to $1/T_{1e}$ within a factor two or three. For great dilution of the impurities, $\langle\mathcal{H}'_{SS}\rangle$, proportional to C^2, is much smaller than $\langle\mathcal{H}'_{IB}\rangle$, proportional to the concentration C, and we can write with good accuracy:

$$\frac{\langle \mathcal{H}'^*_{SS} \rangle}{T^*_{1,SS}} = \frac{\langle \mathcal{H}'_{IB} \rangle}{T_{1e}}.$$

At high temperatures the conclusions are the same for an inhomogeneous broadening of the second type. Equation (6.81) can then be rewritten:

$$\frac{1}{T_{1n}} \approx \frac{1}{T_{1e}} \frac{CD^2}{\omega_n^2} \frac{S(S+1)}{I(I+1)} (1+f), \tag{6.83}$$

where D^2 is now the second moment of the EPR line.

The Provotorov equations become:

$$\begin{aligned} \frac{d\alpha}{dt} &= -W(\alpha-\beta) - \frac{1}{T_{1e}}\left(\alpha - \frac{\omega_S}{\Delta}\beta_L\right) \\ \frac{d\beta}{dt} &= \frac{CS(S+1)}{I(I+1)} \left\{ \frac{\Delta^2}{\omega_n^2} W(\alpha-\beta) - \frac{\beta}{T_{1e}} \frac{D^2}{\omega_n^2}(1+f) \right\}. \end{aligned} \tag{6.84}$$

For $WT_{1e} \gg 1$, that is for complete saturation, it is found for the enhancement of the nuclear polarization:

$$E = \frac{\beta}{\beta_L} = \frac{\omega_S \Delta}{\Delta^2 + (1+f)D^2}, \tag{6.85}$$

which is very similar to eqn (6.78) proposed by Solomon. Apart from the factor $(1+f)$ in (6.85) which takes into account leakage relaxation, the extra factor 2 in the denominator of (6.78) originates in the choice of $2/T_{1e}$ for the relaxation rate of $\langle \mathcal{H}'_{SS} \rangle$ in (6.78) as opposed to $1/T_{1e}$ for $\langle \mathcal{H}'^*_{SS} \rangle \approx \langle \mathcal{H}'_{IB} \rangle$ in (6.85).

Neither the nuclear frequency nor the impurity concentration appear in the limiting enhancement (6.85). This is misleading in the presence of leakage relaxation since the leakage relaxation rate is by definition independent of C and therefore the leakage coefficient f is proportional to C^{-1}.

On the other hand the polarization rate is, in the absence of leakage (and for $I = S$) given by:

$$\left(\frac{1}{\tau}\right)_{\text{pol}} = \frac{C}{T_{1e}} \left\{ \frac{WT_{1e}}{1+WT_{1e}} \frac{\Delta^2}{\omega_I^2} + \frac{D^2}{\omega_I^2} \right\}. \tag{6.86}$$

For $WT_{1e} \gg 1$

$$\left(\frac{1}{\tau}\right)_{\text{pol}} = \frac{C}{T_{1e}} \frac{\Delta^2 + D^2}{\omega_I^2}. \tag{6.87}$$

Equation (6.87) has a simple interpretation: the polarization rate is equal to the electron relaxation rate, times $C(\Delta^2+D^2)/\omega_I^2 = \langle \mathcal{H}'^*_{SS} + Z^*_S \rangle / \langle Z_I \rangle$ where $Z^*_S = \Delta S_Z$ is the electronic Zeeman Hamiltonian in the rotating frame.

The maximum enhancement is obtained from eqn (6.85):

$$E_{\max} = \frac{\omega_S}{2D(1+f)^{1/2}}. \tag{6.88}$$

In the absence of leakage relaxation it is independent of impurity concentration, nuclear gyromagnetic ratio, and even nuclear spin I.

We shall not pursue further the discussion of the high temperature case, our main interest in this chapter being high nuclear polarizations.

(c) Low temperature case

1. *Inhomogeneous broadening of first type*

There exists no general theory of DNP similar to that outlined in **F**(*b*) and making use of Provotorov's equations, for very low temperatures such that the linear expansion of Boltzmann's exponentials $\exp(-\beta\mathcal{H})$ into $(1-\beta\mathcal{H})$ is invalid.

The root of the problem is the lack of a theory capable to predict the temperature of the non-Zeeman electronic Hamiltonian upon off-centre saturation of the ESR resonance line.

A special solution has been proposed by Borghini (1968) for a simplified model of the ESR line. We shall outline in some detail the derivation and the discussion of the formula arrived at by Borghini.

The assumptions of the model are the following. The ESR line is made of narrow individual spin packets, whose frequencies ω_i differ either because of a spread in the electronic g factors or because of hyperfine coupling with neighbouring nuclei. It is assumed that electronic dipolar interactions bring a negligible contribution to the ESR linewidth. However, they play an important role by inducing between the various spin packets a spectral diffusion that is fast enough to maintain at all times a single spin temperature inside the non-Zeeman Hamiltonian $\mathcal{H}'^*_{SS}$ which describes the distribution of spin packets.

The coupling between $\mathcal{H}'^*_{SS}$ and the nuclear Zeeman energy Z_n is assumed strong enough for $\mathcal{H}'^*_{SS}$ and Z_n to be at all times at the same temperature. Finally a spin–lattice relaxation coupling is assumed between the electronic Zeeman energy Z_e and the lattice.

It turns out that the behaviour of DNP is quite different depending on the type of inhomogeneous broadening (in contrast with the assumptions of Borghini (1968)). Only the first type is considered in this section the second type being deferred till **F**(*c*)2.

The Hamiltonian of the system can be written:

$$\mathcal{H} = \omega_e S_z - \sum_i \Delta_i S^i_z + \omega_n I_z. \tag{6.89}$$

Let the relative weight of the packets of frequency Δ_i (with respect to the central frequency ω_e) be f_i with:

$$\sum_i f_i = 1; \qquad \sum_i f_i \Delta_i = 0. \tag{6.90}$$

A strong microwave field is applied at a frequency $\omega = \omega_e - \Delta_0$ saturating a particular spin packet Δ_0 of statistical weight f_0.

Let P_n and T_{1n} be the nuclear polarization and spin–lattice relaxation time. The assumption of two different spin temperatures α^{-1} and β^{-1} for the electronic Zeeman and non-Zeeman Hamiltonians leads to the expression:

$$P_{ei} = -\tanh[\tfrac{1}{2}(\alpha\omega_e - \beta\Delta_i)], \tag{6.91}$$

for the polarization of each spin packet i. P_{ei} is defined here as $2\langle S_z^i \rangle$. (It is negative at thermal equilibrium because of the negative sign of the electronic moment.) Equation (6.91) is obtained by writing:

$$P_{ei} = \frac{h_i - l_i}{h_i + l_i} = \frac{(h_i/l_i) - 1}{(h_i/l_i) + 1} \quad \text{and} \quad h_i/l_i = \exp - (\alpha\omega_e - \Delta\beta), \tag{6.91'}$$

where h_i and l_i are the populations of the higher and lower electronic energy levels of the spin packet. The electronic non-Zeeman energy is given by:

$$\langle \mathcal{H}'^*_{SS} \rangle = -\tfrac{1}{2} N_e \sum_i f_i \Delta_i P_{ei}, \tag{6.92}$$

where N_e is the number of electronic spins. The nuclear Zeeman energy is given by:

$$E_n = \langle Z_n \rangle = \tfrac{1}{2} N_n \omega_n P_n = \frac{1}{2C} N_e \omega_n P_n, \tag{6.93}$$

where $C = N_e/N_n$ is the relative concentration of paramagnetic impurities. The electronic Zeeman energy $\langle Z_e \rangle$ is given by:

$$\langle Z_e \rangle = \tfrac{1}{2} N_e \omega_e \sum_i f_i P_{ei}. \tag{6.94}$$

The electronic Zeeman energy $\langle Z_e \rangle$ is not on 'speaking terms' with the much smaller energies $\langle Z_n \rangle + \langle \mathcal{H}'^*_{SS} \rangle$ unable to absorb or emit the large electronic quanta $\hbar\omega_e$, and we can write *separate* rate equations for $\langle Z_e \rangle$ and $\langle Z_n \rangle + \langle \mathcal{H}'^*_{SS} \rangle$:

$$\frac{d}{dt}\langle Z_e \rangle = \tfrac{1}{2} N_e \omega_e \sum_i f_i \left(\frac{dP_{ei}}{dt} \right). \tag{6.95}$$

Several effects contribute to the rate of change of the polarization P_{ei}: spin–lattice relaxation, spectral diffusion among the various packets and, for the packet Δ_0, saturation by the microwave field. In the summation (6.95), the spectral diffusion terms cancel out and we obtain:

$$\frac{\mathrm{d}}{\mathrm{d}t}\langle Z_e\rangle = -\tfrac{1}{2}N_e\omega_e\left\{\sum_i \frac{1}{T_{1e}} f_i(P_{ei}-P_0) + Uf_0P_{e0}\right\}, \tag{6.95'}$$

where $P_0 = -\tanh(\frac{1}{2}\beta_L\omega_e)$ is the electronic thermal equilibrium polarization, T_{1e} is the electronic spin–lattice relaxation time, U is the transition probability induced by the microwave field irradiating the spin packet Δ_0, and $P_{e0} \neq P_0$ is the polarization of the saturated packet Δ_0. Similarly:

$$\frac{\mathrm{d}}{\mathrm{d}t}\langle Z_n + \mathscr{H}'^{*}_{SS}\rangle = -\frac{N_eI}{C}\omega_n\frac{P_n}{T_{1n}} + \frac{N_e}{2}\left(\sum_i \frac{f_i\Delta_i}{T_{1e}}(P_{ei}-P_0) + Uf_0\Delta_0P_{e0}\right)$$

$$= -\frac{IN_e}{C}\omega_n\frac{P_n}{T_{1n}} + \frac{N_e}{2}\left(\sum_i \frac{f_i\Delta_i}{T_{1e}}P_{ei} + Uf_0\Delta_0P_{e0}\right) \tag{6.96}$$

since $\sum f_i\Delta_i = 0$.

It should be noted that the relaxation rate $1/T_{1n}$ in (6.96) and onwards is a 'leakage' nuclear relaxation rate, distinct from that due to the electronic spins S. The latter is automatically taken into account by assigning the same temperature to $\langle Z_n\rangle$ and $\langle \mathscr{H}^{*}_{SS}\rangle$.

Under steady-state conditions the right-hand sides of (6.95′) and (6.96) both vanish. Multiplying that of (6.95′) by Δ_0/ω_e and adding it to the right-hand side of (6.96) we get rid of the microwave transition probability U and obtain the relation:

$$\sum_i f_i(\Delta_i-\Delta_0)P_{ei} + \Delta_0P_0 - \frac{2I}{C}\omega_n\frac{T_{1e}}{T_{1n}}P_n = 0, \tag{6.97}$$

where P_{ei} is given by (6.91).

In the limit of very strong microwave power the spin packet Δ_0 is saturated, $P_{e0}=0$, $\alpha\omega_e - \beta\Delta_0 = 0$ and:

$$P_{ei} = -\tanh(\tfrac{1}{2}\beta(\Delta_0-\Delta_i)). \tag{6.98}$$

By carrying (6.98) into (6.97) we obtain Borghini's relation:

$$\sum_i f_i(\Delta_0-\Delta_i)\tanh[\tfrac{1}{2}\beta(\Delta_0-\Delta_i)] = -\Delta_0P_0 + \frac{2I}{C}\omega_n\frac{T_{1e}}{T_{1n}}P_n. \tag{6.99}$$

If we replace the sum $\sum_i$ by an integral $\int \mathrm{d}\Delta$, equation (6.99) can be rewritten:

$$-\Delta_0P_0 + \frac{2I}{C}\omega_n\frac{T_{1e}}{T_{1n}}\mathrm{P}_n = \int_{-\infty}^{\infty}(\Delta_0-\Delta)f(\Delta)\tanh[\tfrac{1}{2}\beta(\Delta_0-\Delta)]\,\mathrm{d}\Delta, \tag{6.100}$$

or writing $|P_0| = -P_0$:

$$\Delta_0|P_0| + \frac{2I}{C}\omega_n\frac{T_{1e}}{T_{1n}}P_n = \int_{-\infty}^{\infty}\tanh(\tfrac{1}{2}\beta\Delta)\Delta f(\Delta+\Delta_0)\,d\Delta. \tag{6.101}$$

In most cases $(T_{1e}/CT_{1n}) \ll 1$ (moderate leakage) and can be neglected, leaving:

$$\Delta_0|P_0| = \int_{-\infty}^{\infty}\tanh(\tfrac{1}{2}\beta\Delta)\,\Delta f(\Delta+\Delta_0)\,d\Delta. \tag{6.102}$$

Equation (6.102) is an integral equation for β that can be solved numerically if the distribution function $f(\Delta)$, which is nothing but the shape of the unsaturated EPR line, is known. One can note that in this model, in the absence of leakage, the limiting inverse temperature β and the optimum saturation frequency Δ_0 which maximizes β, depend only on the EPR lineshape $f(\Delta)$. Neither the impurity concentration C, nor their relaxation rate $1/T_{1e}$, nor the nuclear frequency ω_n appear in eqn (6.102). On the other hand all these parameters have a strong influence on the polarization rate as for the high temperature case (eqn (6.87)).

If the equilibrium polarization $|P_0| = \tanh(\frac{1}{2}\beta_L\omega_e)$ is small (6.102) should reduce to equation (6.85) derived for the high-temperature case. We can write $|P_0| \simeq \frac{1}{2}\beta_L\omega_e$, $\tanh(\frac{1}{2}\beta\Delta) \simeq \frac{1}{2}\beta\Delta$ whence, changing $(\Delta+\Delta_0)$ into Δ in (6.102):

$$\begin{aligned}\beta &= \beta_L\omega_e\Delta_0\left[\int_{-\infty}^{\infty}(\Delta-\Delta_0)^2 f(\Delta)\,d\Delta\right]^{-1}\\ &= \beta_L\omega_e\Delta_0(\Delta_0^2+D^2)^{-1},\end{aligned} \tag{6.103}$$

where:

$$D^2 = \int_{-\infty}^{+\infty}\Delta^2 f(\Delta)\,d\Delta, \tag{6.104}$$

a formula identical to (6.85) if we take the leakage factor f equal to zero. (Remember $\int \Delta f(\Delta)\,d\Delta = \sum f_i\Delta_i = 0$.)

As Δ_0, the position of the microwave frequency with respect to the Larmor frequency ω_e, goes from zero to infinity, we expect, in analogy with the high-temperature case of eqn (6.85) that $|\beta|$ will go through a maximum.

Indeed, if Δ_0 is small, i.e. $\Delta_0 \ll D$, the mean square width of $f(\Delta)$, we expect β as given by (6.102) to be small, since the left-hand side $\Delta_0|P_0|$ is small. Replacing in (6.102) $\tanh(\frac{1}{2}\beta\Delta)$ by $\frac{1}{2}\beta\Delta$ we find:

$$\beta = 2\Delta_0|P_0|(\Delta_0^2+D^2)^{-1}$$

which is an *increasing* function of Δ_0 when $\Delta_0 \ll D$. On the other hand, if $\Delta_0 \gg D$ the product $\tanh(\frac{1}{2}\beta\Delta)\Delta f(\Delta_0+\Delta)$ will be significantly different from zero only for $\Delta \simeq -\Delta_0$ and with reasonable accuracy we can write:

$$\int_{-\infty}^{\infty} \tanh(\tfrac{1}{2}\beta\Delta)\Delta f(\Delta_0+\Delta)\,\mathrm{d}\Delta \approx \Delta_0 \tanh(\tfrac{1}{2}\beta\Delta_0),$$

whence:

$$\beta = \frac{2}{\Delta_0}\tanh^{-1}(|P_0|) = \frac{2}{\Delta_0}\times\tfrac{1}{2}\beta_L\omega_e;$$

$$\beta = \beta_L\frac{\omega_e}{\Delta_0}; \tag{6.105}$$

which is a *decreasing* function of Δ_0. It is thus reasonable to expect β to go through a maximum for an intermediate value of Δ_0.

As soon as $\tfrac{1}{2}\beta_L\omega_e \gg 0$, $|P_0| = \tanh(\tfrac{1}{2}\beta_L\omega_e)$ is very nearly unity and it would seem that if $\varepsilon = 1-|P_0|$ is, say, smaller than 1 per cent, there is nothing to be gained for the nuclear polarization by lowering further the lattice temperature β_L^{-1} or increasing the Larmor frequency ω_e. That this is not so can be seen rather simply if we take $\Delta_0 \gg D$. The inverse temperature is then given by (6.105) and the nuclear polarization is:

$$P_n = -\tanh(\tfrac{1}{2}\beta\omega_n) = -\tanh\left(\tfrac{1}{2}\beta_L\omega_e\frac{\omega_n}{\Delta_0}\right). \tag{6.106}$$

If $\omega_n \leqslant D$, $(\omega_n/\Delta_0) \ll 1$, $\tfrac{1}{2}\beta_L\omega_e(\omega_n/\Delta_0)$ is not necessarily large and there may be a definite advantage in increasing β_L further even though the electronic equilibrium polarization is practically unchanged. A numerical calculation to be given in **F**(c)3 confirms this statement in greater detail.

2. *Inhomogeneous broadening of second type*

This case is more complicated to deal with than the previous one, for it requires not only the knowledge of the ESR lineshape but also that of the various hyperfine couplings responsible for that shape. Several features of DNP are qualitatively different in the two cases.

The Hamiltonian of the system can be written as:

$$\mathcal{H} = \omega_e S_z + \sum_{\mu\nu} A_{\mu\nu} S_z^\nu I_z^\mu + \omega_n I_z, \tag{6.107}$$

where the $A_{\mu\nu}$ are the various hyperfine coupling constants responsible for the ESR broadening. Because of the low concentration of impurities no nucleus will interact effectively with more than one electron.

The shift $\Delta_i = \omega_e - \omega_i$ in the frequency of an electronic spin is obtained from the configuration $\{m^\mu\}$ of the quantum numbers $m_\mu = I_z^\mu$ of the nuclei that surround it by:

$$\Delta(\{m^\mu\}) = -\sum_\mu A_\mu m_\mu = \Delta_i. \tag{6.108}$$

(We have dropped the unnecessary index ν in the HFS coupling $A_{\mu\nu}$.) The subset of all configurations $\{m^\mu\}$ such that $\Delta(\{m^\mu\}) = \Delta_i$ is the analogue of the spin packet Δ_i considered previously. Its statistical weight $f(\Delta_i)$ can be obtained experimentally from the shape of the unsaturated ESR line with zero nuclear polarization, or else computed if all hyperfine interactions are known, say from Endor experiments.

The main qualitative difference with **F**(c)1 is that the populations of the hyperfine spin packets are not bound to be constant and can vary in the course of the DNP. Indeed, as the nuclei close to impurities become polarized (as well as the distant nuclei) the hyperfine spin packets produced by configurations with high nuclear polarization become more populated at the expense of those with low nuclear polarization.

The formal treatment is based on the same assumptions as in **F**(c)1: we assume that the state of the system can be described by two inverse temperatures, α for the electronic Zeeman interaction and β common to the electronic non-Zeeman and the nuclear Zeeman interactions; we saturate the electronic spins of resonance frequency $\omega_e - \Delta_0$; we set equal to zero the rates of change of the energies of the two reservoirs, electronic Zeeman and electronic non-Zeeman plus nuclear Zeeman.

As the simplest possible example we consider the case when each electronic spin is surrounded by z nuclear spins I coupled to it by identical scalar couplings:

$$A_\mu = A = \frac{8\pi}{3} \hbar\gamma_n|\gamma_e| |\psi(r_\mu)|^2. \tag{6.109}$$

In that particular case, each set of configurations with,

$$\sum_\mu m_\mu = M,$$

corresponds to a spin packet with frequency shift,

$$\Delta_i = -AM. \tag{6.110}$$

Its statistical weight $f(\Delta_i)$ is proportional to the coefficient of x^M in the expansion of $(x^I + x^{I-1} + \cdots + x^{-I})^z$. The distribution of these coefficients approximates a Gaussian shape with a cut off at $M = \pm zI$.

The *number* of electronic spins in a given packet Δ_i depends on the inverse temperatures α and β of the electronic Zeeman and non-Zeeman reservoirs and so changes in the course of DNP. To use the notations of eqn (6.91′), the *sums*:

$$n_i = n(\Delta_i) = h(\Delta_i) + l(\Delta_i), \tag{6.111}$$

of the upper and lower populations of the electronic spins in the packet are functions of Δ_i instead of being equal to unity, as in the former case, when only the polarizations P_{ei} given by (6.91) were allowed to vary from packet to packet. On the other hand the relaxation mechanisms of both $\langle Z_e \rangle$ and $\langle \mathcal{H}'^{*}_{SS} \rangle$ are the same as before. We replace in (6.95) and (6.96), f_i by $f_i n_i$ and the normalization condition (6.90) by:

$$\sum_i f(\Delta_i) = 1; \qquad \sum_i f(\Delta_i) n(\Delta_i) = 1; \qquad \sum_i f(\Delta_i) \Delta_i n(\Delta_i) \neq 0. \quad (6.112)$$

The combination of the eqns (6.95) and (6.96) thus modified, yields:

$$\sum_i f(\Delta_i) n(\Delta_i)(\Delta_i - \Delta_0)(P_{ei} - P_0) = \frac{2I}{C} \omega_n \frac{T_{1e}}{T_{1n}} P_n, \quad (6.113)$$

or for strong microwave saturation, when as before: $P_{ei} = \tanh(\beta/2)(\Delta_i - \Delta_0)$;

$$\sum_i f(\Delta_i) n(\Delta_i)(\Delta_i - \Delta_0)\left\{\tanh\left\{\frac{\beta}{2}(\Delta_i - \Delta_0)\right\} - P_0\right\} = \frac{2I}{C} \omega_n \frac{T_{1e}}{T_{1n}} P_n. \quad (6.114)$$

Equation (6.114) has a form different from (6.99): because of the last relation (6.112) we may not replace

$$\sum_i f(\Delta_i) n(\Delta_i)(\Delta_0 - \Delta_i) P_0 \quad \text{by } \Delta_0 P_0 \text{ in (6.114).}$$

In order to be able to extract β from (6.114) we must express $n_i = n(\Delta_i)$ as a function of β (for complete saturation $\Delta_0 \beta = \omega_e \alpha$).

$E_h(\Delta_i)$ and $E_l(\Delta_i)$, the higher and lower energy levels of the spin packet Δ_i are given by:

$$\begin{aligned} E_h(\Delta_i) &= (\omega_e - \Delta_i)/2 + M\omega_n = \frac{\omega_e}{2} - \left(\frac{\omega_n}{A} + \frac{1}{2}\right)\Delta_i; \\ E_l(\Delta_i) &= -(\omega_e - \Delta_i)/2 + M\omega_n = -\frac{\omega_e}{2} - \left(\frac{\omega_n}{A} - \frac{1}{2}\right)\Delta_i. \end{aligned} \quad (6.115)$$

Their populations are:

$$\begin{aligned} h(\Delta_i) &= \xi \exp\left(-\frac{\alpha\omega_e}{2}\right) \exp\left[\beta\left(\frac{\omega_n}{A} + \frac{1}{2}\right)\Delta_i\right]; \\ l(\Delta_i) &= \xi \exp\left(\frac{\alpha\omega_e}{2}\right) \exp\left[\beta\left(\frac{\omega_n}{A} - \frac{1}{2}\right)\Delta_i\right]; \end{aligned} \quad (6.116)$$

whence, using $\omega_e \alpha = \Delta_0 \beta$:

$$n(\Delta_i) = h(\Delta_i) + l(\Delta_i) = 2\xi \exp\left(\beta \frac{\Delta_i \omega_n}{A}\right) \cosh\left\{\frac{\beta}{2}(\Delta_i - \Delta_0)\right\}. \quad (6.117)$$

The constant ξ is obtained from $\sum_i f(\Delta_i)n(\Delta_i)=1$ or:

$$\xi^{-1}=2\int f(\Delta)\exp\left(\frac{\beta\Delta\omega_n}{A}\right)\cosh\left\{\frac{\beta}{2}(\Delta-\Delta_0)\right\}d\Delta. \tag{6.117'}$$

The dependence of $n(\Delta)$ on β has the following consequences. If Δ_0 is, say, positive, i.e. if the microwave power is applied *below* the central frequency ω_e, $\beta=\omega_e\alpha/\Delta_0$ is positive. In the present model the ratio,

$$\omega_n/A=-\gamma_n H_0\left(\frac{8\pi}{3}|\gamma_e|\gamma_n|\psi|^2\right)^{-1},$$

is *negative* and $\beta\Delta_i\omega_n/A$ is *positive* for Δ_i *negative*, i.e. for spin packets with resonance frequencies *greater* than ω_e. These packets, according to eqn (6.117) are thus more populated than those *below* ω_e. The ESR line is shifted away from the saturating microwave or to use a more familiar language: it slips away when one treads on its tail.

A consequence of this slipping away of the ESR line is that for optimum DNP the irradiation frequency must not be kept constant but must rather be varied so as to follow the line in the course of polarization. Suppose for instance that the DNP is started with Δ_0 positive, i.e. with an irradiation frequency below ω_e. The nuclear polarization will then grow positive, the ESR will slip away and Δ_0 must be decreased in order to follow it. It can even be made negative and still increase the positive nuclear polarization (roughly as long as the irradiation frequency is lower than the centre of gravity of the shifted ESR line). These arguments, physically plausible, are substantiated by numerical calculations on a simple model.

For high temperatures (6.117′) yields $\xi=\frac{1}{2}+0(\beta^2)$ and (6.117): $n(\Delta_i)\sim 1-\beta(\omega_n/A)\Delta_i$. This carried into (6.113) yields to the lowest order in β the same result as for broadening of the first type and explains why no distinction between these two types of broadening was made in section **F**(*b*).

A similar shift of populations can be observed on the *unsaturated* ESR signal when the nuclear polarization has a sizeable value. After suppressing the microwave power used for DNP, the electronic Zeeman energy snaps back to its thermal equilibrium value at the lattice temperature, because of the short electronic relaxation time. However, the relaxation time of the combined system, non-Zeeman plus nuclear Zeeman, is much longer and its energy keeps for a long time the value achieved by the DNP.

As explained in chapter 5, $\langle Z_e\rangle$ and $\langle\mathcal{H}'^{*}_{SS}\rangle$ depend on both inverse temperatures α and β. On the other hand $\langle Z_I\rangle$ depends only on β and since $\langle Z_I\rangle\gg\langle\mathcal{H}'^{*}_{SS}\rangle$, the constancy of $\langle Z_I\rangle+\langle\mathcal{H}'^{*}_{SS}\rangle$ results in the constancy of the inverse temperature β, equal to that achieved by the DNP. The total system is thus in a metastable state where $\langle Z_e\rangle$ is in thermal equilibrium with the lattice whilst $\langle Z_I\rangle$ keeps the inverse temperature β. The value α_0 corresponding to this metastable state is obtained from the condition:

$$\bar{P}_e = \sum_i f(\Delta_i) n(\Delta_i) P_e(\Delta_i) = P_0.$$

According to (6.116) and the second condition (6.112), the polarization and the population of the packet Δ_i are:

$$P_e(\Delta_i) = \tanh\left(\frac{\beta}{2}\Delta_i - \frac{\alpha}{2}\omega_e\right),$$

and, (for our special model of broadening of the second type):

$$n(\Delta_i) = \frac{\exp\left(\beta \frac{\omega_n}{A}\Delta_i\right)\cosh\left(\frac{\beta}{2}\Delta_i - \frac{\alpha}{2}\omega_e\right)}{\sum_i f(\Delta_i)\exp\left(\beta \frac{\omega_n}{A}\Delta_i\right)\cosh\left(\frac{\beta}{2}\Delta_i - \frac{\alpha}{2}\omega_e\right)}.$$

If we call $\bar{P}_e(\alpha, \beta)$ the average polarization,

$$\bar{P}_e(\alpha, \beta) = \sum_i f(\Delta_i) n(\Delta_i, \alpha, \beta) P_e(\Delta_i, \alpha, \beta),$$

the condition for metastable equilibrium is:

$$\bar{P}_e(\alpha_0, \beta) = P_0 = \tanh\left(-\beta_L \frac{\omega_e}{2}\right),$$

or else:

$$\bar{P}_e(\alpha_0, \beta) = \bar{P}_e(\beta_L, 0). \tag{6.118}$$

In general, the steady-state value α_0 is *not* equal to the inverse lattice temperature β_L, and the polarizations P_{ei} of the different packets are not equal to P_0: they differ from one another and it is only the average polarization $\bar{P}_e(\alpha_0, \beta)$ which is equal to P_0.

The ESR signal at Δ_i, proportional to:

$$f(\Delta_i) n(\Delta_i) P_e(\Delta_i),$$

depends on the nuclear polarization, through the dependence of $n(\Delta_i)$ and $P_e(\Delta_i)$ on β, both explicitly and implicitly through $\alpha_0 = \alpha_0(\beta_L, \beta)$.

Take, for instance, the case when $\beta_L\omega_e \gg 1$, i.e. $P_0 \simeq -1$; the populations of the electronic levels of higher energy are then practically zero, the polarizations $P_e(\Delta_i) \simeq -1$, and the populations of the packets are approximately $n(\Delta_i) = l(\Delta_i)$. In the present model this yields:

$$n(\Delta_i) \propto \exp\left[\beta \Delta_i \left(\frac{\omega_n}{A} - \frac{1}{2}\right)\right],$$

and the ESR signal at Δ_i is proportional to:

$$f(\Delta_i)n(\Delta_i)P_e(\Delta_i) \propto f(\Delta_i)n(\Delta_i).$$

The ratio ω_n/A being negative, it is the packets with Δ_i negative, i.e. with resonance frequencies *above* ω_e which have large populations when β is positive. The ESR slips away in the same direction as in the presence of a saturating microwave.

The general case when all HFS couplings are not equal is fairly complicated and we shall not discuss it here.

In contrast to the simple case considered above, its treatment requires a knowledge of all the hyperfine couplings, that is detailed study and a complete interpretation of the Endor spectrum of the electronic spins.

(d) Numerical results

In order to illustrate the behaviour of DNP described in the previous section we present some results obtained for a simple model.

In the case of hyperfine broadening (case (b)) we assume that each electron spin is surrounded by six neighbouring nuclear spins $I = \frac{3}{2}$ with identical scalar couplings A. We assume for definiteness sake $A > 0$ and therefore according to (6.109)$\gamma_n > 0$ and $\omega_n = -\gamma_n H_0 < 0$. We take in the numerical calculation $|\omega_n| = 3A/2$. The unsaturated ESR spectrum for zero nuclear polarization is a set of 19 discrete lines that we replace by a continuous curve drawn between the lines, which is very nearly a truncated Gaussian (Fig. 6.7). Its half-width $\Delta\omega_e$ at half-intensity is very nearly $3A$ and $|\omega_n| \simeq \frac{1}{2}\Delta\omega_e$.

For the case of a broadening due to a distribution of g factors (case (a)) we assume somewhat unrealistically the *same* lineshape for the ESR line. This choice is made in order to exhibit clearly the difference in the behaviour of DNP in the two cases as due to the different dynamics of the DNP process rather than to a difference in lineshape.

We present the results for case (a) first. Figure 6.8 is a plot of the inverse temperature β as a function of the irradiation frequency. It is obtained by solving numerically eqn (6.102) in the case when T_{1n} is infinite. Table 6.1 shows for three values of P_0 the values of $\beta_L\omega_e$, $(\beta\Delta\omega_e)_{max}$ and $(P_n)_{max}$. These results are in agreement with the qualitative discussion following eqn (6.102), in particular with the fact that there is an appreciable increase in the limiting nuclear polarization when the electronic polarization varies by barely 1 per cent.

For case (b) the inverse temperature β calculated numerically from eqn (6.114) is plotted in Figs 6.9 and 6.10 for various values of P_0. We have again assumed $T_{1n}^{-1} = 0$. These figures exhibit some remarkable features of DNP in case (b). Whilst for small values of P_0, these plots are very similar to those represented in Fig. 6.8 for case (a), their behaviour changes drastically as P_0

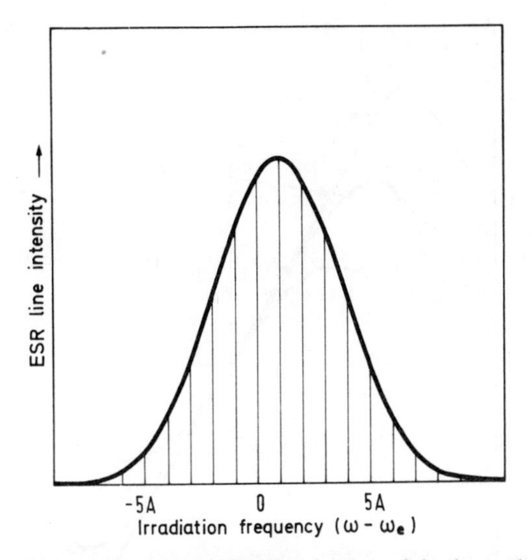

FIG. 6.7. Shape of an unsaturated ESR line in a model where the broadening is caused by equal scalar hyperfine couplings with six nuclear neighbours of spins $\frac{3}{2}$. The smooth curve is the envelope of the 19 discrete lines.

increases, and exhibits a discontinuity for $\Delta_0 = \omega_e - \omega = 0$: an irradiation at a distance Δ_0 from resonance, infinitely small, leads to a *finite* value for the limiting value of β, of the same sign as Δ_0.

The threshold value of P_0 for which this discontinuity appears can be obtained from eqn (6.114), which for $T_{1n}^{-1} = 0$ can be rewritten as:

$$P_0 = \frac{\sum_i f(\Delta_i) n(\Delta_i)(\Delta_i - \Delta_0) \tanh\{(\beta/2)(\Delta_i - \Delta_0)\}}{\sum_i f(\Delta_i) n(\Delta_i)(\Delta_i - \Delta_0)}. \tag{6.119}$$

We replace in (6.119) $n(\Delta_i)$ by its expression (6.117) which yields:

$$P_0 = \frac{\sum_i f((\Delta_i) \sinh\{(\beta/2)(\Delta_i - \Delta_0)\}(\Delta_i - \Delta_0) \exp(\beta(\omega_n/A)\Delta_i)}{\sum_i f(\Delta_i) \cosh\{(\beta/2)(\Delta_i - \Delta_0)\}(\Delta_i - \Delta_0) \exp(\beta(\omega_n/A)\Delta_i)}, \tag{6.120}$$

Table 6.1.
DNP for case (a) according to Fig. 6.8

P_0	$\beta_L\omega_e$	$(\beta\Delta\omega_e)_{max}$	$(P_n)_{max}$
0·9	2·94	2·04	0·62
0·99	5·29	3·85	0·85
0·999	7·60	5·84	0·95

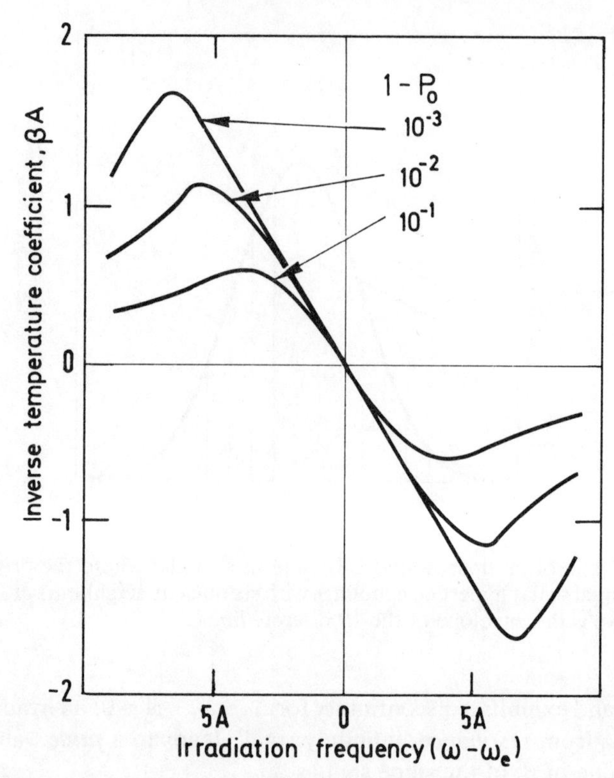

FIG. 6.8. Theoretical variation of the inverse temperature of the non-Zeeman plus nuclear Zeeman reservoir against irradiation frequency for various values of the electronic thermal equilibrium polarization. The ESR shape, identical with that of Fig. 6.7, arises from a distribution of g values. (After Abragam *et al.* 1978.)

and multiplying numerator and denominator by $\exp(-\beta(\omega_n/A)\Delta_0)$

$$P_0 = \frac{\int f(\Delta)(\Delta-\Delta_0)\sinh\{(\beta/2)(\Delta-\Delta_0)\}\exp\{\beta(\omega_n/A)(\Delta-\Delta_0)\}\,d\Delta}{\int f(\Delta)(\Delta-\Delta_0)\cosh\{(\beta/2)(\Delta-\Delta_0)\}\exp\{\beta(\omega_n/A)(\Delta-\Delta_0)\}\,d\Delta}. \tag{6.121}$$

For small β (6.121) reduces to:

$$P_0 = \frac{\frac{\beta}{2}(\Delta_0^2+D^2)}{-\Delta_0+\beta\frac{\omega_n}{A}(\Delta_0^2+D^2)} \quad \text{where } D^2 = \int f(\Delta)\Delta^2\,d\Delta, \tag{6.122}$$

or

$$\beta = \frac{-2P_0\Delta_0}{(\Delta_0^2+D^2)\{1-2P_0(\omega_n/A)\}}. \tag{6.123}$$

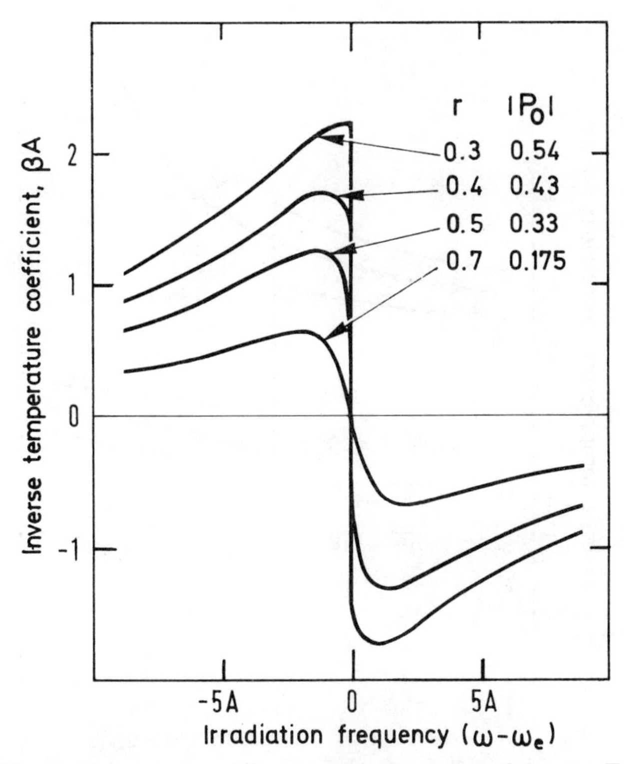

FIG. 6.9. Theoretical variation of the inverse temperature of the non-Zeeman plus nuclear Zeeman reservoir against irradiation frequency for various values of the electronic Boltzmann ratio r. The ESR line is broadened by hyperfine interactions (Fig. 6.7). (After Abragam *et al.* 1978.)

When P_0 is small, eqn (6.123) is equivalent to the high lattice temperature eqn (6.85) (with zero leakage) and the limiting inverse nuclear temperature β is vanishingly small if the irradiation is performed for Δ_0 very small that is very near the electronic Larmor frequency. However if $|P_0|$ is increased beyond the value $P_0 = A/2\omega_n = -1/3$ in our example, it can be shown that the solution approximated by (6.123) for $|\Delta_0|$ small becomes unstable and two stable solutions appear, as represented on Figs. 6.9 and 6.10.

It is important to understand the physics of this problem where one cannot separate the study of the steady state and of the dynamics of the process. The important fact is that as the nuclear inverse temperature and the nuclear polarization increase, the shape and even the position of the ESR line change. Instead of being described by the shape function $f(\Delta)$ for $\beta = 0$ it is described during the irradiation by $f(\Delta)n(\Delta)P(\Delta)$ that is:

$$2\xi f(\Delta)\exp\left(\beta\frac{\Delta\omega_n}{A}\right)\sinh\frac{\beta}{2}(\Delta - \Delta_0). \tag{6.124}$$

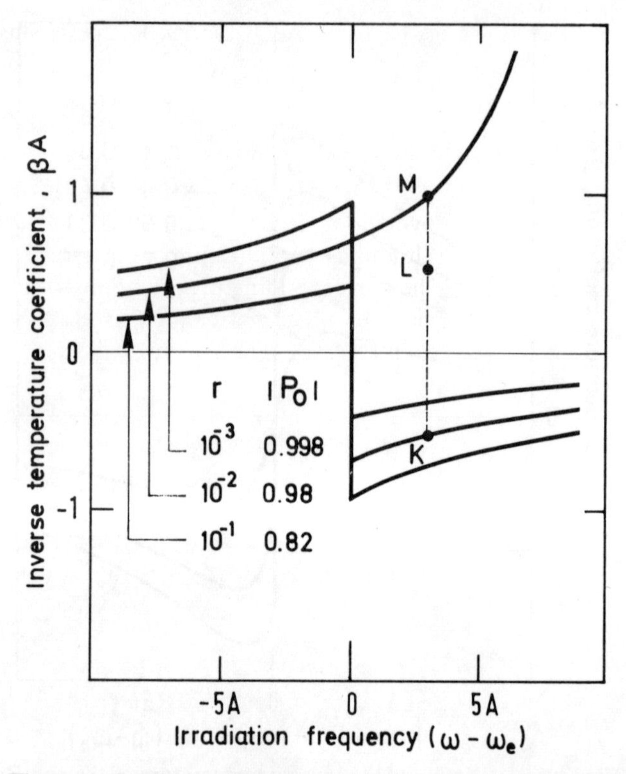

FIG. 6.10. Theoretical variation of the inverse temperature of the non-Zeeman plus nuclear Zeeman reservoir against irradiation frequency for Boltzmann ratios r smaller than in Fig. 6.9. The branch of positive values of βA, for $r = 10^{-2}$ and $\omega - \omega_e > 0$ is reached only by 'pushing' the line in the course of DNP. For an irradiation at a frequency corresponding to the broken line, the inverse temperature coefficient evolves toward K or M, depending on whether its initial value lies between L and K or between L and M. (After Abragam *et al.* 1978.)

If a high inverse temperature β_f is thus compatible with an irradiation at $\Delta_0 = 0$ as shown in Fig. 6.9, provided the equilibrium electron polarization is sufficiently high, the physical reason for it is the following: for $\beta = \beta_f$ the ESR lineshape $n(\beta, \Delta)f(\Delta)P(\Delta)$ is not centred at ω_S and an irradiation at $\Delta_0 = 0$ is actually an off-*centre* irradiation able to cool the electronic non-Zeeman energy including the nuclear spins. On the other hand, if we *start* irradiating at $\Delta_0 = 0$, the nuclear spins being unpolarized, β will *not* grow since we irradiate at the centre of the line where the rate of growth is zero. This means that the steady state value β_f, for a given lattice temperature, depends on the past history of the irradiation. As shown in Fig. 6.10 two equilibrium nuclear temperatures of opposite sign are compatible with an irradiation at the same position. This possibility of maintaining one or the other of two different steady-state values β_{f} under the same steady-state irradiation conditions is a characteristic of many non-linear problems and is by no means limited to high nuclear polarization. It is clear that if the ESR line is going to distort and slip away, the optimum irradiation position Δ_0 at the beginning of DNP will have to change during the process. As explained earlier the line slips away as one treads on its tail. It should be pursued carefully in order not to overshoot and to get on the side where the polarization changes signs. Although a computation of the time-dependence of the phenomenon and of the corresponding programming of the irradiation frequency is by no means unfeasible, in practice the adjustment of the frequency $\Delta_0(t)$ is performed empirically. The limiting values β_{f} obtainable in this manner are shown in Fig. 6.10.

We have given in (6.124) the shape of the ESR line during irradiation. The shape of the unsaturated ESR line when β is large but α has an equilibrium value α_0 is obtained by replacing $\beta\Delta_0$ by $\alpha_0\omega_S$ in (6.124):

$$2\xi f(\Delta)\exp\left(\beta\frac{\omega_{\mathrm{n}}\Delta}{A}\right)\sinh\left(\frac{\beta\Delta-\alpha_0\omega_S}{2}\right). \qquad (6.125)$$

Figure 6.11 shows the shapes computed for several values of β, for the unperturbed lineshape of Fig. 6.6.

(*e*) *Experimental results*

In this section we give a few experimental results, as an illustration of the theory relative to the thermal contact between the electronic spin–spin reservoir and the nuclear Zeeman energy. A few have already been given in section **E**(*b*).

(a) An experimental proof of the existence of a strong thermal coupling between nuclear Zeeman and electronic non-Zeeman energies results from a comparison between the polarizations of several nuclear species present in the same sample throughout the whole duration of the DNP process. The nuclear species were ^{1}H, ^{2}D, and ^{13}C in a sample of deuterated ethanediol (De Boer, Borghini, Morimoto, and Niinikoski, 1974). The ratios of their polarizations do not remain constant through the course of the DNP. However, the Zeeman spin temperatures deduced for each species from the measured polarizations turn out to be the same all through the DNP. This is a natural consequence of the existence between the nuclear species of a strong coupling mediated by the electronic non-Zeeman energy. This is illustrated in Fig. 6.12 for H and D.

(b) As another illustration of the spin-temperature theory of DNP, Fig. 6.13 displays the maximum polarizations of protons and deuterons, obtained in a field of 25 kOe in propanediol, doped with complexes of Cr, plotted against the lattice temperature (De Boer and Niinikoski, 1974; De

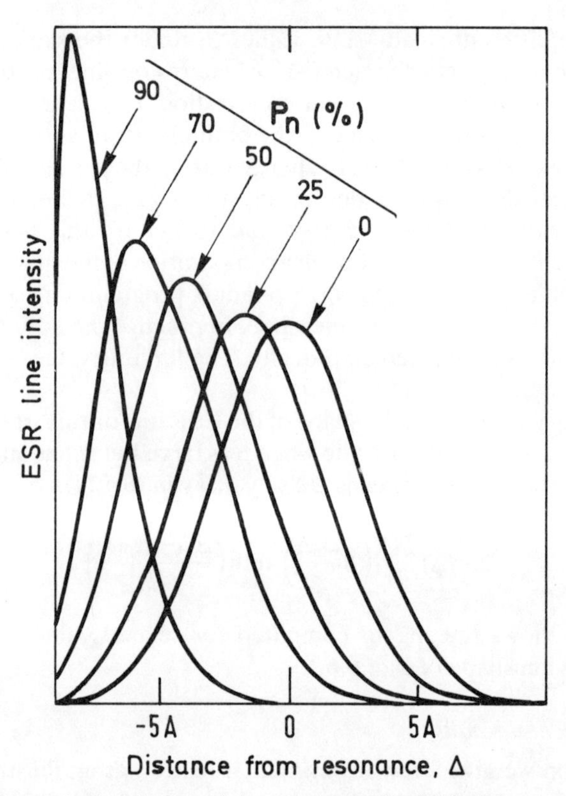

FIG. 6.11. Theoretical variation of the unsaturated ESR lineshape against nuclear polarization in a model system: the electronic spins have a scalar coupling A with six nuclei of spin $\frac{3}{2}$ and the nuclear Larmor frequency is $\omega_n = \frac{3}{2}A$. (After Abragam *et al.* 1978.)

Boer, 1974*). The linewidth of the order of 400 MHz is due to the anisotropy of the g factor. The full curves are theoretical results, computed from eqn (6.101) where $f(\Delta)$ is the experimental ESR lineshape. The difference between maximum positive and negative polarizations results from the asymmetry of the ESR line.

The agreement with experiment is remarkably good and underlines the importance of a lattice temperature as low as possible, even if the corresponding changes in the electronic polarization P_0 seem negligible. (This was already pointed out in connection with the numerical solution for a model lineshape.) As T_L goes from 0·8 to 0·5K, P_0 varies only from 0·97 to 0·998 but for protons the measured values of $(P_n)_{max}$ vary from 0·75 to unity (within experimental error). The increase in $(P_n)_{max}$ is even more spectacular for deuterons which have a smaller Larmor frequency.

* W. De Boer. Thesis, University of Technology, Delft, The Netherlands (CERN 74-11).

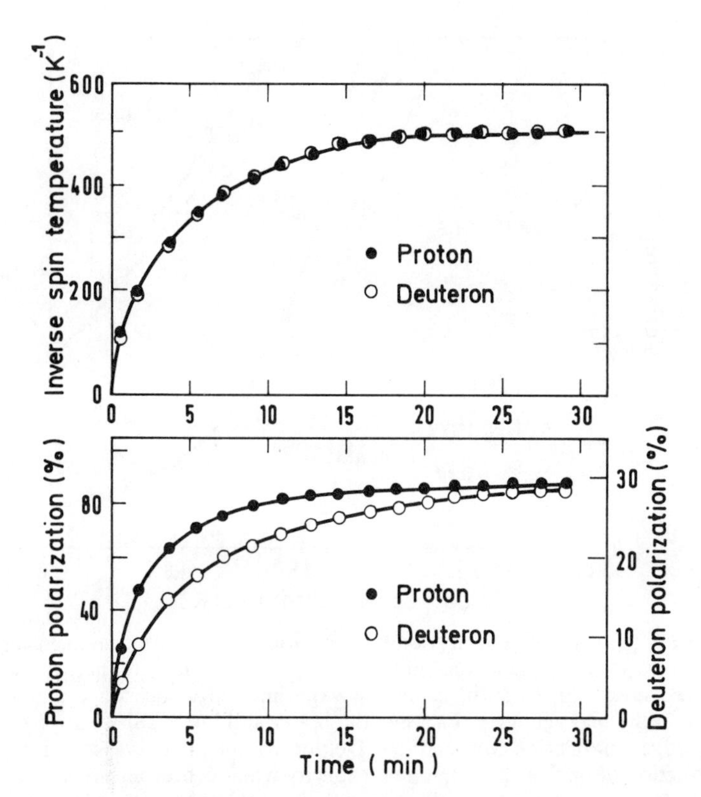

FIG. 6.12. Dynamic polarization of protons and deuterons in ethanediol. Lower curve: variation of the polarizations. Upper curve: variation of the inverse spin temperatures. (After De Boer *et al.* 1974.)

(c) The theory of DNP for ESR lines broadened by hyperfine interactions outlined in section **F**(c)2, which predicts a displacement and a deformation of the ESR line when the nuclear polarization grows, is on this point in qualitative agreement with experimental results obtained in LiH where it is known from Endor studies that the ESR width of F-centres results from coupling with neighbouring nuclei (Roinel and Bouffard, 1976, 1977). This is shown in Fig. 6.14. The behaviour of these curves is comparable to that of the theoretical curves of Fig. 6.11. It should be noted that the experimental electronic lineshape of Fig. 6.14 has not been obtained by ESR but a technique named Nedor (nuclear–electron double resonance) to be described in **G**(a).

(d) DNP through thermal contact between electronic non-Zeeman and nuclear Zeeman reservoirs may occur even when the conditions for a well-separated solid effect exist, i.e. when $\omega_n \gg \Delta\omega_e$. This is observed with a free radical, BDPA, dissolved in m-xylene (Borghini, De Boer, and Mori-

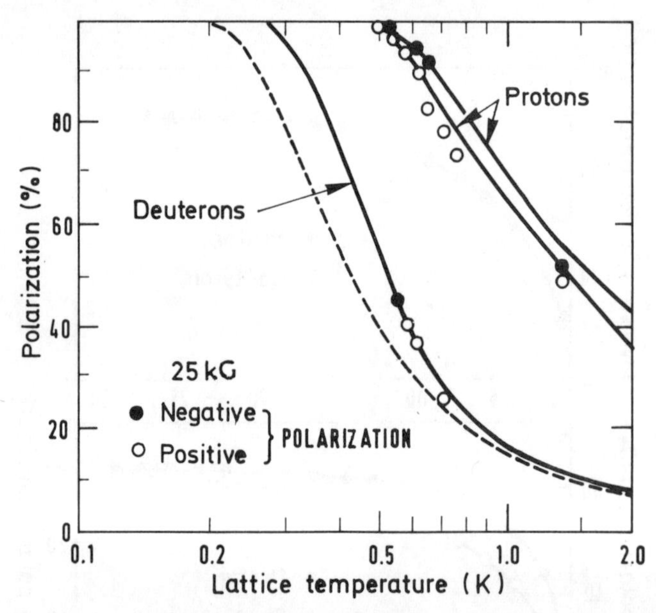

FIG. 6.13. Maximum polarizations of protons and deuterons against lattice temperature in 1,2-propanediol doped with CrV complexes. The full curves are calculated after eqn (6.100), using the experimental values of T_{1e} and T_{1n} for computing the leakage term. Protons: the upper and lower curves are for negative and positive polarizations respectively. Deuterons: the full curve takes into account the reduction of nuclear Zeeman heat capacity when deuterons are substituted for protons. For the broken curve this correction is not made. (After De Boer, 1976.)

moto, 1974). Figure 6.15 shows the variation of the proton polarization as a function of the microwave frequency observed in a field of 25 kOe where the proton frequency is $\omega_n \simeq 106$ MHz, to be compared to $\Delta\omega_e \sim 20$ MHz only. *Near* the electronic Larmor frequency two nuclear polarization peaks of opposite sign are observed, characteristic of DNP by thermal contact of the nuclear Zeeman reservoir with the electronic non-Zeeman energy. The latter is cooled in turn by off-centre saturation of the ESR line by a microwave field which drives *allowed* electronic transitions. At a distance $\pm\omega_n$ from ω_e are observed the two peaks characteristic of a well-separated solid effect, driven by *forbidden* transitions. Smaller peaks at $\omega_e \pm 2\omega_n$, characteristic of a double nuclear flip, can also be seen in Fig. 6.15. They have the same origin as those observed for a double flip of ^{19}F in CaF_2, mentioned in **C**(*b*). The existence of DNP by thermal mixing with the electronic non-Zeeman energy under conditions where $\omega_n \gg \Delta\omega_e$ can be contrasted with the DNP observed under similar conditions in LMN in 20 kOe (Fig. 6.3). Whilst the widths of the solid-effect polarization peaks are comparable in the two substances, there is no trace in LMN of DNP by

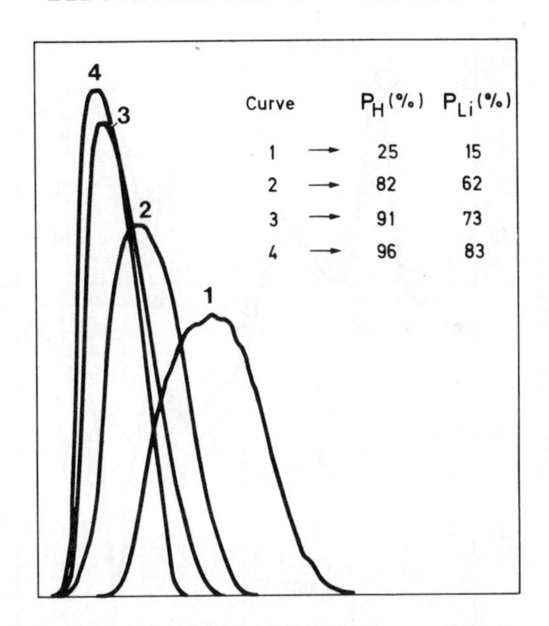

FIG. 6.14. Unsaturated ESR absorption lines in LiH with $H_0 \parallel [100]$, for various nuclear polarizations. (By courtesy of Y. Roinel and V. Bouffard.)

thermal mixing near the Larmor frequency, observed in m-xylene. A possible explanation is that in LMN the Nd^{3+} ions which replace the La^{3+} ions on a regular lattice cannot come nearer each other than the lattice spacing of LMN which is rather large (6·3 Å) and could actually be kept even further away by repulsive interactions between Nd^{3+} ions of which there seems to be some evidence (Van den Heuvel, Swanenberg, and Poulis, 1971). The spectrum of the electronic spin–spin interaction in LMN would then have a sharp cut off at a frequency smaller than ω_n and could not come 'on speaking terms' with the nuclear Zeeman energy.

The same remark applies to $Ca(OH)_2$ doped with (O_2^-) (Marks, Wenckebach and Poulis, to be published, and Fig. 6.4).

There is a second fact which demonstrates rather strikingly that one deals there with a well-resolved solid effect rather than with DNP by thermal mixing.

In a sample containing 75 per cent D and 25 per cent H, heavily doped (1.6×10^{-3} of O_2^-) the saturation of one of the nuclear species had no observable effect on the polarization of the other. This effect was even observed in fields as low as 2 kG where one would have expected the electronic non-Zeeman spectrum to overlap with the Larmor frequency of the protons. This is to be contrasted with, say, the behaviour of LiF in a much higher field as described in section **E**(*b*) and illustrated in Fig. 6.5.

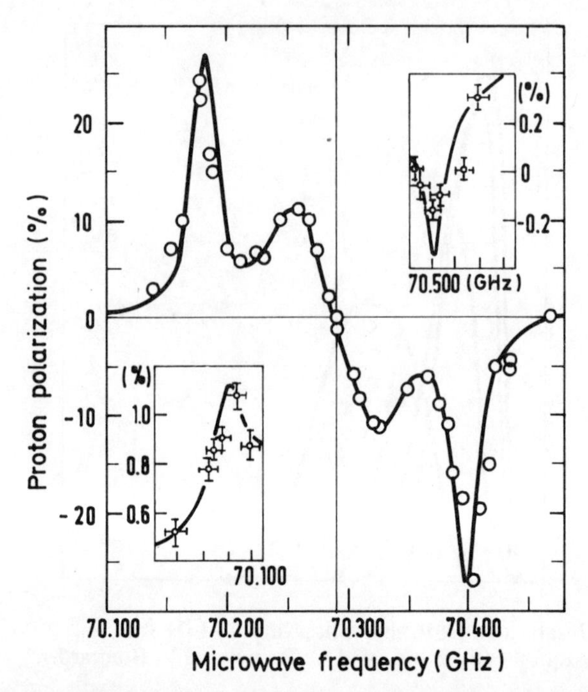

FIG. 6.15. Proton polarization as a function of microwave frequency in m-xylene doped with the radical BDPA, exhibiting both the thermal mixing with the non-Zeeman reservoir and the solid effect. The insets show the double solid effect on an enlarged scale. (After De Boer, 1976.)

There is however a feature of DNP in $Ca(OH)_2$, doped with O_2^-, which is not understood at present. The width of the polarization peaks is such that a change of the applied field by as little as 1 gauss affects the nuclear polarization during DNP in an observable way. The samples of $Ca(OH)_2$ are thin plates, where the Weiss fields produced by the protons are of the order of a few gauss. It was found, nevertheless, that the optimum positions of polarization peaks remained unchanged throughout the whole DNP process.

Whatever the mechanism assumed for DNP in $Ca(OH)_2$, pure solid effect or thermal contact of the non-Zeeman electronic reservoir with the nuclear Zeeman reservoir, this behaviour is unexpected.

(e) Figure 6.16 shows the dependence on the microwave frequency of the deuteron DNP in m-xylene (Borghini *et al.* 1974). The Larmor frequency of the deuteron $\omega_n \sim 16$ MHz is now comparable to $\Delta\omega_e$ and near ω_e two large polarization peaks of opposite sign, centred on ω_e, characteristic of nuclear Zeeman cooling by thermal contact with a cold non-Zeeman electronic reservoir are observed as expected. However, two other pairs of peaks

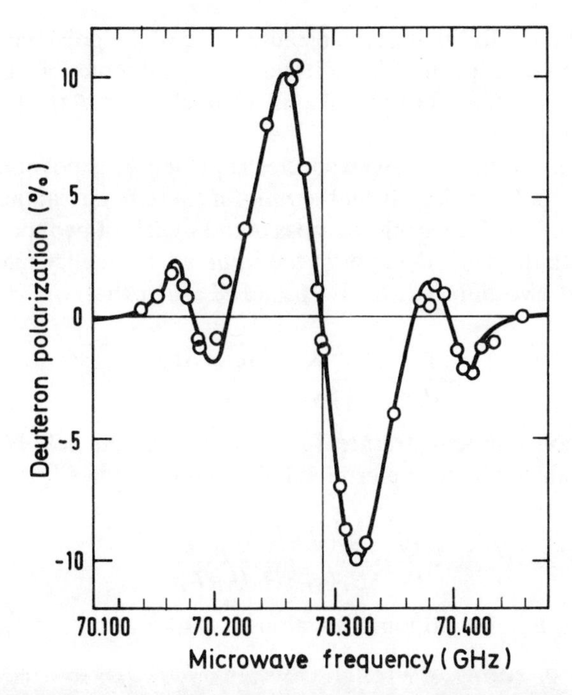

FIG. 6.16. Deuteron polarization as a function of microwave frequency in partially deuterated m-xylene doped with BDPA. The deuterons are polarized by thermal mixing with the non-Zeeman reservoir. The latter is cooled either by allowed electronic transitions (central part) or by forbidden transitions involving the simultaneous flips of an electron and a proton. (After De Boer, 1976.)

centred at the frequencies $\omega_e \pm 106$ MHz, i.e. centred around the solid-effect frequencies for the *proton*, are also observed. These pairs of peaks also originate in the thermal contact of the Zeeman deuteron reservoir with the non-Zeeman energy. The only difference with the central pair of peaks is that now the electronic non-Zeeman reservoir is cooled by off-centre saturation of a forbidden transition that involves, besides an electron flip, a proton flip.

(f) An interesting example of 'spin–temperature' methods is provided by the polarization of nuclei of spin $I > \frac{1}{2}$ with a quadrupole moment. In an environment of non-cubic symmetry the resonance line observed in high field is split by the quadrupole interaction into several components corresponding to the various transitions $m \rightarrow m-1$. If all the levels E_m are populated according to a single spin temperature T_S, the latter can be deduced from the ratios of the various components and the polarization is then given by $P = B_I(\hbar\omega_0/kT_S)$ where ω_0 is the Larmor frequency (supposedly much larger than the quadrupole splittings). This method has been applied to LMN, giving the absolute value of the polarization of ^{139}La and also as a by-product the sign of its quadrupole interaction, unobtainable for small polarizations (Abragam and Chapellier, 1964). Typical ^{139}La signals are shown in Fig. 6.17.

(g) When more than one spin species is present it is possible to extract their *absolute* polarizations from the *ratio* of these polarizations. The ratio can be obtained by comparing the areas of the signals of two species I and I' measured with the same detector at the *same* r.f. frequency (and therefore necessarily at two different fields H and H' such that $\gamma_I H = \gamma_{I'} H'$). We have:

$$\frac{P_I}{P_{I'}} = \frac{(I'\gamma_{I'}^2 N_{I'})}{(I\gamma_I^2 N_I)} \frac{\int v_I(\omega)\,\mathrm{d}\omega}{\int v_{I'}(\omega)\,\mathrm{d}\omega}, \tag{6.126}$$

The common spin temperature T_S of the spins at the field H_0 where the DNP was produced is then determined by solving graphically or numerically the equation for T_S:

$$\frac{P_I}{P_{I'}} = \frac{B_I(\gamma_I \hbar H_0/k_B T_S)}{B_{I'}(\gamma_{I'} \hbar H_0/k_B T_S)}. \tag{6.127}$$

where B_I and $B_{I'}$ are Brillouin functions.

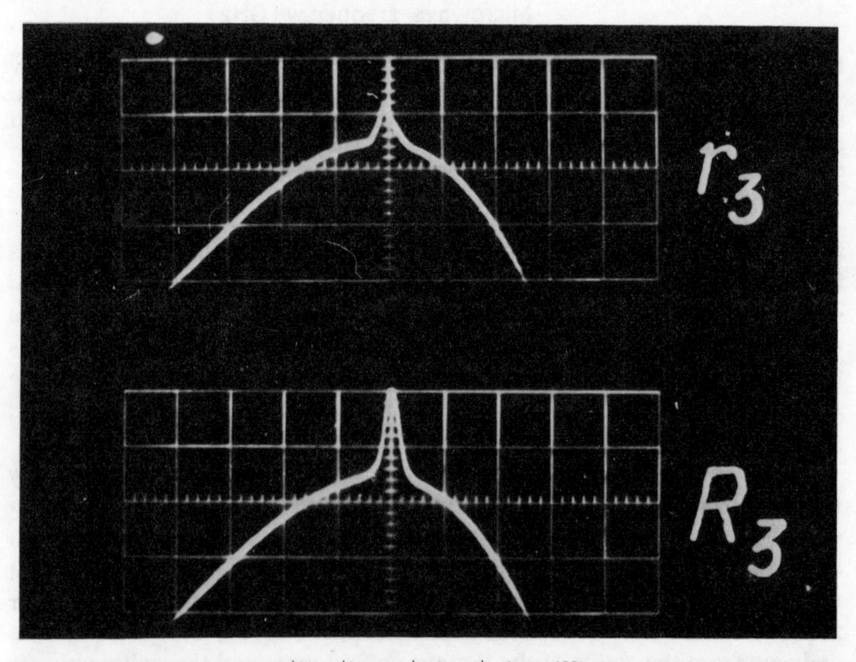

FIG. 6.17. Resonance lines $|\frac{3}{2}\rangle \rightleftarrows |\frac{1}{2}\rangle$ and $|-\frac{3}{2}\rangle \rightleftarrows |-\frac{1}{2}\rangle$ of ^{139}La in polarized LMN. The intensity ratio yields both the value of the spin temperature and the sign of the quadrupole interaction. (After Abragam and Chapellier, 1964.)

The absolute values of P_I and $P_{I'}$ are then given by $B_I(\gamma_I \hbar H_0/k_B T_S)$ and $B_{I'}(\gamma_{I'} \hbar H_0/k_B T_S)$ respectively. Apart from the assumption of common spin temperature it is necessary to assume the relaxation times sufficiently long for the polarizations P and P' to keep the same values in the fields of measurement H and H' as those reached at the end of the DNP in the field H_0 (which can be one of the two fields H or H'). Actually it is experimentally easier to compare at various stages of the DNP (i.e. for different values of T_S) the areas of the NMR signals of spins I and I', each one recorded with a different spectrometer. The gains of the two receivers have to be suitably adjusted.

The validity of this procedure has been checked with ^1H, ^2H, and ^{13}C in some target materials (De Boer *et al.* 1974), and also with ^{19}F and ^7Li in LiF (Roinel and Bouffard, 1975). It is practical for large polarizations only ($P \geq 0{\cdot}4$) and preferably for nuclei with widely different magnetic moments.

It should be noted that if the ratio $N_{I'}/N_I$ is not known beforehand as can happen with enriched isotopes it can be obtained from the two eqns (6.126) and (6.127) by comparing the signals of the two species at low polarizations. The ratio $P_I/P_{I'}$ is then given by a linear expansion of (6.127) where $1/T_S$ factors out, leaving the well-known relation:

$$\frac{P_I}{P_{I'}} = \frac{I+1}{I'+1}\frac{\gamma}{\gamma'}. \tag{6.128}$$

(h) The previous method was used to measure absolute polarizations and relative concentrations of ^6Li, ^7Li, and D in a sample of enriched ^6Li D (Abragam, Bouffard, and Roinel, 1980). The main interest of this experiment was to obtain a high deuterium polarization, which is difficult because of its small magnetic moment. It can be read on Fig. 6.13 that in some materials the maximum polarization of deuterium is 42 per cent under conditions where that of the protons is practically 100 per cent. Similarly in lithium hydride (Roinel, Bouffard, and Roubeau, 1978), for an inverse spin temperature of 360 K^{-1}, the following polarizations were obtained: 95 per cent for ^1H, 80 per cent for ^7Li, and only 35 per cent for ^6Li which has the spin $I = 1$ and practically the same gyromagnetic ratio as deuterium.

Higher polarizations of D or ^6Li can be obtained only by lowering drastically their spin temperature. This can be achieved by using paramagnetic impurities with a narrower ESR line. For the simple model of a line broadened by h.f.s. couplings of strength A with six nearest neighbours of spin $\frac{3}{2}$, used in sections **F**(c)2 and **F**(d), Fig. 6.9 exhibits a plot of βA against the irradiation frequency for various lattice temperatures. Thus, all other things being equal, the smaller A, the larger the expected inverse spin temperature, and replacing ^7LiH by ^6LiD should yield a lower spin temperature. If ^6Li had spin 3/2 as ^7Li, the increase of β, in the framework of this model, would be exactly in the ratio $A_7/A_6 = \gamma(^7Li)/\gamma(^6Li) = 2{\cdot}64$.

Actually $I(^6\mathrm{Li}) = 1$, and the comparison is less direct. However the decrease in spin temperature and the increase in the polarizations of ^{6}Li and D actually observed are spectacular although slightly less than expected. Under conditions very similar to those which gave $P(^6\mathrm{Li}) = 35$ per cent and $\beta = 360\ \mathrm{K}^{-1}$ in ^{7}LiH, one obtains $P(^6\mathrm{Li}) \cong P(^1\mathrm{D}) \cong 70$ per cent and $\beta = 825\ \mathrm{K}^{-1}$ in Li^{6}D (Abragam *et al.* 1980). We shall come back to ^{6}LiD in section **G**(*b*).

G. Applications of DNP

Legend has it that when Kammerling Onnes liquefied helium, he predicted that henceforth every experiment in physics could be repeated by adding the condition 'at low temperature'. The same had been said of Bridgman with 'low temperature' changed to 'high pressure'.

For experiments where atomic nuclei play a part, a similar addition could be 'with polarized nuclei' and in the field of nuclear and particle physics, starting in the early sixties, some, if by no means all, experiments have indeed been renewed by the use of dynamically polarized targets.

During the seventies, other directions of research have been opened up by using the methods of dynamic nuclear polarization, such as studies of ordered nuclear states, ferromagnetic or antiferromagnetic or new features of the spin-dependent scattering of slow neutrons by polarized nuclei, sometimes referred to as pseudomagnetism.

Chapters 7 and 8 of this book deal with the last two subjects in considerable detail and we shall be content to describe briefly in this section a few of the other applications of DNP.

(*a*) *The Nedor method*

The design of the best procedure for DNP in a particular sample often requires an investigation of the mechanisms of DNP in that sample, which is performed through the observation of the nuclear spins *and* of the electronic spins. As shown in sections **C** and **F** useful information on the electronic spin system comprises their spin–lattice relaxation time, their polarization in the course of DNP and their ESR lineshape as a function of nuclear polarization. Very often the microwave irradiation system, primarily designed for performing the DNP, is poorly suited to the observation of the ESR line and has a bad signal to noise ratio.

The Nedor method described in this section allows an indirect detection of the ESR through the observation of the *nuclear* resonance signal in analogy with the well-known Endor method which allows an investigation of the nuclear spins through the observation of the electronic resonance line. The principle of Nedor is to observe in a non-spherical sample the shift of the nuclear resonance frequency produced by the dipolar field of the *electronic* spins.

In principle the electronic dipolar field affects both the position and the shape of the NMR line. However, with the electronic concentrations commonly used for DNP, the shape of the nuclear signal has a negligible dependence on the electronic polarization P_S and its only effect is to shift the NMR line by an amount equal to the first moment of the NMR line M_1^{IS}, as given by eqn (5.66). The field shift is typically a fraction of a Gauss for $P_S = 1$. In practice, one observes the change in the NMR lock-in signal produced by a change of P_S. This signal is calibrated by comparing it with the signal produced by a well-defined variation of the external field. The NMR signal is observed with polarized nuclei, which greatly enhances the sensitivity of the method. Several types of measurements are possible.

(a) The shift produced by the complete saturation of the electronic spins yields, according to eqn (5.66), the concentration N_S of these spins.

(b) The electronic spin–lattice relaxation time T_{1e} is obtained as the recovery time constant of the NMR signal after removing the saturating microwave from the electronic resonance frequency. An example is shown in Fig. 6.18. An interesting application of this type of measurement is the study of the phonon bottleneck in CaF_2 doped with Tm^{2+} ions, referred to in section **C**(*b*) (Abragam, Bouffard, and Roinel, 1976). It was stated in that section that one could change at will the populations of the spins belonging

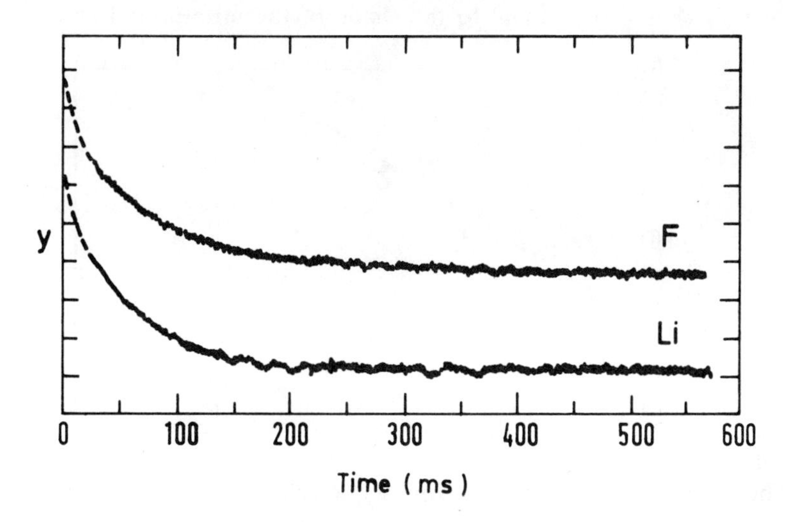

FIG. 6.18. NEDOR signals of ^{19}F and ^{7}Li in LiF, following the removal of a saturating microwave at the electronic frequency. The time constant of the variations of the signals is the electronic spin–lattice relaxation time T_{1e}. (After Abragam, Bouffard, and Roinel, 1976.)

to one of the two hyperfine lines of Tm^{2+} by a transfer of populations between these lines. It is then possible to saturate rapidly one of the lines and to observe by Nedor the recovery of its polarization in a time short enough for the populations of the two lines to remain constant. The signal produced by the saturation yields the population of the line, and its recovery time the apparent electronic relaxation time T_1^*. The latter is equal to σT_{1e} (eqn (6.14)) where σ (the phonon bottleneck coefficient) is, according to eqn (6.13), proportional to the population N_S of the line. A plot of T_1^* as a function of N_S is indeed a straight line whose slope yields the value of σ/N_S and whose extrapolation to $N_S = 0$ yields the value of T_{1e} (Fig. 6.19).

(c) The electronic polarization in the course of DNP is measured as follows. One observes the NMR signal when the polarizing microwave source is on and the electronic polarization is P_S. One then switches off the microwave power: the electronic polarization snaps back quickly to its thermal equilibrium value P_0. The change of NMR signal, which is proportional to $(P_0 - P_S)$, yields the value of P_S. The measurement of P_S would be very difficult by standard ESR detection, at least with a single microwave source.

(d) The ESR lineshape is determined by observing the change of NMR signal produced by applying a microwave field at various distances from the electronic Larmor frequency. The microwave power is small enough to produce only a small electronic saturation. Under these conditions the decrease of electronic polarization $(P_0 - P_S)$ produced by an irradiation of frequency ω is proportional to the shape of the unsaturated absorption

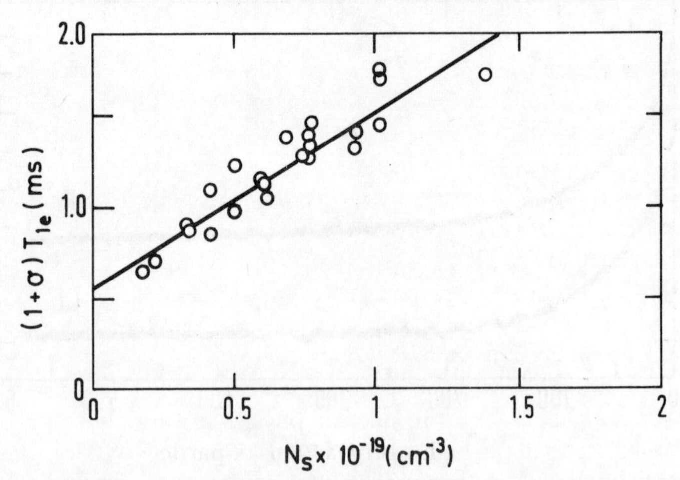

FIG. 6.19. Apparent electronic relaxation time of one of the hyperfine lines of Tm^{2+} in CaF_2 as a function of the electronic concentration of that line. Both T_1^* and N_S are measured by the NEDOR method. The linear variation of T_1^* with N_S is as expected from phonon bottleneck theory. (After Abragam, Bouffard, and Roinel, 1976.)

signal observable with a standard ESR spectrometer. The determination of this shape is performed point by point at discrete frequency values. An example of ESR lineshapes observed by this method in lithium hydride is shown in Fig. 6.14.

(b) Polarized targets for nuclear and particle physics

All billiards players know that if they give a spin to their ball, the way it collides with another ball will depend on the direction of that spin. The same would naturally be true of the target ball if the player knew how to spin it without knocking it off from its position. This is precisely what is achieved in polarized nuclear targets bombarded with other nuclei or with elementary particles.

The result of the scattering of a fast incoming beam of particles on an *unpolarized* target is an average of scatterings for all the orientations of the spins of the target nuclei and represents a considerable loss of information with respect to scatterings on a polarized target, especially by DNP where the sign of the polarization is easily reversed.

If the particles of the incoming beam have spins, it is possible to polarize them also by various methods. These are essentially of two types: atomic physics methods whereby the beam is polarized at low energy prior to its acceleration, and nuclear and particle physics methods where a fast beam is polarized by a primary collision on an auxiliary target before striking the main target. As in the billiards game, the production of polarized beams has turned out to be easier than that of polarized targets and was achieved somewhat earlier. The study of polarized beams is outside of the scope of this book.

Although more relevant to the problem of DNP our description of polarized targets will be very brief. Too many aspects of this problem, such as theory of elementary particles or nuclear physics, particle detection techniques (in particular, counting rates, angular and energy resolution, particle discrimination, etc.), radiation damage and also plain cryogenic engineering are alien to the subject of DNP proper. In dynamically polarized targets for nuclear and elementary particle physics there has been practically no demand for nuclei other than protons and occasionally deuterons. (An exception is ^{3}He with gaseous targets polarized by optical pumping.)

We outline below some of the main features of targets for low-energy nuclear physics and for high-energy elementary particle physics.

Low-energy beams used in nuclear physics are usually produced with higher intensity than the high-energy beams of particle physics, especially if the latter are secondary beams of mesons or leptons rather than primary beams of accelerated protons. Furthermore, cross sections are usually greater at low energy than at high energy. The consequence of these two facts is that far more events occur per unit volume in a low-energy target, making it possible to use much smaller targets in nuclear physics than in high-energy physics. For very low energies, say below 20 MeV, this possibility becomes a necessity because of the spread of energy of the beam inside a thick target. Thus in the first scattering experiment of polarized protons of 20 MeV on a polarized proton target (Abragam, Borghini, Catillon, Coustham, Roubeau, and Thirion, 1962) the thickness of the target was 0·1 mm, to be compared with targets as thick as 150 mm or more in particle physics (Vermeulen, 1972; Niinikoski and Udo, 1976).

The presence in the target of nuclei other than protons is not a serious drawback for low-energy scattering: because of the existence of an internal Fermi energy for the nucleons inside these nuclei, scattering from them occurs with an energy different from scattering by the free (and polarized) protons of the target and the two types of events are easily discriminated.

On the contrary, in high-energy physics this discrimination, based on differences in energy much smaller than that of the beam, is far less efficient and leads to a large background to be subtracted with more or less certainty by comparison with a 'dummy' target. It is clear that the higher the ratio of free polarized protons to the total number of protons, free or bound in nuclei, the better the target.

Progress in high-energy targets has evolved in several directions.

Replacing hydrated rare earth salts such as double nitrates or ethylsulphates by proton-rich organic substances has increased the ratio of free protons. For instance, this ratio has gone from 0·03 in LMN to 0·18 in ethanediol and 0·23 in butanol, doped respectively with complex paramagnetic compounds such as chromium V and a free radical such as porphyrexide, two common target materials. Substances with an even higher free proton ratio, such as solid hydrogen H_2 which offers a free proton ratio of 0·75 (the proportion of metastable orthomolecules), deuterated hydrogen HD with 0·5, methane with 0·4, have never yielded any significant dynamic polarization.

Increasing the polarizing field to 25 and even 50 kOe and decreasing the temperature during DNP down to 0·4 K, polarizations very near 100 per cent have been achieved.

With heavy paramagnetic doping up to concentrations of the order of $c \sim 10^{-3}$, short polarization times have been achieved enabling the polarization to be completely reversed in a few minutes.

A refinement in the technique of polarized targets relevant to the theory of DNP is worth mentioning. The DNP requirement of a large homogeneous magnetic field H_0 puts severe limitations on the geometry of high-energy experiments such as the solid angle into which the products of the collision are emitted. This problem is solved by the device of the 'frozen target'. After the polarization has been brought to the required value P_0 in the polarizing field H_0 at a polarizing temperature T, of the order of, say, 0·4 K, the target is moved into a different position. There a 'holding' field H_h much smaller and not necessarily homogeneous, which 'holds' the polarization to the value P_0, can be produced in a magnet with a much larger gap that does not interfere with the trajectories of the outgoing particles. Solid angles of the order of 4π can thus be available for these particles. The gist of the method is that the nuclear relaxation time in the holding field H_h remains very long in spite of the relative weakness of that field. This is achieved by lowering the temperature of the target by means of dilution refrigeration down to a holding value T_h of the order of, say, 0·04 K. This is done *after* dynamically polarizing the target to P_0 but *before* removing it into the holding field. It is the factor $(1-P_e^2)$ in $1/T_{1n}$, where $|P_e| = \tanh(\hbar\omega_e/2k_B T_n)$, which gives to T_{1n} values of the order of thousands of hours.

For deuterium, polarizations of the order of 40 per cent have been achieved in standard organic target materials. The main interest of polarized deuteron targets in high-energy physics is that if one neglects the weak binding energy of the deuteron, it can be considered as a target of polarized neutrons. Detailed information on the realization of polarized targets can be found in the proceedings of the two conferences held on this subject and in some more recent papers (Vermeulen, 1972; Niinikoski and Udo, 1976; Borisov, Glonti, Kazarinov, Kazarinov, Kiselev, Kiselev, Matafonov, Macharashvili, Neganov, Strakhota, Trofimov, Usov, and Khachaturov, 1977).

Two new problems have arisen in the late seventies.

Firstly an interest in scattering from polarized targets at large angles. The cross sections at these angles are small and very intense beams of incoming particles have to be used. Under these conditions the radiation damage of the target is important and its polarization falls off in the course of the experiment. The phenomenon is not fully understood and no good remedy exists.

Secondly there is a trend toward so-called inclusive reactions where it is impossible to discriminate not only between scattering events on bound and free protons but even between scatterings on protons and neutrons. This implies that the effective ratios of polarized nucleons such as 0·18 in ethanediol and 0·23 in butanol have to be divided by a further factor two.

In this connection the dynamic polarization of ^{6}LiD where polarizations up to 70 per cent have been obtained for both ^{6}Li and D looks promising. Its interest is enhanced by the fact that, with respect to nucleon polarization, ^{6}Li can be considered, with reasonable accuracy as a deuterium nucleus plus an alpha particle. This means that in polarized Li^{6}D, out of eight nucleons, four are polarized, two neutrons and two protons, a far higher ratio than in any other target.

H. Other examples of coupling between the electronic spin–spin reservoir and the nuclear Zeeman reservoir

(a) *The rotating crystal*

The principle of the DNP methods discussed in detail in sections **C** and **F** can be summarized as follows: in order to cool the nuclear Zeeman reservoir far below the lattice temperature by thermal contact with the electronic Zeeman reservoir the latter must undergo two changes:

(i) it must itself be cooled far below the lattice temperature;

(ii) it must be brought onto 'speaking terms' with the nuclear Zeeman reservoir.

It is rather remarkable that by applying to the electron spin system a microwave field at a distance Δ from resonance both conditions can be fulfilled at once:

(i) seen in the rotating frame the electronic Zeeman reservoir $\langle Z_S^* \rangle$ is indeed colder than $\langle Z_I \rangle$ of the laboratory frame by the large ratio (ω_S/Δ);

(ii) in some cases $\langle Z_S^* \rangle$ and $\langle Z_I \rangle$ can be brought onto speaking terms by matching their frequencies Δ and ω_n (or rather $|\Delta|$ and $|\omega_n|$). This is the well separated solid effect of section **C**. Another situation, described in section **F** is that when $\langle Z_I \rangle$ and the electronic spin–spin reservoir $\langle \mathcal{H}'_{SS} \rangle$ are at all times on speaking terms and at the same temperature and the effect of the microwave field is to bring the cold $\langle Z_S^* \rangle$ into contact with $\langle \mathcal{H}'^*_{SS} \rangle$ and through it, with $\langle Z_I \rangle$. In either scheme, the heat capacity of $\langle Z_S^* \rangle$ is much smaller than that of $\langle Z_I \rangle$. If $\langle Z_S^* \rangle$ is nonetheless capable of cooling appreciably $\langle Z_I \rangle$, either directly or through $\langle \mathcal{H}'^*_{SS} \rangle$, as the case may be, it is because the electronic relaxation keeps $\langle Z_S \rangle$ reasonably cold and $\langle Z_S^* \rangle$ even colder, in the ratio (ω_S/Δ).

We outline in this section a different scheme, capable of cooling the electronic Zeeman reservoir and of bringing it into thermal contact with the nuclear Zeeman reservoir. It is based on the large magnetic anisotropy of certain magnetic centres and was proposed simultaneously and independently by two authors (Jeffries, 1963; Abragam, 1963).

Consider a crystal containing magnetic centres with an anisotropic tensor **g**. The Larmor frequency of such a centre is proportional to the magnitude of the vector (**g** . **H**) with components:

$$(\mathbf{g} \,.\, \mathbf{H})_i = \sum_k g_{ik} H_k. \tag{6.129}$$

If we take as co-ordinate axes the principal axes of the tensor **g** with principal values g_i the electron Larmor frequency is proportional to:

$$\omega_S \propto \left[\sum_i (g_i H_i)^2 \right]^{1/2} \tag{6.130}$$

There exist magnetic centres for which one or two of the principal values g_i vanish (or at least are very small). Then if the field is in the vicinity of such an axis, or of the plane of two such axes, ω_S is very small. If the crystal is rotated with respect to the field (or the field with respect to the crystal) around an axis of rotation **n**, chosen so as to bring the field along an axis or into a plane where ω_S becomes very small, we shall have achieved the two goals outlined above (i) cooling $\langle Z_S \rangle$ (ii) bringing it onto speaking terms with $\langle Z_I \rangle$, directly or indirectly. This demagnetization by rotation differs from ordinary demagnetization through the fact that the isotropic nuclear Zeeman reservoir $\langle Z_I \rangle$ takes no part in it. In that sense demagnetization by rotation bears a great analogy to ADRF.

As in standard DNP, the heat capacity of $\langle Z_S \rangle$ when on 'speaking terms' with $\langle Z_I \rangle$ is very much smaller than that of $\langle Z_I \rangle$ and a 'one shot' cooling of $\langle Z_I \rangle$ by a single rotation of the crystal is impossible. The crystal is then rotated continuously. For relative orientations of field and crystal such that $|\mathbf{g} \cdot \mathbf{H}|$ is large, $\langle Z_S \rangle$ relaxes to the lattice and recovers the thermal equilibrium value $\langle Z_S \rangle_{\mathrm{L}}$ which enables the cooling to start again, and again. A favorable feature is that for relaxation by the direct process, large values of the electronic Larmor frequency favour short relaxation times: thus $\langle Z_S \rangle$ is poorly coupled to the lattice while cooling the nuclei, and strongly coupled to it whilst uncoupled from nuclei.

The first experiment which demonstrated the feasibility of the rotational nuclear cooling (Robinson, 1963) was performed in LMN doped with Ce^{3+} where rotational enhancements of the proton polarization up to a factor 15 were obtained. For this ion: $g_{\perp} \simeq 1{\cdot}8$; $g_{\parallel} \approx 0{\cdot}023$.

The smallness of $g_{\parallel}$ does not result from the symmetry of the crystal but rather from an accidental cancellation of some matrix elements.

All subsequent studies were performed on Yttrium ethyl sulphate (abbreviated as YES) doped with Yb^{3+} for which $g_{\parallel} = 3{\cdot}35$; $g_{\perp} \cong 0$. The difference with Ce^{3+} in LMN is twofold: g vanishes in a plane rather than along an axis; the vanishing or at any rate the smallness of $g_{\perp}$ results from symmetry considerations rather than from accident.

The fact that g vanishes in a plane makes it much easier to bring the magnetic field into that plane, by rotation of either the crystal or the field, than in the case when $g_{\parallel}$ vanishes along an axis, which, as for Ce^{3+} in LMN, must be made to coincide with the field.

Furthermore when $g_{\perp}$ is zero rather than $g_{\parallel}$, it is possible to polarize a polycrystalline sample by rotation, of either the sample or the field, around an axis **n** perpendicular to the field: it is clear that only crystallites for which $g_{\parallel}$ is along **n**, will not polarize at all. For all the others the limiting polarization will depend on the orientation of the crystallite. Its average value will be smaller than the maximum value for a single crystal with $g_{\parallel}$ perpendicular to the axis of rotation.

Proton polarizations of up to 85 per cent have been observed in polycrystalline samples of YES doped with Yb^{3+} (Button-Shafer, Lichti, and Potter, 1977; Button-Shafer, 1979).

A detailed discussion of rotational nuclear cooling has been given by Jeffries (1972). It ignores however the role of the electronic spin–spin interaction in the cooling process. In a substance such as LMN doped with Ce^{3+} the smallest value of the electronic frequency $\Delta_0 = (g_{\min}/2)\gamma_S H_0$ (where γ_S is the gyromagnetic ratio of the free electron) exceeds the proton frequency $\gamma_I H_0$ by a factor eight. It is clear that in that case at least the thermal contact between $\langle Z_S \rangle$ and $\langle Z_I \rangle$ passes through the spin–spin reservoir $\langle \mathcal{H}'^*_{SS} \rangle$ and it is interesting to discuss its role at least qualitatively. We define besides Δ_0, and somewhat loosely, two cut off frequencies for $\mathcal{H}'^*_{SS}$; an electronic Zeeman frequency Δ_S above which $\mathcal{H}'^*_{SS}$ and Z_S do not communicate and a nuclear Zeeman frequency Δ_I above which $\mathcal{H}'^*_{SS}$ and Z_I do not communicate. It is reasonable to assume that $\Delta_S > \Delta_I$: it is easier for an electronic spin–spin interaction to communicate with its own Zeeman reservoir than with that of a system of nuclear spins. Since we know from experiment, and from theory that even in a field of a few teslas $\mathcal{H}'^*_{SS}$ and Z_I do communicate with each other, unless the electronic concentration is exceedingly small, we conclude that $\Delta_I > \omega_n$ whence the sequence: $\Delta_S > \Delta_I > \omega_n$. It remains to place Δ_0 in this sequence: if Δ_0 were larger than Δ_S there would be no communication between $\langle Z_S \rangle$ and $\langle Z_I \rangle$ through $\langle \mathcal{H}'^*_{SS} \rangle$ and no nuclear polarization whatsoever. The final inequalities are:

$$\Delta_S > \Delta_I, \Delta_0; \qquad \Delta_I > \omega_n.$$

We can now visualize the rotational demagnetization process as follows. When the electronic frequency reaches the value Δ_S, $\mathcal{H}'^*_{SS}$ and Z_S come into contact. We saw in section **E** that the spectral density of $\mathcal{H}'^*_{SS}$ has a quasi-Lorentzian shape with a long tail and for $\omega_S = \Delta_I$ the Zeeman energy $\langle Z_S \rangle$ is probably very much larger yet than $\langle \mathcal{H}'^*_{SS} \rangle$. However $\langle \mathcal{H}'^*_{SS} \rangle$ is tightly coupled to $\langle Z_1 \rangle$ and therefore, as soon as $\omega_S = \Delta_I$, $\langle Z_S \rangle$ begins to couple to $\langle Z_I \rangle$. The ratio $\langle Z_S(\Delta_I) \rangle / \langle Z_I \rangle = C(\Delta_I/\omega_n)$ is a small number and $\langle Z_S \rangle$, through $\langle \mathcal{H}'^*_{SS} \rangle$, comes rapidly into thermal equilibrium with $\langle Z_I \rangle$. If the time τ required for this process is short compared to the time constant $(\omega_S/\dot{\omega}_S)$ for the change of the electron frequency due to rotation, the final temperature of $\langle ZI \rangle$ will not be lower than $T_L(\Delta_I/\omega_S^0)$ (in the linear approximation), where ω_S^0 is the maximum electronic Larmor frequency. Since, as stated earlier, Δ_I is large, the cooling will be poor.

The answer is twofold: firstly the obvious one of going to high fields and low lattice temperature which reduces (T_L/ω_S^0), but secondly an increase of the speed of rotation: this would permit to $\langle Z_S \rangle$ to reach by demagnetization a lower temperature than $(\omega_S^0/\Delta_I)T_L$, before coming into thermal equilibrium with $\{\langle\mathcal{H}'^{*}_{SS}\rangle+\langle Z_I\rangle\}$. In the experiment of Button-Shafer *et al.* the speed of rotation exceeded 200 r.p.s.

If the rotation is sufficiently fast for that, there is another lower limit to the final temperature:

$$T_f \cong T_L \frac{\omega_S^0}{\Delta_0} = T_L \frac{g_{max}}{g_{min}}.$$

This for Ce-doped LMN gives a maximum cooling of 80.

In this discussion we have disregarded the problems of spin–lattice relaxation. Too fast a rotation, would not leave time for $\langle Z_S\rangle$ to come to thermal equilibrium with the lattice when ω_S is large. Fortunately, as stated earlier, T_{1e} is short when ω_S is large.

In spite of its attractive features, rotational nuclear cooling has not been a successful competitor to microwave DNP. It has been promoted as a device for polarized targets on the ground that it requires no microwaves and no homogeneous field. Its main drawbacks in this respect however are the necessity of a fast rotation (200 r.p.s.), the impossibility of reversing the polarization without reversing the field, and the very limited choice of working materials; in fact no choice at all since YES seems to be the only successful candidate, with a ratio of free protons to all protons of 0·057.

(*b*) *The bootstrap effect: cooling of electron spins by nuclei in low fields*

1. *The triplet*

In section **F** we saw how the nuclear Zeeman reservoir $\langle Z_I\rangle$ could be cooled by thermal contact with the non-Zeeman electronic reservoir itself cooled by off-centre irradiation of the ESR line. As explained in section **H**(*a*) this irradiation can be described as a thermal contact in the rotating frame between $\langle\mathcal{H}'^{*}_{SS}\rangle$ and the electronic Zeeman Hamiltonian as it appears in the rotating frame. Order can thus be transferred from $\langle Z_S^*\rangle$ into $\langle Z_I\rangle$. The reason why the order of $\langle Z_S^*\rangle$ is not destroyed in the process in spite of its heat capacity being much smaller than that of $\langle Z_I\rangle$, is the close link of $\langle Z_S\rangle$ with a cold lattice through the electronic spin–lattice relaxation, which maintains $\langle Z_S^*\rangle$ at a low temperature. In this section we describe the opposite situation where a very cold nuclear Zeeman reservoir $\langle Z_I\rangle$ cools an electronic Zeeman reservoir $\langle Z_S\rangle$ of much smaller heat capacity through a thermal contact across an electronic non-Zeeman Hamiltonian $\langle\mathcal{H}'^{*}_{SS}\rangle$. The experiment (Urbina and Jacquinot, 1980) was performed on CaF_2 doped with Tm^{2+}. As indicated in section **C**(*b*) the nuclei of ^{19}F were polarized in a field $H_0=2{\cdot}7$ T at a frequency $\omega_S\simeq 130$ GHz up to a polarization of 90 per cent. The NMR signal of ^{19}F was observed at a frequency of 107 mHz. The same receiver could be used to observe the ESR signal of Tm^{2+} in low field.

The spin Hamiltonian of Tm^{2+} is:

$$\mathscr{H} = g\frac{\mu_B}{\hbar}\mathbf{H}\,.\,\mathbf{s} + A\mathbf{K}\,.\,\mathbf{s} + \gamma_n\mathbf{K}\,.\,\mathbf{H}, \tag{6.131}$$

where we represent the nuclear spin of ^{169}Tm by the symbol **K** to avoid confusion with the spins **I** of ^{19}F. The isotropic g factor is 3·45 and the isotropic h.f.s. constant $A = -1100$ MHz. In zero field the Hamiltonian (6.131) gives rise to a triplet and to a singlet 1100 MHz higher. In order to be able to observe the ESR signal at 107 MHz, fields of the order of a few milliteslas are used. The energy levels of Tm^{2+} are given by the Breit–Rabi formula: they are represented in Fig. 6.20. The coupling between Z_I and Z_S

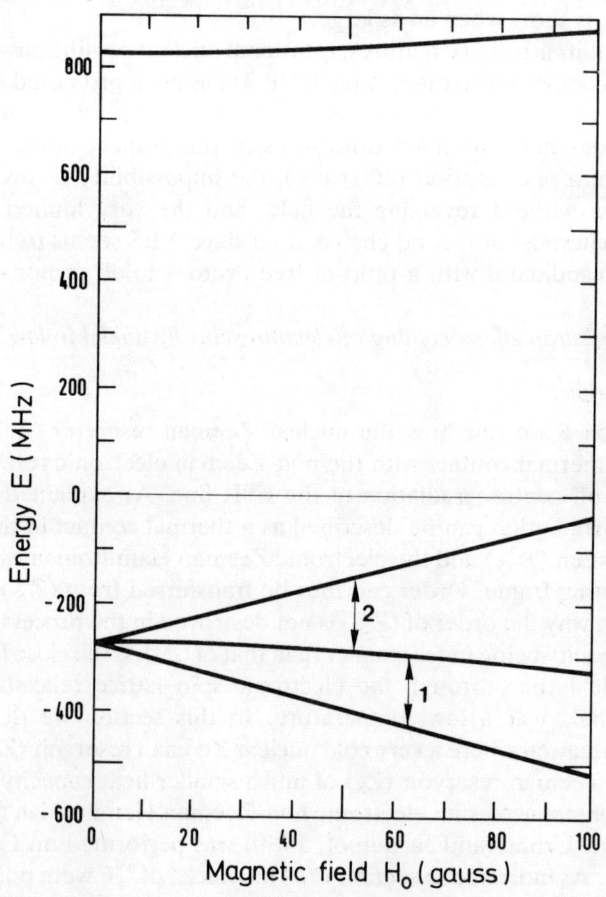

FIG. 6.20. Tm^{2+} energy levels in CaF_2:Tm^{2+} as a function of applied magnetic field H_0. Also shown are the two allowed triplet transitions. (After Urbina and Jacquinot, 1980).

is demonstrated dramatically in Fig. 6.21. Upon demagnetization down to a field H_f^0, low enough to permit the observation of the ESR resonance of Tm^{2+} at 107 MHz, a strong absorption signal is observed if the initial fluorine polarization P_F is high and positive, a strong emission signal if P_F is high and negative and a weak signal if P_F is low.

The interpretation is as follows: the polarized nuclear spins of ^{19}F can be cooled by the demagnetization, from their initial Zeeman temperature of a few mK down to, say, 5 μK. If the Zeeman triplet of Tm^{2+} comes to thermal equilibrium with the fluorine spins, it must be 100 per cent polarized, with only the lowest or the highest level of the triplet being occupied, depending on the sign of P_F. The coupling between Z_S and Z_I, with frequencies of the order of 100 MHz and 100 kHz respectively, occurs through $\mathscr{H}'^*_{SS}$. The difference with the high field case is that in low fields $\mathscr{H}'^*_{SS}$ is on speaking terms not only with Z_I but also with Z_S, directly, rather than with the help of microwaves as in high field.

In order to check this interpretation more quantitatively it is necessary to understand the position and the shape of the ESR signal. The essential point is that the electron spin **s** (coupled to the nuclear spin **K**) 'sees' not only the applied field H_f^0 of the order of a few milliteslas but also the local field of the fluorines. The latter behaves in a way very different from that in high field. For nuclei of ^{19}F, near neighbours of Tm^{2+}, the local field H_e that they

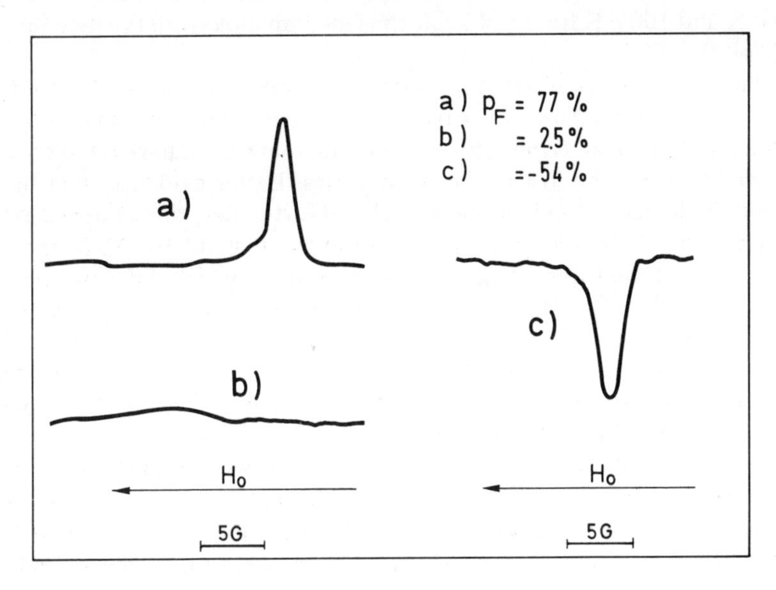

FIG. 6.21. ESR low field signals of Tm^{2+} in CaF_2, for different ^{19}F nuclear polarizations, showing a strong thermal mixing between electronic and nuclear spins. (a): $p_F = 0{\cdot}77$; (b): $p_F = 0{\cdot}025$; (c): $p_F = -0{\cdot}54$. (After Urbina and Jacquinot, 1980.)

experience from it, is much higher than the applied field H_f^0 (it is of the order of a thousand Oersteds for nearest neighbours).

As long as $H_e \gg H_f^0$, the ^{19}F spins are quantized along $\mathbf{H}_e$. The local field $\mathbf{H}_n$ that they produce in turn at the spin of Tm^{2+} and which shifts and broadens its ESR line is quite different from that observed in a high field H_0. The calculation of H_n is far from simple since it is in fact a many-body problem involving the interactions of the spins of ^{19}F among themselves and with that of Tm^{2+}. The reader is referred for a discussion to the paper of Urbina and Jacquinot (1980).

Suffice it to say here that as the ^{19}F spin temperature varies from say, 5 μK to 100 μK, the shift δH of the ESR line varies from 25 to 10 Gauss and its width ΔH from 1 to 6 Gauss. This has an important bearing on the discussion of the Tm^{2+} cooling by the fluorines.

If we assume that the Tm^{2+} triplet is at the same spin temperature as the ^{19}F spins, since $\hbar\omega_S/k_B \sim 5$ mK, its polarization should be 100 per cent at 10 μK as well as at 100 μK and the areas of the ESR signal should be the same for both. Furthermore, an absolute calibration of the ESR signal, which is possible since it is the same receiver that is used for NMR in high field, should yield an absolute polarization of 100 per cent. This is indeed what is observed within experimental error for the lowest ^{19}F spin temperature. On the other hand, as the ^{19}F temperature goes up, the measured area of the Tm^{2+} ESR signal seems to decrease by a factor 2 to 3 between say 10 μK and 100 μK for $T(^{19}F)$, which is incompatible with the assumption: $T(Tm^{2+}) = T(^{19}F)$.

One can talk oneself out of this predicament by noticing that the width of the ESR signal increases by a factor 5 between these two limits, that the signal has a peculiar asymmetrical shape and that a large part of the broader signal could well be lost in the noisy wings. Better evidence than that is needed to lend credibility to the model, but fortunately this evidence exists.

It follows from the Breit–Rabi formulae, that at 107 MHz the two transitions of the triplet (Fig. 6.20) occur at fields which differ by approximately 5 Gauss. Therefore, unless the triplet polarization is rigorously 100 per cent, *two* lines of unequal strength should be observed. This does *not* occur, showing that the middle level of Fig. 6.20 is completely empty. Even stronger evidence is provided by the following experiment. If one passes through the first line with a strong saturating r.f. field populating to a certain extent the middle level, the second line *is* observed, and the stronger the saturating field (Fig. 6.22). This shows convincingly that the model is at least qualitatively correct and that the Tm^{2+} triplet is indeed cooled by the ^{19}F spin system to a temperature which upon the existing evidence could be equal to its own.

It is clear that an electron–spin polarization higher than the initial thermal equilibrium polarization used in high field for nuclear polarization, can be

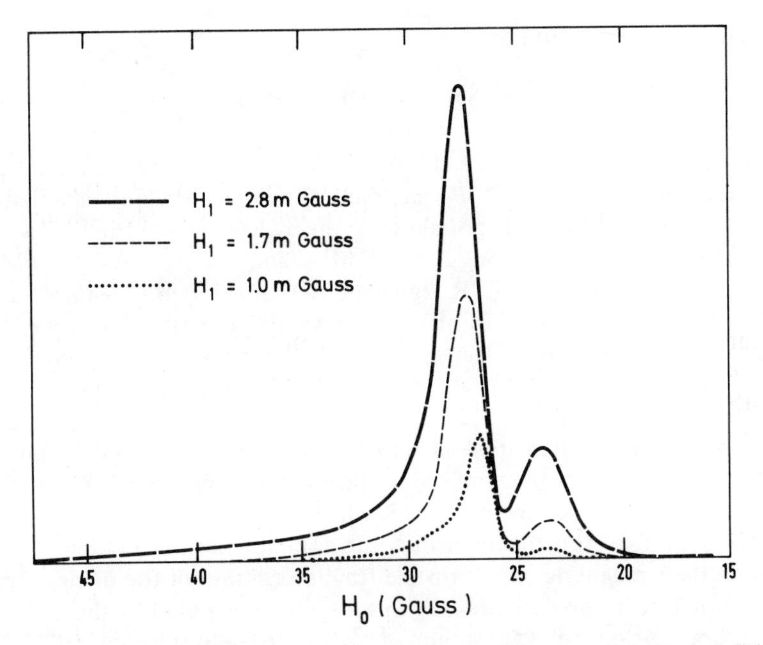

FIG. 6.22. ESR low field signals of Tm^{2+} in CaF_2, for different saturation conditions, exhibiting the two triplet transition lines. (After Urbina and Jacquinot, 1980.)

obtained by this procedure of first cooling the nuclei by the electrons and second the electrons by the nuclei, not unlike lifting oneself in the air by the bootstraps, whence the title of the section.

2. *The singlet*

It is also possible to create a process which is the reverse of the cooling of a nuclear Zeeman reservoir Z_I by an electronic Zeeman reservoir Z_S^* in the rotating frame (with as always $\mathcal{H}_{SS}^{\prime *}$ as an intermediary). Assume for definiteness sake that the spins of ^{19}F have been demagnetized at positive temperature so that in low field only the lowest level of the electronic triplet that we represent by the symbol $|-1\rangle$, is populated. The singlet level $|S\rangle$ is higher than $|-1\rangle$ by $(1100+\delta)$ MHz where $\delta \sim (\gamma_S/2)H_f^0$ depends on the low applied field H_f^0. $|S\rangle$ is also empty after the demagnetization. The two levels $|S\rangle$ and $|-1\rangle$ of Tm^{2+} can be considered as the two levels $|\pm\frac{1}{2}\rangle$ of a fictitious spin S' with a Larmor frequency, in megahertz: $\omega_S' \simeq 1100+\delta$. In the actual experiment $H_f^0 \sim 18$ Gauss, and $\delta \sim 45$ MHz. If we irradiate the system at a frequency $\omega = \omega_S' - \Delta$, in the frame rotating at that frequency the effective Larmor frequency of the fictitious spin is Δ.

If this frequency is on speaking terms with the ^{19}F spin system, the fictitious electronic Zeeman reservoir will reach the temperature $T(^{19}\text{F})$ and its fictitious polarization will be:

$$P_{S'} = 2\langle S'_z \rangle = n(|S\rangle) - n(|-1\rangle) = -\tanh\left(\frac{\Delta}{2k_B T(^{19}\mathrm{F})}\right), \qquad (6.132)$$

where $n(|S\rangle)$ and $n(|-1\rangle)$ are the populations of these two states. For Δ negative and $|\Delta| \gg k_B T(^{19}\mathrm{F})$, $P_{S'}$ is practically unity, which means that the singlet has been completely populated at the expense of the triplet. This was achieved in practice by taking $\Delta \approx -69$ MHz and $\omega = \omega_{S'} - \Delta \simeq 1214$ MHz.

Figure 6.23 shows an ESR signal of the triplet before and after the application of the r.f. at 1214 MHz. It shows that at least 80 per cent of the electronic spins have been transferred from the triplet into the singlet.

It is worth pointing out that the transfer of the electronic spins into the singlet state was the prime motivation of the experiment (Abragam, 1976). The paramagnetic impurities introduced into a sample for DNP can be a nuisance for certain applications, once the nuclei are polarized. We shall see examples of this fact in chapter 8 which deals with nuclear magnetic ordering. Transferring the impurities into a singlet non-magnetic state could quench their magnetic effects on the fragile ordering of the nuclear spins. Unfortunately in the case of CaF_2 this attempt was foiled by the unexpectedly short nuclear relaxation time of ^{19}F in zero field (Urbina, 1977*).

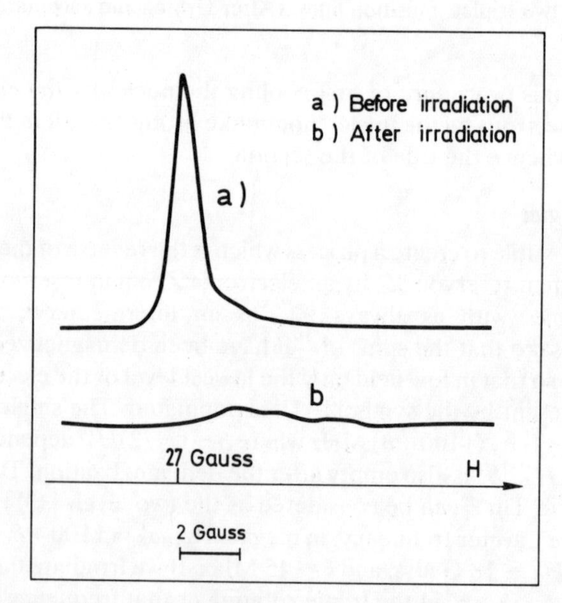

FIG. 6.23. Triplet ESR signal of Tm^{2+} in CaF_2: (a) before and: (b) after selective population of the singlet state. (After Urbina and Jacquinot, 1980.)

* C. Urbina (1977). Thèse de 3e cycle, Paris-Sud, Orsay, No 2343.

(*c*) *Cooling of a gas by laser irradiation: an analogue of DNP*

It has been suggested (Hänsch and Schawlow, 1975) and demonstrated by several authors (for instance Wineland, Drullinger, and Walls, 1978) that irradiation by laser light of an atomic vapour at a frequency: $\omega = \Omega - \Delta_0$ slightly *below* a sharp atomic absorption line of frequency Ω, would lead to an appreciable cooling of the translational energy of the vapour. The microscopic mechanism of this phenomenon can be understood qualitatively as follows: only atoms moving *toward* the source, and whose frequency in the laboratory frame is Doppler-shifted *downwards* can absorb resonantly the laser light. The conservation of momentum implies that they lose kinetic energy, whence the cooling.

It is easy to see that this phenomenon is thermodynamically identical to the cooling of a non-Zeeman reservoir by off-centre saturation of its Zeeman energy.

Let us call Z the Hamiltonian of the ensemble of two-level atomic systems driven by the laser and $\mathcal{H}_d$ the Hamiltonian with a continuous spectrum representing the translational energy of the atoms. The populations n_+ and n_- of the upper and lower level of an atom, under the combined effect of the laser light and of the finite lifetime of the upper level can be described by an inverse 'Zeeman' temperature β_Z with:

$$n_+/n_- = \exp(-\beta_Z\Omega).$$

Under steady state conditions, between the rate $\mathscr{E}_Z$ and $\mathscr{A}_Z$ for emission and absorption of light by each atom, there is a ratio:

$$\frac{\mathscr{E}_Z}{\mathscr{A}_Z} = \frac{n_-}{n_+} = \exp(\beta_Z\Omega). \tag{6.133}$$

If energy is to be conserved, each absorption (emission) of Ω by the 'Zeeman' system is accompanied by an emission (absorption) of an energy quantum Δ_0 by the translational energy reservoir $\mathcal{H}_d$ and $(\mathscr{A}_{\mathcal{H}_d}/\mathscr{E}_{\mathcal{H}_d}) = (\mathscr{E}_Z/\mathscr{A}_Z) = \exp(\beta_Z\Omega)$. If collisions between atoms maintain at all times the system $\mathcal{H}_d$ in a state of internal thermal equilibrium, under steady state conditions its inverse temperature β is given by:

$$\exp(\beta\Delta_0) = \frac{\mathscr{A}_{\mathcal{H}_d}}{\mathscr{E}_{\mathcal{H}_d}} = \frac{\mathscr{E}_Z}{\mathscr{A}_Z} = \exp(\beta_Z\Omega), \tag{6.134}$$

whence:

$$\beta = \beta_Z \frac{\Omega}{\Delta_0}, \tag{6.135}$$

and a cooling which can be very appreciable. The analogy with DNP is indeed very far reaching except for one point. The energy spectrum of the non-Zeeman Hamiltonian has an upper bound and can be given negative temperatures. There is symmetry between the effects of detuning the saturating frequency on either side of the resonance frequency Ω. There is nothing like it in the atomic vapour. An irradiation at $\omega = \Omega + \Delta$ leads to an unlimited heating of its translational energy.

The analogy can be pursued beyond thermodynamics: one can assimilate all the atoms with a given radial velocity v_r^i with respect to the light source to a spin-packet Δ^i, those for which the Doppler shift Δ^i is equal to Δ_0 correspond to the spin packet Δ_0^i which is being saturated. Radiationless collision between an atom in the ground state and another in the excited state is the equivalent of an electronic flip-flop. Collisions between two atoms, both in the excited state or in the ground state are equivalent in inhomogeneous broadening of the second type to the flip-flop of two nuclear spins. The lifetime of the excited state against radiation is the equivalent of the electronic relaxation time T_{1e}. The narrowing of the optical line by suppression of the Doppler shift, corresponds to the narrowing of the ESR line (accompanied by a shift), through driving of all the nuclear spins into their lowest energy level and thus of all the electronic spins into the spin packet of lowest energy.

7

NUCLEAR MAGNETISM AND NEUTRONS: NUCLEAR PSEUDOMAGNETISM

I am not what I am.

SHAKESPEARE (*Othello I, 1*)

THE usefulness of beams of slow neutrons as a tool for the study of condensed matter and in particular of its magnetic properties has been known for a long time. However its application to research in nuclear magnetism is much more recent since the very first experiments with neutron beams to yield new information in nuclear magnetism were performed in the early seventies. The reason for this lateness is the following. The superiority of the neutron beams over the classical magnetic measurements resides in the ability of the neutron to sense and to measure correlations between the orientations of spins, be they short range as in the paramagnetic state or long range as in the ordered states of magnetic moments, say, ferromagnetic and antiferromagnetic.

States of nuclear spins exhibiting long range order were first produced and observed by purely magnetic methods in 1969 (Chapellier, Goldman, Vu Hoang Chau, and Abragam, 1969) and observed by neutron diffraction in 1978 (Abragam, Bacchella, Bouffard, Goldman, Meriel, Pinot, Roinel, and Roubeau, 1978). Chapter 8 of this book is devoted to their study.

As for the short range correlations that exist in the paramagnetic state, the only short range correlations between the neighbouring nuclear spins revealed by neutron scattering have been those due to symmetry constraints stemming from the Pauli principle. The best known and fairly ancient example is the difference in neutron scattering by molecules of ortho- and para-hydrogen. On the other hand, short range correlations between nuclear spins induced by their dipolar interactions and evidenced by the existence of an NMR signal in the ADRF state, are at most of the order of $\beta\gamma H_L$, (actually $(\beta\gamma H_L)^2$) where β is the inverse dipolar temperature in the demagnetized state and H_L the dipolar local field. They are in general very small. (See however section **F**(c).)

The mechanism whereby the neutron senses the orientation of a magnetic atom is the electromagnetic interaction of its own magnetic moment with that of the atom. If the interaction between a neutron spin and a nuclear spin were also purely magnetic, it would be hopelessly inadequate for a study of nuclear magnetism, because of the smallness of the nuclear magnetic moments. Fortunately there exists another much stronger interaction between the spin of the neutron and that of a nucleus, which is a purely nuclear force. Although it has been known for many years, as witnessed for instance by the difference in the neutron scattering by ortho- and para-hydrogen, its systematic use for the study of nuclear magnetism is fairly recent as mentioned earlier. For reasons that will appear later we have found it convenient to refer to the manifold aspects of the interaction of the spin of the neutron with a system of nuclear spins as nuclear pseudomagnetism. The present chapter is devoted to the study of the various features of nuclear pseudomagnetism and of their applications to the study of nuclear magnetism proper. In order to make this chapter reasonably self-contained we have included a few developments perhaps less well known to students of nuclear magnetism than to those of neutron scattering. The latter will do well to skip these developments.

A. Slow neutron scattering by an isolated nucleus

(*a*) *Scattering on a spinless nucleus*

We consider first the elastic scattering of a neutron by a fixed nucleus placed at the origin. There is an apparent contradiction in studying the scattering of a neutron by a nucleus that is both isolated and fixed in space. Actually this section is an introduction to a situation where the nucleus is embedded in a solid target and maintained more or less fixed by its environment. The wave function of the neutron obeys the Schrödinger equation:

$$-\frac{\hbar^2}{2M}\nabla^2\psi + V(\mathbf{r})\psi = \frac{\hbar^2 k^2}{2M}\psi, \tag{7.1}$$

where $V(\mathbf{r})$ is the interaction potential between the neutron and the nucleus, M the mass of the scattered neutron and k its wave number. The incoming neutron is represented by the plane wave $\psi_0 = \exp(i\mathbf{k}\,.\,\mathbf{r})$ and the scattering process is described by a solution $\psi_+(\mathbf{r})$ of (7.1) which at a large distance from the scatterer is the sum of the incoming plane wave and of an outgoing spherical wave, that is with an asymptotic behaviour:

$$\psi_+(\mathbf{r}) \to \exp(i\mathbf{k}\,.\,\mathbf{r}) + f(\theta)\frac{\exp(ikr)}{r}, \tag{7.2}$$

where $f(\theta)$, called the scattering amplitude, depends in general on the angle θ between the incoming and outgoing particle. We rewrite eqn (7.1) as:

$$(\nabla^2 + k^2)\psi = \frac{2M}{\hbar^2}V(\mathbf{r})\psi. \tag{7.3}$$

The solution of (7.3) that has the asymptotic behaviour (7.2) is given by the integral equation:

$$\psi_+(\mathbf{r}) = \exp(i\mathbf{k}\,.\,\mathbf{r}) - \frac{2M}{4\pi\hbar^2}\int \psi_+(\mathbf{r}')\,V(\mathbf{r}')\frac{\exp(ik|\mathbf{r}-\mathbf{r}'|)}{|\mathbf{r}-\mathbf{r}'|}\,d^3r', \qquad (7.4)$$

where the Green function,

$$G(\mathbf{r}-\mathbf{r}') = -\frac{1}{4\pi}\,\frac{\exp(i\mathbf{k}|\mathbf{r}-\mathbf{r}'|)}{|\mathbf{r}-\mathbf{r}'|},$$

is the solution of the equation,

$$(\nabla^2 + k^2)G(\mathbf{r}-\mathbf{r}') = \delta(\mathbf{r}-\mathbf{r}'), \qquad (7.5)$$

which for $r \gg r'$ behaves like $\exp(ikr)/r$. When $r \gg r'$, $|r-r'| \sim r - (\mathbf{r}\,.\,\mathbf{r}'/r)$ and (7.4) takes the asymptotic form:

$$\psi_+(\mathbf{r}) \rightarrow \exp(i\mathbf{k}\,.\,\mathbf{r}) - \frac{2M}{\hbar^2}\,\frac{1}{4\pi}\,\frac{\exp(ikr)}{r}\int \psi_+(\mathbf{r}')V(\mathbf{r}')\exp(-i\mathbf{k}'\,.\,\mathbf{r}')\,\mathrm{d}^3r', \qquad (7.6)$$

where $\mathbf{k}' = k(\mathbf{r}/r)$ is the wave vector of the wave scattered in the direction $\mathbf{r}$. The scattering amplitude $f(\theta)$ of eqn (7.2) becomes:

$$f(\theta) = -\frac{M}{2\pi\hbar^2}\int \psi_+(\mathbf{r}')V(\mathbf{r}')\exp(-i\mathbf{k}'\,.\,\mathbf{r}')\,\mathrm{d}^3r', \qquad (7.7)$$

where θ is the angle between $\mathbf{k}$ and $\mathbf{k}'$. There are two cases, both of interest to us, when eqn (7.7) is considerably simplified.

The first is when the scattering potential $V(\mathbf{r})$ is weak and the solution $\psi_+(\mathbf{r})$ of (7.4) differs little from the incoming wave $\exp(i\mathbf{k}\,.\,\mathbf{r})$, and can be replaced by it in the integral of eqn (7.7). This is known as the Born approximation. The scattering amplitude $f(\theta)$ is then given explicitly by:

$$f(\theta) = -\frac{M}{2\pi\hbar^2}\int \exp(i(\mathbf{k}-\mathbf{k}')\,.\,\mathbf{r}')V(\mathbf{r}')\,\mathrm{d}^3r'. \qquad (7.8)$$

The second simplification occurs when the range of the interaction $V(\mathbf{r})$ is very much smaller than the wavelength $\lambda = 2\pi/k$ of the neutron. We can then replace by unity in the integral (7.7), the exponential $\exp(-i\mathbf{k}'\,.\,\mathbf{r}')$, which through the vector $\mathbf{k}' = k(\mathbf{r}/r)$ defines the direction of the scattered neutron. Then the scattering amplitude,

$$f = -\frac{M}{2\pi\hbar^2}\int \psi_+(\mathbf{r}')V(\mathbf{r}')\,d^3\mathbf{r}', \qquad (7.9)$$

does not depend on the angle θ between $\mathbf{k}$ and $\mathbf{k}'$ and the scattering becomes isotropic. This is the situation which prevails in nuclear scattering of slow neutrons. An isotropic scattering corresponds to an orbital momentum $l = 0$ of the scattered neutrons, called for that reason s-neutrons.

Neutron physicists prefer to define a nuclear scattering amplitude $a = -f$. We shall abide by their custom and write the isotropic scattered wave as:

$$-a\frac{\exp(ikr)}{r}.$$

In contrast to the Born approximation (7.8) it would be profoundly incorrect to replace $\psi_+(\mathbf{r}')$ by $\exp(i\mathbf{k}\,.\,\mathbf{r}')$ in (7.9). The nuclear interaction V is strong and distorts considerably the wave function within its range. It is not possible to calculate the nuclear scattering amplitude a from first principles. On the other hand it is often convenient, as we shall see later in a particular example, to be able to treat on the same footing the weak interactions (of possibly long range) describable by the Born approximation and the strong short range nuclear interaction which results in an isotropic scattering amplitude a. This is done by introducing the so-called fictitious Fermi potential:

$$V_F(\mathbf{r}) = \frac{2\pi\hbar^2}{M}a\,\delta(\mathbf{r}). \tag{7.10}$$

$V_F(\mathbf{r})$ is *not* the actual interaction between the nucleus and the neutron. However, by pretending that it is, and also that it is weak and using it in the Born approximation (7.8) we find indeed:

$$a = -f = \frac{M}{2\pi\hbar^2}\int \exp(i(\mathbf{k}-\mathbf{k}')\,.\,\mathbf{r}')V_F(\mathbf{r}')\,d^3r' = a. \tag{7.11}$$

In other words $V_F(\mathbf{r})$ is the wrong potential which, used in the wrong (Born) approximation, gives the right answer. As was shown by Fermi and others (see for instance Blatt and Weisskopf, 1952, p. 71) the use of $V_F(\mathbf{r})$ is justified in all problems where the wavelength of the neutron and also the wavelength associated with a possible motion of the scattering nucleus within the target are both much larger than the nuclear scattering amplitude a. This makes the Fermi potential (7.10) particularly convenient in a realistic description of the scattering of thermal neutrons by nuclei embedded and moving in bulk matter at thermal velocities.

The scattering amplitude $a = a' + ia''$ may have an imaginary part a''. The differential scattering cross section is: $d\sigma/d\Omega = |a|^2 = a'^2 + a''^2$ and since the nuclear scattering is isotropic the integrated scattered cross section is $\sigma^{sc} = 4\pi|a|^2$. The magnitude of the imaginary part a'' is given by the optical theorem which states quite generally (and not only for s-scattering):

$$\operatorname{Im} f(0) = \frac{k}{4\pi}(\sigma^{sc} + \sigma^{c}) = \frac{k}{4\pi}\sigma^{T} \tag{7.12}$$

where $f(0)$ is the forward scattering amplitude (see for instance Cohen-Tannoudji, Diu, and Laloe, 1973, p. 944).

σ^c (c for capture) is the absorption cross section relative to all the processes different from an elastic scattering of the neutron. For thermal neutrons, which have practically zero kinetic energy, an inelastic scattering which would leave the target nucleus in an excited state is excluded. The neutron is actually absorbed by the target nucleus which then disintegrates into various products. A typical example is:

$$\mathrm{n} + {}^6\mathrm{Li} \to {}^3\mathrm{H} + {}^4\mathrm{He}. \tag{7.13}$$

(Strictly speaking this example does not belong in this section for ^{6}Li does have a spin $I = 1$.)

The optical theorem expresses simply a conservation of flux: the total flux of incoming particles (neutrons in our case) through a sphere of radius r is equal to the total flux of outgoing particles plus those absorbed: this is particularly easy to check for s-neutrons. The s-part of the incoming plane wave $\exp(i\mathbf{k}\,.\,\mathbf{r})$ is:

$$\frac{1}{4\pi}\int \exp(i\mathbf{k}\,.\,\mathbf{r})\,d\Omega = \tfrac{1}{2}\int_{-1}^{1} \exp(ikru)\,du = \frac{1}{2ikr}\{\exp(ikr) - \exp(-ikr)\}. \tag{7.14}$$

The outgoing wave being $-[(a' + ia'')/r]\exp(ikr)$ the total incoming and outgoing fluxes are:

$$\Phi_{\text{in}} = v4\pi r^2 \frac{1}{4k^2r^2}; \qquad \Phi_{\text{out}} = v4\pi r^2 \left| a' + ia'' - \frac{1}{2ik} \right|^2 r^{-2}, \tag{7.15}$$

where $v = \hbar k/M$ is the velocity of the neutrons. By writing that $\Phi_{\text{in}} - \Phi_{\text{out}} = v\sigma^c$ one gets:

$$\frac{1}{4k^2} - \left(a'^2 + \left(a'' + \frac{1}{2k}\right)^2\right) = \frac{\sigma^c}{4\pi},$$

or:

$$a'' = -\frac{k}{4\pi}\{4\pi(a'^2 + a''^2) + \sigma^c\} = -\frac{k}{4\pi}(\sigma^{\text{sc}} + \sigma^c) = -\frac{k}{4\pi}\sigma^{\text{T}}, \tag{7.16}$$

which for s-neutrons is the same as (7.12). As the velocity v of the neutron goes toward zero the scattering cross section $4\pi|a|^2$ tends to a fixed limit whereas σ^c is known to go up as $1/v$. It follows that for slow neutrons in the absence of capture the imaginary part a'' is very small with:

$$\left|\frac{a''}{a'}\right| \simeq |ka'| \simeq \left|\frac{2\pi a'}{\lambda}\right| \simeq 2\pi\frac{|a|}{\lambda}. \tag{7.17}$$

Typical values are $a \approx 10^{-12}$ cm, $\lambda \sim 10^{-8}$ cm whence $|a''/a| \leqslant 10^{-3}$, and a can be taken as real.

Even for a nucleus with a very large capture cross section like ^{6}Li where for $\lambda = 10^{-8}$ cm, $\sigma^c = 520$ Barns, and $a = 1{\cdot}8 \times 10^{-13}$ cm,

$$\left|\frac{a''}{a'}\right| = \frac{\sigma^c}{2\lambda a} = \frac{520 \times 10^{-24}}{2 \times 10^{-8} \times 1{\cdot}8 \times 10^{-13}} = 0{\cdot}14, \tag{7.18}$$

and the neglect of a'' in the scattering amplitude $a' + ia''$ affects the scattering cross section only by 2 per cent.

(b) Slow neutron scattering by a nucleus with a spin I

In the scattering of a neutron by a nucleus with a spin I, the magnitude J of the total angular momentum $\mathbf{J} = \mathbf{I} + \mathbf{s} + \mathbf{l}$, where $\mathbf{s}$ is the neutron spin and $\mathbf{l}$ its orbital momentum with respect to the nucleus, is a good quantum number. As we saw earlier, for slow neutrons the scattering is negligible for $l \neq 0$. There are then only two possible values for J, $J_\pm = I \pm \frac{1}{2}$ to which correspond two scattering amplitudes, that, to conform to current notations, we denote by b_+ and b_-. It is convenient to express the scattering amplitude in an operator form as:

$$a = b_0 + b\mathbf{I}\,.\,\mathbf{s}. \tag{7.19}$$

The constants b_0 and b are determined by noticing that the operator $(\mathbf{I}\,.\,\mathbf{s})$ has different eigenvalues in the states $J_\pm = I \pm \frac{1}{2}$:

$$(\mathbf{I}\,.\,\mathbf{s})_\pm = \tfrac{1}{2}\{J_\pm(J_\pm + 1) - I(I+1) - s(s+1)\} = \tfrac{1}{2}\{(I \pm \tfrac{1}{2})(I \pm \tfrac{1}{2} + 1) - I(I+1) - \tfrac{3}{4}\};$$

$$(\mathbf{I}\,.\,\mathbf{s})_+ = \frac{I}{2}; \qquad (\mathbf{I}\,.\,\mathbf{s})_- = -\frac{(I+1)}{2}; \tag{7.20}$$

$$b_+ = b_0 + b(\mathbf{I}\,.\,\mathbf{s})_+ = b_0 + b\frac{I}{2}; \qquad b_- = b_0 + b(\mathbf{I}\,.\,\mathbf{s})_- = b_0 - b\left(\frac{I+1}{2}\right);$$

$$b_0 = \frac{(I+1)b_+ + Ib_-}{(2I+1)}; \qquad b = \frac{2(b_+ - b_-)}{(2I+1)}. \tag{7.21}$$

The notation (7.19) for the scattering amplitude has been introduced simply as a shorthand reminder of the fact that the scattering channels $J_\pm = I \pm \frac{1}{2}$ have different amplitudes $b_\pm$. We can however consider the spin-dependent part $b(\mathbf{I}\,.\,\mathbf{s})$ as resulting from an interaction between the spins of the neutron and the nucleus, represented in accordance with (7.10) by a Fermi potential:

$$V_F(\mathbf{r}) = \frac{2\pi\hbar^2}{M} b(\mathbf{I}\,.\,\mathbf{s})\,\delta(\mathbf{r}). \tag{7.22}$$

As stated earlier, we know that the interaction (7.22) used in the Born approximation gives the correct value for the spin-dependent part of the nuclear scattering of slow neutrons. It is interesting to compare the amplitude (7.19) and the (pseudo) interaction (7.22) with the *magnetic* interaction, between the magnetic moment of the neutron and that of the target nucleus and with the corresponding scattering amplitude.

$$\boldsymbol{\mu}_{\rm N} = \gamma_{\rm N}\hbar\mathbf{s} = 2\mu_{\rm N}\mathbf{s} = 2g_{\rm N}\frac{e\hbar}{2Mc}\mathbf{s} \tag{7.23}$$

represents the magnetic moment of the neutron. The constant $g_{\rm N}$ expresses its value in units of nuclear Bohr magnetons: $g_{\rm N} = -1{\cdot}91$. Similarly $\boldsymbol{\mu}_{\rm n} = \gamma_{\rm n}\hbar\mathbf{I} = (\mu_{\rm n}/I)\mathbf{I}$ is the magnetic moment of the target nucleus. As is well known, $\mu_{\rm n}$ produces a vector potential $\mathbf{A}_{\rm n} = {\rm curl}(\boldsymbol{\mu}_{\rm n}/r) = \boldsymbol{\nabla}\wedge(\boldsymbol{\mu}_{\rm n}/r)$ from which derives a magnetic field:

$$\mathbf{H}_{\rm n} = \boldsymbol{\nabla}\wedge\mathbf{A}_{\rm n}(\mathbf{r}) = \boldsymbol{\nabla}\wedge\left(\boldsymbol{\nabla}\wedge\left(\frac{\boldsymbol{\mu}_{\rm n}}{r}\right)\right). \tag{7.24}$$

The magnetic interaction of that field with the magnetic moment $\mu_{\rm N}$ of the neutron is:

$$\begin{aligned} V_{\rm M}(\mathbf{r}) &= -\boldsymbol{\mu}_{\rm N}\,.\,\mathbf{H}_{\rm n} = -\boldsymbol{\mu}_{\rm N}\,.\left\{\nabla\wedge\left(\nabla\wedge\left(\frac{\boldsymbol{\mu}_{\rm n}}{r}\right)\right)\right\} \\ &= -(\boldsymbol{\mu}_{\rm N}\,.\,\boldsymbol{\nabla})(\boldsymbol{\mu}_{\rm n}\,.\,\boldsymbol{\nabla})\left(\frac{1}{r}\right) + (\boldsymbol{\mu}_{\rm N}\,.\,\boldsymbol{\mu}_{\rm n})\nabla^2\left(\frac{1}{r}\right). \end{aligned} \tag{7.25}$$

Using the Born approximation we find for the magnetic scattering amplitude:

$$a_{\rm M} = \frac{M}{2\pi\hbar^2}\int \exp({\rm i}(\mathbf{k}-\mathbf{k}')\,.\,\mathbf{r})V_{\rm M}(\mathbf{r})\,{\rm d}^3r, \tag{7.26}$$

which after integration by parts yields:

$$a_{\rm M} = -\frac{2M}{\hbar^2}\{(\boldsymbol{\mu}_{\rm N}\,.\,\boldsymbol{\mu}_{\rm n}) - (\boldsymbol{\mu}_{\rm N}\,.\,\hat{\mathbf{u}})(\boldsymbol{\mu}_{\rm n}\,.\,\hat{\mathbf{u}})\} \tag{7.27}$$

where $\hat{\mathbf{u}} = (\mathbf{k}-\mathbf{k}')/|\mathbf{k}-\mathbf{k}'|$, is a unit vector along the impulse transfer of the neutron.

There is a great similarity between the nuclear amplitude $b(\mathbf{I}\,.\,\mathbf{s})$ and the magnetic amplitude (7.27), both bilinear with respect to the spins of the neutron and of the nucleus although the latter is anisotropic with respect to the impulse transfer. (It can be written $a_{\rm M} = -(2M/\hbar^2)(\boldsymbol{\mu}_{\rm N}\cdot\boldsymbol{\mu}_{{\rm n}\perp})$ where $\boldsymbol{\mu}_{{\rm n}\perp}$ is the component of $\boldsymbol{\mu}_{\rm n}$ perpendicular to the impulse transfer.) It is illuminating to assign to the nucleus a fictitious pseudomagnetic moment

$\boldsymbol{\mu}_n^* = (\boldsymbol{\mu}_n^*/I)\mathbf{I}$ such that in analogy with (7.27) the nuclear scattering amplitude $b(\mathbf{I}\,.\,\mathbf{s})$ be written $-(2M/\hbar^2)(\boldsymbol{\mu}_N\,.\,\boldsymbol{\mu}_n^*)$. Using (7.23) we find:

$$-\frac{2M}{\hbar^2}(\boldsymbol{\mu}_N\,.\,\boldsymbol{\mu}_n^*) = -\frac{2M}{\hbar^2}2g_N\frac{e\hbar}{2Mc}\frac{\mu_n^*}{I}(\mathbf{I}\,.\,\mathbf{s}) = b(\mathbf{I}\,.\,\mathbf{s}), \qquad (7.28)$$

whence:

$$\mu_n^* = -\frac{Ib}{2}\frac{\hbar c}{e}\frac{1}{g_N} = \frac{Ib}{2}\frac{\hbar c}{e}\frac{1}{|g_N|}. \qquad (7.29)$$

A more interesting way of writing μ_n^* is by introducing the classical radius of the electron $r_0 = e^2/mc^2 = 2{\cdot}8\times10^{-13}$ cm and the Bohr magneton (electronic not nuclear) $\mu_B = e\hbar/2mc$ where m is the mass of the electron. Equation (7.29) can then be given the dimensionless form:

$$\frac{\mu_n^*}{\mu_B} = \frac{I}{|g_N|}\left(\frac{b}{r_0}\right) \quad \text{with } |g_N| = 1{\cdot}91. \qquad (7.30)$$

Since for some nuclei the spin-dependent nuclear amplitude b turns out to be comparable to, and even sometimes a good deal larger than r_0, eqn (7.30) shows that the pseudomagnetic moment μ_n^* can be of the order of a Bohr magneton and that the nuclear spin-dependent scattering can be comparable in magnitude to the magnetic scattering on electronic magnetic moments. Thus for the proton $b = 5{\cdot}8\times10^{-12}$ and $\mu_P^* = \mu^*(^1\text{H}) = 5{\cdot}4\mu_B = 3600\mu_P$ where μ_P is the *true* magnetic moment of the proton. It is then reasonable to expect that if polarizations and/or spatial correlations comparable to those that exist for electron spins can be produced for nuclear spins, they can be investigated in a similar manner by neutron beams. Actually as we shall see shortly there is more to the definition (7.29) of a nuclear pseudo-magnetic moment than a simple comparison of magnitude with electron magnetic scattering.

B. Slow neutron scattering by a macroscopic target: Bragg scattering

The four features which have made the slow neutron such a marvellous tool for the study of bulk matter are the following.

(i) Its absence of electric charge enables it to penetrate deeply into matter in contrast with charged particles.

(ii) Its wavelength, comparable to the size of atoms or molecules enables it to probe spatial correlations on that scale.

(iii) Its energy E, of the order of the thermal energy of the atoms or molecules and what is even more important, the resolution in energy δE, which, thanks to the progress of techniques, is a very small fraction of E, enables it, through inelastic scattering, to probe time correlations on a time scale of the order of $(\hbar/\delta E)$.

(iv) Because of its magnetic moment, the neutron can probe such correlations not only between positions of nuclei but also between orientations of electronic magnetic moments.

An enormous literature exists on the subject that we shall not attempt even to skim. Our purpose is to consider what these four basic features of the neutron can bring to the study of nuclear magnetism.

Features (i) and (ii) clearly play no new roles. For feature (iv) the magnetic interaction of the neutron magnetic moment with the magnetic moments of the electrons is replaced by the pseudo-magnetic interaction of the neutron spin with the nuclear spins, and spatial correlations between nuclear spins can be probed in a manner very similar to that used for electronic spins. For feature (iii) the situation at least in the present state of techniques is somewhat different. The spread ΔE in the spectrum of the interaction energy of a system of nuclear spins, smaller than a microkelvin, is far less than the resolution δE of present-day spectrometers. This means that the time-dependence of the correlation of a nuclear spin with a neighbouring spin or with itself cannot be measured by neutron scattering in the present state of the art. The observed differential cross sections $d\sigma/d\Omega$ are perforce integrals over all the energies transferred by the motion of the spins to the scattered neutrons. This loss of information results in a considerable simplification of the formalism since the cross section $d\sigma/d\Omega$ over a macroscopic target is then simply the squared modulus of the scattering amplitude, over a rigid system of spins. The foregoing does not take into account the exchange of energy between the neutron and the nuclei due to the motions of the atoms (rather than of the nuclear spins). Those we disregard not because they are very small but because they are of no interest for the study of nuclear magnetism.

(a) *Spinless nuclei*

If a scatterer is placed at the point r_i rather than at the origin the asymptotic expression of $\psi_+(\mathbf{r})$ namely $\exp i(\mathbf{k}\,.\,\mathbf{r})-(a/r)\exp(ikr)$ becomes through a change in the origin of the co-ordinates:

$$\psi_+(\mathbf{r})\rightarrow\exp(i\mathbf{k}\,.\,(\mathbf{r}-\mathbf{r}_i))-\frac{a}{|\mathbf{r}-\mathbf{r}_i|}\exp(ik\,.\,|\mathbf{r}-\mathbf{r}_i|). \tag{7.31}$$

We can use instead the wave function:

$$\psi_+(\mathbf{r})=\exp(i\mathbf{k}\,.\,\mathbf{r})-a\exp(i(\mathbf{k}\,.\,\mathbf{r}_i))\frac{\exp(ik|\mathbf{r}-\mathbf{r}_i|)}{|\mathbf{r}-\mathbf{r}_i|}, \tag{7.31$'$}$$

which differs from (7.31) only by an overall phase factor. It is then clear that the scattering by a large number of nuclei will be represented by:

$$\psi_+(\mathbf{r})=\exp(i\mathbf{k}\,.\,\mathbf{r})-\sum_i a_i\exp(i(\mathbf{k}\,.\,\mathbf{r}_i))\frac{\exp(ik|\mathbf{r}-\mathbf{r}_i|)}{|\mathbf{r}-\mathbf{r}_i|}. \tag{7.32}$$

For $r \gg r_i$,

$$|\mathbf{r}-\mathbf{r}_i| \approx r - \frac{(\mathbf{r}\,.\,\mathbf{r}_i)}{r},$$

and writing $\mathbf{k}' = k(\mathbf{r}/r)$, (7.32) becomes:

$$\psi_+(\mathbf{r}) = \exp \mathrm{i}(\mathbf{k}\,.\,\mathbf{r}) - \frac{\exp(\mathrm{i}kr)}{r} \sum_i a_i \exp(-\mathrm{i}\boldsymbol{\kappa}\,.\,\mathbf{r}_i) \tag{7.33}$$

where $\boldsymbol{\kappa} = \mathbf{k}' - \mathbf{k}$ is the momentum transfer. There is a striking difference between the scattering by a single nucleus and by an assembly of nuclei: whereas the former is isotropic, the latter with an amplitude,

$$\mathscr{A} = \sum_i a_i \exp(-\mathrm{i}\boldsymbol{\kappa}\,.\,\mathbf{r}_i), \tag{7.34}$$

clearly is not. The differential cross section resulting from (7.34) is

$$\frac{\mathrm{d}\sigma}{\mathrm{d}\Omega} = |\mathscr{A}|^2 = \sum_{i,j} a_i a_j^* \exp[-\mathrm{i}\boldsymbol{\kappa}\,.\,(\mathbf{r}_i - \mathbf{r}_j)]. \tag{7.35}$$

Assume first for simplicity that there is only one species of nuclei in the target and that this species has spin zero. Then all the a_i have the same value a and (7.35) can be rewritten:

$$\frac{\mathrm{d}\sigma}{\mathrm{d}\Omega} = |a|^2 \sum_{i,j} \exp[-\mathrm{i}\boldsymbol{\kappa}\,.\,(\mathbf{r}_i - \mathbf{r}_j)]. \tag{7.36}$$

Assume further that the target is a crystal with a single atom per unit cell. The double sum in (7.36) can be written $N \sum_i \exp(-\mathrm{i}\boldsymbol{\kappa}\,.\,\mathbf{r}_i)$ where N is the number of scattering nuclei and the single sum is to be taken over all the cells. It is well known that this sum vanishes unless the vector $\boldsymbol{\kappa} = \mathbf{k}' - \mathbf{k}$ is a lattice vector $\boldsymbol{\tau}$ of the reciprocal lattice. It can be shown (see for instance Squires, 1978, p. 32) that:

$$\frac{\mathrm{d}\sigma}{\mathrm{d}\Omega} = N|a|^2 \sum_i \exp(-\mathrm{i}\boldsymbol{\kappa}\,.\,\mathbf{r}_i) = \frac{N|a|^2}{v_0}(2\pi)^3 \sum_{\boldsymbol{\tau}} \delta(\boldsymbol{\tau} - \boldsymbol{\kappa}), \tag{7.37}$$

where v_0 is the volume of the unit cell. We see that the scattering is violently anisotropic and that it occurs only for special orientations of the incoming wave vector $\mathbf{k}$ and of the outgoing wave vector $\mathbf{k}'$ with $|\mathbf{k}'| = |\mathbf{k}|$, such that $\mathbf{k}' - \mathbf{k} = \boldsymbol{\tau}$. The various peaks in the scattering along $\mathbf{k}'$ which occur when $\mathbf{k}$ has the right orientation with respect to the crystal are called Bragg peaks. No scattering whatsoever occurs outside of the Bragg peaks. One says that the scattering is entirely coherent. The conditions: $\mathbf{k}' - \mathbf{k} = \boldsymbol{\tau}$; $|\mathbf{k}'| = |\mathbf{k}|$ require that $\mathbf{k}$ and $\mathbf{k}'$ make with the lattice plane perpendicular to $\boldsymbol{\tau}$ the same angle θ and that $|\boldsymbol{\tau}| = 2k \sin\theta$ or $|\boldsymbol{\tau}| = (4\pi \sin\theta)/\lambda$. We see that if $\lambda > 4\pi/\tau_0$, where τ_0 is the smallest of the vectors of the reciprocal lattice, no Bragg scattering can occur. The length $\lambda = 4\pi/\tau_0$ is called the cut-off wavelength.

In practice the expression (7.37) of infinitely narrow and infinitely high Bragg peaks for the scattering by a perfect crystal of perfectly mono-energetic and perfectly collimated neutrons is unrealistic. Imperfections in any of the parameters listed above result in finite width and finite height for the Bragg peaks.

Assume for simplicity that the only imperfection is the collimation of the incident beam. This beam has a small but finite aperture and, as the crystal is slowly rotated, different neutrons from this beam are scattered coherently satisfying in turn the condition: $\boldsymbol{\kappa} = \mathbf{k}' - \mathbf{k} = \boldsymbol{\tau}$.

The counting rate $I(\theta)$ of the scattered neutrons plotted against the orientation of the crystal is what is known as the rocking curve. Its finite width reflects the angular spread of the incident beam but in practice it gets contributions from the spread in energy and from imperfections in the crystal. Let us call P the number of neutrons scattered per unit time into the Bragg peak integrated over the rocking curve. It is a simple matter to establish that P is related to the incident flux Φ over unit surface, by:

$$\frac{P}{\Phi} = \frac{V}{v_0^2} \frac{\lambda}{\sin 2\theta} |a|^2, \tag{7.38}$$

where V is the volume of the sample and v_0 that of the unit cell (see for instance Squires, 1978, p. 42).

The restriction that the crystal contains a single atom per unit cell can be lifted. Assume that the unit cell contains ρ atoms, whose nuclei have scattering amplitudes $a_1, \ldots a_\alpha, \ldots a_\rho$ which occupy within the unit cell positions $d_1, \ldots d_\sigma, \ldots d_\rho$. We also assume that for each atomic species there is only one isotope. It is easily shown that the eqns (7.37) and (7.38) are still valid provided the scattering amplitude a is replaced by the scattering amplitude of the unit cell relative to the impulse transfer $\boldsymbol{\kappa}$:

$$F(\boldsymbol{\kappa}) = \sum_{\alpha=1}^{\rho} a_\alpha \exp(-i\boldsymbol{\kappa} \cdot \mathbf{d}_\alpha). \tag{7.39}$$

We consider now a situation where the crystal, with still a single atom per unit cell, is made of several isotopes distributed at random. Let $a_1, a_2, \ldots a_\nu$ be their different scattering amplitudes and $C_1, C_2, \ldots C_\nu$ the relative isotopic concentrations with $\sum C_\nu = 1$. (It should be understood that in eqn (7.35) a_i means scattering amplitude of the nucleus at site i, whereas a_ν means the scattering amplitude of a nucleus belonging to the isotope ν.) Since the isotopes are distributed at random there is no correlation between the relative positions $(\mathbf{r}_i - \mathbf{r}_j)$ and the products $a_i^* a_j$ of the scattering amplitudes. The nuclear target which contains a very large number of nuclei can be considered as a statistical ensemble and we can replace on the right-hand side of (7.35) $a_i a_j^* \exp[-i\boldsymbol{\kappa} . (\mathbf{r}_i - \mathbf{r}_j)]$ by:

$$\langle a_i a_j^* \exp(-i\boldsymbol{\kappa} . (\mathbf{r}_i - \mathbf{r}_j))\rangle = \langle a_i a_j^* \rangle \exp(-i\boldsymbol{\kappa} . (\mathbf{r}_i - \mathbf{r}_j)).$$

Equation (7.35) can be rewritten:

$$\frac{d\sigma}{d\Omega} = \sum_i \langle |a_i|^2 \rangle + \langle a_i a_j^* \rangle_{i \neq j} \sum_{i \neq j} \exp(-i\boldsymbol{\kappa} . (\mathbf{r}_i - \mathbf{r}_j)); \qquad (7.40a)$$

$$\langle a_i a_j^* \rangle_{i \neq j} = \sum_{\nu,\nu'} C_\nu C_{\nu'} a_\nu a_{\nu'}^* = |\langle a \rangle|^2;$$
$$\langle |a_i|^2 \rangle = \langle |a|^2 \rangle = \sum_\nu C_\nu |a_\nu|^2. \qquad (7.40b)$$

We can add and subtract on the right hand side of (7.40*a*):

$$N\langle a_i a_j^* \rangle_{i \neq j} = N|\langle a \rangle|^2;$$

$$\frac{d\sigma}{d\Omega} = N\{\langle |a|^2 \rangle - |\langle a \rangle|^2\} + |\langle a \rangle|^2 \sum_{i,j} \exp(-i\boldsymbol{\kappa} . (\mathbf{r}_i - \mathbf{r}_j))$$
$$= N\{\langle |a|^2 \rangle - |\langle a \rangle|^2\} + N|\langle a \rangle|^2 \sum_i \exp(-i\boldsymbol{\kappa} . \mathbf{r}_i). \qquad (7.41)$$

The second term in (7.41) is the same as in (7.36) for a single isotope. It gives rise to Bragg scattering with a coherent amplitude $\langle a \rangle = \sum_\nu C_\nu a_\nu$. There is now however an isotropic scattering, called incoherent with an incoherent cross-section:

$$\left(\frac{d\sigma}{d\Omega}\right)_{\text{inc}} = N\{\langle |a|^2 \rangle - |\langle a \rangle|^2\} = N\sum_\nu C_\nu |a_\nu|^2 - \left|\sum_\nu C_\nu a_\nu\right|^2. \qquad (7.41')$$

The generalization to a lattice with more than one atom per unit cell is straightforward: in the scattering amplitude (7.39), for each atomic site $\mathbf{d}_\alpha$ in the unit cell, the amplitude a_α is to be replaced by the average $\langle a_\alpha \rangle$ over the various isotopes.

(b) Nuclei with spin

This is the only situation of interest for nuclear magnetism. Its handling is greatly simplified by the operator notation (7.19) for the scattering amplitude. We can write, using standard commutation relations of spin components:

$$|a_i|^2 = [b_0 + b(\mathbf{s} . \mathbf{I}_i)]^2 = b_0^2 + 2bb_0(\mathbf{s} . \mathbf{I}_i) + b^2(\mathbf{s} . \mathbf{I}_i)^2;$$
$$(\mathbf{s} . \mathbf{I}_i)^2 = \tfrac{1}{4}(\mathbf{I}_i . \mathbf{I}_i) + \frac{i\mathbf{s}}{2}(\mathbf{I}_i \wedge \mathbf{I}_i) = \frac{I(I+1)}{4} - \frac{(\mathbf{s} . \mathbf{I}_i)}{2}; \qquad (7.42)$$

whence:

$$|a_i|^2 = b_0^2 + 2bb_0(\mathbf{s}\,.\,\mathbf{I}_i) + b^2\left\{\frac{I(I+1)}{4} - \frac{(\mathbf{s}\,.\,\mathbf{I}_i)}{2}\right\};$$

$$a_i a_j^* = [b_0 + b(\mathbf{s}\,.\,\mathbf{I}_i)][b_0 + b(\mathbf{s}\,.\,\mathbf{I}_j)] \tag{7.43}$$

$$= b_0^2 + b_0 b\{(\mathbf{s}\,.\,\mathbf{I}_i) + (\mathbf{s}\,.\,\mathbf{I}_j)\} + b^2\left\{\frac{(\mathbf{I}_i\,.\,\mathbf{I}_j)}{4} + \frac{i\mathbf{s}}{2}\,.\,(\mathbf{I}_i \wedge \mathbf{I}_j)\right\}.$$

The equations (7.42) and (7.43) greatly facilitate the calculations of average values in the formula (7.40a) for the differential cross section:

$$\frac{d\sigma}{d\Omega} = \sum_i \langle |a_i|^2 \rangle + \sum_{i \neq j} \langle a_i a_j^* \exp(-i\boldsymbol{\kappa}\,.\,(\mathbf{r}_i - \mathbf{r}_j)) \rangle. \tag{7.44}$$

In this section we shall assume that there is no correlation between the relative orientations of two nuclear spins $\mathbf{I}_i$ and $\mathbf{I}_j$ on the one hand and their relative positions on the other. This is certainly not true for antiferromagnetic nuclear states nor even for a demagnetized paramagnetic state with a cold dipolar energy. These situations will be dealt with in section **F**. On the other hand there is no correlation between $(\mathbf{I}_i\,.\,\mathbf{I}_j)$ and $\mathbf{r}_i - \mathbf{r}_j$ in an unpolarized target or in a polarized target with Zeeman order only. We can then rewrite (7.45) as:

$$\frac{d\sigma}{d\Omega} = \sum_i \langle |a_i|^2 \rangle + \langle a_i a_j^* \rangle_{i \neq j} \sum_{i \neq j} \exp(-i\boldsymbol{\kappa}\,.\,(\mathbf{r}_i - \mathbf{r}_j)) \tag{7.45}$$

that is in a form identical with (7.40a).

We assume a polarized target with a polarization P along an axis and a neutron beam polarized along the same axis with polarization p. Then:

$$\langle \mathbf{I}_i\,.\,\mathbf{I}_j \rangle = \langle \mathbf{I}_i \rangle\,.\,\langle \mathbf{I}_j \rangle = P^2 I^2;$$

$$\langle \mathbf{I}_i \wedge \mathbf{I}_j \rangle = 0 \qquad \langle \mathbf{s}\,.\,\mathbf{I}_i \rangle = \frac{pPI}{2};$$

using (7.43):

$$\langle |a_i|^2 \rangle = b_0^2 + bb_0 IpP + \frac{b^2}{4}\{I(I+1) - pPI\}; \tag{7.46}$$

$$\langle a_i a_j^* \rangle_{i \neq j} = b_0^2 + bb_0 IpP + \frac{b^2 I^2 P^2}{4}. \tag{7.47}$$

We can rewrite (7.45′) as:

$$\frac{\mathrm{d}\sigma}{\mathrm{d}\Omega}=\sum_i\{\langle|a_i|^2\rangle-\langle a_ia_j\rangle_{i\neq j}\}+\langle a_ia_j^*\rangle_{i\neq j}\sum_{i,j}\exp(-\mathrm{i}\boldsymbol{\kappa}\,.\,(\mathbf{r}_i-\mathbf{r}_j))$$

$$=\frac{Nb^2}{4}\{I(I+1)-pPI-P^2I^2\}+N\left\{b_0^2+bb_0IpP+\frac{b^2I^2P^2}{4}\right\}\sum_i\exp(-\mathrm{i}\boldsymbol{\kappa}\,.\,\mathbf{r}_i)$$

$$=N\left(\frac{\mathrm{d}\sigma}{\mathrm{d}\Omega}\right)_{\mathrm{inc.}}+N\left(\frac{\mathrm{d}\sigma}{\mathrm{d}\Omega}\right)_{\mathrm{coh}}\sum_i\exp(-\mathrm{i}\boldsymbol{\kappa}\,.\,\mathbf{r}_i),\tag{7.48}$$

with:

$$\left(\frac{\mathrm{d}\sigma}{\mathrm{d}\Omega}\right)_{\mathrm{inc.}}=\frac{b^2}{4}\{I(I+1)-pPI-P^2I^2\};$$
$$\left(\frac{\mathrm{d}\sigma}{\mathrm{d}\Omega}\right)_{\mathrm{coh}}=\left\{b_0^2+bb_0IpP+\frac{I^2P^2b^2}{4}\right\}.\tag{7.49}$$

We see that $(\mathrm{d}\sigma/\mathrm{d}\Omega)_{\mathrm{inc}}$ reduces to zero for $pP=1$ but *not* for $pP=-1$. This is understandable: for $pP=-1$ an incoming neutron can undergo a flip-flop with an individual nucleus through the s_+I_- part of the scattering amplitude. This process does not interfere with a flip-flop on another nucleus since the two processes leave the target in different final states. A scattering on an individual nucleus is thus possible, giving rise to an isotropic incoherent cross-section.

We can also notice that in contrast with isotopic incoherence where according to (7.41) $(\mathrm{d}\sigma/\mathrm{d}\Omega)_{\mathrm{coh}}$ and $(\mathrm{d}\sigma/\mathrm{d}\Omega)_{\mathrm{inc}}$ are given by $|\langle a\rangle|^2$ and $\{\langle|a|^2\rangle-|\langle a\rangle|^2\}$, in the present case: $\langle a\rangle=b_0+bpPI/2$;

$$|\langle a\rangle|^2=\left[b_0+\frac{bpPI}{2}\right]^2=b_0^2+bb_0pPI+\frac{b^2p^2P^2I^2}{4};\tag{7.50a}$$

$$\langle|a|^2\rangle-|\langle a\rangle|^2=\frac{b^2}{4}\{I(I+1)-pPI-p^2P^2I^2\}.\tag{7.50b}$$

We can see that the two expressions (7.50) differ from the corresponding expressions (7.49) with which they coincide only for $|p|=0$ or 1. These differences originate in the fact that $\langle a_ia_j^*\rangle$ given by (7.47) differs from $|\langle a\rangle|^2$ given by (7.50*a*).

For a lattice with more than one atom per unit cell the coherent cross-section is obtained as follows (we assume for each atom a single isotope).

Define in the unit cell:

$$B_0=\sum_{\alpha=1}^{\rho}b_\alpha^0\exp(-\mathrm{i}\boldsymbol{\kappa}\,.\,\mathbf{d}_\alpha);\qquad B=\sum_{\alpha=1}^{\rho}P_\alpha I_\alpha b_\alpha\exp(-\mathrm{i}\boldsymbol{\kappa}\,.\,\mathbf{d}_\alpha),\tag{7.51}$$

where $b_\alpha^0 + b_\alpha(\mathbf{s}\,.\,\mathbf{I}_\alpha)$ is the scattering amplitude for the nucleus of the atom which has the position $\mathbf{d}_\alpha$ in the unit cell, whose spin is $\mathbf{I}_\alpha$ and the polarization P_α,

$$\left(\frac{\mathrm{d}\sigma}{\mathrm{d}\Omega}\right)_{\mathrm{coh}} = |B_0|^2 + \tfrac{1}{4}|B|^2 + p\ \mathrm{Re}(B_0 B^*). \tag{7.52}$$

C. Pseudomagnetic nuclear field and pseudomagnetic resonance

Pursuing the analogy between nuclear and magnetic scattering, emphasized by the definition (7.29) of a pseudomagnetic nuclear moment, leads to interesting consequences.

It is a well-known result of classical magnetostatics that the average magnetic field inside a magnetic medium is the induction $\mathbf{B} = \mathbf{H} + 4\pi\mathbf{M}$ where $\mathbf{M}$ is the magnetization. Some caution must be exercised in using this statement because of various problems, including the shape of the sample, due to the long range of magnetic interactions, but let us not worry about these for the time being. Arguing that a neutron going through such a medium would sample this average field, one would expect its Larmor frequency to be $-\gamma_N B$ rather than $-\gamma_N H$. Going one step further, it is tempting to treat a polarized nuclear target as a pseudomagnetic medium with a pseudomagnetization $M^* = N_0\mu^* P$ (N_0 being the density of nuclei and P their polarizations) and to speculate that a neutron passing through such a target would see a nuclear pseudomagnetic field $H^* = 4\pi M^* = 4\pi N_0\mu^* P$ resulting in a shift of its Larmor frequency inside the sample with respect to vacuum:

$$\Delta\omega_N = -\gamma_N H^* = -4\pi\gamma_N N_0 \mu^* P. \tag{7.53}$$

This heuristic argument is easily made quantitative. The Fermi pseudopotential introduced in eqn (7.22) can be rewritten using (7.28) as:

$$V_F(\mathbf{r}) = -\boldsymbol{\mu}_N\,.\,\boldsymbol{\mu}_n^* 4\pi\,\delta(\mathbf{r}-\mathbf{r}_n) = -(\boldsymbol{\mu}_N\,.\,\mathbf{I})\frac{4\pi\mu_n^*}{I}\,\delta(\mathbf{r}-\mathbf{r}_n), \tag{7.54}$$

where $\mathbf{r}_n$ is the position of the nucleus.

For a target containing many identical nuclei at positions r_j and with spins $\mathbf{I}_j$, $V_F(\mathbf{r})$ becomes:

$$V_F(\mathbf{r}) = -\boldsymbol{\mu}_N\,.\left(\sum_j \frac{4\pi\mu_n^*}{I}\mathbf{I}_j\,\delta(\mathbf{r}-\mathbf{r}_j)\right) = -\boldsymbol{\mu}_N\,.\,H^*(\mathbf{r}), \tag{7.55}$$

where:

$$\mathbf{H}^*(\mathbf{r}) = \frac{4\pi\mu_n^*}{I}\sum_j \mathbf{I}_j\,\delta(\mathbf{r}-\mathbf{r}_j). \tag{7.55$'$}$$

We call $\mathbf{H}^*(r)$ the pseudomagnetic nuclear field. We rewrite $\mathbf{H}^*(\mathbf{r})$ as:

$$\mathbf{H}^*(\mathbf{r}) = \mathbf{H}^* + \mathbf{h}^*(\mathbf{r})$$

where $\mathbf{H}^*$, the spatial average of $\mathbf{H}^*(\mathbf{r})$ is given by:

$$\mathbf{H}^* = \frac{1}{V}\int \mathbf{H}^*(\mathbf{r})\,\mathrm{d}^3r = \frac{4\pi}{I}\frac{\mu_{\mathrm{n}}^*}{V}\sum_j \mathbf{I}_j = 4\pi N_0\mu_{\mathrm{n}}^* P\hat{\mathbf{n}} = \alpha P\hat{\mathbf{n}}, \qquad (7.56)$$

where P is the nuclear polarization along a unit vector $\hat{\mathbf{n}}$ and N_0 the number of nuclei per unit volume.

The effect of $\mathbf{H}^*$ is to shift the Larmor frequency of a neutron crossing the target by $\Delta\omega_{\mathrm{N}} = -\gamma_{\mathrm{N}}H^*$. The spatially varying part $\mathbf{h}^*(r)$ is seen by the incoming neutron as a time-dependent field, of zero average value, fluctuating with a frequency of the order of $\Omega = 2\pi v/a$ where a is the internuclear distance and v the neutron velocity. For $\Omega \gg |\omega_{\mathrm{N}}|$, which is the case in most experimental situations, the effect of $\mathbf{h}^*(\mathbf{r})$ is completely averaged out and there remains the shift: $\Delta\omega_{\mathrm{N}} = -\gamma_{\mathrm{N}}H^* = -\gamma_{\mathrm{N}}4\pi M^*$ as predicted by the magnetic analogy, but, thanks to the short range of nuclear forces (as expressed by the δ-function), without any sample shape problem.

Going back to the expression (7.29) of μ_{n}^*, and to the expression (7.23) for the magnetic moment μ_{N} of the neutron we find:

$$\Delta\omega_{\mathrm{N}} = -\gamma_{\mathrm{N}}H^* = \frac{4\pi\hbar}{M}N_0\frac{bPI}{2} = \frac{4\pi\hbar}{M}N_0\frac{I}{2I+1}(b_+ - b_-)P. \qquad (7.57)$$

This equation was first derived (Baryshevskii and Podgoretskii, 1964) using the somewhat different approach of the refractive index of neutron optics, that we shall describe later in section **E**, rather than the magnetic analogy outlined above.

We describe now the experimental proof of the existence of the pseudomagnetic field which can be found in Abragam, Bacchella, Glättli, Meriel, Piesvaux, and Pinot (1972).

(a) A simple scheme

Under favourable circumstances the magnitude of the nuclear pseudomagnetic field $H^* = 4\pi N\mu^*P = \alpha P$ can be very large. In LMN, that is the double nitrate of lanthanum and magnesium, where the protons of the water of crystallization are polarized, the coefficient α in $H^* = \alpha P$ is 25 000 Gauss!

The measurement of the Larmor frequency ω_{N} of the neutron is straightforward, in principle. Neutrons coming out of a reactor through a polarizer emerge with a polarization p higher than 99 per cent. They cross a dynamically polarized target of LMN placed in the static field H of a magnet, then an analyser, before striking a counter. An r.f. coil wrapped around the target produces inside it a linearly polarized r.f. field $2H_1 \cos\omega t$ with a component of amplitude H_1 rotating at the frequency ω (Fig. 7.1). This field

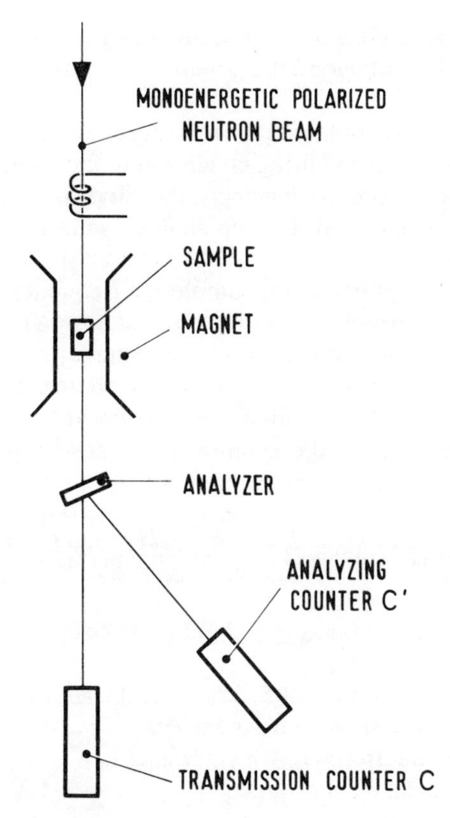

FIG. 7.1. Experimental arrangement for the observation of the neutron pseudo-magnetic resonance in a target of polarized protons.

may flip the neutron if the resonance condition $\omega = \omega_N$ is fulfilled. If the analyser has been set to let through only neutrons of polarization opposite to that prepared by the polarizer, the counting rate will be vanishingly small unless $\omega \approx \omega_N$.

(*b*) *Rotating pseudomagnetic field*

An appreciable reversal of the neutron spin will only occur if the time of flight through the target, $t = l/v$ is sufficiently long for the quantity $\phi = \gamma_N H_1 t$ not to be very small. With experimental conditions $l = 4$ mm, $v = 4 \times 10^5$ cm^{-1}, $t = 1\ \mu$s, the condition $\phi \approx 1$ requires $2H_1 \sim 100$ Gauss. Dynamically polarized targets are cooled by liquid helium to temperatures of 1 K and below, and amplitudes $H_1 \gg 1$ Gauss are difficult to produce under these conditions.

An obvious way out would have been to decrease the velocity of the neutrons by, say, a factor 10 and to increase by the same factor the size of the sample thus making t and ϕ a hundred times larger. These expensive brute force solutions were far beyond the possibilities of the reactor, the cryostat and the magnet.

Instead a method for amplifying the rotating field H_1 was used, which had the twofold advantage of requiring no change in the equipment and also of demonstrating even more convincingly the physical reality of the pseudomagnetic nuclear field and its deep analogy with an ordinary magnetic field.

If an r.f. field H_1 is applied to the sample at a frequency $\omega = \omega_P + \Delta$ where ω_P is the proton Larmor frequency and Δ a small shift, the proton magnetization will reach in a very short time a steady state where it is tilted with respect to the d.c. field H by an angle θ such that: $\tan\theta = \gamma_P H_1/\Delta \approx \theta$ (assuming $|\Delta| \gg \gamma_P H_1$). The pseudomagnetization M^* and $H^* = 4\pi M^*$ are tilted by the same angle. The d.c. component $H^* \cos\theta$ is practically equal to H^*, but there appears a rotating component:

$$H_1^* = H^* \sin\theta \simeq H^*\theta \approx \frac{H^*\gamma_P H_1}{\Delta} = H_1\left(\gamma_P \frac{\alpha P}{\Delta}\right), \tag{7.58}$$

precessing at the frequency ω (Fig. 7.2). H_1^* can be larger than H_1 by a huge factor.

For instance in LMN: for $P = 0{\cdot}5$, $\Delta/\gamma_p = 100$ Gauss, $H_1^* = 125 H_1$. It still remains to make the precession frequency of H_1^*: $\omega = \omega_P + \Delta \approx \omega_P = -\gamma_P H$ equal to that of the neutron: $\omega_N = -\gamma_N(H + \alpha P)$.

This can be achieved by choosing:

$$H \approx \left(\frac{\gamma_P}{\gamma_N} - 1\right)^{-1} \alpha P \quad \text{or since } \frac{\gamma_P}{\gamma_N} \approx -\tfrac{3}{2};$$
$$H \approx -\tfrac{2}{5}\alpha P \approx -10\,000\, P \text{ Gauss.} \tag{7.59}$$

This requires a negative value of the proton polarization which is possible with DNP by the solid effect. For $P = 50$ per cent, (7.59) gives $H \approx$ 5000 Gauss.

(c) *Experimental procedure and results*

In order to produce a high proton polarization, the sample of LNM, doped with Nd^{3+} is placed in a high static field H_0 and irradiated by a microwave field at a frequency slightly above or slightly below the Larmor frequency $\omega_S = -\gamma_S H_0$ of the paramagnetic impurities (depending on the sign sought for the nuclear polarization). The microwave source used in the experiment had a frequency of 70 GHz and required a field $H_0 = -\omega/\gamma_S \approx 18\,500$ Gauss for the type of polarizing impurities used in LMN, namely Nd^{3+} ions. This field cannot be made to coincide with $H = \frac{2}{5}\alpha|P| = 10\,000\, P$.

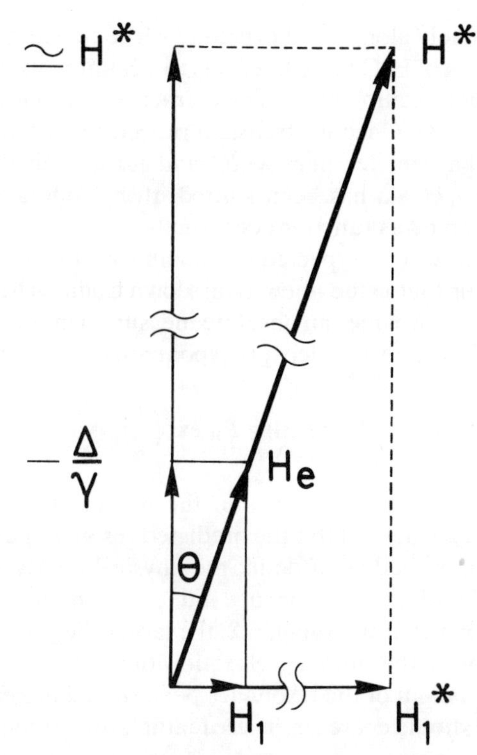

FIG. 7.2. Pseudomagnetic resonance of neutrons in a target of polarized protons. Effective field experienced by the protons, and effective pseudomagnetic field experienced by the neutrons in the rotating frame.

The experiment could still be done however in a transient two step procedure. First the protons are polarized in the high field H_0 to a negative polarization P_i. The field is then reduced to a value H somewhat below $\frac{2}{5}\alpha|P_i|$, and an r.f. field is applied, as explained earlier, at a frequency $\omega = -\gamma_P H + \Delta(|\Delta| \ll |\gamma_P H|)$. In the presence of the r.f. field the nuclear polarization decreases with a time constant τ, (which depends on Δ and H_1), sweeps through the resonant value: $P_R \approx (-5H/2\alpha) \approx (-H/10\,000)$ where $\omega \approx \omega_N$ and the resonant reversal of the neutron spin occurs, and eventually vanishes.

The smaller Δ and the larger H_1, the shorter τ. For a given H_1 there is an optimum value of Δ below which τ is too short and the passage through the resonance too fast for enough neutrons to be counted, and above which the amplified flipping field H_1^* given by (7.58) is too weak to flip effectively the neutron spins.

In the actual experiment the incoming neutron beam has a negative polarization $p \sim -100$ per cent and the target nuclei a negative polarization

P_i (counted positively along the d.c. field H). The transmitted beam strikes an analysing crystal of FeCo which reflects to a counter none of the neutrons with a negative polarization but a certain fraction f of those with a positive polarization (Fig. 7.1). Once the transient procedure of setting the field at a value H somewhat smaller than $\frac{2}{5}\alpha|P_i|$ and turning on the r.f. field at a frequency $\omega = -\gamma_P H + \Delta$ has been started, the counting rate $N(t)$ of the counter C is observed as functions of time.

This counting rate N is expected to remain vanishingly small until P_R is reached, to rise for that value, then to fall down again. When the resonance occurs at a time t, it provides an absolute measurement of the polarization: $P(t) = P_R = -5H/2\alpha$ and if its decay is exponential the initial polarization is:

$$P_i = P(0) = P_R \exp\left(\frac{t}{\tau}\right). \tag{7.60}$$

Figure 7.3 shows for certain values of the parameters Δ, H, and P_i the counting rates $I(t)$ which exhibit the predicted resonant peaks thus demonstrating beyond any shadow of doubt the physical reality of the d.c. pseudomagnetic field and of the resonance induced by its rotating component.

We stated earlier that the smaller Δ, the larger the pseudomagnetic field H_1^* and the shorter the nuclear relaxation time. As a consequence, with decreasing Δ the height of the resonance peak should increase and its width on the time scale should decrease. These features are brought out by the data of Fig. 7.3.

The reader is referred to the literature for further details on this experiment (Abragam, 1973).

(*d*) *Methodological digressions*

Let us first point out that the concept of a nuclear pseudomagnetic field has a very respectable ancestor in the person of the molecular or Weiss field. Long before the advent of quantum mechanics, Pierre Weiss had described the interactions that give rise to the ordering of atomic magnetic moments in ferromagnetic substances as a very large molecular field experienced by each magnetic moment and produced by its neighbours. We know now that these interactions are quantum mechanical exchange spin–spin couplings of the form $J\mathbf{S}_1 . \mathbf{S}_2$ and that they originate from electric rather than magnetic forces. Nevertheless we still often find it convenient to treat the Weiss field as a pseudomagnetic field. In so doing we pretend to ourselves that an interaction we *know* to be a coupling between spins, and of *electrostatic* origin, is a *magnetic* coupling. This is very similar to our pretence of the neutron seeing a pseudomagnetic nuclear field, thus describing as a *magnetic* coupling what we *know* to be a coupling between spins, originating in *nuclear* forces. The reason why both pseudomagnetic fields, Weiss and

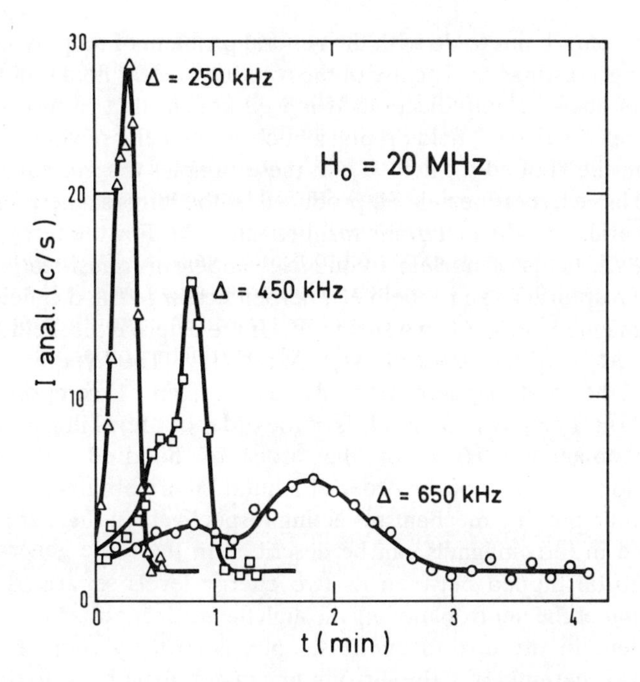

FIG. 7.3. Pseudomagnetic resonance of neutrons in a LMN target of polarized protons. Influence of the proton longitudinal effective frequency Δ on the neutron resonance shape at constant field H_1. (After Abragram, 1973.)

nuclear, so far exceed those produced by the true magnetic moments present is that electric as well as nuclear forces are very much stronger than magnetic forces.

A second remark has to deal with the terminology of the two usual descriptions of the magnetic resonance phenomenon: classical and quantum mechanical. In the classical description, at resonance the rotating field H_1 precesses in a plane perpendicular to the d.c. field H_0 with an angular velocity $\omega = \omega_0 = -\gamma H_0$ while the resonating spin precesses around H_1 with an angular velocity $\omega_1 = -\gamma H_1$. If parallel to H_0 at $t = 0$, the spin will be antiparallel to it $|\pi/\gamma H_1|$ seconds later.

In the quantum mechanical description, the r.f. field is a coherent set of photons of energy $\hbar\omega$ which are absorbed or emitted when the spin makes an upward or downward transition of energy $\hbar\omega_0$ and the resonance condition $\omega = \omega_0$ is that of the conservation of energy.

When we go over to a transition induced by the rotating nuclear pseudomagnetic field, there is nothing to change in the classical description but in the quantum mechanical one, what is the field quantum, emitted or absorbed by the neutron? (See Odehnal, 1975.)

Our last remark has to do with the general problem of amplifying the r.f. field driving a resonance. The use of the rotating nuclear field is not the first example of such an amplification. It is well known that in ferromagnetic samples the effective r.f. field driving a nuclear magnetic resonance is much larger than the applied r.f. field H_1. In these samples the nuclear moments see a very large hyperfine field H_i produced by the ferromagnetic electrons, which is parallel to the electronic magnetization M. For the ferromagnetic electrons, r.f. fields of nuclear frequencies appear as quasi-static and the electronic response to an r.f. field H_1, perpendicular to the d.c. field H_0 is a small r.f. magnetization M_1 parallel to H_1. If the external d.c. field H_0 is not too small, M_1 is of the order of: $M_1 \approx M(H_1/H_0)$. The hyperfine field H_i, parallel to the total magnetization, also acquires an r.f. component $H_{i_1} \approx H_i(M_1/M) \approx H_1(H_i/H_0)$. Since H_0 is of the order of a few kilogauss while in many ferromagnets, H_i is of the order of hundreds of kilogauss, amplifications (H_{i_1}/H_1) of the order of a hundred are obtained.

The two amplifying mechanisms acting respectively in nuclear polarized targets and in ferromagnets can be described in the same general terms: the spin to be flipped between its two energy levels separated by $\hbar\omega_0$, whether that of the neutron moving through the nuclear target, or that of the fixed nucleus in the ferromagnetic sample, is strongly coupled to some characteristic parameter of the surrounding material medium: to the proton polarization in the nuclear target by means of the pseudomagnetic nuclear field H^*, to the electron magnetization in the ferromagnetic sample by means of the hyperfine field H_i. An applied magnetic r.f. field H_1 of frequency $\omega \approx \omega_0$ may modulate at the same frequency the proton polarization (the electron magnetization) giving an r.f. component H_1^* (H_{i_1}) to the pseudomagnetic nuclear field H^* (the hyperfine field H_i). Under favourable circumstances the amplified r.f. field H_1^* (H_{i_1}) can be much stronger than the applied field H_1. We will call this type of resonance resulting from a forced r.f. modulation of a certain parameter, characteristic of a material medium, parametric resonance.

D. Systematic measurements of nuclear pseudomagnetic moments

It has been made abundantly clear in the previous sections that the pseudomagnetic nuclear moment, μ^* (related to the spin-dependent scattering amplitude b by eqn (7.29) or (7.30)) is the handle whereby the neutron can be used as a probe of nuclear magnetism. The knowledge of the pseudomagnetic moments of the various nuclear species is of primary importance for this task. Besides, the knowledge of these quantities is of some interest for other studies of condensed matter with neutrons, and potentially for a better understanding of nuclear structures. It turns out however that whilst the spin-independent part b_0 of the scattering amplitude has been known with good accuracy for a long time and for most nuclei, the μ^* of all the nuclei but a few were very poorly known until the middle seventies.

The methods of measuring b_0 are numerous; they are described in detail in many books on neutron physics (see for instance Koester and Steyerl, 1977). All these methods give directly the scattering amplitude b_0. On the other hand access to the spin-dependent part b passed traditionally through a measurement of the total transmission cross section of a target, from which the elastic scattering cross section of a single nucleus, $\sigma^{sc} = 4\pi(b_0^2+(b^2/4)I(I+1))$ could be extracted. This extraction is far from simple: it requires a knowledge of the capture cross section, of the inelastic scattering processes that can take place inside the target, of possible Bragg scatterings at various angles, etc. Furthermore an accurate absolute measurement of the transmission cross section itself is not easy. It is not so surprising therefore that until 1972 the amplitude b was known with reasonable accuracy for four nuclei only: ^{1}H, ^{2}D, ^{51}V, ^{59}Co for all of which it is fairly large. Since then the procedures for measuring the total cross section and for extracting b^2 have improved considerably for some nuclei (see Koester and Steyerl, 1977) but even so they do not yield the sign of b. The determination of b_+ and b_- by the eqns (7.21) suffers from this ambiguity by offering two sets of values for b_+ and b_-. Until the use of polarized targets the sign of b was known with certainty for one nucleus only, ^{1}H, thanks to the striking difference in scattering by ortho- and para-hydrogen molecules. Finally the quadratic dependence of σ_{sc} with respect to b makes very difficult the determination of b when $|b| \ll |b_0|$.

(a) Bragg scattering on polarized targets

The availability of polarized neutron beams and of polarized targets offers direct access to the sign of b.

This follows from the second eqn (7.49) for $(d\sigma/d\Omega)_{coh}$ which contains the term bb_0IPp linear with respect to b.

An experiment based on this fact was performed on a single crystal of CaF_2 in order to measure b (^{19}F) (Abragam, Bacchella, Long, Meriel, Piesvaux, and Pinot, 1972).

At the time of this experiment it was already known (see chapter 8) that the spins of ^{19}F could be brought in that crystal to an antiferromagnetic state and it was highly desirable to know whether the magnitude of $\mu^*(^{19}\text{F})$ was sufficient to allow the observation of this antiferromagnetism by neutron diffraction. From former measurements of the total scattering cross section, $b(^{19}\text{F})$ was expected to be small: $0 \leq |b| \leq 3$F (where F, a Fermi, is 10^{-13} cm).

A theoretical estimate (Gillet and Normand, 1971) predicted: $b = -1{\cdot}4$ F.

The principle of the experiment was as follows. A monochromatic neutron beam with a polarization $p \approx \pm 100$ per cent was coherently diffracted by a single crystal of CaF_2. The polarization of the beam could be reversed rapidly by a flipping coil. The nuclei of ^{19}F were polarized dynamically up to $|P| \approx 40$ per cent. In view of the smallness of $b(^{19}\mathrm{F})$ the term proportional to b^2 could be neglected. If ^{19}F had been the only nucleus in the crystal, the change in the diffracted Bragg intensity upon reversal of the neutron polarization p would have been:

$$I_+/I_- \cong 1 + \frac{2bIPp}{b_0}. \tag{7.61}$$

In a CaF_2 crystal, the scattering amplitude $b_0(^{40}\mathrm{Ca})$ of the spinless nucleus ^{40}Ca contributes to the Bragg scattering amplitude in accordance with eqns (7.51) and (7.52). In order to enhance the effect of the spin-dependent term a suitable Bragg reflection was chosen, namely 200, and (7.61) must then be replaced by:

$$\frac{I_+}{I_-} = 1 - \frac{4b(^{19}\mathrm{F})I|p|P}{b_0(^{40}\mathrm{Ca}) - 2b_0(^{19}\mathrm{F})}. \tag{7.62}$$

The values extracted from the experimental values of (I_+/I_-) and the known values $b_0(^{40}\mathrm{Ca}) = 4{\cdot}9$ F, $b_0(^{19}\mathrm{F}) = 5{\cdot}7$ F was:

$$b(^{19}\mathrm{F}) = (-0{\cdot}15 \pm 0{\cdot}02)\ \mathrm{F}.$$

This value is so small that a correction had to be made for the magnetic scattering by the *true* nuclear magnetic moment $\mu(^{19}\mathrm{F})$, which by itself would have led to a value: $b(^{19}\mathrm{F})_{\mathrm{magn}} \cong 0{\cdot}015$ F. This yields for the nuclear $b(^{19}\mathrm{F})$ a corrected value $b(^{19}\mathrm{F}) \cong -0{\cdot}165$ F and a value $\mu^*(^{19}\mathrm{F})/\mu_{\mathrm{B}} \cong -0{\cdot}016$. (This correction was made with the wrong sign in Abragam, Bacchella, Long, Meriel, Piesvaux, and Pinot (1972).)

The theoretical overestimate of $b(^{19}\mathrm{F})$ by a factor ten turned out to be a blessing in disguise. It is doubtful that the experiment would have been undertaken at all if a value as small as that actually found, had been expected.

The measurement of $b(^{19}\mathrm{F})$ was not quite the first of its kind: the amplitude $b(^{51}\mathrm{V})$ (Schull and Ferrier, 1963) and $b(^{59}\mathrm{Co})$ (Ito and Schull, 1969) had been obtained before, by Bragg scattering of polarized neutrons on targets with finite nuclear polarizations.

Vanadium ^{51}V is exceptional among nuclei insofar as $b_0(^{51}\mathrm{V})$ is very small, much smaller than $b(^{51}\mathrm{V})$ which is very large. The total scattering cross section is thus a measure of b^2 and the experiment of Schull and Ferrier was really a determination of the *sign* of $b(^{51}\mathrm{V})$. In the counting rate of the neutrons scattered from a Bragg reflection one looks for contributions bilinear with respect to the neutron polarization p and the nuclear polarization P. The latter is exceedingly small in this experiment ($P \leqslant 4 \times 10^{-4}$) and proportional to $1/T$. A ratio of the order of 120 between $\mu^*(^{51}\mathrm{V})$ and

$\mu^*(^{19}\text{F})$ explains why $\mu^*(^{51}\text{V})$ (or rather its sign, since its magnitude was already known) could be determined in spite of a nuclear polarization a thousand times smaller than in the experiment on ^{19}F.

A similar experiment performed on a single crystal of ferromagnetic cobalt (Ito and Shull, 1969) gave $\mu^*(^{59}\text{Co})$. A favourable feature there is a nuclear polarization up to 7×10^{-3} that is 17 times higher than in the preceding experiment, thanks to the hyperfine field H_i, seen by the nuclei of ^{59}Co, much higher than the applied field $H_0(|H_i|\approx 200$ kG; $H_0 = 15$ kG). Even so, what made the experiment feasible was the high value of $\mu^*(^{59}\text{Co})$, $(\mu^*(^{59}\text{Co})/\mu^*(^{19}\text{F})\approx 120)$. In either of the two experiments there are in (I_+/I_-) other terms linear in p and due to the coupling of the neutron spin with electronic magnetic moments. They are discriminated against through their lack of temperature dependence. The electronic magnetization is either temperature independent as in metallic vanadium or saturated as in ferromagnetic cobalt, in contrast with the nuclear polarization proportional to $(1/T)$.

An experiment (Herpin and Meriel, 1973) which gave the pseudomagnetic moment of ^{165}Ho deserves to be cited if only for its elegance and economy of means since it requires neither a single crystal nor a polarized beam nor very low temperatures. In the scattering amplitude of a neutron by a magnetized target there are two terms proportional respectively to $(\mathbf{s}\,.\,\mathbf{M}_\perp)$ and $(\mathbf{s}\,.\,\boldsymbol{\mu}_n^*)$ where $\mathbf{M}_\perp$ is the component of the atomic magnetic moment at right angle to the impulse transfer $\boldsymbol{\kappa}$, $\boldsymbol{\mu}_n^*$ is the pseudomagnetic moment of the nucleus, and $\mathbf{s}$ the spin of the neutron. In the Bragg cross section, the cross product of these two terms yields a contribution, independent of the neutron polarization, and proportional to $M_\perp$ (which is temperature independent because of magnetic saturation, and known), to $\mu_n^*(^{165}\text{Ho})$ (to be measured) and to the nuclear polarization P (varying as $1/T$ and sizeable in spite of the smallness of $1/T$ because of the huge hyperfine field $H_i\approx 800$ Teslas, seen by the nucleus ^{165}Ho). The correct sign and a correct magnitude for $\mu^*(^{165}\text{Ho})\approx -0.34\mu_B$ could be extracted from this elegant experiment, done 'on the cheap'.

Since then a similar experiment performed on the antiferromagnetic intermetallic compound AgTb gave $\mu^*(^{159}\text{Tb})$ (Akopyan, Alfimenkov, Lason, Ovchinnikov, and Shapiro, 1975).

(*b*) *Pseudomagnetic precession and the two-coils methods*

1. *The principle*

The physical reality of the pseudomagnetic field was demonstrated by the resonance experiment, described in section **C**, on a sample of LMN polarized dynamically up to $|P|\approx 50$ per cent. This experiment, which would have yielded the value of $\mu^*(^1\text{H})$, if it had not been known beforehand, presents, however, certain features which make difficult its extension to nuclei with smaller μ^* and/or smaller polarizations.

An accurate resonance measurement of the pseudomagnetic precession frequency requires that the precession phase angle $\psi = -\gamma_N H^* t$ accumulated during the transit time t in the sample be much larger than 2π (it was of the order of $37 \times 2\pi$ for $|P| \approx 50$ per cent, in that experiment).

For the weak nuclear fields produced by nuclei other than protons, ψ is much smaller and it is preferable to replace the resonance experiment by a precession experiment where the phase angle ψ is measured directly.

This is achieved in the two-coils method invented many years ago (Ramsey, 1949).

The experimental arrangement is shown in Fig. 7.4. Two r.f. coils, C′ and C″, of length l placed in the gap of a magnet and separated by a distance $L \gg l$ produce two magnetic r.f. fields, $2H_1' \cos(\omega t)$ and $2H_1'' \cos(\omega t - \phi)$, respectively, perpendicular to the steady magnetic field H_0 of the magnet. The frequency ω can be adjusted to the resonant value $\omega = |\omega_0| = |(\gamma_N H_0)|$ of the Larmor frequency of the neutron. A beam of monochromatic neutrons of velocity v, with a polarization p_0 of nearly 100 per cent, parallel to or antiparallel to H_0 when it enters the first coil C′, crosses the two coils and the sample placed between them before striking an analyser A which measures its polarization along H_0.

What happens to the neutron spin is best described in a frame rotating at the frequency ω of the r.f. field. In that frame the neutron 'sees' at resonance a static field in each coil of respective magnitudes H_1' and H_1''. These two fields are perpendicular to $0z$ and make between them the angle ϕ. In the space between the coils the spins see only the pseudomagnetic field H^* of the sample parallel to H_0. An elementary calculation shows that when the beam strikes the analyser, its polarization p is given by:

$$p = p_0(\cos\theta' \cos\theta'' - \sin\theta' \sin\theta'' \cos(\psi + \phi)). \qquad (7.63)$$

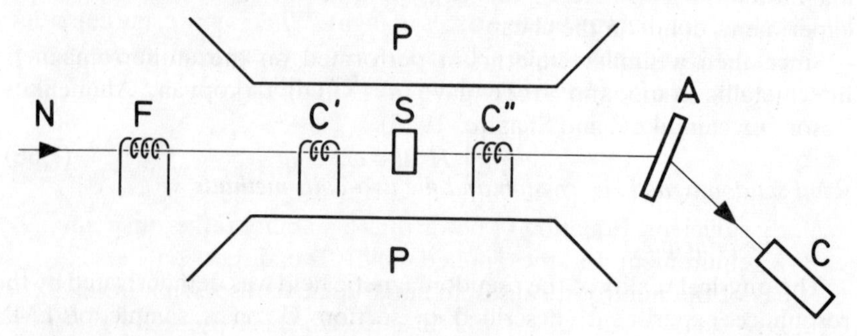

FIG. 7.4. Experimental arrangement for the Ramsey type measurement of the neutron pseudomagnetic precession angle after crossing a nuclear polarized target.

In this formula $\theta' = -\gamma_N H_1' l/v$ and $\theta'' = -\gamma_N H_1'' l/v$ are the spin nutation angles in each coil and $\psi = -\gamma_N H^* d/v$ is the extra precession angle due to the pseudomagnetic field of the sample of thickness d. Formula (7.63) is greatly simplified if $\theta' = \theta'' = \pi/2$. In that case:

$$H^* = -\frac{v}{\gamma_N d}\left[\cos^{-1}\left(-\frac{p}{p_0}\right) - \phi\right]. \tag{7.64}$$

For small nuclear polarization P, H^* is proportional to the inverse temperature of the sample, and a plot of $\cos^{-1}(p/p_0)$ against $1/T$ should be a straight line whose slope gives access to μ^*. If the condition $\theta' = \theta'' = \pi/2$ is not fulfilled, this plot is distorted and the more general formula (7.63) should be used.

An alternative procedure is to keep the final neutron polarization constant by compensating the change with temperature in the precession angle $\psi = -\gamma_N H^*(d/v)$ by a change ΔH_0 in the applied field, which introduces a precession angle $-\gamma_N \Delta H_0 L/v$.

This is not entirely correct since a change in H_0 breaks the resonance condition $\omega = \omega_0$. However, if θ' and θ'' differ little from $(\pi/2)$ and if $|\Delta H_0| \ll H_1', H_1''$ this effect can be taken into account by using the easily established formula:

$$H^* = -\frac{\Delta H_0}{d}\left(L + \frac{4l}{\pi}\right), \tag{7.65}$$

where the restrictive assumptions θ', $\theta'' = \pi/2$ are required for the validity of the corrective term $4l/\pi$ only.

The beauty of the two-coils method resides in the fact that to a large extent it is insensitive to the dispersion of the velocities and thus of the transit times of the neutrons in the space between the coils. Again this is easiest to see in the rotating frame: if the condition $\omega = |\gamma_N H_0|$ is fulfilled, in the rotating frame the spins of all the neutrons see the same static fields H_1' and H_1'', independently of their respective times of flight. What is more, the condition $\omega = |\gamma_N H_0|$ does not have to be satisfied locally in every point of the path of the neutrons. The far less stringent average condition,

$$\frac{|\gamma_N|}{L}\int H \, \mathrm{d}l = \omega, \tag{7.66}$$

is clearly sufficient. Equation (7.66) is thus a stability rather than a homogeneity requirement for the applied field. The dispersion $\delta v/v$ in the velocities of the neutrons, causes an equal relative dispersion in the extra phase angles $\psi = -\gamma_N H^* \tau = -\gamma_N H^* d/v$ due to the presence of the pseudomagnetic field of the polarized sample, but not in the much larger angles of precession between the coils: $\psi_0 = -\gamma_N H_0 L/v$.

Last but not least, in contrast to the resonance experiment of section **C**, the r.f. fields which flip the neutron spins are now outside of the cold sample and can be given much larger values without any cryogenic problems.

Before using the method as a search for new μ^* its soundness was checked on two nuclei whose μ^* were known, ^1H and ^{51}V (Abragam, Bacchella, Glättli, Meriel, Pinot, and Piesvaux, 1973). The samples were respectively a single crystal of LMN, 6 mm thick, doped with a small amount of Praseodymium to shorten the nuclear relaxation times and a polycrystalline sample of metallic vanadium, 10 mm thick. The neutron wavelength was 1.075 Å. The temperature of the samples could be varied between roughly 1 and 4 Kelvins. The applied field was 25 kG. The precession angle of the neutron spin, related to the measured polarization of the outgoing neutron by (7.63) was plotted against the inverse temperature and gave as expected a straight line. Its slope is obtained from the following relations:

$$\psi = -\gamma_N H^* \tau = -\gamma_N H^* \frac{d}{v} = -\gamma_N 4\pi N_0 \mu^* P \frac{d}{v};$$
$$P \cong \frac{\gamma_n \hbar H_0}{k_B T} \frac{I+1}{3}. \tag{7.67}$$

In (7.67) γ_N and γ_n are the gyromagnetic ratios of the neutron and of the nucleus of spin I. The values of $\mu^*(^1H)$ and $\mu^*(^{51}V)$, extracted from these measurements coincided within experimental errors with the known values.

As an order of magnitude, the change in the pseudomagnetic precession angle of the LMN sample was of the order of 80° between 1 and 4 Kelvins.

It is interesting to compare the sensitivity of the precession method with the Bragg scattering on a polarized target of the type described in section **D**(a). We consider an experiment of that type, actually performed on a single crystal of LiH for the most favourable reflexion (111) (Abragam, Bacchella, Long, Meriel, Pinot, and Piesvaux, 1972). In a field of 20 kG and at a temperature of 1·15 K, the equilibrium proton polarization is ~0·18 per cent. The change (I_+/I_-) in the counting rate, upon reversal of the neutron polarization was predicted, and observed to be approximately +6 per cent. In the precession method for the same field, the same neutron beam and for a sample of LiH 6 mm thick, the change in the precession angle between 1 and 4 Kelvins is of the order of 90°. This corresponds to a change in the final neutron polarization of the order of 100 per cent.

The precession method is clearly far more sensitive than the Bragg method. The precession method has yet another important advantage: it can be applied equally well to a polycrystalline or amorphous sample.

In the next section we list some results obtained with this method.

2. *The results*

By the end of 1979 the sign and magnitude of pseudomagnetic moments, had been determined for more than thirty nuclear species, by pseudomagnetic precession (Glättli, Bacchella, Fourmond, Malinovski, Meriel, Pinot, Roubeau, and Abragam, 1979; Malinovski, Coustham, and Glättli, to be published, where references to earlier work can be found). For a few more μ^*, very small, an upper limit was set.

The accuracy varies in general between 10 and 30 per cent. In many cases far greater accuracy (down to 1 or 2 per cent) could be achieved in a more elaborate experiment, if there had been a definite call for it. Table 7.1 sums up the results which are discussed in Glättli *et al.* (1979).

The nuclei which produce the pseudomagnetic fields in the target can be polarized either statically (brute force method) or by DNP. Each method has its good points and its problems.

Static method

The precession angle ψ is according to (7.67) proportional to:

$$\mu^*\mu_n \frac{I+1}{3}\frac{H_0}{T}.$$

The ratio of these quantities for ^1H and say ^{197}Au is of the order of 500 which shows in passing that contamination of the sample by hydrogen is a serious cause of error.

The necessity of going to low temperatures to obtain observable precessions (down to 20 mK) rules out non-metallic samples for which the relaxation times T_1 are prohibitively long. With the exception of the first precession experiment on ^1H, mentioned in **D**(b)1 where a sample of LMN was used and of the ethylsulphate for ^{165}Ho, all static experiments were done on metals.

The main advantage of the static polarization method is that once the cryogenic problems have been solved, the samples containing the pseudomagnetic moments to be measured can be processed, one after the other, without creating each time a new problem.

The main disadvantages are: relatively low nuclear polarizations (it is impractical to work below 20 mK), and for metallic samples the impossibility of an r.f. field at nuclear Larmor frequencies penetrating inside the sample. The first feature requires relatively thick samples in order to obtain measurable precession angles. This in turn rules out nuclei with strong capture because of excessive loss in transmission. The second feature has two consequences: (i) it is impossible to monitor the polarization by an NMR signal; (ii) if the sample contains more than one nuclear species it is not possible to saturate their resonances selectively. If C_α are the proportions of

Table 7.1.

Isotope	I	μ^*/μ_B	$b^+ - b^-$ ($\times 10^{-12}$ cm)
^{1}H	$\frac{1}{2}$	+5·4	+5·82
^{6}Li	1	−0·47 ±0·07	−0·38 ±0·05
^{7}Li	$\frac{3}{2}$	−0·63 ±0·03	−0·45 ±0·02
^{13}C	$\frac{1}{2}$	−0·11 ±0·02	−0·12 ±0·02
^{19}F	$\frac{1}{2}$	−0·018±0·002	−0·019±0·002
^{23}Na	$\frac{3}{2}$	+0·99 ±0·04	+0·71 ±0·03
^{27}Al	$\frac{5}{2}$	0·081±0·003	0·052±0·002
^{39}K	$\frac{3}{2}$	+0·42	+0·30
^{45}Sc	$\frac{7}{2}$	−2·20 ±0·15	−1·36 ±0·09
^{51}V	$\frac{7}{2}$	+2·1	+1·294
^{59}Co	$\frac{7}{2}$	−2·03 ±0·06	−1·25 ±0·04
^{63}Cu	$\frac{3}{2}$	+0·063±0·006	+0·045±0·005
^{65}Cu	$\frac{3}{2}$	+0·51 ±0·02	+0·37 ±0·02
^{75}As	$\frac{3}{2}$	−0·199±0·017	−0·143±0·012
^{89}Y	$\frac{1}{2}$	+0·24 ±0·06	+0·26 ±0·07
^{91}Zr	$\frac{5}{2}$	−0·34 ±0·05	−0·22 ±0·03
^{93}Nb	$\frac{9}{2}$	−0·046±0·004	−0·028±0·002
^{107}Ag	$\frac{1}{2}$	+0·21 ±0·03	+0·23 ±0·03
^{109}Ag	$\frac{1}{2}$	−0·34 ±0·03	−0·37 ±0·03
^{133}Cs	$\frac{7}{2}$	+0·42 ±0·04	+0·26 ±0·03
^{139}La	$\frac{7}{2}$	+0·99 ±0·07	+0·61 ±0·04
^{165}Ho	$\frac{7}{2}$	+0·57 ±0·06	−0·35 ±0·04
^{181}Ta	$\frac{7}{2}$	−0·096±0·01	−0·059±0·006
^{195}Pt	$\frac{1}{2}$	−0·21 ±0·04	−0·23 ±0·04
^{197}Au	$\frac{3}{2}$	−0·49 ±0·04	−0·35 ±0·03
^{207}Pb	$\frac{1}{2}$	+0·02 ±0·04	+0·02 ±0·04
^{209}Bi	$\frac{9}{2}$	+0·08 ±0·02	+0·044±0·099

the various nuclear species α with spin, only the weighted average:

$$H^* = \sum_\alpha C_\alpha H^*_\alpha \propto \sum_\alpha C_\alpha \mu^*_\alpha P_\alpha, \tag{7.68}$$

can be measured. If a sample has more than one isotope with spin, more than one measurement must be made, on samples with different isotopic concentrations. A reliable thermometer is an essential feature. In an early experiment it was calibrated from the known $\mu^*(^{51}V)$ and the measure of its precession angle.

Nuclei of magnetic atoms are a special case. They yield relatively large nuclear polarizations because of the huge hyperfine fields $H_i \gg H_0$ created by the magnetic electrons at the nucleus. On the other hand one must consider the purely magnetic precession of the neutron spins, caused by the electronic magnetization. Below a temperature where this magnetization is completely saturated, its contribution ψ_M to the phase angle, although much larger than the nuclear precession angle ψ, remains constant and can easily be separated from ψ, proportional to $1/T$ down to the lowest temperatures.

The effects of the magnetic precession could be disregarded altogether if ψ_M were homogeneous through a cross section of the target and if the neutron beam were strictly monoenergetic. In the presence of a velocity spread δv, the precessing neutron spins will fan out in the plane perpendicular to the magnetization through an angle of the order of $\psi_M(\delta v/v)$ and their vector polarization will be decreased. For example assuming a Gaussian velocity distribution of half-width Δv, the neutron polarization is reduced in the ratio $\rho = \exp -\{(\psi_M^2/2)(\Delta v/v)^2\}$. This limits severely ψ_M and therefore the thickness d of the sample and the magnitude of the pseudomagnetic angle ψ. The inhomogeneity of ψ_M produces similar depolarization effects.

DNP method

The main drawback of DNP as a tool in the measurement of pseudomagnetic moments, is that each nucleus offers a new problem. A suitable compound must be found, associated to a suitable type of paramagnetic impurities, introduced into the sample chemically or by irradiation. It is usually impossible to predict beforehand whether the combination will lead to sizeable polarizations. Another problem is the absolute calibration of the polarization, for which the reader is referred to chapter 5.

On the other hand once these problems have been solved, the advantages of DNP are impressive. The polarizations are sometimes two orders of magnitude higher than those obtainable statically and the range of precession angle is yet doubled by going over to negative polarizations. Nuclei of atoms not found in the metallic state can be polarized in reasonable times in insulating compounds. For nuclei with large capture cross sections, the high polarizations, by allowing acceptable precession angles in thin targets, improve the intensity of the transmitted beam by keeping down the absorption.

Last but not least, the possibility of saturating selectively the resonances of various nuclear species, and reducing to zero the polarizations of all but one, makes it possible to measure in the same sample the μ^* of various isotopes, a most desirable feature. This procedure requires that, as discussed in chapter 6, in the absence of microwave irradiation the various polarizations be very weakly coupled with each other by the electronic spin–spin interaction. This is achieved by going to rather low lattice temperatures, higher than what would be necessary in static methods but lower than strictly necessary for the DNP itself.

If the μ^* of one of the species, say ^1H, is well known, the selective measurement of the precession it produces, provides an absolute calibration for its polarization and thus for the absolute polarizations of the other species if their *relative* polarizations are known (either by the assumption of a common spin temperature or by direct comparison of NMR signals). Actually the ratio of two pseudomagnetic moments μ^* and $\mu^{*\prime}$ is given by:

$$\frac{\mu^{*\prime}}{\mu^*}=\frac{\psi' PN}{\psi P'N'}, \tag{7.69}$$

which shows that if μ^* is known and the precession angles ψ and ψ' relative to the two species are measured separately, all that is needed are the *relative* concentrations and the *relative* polarizations of the two species. Even these ratios can be dispensed with, the ratio $PN/P'N'$ being directly related to the ratio S/S' of the areas of the absorption signals for the two species. The velocity of the neutron, the thickness of the sample and the absolute nuclear densities factor out.

Dynamic polarization has so far made possible the measurement of the μ^* of ^{6}Li and ^{7}Li in LiF, ^{13}C in $BaCO_3$, ^{17}O in CaO, ^{19}F in CaF_2, ^{35}Cl in KCl, ^{79}Br and ^{81}Br in KBr. Upper limits were obtained for ^{31}P and ^{37}Cl.

Some species such as ^{7}Li and ^{27}Al were measured both statically in the metal and dynamically in LiF and Al_2O_3.

3. *Fall out for nuclear magnetism*

In the course of measurements of pseudomagnetic nuclear moments, new results of some interest for the study of nuclear magnetism proper have come to light. We refer specifically to information obtained from the study of pseudomagnetic precession, leaving aside the far more important contribution of neutrons to the study of nuclear magnetic ordering, to be discussed in section **F** of this chapter, and, in greater detail, in Chapter 8.

The pseudomagnetic precession angle is proportional to a weighted sum $\sum_\alpha C_\alpha \mu_\alpha^* P_\alpha$ of nuclear polarizations. If the C_α and μ_α^* are known, it gives a *direct, absolute* measure of this average of nuclear polarizations. Combined if necessary with NMR, which through selective saturation, can single out a given polarization, it provides a tool for measuring spin–lattice relaxation of a single species, or cross-relaxation between two species.

However the main interest of the method is to allow polarization and spin–lattice relaxation measurements, when for some reason, (lines too broad, bulk metallic samples, inconvenient frequencies) NMR is not feasible. We consider first the example of ^{35}Cl$(I=3/2)$ in KCl. It is found that $\mu^*(^{35}\text{Cl})$ is overwhelmingly larger than $\mu^*(^{37}\text{Cl})$ and $\mu^*(^{39}\text{K})$. ^{35}Cl is thus responsible for the total precession observed in KCl, polarized dynamically by means of F-centres (Glättli and Coustham, 1980). It is found that the saturation of the resonance of either ^{37}Cl or ^{39}K has no observable effect on the precession angle and that the saturation of the resonance of ^{35}Cl leaves an appreciable fraction f of this angle. This is interpreted as the existence of a certain proportion of anomalous nuclei in non-cubic sites. For those, only the $-\frac{1}{2} \rightarrow \frac{1}{2}$ transition is visible by NMR. The contribution of the anomalous nuclei to precession can only be destroyed by r.f. saturation in a broad range of ± 300 kHz around the NMR line of ^{35}Cl. From the knowledge of the fraction f and the assumption that the NMR signal is made of *all* the transitions for nuclei in normal sites and the $-\frac{1}{2} \rightarrow \frac{1}{2}$ transition in the anomalous sites the proportion of those sites can be determined to be ≈ 15 per cent. Once this is known, the NMR signal can be used to measure the polarization of ^{35}Cl from which $\mu^*(^{35}\text{Cl})$ is deduced.

By far the most interesting results came from the measurement by pseudomagnetic precession of the relaxation rates of ^{165}Ho and of protons in holmium ethylsulphate $Ho(C_2H_5SO_4)_3$, $9H_2O$ (Abragam *et al.* 1976*a*) due to mechanisms that had not been observed previously. These results and these mechanisms have been analysed in chapter 6 and we shall be content to describe here the main features of the experimental procedure.

The ion Ho^{3+} is sited in a crystal field of trigonal symmetry. Its three lowest electronic levels can be described by a spin Hamiltonian with a fictitious spin $S = 1$. The nuclear spin of ^{165}Ho is $I = \frac{7}{2}$. The experiment is performed in a field $H_0 = 2{\cdot}5$ T parallel or perpendicular to the c-axis and the temperature goes from 1 K to 30 mK.

There are three contributions to the neutron precession angle: the electronic magnetism of Ho^{3+}, the pseudomagnetism of the nucleus ^{165}Ho and that of the 33 protons per molecule of ethylsulphate. The first of these, although very large is completely frozen out even at the highest temperature of 1 K; its variation with temperature is negligible although, as explained in section **D**(b)2, the existence of a large ψ_M is still a limitation on the accuracy because of its inhomogeneity and of the velocity spread in the neutron beam. Although the pseudomagnetic $\mu^*(^{165}\text{Ho})$ is 300 times smaller than that of the 33 protons, this is compensated (except at the lowest temperatures when the *nuclear* polarization of ^{165}Ho saturates also), by the fact that ^{165}Ho 'sees' a hyperfine field $H_i \approx 800$ Teslas, for $H \parallel c$ and 400 Teslas for $H \perp c$, that is 300 to 150 times larger than that seen by the protons. The two angles $\psi(^{165}\text{Ho})$ and $\psi(^1\text{H})$ are thus comparable over a wide range. The relaxation rates of ^{165}Ho and ^{1}H are obtained from the time-dependence of the precession angles. No selective saturation of the resonance of ^{165}Ho or ^{1}H is possible in this experiment for practical reasons. Fortunately the contributions to precession of the two nuclear species can be separated because of their very different relaxation rates. One favourable feature is that the precession gives *absolute* values of the proton polarizations at all times of the relaxation process and that one need not wait a very long time to know the equilibrium value, in contrast with NMR measurements.

As stated earlier the results of these measurements of $T_1(^{165}\text{Ho})$ and $T_1(^1\text{H})$ and their theoretical interpretation have been given in chapter 6. For experimental details the reader is referred to Abragam *et al.* (1976*a*).

E. Wave-like description of the neutron

(a) *Refractive index*

In our treatment of pseudomagnetic precession beginning with the eqn (7.54) for the interaction between the magnetic moment of the neutron and the pseudomagnetic moment of the nucleus, we have systematically described the neutron as a particle rather than as a wave. In the present section we adopt for the neutron, the opposite, or rather complementary, description (in the sense of Niels Bohr) of a wave propagating through matter. All the theoretical results relative to the pseudomagnetic precession can be derived using this approach. It is only fair to say that not only is this how the idea of this precession was first discovered (Baryshevskii and Podgoretskii, 1964) but also that it provides a natural description for the damping of the precession, absent from the former approach.

We begin by introducing the concept of a refractive index for a neutron propagating through a material medium. The derivation is borrowed from Glauber (1962). It is well known in optics that the existence of a refractive index is the result of interferences between the incident wave and the waves scattered by the various parts of the medium in the forward direction. It is clear therefore that it is connected with multiple scattering.

A single scattering on several nuclei is described by the wave function (7.32):

$$\psi(\mathbf{r}) = \exp(i\mathbf{k}\,.\,\mathbf{r}) - \sum_i a_i \exp(i\mathbf{k}\,.\,\mathbf{r}_i)\frac{\exp(ik|\mathbf{r}-\mathbf{r}_i|)}{|\mathbf{r}-\mathbf{r}_i|}.$$

We assume no correlation between the scattering amplitudes a_i and the positions $\mathbf{r}_i$ and rewrite (7.32) as:

$$\psi(\mathbf{r}) = \exp(i\mathbf{k}\,.\,\mathbf{r}) - \langle a\rangle \sum_i \exp(i\mathbf{k}\,.\,\mathbf{r}_i)\frac{\exp(ik|\mathbf{r}-\mathbf{r}_i|)}{|\mathbf{r}-\mathbf{r}_i|}, \qquad (7.70)$$

where $\langle a\rangle$ is an average scattering amplitude (see the remark at the end of section **E**(a)). If we replace the sum over the discrete scatterers i by an integral over a nuclear density $\rho(\mathbf{r})$, (7.70) is replaced by:

$$\psi(\mathbf{r}) = \exp(i\mathbf{k}\,.\,\mathbf{r}) - \langle a\rangle \int \rho(\mathbf{r}')\exp(i\mathbf{k}\,.\,\mathbf{r}')\frac{\exp(ik|\mathbf{r}-\mathbf{r}'|)}{|\mathbf{r}-\mathbf{r}'|}\,\mathrm{d}^3 r'. \qquad (7.71)$$

In order to describe multiple scattering we replace in the integral, $\exp(i\mathbf{k}\,.\,\mathbf{r}')$ which is the incident wave, by the expression (7.71), which is that of the incident wave modified by a single scattering, and then iterate again and again.

We can short-circuit the successive steps by replacing in the integral, $\exp(i\mathbf{k}\,.\,\mathbf{r}')$ by $\psi(\mathbf{r}')$ straightaway, obtaining the integral equation:

$$\psi(\mathbf{r}) = \exp(i\mathbf{k}\,.\,\mathbf{r}) - \langle a\rangle \int \rho(\mathbf{r}')\psi(\mathbf{r}')\frac{\exp(ik|\mathbf{r}-\mathbf{r}'|)}{|\mathbf{r}-\mathbf{r}'|}\,\mathrm{d}^3r'. \tag{7.72}$$

Applying to both sides of (7.72) the operator (∇^2+k^2) and using equation (7.5) for $[\exp(ik|\mathbf{r}-\mathbf{r}'|)/|\mathbf{r}-\mathbf{r}'|]$ we find:

$$(\nabla^2+k^2)\psi = 4\pi\rho\langle a\rangle\psi. \tag{7.73}$$

If we introduce the refractive index n by looking for a solution of (7.73) of the form $\exp(inkz)$ we find, for constant $\rho(\mathbf{r}) = N_0$:

$$k^2(1-n^2) = 4\pi N_0\langle a\rangle;$$

$$n^2 = 1 - \frac{4\pi N_0\langle a\rangle}{k^2} = 1 - \frac{N_0\langle a\rangle\lambda^2}{\pi}. \tag{7.74}$$

(n^2-1) is very small: for $\langle a\rangle \approx 10^{-12}$, $\lambda \approx 10^{-8}$, $N_0 \approx 3\,.\,10^{22}$, $|n^2-1| \sim 10^{-6}$ and we can write:

$$n \approx 1 - \frac{2\pi N_0\langle a\rangle}{k^2} = 1 - \frac{N_0\langle a\rangle\lambda^2}{2\pi}. \tag{7.75}$$

A word of caution about the use of eqn (7.73) and of its solution $\psi = \exp(iknz)$ is in order. In eqn (7.72) we use the average amplitude $\langle a\rangle$. This is to be contrasted with the procedures used in **A** and **B** where averages were taken on the cross sections, *after* squaring the amplitudes. Taking the average amplitude is permissible here only because we consider the *forward* scattering for which all the scattered waves are in phase and the amplitudes can be added. The use of the form $\psi = \exp(iknz)$ is thus to be reserved for the transmitted beam.

(b) Transmission and absorption

$\langle a\rangle$ and thus also n, have real and imaginary parts: $\langle a\rangle = a' + ia''$, $n = n' + in''$. a'' is related to the total cross-section by the optical theorem:

$$a'' = -\frac{k}{4\pi}\sigma^{\mathrm{T}} \tag{7.76}$$

As a rule $|a''| \ll |a'|$, $|n''| \ll |n'-1|$.

It is worth pointing out that (7.76) follows here from the general optical theorem rather than from the special case that we have derived for isotropic s-scattering in **A**(*a*): the average amplitude $\langle a\rangle$ corresponds to scattering from a target containing many nuclei and the latter is not isotropic in general as stated in section **B**. The total cross-section in (7.76) is not the total cross-section, capture plus scattering for a single nucleus, but the sum of the capture cross-section plus the cross-sections for all the scattering processes, coherent and incoherent, elastic and inelastic (in the latter case exciting or de-exciting the lattice, not the nucleus, which is impossible for slow neutrons), that is for all the processes which in one way or another remove a neutron from the transmitted beam. This becomes apparent if we rewrite the transmitted wave function:

$$\psi_t = \exp(iknz) = \exp(ikn'z)\exp\left(ikz\left(-ia''\frac{2\pi N_0}{k^2}\right)\right)$$

$$= \exp(ikn'z)\exp-\left(\frac{N_0\sigma^{\mathrm{T}}z}{2}\right). \tag{7.77}$$

The interpretation of (7.77) is straightforward. Per unit length ψ_t is attenuated in the ratio $\exp-(N_0\sigma^{\mathrm{T}}/2)$ and the number of particles transmitted proportional to $|\psi_t|^2$ is reduced in the ratio $\exp-(N_0\sigma^{\mathrm{T}})$.

Spin-dependence of absorption cross sections. The measurement of the spin dependence of a total cross section is straightforward: the transmitted intensity of a beam of polarized neutrons is measured as a function of the polarization of the beam and of the target. What is less straightforward is the interpretation of the results.

There is no difficulty for the capture cross section. The channel spin $(I\pm\frac{1}{2})$ is a good quantum number for the capture, with two cross-sections σ_+^{c} and σ_-^{c}. It is convenient, as for the scattering amplitude, to express this fact in a shorthand operator notation:

$$\sigma^{\mathrm{c}} = \sigma_0^{\mathrm{c}} + \frac{2}{I}(\mathbf{I}\,.\,\mathbf{s})\sigma_{\mathrm{pol}}^{\mathrm{c}}. \tag{7.78}$$

The equation (7.78) is a *definition* of σ_0^{c} and $\sigma_{\mathrm{pol}}^{\mathrm{c}}$. Remembering that $(\mathbf{I}\,.\,\mathbf{s})_+ = I/2$ and $(\mathbf{I}\,.\,\mathbf{s})_- = -\{(I+1)/2\}$ we get:

$$\sigma_0^{\mathrm{c}} = \frac{1}{2I+1}\{(I+1)\sigma_+^{\mathrm{c}} + I\sigma_-^{\mathrm{c}}\}, \qquad \sigma_{\mathrm{pol}}^{\mathrm{c}} = \frac{I}{2I+1}(\sigma_+^{\mathrm{c}} - \sigma_-^{\mathrm{c}}). \tag{7.79}$$

For a beam of polarization p and a target of polarization P it follows from (7.78):

$$\sigma^{\mathrm{c}} = \sigma_0^{\mathrm{c}} + pP\sigma_{\mathrm{pol}}^{\mathrm{c}}, \tag{7.80}$$

where σ_0^{c} and $\sigma_{\mathrm{pol}}^{\mathrm{c}}$ are independent of p and P. This is to be contrasted with the eqns (7.49) for the *scattering* cross-sections which contain terms independent of p and proportional to P^2. The terms proportional to P^2 result from interference between spin-dependent scatterings from two different nuclei: there is no interference between *capture* from two different nuclei.

Polarized beams and targets make it possible to go beyond the knowledge of σ_0^c and to reach σ_{pol}^c by measuring the dependence of σ^T on the product pP. We can write for the total cross-section σ^T an equation similar to (7.80):

$$\sigma^T = \sigma_0^T + pP\sigma_{pol}^T. \tag{7.81}$$

It should be noted again that because of the existence in the scattering cross-section of terms proportional to P^2, in contrast to σ_0^c, σ_0^T is *not* the total cross-section for an unpolarized target. However in an experiment where one compares the transmissions for two opposite neutron polarizations, σ_0^T cancels out:

$$\sigma_\uparrow^T - \sigma_\downarrow^T = 2pP\sigma_{pol}^T. \tag{7.82}$$

The problem is to separate in σ_{pol}^T the part $(\sigma_{pol}^T - \sigma_{pol}^c)$ due to scattering. As we stated earlier this is in general a difficult problem which is greatly simplified in two special cases.

(i) If the target is such that no Bragg scattering is possible, either because the neutron wavelength is greater than the cut off value defined in **A**(a) or because the target is a single crystal so oriented that no Bragg scattering can occur, the scattering cross-section reduces to the incoherent cross-section. According to the first eqn (7.49) its contribution to σ_{pol}^T is:

$$\sigma_{pol}^T - \sigma_{pol}^c = -4\pi\frac{b^2}{4}I = -\frac{4\pi I}{(2I+1)^2}(b_+ - b_-)^2. \tag{7.83}$$

(ii) If the wavelength is much shorter than the cut-off value and if the sample is a crystalline powder, many Bragg scatterings do occur. Then the separation of the differential cross-section into coherent and incoherent parts as is done in the last eqn (7.48) is not the best procedure. It is preferable to use the expression (7.45). A term such as $\exp[-i\boldsymbol{\kappa}\,.\,(\mathbf{r}_i - \mathbf{r}_j)]$ with $i \neq j$ in (7.45) referring to scattering by a given crystallite with a given impulse transfer $\boldsymbol{\kappa}$ can be written $\exp(-i|\boldsymbol{\kappa}|l\cos\phi)$ where l is the length of a lattice vector of the crystallite and ϕ its angle with $\boldsymbol{\kappa}$. The contribution to this particular scattering from all the crystallites, corresponds to an average of the exponential over all the values of the angle ϕ, which for $|\boldsymbol{\kappa}|l \gg 1$ is vanishingly small. We are thus left for the total scattering cross-section with the first term in (7.45): $4\pi\langle|a_i|^2\rangle$. According to (7.46) the part of $4\pi\langle|a_i|^2\rangle$ proportional to pP is $4\pi I\{bb_0 - b^2/4\}$ and in this case:

$$\sigma_{pol}^T - \sigma_{pol}^c = 4\pi bI\left(b_0 - \frac{b}{4}\right) = \frac{4\pi I}{2I+1}(b_+^2 - b_-^2). \tag{7.84}$$

The comparison of eqns (7.83) and (7.84) reveals an interesting fact: assuming no capture, then two polarized targets containing the *same* nuclei, with the *same* polarization can transmit preferentially neutrons with *opposite* polarizations if, for these nuclei $b_+^2 > b_-^2$.

The attenuation of a polarized beam by a polarized target has permitted the measure of σ^{c}_{pol} for the strongly absorbing isotope ^{6}Li (Glättli *et al.* 1978). Using a polycrystalline metallic target of Li with natural isotopic abundance polarized statically down to 40 mK, it was found, for $\lambda =$ 1·074 Å,

$$(\sigma^{c}_{+} - \sigma^{c}_{-}) = -1160 \pm 50b. \tag{7.85}$$

The effects of spin-dependent scattering by ^{7}Li were taken into account by (7.84). Spin-dependent scattering of ^{6}Li and spin-dependent capture of ^{7}Li could be neglected. In a second experiment the ^{7}Li nuclei were polarized dynamically in a single crystal of LiF. An analysis of possible Bragg reflections showed that only one, rather weak, Bragg reflection could occur and that it was safe to correct for the spin-dependent ^{7}Li contribution by using eqn (7.83).

The result,

$$\sigma^{c}_{+} - \sigma^{c}_{-} = -1300 \pm 200b, \tag{7.86}$$

is compatible with, although rather less precise than (7.85). The main uncertainty in (7.86) comes from the measure of the nuclear polarization. The weighted average of (7.85) and (7.86), combined with the known value $\sigma^{c}_{0} = 560 \pm 4b$ for $\lambda = 1.074$ Å (Uttley and Diment, 1968, 1969), yields for ^{7}Li:

$$\sigma^{c}_{+} = 170 \pm 20b; \qquad \sigma^{c}_{-} = 1340 \pm 40b. \tag{7.87}$$

The main motivation for the DNP experiment in LiF was the measurement by pseudomagnetic precession of $\mu^{*}(^{6}\mathrm{Li})$, the polarizations of ^{7}Li and ^{19}F being destroyed by selective saturation, with the result $\mu^{*}(^{6}\mathrm{Li})/\mu_{B} =$ $-0{\cdot}47 \pm 0{\cdot}07$.

(*c*) *Pseudomagnetic precession*

Consider for simplicity a polarized target with a single species of nuclei *I*. For a neutron with a spin up travelling through the target the scattering amplitude $\langle a \rangle_{\uparrow} = \langle b_0 + b(\mathbf{I}\,.\,\mathbf{s}) \rangle_{\uparrow} = b_0 + (bPI/2)$. It is $\langle a \rangle_{\downarrow} = b_0 - (bPI/2)$ for a neutron with spin down. They have different refractive indices, $n_{\uparrow}$ and $n_{\downarrow}$ with: $n = 1 - (2\pi N_0/k^2)\langle a \rangle$:

$$\frac{n_{\uparrow} + n_{\downarrow}}{2} = n_0 = 1 - \frac{2\pi N_0}{k^2} b_0; \qquad n_{\uparrow} - n_{\downarrow} = n_1 = -\frac{2\pi N_0}{k^2} bPI. \tag{7.88}$$

The wave function of the transmitted neutron can be written:

$$|\psi(z)\rangle = \alpha_{\uparrow}|+\rangle \exp(in_{\uparrow}kz) + \alpha_{\downarrow}|-\rangle \exp(in_{\downarrow}kz), \tag{7.89}$$

where $\alpha_\uparrow|+\rangle+\alpha_\downarrow|-\rangle$ describes the spin state of the neutron as it enters the target for $z=0$. Equation (7.89) can be rewritten as:

$$|\psi(z)\rangle=\exp(\mathrm{i}n_0kz)\left\{\alpha_\uparrow|+\rangle\exp\left(\frac{\mathrm{i}n_1}{2}kz\right)+\alpha_\downarrow|-\rangle\exp\left(-\frac{\mathrm{i}n_1}{2}kz\right)\right\}. \quad (7.90)$$

All the information relative to the spin of the neutron is contained in the curly bracket of (7.90), that we represent by the ket $|\psi_1\rangle$

$$|\psi_1\rangle=\alpha_\uparrow|+\rangle\exp\left(\mathrm{i}\frac{n_1}{2}kz\right)+\alpha_\downarrow|-\rangle\exp\left(-\mathrm{i}\frac{n_1}{2}kz\right). \quad (7.91)$$

The expectation value of the transverse component $s_+=s_x+\mathrm{i}s_y$ of the neutron spin is given at each point z by:

$$\langle s_+\rangle=\langle\psi_1|s_+|\psi_1\rangle=\alpha_\uparrow^*\alpha_\downarrow\exp(-\mathrm{i}kn_1z). \quad (7.92)$$

For a neutron wave packet moving with a velocity $v=\hbar k/M$, $z=\hbar kt/M$ and:

$$\langle s_+\rangle=\alpha_\uparrow^*\alpha_\downarrow\exp\left(-\mathrm{i}\frac{\hbar k^2}{M}n_1t\right)=\alpha_\uparrow^*\alpha_\downarrow\exp(\mathrm{i}\omega t). \quad (7.93)$$

Equation (7.93) states that the transverse neutron spin precesses at a frequency $\omega=-(n_1\hbar k^2/M)$ which can be interpreted as a Larmor precession in a (pseudo) magnetic field H^* given by:

$$H^*=-\frac{\omega}{\gamma_\mathrm{N}}=-\frac{\omega\hbar}{2\mu_\mathrm{N}}=\frac{n_1\hbar^2k^2}{2M\mu_\mathrm{N}}. \quad (7.94)$$

Replacing in (7.94) n_1 by its value (7.88), μ_N by $g_\mathrm{N}(e\hbar/2Mc)$ and introducing μ_n^* by its expression (7.29) we find: $H^*=4\pi N_0\mu_\mathrm{n}^*P$ that is again the value derived in **C** by a somewhat different approach for the pseudo magnetic field. In the foregoing we have disregarded the fact that even in the absence of capture the spin dependent scattering amplitude b and therefore by (7.88) the refractive index n_1 have small imaginary parts, b'' and n_1'', with $b''/b=n_1''/n_1\equiv-\varepsilon$, where $|\varepsilon|\ll 1$.

According to the definition (7.81), the part of the total cross-section proportional to the neutron polarization p, is $pP\sigma_\mathrm{pol}^\mathrm{T}$. According to the optical theorem, the imaginary part b'' of b is given by

$$b''\langle\mathbf{I}\cdot\mathbf{s}\rangle=\frac{b''pPI}{2}=-\frac{k}{4\pi}pP\sigma_\mathrm{pol}^\mathrm{T}$$

or

$$b''=-\frac{k}{4\pi}\frac{2}{I}\sigma_\mathrm{pol}^\mathrm{T}, \qquad \varepsilon=+\frac{\sigma_\mathrm{pol}^\mathrm{T}}{b\lambda I} \quad (7.95)$$

where λ is the neutron wavelength.

The ket $|\psi_1\rangle$ defined by (7.91) becomes:

$$|\psi_1\rangle = \exp\left(\mathrm{i}\frac{k}{2}n_1 z\right)\exp\left(\varepsilon\frac{k}{2}n_1 z\right)\alpha_\uparrow|+\rangle + \exp\left(-\mathrm{i}\frac{k}{2}n_1 z\right)\exp\left(-\varepsilon\frac{k}{2}n_1 z\right)\alpha_\downarrow|-\rangle \tag{7.96}$$

$|\psi_1\rangle$ is not normalized anymore. Assuming for simplicity $\alpha_\uparrow = \alpha_\downarrow = 1/\sqrt{2}$, that is the neutron spin along $0x$ for $z = 0$, we find

$$\langle\psi_1|\psi_1\rangle = \cosh(\varepsilon k n_1 z) = \cosh(\varepsilon\psi) \tag{7.97}$$

where

$$\psi = -kn_1 z = \frac{2\pi N_0}{k}bPIz \tag{7.98}$$

is the angle of pseudomagnetic precession of the neutron spin between the origin and z.

We can now easily compute the change, due to damping, in the longitudinal and transverse polarization of the neutron. From the expression (7.96) of $|\psi_1\rangle$, with $\alpha_\uparrow = \alpha_\downarrow = 1/\sqrt{2}$, we find:

$$2\langle s_z\rangle = 2\frac{\langle\psi_1|s_z|\psi_1\rangle}{\langle\psi_1|\psi_1\rangle} = \tanh(\varepsilon k n_1 z) = -\tanh(\varepsilon\psi) \tag{7.99}$$

$$2\langle s_+\rangle = 2\frac{\langle\psi_1|s_+|\psi_1\rangle}{\langle\psi_1|\psi_1\rangle} = \frac{\exp(\mathrm{i}\psi)}{\cosh(\varepsilon\psi)}. \tag{7.99'}$$

We see on eqn (7.99) that as the neutron progresses through the target, its longitudinal polarization, which is zero upon entering the target, increases in absolute value and for large $|\varepsilon\psi|$, tends towards ∓ 1, depending on the sign of $\varepsilon\psi$ which by (7.95) and (7.98) is that of $\sigma^{\mathrm{T}}_{\mathrm{pol}}$. This follows immediately from the definition (7.81) of $\sigma^{\mathrm{T}}_{\mathrm{pol}}$. If, say, $\sigma^{\mathrm{T}}_{\mathrm{pol}} > 0$, neutrons with positive polarizations have a larger total cross-section and are removed preferentially from the transmitted beam whose polarization then tends toward -1. At the same time the total intensity of the beam decreases toward zero. This trend of the neutron polarization towards unity is the basis of the method for producing polarized neutrons by sending an unpolarized beam through a polarized target (Lushchikov, Taran, and Shapiro, 1969).

The amplitude of the transverse polarization is also damped as the neutron progresses through the target, an effect we had neglected in **C** and **D**.

It is interesting to evaluate the amount of damping of this amplitude for a given precession angle $\omega t = \psi$.

According to (7.99′) it is reduced by:

$$\frac{1}{\cosh(\varepsilon\psi)} \cong 1 - \frac{\varepsilon^2\psi^2}{2}. \tag{7.100}$$

As an example, we evaluate ε, given by eqn (7.95) for three different situations.

(i) $\sigma^{\mathrm{T}}_{\mathrm{pol}}$ is the spin-dependent part of the incoherent cross-section (eqn (7.83)):

$$\varepsilon = \frac{1}{b\lambda I}\sigma^{\mathrm{T}}_{\mathrm{pol}} = \frac{1}{b\lambda I}(-\pi b^2 I) = -\frac{b\pi}{\lambda} \ll 1. \tag{7.101}$$

For instance for protons which have the largest $b = 5{\cdot}4\times 10^{-12}$ cm and $\lambda = 2\pi/k = 10^{-8}$ cm, $\varepsilon = -1{\cdot}7\times 10^{-3}$ and the correction (7.100), quadratic in ε is negligible.

(ii) For a polycrystalline target with numerous Bragg reflections and no capture, $\sigma^{\mathrm{T}}_{\mathrm{pol}}$ is given by eqn (7.84) and ε is given by:

$$\varepsilon = \frac{\pi}{\lambda}(4b_0 - b). \tag{7.102}$$

For the proton $b_0 = -0{\cdot}375\times 10^{-12}$ cm and $\varepsilon = -2{\cdot}16\times 10^{-3}$, again quite small.

(iii) For a nucleus with a large spin-dependent capture $\sigma^{\mathrm{T}}_{\mathrm{pol}} = \sigma^{\mathrm{c}}_{\mathrm{pol}}$. By far the largest value of ε is obtained for ^{6}Li:

$$\sigma^{\mathrm{T}}_{\mathrm{pol}} = \sigma^{\mathrm{c}}_{\mathrm{pol}} = -390 \text{ Barns} \quad \text{for } \lambda \cong 10^{-8} \text{ cm}$$

$$b = -0{\cdot}25\times 10^{-12} \text{ cm}$$

$$\varepsilon = \frac{\sigma^{\mathrm{c}}_{\mathrm{pol}}}{(b\lambda I)} \cong 0{\cdot}15.$$

The damping $-\varepsilon^2\psi^2/2$ is negligible even in this extreme case especially in view of the fact that the smallness of b and the strong overall absorption, enforce a small precession angle ψ.

The neglect of the damping in the treatment given in **C** and **D** is thus justified *a posteriori*.

F. Neutron scattering and spatial correlations between nuclear spins; domain size

(*a*) *The correlation function*

It has already been stated at the beginning of this chapter that states of nuclear spins exhibiting long range order could be, and had been, studied by neutron diffraction. The whole problem of nuclear ordering is treated in detail in chapter 8.

We begin by a brief discussion of antiferromagnetic nuclear states where the usefulness of neutron diffraction is particularly striking. It is well known that an antiferromagnetic structure with its two sublattices of opposite polarizations is 'seen' by the neutron spin, as a 'magnetic' lattice, with a periodicity different from that of the crystal lattice. This follows from the fact that two sites with opposite spins are seen by the neutron as 'different'. As a result, new Bragg peaks, the so-called superstructure peaks, appear in the antiferromagnetic state, that is below the Neel temperature.

The formalism necessary to establish the existence of and find the positions of these new peaks, requires some modifications of procedures used earlier in this chapter.

In the previous derivation of the coherent scattering cross-section of a polarized target (eqn (7.49)), an essential point was the lack of correlation between the average value $\langle(\mathbf{I}_i \,.\, \mathbf{I}_j)\rangle$ of the scalar product of two nuclear spins, and their relative position $(\mathbf{r}_i - \mathbf{r}_j)$. This is what made possible the transition from eqns (7.44) to (7.45) and then to (7.49). This assumption must be given up for an antiferromagnetic state where for each spin: $\langle\mathbf{I}_i\rangle = \pm PI$ depending on the sublattice to which it belongs. In order to take into account the alternation in the orientations of the spin in an antiferromagnet through successive lattice planes, we must replace the former relation:

$$\langle(\mathbf{I}_i \,.\, \mathbf{I}_j)\rangle = P^2I^2 \quad \text{by:} \quad \langle(\mathbf{I}_i \,.\, \mathbf{I}_j)\rangle = (\langle\mathbf{I}_i\rangle \,.\, \langle\mathbf{I}_j\rangle) = \pm P^2I^2, \tag{7.103}$$

with the minus sign for spins belonging to opposite sublattices. This is done by writing:

$$\langle(\mathbf{I}_i \,.\, \mathbf{I}_j)\rangle = P^2I^2 \exp\{\mathrm{i}\boldsymbol{\tau}_\mathrm{a} \,.\, (\mathbf{r}_i - \mathbf{r}_j)\} \tag{7.104}$$

where $\boldsymbol{\tau}_\mathrm{a}$ (a for antiferromagnetic) is a vector of the magnetic reciprocal lattice. The scalar product of the vector $\boldsymbol{\tau}_\mathrm{a}$ with a lattice vector $\mathbf{l}$ of the crystal lattice is $(\boldsymbol{\tau}_\mathrm{a} \,.\, \mathbf{l}) = n\pi$ (rather than $(\boldsymbol{\tau} \,.\, \mathbf{l}) = 2n\pi$ if $\boldsymbol{\tau}$ is a vector reciprocal to the crystal lattice). From there on, the calculation of the spin-dependent part of the coherent cross-section follows the same lines as before:

$$\begin{aligned}
\sum_{i\neq j} \langle a_i a_j^* \exp(-\mathrm{i}\boldsymbol{\kappa} \,.\, (\mathbf{r}_i - \mathbf{r}_j))\rangle &= \frac{b^2}{4} \sum_{i\neq j} \langle(\mathbf{I}_i \,.\, \mathbf{I}_j) \exp(-\mathrm{i}\boldsymbol{\kappa} \,.\, (\mathbf{r}_i - \mathbf{r}_j))\rangle \\
&= \frac{b^2}{4} P^2I^2 \sum_{i\neq j} \exp(-\mathrm{i}(\boldsymbol{\kappa} - \boldsymbol{\tau}_\mathrm{a}) \,.\, (\mathbf{r}_i - \mathbf{r}_j)) \\
&= -\frac{NB^2}{4} P^2I^2 + \frac{Nb^2}{4} P^2I^2 \sum_i \exp(-i(\boldsymbol{\kappa} - \boldsymbol{\tau}_\mathrm{a}) \,.\, \mathbf{r}_i) \\
&= -\frac{Nb^2}{4} P^2I^2 + \frac{Nb^2}{4} P^2I^2 \frac{(2\pi)^3}{v_0} \sum_{\boldsymbol{\tau}} \delta(\boldsymbol{\kappa} - \boldsymbol{\tau}_\mathrm{a} - \boldsymbol{\tau}).
\end{aligned} \tag{7.105}$$

We can notice that a vector $\boldsymbol{\tau}_a+\boldsymbol{\tau}$ is also a vector of the 'magnetic' reciprocal lattice and replace in (7.105)

$$\sum_{\boldsymbol{\tau}} \delta(\boldsymbol{\kappa}-\boldsymbol{\tau}_a-\boldsymbol{\tau}) \quad \text{by} \quad \sum_{\boldsymbol{\tau}_a} \delta(\boldsymbol{\kappa}-\boldsymbol{\tau}_a).$$

In the foregoing we have not written out the cross term between the neutron spin $\mathbf{s}$ and $\mathbf{I}_i$. The reader will easily convince himself that for an antiferromagnetic state it gives zero, whatever the polarization of the neutron beam.

What we have described so far is an ideal antiferromagnet where the alternation of spin polarizations is perfectly regular through the whole crystal; according to (7.104) the product:

$$(\langle(\mathbf{I}_i \,.\, \mathbf{I}_j)\rangle/P^2I^2)\times\exp\{-\mathrm{i}\boldsymbol{\tau}_a \,.\, (\mathbf{r}_i-\mathbf{r}_j)\},$$

is equal to unity, whatever the distance $|\mathbf{r}_i-\mathbf{r}_j|$. Such a view is unrealistic. Some faults are bound to occur in the crystal, which will alter the regular succession of antiferromagnetic planes. For $|\mathbf{r}_i-\mathbf{r}_j|$ sufficiently large it is reasonable to expect that the correlation between $\langle\mathbf{I}_i\rangle$ and $\langle\mathbf{I}_j\rangle$ vanishes altogether and that:

$$\overline{\langle(\mathbf{I}_i \,.\, \mathbf{I}_j)\rangle \exp\{-\mathrm{i}\boldsymbol{\tau}_a \,.\, (\mathbf{r}_i-\mathbf{r}_j)\}}=\phi(\mathbf{r}_i-\mathbf{r}_j)P^2I^2, \tag{7.106}$$

where $\phi(\mathbf{r})$ is a slowly varying function, with $\phi(-\mathbf{r})=\phi(\mathbf{r})$, which is equal to 1 when $|\mathbf{r}|$ is small (of the order of a few lattice spacings) and decays to zero when $|\mathbf{r}|$ becomes very large. $\phi(\mathbf{r}_i-\mathbf{r}_j)$ is not defined for $\mathbf{r}_i=\mathbf{r}_j$. In order to be coherent with the formalism for a perfect arrangement of the spins, we make the convention $\phi(0)=1$. In (7.104) a symbol such as $\langle(\mathbf{I}_i \,.\, \mathbf{I}_j)\rangle$ represents an average value in a perfect crystal. According to (7.104) it is equal to: $P^2I^2\exp\{-\mathrm{i}\boldsymbol{\tau}_a \,.\, (\mathbf{r}_i-\mathbf{r}_j)\}$. The symbol $\overline{\langle(\mathbf{I}_i \,.\, \mathbf{I}_j)\rangle}$ in (7.106) represents in an imperfect crystal where defects which destroy the regular arrangement of the spins, such as, say, paramagnetic impurities, occur in a random way, a further average over these imperfections.

Regions of the crystal through which the arrangement of spins is regular and the relation (7.104) verified, are called domains. Since they are interrupted by random defects, the size of domain is to be considered as a random function with an average value and a probability distribution. The coherent cross-section can be written:

$$\frac{b^2}{4}\sum_{i,j}\overline{\langle(\mathbf{I}_i \,.\, \mathbf{I}_j)\rangle}\exp(-\mathrm{i}\boldsymbol{\kappa} \,.\, (\mathbf{r}_i-\mathbf{r}_j))=\frac{Nb^2}{4}P^2I^2\sum_i\exp\{\mathrm{i}(\boldsymbol{\tau}_a-\boldsymbol{\kappa}) \,.\, \mathbf{r}_i\}\phi(\mathbf{r}_i). \tag{7.107}$$

We have so far considered antiferromagnetic states only. We shall see in chapter 8 that after an ADRF, under certain conditions, ordered ferromagnetic states with zero net magnetization can be produced. Clearly this is only possible if there are ferromagnetic domains with magnetizations of opposite sign. In a certain sense, the parallel with antiferromagnetic domains is not perfect: in a demagnetized ferromagnetic state with no net magnetization, domains are a necessity even in a perfect crystal; it is not obvious that their size is a random function. The existence of ferromagnetic domains implies an equation analogous to (7.106):

$$\overline{\langle(\mathbf{I}_i \,.\, \mathbf{I}_j)\rangle} = P^2 I^2 \phi(\mathbf{r}_i - \mathbf{r}_j), \tag{7.108}$$

where $\phi(\mathbf{r}_i - \mathbf{r}_j)$ is a correlation function of the same type as in (7.106).

The Bragg peak for a ferromagnet occurs naturally for $\boldsymbol{\kappa} = \boldsymbol{\tau}$ rather than $\boldsymbol{\kappa} = \boldsymbol{\tau}_\mathrm{a}$ as for an antiferromagnet, that is at the same position as for a spin-disordered crystal.

A simple minded argument suggests that between the width Δk of the rocking curve and the average domain size l_0 exists a relationship: $\Delta k \,.\, \Delta l_0 \sim 1$. The purpose of the next section is to sharpen this relationship in order to gain more information about the size and the shape of the domains.

(*b*) *Mapping of the correlation function and domain size*

The measurement of the correlation function has been renewed by the introduction of the two-dimensional counter which permits a far more detailed analysis of the Bragg scattering than that afforded by the study of the rocking curve. Figure 7.5 displays the geometry of the method. We

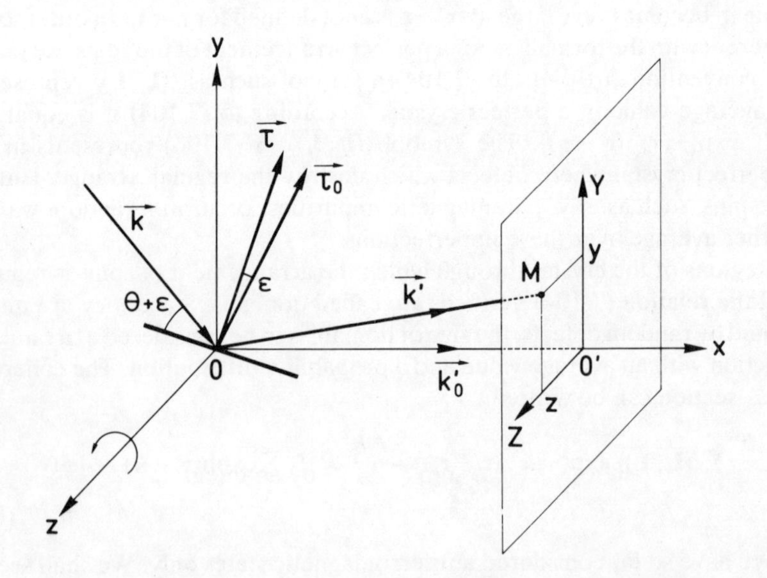

FIG. 7.5. Geometry of the 2-dimensional detection of neutrons diffracted by a nuclear magnetic ordered sample.

assume, in accordance with experiment (see chapter 8) that the random distribution of the domains dominates all the other imperfections of the crystal and of the incoming beam.

The latter, of wave vector $\mathbf{k}$ is then assumed perfectly collimated and perfectly monochromatic. The crystal is first positioned at the exact Bragg angle θ_0. It can be rotated around an axis z parallel to the crystal planes on which the reflection takes place and perpendicular to the incoming beam. In the absence of domains the scattered beam is also perfectly collimated with a wave vector $\mathbf{k}'_0$ and gives a single point of impact 0 on a screen, normal to the scattered beam, situated at a large distance D from the crystal.

The plane $0xy$ thus contains the wave vector $\mathbf{k}$ of the incoming beam, the vector $\boldsymbol{\tau}_a$ (or $\boldsymbol{\tau}$, depending on whether the reflexion is antiferromagnetic or ferromagnetic), and the wave vector $\mathbf{k}'_0$ of the scattered beam, along the $0x$ axis. We have the Bragg condition

$$\mathbf{k}'_0 - \mathbf{k} = \boldsymbol{\tau}_a. \tag{7.109}$$

Actually because of the randomness of the domains, there will be a manifold of orientations for the scattered wave vectors $\mathbf{k}' = \mathbf{k}'_0 + \delta\mathbf{k}'$ where $|\delta\mathbf{k}'| \ll k$. To each vector $\delta\mathbf{k}'$ corresponds an impact M, of co-ordinates Y and Z on the screen with:

$$\delta k'_y = k\frac{Y}{D}; \qquad \delta k'_z = k\frac{Z}{D}; \qquad \delta k'_x = 0. \tag{7.110}$$

The condition $\delta k'_x = 0$ follows from $|\mathbf{k}'| = |\mathbf{k}'_0| = k$ whence $\delta k'_x = (\delta\mathbf{k}' \,.\, \mathbf{k}'_0)/k \approx 0$. The two-dimensional counter is able to measure the number of neutrons impinging on each point (Y, Z) of the screen. This number is proportional to the cross-section for scattering along $\mathbf{k}'$ that is, according to (7.107), to:

$$I(Y, Z) \propto \sum_i \exp \mathrm{i}\{(\boldsymbol{\tau}_a - \mathbf{k}' + \mathbf{k}) \,.\, \mathbf{r}_i\}\phi(\mathbf{r}_i), \tag{7.111}$$

that is, because of the condition (7.109):

$$I(Y, Z) \propto \sum_i \exp-\{\mathrm{i}(\delta\mathbf{k}' \,.\, \mathbf{r}_i)\}\phi(\mathbf{r}_i). \tag{7.111$'$}$$

We approximate the discrete sum (7.111′) by an integral,

$$I(Y, Z) \propto N_0 \int \exp\{-\mathrm{i}(\delta\mathbf{k}' \,.\, \mathbf{r})\}\phi(\mathbf{r})\, \mathrm{d}^3 r, \tag{7.112}$$

where N_0 is the nuclear density per unit volume.

We introduce the Fourier transform of $\phi(\mathbf{r})$ by:

$$A(\mathbf{q})=\frac{(2\pi)^3}{V}\int\phi(\mathbf{r})\exp(-\mathrm{i}\mathbf{q}\,.\,\mathbf{r})\,\mathrm{d}^3r;$$
$$\phi(\mathbf{r})=V\int A(\mathbf{q})\exp(\mathrm{i}\mathbf{q}\,.\,\mathbf{r})\,\mathrm{d}^3q; \tag{7.113}$$

whence:

$$I(Y,Z)\propto N_0V\iint A(\mathbf{q})\exp\{\mathrm{i}(\mathbf{q}-\delta\mathbf{k}')\,.\,\mathbf{r}\}\,\mathrm{d}^3r\,\mathrm{d}^3q$$
$$=N_0V^2(2\pi)^3A(\delta\mathbf{k}')=N_0V(2\pi)^3A\left(0,\frac{kY}{D},\frac{kZ}{D}\right). \tag{7.114}$$

We see that the two-dimensional counter gives a map of the Fourier transform of $\phi(r)$ in a plane perpendicular to $0x$. To complete the information we rotate the crystal by a small angle ε which changes $\boldsymbol{\tau}_\mathrm{a}$ into $\boldsymbol{\tau}_\mathrm{a}+\delta\boldsymbol{\tau}_\mathrm{a}$ with:

$$\delta\boldsymbol{\tau}_\mathrm{a}=\varepsilon\tau_\mathrm{a}(-\hat{x}\cos\theta_0+\hat{y}\sin\theta_0). \tag{7.115}$$

Equation (7.111) becomes:

$$\sum_i\exp\mathrm{i}\{(\boldsymbol{\tau}_\mathrm{a}-\mathbf{k}_0'+\mathbf{k}+\delta\boldsymbol{\tau}_\mathrm{a}-\delta\mathbf{k}')\,.\,\mathbf{r}_i\}\phi(\mathbf{r}_i), \tag{7.116}$$

and the intensity of impacts on the screen which now depends on the angle ε through which the crystal has rotated, is:

$$I(Y,Z,\varepsilon)\propto N_0V\iint A(\mathbf{q})\exp\mathrm{i}\{(\mathbf{q}-\delta\mathbf{k}'+\delta\boldsymbol{\tau}_\mathrm{a})\,.\,\mathbf{r}\}\,\mathrm{d}^3r\,\mathrm{d}^3q$$
$$=N_0V(2\pi)^3A(\delta\mathbf{k}'-\delta\boldsymbol{\tau}_\mathrm{a}) \tag{7.117}$$
$$=N_0V(2\pi)^3A\left(\varepsilon\tau_\mathrm{a}\cos\theta_0,\frac{kY}{D}-\varepsilon\tau_\mathrm{a}\sin\theta_0,\frac{kZ}{D}\right).$$

The collection of the two-dimensional intensity plots for a succession of orientations of the crystal provides according to (7.117) a complete three-dimensional mapping of $A(\mathbf{q})$ and, at least in principle, through (7.113) a complete determination of $\phi(\mathbf{r})$.

In accordance with eqn (7.113), eqn (7.117) can be rewritten:

$$I(Y,Z,\varepsilon)\propto N_0(2\pi)^6\iiint\phi(x,y,z)\exp\left[-\mathrm{i}\left\{x\varepsilon\tau_\mathrm{a}\cos\theta+y\left(\frac{kY}{D}-\varepsilon\tau_\mathrm{a}\sin\theta_0\right)+z\frac{kZ}{D}\right\}\right]\mathrm{d}x\,\mathrm{d}y\,\mathrm{d}z. \tag{7.118}$$

If a two-dimensional counter is not available we fall back on the traditional rocking-curve measurement: only the total number of neutrons scattered coherently for each value of the angle ε is available, an appreciable loss of information.

We define:

$$I(\varepsilon) \propto \iint I(Y, Z, \varepsilon)\, \mathrm{d}\left(\frac{kY}{D}\right) \mathrm{d}\left(\frac{kZ}{D}\right). \tag{7.119}$$

From (7.118) and (7.119) we obtain:

$$I(\varepsilon) \propto N_0 (2\pi)^6 \int \phi(x, 0, 0) \exp\{-\mathrm{i}x\varepsilon\tau_\mathrm{a} \cos\theta_0\}\, \mathrm{d}x. \tag{7.120}$$

The rocking curve is the Fourier transform with wave number $\varepsilon\tau_\mathrm{a}\cos\theta_0$ of the domain correlation function in the direction of the diffracted beam. Having assumed that the size of domains is a random function we have to choose a mathematical model to represent it. Consider first the case of ferromagnetic domains. We assume that the variation of the nuclear polarization $\langle I_j \rangle_z$ along a line, across successive magnetic planes has a square shape, that is a value PI for a certain number n_1 of lattice sites, then abruptly $-PI$ for n_2 sites, and so on.

The antiferromagnetic case is the same if one replaces the polarization $\langle I_j \rangle_z$ by the 'staggered' polarization $\langle I_j \rangle_z \exp -\mathrm{i}(\boldsymbol{\tau}_\mathrm{a} \cdot \mathbf{r}_j)$. We shall see in chapter 8 that this very simple model is probably a realistic description of the physical situation.

We shall call the domain wall that narrow region (of the order of a lattice spacing) over which occurs the reversal of the polarization (staggered or not as the case may be). Let $P(l)$ be the probability for a domain to have an extent l, with an average value $l_0 = \int_0^\infty lP(l)\, \mathrm{d}l$.

Consider the distribution of domain walls in an interval $r \ll l_0$. The probability for there being more than one wall is negligible, that for one wall is r/l_0 and for no wall $(1 - r/l_0)$.

The correlation between polarizations (staggered or not) at the points 0 and r is:

$$\phi(r) = (+1)\left(1 - \frac{r}{l_0}\right) + (-1)\left(\frac{r}{l_0}\right) = 1 - \frac{2r}{l_0}.$$

The expansion of $\phi(r)$ for r small is thus:

$$\phi(r) = 1 - \frac{2r}{l_0}. \tag{7.121}$$

The following relation can be shown to exist between $\phi(r)$ and $P(l)$ (Goldman, 1980): if $\Phi(z)$ and $\mathscr{P}(z)$ are the Laplace transforms of $\phi(r)$

and $P(l)$:

$$\mathscr{P}(z)=\frac{2-zl_0[1-z\Phi(z)]}{2+zl_0[1-z\Phi(z)]}. \tag{7.122}$$

An interesting special case is: $\phi(r)=\exp(-2r/l_0)$ which according to (7.120) corresponds to a rocking curve of Lorentzian shape. Its Laplace transform is: $\Phi(z)=1/z+(2/l_0)$ whence from (7.122):

$$\mathscr{P}(z)=\frac{(1/l_0)}{z+(1/l_0)}, \tag{7.123}$$

and:

$$P(l)=\frac{1}{l_0}\exp(-l/l_0). \tag{7.124}$$

(*c*) *Short-range spin correlations in a paramagnetic dipolar state*

Following an ADRF from a state with polarization P, there appears a short range correlation $\langle(\mathbf{I}_i\,.\,\mathbf{I}_j)\rangle$ between neighbouring spins, to which the neutron is sensitive, in spite of the fact that at each site the polarization $\langle\mathbf{I}_j\rangle$ vanishes.

This correlation is given by:

$$\langle(\mathbf{I}_i\,.\,\mathbf{I}_j)\rangle=\frac{\mathrm{Tr}\{\exp(-\beta\mathscr{H}'_{\mathrm{D}})(\mathbf{I}_i\,.\,\mathbf{I}_j)\}}{\mathrm{Tr}\,\{\exp(-\beta\mathscr{H}'_{\mathrm{D}})\}}, \tag{7.125}$$

where $\mathscr{H}'_{\mathrm{D}}$ is the truncated dipolar Hamiltonian with inverse temperature β:

$$\begin{aligned}\mathscr{H}'_{\mathrm{D}}&=\tfrac{1}{2}\sum_{i,j}A_{ij}\{3I_iI_j-(\mathbf{I}_i\,.\,\mathbf{I}_j)\};\\ A_{ij}&=\frac{\gamma^2\hbar}{2}(1-3\cos^2\theta_{ij})r_{ij}^{-3}.\end{aligned} \tag{7.126}$$

If the temperature is sufficiently high to allow a limited expansion of $\exp(-\beta\mathscr{H}'_{\mathrm{D}})$ in powers of β, the lowest term of (7.125) is proportional to:

$$\begin{aligned}\langle(\mathbf{I}_i\,.\,\mathbf{I}_j)\rangle&\cong\frac{\beta^2}{2}\mathrm{Tr}\{\mathscr{H}'^2_{\mathrm{D}}(\mathbf{I}_i\,.\,\mathbf{I}_j)\}\\ &=\frac{3\beta^2}{4}\left[\frac{I(I+1)}{3}\right]^2\left\{A_{ij}^2+\frac{8I(I+1)}{3}\sum_k A_{ik}A_{kj}\right\},\end{aligned} \tag{7.127}$$

or, for $I=\frac{1}{2}$:

$$\langle(\mathbf{I}_i\,.\,\mathbf{I}_j)\rangle\cong\frac{\beta^2}{16}\left\{\tfrac{3}{4}A_{ij}^2+\tfrac{3}{2}\sum_k A_{ik}A_{kj}\right\}. \tag{7.127'}$$

The contribution of the spin-correlation terms to the differential cross-section is:

$$\frac{d\sigma'}{d\Omega}=\frac{b^2}{4}\sum_{i\neq j}\langle(\mathbf{I}_i\,.\,\mathbf{I}_j)\rangle_{i\neq j}\exp-(i\boldsymbol{\kappa}\,.\,(\mathbf{r}_i-\mathbf{r}_j)).\qquad(7.128)$$

In the paramagnetic dipolar state, $\langle(\mathbf{I}_i\,.\,\mathbf{I}_j)\rangle$ falls off with the distance r_{ij} over a few atomic distances. Its Fourier transform has a large spread in k space and will contribute to a Bragg peak only very far in the wings where it is difficult to separate the peak from the background.

We estimate below, in order of magnitude, the contribution of the spin correlation terms to the total cross-section in the absence of Bragg scattering for a wave number $k \ll 1/a$ the inverse lattice spacing. The effects should be apparent in a transmission experiment.

We introduce the Fourier transform $A(\mathbf{q})$ of A_{ij} through:

$$A_{ij}=\sum_{\mathbf{q}}A(\mathbf{q})\exp[-i\mathbf{q}\,.\,(\mathbf{r}_i-\mathbf{r}_j)],\qquad(7.129)$$

which for $qa \ll 1$ can be approximated by:

$$A(\mathbf{q})=\frac{2\pi}{3}N_0\gamma^2\hbar(3\cos^2\theta_{\mathbf{q}}-1),\qquad(7.130)$$

where N_0 is the number of spins per unit volume, $\theta_{\mathbf{q}}$ is the angle of the vector $\mathbf{q}$ with the applied field $\mathbf{H}_0$, a formula already used in chapter 3, eqn (3.162).

The validity of (7.130) is independent of the lattice structure (see chapter 8).

Using (7.127′) eqn (7.128) can be rewritten:

$$\left(\frac{d\sigma'}{d\Omega}\right)\Big/\left(\frac{b^2\beta^2}{64}\right)=\sum_{i,j}\tfrac{3}{4}A_{ij}^2\exp(-i\boldsymbol{\kappa}\,.\,(\mathbf{r}_i-\mathbf{r}_j))$$
$$+\sum_{i,j,k}\tfrac{3}{2}A_{ik}A_{jk}\exp(-i\boldsymbol{\kappa}\,.\,(\mathbf{r}_i-\mathbf{r}_j))-\sum_{i,k}\tfrac{3}{2}A_{ik}^2.\qquad(7.131)$$

In the first term of (7.131) which converges rapidly (as r_{ij}^{-6}) we approximate the exponential by unity ($ka \ll 1$), whence using (7.129):

$$\left(\frac{d\sigma'}{d\Omega}\right)\Big/\left(\frac{b^2\beta^2}{64}\right)=-\tfrac{3}{4}N\sum_k A_{ik}^2+\frac{3N}{2}[A(\boldsymbol{\kappa})]^2.\qquad(7.132)$$

The total cross-section σ' relative to spin–spin correlations is:

$$\sigma'=4\pi\,.\,\frac{1}{4\pi}\int\left(\frac{d\sigma'}{d\Omega}\right)d\Omega=4\pi\overline{\left(\frac{d\sigma'}{d\Omega}\right)};$$
$$\sigma'\Big/\left(\frac{4\pi b^2\beta^2}{64}\right)=-\tfrac{3}{4}N\sum_k A_{ik}^2+\frac{3N}{2}\overline{[A(\boldsymbol{\kappa})]^2};\qquad(7.133)$$

where $A(\boldsymbol{\kappa})$ is approximated by (7.130).

The angular average $\overline{[A(\boldsymbol{\kappa})]^2}$ is taken, assuming an incident beam $\mathbf{k}$ perpendicular to the applied field $\mathbf{H}_0$ and averaging $(3\cos^2\theta_\kappa - 1)^2$ over all orientations $\mathbf{k}'$ of the scattered beam.

With a little algebra we get $\overline{(3\cos^2\theta_\kappa - 1)^2} = 5/8$ and:

$$\sigma' = 4\pi N \frac{b^2\beta^2}{64}\left\{\frac{15}{16}\left(\frac{N_0\gamma^2\hbar 2\pi}{3}\right)^2 - D^2\right\}, \tag{7.134}$$

where:

$$D^2 = \gamma^2 H_L'^2 = \frac{\mathrm{Tr}\{\mathcal{H}_D'^2\}}{\mathrm{Tr}\{I_z^2\}} = \frac{3}{4}\sum_j A_{ij}^2.$$

To proceed, we consider as an example, a target, which is a single crystal with a simple cubic lattice and, for definiteness sake, the field $\mathbf{H}_0$ along the [100] axis. $N_0 = a^{-3}$ and $D^2\,(100) = M_2(100)/3 \cong 2\cdot5(\gamma^2\hbar/a^3)^2$:

$$\sigma' = \frac{4\pi N b^2\beta^2}{64}\left(\frac{\gamma^2\hbar}{a^3}\right)^2\left\{\frac{5\pi^2}{12} - 2\cdot5\right\}$$

$$= \frac{4\pi N b^2}{16}\left(\frac{\beta D(100)}{2}\right)^2\left\{\frac{5\pi^2}{12\times 2\cdot5} - 1\right\} = \frac{4\pi N b^2}{16}\left(\frac{\beta D}{2}\right)^2 \times 0\cdot64. \tag{7.135}$$

We write σ' in this form for the following reason: the magnetized state with the large inverse dipolar temperature is obtained by ADRF from a state with a polarization $P = \beta D/2$ (for small polarizations which is the only situation when the expansion (7.127) is valid). We can rewrite:

$$\sigma'(100) = \frac{4\pi N b^2}{16} P^2 \times 0\cdot64. \tag{7.136}$$

This is to be compared to the contribution of the term quadratic in P, to the incoherent scattering cross-section of a target of polarization P which according to (7.49) is $-4\pi N b^2 P^2/16$ (for $I = \frac{1}{2}$).

Performing an ADRF on the target changes the cross-section by an amount:

$$\frac{4\pi N b^2 P^2}{16} \times 1\cdot64.$$

The change is far more striking if the field is along the [111] direction: Then:

$$D^2(111) = 0\cdot449\left(\frac{\gamma^2\hbar}{a^3}\right)^2;$$

$$\sigma'(111) = \frac{4\pi N b^2}{16}\left(\frac{\beta D(111)}{2}\right)^2\left\{\frac{5\pi^2}{12\times 0\cdot449} - 1\right\}$$

$$= \frac{4\pi N b^2}{16} P^2 \times 8\cdot16; \tag{7.137}$$

which is more than 8 times the term in P^2, before the ADRF. Similarly:

$$\sigma'(110) = \frac{4\pi N b^2 P^2}{16} \times 3{\cdot}27. \tag{7.138}$$

Higher order terms in the expansion of (7.125) are needed for higher values of the starting polarization P. The expansion breaks down when one gets near to the critical inverse temperature β_c, where the correlation length is expected to become very large.

It may appear surprising that a short range correlation $\langle(\mathbf{I}_i \,.\, \mathbf{I}_j)\rangle$ in the demagnetized state, given by (7.127), results in a larger effect on the total cross-section than the long range correlation

$$\langle(\mathbf{I}_i \,.\, \mathbf{I}_j)\rangle = P^2 I^2,$$

which exists before ADRF.

The reason is the following: in the polarized state (7.128) can be rewritten:

$$-\frac{Nb^2}{4} P^2 I^2 + \frac{Nb^2 P^2}{4} \sum_i \exp -(i\mathbf{k} \,.\, \mathbf{r}_i),$$

and the second term vanishes in the absence of Bragg scattering. On the contrary, in the demagnetized state, $\langle(\mathbf{I}_i \,.\, \mathbf{I}_j)\rangle$ is a short range function $P(\mathbf{r}_i - \mathbf{r}_j)$ which cuts-off in (7.128) the destructive interference of the sum of exponentials.

8

NUCLEAR DIPOLAR MAGNETIC ORDERING

Order is Heaven's first law.

ALEXANDER POPE

A. Introduction

WHEN a system of interacting nuclear spins is cooled to a sufficiently low temperature, it may undergo a transition to a magnetic ordered phase, in complete analogy with electronic spin systems, whose magnetic ordering has been extensively studied for a long time.

This chapter is devoted to the study of ordering in nuclear spin systems with purely dipole–dipole interactions. This is a clean problem of phase transitions since dipole–dipole interactions are known with certainty, without any adjustable parameter, and the experimental properties of ordering provide a direct test for the validity of the approximate theories used for their prediction. There is no room, as with electronic spin systems, for an adjustment of the form and the magnitude of the interactions to improve the agreement between experiment and theory.

Besides their dipole–dipole interactions, nuclear spins experience indirect interactions mediated by the surrounding electrons (Abragam, 1961, chapter VI), whose form was recalled in chapter 2. The indirect interactions increase with atomic number and are negligible compared with the dipolar interactions for light nuclei, at least in diamagnetic insulators. They may be somewhat larger in substances with low-lying excited states: metals, semiconductors and so-called weak or Van Vleck paramagnets. Therefore the choice of purely dipolar interactions suggests the use of systems of light nuclei in insulating materials.

The production and study of nuclear magnetic ordering raise problems and use methods that differ widely from those encountered in the study of the magnetic ordering of electronic spin systems. They are briefly sketched in the rest of this introduction.

(*a*) *Production of nuclear magnetic ordering: the principle*

The first, and most obvious problem in producing nuclear magnetic ordering is that it requires exceedingly low temperatures, because of the smallness of nuclear dipole–dipole interactions. As a rough order of magnitude, the critical temperature T_c for the onset of ordering is such that the thermal energy $k_B T_c$ is comparable with the interaction energy of a nuclear spin with its neighbours. The latter, of the order of the local frequency $D = \gamma H_L$ is typically a few kHz, and since:

$$k_B \simeq 2 \,.\, 10^4 \text{ MHz K}^{-1}$$

T_c is expected to be in the range 10^{-6} to 10^{-7} K. The cooling technique for achieving such low temperatures must be entirely different from those currently used in cryogenics.

The best known method for producing low spin temperatures is adiabatic demagnetization from a high external field. With electronic spins in insulators or nuclear spins in metals, whose spin–lattice relaxation is relatively fast, their adiabatic demagnetization produces a cooling of the lattice and is commonly used for that very purpose. On the other hand, in insulators the nuclear spins are very loosely coupled to the lattice, the less so the lower the lattice temperature (chapter 6), and nuclear adiabatic demagnetization produces a cooling of the nuclear spins only, leaving the lattice temperature unaltered. This is immaterial as long as one is interested in the properties of the nuclear spins only: it is well established through numerous experiments that a nuclear spin system in low field, isolated from the lattice, reaches in a time of the order of T_2, much shorter than the spin–lattice relaxation time, a state of internal equilibrium characterized by a temperature (Goldman, 1970, chapters 1 and 5).

Furthermore, if the coupling with the lattice can be neglected, it is possible to produce the nuclear demagnetization either by actually reducing the external field to zero, as stated earlier or, remaining in high field, by an ADRF stopped at resonance: this, as will appear shortly, offers a wealth of possibilities for investigating nuclear magnetic ordering which has no equivalent in electronic spin systems.

According to sections **A**(b)1 and **A**(b)3 of chapter 1, the ratio T_f/T_i of final to initial spin temperatures following an adiabatic demagnetization is of the order of H_L/H_0, where H_0 is the initial field and H_L is the local field (in the laboratory frame or in the rotating frame). With easily available fields H_0, this ratio is typically of the order of 10^{-4}. In order for the spin temperature in low field to be in the μK range required for the production of ordering, it is then necessary that the initial spin temperature in high field be in the mK range.

Lattice temperatures in this range can be produced with dilution refrigerators. However, in high field and at low temperature, the nuclear Zeeman spin–lattice relaxation in insulators is so fantastically slow that under its sole effect the nuclear spins would *never* reach the lattice temperature. The way out is to increase the polarization of the spins through the dynamic polarization method described in chapter 6, which requires the doping of the sample with a small amount of paramagnetic centres.

The polarization and the spin temperature are related through the Brillouin function. For spins $\frac{1}{2}$ for instance:

$$p = \tanh(-\hbar\omega_0/2k_B T)$$

where ω_0 is the Larmor frequency. As an example, a proton polarization of 90 per cent in a field of 5 Teslas corresponds to a temperature, $T \simeq 3{\cdot}5$ mK, which is a suitable value prior to the adiabatic demagnetization intended to produce nuclear ordering.

To sum up, the cooling of the nuclear spins below the transition temperature is performed in two successive steps: nuclear dynamic polarization in high field followed by nuclear adiabatic demagnetization either in low field or in high field (Abragam, 1960, 1962).

It is useful to give an alternative description of the production of nuclear ordering, in terms of entropy rather than temperature. What is required for nuclear ordering to take place is that the entropy of the system be low. The result of the DNP in high field is to reduce the entropy of the system to the required value. However, the Zeeman order of the nuclear spins produced by DNP in high field is paramagnetic in nature: there is no correlation between their orientations, other than their common alignment along the applied field. The role of the adiabatic demagnetization is to transfer this order to the dipolar reservoir. In the process the Zeeman order is destroyed. This is evident in the laboratory frame but it is also true in the ADRF.

It may at first sight appear surprising that, after an ADRF, an ordered state of nuclear spins where the local ordering fields are of the order of a few Gauss only, could survive in the presence of an applied field H_0 of several kilogauss. The answer, already known from chapter 1, is that as soon as the demagnetizing r.f. field is cut off there is no way for the Zeeman energy, quantized in units of hundreds of MHz, to flow into the dipolar reservoir whose energy spread is at most a few tens of kHz per spin.

This process of production of nuclear ordering calls for several comments. Firstly, the temperature T_f of the system after adiabatic demagnetization is determined by its entropy, which is the same as that achieved in high field by the DNP. The entropy in high field depends only on the polarization p_i, which is a well defined function of the ratio $(\gamma H_0/k_B T_i)$. It is the initial polarization p_i which turns out to be the important quantity, and not the initial temperature T_i. The choice of a high initial field is not made for obtaining a lower initial temperature, but is dictated by a practical reason, that of increasing the efficiency of the DNP process (chapter 6).

Secondly, the value of the final temperature is not significant by itself either. The final physical state of the system depends only on its entropy or rather on the dimensionless quantity S/k_B. The latter is a well-defined function of $\langle\mathcal{H}_D\rangle/k_B T_f$ (in zero field) or $\langle\mathcal{H}'_D\rangle/k_B T_f$ (in high field). The dipolar interaction is proportional to γ^2. As a consequence, if we consider two spin systems with identical crystalline structures but different gyromagnetic ratios γ_a and γ_b, adiabatic demagnetizations with equal initial polarizations p_i (that is equal entropies) will yield different final temperatures T_f^a and T_f^b for these two systems, in the ratio:

$$T_f^a/T_f^b = (\gamma_a/\gamma_b)^2,$$

but their properties would be the same (except for a scaling of the energies and the quantitities that depend on the gyromagnetic ratio).

Extremely low temperatures could be achieved with nuclear spins having an extremely small γ, but the physics would be the same as with a larger γ and a higher final temperature.

(b) General features of nuclear magnetic ordering

When a nuclear spin system is demagnetized in the laboratory frame, that is by actually lowering the external field to zero, the final Hamiltonian is the full dipole–dipole Hamiltonian $\mathcal{H}_D$, whose form for two spins is given by eqn (1.1). When the adiabatic demagnetization is performed in high field, by an ADRF stopped at resonance, the final Hamiltonian is the secular dipolar Hamiltonian $\mathcal{H}'_D$ to which contributions from like and unlike spins are given by eqns (5.43) and (5.45), respectively.

The secular dipolar Hamiltonian depends on the orientation of the external field with respect to the crystalline axes of the sample, as witnessed by the angular variations of the coefficients A_{ij} and $B_{i\mu}$ (eqns (5.44) and (5.46)). By performing the demagnetization either to zero field or in high field, and in the latter case by varying the orientation of the field it is thus possible, with one single crystal, to study the low temperature properties of a whole variety of different Hamiltonians, leading to a variety of different ordered structures.

Furthermore, since the demagnetization acts only on the nuclear spins, whose energy spectrum has an upper bound, it is possible to choose the sign of the spin temperature to be either positive or negative. For a demagnetization to zero field the sign of the temperature is determined by the sign of the polarization achieved by the DNP: positive if the polarizing microwave frequency is below the electronic Larmor frequency, and negative if it is above the electronic Larmor frequency (chapter 6). For a demagnetization in high field, whatever the sign of the polarization, either sign of temperature can be produced by choosing properly the sign of the effective field at the start of the ADRF, i.e. by applying the r.f. field at a frequency either above or below the nuclear Larmor frequency.

When the spin temperature tends to $+0$, the state of the system is that of lowest energy, whereas when T tends to -0, the state of the system is that of highest energy. For a given Hamiltonian, the ordered phases corresponding to these states will be different. In this connection, a word of caution against a too hasty comparison with electronic spin systems is in order.

It is possible to conceive electronic spin systems at negative temperature, despite the practical difficulty of keeping them at negative temperature during a substantial time, because of their short spin–lattice relaxation time. In most such systems at normal spin concentrations, the spin–spin interactions are short-range Heisenberg interactions which can often be limited to nearest neighbours. In such a case, ferromagnetism for one sign of the temperature corresponds to antiferromagnetism for the opposite sign of temperature.

This is not necessarily so for nuclear spins whose dipolar interactions are long-range, since they decrease as r_{ij}^{-3} and, in the case of the truncated dipolar interactions $\mathscr{H}'_{\mathrm{D}}$, have a sign which depends on the orientation of the vector $\mathbf{r}_{ij}$ with respect to the applied field. We will see in the next section examples where the ordered structures for a given Hamiltonian, and for opposite signs of the temperature, are either both antiferromagnetic or both ferromagnetic, albeit different.

The field of study of nuclear ordering is further enlarged by the following feature: when producing ordering by ADRF in a system consisting of two different nuclear species, one can adjust independently the effective field seen by each species. The ADRF with two spin species is performed by applying two r.f. fields of frequencies ω_I and ω_S close to the Larmor frequencies $-\gamma_I H_0$ and $-\gamma_S H_0$ of the two species, and then sweeping the external field. The effective Hamiltonian in the rotating frame is:

$$\mathscr{H} = (-\gamma_I H_0 - \omega_I) I_z + (-\gamma_S H_0 - \omega_S) S_z + \mathscr{H}'_{\mathrm{D}},$$

and the effective fields:

$$h_I = H_0 + \omega_I/\gamma_I;$$

$$h_S = H_0 + \omega_S/\gamma_S;$$

can be chosen at will. It is only when $(\omega_I/\omega_S) = (\gamma_I/\gamma_S)$ that these effective fields are equal. The problem of ordering in a system where different spins experience different external fields is a new problem, with no equivalent in electronic spin systems.

The experimental methods for studying nuclear magnetic ordering are of two types: magnetic measurements and neutron diffraction.

For magnetic measurements, there is a definite advantage in studying magnetic ordering in high field produced by ADRF. The large external field, while it does not interfere with the ordering, makes it possible to use magnetic resonance and to benefit from its sensitivity and versatility as compared with static measurements that would be the only possible method in actual zero field. As described in chapter 5, it is possible to measure by

NMR the entropy of the system from the initial polarization (or polarizations when several spin species are present), the various contributions to the dipolar energy, the transverse and longitudinal susceptibilities, and the temperature from the resonance signal of auxiliary rare spins. Still other types of measurement will be described in the following. The rest of this chapter will be restricted as a rule to nuclear ordering produced in high field by ADRF, so-called nuclear ordering in the rotating frame.

Nuclear ordering by ADRF implies however two constraints on the samples which do not apply to demagnetization to zero field. Both the final spin temperature and the final structure depend on the orientation of the field with respect to the crystal axes. The samples must therefore be single crystals. In order for the demagnetizing r.f. field to penetrate inside the samples they must also be insulators.

Thanks to the spin-dependence of the strong neutron–nucleus interactions, described at length in chapter 7, neutron diffraction by suitably chosen nuclear spins can be used to determine the exact structure of antiferromagnetic states, to measure the nuclear sublattice polarizations, and to study the distribution of sizes and shapes of the ordered domains.

In spite of its many attractive and highly original features the study of nuclear magnetic ordering is complicated by several circumstances which render the problem both more difficult and less 'clean' than ideally described above.

It is a difficult and time-consuming task to find the proper way of doping the samples with suitable impurities for an efficient dynamic polarization.

Once the polarization is achieved, the paramagnetic centres cannot be ignored because of their large magnetic moment, about three orders of magnitude larger than the nuclear ones. Nuclei close to the impurities experience a large shift of their resonance frequency, and they participate neither in the nuclear demagnetization nor in the nuclear ordering. Outside these 'neutral' pockets, the impurities produce at the sites of the bulk nuclei a distribution of dipolar fields. The energy reservoir represented by the nuclear Zeeman interactions with this field distribution is thermally coupled to the dipolar interactions, and they form together the 'non-Zeeman' reservoir described in section **D**(c) of chapter 5. In order to minimize this perturbation, it is necessary to keep the impurity concentration as low as possible. This conflicts with the needs of dynamic polarization which, as a general rule, is faster and more efficient the larger the impurity concentration. In the necessary compromise between these requirements, the perturbation of the nuclear system by the impurities cannot be made entirely negligible.

Onto this electronic dipolar field inhomogeneity is superimposed the inhomogeneity of the external field over the sample and, more important, the inhomogeneity of the nuclear dipolar fields in the course of the demagnetization when the shape of the sample is not a perfect ellipsoid.

Another problem is that of the homogeneity of the spin temperature in the demagnetized state. Several factors concur in producing an inhomogeneity of this temperature. Firstly, the initial temperature in high field (and subsequently the final temperature) may be inhomogeneous as a result of an inhomogeneity of the nuclear polarization produced either by an inhomogeneity of the polarizing microwave power, an inhomogeneity of the impurity concentration, or an inhomogeneity of the lattice temperature. Secondly, the ADRF is not rigorously isentropic. The irreversible increase of entropy in the course of the ADRF is larger for smaller r.f. fields. An inhomogeneity of r.f. field produces an inhomogeneity of final entropy per unit volume, that is again an inhomogeneity of temperature. Finally, the spin–lattice relaxation may be inhomogeneous as a result of inhomogeneities in the impurity concentration or lattice temperature, and the warming up of the system by spin–lattice relaxation will broaden the distribution of spin temperatures. The variation of spin temperature produced by these factors takes place over macroscopic distances and cannot therefore be neutralized by spin diffusion, which is too slow.

The final difficulty arises from the finite dipolar spin–lattice relaxation time. Once cooled by ADRF, the nuclear dipolar system warms up by relaxation. Measurements on a system whose temperature increases continuously must be performed on the run over times short compared with the relaxation time, which may put a severe limitation on their accuracy. In particular, the study of the critical properties in the vicinity of the transition temperature is made very difficult. The dipolar relaxation is slowed down by increasing the external field and lowering the lattice temperature, which makes it easier to study nuclear ordering in the rotating frame than in the laboratory frame. In practice, nuclear dipolar relaxation is often caused by spurious paramagnetic impurities other than those used for the DNP and the decrease of its rate when lowering the lattice temperature is not as large as expected theoretically, so far an unsolved problem.

As a rule, the dynamic polarization time is much longer than the dipolar nuclear relaxation time, which results in a poor duty ratio for the experimental investigation of nuclear ordering.

Much time and effort are needed to overcome, even partially, all the difficulties listed above.

B. Prediction of the ordered structures

The exact prediction of the ordered structures arising in a system of interacting spins is as yet an unsolved theoretical problem. A prediction of such structures can be made only through approximate methods whose validity must be tested by experiment. We describe below two closely related approximation methods based on a local version of the Weiss field approximation, which have proved very successful (Goldman, Chapellier, Vu Hoang Chau, and Abragam, 1974; Goldman, 1977).

(a) The method of Villain

The method invented by Villain (1959) was initially used to investigate ordering in electronic spin systems with short-range interactions, and led to the remarkable prediction of helical structures.

The description of this very general method is specialized here to the case of a system of spins $\frac{1}{2}$ with a truncated dipolar Hamiltonian. Its use for a system with two spin species will be briefly considered afterwards.

The gist of the Weiss field approximation for magnetic ordering is to neglect short-range correlations between spins. For a Hamiltonian of the form (5.43), it amounts to replacing the actual dipolar energy:

$$\langle \mathcal{H}'_{\mathrm{D}} \rangle = \tfrac{1}{2} \sum_{i,j} A_{ij} (2\langle I^i_z I^j_z \rangle - \langle I^i_x I^j_x \rangle - \langle I^i_y I^j_y \rangle), \tag{8.1}$$

by the approximation:

$$\langle \mathcal{H}'_{\mathrm{D}} \rangle \simeq \tfrac{1}{2} \sum_{i,j} A_{ij} (2\langle I^i_z \rangle \langle I^j_z \rangle - \langle I^i_x \rangle \langle I^j_x \rangle - \langle I^i_y \rangle \langle I^j_y \rangle), \tag{8.2}$$

which can be written:

$$\langle \mathcal{H}'_{\mathrm{D}} \rangle = \tfrac{1}{2} \sum_{i} \boldsymbol{\omega}_i \, . \, \langle \mathbf{I}_i \rangle \tag{8.3}$$

where $\boldsymbol{\omega}_i$ is the Larmor frequency corresponding to the local Weiss field experienced by the spin *i*, defined as the truncated dipolar field produced by the other spins of the sample in their state of average polarization. This frequency, called the local Weiss frequency, is equal to:

$$\boldsymbol{\omega}_i = \partial \langle \mathcal{H}'_{\mathrm{D}} \rangle / \partial \langle \mathbf{I}_i \rangle$$

that is, according to (8.2):

$$\begin{aligned} \omega^i_x &= -\sum_j A_{ij} \langle I^j_x \rangle; \\ \omega^i_y &= -\sum_j A_{ij} \langle I^j_y \rangle; \\ \omega^i_z &= 2 \sum_j A_{ij} \langle I^j_z \rangle. \end{aligned} \tag{8.4}$$

The factor $\frac{1}{2}$ in front of the sum on the right-hand side of (8.3) originates from the fact that $\langle \mathcal{H}'_{\mathrm{D}} \rangle$ is a self-energy: a dipolar term such as $\mathcal{H}'_{\mathrm{D}ij}$ contributes to both local frequencies $\boldsymbol{\omega}_i$ and $\boldsymbol{\omega}_j$, and is therefore counted twice in the sum.

The next step is to assume that the expectation value $\langle \mathbf{I}_i \rangle$ of an individual spin i is equal to its thermal equilibrium value at the temperature β^{-1} in its local Weiss field, that is, for a spin $\frac{1}{2}$:

$$\langle \mathbf{I}_i \rangle = -\frac{1}{2} \frac{\boldsymbol{\omega}_i}{|\boldsymbol{\omega}_i|} \tanh(\tfrac{1}{2}\beta |\boldsymbol{\omega}_i|). \tag{8.5}$$

Equations (8.5) supplemented with the definitions (8.4) represent a system of $3N$ self-consistent equations for the components of the N spins of the sample. For every value of β it has a number of solutions among which one must select those for which the components $\langle I_\alpha^i \rangle$ are real. Among these solutions, the stable one must be selected according to a thermodynamic stability criterion. A characteristic feature of phase transitions is the fact that there is an apparent inconsistency in using thermodynamics for comparing a collection of possible states whereas the very existence of a temperature implies that the statistical state of the system is unique and well defined.

Each of the solutions to be derived has a lower symmetry than the Hamiltonian. It can be considered as a metastable state with given built-in symmetry breaking, which can be assigned all usual thermodynamic properties: temperature, energy, entropy, free energy, etc. .... We have used this procedure in section **F** of chapter 4 to find the stable phases of superfluid ^{3}He.

The criterion of thermodynamic stability is slightly more elaborate than usual, because of the possibility of negative temperatures. It can be given three different forms, derived as follows.

(i) The use of a traceless Hamiltonian is akin to choosing the zero of energy to be that of the system at infinite temperature. The energy of the system is negative or positive when the temperature is respectively positive or negative.

(ii) The stable state has to be determined independently for positive temperatures among the states with negative energies, and at negative temperature among the states with positive energies.

(iii) The first and most elementary form of stability criterion is that at constant energy, the entropy of the system is maximum.

The second form of stability criterion is derived as follows. In the Lagrange multiplier formalism, maximizing the entropy at constant energy amounts to looking for the unrestricted maximum of:

$$S - \beta E = -\beta F, \tag{8.6}$$

where β is the inverse temperature and F the free energy. When comparing several structures at the same temperature, it is possible to divide both sides of (8.6) by $-\beta$, which yields the second form of stability criterion.

At constant temperature the stable state is that for which F is minimum if the temperature is positive, or maximum if the temperature is negative.

Finally, using the general property that for either sign of the energy E, the entropy is a monotonic decreasing function of the absolute value $|E|$, we obtain the third form of stability criterion:

At constant entropy the stable state is that of lowest energy at positive temperature, and of highest energy at negative temperature.

For the experimental procedure described in **A**(a) where the ordering is produced by isentropic demagnetization, this is the more convenient form of the stability criterion.

The energy is given by (8.3). Insofar as short-range correlations are ignored, the entropy is the same as that of a collection of independent spins. For spins I_i with substates m and populations $n_{m,i}$ the entropy is given by the well-known formula:

$$S/k_{\mathrm{B}} = -\sum_{m,i} n_{m,i} \ln(n_{m,i}). \tag{8.7}$$

For spins $\frac{1}{2}$:

$$n_{\pm,i} = \tfrac{1}{2}(1 \pm p_i),$$

where p_i is the polarization of the spin i, and (8.7) becomes:

$$S/k_{\mathrm{B}} = \sum_i \{\ln 2 - \tfrac{1}{2}[(1+p_i)\ln(1+p_i) + (1-p_i)\ln(1-p_i)]\}. \tag{8.7'}$$

The only trouble is that solving the system of eqns (8.5) is so formidably complicated as to be well-nigh impossible. The 'second best' procedure adopted by Villain is to solve eqns (8.5) in the immediate vicinity of the transition. The polarizations and the corresponding local Weiss frequencies being all very small, one can use a linear expansion of the Brillouin function, which yields a set of linear equations; (8.5) become:

$$\langle \mathbf{I}_i \rangle = -\tfrac{1}{4}\beta_{\mathrm{c}} \boldsymbol{\omega}_i; \tag{8.8}$$

or else:

$$\boldsymbol{\omega}_i = \lambda \langle \mathbf{I}_i \rangle; \tag{8.9}$$

with:

$$\lambda = -4/\beta_{\mathrm{c}} = -4k_{\mathrm{B}}T_{\mathrm{c}}/\hbar. \tag{8.10}$$

In this limit we have (within k_{B}):

$$S = N \ln 2 - \tfrac{1}{2}\sum_i p_i^2 \tag{8.11}$$

and:

$$\begin{aligned} E &= \tfrac{1}{2}\sum_i \boldsymbol{\omega}_i \,.\, \langle \mathbf{I}_i \rangle \\ &= \tfrac{1}{8}\lambda \sum_i p_i^2 \\ &= \tfrac{1}{4}\lambda (N \ln 2 - S). \end{aligned} \tag{8.12}$$

According to the third form of the stability criterion, the stable structures for the two signs of the temperature are:

at $T>0$ the one for which λ is minimum,
at $T<0$ the one for which λ is maximum.

1. *General form of the solutions*

The system of eqns (8.9) is solved by using space Fourier transforms. We suppose for simplicity that the spins are located at the lattice sites of a Bravais lattice. We define:

$$\mathbf{I}(\mathbf{k}) = N^{-1/2} \sum_i \langle \mathbf{I}_i \rangle \exp(i\mathbf{k} \cdot \mathbf{r}_i) \tag{8.13}$$

$$A(\mathbf{k}) = \sum_i A_{ij} \exp[i\mathbf{k} \cdot (\mathbf{r}_i - \mathbf{r}_j)], \tag{8.14}$$

whence $A(\mathbf{k}) = A(-\mathbf{k}^*)$ and, since each lattice site of a Bravais lattice is a centre of symmetry, $A(\mathbf{k})$ is real and equal to $A(-\mathbf{k})$.

The last equation is the same as eqn (5.174) in section **D**(*b*) of chapter 5. As in that section, we neglect the small 'pathologic' region of the first Brillouin zone where $|\mathbf{k}|^{-1}$ is comparable with the sample dimensions.

According to (8.4), (8.13), and (8.14), the system (8.9) becomes:

$$\begin{aligned} [-A(\mathbf{k}) - \lambda] I_x(\mathbf{k}) &= 0; \\ [-A(\mathbf{k}) - \lambda] I_y(\mathbf{k}) &= 0; \\ [2A(\mathbf{k}) - \lambda] I_z(\mathbf{k}) &= 0. \end{aligned} \tag{8.15}$$

The solutions are found by direct inspection. They are of two types: longitudinal structures for which $I_x(\mathbf{k}) = I_y(\mathbf{k}) = 0$, for all $\mathbf{k}$ whence $\langle I_x^i \rangle = \langle I_y^i \rangle = 0$, and transverse structures for which $I_z(\mathbf{k}) = 0$ for all $\mathbf{k}$, whence $\langle I_z^i \rangle = 0$.

Let us consider a longitudinal structure, for which the first two eqns (8.15) are automatically satisfied. We look for non-trivial solutions where the $\langle I_z^i \rangle$ do not all vanish, that is where at least one $I_z(\mathbf{k})$ is non-zero, say $I_z(\mathbf{k}_0)$. Since for a solution to be acceptable all $\langle I_z^i \rangle$ must be real, one has according to (8.13):

$$I_z(-\mathbf{k}_0) = I_z(\mathbf{k}_0)^* \neq 0.$$

The system (8.15) is satisfied with:

$$\lambda = 2A(\mathbf{k}_0) = 2A(-\mathbf{k}_0).$$

An acceptable longitudinal structure is then built from a couple of vectors $\pm\mathbf{k}_0$. It corresponds to:

$$\lambda = 2A(\mathbf{k}_0), \tag{8.16}$$

and the only non-vanishing spin components:

$$I_z(\mathbf{k}_0) = I_z(-\mathbf{k}_0)^*. \tag{8.17}$$

If $2\mathbf{k}_0$ is *not* a vector of the reciprocal lattice, the last condition can be fulfilled in two independent ways, and we obtain two solutions.

There are as many longitudinal solutions as there are independent vectors $\mathbf{k}$ in the first Brillouin zone, that is N.

Transverse solutions are built in a similar way. For each couple of distinct vectors $\pm\mathbf{k}_0$, there are four independent solutions. They correspond to:

$$\lambda = -A(\mathbf{k}_0). \tag{8.18}$$

The only non-vanishing spin components are either:

$$I_x(\mathbf{k}_0) = I_x(-\mathbf{k}_0)^*, \tag{8.19}$$

which can be achieved in two independent ways, or:

$$I_y(\mathbf{k}_0) = I_y(-\mathbf{k}_0)^*, \tag{8.19$'$}$$

which can also be achieved in two independent ways.

There are twice as many transverse structures as independent vectors $\mathbf{k}$, that is $2N$.

Many of these solutions are degenerate, i.e. they correspond to the same value of λ. New solutions can be obtained by linear combinations of degenerate solutions.

The determination of the stable solution in the limit of small polarizations goes as follows. We first determine the vectors $\mathbf{k}_1$ and $\mathbf{k}_2$ such that $A(\mathbf{k}_1)$ is the minimum, and $A(\mathbf{k}_2)$ the maximum of $A(\mathbf{k})$.

The stable structure at positive temperature is either longitudinal with $\mathbf{k}_0 = \mathbf{k}_1$, $\lambda = \lambda_1 = 2A(\mathbf{k}_1)$, or transverse with $\mathbf{k}_0 = \mathbf{k}_2$, $\lambda = \lambda_2 = -A(\mathbf{k}_2)$, depending on which λ is the smaller.

The stable structure at negative temperature is either transverse with $\mathbf{k}_0 = \mathbf{k}_1$, $\lambda = \lambda_1' = -A(\mathbf{k}_1)$, or longitudinal with $\mathbf{k}_0 = \mathbf{k}_2$, $\lambda = \lambda_2' = 2A(\mathbf{k}_2)$, depending on which λ is the larger.

The solutions obtained at the transition for infinitely small polarizations may be different from those valid at lower temperatures, for the linear system (8.8) is then replaced by the non-linear system (8.5). The only tractable cases are those when the polarizations of all spins have the same magnitude $|p_i| = p$. Such structures satisfy (8.5) at all temperatures within a scaling of the polarizations, and the relation between Weiss frequencies and polarizations is of the form (8.9) where λ, identical for all spins, is independent of temperature. We call these structures 'permanent' structures.

The energy and entropy of a permanent structure are of the form:

$$E = \tfrac{1}{8} N \lambda p^2 \tag{8.20}$$

$$S = N\{\ln 2 - \tfrac{1}{2}[(1+p)\ln(1+p) + (1-p)\ln(1-p)]\}. \tag{8.21}$$

All permanent structures with the same entropy have the same spin polarization and, since λ in (8.20) is temperature-independent, the relative stability of permanent structures is independent of temperature.

As will be seen shortly, this result is valid only for systems with one spin species. Even then, it is valid only within the Weiss field approximation and may be disproved when short-range correlations are taken into account.

Longitudinal permanent structures are of four types:

(i) When $\mathbf{k}_0 = 0$, the structure is ferromagnetic: all spins are parallel.

(ii) When $\mathbf{k}_0$ is at the boundary of the Brillouin zone, the system is antiferromagnetic:

$$\langle I_z^i \rangle \propto \exp(-\mathrm{i}\mathbf{k}_0 \cdot \mathbf{r}_i) = \pm 1$$

is alternatively parallel and antiparallel to H_0 in successive crystalline planes perpendicular to $\mathbf{k}_0$.

(iii) When $2\mathbf{k}_0$ is at the boundary of the Brillouin zone, the choice

$$I_z(\mathbf{k}_0) \propto 1 + i, \qquad I_z(-\mathbf{k}_0) \propto 1 - i,$$

yields an antiferromagnetic structure with pairs of adjacent planes alternately parallel and antiparallel to H_0. It is such a structure which is postulated for the ordered spin phase in solid b.c.c. ^{3}He (see section **I**(b) of this chapter).

Another, quasi-permanent, structure in the form of a ferromagnet with domains will be described in section **B**(c).

Transverse permanent structures can be obtained for any vector $\mathbf{k}_0$ by choosing:

$$I_x(\mathbf{k}_0) = I_x(-\mathbf{k}_0) \quad \text{real},$$

and:

$$I_y(\mathbf{k}_0) = -I_y(-\mathbf{k}_0) = \mathrm{i}I_x(\mathbf{k}_0),$$

whence, by the inverse Fourier transform of (8.13):

$$\begin{aligned} \langle I_x^i \rangle &= \tfrac{1}{2}p \cos(\mathbf{k}_0 \cdot \mathbf{r}_i); \\ \langle I_y^i \rangle &= \tfrac{1}{2}p \sin(\mathbf{k}_0 \cdot \mathbf{r}_i). \end{aligned} \tag{8.22}$$

This corresponds to a helical structure where, as one moves in space along the direction of $\mathbf{k}_0$, the magnetization rotates in a plane perpendicular to H_0.

The great simplicity of the solutions is a direct consequence of the use of a truncated dipolar interaction in a system with only one spin per unit cell of the Bravais lattice. When the unit cell contains several spins or when the full dipolar interactions are used, that is for the study of magnetic ordering in actual zero field, the linear system (8.15) is replaced by a more complicated one and the form of the solutions is not so simple.

2. *Systems with two spin species*

The theory is not much more complicated when the system contains several spin species. We sketch its derivation when there are two spin species I and S located in a Bravais lattice, with only one spin of each species per unit cell. We drop the simplifying assumption made in **B**(a)1 of spins $\frac{1}{2}$ only.

The truncated dipolar Hamiltonian is written in the form:

$$\mathcal{H}'_{\mathrm{D}} = \tfrac{1}{2}\sum_{i,j} A_{ij}(2I^i_z I^j_z - I^i_x I^j_x - I^i_y I^j_y) + \tfrac{1}{2}\sum_{\mu,\nu} B_{\mu\nu}(2S^\mu_z S^\nu_z - S^\mu_x S^\nu_x - S^\mu_y S^\nu_y) + \sum_{i,\mu} C_{i\mu} \times 2I^i_z S^\mu_z. \qquad (8.23)$$

In the local Weiss-field approximation, the Weiss frequencies are:

$$\begin{aligned} \omega^i_{I_x} &= -\sum_j A_{ij}\langle I^j_x\rangle; \\ \omega^i_{I_y} &= -\sum_j A_{ij}\langle I^j_y\rangle; \\ \omega^i_{I_z} &= 2\sum_j A_{ij}\langle I^j_z\rangle + 2\sum_\mu C_{i\mu}\langle S^\mu_z\rangle; \end{aligned} \qquad (8.24)$$

and:

$$\begin{aligned} \omega^\mu_{S_x} &= -\sum_\nu B_{\mu\nu}\langle S^\nu_x\rangle; \\ \omega^\mu_{S_y} &= -\sum_\nu B_{\mu\nu}\langle S^\nu_y\rangle; \\ \omega^\mu_{S_z} &= 2\sum_\nu B_{\mu\nu}\langle S^\nu_z\rangle + 2\sum_i C_{i\mu}\langle I^i_z\rangle. \end{aligned} \qquad (8.25)$$

The individual polarizations are related to the temperature and the local Weiss frequencies by Brillouin functions. In the vicinity of the transition, this yields:

$$\langle \mathbf{I}_i\rangle = -\beta_{\mathrm{c}}\frac{I(I+1)}{3}\boldsymbol{\omega}^i_I; \qquad (8.26)$$

$$\langle \mathbf{S}_\mu\rangle = -\beta_{\mathrm{c}}\frac{S(S+1)}{3}\boldsymbol{\omega}^\mu_S; \qquad (8.27)$$

or else:

$$\boldsymbol{\omega}^i_I = \lambda_I\langle \mathbf{I}_i\rangle \qquad (8.26')$$

$$\boldsymbol{\omega}^\mu_S = \lambda_S\langle \mathbf{S}_\mu\rangle \qquad (8.27')$$

with:

$$\lambda_I = -\frac{3\beta_c^{-1}}{I(I+1)};\tag{8.28}$$

$$\lambda_S = -\frac{3\beta_c^{-1}}{S(S+1)} = \frac{I(I+1)}{S(S+1)}\lambda_I.\tag{8.29}$$

By using the space-Fourier transforms of (8.26) and (8.27) in accordance with (8.24) and (8.25) we obtain for each vector **k** of the first Brillouin zone a system of six linear equations for the components of **I**(**k**) and **S**(**k**). It is easy to solve and yields the values of λ_I (and λ_S) for the different solutions.

The energy is of the form:

$$E = \frac{1}{2}\left\{\sum_i \boldsymbol{\omega}_I^i \cdot \langle \mathbf{I}_i \rangle + \sum_\mu \boldsymbol{\omega}_S^\mu \cdot \langle \mathbf{S}_\mu \rangle\right\}$$

$$= \frac{1}{2}\left\{I^2 \lambda_I \sum_i p_i^2 + S^2 \lambda_S \sum_\mu p_\mu^2\right\},\tag{8.30}$$

where $p_i = |\langle \mathbf{I}_i \rangle|/I$ and $p_\mu = |\langle \mathbf{S}_\mu \rangle|/S$ are the individual polarizations.

The entropy per spin of a system of independent spins I at low polarization p is:

$$S_I = \ln(2I+1) - \frac{3}{2}\frac{I}{I+1}p^2.\tag{8.31}$$

This result is obtained by writing:

$$n_m = \exp(-\beta\omega_0 m)\Big/\sum_{m'} \exp(-\beta\omega_0 m'),$$

expanding (8.7) to the lowest (second) order in $\beta\omega_0$, and introducing the polarization p by its high temperature value:

$$p = -\beta\omega_0(I+1)/3.$$

According to (8.31) the entropy of the system is:

$$S = S_0 - \frac{3}{2}\left\{\frac{I}{I+1}\sum_i p_i^2 + \frac{S}{S+1}\sum_\mu p_\mu^2\right\}\tag{8.32}$$

where S_0 is the entropy at infinite temperature.

Using (8.29), (8.30) and (8.32), a little algebra yields;

$$E = \tfrac{1}{3}I(I+1)\lambda_I(S_0 - S).\tag{8.33}$$

Then, according to the third form of the stability criterion, the stable state at positive temperature is that for which λ_I is minimum, and at negative temperature that for which λ_I is maximum.

Permanent structures are those for which $p_i = p_I$ is the same for all spins I and $p_\mu = p_S$ is the same for all spins S. The form of these solutions remains unchanged when the temperature is lowered below T_c. However, there is no reason for the ratio p_S/p_I to remain constant. Its variation as a function of temperature will be different for different permanent structures, and the relative stability of two permanent structures may be reversed by lowering the temperature.

(*b*) *The method of Luttinger and Tisza*

This method is very similar to that of Villain in that it uses the same local Weiss field approximation. The difference is that instead of considering the vicinity of the transition temperature where the polarizations are infinitely small, one looks for the stable structures in the limit when the temperature goes to zero. Disregarding the hypothetical case when the dipolar field at the sites of some nuclei would be zero, the polarization of each spin is equal to unity.

Since at zero temperature, the entropy of the system is zero, the stability of the structures is determined by the sole consideration of their energy: at positive temperature, the energy is minimum, and at negative temperature it is maximum.

We first consider the case when there is only one spin species I. The energy is of the form:

$$E = \tfrac{1}{2}\sum_i \boldsymbol{\omega}_i \,.\, \langle \mathbf{I}_i \rangle \tag{8.34}$$

where the Weiss frequency $\boldsymbol{\omega}_i$ is given by eqns (8.4). The fact that each spin polarization is unity is described by the system of N equations:

$$\langle \mathbf{I}_i \rangle \,.\, \langle \mathbf{I}_i \rangle = I^2, \tag{8.35}$$

which will be called the 'strong constraint'.

In order to determine say, the state of lowest energy (8.34) subjected to the strong constraint (8.35), we can use the Lagrange multiplier method and look for the unrestricted minimum of:

$$\sum_i \tfrac{1}{2}\boldsymbol{\omega}_i \,.\, \langle \mathbf{I}_i \rangle - \tfrac{1}{2}\lambda_i \langle \mathbf{I}_i \rangle \langle \mathbf{I}_i \rangle,$$

which yields the system of N equations:

$$\boldsymbol{\omega}_i = \lambda_i \langle \mathbf{I}_i \rangle. \tag{8.36}$$

We shall call for brevity the solutions of the system (8.36) the 'strong' solutions.

Solving this system with *a priori* different Lagrange multipliers for each spin is an impossible task. This difficulty is by-passed by using an artifice due to Luttinger and Tisza, which consists in looking for the structure of lowest energy subjected to a condition much less stringent than (8.35):

$$\sum_i \langle \mathbf{I}_i \rangle \cdot \langle \mathbf{I}_i \rangle = NI^2 \tag{8.37}$$

which is called the 'weak constraint'.

The unrestricted minimum of

$$\sum_i \tfrac{1}{2}\boldsymbol{\omega}_i \cdot \langle \mathbf{I}_i \rangle - \tfrac{1}{2}\lambda \langle \mathbf{I}_i \rangle \cdot \langle \mathbf{I}_i \rangle,$$

with only one Lagrange multiplier λ, corresponds to:

$$\boldsymbol{\omega}_i = \lambda \langle \mathbf{I}_i \rangle \tag{8.38}$$

which is identical with (8.9). The structures satisfying the weak constraint (8.37) and for which the energy is stationary are therefore the same as in the method of Villain. We shall call them for brevity the 'weak' solutions.

The next step is to note that the strong constraint (8.35) also satisfies the weak constraint (8.37), whereas the converse is not necessarily true. As a consequence, the energy of a 'strong' solution cannot be lower (higher) than that of a 'weak' solution of lowest (highest) energy.

Structures satisfying the strong constraint are 'permanent' structures as defined in the preceding section since all the spins have the same polarization (namely unity). We then obtain the following rule.

The stable structure at zero positive temperature is the structure of lowest energy, provided it is permanent.

Similarly, the stable structure at zero negative temperature is the 'weak' structure of highest energy, if it is permanent.

The only tractable case is that when the 'weak' solutions of lowest and highest energy are permanent. In that case it is the same structures which are stable at the transition and at zero temperature. It is not unreasonable to assume that they remain stable in the intermediate temperature range.

Let us now consider the case when the system contains two spin species I and S. The energy is of the form:

$$E = \tfrac{1}{2}\left\{\sum_i \boldsymbol{\omega}_I^i \cdot \langle \mathbf{I}_i \rangle + \sum_\mu \boldsymbol{\omega}_S^\mu \cdot \langle \mathbf{S}_\mu \rangle\right\}, \tag{8.39}$$

where the local Weiss frequencies are given by (8.24) and (8.25).

The strong constraint corresponds to:

$$\begin{aligned} \langle \mathbf{I}_i \rangle \cdot \langle \mathbf{I}_i \rangle &= I^2; \\ \langle \mathbf{S}_\mu \rangle \cdot \langle \mathbf{S}_\mu \rangle &= S^2. \end{aligned} \tag{8.40}$$

For the weak constraint we use two conditions:

$$\sum_i \langle \mathbf{I}_i \rangle \cdot \langle \mathbf{I}_i \rangle = N_I I^2; \tag{8.41}$$

$$\sum_\mu \langle \mathbf{S}_\mu \rangle \cdot \langle \mathbf{S}_\mu \rangle = N_S S^2. \tag{8.41'}$$

The 'weak' solutions satisfying these last conditions and for which the energy is stationary correspond to:

$$\boldsymbol{\omega}_I^i = \lambda_I \langle \mathbf{I}_i \rangle; \tag{8.42}$$

$$\boldsymbol{\omega}_S^\mu = \lambda_S \langle \mathbf{S}_\mu \rangle; \tag{8.42$'$}$$

which are of the same form as eqns (8.26′) and (8.27′), so that the stationary 'weak' solutions at zero temperature are the same as the stationary solutions near the transition.

The energies of the 'weak' and 'strong' solutions are related in the same way as when there is one spin species. The only tractable case is again that when the 'weak' solutions of (8.42) and (8.42′) with lowest and highest energies correspond to permanent structures. These structures, stable at $T=+0$ and $T=-0$, respectively, may then remain stable when $|T|$ increases. In contrast to the one spin species case, the stable structures at the transition (which we suppose to be permanent) are not determined in the same way as at zero temperature, for the following reason: near the transition, the ratio λ_I/λ_S is well-defined (eqn (8.29)) and the ratio p_I/p_S which it determines through (8.26′) and (8.27′) is in general different for different structures. At zero temperature, it is the ratio $p_I/p_S=1$ which is defined beforehand and the ratio λ_I/λ_S, as obtained from (8.42) and (8.42′), depends on the structure and has no reason to be given by (8.29). As a consequence, the stable structures may be different at the transition and at zero temperature. When this happens, one may expect a first-order transition between these structures when the temperature is lowered below T_c.

(c) *The stable structures in CaF_2*

In calcium fluoride, the ^{19}F nuclei, of spin $\frac{1}{2}$, form a simple cubic lattice. The calcium atoms, which form an f.c.c. lattice, occupy one half of the centres of the fluorine cubes. Most calcium nuclei belong to the spinless isotope ^{40}Ca, and only 0·13 per cent of them to the isotope ^{43}Ca. The latter has spin $\frac{7}{2}$ and a gyromagnetic ratio about 14 times smaller than that of ^{19}F. Both because of its low abundance and its small gyromagnetic ratio, it plays a very small role in the ordering, which is essentially due to the ^{19}F spins.

The Fourier transforms of the dipole–dipole interactions have been computed by Cohen and Keffer (1955) for various cubic lattices (s.c., b.c.c., and f.c.c.). They have tabulated, for a series of discrete values of k regularly distributed in the first Brillouin zone, the following quantities:

$$S_3(\mathbf{k}) = \rho^{-1} \sum_l |r_l|^{-3} \exp(\mathrm{i}(\mathbf{k} \cdot \mathbf{r}_l))$$

$$S_5^{ij}(\mathbf{k}) = \rho^{-1} \sum_l r_l^i r_l^j |r_l|^{-5} \exp(\mathrm{i}(\mathbf{k} \cdot \mathbf{r}_l))$$

where the indices i, j stand for the fourfold axes X, Y, Z of the cubic system, ρ is the number of lattice points per unit volume, and the summations are over lattice vectors $\mathbf{r}_l$, except $\mathbf{r}_l = 0$. The knowledge of S_3 and S_5 determines $A(\mathbf{k})$ (eqn (8.14)), with A_{ij} given by eqn (5.44). Let α, β, γ be the cosines of the direction z of the external field H_0 with respect to the fourfold crystalline axes. The dipolar sum $A(\mathbf{k})$ is equal to:

$$A(\mathbf{k}) = n\frac{\gamma^2\hbar}{2a^3}\{S_3(\mathbf{k}) - 3[\alpha^2 S_5^{XX}(\mathbf{k}) + \beta^2 S_5^{YY}(\mathbf{k}) + \gamma^2 S_5^{ZZ}(\mathbf{k}) + 2\alpha\beta S_5^{XY}(\mathbf{k}) + 2\beta\gamma S_5^{YZ}(\mathbf{k}) + 2\gamma\alpha S_5^{ZX}(\mathbf{k})]\}, \tag{8.43}$$

where a is the lattice parameter and n the number of spins per unit cell.

In a simple cubic lattice of spins $\frac{1}{2}$, such as ^{19}F in CaF_2, the preceding theory predicts, depending on the external field orientation and on the sign of the temperature, the occurrence of five different ordered structures. Three are antiferromagnetic, one is ferromagnetic with domains, and the fifth one is also probably ferromagnetic, although its exact nature is not yet firmly established. They are listed below for simple orientations of the external field, and shown in Fig. 8.1.

Positive temperature

Structure (I), longitudinal antiferromagnetic. $\mathbf{H}_0 \| [001]$.

$$p_{iz} = p\exp(i\mathbf{k}_I \cdot \mathbf{r}_i),$$

with:

$$\begin{aligned} \mathbf{k}_I &= (\pi/a, \pi/a, 0); \\ A(\mathbf{k}_I) &= -5{\cdot}352\gamma^2\hbar/(2a^3) = \tfrac{1}{2}\lambda_I; \\ T_c &= -\hbar\lambda_I/(4k_B) \\ &= 0{\cdot}34\ \mu\text{K} \quad \text{in } CaF_2, \end{aligned} \tag{8.44}$$

where a is the lattice parameter.

It is a two-sublattice antiferromagnetic structure where successive planes perpendicular to [110] carry magnetizations alternatively parallel and antiparallel to [001], the direction of the external field (Fig. 8.1I).

Structure (II), longitudinal antiferromagnetic. $\mathbf{H}_0 \| [110]$.

$$\begin{aligned} \mathbf{k}_{II} &= (0, 0, \pi/a) \\ A(\mathbf{k}_{II}) &= -4{\cdot}843\gamma^2\hbar/(2a^3) \\ T_c &= 0{\cdot}306\ \mu\text{K} \quad \text{in } CaF_2. \end{aligned} \tag{8.45}$$

Successive planes perpendicular to [001] carry magnetizations alternatively parallel and antiparallel to [110] (Fig. 8.1II).

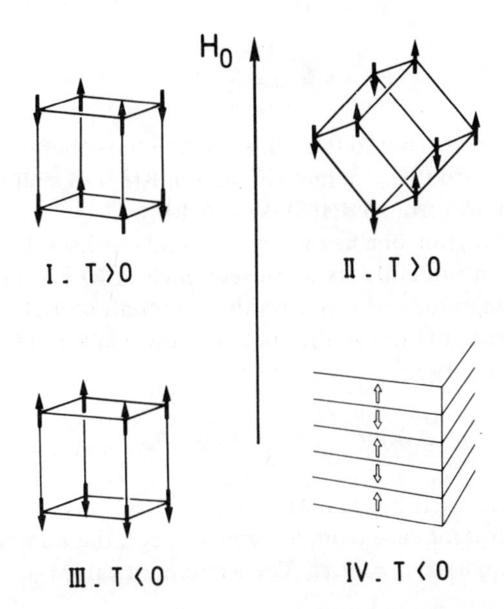

FIG. 8.1. Predicted ordered structures in a simple cubic lattice of spins $\frac{1}{2}$ subjected to truncated dipole–dipole interactions. (After Goldman, Chapellier, Vu Hoang Chau, and Abragam, 1974.)

Negative temperature

Structure (III), longitudinal antiferromagnetic. $\mathbf{H}_0 \| [001]$.

$$\begin{aligned} \mathbf{k}_{III} &= \mathbf{k}_{II} = (0, 0, \pi/a); \\ A(\mathbf{k}_{III}) &= 9{\cdot}687\gamma^2\hbar/(2a^3); \qquad (8.46) \\ T_c &= -0{\cdot}613\ \mu\mathrm{K} \quad \text{in } \mathrm{CaF_2}. \end{aligned}$$

Successive planes perpendicular to [001] carry magnetizations alternatively parallel and antiparallel to [001] (Fig. 8.1III).

Structure (IV), longitudinal ferromagnetic. $\mathbf{H}_0 \| [111]$ or $\mathbf{H}_0 \| [110]$.

This is a quasi-permanent structure derived as follows. It turns out that for these orientations of the field $\mathbf{H}_0$, the maximum and minimum values of $A(\mathbf{k})$ correspond to small values of $\mathbf{k}$, i.e. $|\mathbf{k}|a \ll 1$. We still assume $|\mathbf{k}|^{-1}$ much smaller than the sample dimensions in order that $A(\mathbf{k})$ be well-defined.

When $k = |\mathbf{k}|$ is small, the summation (8.14) over the index i can be performed separately over two regions: an internal region inside a sphere of radius R centred on the spin j, with $a \ll R \ll k^{-1}$, and an external region consisting of the rest of the sample. Since in the internal region:

$$\mathbf{k}\,.\,(\mathbf{r}_i - \mathbf{r}_j) \ll 1,$$

one has:

$$A(\mathbf{k})_{\text{int}} \simeq \sum_i{}' A_{ij}, \tag{8.47}$$

where the sum is restricted to the spins i inside the sphere of radius R. The sum (8.47) is identical to a_{int} defined in section $\mathbf{B}(b)1$ of chapter 5. In a cubic system it is equal to zero, so that $A(\mathbf{k}) = A(\mathbf{k})_{\text{ext}}$.

In the external region, one has $|\mathbf{r}_i - \mathbf{r}_j| \gg a$, and the discrete summation can be replaced by an integral. As a consequence $A(\mathbf{k})$ for small $\mathbf{k}$ depends neither on the magnitude of k nor on the orientation of $\mathbf{H}_0$ with respect to the crystalline axis, but only on the angle θ_k between $\mathbf{k}$ and $\mathbf{H}_0$. It is given by (Cohen and Keffer, 1955):

$$A(\mathbf{k}) = \frac{\gamma^2 \hbar}{2a^3} \times \frac{4\pi}{3}(3\cos^2\theta_k - 1), \tag{8.48}$$

a formula already used in chapter 3, eqn (3.162) and in chapter 7, eqn (7.130). In the present case (simple cubic lattice), the number of spins per unit volume is equal to: $n = 1/a^3$. The extremum values of $A(\mathbf{k})$ are:

$$A(\mathbf{k})_{\max} = \frac{8\pi}{3} \frac{\gamma^2 \hbar}{2a^3} \quad \text{for } \mathbf{k} \parallel \mathbf{H}_0; \tag{8.49}$$

$$A(\mathbf{k})_{\min} = -\frac{4\pi}{3} \frac{\gamma^2 \hbar}{2a^3} \quad \text{for } \mathbf{k} \perp \mathbf{H}_0. \tag{8.49'}$$

We obtain a longitudinal solution of eqns (8.15) with λ maximum, i.e. acceptable at $T < 0$ near the transition, by taking:

$$\begin{aligned} \langle I_z^i \rangle &= A \sin(\mathbf{k}_0 \cdot \mathbf{r}_i) + B \cos(\mathbf{k}_0 \cdot \mathbf{r}_i); \\ \lambda &= 2A(\mathbf{k})_{\max}; \end{aligned} \tag{8.50}$$

where $\mathbf{k}_0$ is any small wave vector parallel to $\mathbf{H}_0$. This solution is however not 'permanent' insofar as the polarization $p_i = 2|\langle I_z^i \rangle|$ is *not* the same for all spins. This difficulty is by-passed by noting that the solution (8.50) is degenerate with all those corresponding to $\mathbf{k} \parallel \mathbf{H}_0$ and $|\mathbf{k}|$ small. In particular we can use a linear combination of solutions corresponding to multiples of $\mathbf{k}_0$, of the form:

$$\langle I_z^i \rangle \propto \sum_{n=0} \frac{1}{2n+1} \sin[(2n+1)\mathbf{k}_0 \cdot \mathbf{r}_i] \tag{8.51}$$

If the sum were extended to $n \to \infty$, this would correspond to a square-wave modulation of $\langle I_z^i \rangle$, that is to a permanent structure consisting of ferromagnetic domains in the form of slices perpendicular to $\mathbf{H}_0$, of width $d = \pi/k_0$. In fact, the sum must be restricted to values of n such that $(2n+1)k_0 a \ll 1$, in order that the various solutions be degenerate. This results in a rounding-off of the variation of $\langle I_z \rangle$ at the edges of the domains over a distance comparable with the lattice parameter a. Since $d \gg a$, we have thus constructed a quasi-permanent structure satisfying the stability criterion, which is a ferromagnetic structure with domains, whose domain walls are thin.

In order to decide whether a single-domain ferromagnet can exist, it is necessary to consider the value of $A(0)$, which depends on the sample shape. Selected values of $A(0) = \bar{a}_{\text{ext}}$ are given in eqns (5.57) to (5.60) for various ellipsoïds. We conclude from this that at $T < 0$, an infinitely flat disk perpendicular to $\mathbf{H}_0$ might give rise to single-domain ferromagnetism, but that for other shapes the ferromagnetic structure will necessarily be composed of domains.

Equation (8.51), which corresponds to domains of equal thickness, is but an example of the way a permanent structure can be built from small vectors $\mathbf{k} \| \mathbf{H}_0$. In fact, the thickness of the various domains need not be equal. However, except for an infinitely flat sample, the vectors $\mathbf{k}$ from which the structure is built must be different from zero, so that the bulk magnetization of the sample is zero: the volume of the domains with polarization up is equal to that of the domains with polarization down.

The following remark is in order. Let us consider an individual spin I_i in a given domain of polarization p. The local frequency it experiences is, according to (8.9) and (8.50), equal to:

$$\omega_{iz} = pA(\mathbf{k})_{\text{max}}. \tag{8.52}$$

It is the sum of two contributions: that produced by the spins in the same domain as the spin I_i, and that produced by the spins in all other domains. Since the domains are in the shape of quasi-infinitely flat disks, the former contribution is equal to $pA(0)$ where $A(0)$ given by eqn (5.59), is identical with $A(\mathbf{k})_{\text{max}}$ (eqn (8.49)). We obtain therefore the result that the spins of all domains other than that where the spin I_i is located produce a vanishing dipolar frequency at the site of this spin. This result can be obtained by the following intuitive argument. Insofar as the domains are thin, we make the assumption that, for calculating the dipolar frequency at the site of the spin I_i, we can replace the whole sample outside the domain where the spin I_i belongs by a continuum of homogeneous polarization equal to its average polarization. Since the volume of the domain where the spin I_i sits is a negligible fraction of the total volume of the sample, this average polarization is very nearly zero, and the dipolar frequency produced by the spins outside this domain vanishes. The approximation of replacing the part of the sample outside a domain by a continuum of homogeneous polarization will be used in sections **E** and **F**.

An essential difference between the domain structure described above and that of ferromagnetic electronic spin systems is that in the present case it is the *same* dipolar interaction which gives rise to the ordering and to the occurrence of domains. The shape of these domains is an intrinsic property of the ordering, and is not determined by the shape of the sample (except possibly for a single-domain ferromagnet).

The fifth structure concerns the case of positive temperature when $\mathbf{H}_0$ is in the vicinity of [111]. By analogy with the case of negative temperature, it is possible to construct a quasi-permanent longitudinal ferromagnet with domains corresponding to $\lambda = 2A(\mathbf{k})_{\min}$ (eqn (8.50)).

Since $A(\mathbf{k})_{\min}$ for k small is degenerate with respect to the orientation of $\mathbf{k}$ in the plane perpendicular to $\mathbf{H}_0$, the shape of the domains is quite undetermined. All that is required is that at least one transverse dimension of each domain be much smaller than its longitudinal dimension.

It is also possible to build a *transverse* ferromagnet with domains or a helical structure, corresponding to $\langle \mathbf{I}_i \rangle \perp \mathbf{H}_0$, $\mathbf{k} \parallel \mathbf{H}_0$ and small, and $\lambda = -A(\mathbf{k})_{\max}$ (eqn (8.18)). Since in the present case, according to (8.49), (8.49′):

$$A(\mathbf{k})_{\min} = -\tfrac{1}{2}A(\mathbf{k})_{\max}$$

these transverse structures are degenerate with the logitudinal ferromagnet with domains. There is therefore some uncertainty as to the actual ordered structure in that case, which at the time of writing has not yet been experimentally investigated.

The regions of existence of these five structures as a function of the orientation of the field $\mathbf{H}_0$ are derived from the variation of the $A(\mathbf{k})$s with this orientation. They are shown in Fig. 8.2. The boundaries correspond to

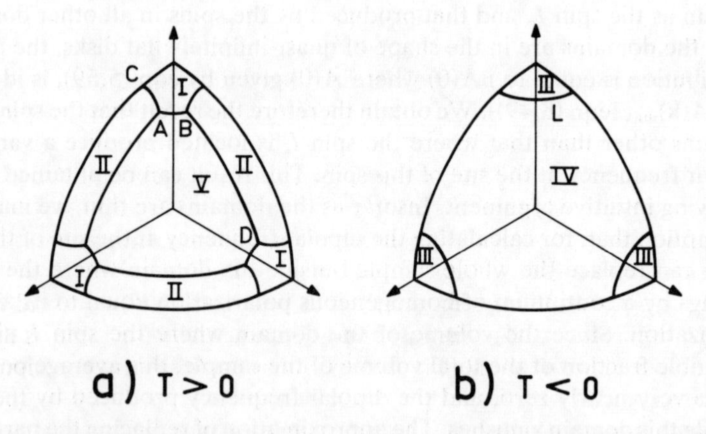

FIG. 8.2. Stable structures in a simple cubic system of spins $\frac{1}{2}$ with truncated dipole–dipole interactions as a function of orientation of the external field with respect to the crystalline axes. Structures (I) to (IV) are those of Fig. 8.1. Structure (V) is not analysed. (After Goldman, Chapellier, Vu Hoang Chau, and Abragam, 1974.)

the following angles:

$$\text{Along AB: } \theta_Z \simeq 22°$$

$$\text{At C: } \theta_Z \simeq 15°$$

$$\text{Along BD: } \theta_X \simeq 78°$$

$$\text{Along L: } \theta_Z \simeq 17{\cdot}5°.$$

Within the local Weiss field approximation, these angles are independent of entropy.

(*d*) *The stable structures in LiF and LiH*

The crystalline structure of these compounds is of the NaCl type: each nuclear species forms an f.c.c. lattice, and these lattices are imbricated so as to form a simple cubic arrangement of nuclei.

The isotopic abundance of ^{19}F is 100 per cent. That of ^{1}H is nearly 100 per cent, since the abundance of deuterium is only $1{\cdot}56 \times 10^{-2}$ per cent. Both ^{19}F and ^{1}H have spin $\frac{1}{2}$. The majority of the lithium nuclei consist of the isotope ^{7}Li, of spin $\frac{3}{2}$, but the abundance of ^{6}Li, of spin 1, is not negligible: 7 per cent in natural lithium, and 3 per cent in the samples of lithium hydride experimentally investigated.

Within a small correction, the system is essentially composed of two spin species: $S = {}^{7}$Li, and $I = {}^{19}$F or ^{1}H.

The magnetic moments $\mu_I = \hbar I \gamma_I$ and $\mu_S = \hbar S \gamma_S$ are close to each other:

$$\mu(^{1}\text{H})/\mu(^{7}\text{Li}) \simeq 0{\cdot}86;$$

$$\mu(^{19}\text{F})/\mu(^{7}\text{Li}) \simeq 0{\cdot}81.$$

It is then not unlikely that the ordered structures will be the same as for a cubic lattice of identical spins, i.e. the structures described in section **B**(*c*) for CaF_2. This is but a reasonable guess, because of the differences between the two problems: although the magnetic moments of the two species are close to each other, their spins and their gyromagnetic ratios are very different, and furthermore the flip-flop terms between the two species are absent from their secular dipolar interactions.

The prediction of the approximations of Villain and Luttinger and Tisza is that the ordered structures in LiH and LiF are actually the same as those in CaF_2 (Fig. 8.3), with roughly the same regions of stability as a function of the orientation of $\mathbf{H}_0$ with respect to the crystalline axes.

As an example, when $\mathbf{H}_0$ is in a plane perpendicular to the [110] crystalline axis, and at negative temperature, the angle between H_0 and the axis [001], at the boundary between the antiferromagnetic structure III and the ferromagnetic structure IV, varies from 17.06° at the transition to 17·28° at zero temperature, to be compared with 17·5° at all temperatures in CaF_2.

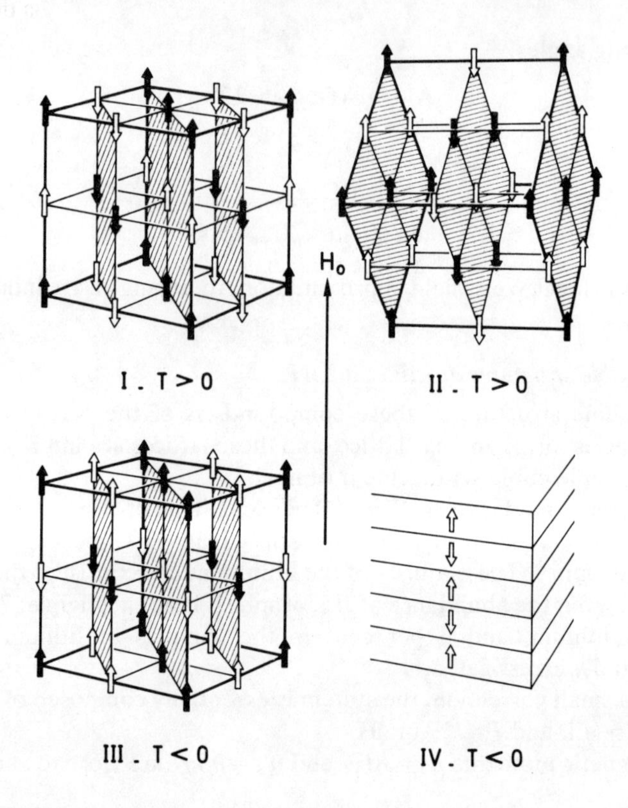

FIG. 8.3. Predicted ordered structures in lithium fluoride and lithium hydride. (After Cox *et al.* 1975)

However, at $T > 0$ and $\mathbf{H}_0$ in the vicinity of [111] the absence of flip-flop terms between the spins of ^{1}H and ^{7}Li precludes the existence of a transverse ferromagnet. The stable structure can only be a longitudinal ferromagnet.

These structures, in which both spin species participate in the ordering, are the stable structures when both spin species are in zero field. They would be different under different field conditions. As a somewhat idealized example, let us consider the case when after the two species have been polarized to $p = 1$, only one of them, say ^{1}H, is demagnetized by ADRF. In a sample whose shape is an ellipsoïd, the only effect of the polarized ^{7}Li spins is to produce at the sites of all protons a dipolar field uniform and constant (it is constant because a 100 per cent polarization precludes any flip-flops). After ADRF of the sole protons, their ordering will be produced by the dipolar interactions between protons only. The ordered structures of a system of like spins in an f.c.c. lattice are very different from those described above. As discussed in section **B**(f), they are of the types IV and V for all orientations of the external field, i.e. ferromagnetic at $T<0$, and ferromagnetic or helical at $T>0$.

(*e*) *Stable structures in* $Ca(OH)_2$

The crystalline structure of $Ca(OH)_2$ is shown in Fig. 8.4. It is a simple hexagonal structure for the calcium atoms, with two OH^- ions per unit cell. Except for a very small percentage of ^{43}Ca, ^{17}O, and D, the only magnetic nuclei are the protons. They are grouped around planes normal to the *c*-axis and 4·88 Å apart. Half of the spins lie above, and half of them below a plane at a distance of 0·355 Å from it, the distance between protons nearest neighbours being 2·186 Å.

The calculation of the stable magnetic ordered structures is slightly more complicated than in section **B**(*a*), **B**(*b*), because the protons are not centres of symmetry. The predicted ordered structures will be described only for simple orientations of the external field (Marks, Wenckebach and Poulis (1979), and private communication).

We define axes $0XYZ$ linked to the crystal. The axis $0Z$ is parallel to the *c*-axis, that is perpendicular to the proton planes. The axes $0X$ and $0Y$ parallel to the proton planes are shown in Fig. 8.5.

1. *Field* H_0 *parallel to* $0Z$

Structure I, $T<0$

Longitudinal ferromagnet with domains in the form of thin slices perpendicular to $0Z$ (same as Fig. 8.1IV):

$$\lambda_{\mathrm{I}} = 16{\cdot}09 n\gamma^2\hbar,$$

$$T_c = -0{\cdot}853\ \mu\mathrm{K},$$

where n is the number of protons per unit volume.

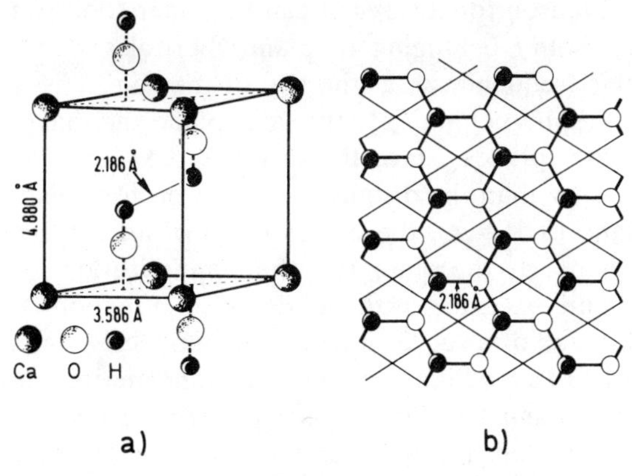

FIG. 8.4. Crystalline structure of $Ca(OH)_2$.

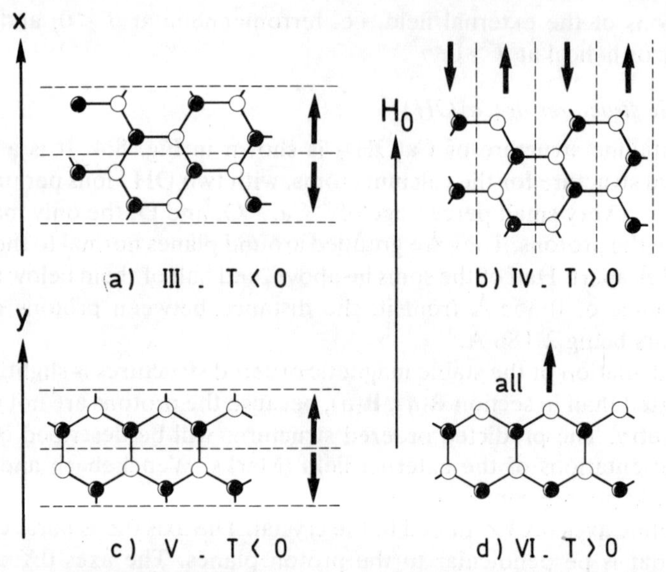

FIG. 8.5. Predicted ordered structures in $Ca(OH)_2$. (After Marks, Wenckebach and Poulis, private communication.)

This structure is nearly degenerate with an antiferromagnetic structure where protons belonging to successive planes would be alternatively parallel and antiparallel to $0Z$. The ratio of the coefficients λ for these two structures is:

$$\lambda_{AF}/\lambda_{ferro} \simeq 1 - 2{\cdot}4 \times 10^{-4},$$

so that there is some uncertainty as to which of these orderings will actually take place.

The physical reason for this result can be understood as follows. Let us consider a given spin I_i belonging to a plane where all spins are, say, pointing up. The dipolar frequency ω_i at the site of the spin I_i is the sum of that produced by the spins belonging to the same plane and that produced by the spins in the other planes. Since the distance between planes (4·88 Å) is substantially larger than the distance between protons nearest neighbours within the plane (2·186 Å), the contribution of an external plane to the frequency ω_i is nearly the same as if the magnetization in this plane was continuous. By an extension of the model used in section $\mathbf{B}(c)$, the dipolar frequency produced by a succession of homogeneously magnetized plane is very close to that of a medium homogeneously magnetized in volume. In the present case, the magnetization of this hypothetical medium is zero, and its contribution to ω_i vanishes, whether the structure is antiferromagnetic or ferromagnetic with thin domains. The dipolar frequency ω_i, overwhelmingly due to the spins belonging to the same plane as the spin I_i, is therefore nearly the same for the two structures.

Structure II, $T>0$

Transverse ferromagnet with domains in the form of thin slices perpendicular to $0Z$, or transverse helical structure with $\mathbf{k}\|0Z$. These degenerate structures correspond to:

$$\lambda_{\text{II}} = -8{\cdot}045 n\gamma^2\hbar,$$

$$T_c = 0{\cdot}427\ \mu\text{K}.$$

By contrast with the case of CaF_2 with $T<0$ and $H_0\|[111]$, these structures are *not* degenerate with a longitudinal ferromagnetic structure with domains that are thin in a direction perpendicular to $0Z$. The reason is that in $Ca(OH)_2$, the dipolar sum (8.47) in a sphere surrounding a spin does not vanish as in CaF_2, so that the dipolar sum for small k is not given by (3.162). The longitudinal ferromagnet would correspond in the present case to:

$$\lambda = -3{\cdot}526 n\gamma^2\hbar.$$

2. *Field H_0 parallel to $0X$*

Structure III, $T<0$

Longitudinal antiferromagnet (Fig. 8.5a):

$$\lambda_{\text{III}} = 6{\cdot}758 n\gamma^2\hbar,$$

$$T_c = -0{\cdot}358\ \mu\text{K}.$$

Structure IV, $T>0$

Longitudinal antiferromagnet (Fig. 8.5b):

$$\lambda_{\text{IV}} = -9{\cdot}945 n\gamma^2\hbar,$$

$$T_c = 0{\cdot}528\ \mu\text{K}.$$

3. *Field H_0 parallel to $0Y$*

Structure V, $T<0$

Longitudinal antiferromagnet (Fig. 8.5c):

$$\lambda_{\text{V}} = 10{\cdot}625 n\gamma^2\hbar,$$

$$T_c = -0{\cdot}564\ \mu\text{K}.$$

For the structures III to V, as for structure I, what is well determined is the relative spin orientations within each plane, but not from one plane to the next, because of a quasi-degeneracy of the ordered state.

Structure VI, $T>0$

Longitudinal ferromagnet with domains where at least one domain dimension in the XY plane is much smaller than along $0Z$. Figure 8.5(d) corresponds to the case when the small domain dimension is parallel to the c-axis, that is when all the spins in a given plane are parallel.

$$\lambda_{\mathrm{VI}} = -8{\cdot}045 n\gamma^2\hbar,$$

$$T_c = 0{\cdot}427\ \mu\mathrm{K}.$$

These structures are nearly degenerate with the antiferromagnetic structures where successive planes carry magnetizations alternatively parallel and antiparallel to the field.

(f) General character of the nuclear ordered structures

The methods of prediction of the ordered structures described above, which are based on the comparison of the various Fourier transforms of the dipolar interaction coefficients, do not provide a simple insight into the reason why the ordered structures are as predicted.

It is possible to obtain a more direct physical feeling for what determines the nature of the ordered structures through a very simple-minded approach, by considering only the coupling of a given spin with its nearest neighbours (Chapellier, private communication).

We take as an example the case of CaF_2, i.e. a simple cubic lattice of spins $\frac{1}{2}$. We suppose that the temperature is zero, and we consider first the case when the external field is along a fourfold axis.

The cluster of 7 spins that we consider is shown in Fig. 8.6. The spins I_0, I_1, I_2 are on a line parallel to $\mathbf{H}_0$, whereas I_0, I_3, I_4, I_5, and I_6 are in a plane normal to H_0.

According to (5.44) the coupling coefficients of the central spin with its neighbours are, in units of $\gamma^2\hbar(2a^3)$:

$$\begin{aligned} A_{01} &= A_{02} = -2 \\ A_{03} &= A_{04} = A_{05} = A_{06} = 1. \end{aligned} \tag{8.53}$$

Suppose for the sake of definiteness that the central spin is pointing up. At positive temperature the energy must be minimum. According to (8.53) this is achieved when the products $I_z^0 I_z^1$ and $I_z^0 I_z^2$ are positive, and the products $I_z^0 I_z^3$ to $I_z^0 I_z^6$ are negative. That is, the spins I_1 and I_2 are pointing up, and the spins I_3 to I_6 are pointing down (Fig. 8.6(a)). By the translational repetition of this pattern, we recover the structure I of Fig. 8.1, theoretically predicted in section $\mathbf{B}(c)$.

At negative temperature, the energy is maximum. The conditions for achieving this are opposite to those at $T>0$, and the orientations of the 6

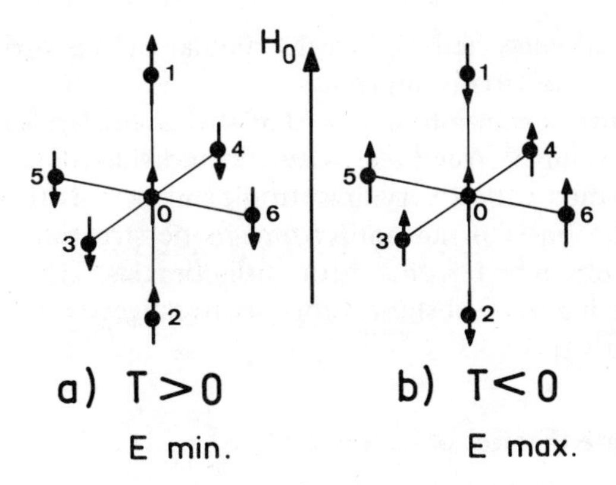

FIG. 8.6. Spin orientations which (a) minimize and (b) maximize the truncated dipolar energy in a cluster of 7 spins.

neighbours of the spin I_0 must be reversed (Fig. 8.6(b)). By translational repetition of this pattern we recover the structure III of Fig. 8.1.

The present crude approach yields the correct antiferromagnetic structures for both signs of the temperature. However, the energies it predicts at $T = \pm 0$ are opposite, whereas when all dipolar couplings are taken into account, the ratio of these energies is, according to (8.44) and (8.46):

$$\frac{E(T=+0)}{E(T=-0)} = \frac{-5{\cdot}352}{9{\cdot}687} \simeq -0{\cdot}55,$$

which shows that the present model is very crude indeed and is at best qualitative.

Let us now consider the case when the field is along an axis [111]. For this orientation, the dipolar couplings of a given spin with its nearest neighbours vanish, and one must consider its next-nearest neighbours. There are 12 of them, whose positions relative to the central spin are the same as in a f.c.c. lattice. It is possible, as before, to orient these twelve spins so as to minimize or maximize their truncated dipolar couplings with the central spin. However, when all dipolar interactions are taken into account, as in section **B**(c), it turns out that the actual stable structures are ferromagnetic with domains (or possibly helical at $T > 0$), i.e. structures in which the dipolar field experienced by a given spin originates not from the spins in its immediate neighbourhood, but from more distant spins.

When the calculation of section **B**(a) is applied to b.c.c. or f.c.c. lattices of identical spins, where each spin is surrounded by 8 and 12 nearest neighbours, respectively, it is found that the maximum and minimum values of $A(\mathbf{k})$ correspond to small $\mathbf{k}$ vectors, whatever the orientation of the external field. The stable structures are therefore always ferromagnetic with domains, or transverse helical, and the dipolar field experienced by each individual spin is due to distant spins.

The qualitative conclusion suggested by this discussion is that it is only in very open crystalline structures, when an individual spin has very few nearest neighbours with a very anisotropic angular distribution, that one may expect the onset of an antiferromagnetic structure. Otherwise the structure is likely to be ferromagnetic with domains. This may be a useful guide for selecting new substances for the investigation of nuclear dipolar magnetic ordering.

C. Approximate theories of ordering

Once the nature of the ordered structures has been determined by the methods of sections **B**(*a*) and **B**(*b*), it is necessary to investigate their properties in their range of existence, in zero as well as in a finite effective field. We sketch below the principle of two standard approximations. Their predictions for specific properties will not be given in this section, but in the sections describing the experimental study of these properties.

These methods are: the Weiss-field approximation occasionally supplemented with the high temperature expansion described in section **D** of chapter 5, and the restricted trace approximation. The case of CaF_2 will be treated as an example.

(*a*) *The Weiss-field approximation*

1. *The Weiss-field equations*

Let us consider for instance an antiferromagnetic structure with two sublattices $\mathscr{A}$ and $\mathscr{B}$. This structure is defined by a vector $\mathbf{k}_0$: when the origin is taken at a lattice site of sublattice $\mathscr{A}$, $\exp(i\mathbf{k}_0 \,.\, \mathbf{r}_i) = +1$ or -1, depending on whether the spin i belongs to the sublattice $\mathscr{A}$ or $\mathscr{B}$, respectively. All spins in $\mathscr{A}$ have the same vector polarization $\mathbf{p}_A$, and all spins in $\mathscr{B}$ the same polarization $\mathbf{p}_B$.

Using (8.4) and (8.14), the Weiss frequencies $\boldsymbol{\omega}_A$ and $\boldsymbol{\omega}_B$ are easily shown to be:

$$\begin{aligned} \omega_A^x &= -\tfrac{1}{4}A(\mathbf{k}_0)(p_x^A - p_x^B) - \tfrac{1}{4}A(0)(p_x^A + p_x^B); \\ \omega_A^y &= -\tfrac{1}{4}A(\mathbf{k}_0)(p_y^A - p_y^B) - \tfrac{1}{4}A(0)(p_y^A + p_y^B); \\ \omega_A^z &= \tfrac{1}{2}A(\mathbf{k}_0)(p_z^A - p_z^B) + \tfrac{1}{2}A(0)(p_z^A + p_z^B); \end{aligned} \tag{8.54}$$

and:

$$\boldsymbol{\omega}_B(\mathbf{p}_A, \mathbf{p}_B) = \boldsymbol{\omega}_A(\mathbf{p}_B, \mathbf{p}_A). \tag{8.54'}$$

Consider for instance the last eqn (8.4):

$$\omega_z^i = \sum_j A_{ij} p_z^j. \tag{8.4'}$$

If $\mathbf{I}_i$, chosen as origin, belongs to the sublattice $\mathscr{A}$, we have:

$$p_z^j = \tfrac{1}{2}[1+\exp(\mathrm{i}\mathbf{k}_0 \cdot \mathbf{r}_j)]p_z^{\mathrm{A}} + \tfrac{1}{2}[1-\exp(\mathrm{i}\mathbf{k}_0 \cdot \mathbf{r}_j)]p_z^{\mathrm{B}}.$$

Carrying this into (8.4′) and using the definition (8.14) of $A(\mathbf{k})$, the last eqn (8.54) results immediately.

If the precession frequency corresponding to the external effective field is $\boldsymbol{\omega}_{\mathrm{e}}$, the total spin frequencies are:

$$\begin{aligned} \boldsymbol{\omega}_{\mathrm{A}}^{\mathrm{T}} &= \boldsymbol{\omega}_{\mathrm{e}} + \boldsymbol{\omega}_{\mathrm{A}} \\ \boldsymbol{\omega}_{\mathrm{B}}^{\mathrm{T}} &= \boldsymbol{\omega}_{\mathrm{e}} + \boldsymbol{\omega}_{\mathrm{B}}. \end{aligned} \tag{8.55}$$

In the Weiss-field approximation, the polarizations are, according to (8.5), equal to:

$$\begin{aligned} \mathbf{p}_{\mathrm{A}} &= -\frac{\boldsymbol{\omega}_{\mathrm{A}}^{\mathrm{T}}}{|\boldsymbol{\omega}_{\mathrm{A}}^{\mathrm{T}}|} \tanh(\tfrac{1}{2}\beta|\boldsymbol{\omega}_{\mathrm{A}}^{\mathrm{T}}|); \\ \mathbf{p}_{\mathrm{B}} &= -\frac{\boldsymbol{\omega}_{\mathrm{B}}^{\mathrm{T}}}{|\boldsymbol{\omega}_{\mathrm{B}}^{\mathrm{T}}|} \tanh(\tfrac{1}{2}\beta|\boldsymbol{\omega}_{\mathrm{B}}^{\mathrm{T}}|). \end{aligned} \tag{8.56}$$

The energy is equal to:

$$E = \tfrac{1}{2}N\{\tfrac{1}{2}\boldsymbol{\omega}_{\mathrm{e}} \cdot (\mathbf{p}_{\mathrm{A}} + \mathbf{p}_{\mathrm{B}}) + \tfrac{1}{4}(\boldsymbol{\omega}_{\mathrm{A}} \cdot \mathbf{p}_{\mathrm{A}} + \boldsymbol{\omega}_{\mathrm{B}} \cdot \mathbf{p}_{\mathrm{B}})\}. \tag{8.57}$$

As for the entropy, it is, according to (8.7):

$$S = \tfrac{1}{2}N\{s(p_{\mathrm{A}}) + s(p_{\mathrm{B}})\}, \tag{8.58}$$

with:

$$s(p) = \ln 2 - \tfrac{1}{2}[(1+p)\ln(1+p) + (1-p)\ln(1-p)]. \tag{8.59}$$

It is easily shown that the eqns (8.56) to (8.59) satisfy the thermodynamic relation:

$$\frac{\mathrm{d}S}{\mathrm{d}\beta} = \beta\frac{\mathrm{d}E}{\mathrm{d}\beta}. \tag{8.60}$$

It will prove useful in the following to use a slightly different approach to the Weiss-field equations, due to Bragg and Williams (1934). In this approach, the form (8.56) for the Weiss-field equations, instead of being *assumed* as before, is shown to result from the condition that the free energy be minimum (or maximum). The starting point is eqn (8.57) for the energy and eqns (8.58) and (8.59) for the entropy, obtained by neglecting short-range correlations between the spins. In the equilibrium state,

$$-\beta F = S - \beta E$$

must be maximum. The polarizations $\mathbf{p}_A$ and $\mathbf{p}_B$ are obtained from the conditions:

$$\frac{\partial}{\partial \mathbf{p}_A}(-\beta F) = \frac{\partial}{\partial \mathbf{p}_B}(-\beta F) = 0. \tag{8.61}$$

It is a matter of straight algebra to show that these conditions yield the Weiss-field equations (8.56).

Equations (8.56) to (8.59) describe all properties of the antiferromagnet, provided that all spins belong to either one of the sublattices $\mathscr{A}$ and $\mathscr{B}$. It will be shown in section **F** that this is true only when the effective frequency $\boldsymbol{\omega}_e$ is not too large. In spite of their usefulness the Weiss-field equations are a rather crude approximation to the problem of magnetic ordering. We list below some of their flaws.

(i) In zero effective field, the polarizations are longitudinal, with:

$$\mathbf{p}_A = p_{Az} = -\mathbf{p}_B = p,$$

and eqns (8.56) reduce to:

$$p = -\tanh[\tfrac{1}{2}\beta A(\mathbf{k}_0)p], \tag{8.62}$$

or else:

$$\tanh^{-1}(p) = -\tfrac{1}{2}\beta A(\mathbf{k}_0)p. \tag{8.62$'$}$$

The transition temperature corresponds to the limit $p \to 0$, that is:

$$p = -\tfrac{1}{2}\beta_c A(\mathbf{k}_0)p,$$

and the inverse temperature is,

$$\beta_c = -2/A(\mathbf{k}_0), \tag{8.63}$$

in accordance with (8.10) and (8.16).

Slightly below the transition, we use a power expansion of $\tanh^{-1}(p)$:

$$\tanh^{-1}(p) \simeq p + \frac{p^3}{3} = -\tfrac{1}{2}\beta A(\mathbf{k}_0)p,$$

whence:

$$\begin{aligned} p^2 &= 3[-\tfrac{1}{2}\beta A(\mathbf{k}_0) - 1] \\ &= \frac{3}{\beta_c}(\beta - \beta_c). \end{aligned} \tag{8.64}$$

The proportionality of p to $(|\beta - \beta_c|)^{1/2} \propto (|T - T_c|)^{1/2}$ slightly below the transition is characteristic of the Weiss-field approximation. It is known that, at least for short-range interactions, the critical exponent $\frac{1}{2}$ is not correct.

(ii) A far more glaring inconsistency resides in its completely false prediction for the transition entropy. To see this let us recall the method of production of the ordering. We start in high field, where all spins have the same polarization p_{in} along the field. The entropy is then:

$$S_{in} = Ns(p_{in}).$$

After adiabatic demagnetization to zero effective field, we have $\mathbf{p}_A = -\mathbf{p}_B$, and since the entropy depends only on the modulus of the polarizations, the final entropy is:

$$S_f = Ns(p_A).$$

Since the demagnetization takes place at constant entropy, we must have $p_A = p_{in}$. The theory predicts incorrectly that, however small the initial polarization, adiabatic demagnetization will produce antiferromagnetism. The flaw that is the origin of this faulty prediction is the neglect of the entropy associated with short range order.

2. *Weiss-field and spin temperature*

In order to have an estimate of the transition entropy, that the Weiss theory is unable to provide, the simplest but rather crude makeshift procedure is the following.

(i) We take for granted the critical inverse temperature predicted by the Weiss-field theory:

$$\beta_c = -2/A(\mathbf{k}_0). \tag{8.63}$$

(ii) We use the high temperature approximation for spin temperature to calculate the initial polarization $p_0 = -\frac{1}{2}\beta_0\Delta_0$ which yields the inverse temperature β_c after ADRF.

According to (1.37):

$$p_0 = -\tfrac{1}{2}\beta_c D. \tag{8.65}$$

By carrying (8.63) into (8.65) we obtain:

$$|p_0| = D/|A(\mathbf{k}_0)|. \tag{8.66}$$

This crude approximation tells us how much initial polarization is required to produce antiferromagnetism by ADRF. The same procedure can be used for the ferromagnet with domains: the only change is that spins with opposite polarizations in zero effective field are located in different domains, rather than in different sublattices.

The predicted values of p_0 for the structures I to IV of section **B**(c) are listed below.

Structure I. $T>0, H_0\|[\bar{0}01]$:

$$A(\mathbf{k}_0)=-5{\cdot}352\gamma^2\hbar/(2a^3);$$

$$D=3{\cdot}16\gamma^2\hbar/(2a^3);$$

$$p_0=0{\cdot}59.$$

Structure II. $T>0, H_0\|[110]$:

$$A(\mathbf{k}_0)=-4{\cdot}843\gamma^2\hbar/(2a^3);$$

$$D=1{\cdot}96\gamma^2\hbar/(2a^3);$$

$$p_0=0{\cdot}405.$$

Structure III. $T<0, H_0\|[001]$:

$$A(\mathbf{k}_0)=9{\cdot}687\gamma^2\hbar/(2a^3);$$

$$D=3{\cdot}16\gamma^2\hbar/(2a^3);$$

$$p_0=0{\cdot}326.$$

Structure IV. $T<0$:

$$A(\mathbf{k}_0)=8{\cdot}378\gamma^2\hbar/(2a^3);$$

$$H_0\|[111]\colon D=1{\cdot}31\gamma^2\hbar/(2a^3);$$

$$p_0=0{\cdot}156;$$

$$H_0\|[110]\colon D=1{\cdot}96\gamma^2\hbar/(2a^3);$$

$$p_0=0{\cdot}234.$$

According to these figures, structures III and IV are the easiest to produce, since they require the lowest initial polarizations.

Remarks The preceding results correspond to ordered structures produced by adiabatic demagnetization in the rotating frame. When the adiabatic demagnetization is produced in the laboratory frame, the theory of sections **B**(a) and **B**(b) predicts that the ordered structures at positive and negative temperatures are identical with structures I and III, respectively, and correspond to the same values of $A(\mathbf{k}_0)$. The local frequency D_L is however larger than in the rotating frame, because the full dipolar Hamiltonian $\mathcal{H}_D$ contains extra terms besides the truncated $\mathcal{H}'_D$.
The calculation yields:

$$D_L=5{\cdot}02\gamma^2\hbar/(2a^3),$$

whence, by analogy with (8.66):

$$p_0 = D_L/|A(\mathbf{k}_0)|$$
$$= 0{\cdot}518 \quad \text{for } T < 0,$$
$$= 0{\cdot}938 \quad \text{for } T > 0.$$

To produce the *same* ordered structure, in zero field as in high field it is necessary to start from a much higher polarization.

At positive temperature, the local frequency D_L is so close to the Weiss-field factor $A(\mathbf{k}_0)$ that it is even possible that short-range correlations would hinder the onset of antiferromagnetism, i.e. that no long-range ordering will ever take place.

(*b*) *The restricted-trace approximation*

The restricted-trace approximation represents an attempt to go beyond the Weiss-field approximation by taking at least partially into account the short-range correlation between the spins (Kirkwood, 1938; Brout, 1965*a*, p. 12). Its principle can be illustrated as follows.

Let us consider a transition to an antiferromagnetic state of known structure, with sublattices $\mathscr{A}$ and $\mathscr{B}$. We call I_i and I_μ the spins belonging to $\mathscr{A}$ and $\mathscr{B}$, respectively. There are $N/2$ spins of each species.

We need to calculate the partition function:

$$\mathscr{Z} = \mathrm{Tr}\{\exp(-\beta\mathscr{H})\}. \tag{8.67}$$

We choose as a basis the eigenstates of both:

$$I_z^A = \sum_i I_j^i \quad \text{and} \quad I_z^B = \sum_\mu I_z^\mu,$$

and we split the Hilbert space of the system into subspaces corresponding to well-defined eigenvalues of I_z^A and I_z^B:

$$\langle I_z^A \rangle = \tfrac{1}{4}Np_A \quad \text{and} \quad \langle I_z^B \rangle = \tfrac{1}{4}Np_B.$$

The trace (8.67) can be written as a sum of partial traces restricted to the various subspaces:

$$\mathscr{Z} = \sum_{p_A, p_B} \mathrm{Tr}'\{\exp(-\beta\mathscr{H})\}$$
$$= \sum_{p_A, p_B} \mathscr{Z}'(\beta, p_A, p_B). \tag{8.68}$$

At this point, the assumption is made that the restricted traces $\mathscr{Z}'(\beta, p_A, p_B)$ exhibit a sharp maximum for those values of p_A and p_B which correspond to the sublattice polarizations in the ordered phase. In the following it is this maximum restricted trace that is called $\mathscr{Z}'(\beta, p_A, p_B)$. The sum (8.67) is made of $2N+2$ restricted traces, and we have:

$$\mathscr{Z}'(\beta, p_A, p_B) < \mathscr{Z} < (2N+2)\mathscr{Z}'(\beta, p_A, p_B),$$

whence:

$$\ln \mathscr{Z} = \ln \mathscr{Z}'(\beta, p_A, p_B) + 0(\ln N). \tag{8.69}$$

Since $\ln \mathscr{Z}$ is proportional to the number N of spins, we have in the thermodynamic limit ($N \to \infty$):

$$\ln \mathscr{Z} \simeq \ln \mathscr{Z}'(\beta, p_A, p_B). \tag{8.70}$$

In fact there will be in general two couples (p_A, p_B) with equal sharp values of the restricted trace. One corresponds to $p_A = p'$, $p_B = p''$, and the other to $p_A = p''$, $p_B = p'$. We choose arbitrarily one of them, thus imposing the symmetry breaking of the ordered phase.

The procedure is then to calculate $\mathscr{Z}'(\beta, p_A, p_B)$ as a function of p_A and p_B, and to determine the latter by the condition that they maximize the restricted trace or, more conveniently its logarithm:

$$\frac{\partial}{\partial p_A} \ln \mathscr{Z}' = \frac{\partial}{\partial p_B} \ln \mathscr{Z}' = 0. \tag{8.71}$$

In practice, we write:

$$\mathscr{Z} = \mathrm{Tr}'(1) \times [\mathrm{Tr}'\{\exp(-\beta(\mathscr{H})\}/\mathrm{Tr}'(1)],$$

where 1 is the unit matrix, or else:

$$\ln \mathscr{Z} = \ln \mathrm{Tr}'(1) + \ln[\mathrm{Tr}'\{\exp(-\beta\mathscr{H})\}/\mathrm{Tr}'(1)]. \tag{8.72}$$

$\mathrm{Tr}'(1)$ is the number of states with the prescribed values of p_A and p_B. Its logarithm is, within the factor k_B, equal to the entropy $S_W(p_A, p_B)$ in the Weiss-field approximation (eqn (8.58)).

When the Hamiltonian is of the form:

$$\mathscr{H} = \Delta(I_z^A + I_z^B) + \mathscr{H}'_D, \tag{8.73}$$

where $\mathscr{H}'_D$ commutes with the effective Zeeman term we have:

$$\begin{aligned}\mathrm{Tr}'\{\exp(-\beta\mathscr{H})\} &= \mathrm{Tr}'\{\exp[-\beta\Delta(I_z^A + I_z^B)] \times \exp(-\beta\mathscr{H}'_D)\} \\ &= \exp[-\tfrac{1}{4}N\beta\Delta(p_A + p_B)] \times \mathrm{Tr}'\{\exp(-\beta\mathscr{H}'_D\},\end{aligned} \tag{8.74}$$

and (8.72) becomes:

$$\ln \mathscr{Z} = S_W(p_A, p_B) - \tfrac{1}{4}N\beta\Delta(p_A + p_B) + \ln[\mathrm{Tr}'\{\exp(-\beta\mathscr{H}'_D)\}/\mathrm{Tr}'(1)]. \tag{8.75}$$

The energy and the entropy are obtained from the equations:

$$E = -\frac{\mathrm{d}}{\mathrm{d}\beta} \ln \mathscr{Z} \tag{8.76}$$

$$\frac{\mathrm{d}S}{\mathrm{d}\beta} = \beta \frac{\mathrm{d}E}{\mathrm{d}\beta}. \tag{8.77}$$

Through p_A and p_B, $\ln \mathscr{Z}$ depends on β both explicitly and implicitly. We can write

$$E = -\frac{\partial \ln \mathscr{Z}}{\partial p_\mathrm{A}} \frac{\mathrm{d}p_\mathrm{A}}{\mathrm{d}\beta} - \frac{\partial \ln \mathscr{Z}}{\partial p_\mathrm{B}} \frac{\mathrm{d}p_\mathrm{B}}{\mathrm{d}\beta} - \frac{\partial \ln \mathscr{Z}}{\partial \beta},$$

whence, since $\mathscr{Z} \simeq \mathscr{Z}'$, and according to (8.71):

$$\begin{aligned} E &= -\frac{\partial \ln \mathscr{Z}}{\partial \beta} \\ &= \tfrac{1}{4} N \Delta (p_\mathrm{A} + p_\mathrm{B}) + \frac{\mathrm{Tr}'\{\mathscr{H}'_\mathrm{D} \exp(-\beta \mathscr{H}'_\mathrm{D})\}}{\mathrm{Tr}'\{\exp(-\beta \mathscr{H}'_\mathrm{D})\}} \\ &= E_\mathrm{Z} + E_\mathrm{D}. \end{aligned} \tag{8.78}$$

The entropy is written in the form:

$$S = S_\mathrm{W}(p_\mathrm{A}, p_\mathrm{B}) + s(\beta, p_\mathrm{A}, p_\mathrm{B}), \tag{8.79}$$

where the first term does not depend explicitly on β. Equations (8.77) and (8.78) then yield:

$$\frac{\partial s}{\partial \beta} = \beta \frac{\partial E_\mathrm{D}}{\partial \beta}, \tag{8.80}$$

with:

$$s(0, p_\mathrm{A}, p_\mathrm{B}) = 0. \tag{8.81}$$

All the difficulty of the theory resides in the calculation of the last term on the right-hand side of (8.75). The approximation made by Kirkwood consists in the use of a limited expansion in powers of β for this term.

When the expansion is limited to the first order in β, one recovers exactly the Weiss-field approximation. The whole procedure is then very close to that of Bragg and Williams. Terms of higher order in β yield corrections to the Weiss-field result, arising from short-range correlations between the spins. Inclusion of terms in β^2 and β^3 are referred to as 1st and 2nd order restricted-trace approximations.

As an example, we give the results of the 1st order restricted-trace approximation for the structure III of $\mathrm{CaF_2}$, at $T < 0$ and $H_0 \| [001]$. The dipolar energy is written in the form:

$$\mathcal{H}'_{\mathrm{D}} = \tfrac{1}{2}\sum_{i,j} A_{ij}[2I_z^i I_z^j - \tfrac{1}{2}(I_+^i I_-^j + I_-^i I_+^j)]$$
$$+\tfrac{1}{2}\sum_{\mu,\nu} A_{\mu\nu}[2I_z^\mu I_z^\nu - \tfrac{1}{2}(I_+^\mu I_-^\nu + I_-^\mu I_+^\nu)]$$
$$+\sum_{i,\mu} B_{i\mu}[2I_z^i I_z^\mu - \tfrac{1}{2}(I_+^i I_-^\mu + I_-^i I_+^\mu)],$$

where the spins I_i and I_μ belong to the sublattices $\mathcal{A}$ and $\mathcal{B}$, respectively. The dipolar energy per spin is:

$$E_{\mathrm{D}}/N = K_0 + \beta K_1 + \cdots \tag{8.82}$$

with:

$$K_0 = A(\mathbf{k}_0)(p_{\mathrm{A}} - p_{\mathrm{B}})^2/16; \tag{8.83}$$

$$K_1 = -\tfrac{1}{32}\{\mathscr{D}[6 + (p_{\mathrm{A}}^2 + p_{\mathrm{B}}^2)^2 - 4(p_{\mathrm{A}}^2 + p_{\mathrm{B}}^2) - \tfrac{1}{2}(p_{\mathrm{A}} + p_{\mathrm{B}})^2]$$
$$+ \mathscr{E}[(p_{\mathrm{A}}^2 - p_{\mathrm{B}}^2)^2 - \tfrac{1}{2}(p_{\mathrm{A}} - p_{\mathrm{B}})^2]\}, \tag{8.84}$$

where:

$$\mathscr{D} = \sum_i (A_{ij}^2 + B_{i\mu}^2)$$
$$= 13{\cdot}357[\gamma^2\hbar/(2a^3)]^2,$$

and:

$$\mathscr{E} = \sum_i (A_{ij}^2 - B_{i\mu}^2)$$
$$= -3{\cdot}396[\gamma^2\hbar/(2a^3)]^2.$$

The entropy per spin is:

$$S/N = S_{\mathrm{W}}/N + \frac{\beta^2}{2}K_1 + \cdots. \tag{8.85}$$

The conditions (8.71) (which are the same as (8.61): maximum of $-\beta F = (S - \beta E)$ yield:

$$\tanh^{-1}(p_{\mathrm{A}}) = -\tfrac{1}{2}\beta\Delta - \tfrac{1}{4}\beta A(\mathbf{k}_0)(p_{\mathrm{A}} - p_{\mathrm{B}}) - \beta^2\frac{\partial K_1}{\partial p_{\mathrm{A}}}; \tag{8.86}$$

$$\tanh^{-1}(p_{\mathrm{B}}) = -\tfrac{1}{2}\beta\Delta + \tfrac{1}{4}\beta A(\mathbf{k}_0)(p_{\mathrm{A}} - p_{\mathrm{B}}) - \beta^2\frac{\partial K_1}{\partial p_{\mathrm{B}}}, \tag{8.86'}$$

when $\Delta = 0$, we have $p_{\mathrm{A}} = -p_{\mathrm{B}} = p$ and eqns (8.82) to (8.86′) become:

$$E/N = \tfrac{1}{4}A(\mathbf{k}_0)p^2 - \tfrac{1}{16}\beta\{\mathscr{D}[3 - 4p^2 + 2p^4] - \mathscr{E}p^2\}; \tag{8.87}$$

$$S/N = S_{\mathrm{W}}/N - \tfrac{1}{32}\beta^2\{\mathscr{D}[3 - 4p^2 + 2p^4] - \mathscr{E}p^2\}; \tag{8.88}$$

$$\tanh^{-1}(p) = -\tfrac{1}{2}\beta A(\mathbf{k}_0)p - \tfrac{1}{16}\beta^2 p[4\mathscr{D}(1 - p^2) + \mathscr{E}]. \tag{8.89}$$

The critical inverse temperature β_c corresponds to:

$$-\tfrac{1}{2}\beta_c A(\mathbf{k}_0) - \tfrac{1}{16}\beta_c^2[4\mathscr{D} + \mathscr{E}] = 1. \quad (8.90)$$

Slightly below the transition, it can be verified that according to (8.89) and (8.90), one has:

$$p \propto |\beta - \beta_c|^{1/2}.$$

The restricted-trace approximation becomes meaningless when the temperature is too low: when $|\beta|$ is large and p tends to unity, according to (8.87) and (8.88), $|E|$ increases beyond limits whereas S becomes negative. This approximation must therefore be limited to values of $|\beta|$ well below that which causes the expression (8.88) for the entropy to vanish.

As will be seen by comparison with experiment, the predictions of the restricted-trace approximation, although qualitatively correct, are still far from satisfactory. Despite its weaknesses, the advantage of this approximation is that it is the simplest one beyond the Weiss-field approximation which yields correct qualitative predictions for the properties of the system in the whole range from paramagnetism to ordering.

As an example, Fig. 8.7 reproduces the variation of the sublattice polarizations in zero effective field as a function of the initial polarization,

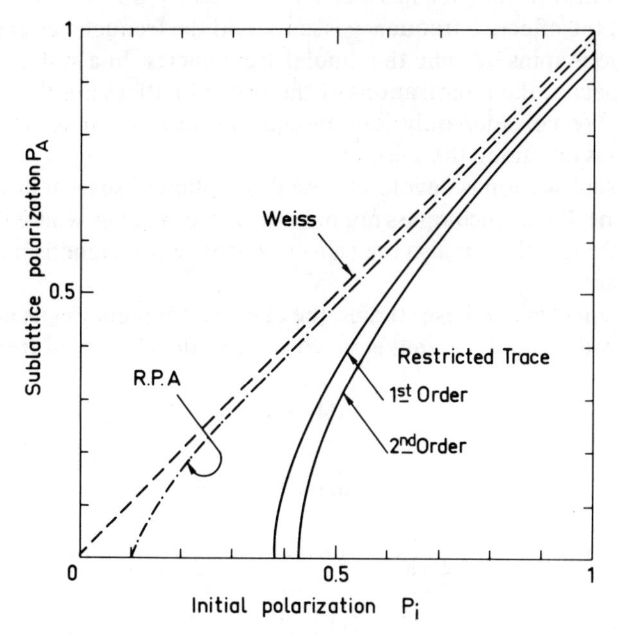

FIG. 8.7. Sublattice polarizations in zero field against initial polarization for the antiferromagnetic structure III of CaF_2, according to Weiss field, RPA and 1st and 2nd order restricted-trace approximations. (After Goldman and Sarma, 1975.)

for the structure III of CaF_2. The predictions of the 1st and 2nd order approximations are compared with those of the Weiss-field approximation, and of the random-phase approximation (described in section **H**(b)). They are much more reasonable than the latter. The predicted transition initial polarizations: 38 per cent and 43 per cent for 1st and 2nd order approximations, respectively, are in reasonable agreement with the value of 32·6 per cent obtained in section **C**(a).

Table 8.1 lists the values of critical temperature and energy in these various approximations for the structure III.

Many more approximation methods could be developed. Most of them are much more complex than the Weiss-field and the restricted-trace methods, and would yield no evident decisive improvement over the latter. Their description is therefore not warranted in this book.

An exception will be made for three methods: spin wave, random phase approximation (RPA) and Monte Carlo simulation, which provide a physical insight into the problem of dipolar ordering and will be described in section **H**.

D. Antiferromagnetism in zero field

When an antiferromagnet has been produced by an ADRF stopped at resonance, the effective frequency is zero, and the frequencies experienced by the various spins become the dipolar frequencies. In a system with only one spin species, the polarizations of the two sublattices are then opposite: $\mathbf{p}_B = -\mathbf{p}_A$. We consider only longitudinal antiferromagnets, where these polarizations are along the z-axis.

The present section is devoted to the description of several properties of such systems. It will encompass not only the entropy range where the system is antiferromagnetic, but also the transition from paramagnetism to antiferromagnetism.

Actually, in some of these studies, the external frequency $\boldsymbol{\omega}_e$ is not exactly zero but always small, i.e. $|\boldsymbol{\omega}_e|$ is much smaller than the local frequency D

Table 8.1

	T_c in units of:		E_c/N in units of:	
Approx.	$\frac{\gamma^2\hbar}{2k_B a^3}$	μK (CaF_2)	$\frac{\gamma^2\hbar}{2k_B a^3}$	kHz (CaF_2)
Weiss	−4·844	−0·614	0	0
RPA	−4·654	−0·590	0·18	0·475
RTA 1st order	−4·078	−0·517	0·615	1·62
RTA 2nd order	−4.034	−0.511	0·725	1·91

and, in the antiferromagnetic state, also much smaller than the Weiss frequency. The bulk polarization $\mathbf{p}$ of the sample is then a linear function of $\boldsymbol{\omega}_e$. Their relationship is of the form

$$\mathbf{p} = \boldsymbol{\chi} \cdot \boldsymbol{\omega}_e \tag{8.91}$$

where $\boldsymbol{\chi}$ is the susceptibility in zero field.

(*a*) *Transverse susceptibility*

When a small r.f. field is applied at exact resonance, the effective frequency $\boldsymbol{\omega}_e$ is purely transverse, with $\omega_{ex} = \omega_1$, and the average polarization per spin, aligned with ω_1, is

$$p_x = \chi_\perp \omega_1. \tag{8.92}$$

As described in section **C**(*f*) of chapter 5, two experimental methods can be used to measure $\chi_\perp$: either observing on the run the dispersion signal at the Larmor frequency during a fast passage, or computing the Kramers–Kronig transform of the absorption signal recorded with a non-saturating r.f. level.

We will consider in turn the cases when the system contains only one spin species (CaF_2), and two spin species (LiF, LiH).

1. *Calcium fluoride*

One of the most characteristic properties of an antiferromagnet with one spin species is that below the transition its transverse susceptibility is, within the Weiss-field approximation, independent of temperature. This prediction is approximately verified in all antiferromagnets, and only a small variation of $\chi_\perp$ is observed in the range of existence of the ordering.

The proof goes as follows. In the Weiss field approximation, the spins are aligned along the total frequency they experience. When the effective frequency is perpendicular to the normal orientation of the sublattice polarizations, one has by symmetry:

$$\begin{aligned} p_z^{\mathrm{A}} &= -p_z^{\mathrm{B}}; \\ p_x^{\mathrm{A}} &= p_x^{\mathrm{B}}; \\ p_y^{\mathrm{A}} &= p_y^{\mathrm{B}} = 0. \end{aligned} \tag{8.93}$$

Equations (8.54) and (8.54′) then yield:

$$\begin{aligned} \omega_{\mathrm{A}z}^{\mathrm{T}} &= A(\mathbf{k}_0) p_z^{\mathrm{A}}; \\ \omega_{\mathrm{A}x}^{\mathrm{T}} &= \omega_1 - \tfrac{1}{2} A(0) p_x^{\mathrm{A}}; \\ \omega_{\mathrm{A}y}^{\mathrm{T}} &= 0; \end{aligned} \tag{8.94}$$

and (8.56) yields:

$$p_x^{\mathrm{A}}/p_z^{\mathrm{A}} = \omega_{\mathrm{A}x}^{\mathrm{T}}/\omega_{\mathrm{A}z}^{\mathrm{T}}, \tag{8.95}$$

whence:

$$\chi_\perp = p_x^{\mathrm{A}}/\omega_1 = [A(\mathbf{k}_0) + \tfrac{1}{2}A(0)]^{-1}, \tag{8.96}$$

which is independent of sublattice polarization, that is of temperature. (We recall once more that $A(0) = \sum_j A_{ij}$ is uniquely defined only if the shape of the sample is an ellipsoid.)

This result can be understood intuitively as follows. We consider for simplicity, a spherical sample, where $A(0) = 0$. Let $p_{\mathrm{A}} = -p_{\mathrm{B}} = p$ be the (longitudinal) sublattice polarizations in zero effective field. When a small transverse field H_1 is applied, these polarizations tilt by a small angle θ so as to become aligned with the total frequency they experience, dipolar plus external (Fig. 8.8), and their transverse polarization is then $p_x = \theta p$. Let us now increase the temperature, so that in zero effective field $p_{\mathrm{A}} = -p_{\mathrm{B}} = p' < p$. The dipolar field experienced by the spins is now smaller than before, in the ratio p'/p, so that if the *same* transverse field H_1 is applied, the tilting angle θ' of the polarizations will be larger than θ:

$$\theta' = \theta p/p'.$$

The decrease in the magnitude of the polarization with temperature is compensated by the increase of tilting angle, so that the transverse polarization:

$$p_x = p'\theta' = p'\theta p/p' = \theta p,$$

is the same as before.

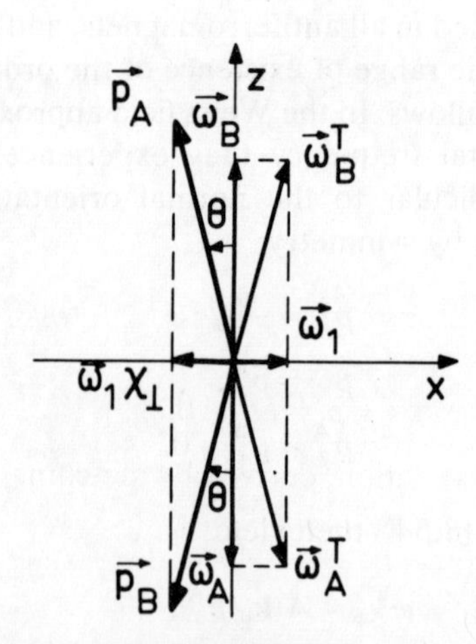

FIG. 8.8. Sublattice polarizations and frequencies for an antiferromagnet in the presence of a small transverse field. The figure corresponds to the case of a positive temperature. (After Goldman, Chapellier, Vu Hoang Chau, and Abragam, 1974.)

Equation (8.96) is valid only below the antiferromagnetic transition since its derivation is based explicitly on the assumption of the existence of two sublattices. The Weiss-field approximation which, as we saw in section **C**(*a*), predicts such a structure for *all* values of the entropy, and its direct consequence (8.96), is naturally invalid above the transition. Instead, in the paramagnetic state we expect for the susceptibility a value predicted by the high temperature approximation of spin temperature theory, corrected for the effect of demagnetizing dipolar fields.

We start from an inverse temperature β_0 in an external frequency Δ_0, with $|\Delta_0| \gg D$. We call p_i the modulus of the initial polarization. In the high temperature limit it is, for spins $\frac{1}{2}$:

$$p_i = \tfrac{1}{2}|\beta_0\Delta_0|. \tag{8.97}$$

We then apply an r.f. frequency ω_1 and perform an ADRF. According to (1.37), and since $\omega_1 \ll D$, the final inverse temperature β is:

$$\begin{aligned} \beta &= \beta_0|\Delta_0|/D \\ &= \pm 2p_i/D, \end{aligned} \tag{8.98}$$

depending on whether the temperature is positive or negative.

All the spins are equally polarized along ω_1, with a polarization:

$$p_x = -\tfrac{1}{2}\beta(\omega_1 + \omega_d), \tag{8.99}$$

where ω_d is the average dipolar frequency. It is obtained from the first eqn (8.54), with $p_x^A = p_x^B = p_x$:

$$\omega_d = \omega_x^A = -\tfrac{1}{2}A(0)p_x. \tag{8.100}$$

By carrying (8.98) and (8.100) into (8.99), we obtain finally:

$$\chi_{\perp\text{para}} = \frac{p_x}{\omega_1} = \frac{\mp p_i}{D \mp \frac{1}{2}A(0)p_i}, \tag{8.101}$$

where the sign $-$ is for positive temperature.

The values (8.96) and (8.101) become equal when the initial polarization is equal to:

$$p_i = D/|A(\mathbf{k}_0)|.$$

This critical initial polarization, derived by matching the transverse susceptibilities $\chi_{\perp AF}$ and $\chi_{\perp\text{para}}$, is identical with that calculated in section **C**(*a*)2, eqn (8.66). One can expect that $\chi_\perp$ will undergo a smooth transition from the behaviour predicted by (8.101) to that predicted by (8.96) when the initial polarization p_i is increased.

The variation of $\chi_\perp$ with p_i according to (8.96) and (8.101) is shown for various sample shapes in Fig. 8.9. The figures correspond to CaF_2, with

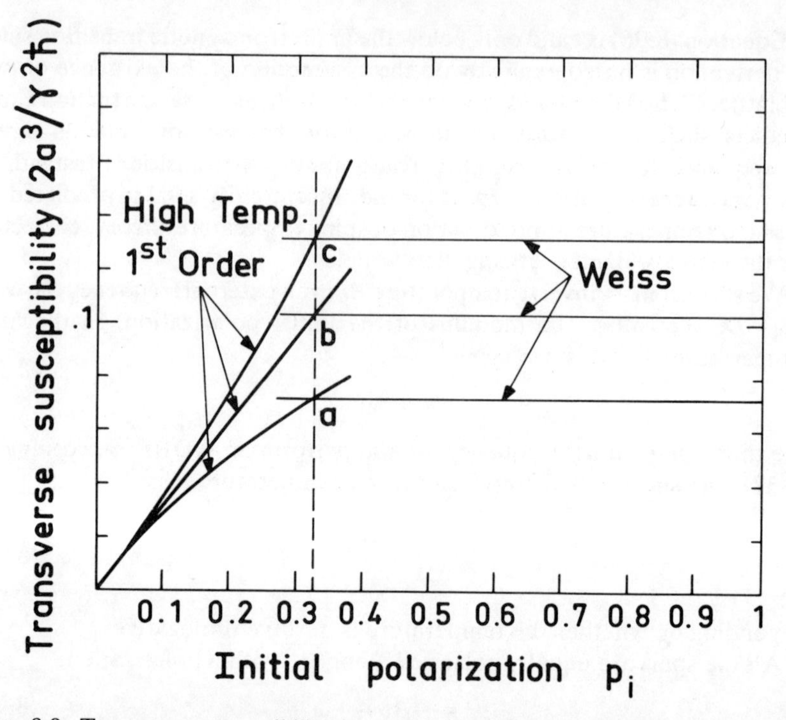

FIG. 8.9. Transverse susceptibility in zero field against initial polarization in the antiferromagnetic structure III of CaF_2, according to high-temperature and Weiss field approximations. (a) Infinitely flat disc perpendicular to H_0. (b) Spherical sample. (c) Infinitely long needle parallel to H_0. (After Goldman, Chapellier, Vu Hoang Chau, and Abragam, 1974.)

$H_0 \| [001]$ and $T<0$. The antiferromagnetic structure is structure III (eqn (8.46)) for which the critical initial polarization is $p_0 = 0{\cdot}326$.

More elaborate calculations have been made for spherical samples (for which $A(0)=0$): expansion up to β^3 in the paramagnetic phase, described in section **D**(b)5 of chapter 5, and 1st order restricted-trace approximation, whose predictions will be described together with the experimental results.

Three different measurements of $\chi_\perp$ have been performed on spherical samples of CaF_2, with $H_0 \| [001]$ and $T<0$.

The first measurement, described in section **D**(c)2 of chapter 5, consisted in observing the dispersion signal at the centre of a fast passage as a function of initial polarization (Chapellier *et al.* 1969). The results are shown in Fig. 8.10, together with the Weiss-field, high temperature and restricted-trace approximations. The occurrence of a plateau at high initial polarizations is clearly visible. When this plateau is adjusted to the Weiss-field value (8.96), the low initial polarization results are in fair agreement with the 3rd order

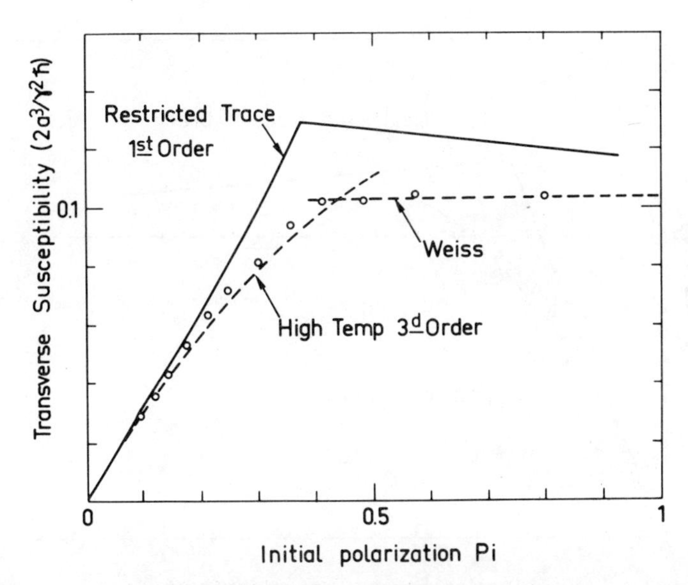

FIG. 8.10. Transverse susceptibility against initial polarization for CaF_2 in the antiferromagnetic structure III. The curves are the predictions of 3rd order high temperature, Weiss field and 1st order restricted-trace approximation. The value of the plateau is adjusted to the Weiss field prediction. (After Goldman and Sarma, 1975.)

high temperature expansion. The agreement is not so good with the restricted-trace prediction. This last approximation does not yield a correct prediction for the non-linearities observed in the paramagnetic region (as expected from a 1st order approximation). In the antiferromagnetic region, an almost constant value of $\chi_\perp$ is predicted, which is however approximately 20 per cent higher than the Weiss-field value.

In the second measurement, described in section **D**(c)2 of chapter 5, both the dispersion and the absorption signals were observed in the course of a square-wave modulation of the external field around the Larmor value in the presence of r.f. irradiation (Jacquinot, Chapellier, and Goldman, 1974). It yielded the variations of both the transverse and longitudinal susceptibilities as a function of the dipolar energy, but we are concerned only with the former in the present section. The results are shown in Fig. 8.11. The plateau is adjusted to the Weiss-field value, which again results in a very good agreement of the low initial polarization data with the high temperature expansion.

Finally, the experimental results shown in Fig. 8.12 were obtained from the absorption signal at low r.f. level: its first moment yields the dipolar energy according to the theory of chapter 5, section **C**(d), and its Kramers–Kronig transform yields the value of $\chi_\perp$ according to the theory of chapter 5,

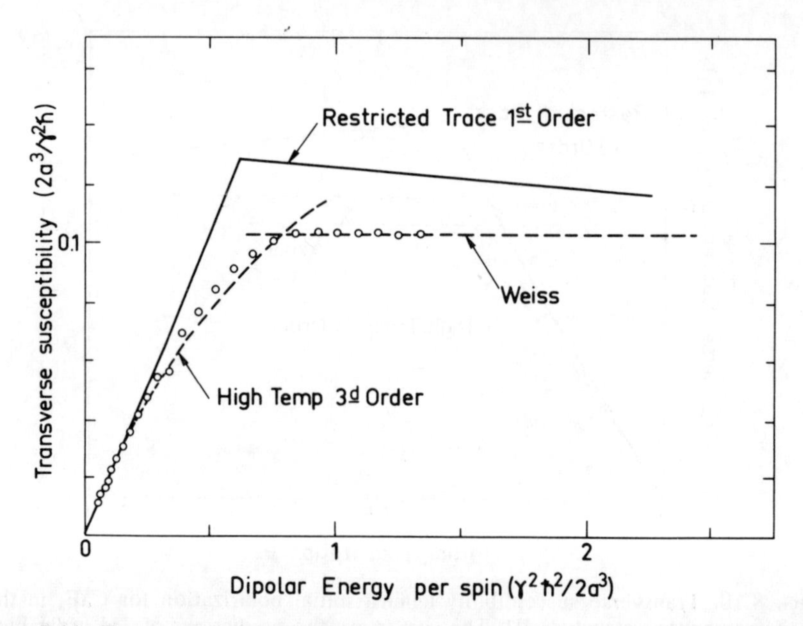

FIG. 8.11. Transverse susceptibility against dipolar energy for CaF_2 in the antiferromagnetic structure III. The curves are the predictions of 3rd order high temperature, Weiss field and 1st order restricted-trace approximations. The susceptibility is measured as a function of time and the variation of the energy with time is computed afterwards (see text). $\chi_\perp$ in the plateau is adjusted to the Weiss field value. In CaF_2, $\gamma^2\hbar/(2a^3) = 2{\cdot}72$ kHz. (After Goldman and Sarma, 1975.)

section **C**(f). Both quantities are calibrated as explained in chapter 5, section **C**(a). This is in principle an *absolute* measurement of $\chi_\perp$, and the experimental value in the plateau is about 20 per cent higher than the Weiss-field value and thus in very good agreement with the restricted-trace prediction. On the other hand, in the paramagnetic region the agreement with the high temperature expansion is rather poor. It is not known whether this is due to a real effect, such as the perturbation due to the paramagnetic impurities, or a consequence of systematic errors in the calibrations.

The net result of this study is that one does observe a plateau of $\chi_\perp$ as expected for an antiferromagnet, and that the value of $\chi_\perp$ in the plateau, as well as the values of the energy and the entropy at the beginning of the plateau are qualitatively as expected.

2. *Lithium fluoride and hydride*

When the system contains two spin species I and S it is possible to measure separately the transverse susceptibilities of both spin species by observing their resonance signals.

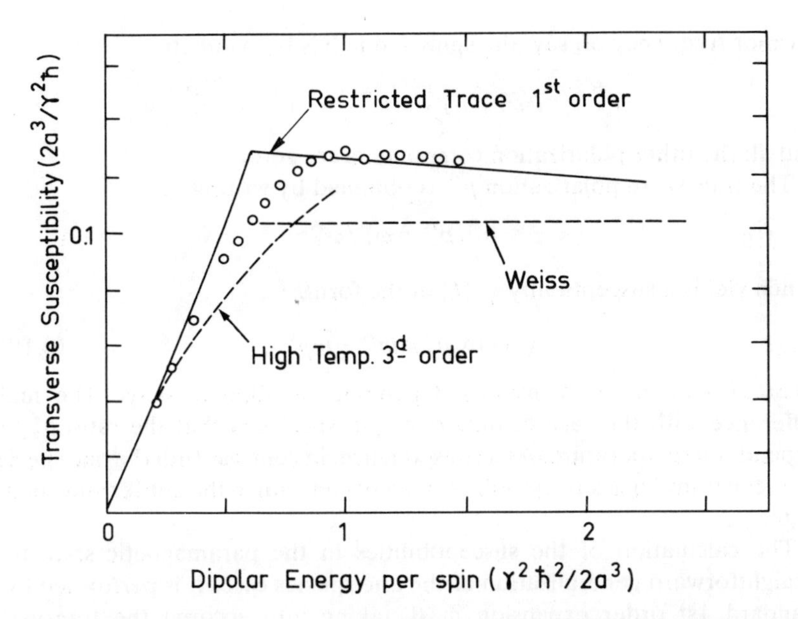

FIG. 8.12. Calibrated transverse susceptibility against dipolar energy in the antiferromagnetic structure III of CaF_2. The susceptibility is obtained from the Kramers–Kronig transform of the absorption signal, and the energy from its first moment. Same theoretical approximations as in Fig. 8.11. (After Goldman, 1977.)

The calculation of these susceptibilities in the Weiss-field approximation is slightly more complicated than when there is only one spin species (sections **C**(a) and **D**(a)1). We outline the principle of calculation of these susceptibilities as a function of the entropy.

We saw in section **B**(d) that in the antiferromagnetic structure predicted for LiH and LiF there are still two sublattices $\mathscr{A}$ and $\mathscr{B}$ of opposite polarizations, but with spins of either species in each sublattice. We thus distinguish four types of spins, labelled as follows:

A and B for spins I in sublattices $\mathscr{A}$ and $\mathscr{B}$, respectively;

C and D for spins S in sublattices $\mathscr{A}$ and $\mathscr{B}$, respectively.

Four steps are involved in the theory. They consist in calculating the following quantities:

(i) Weiss frequencies as a function of spin polarizations;
(ii) Transverse susceptibilities as a function of spin polarizations;
(iii) Spin polarizations as a function of temperature;
(iv) Entropy.

The Weiss frequency experienced by each type of spin is a linear combination of the four polarizations. In the presence of an r.f. field at the

Larmor frequency of, say, the spins I, one has by symmetry:

$$p_z^{\mathrm{B}} = -p_z^{\mathrm{A}}; \qquad p_x^{\mathrm{B}} = p_x^{\mathrm{A}}; \qquad p_z^{\mathrm{D}} = -p_z^{\mathrm{C}},$$

and all the other polarization components are zero.

The transverse polarization p_x^{A} is obtained by writing:

$$p_x^{\mathrm{A}}/p_z^{\mathrm{A}} = \omega_x^{\mathrm{A}}/\omega_z^{\mathrm{A}},$$

which yields a susceptibility $\chi_\perp(I)$ of the form:

$$\chi_\perp(I) \propto p_z^{\mathrm{A}}/(lp_z^{\mathrm{A}} + mp_z^{\mathrm{C}}), \tag{8.102}$$

where l and m are constants depending on dipolar sums. The main difference with the case of only one spin species is that the ratio $p_z^{\mathrm{A}}/p_z^{\mathrm{C}}$ depends on temperature. As a consequence, in contrast with the one species case, the transverse susceptibilities are *not* constant in the antiferromagnetic state.

The calculation of the susceptibilities in the paramagnetic state is a straightforward generalization of the one species case. It is performed by a standard 1st order expansion in β, taking into account the transverse Weiss-field associated with the transverse polarizations.

Figure 8.13 is a plot of the transverse susceptibilities according to these approximations, in an infinitely flat sample of LiF parallel to the external field, with $H_0 \| [001]$ and $T < 0$. The antiferromagnetic structure is structure III of Fig. 8.3.

The entropy parameter used for the abscissa of Fig. 8.13 is:

$$R = [(S_0 - S)/2Nk_{\mathrm{B}}]^{1/2},$$

where S_0 is the entropy at infinite temperature and N the number of spins of each species. This choice of abscissa has the advantage that at high temperature, R is proportional to the inverse temperature β, so that the susceptibilities are linear functions of R. The transition is expected to occur for $R \simeq 0{\cdot}2$, that is:

$$p_I \simeq 30 \text{ per cent} \quad \text{and} \quad p_S \simeq 20 \text{ per cent}.$$

In Fig. 8.14 are shown the experimental results in a sample of LiF, obtained by observing the heights of the fast passage signals at the centre of the resonance (Cox *et al.* 1975). The levelling-off of the susceptibilities at low entropy, in accordance with expectation, was the first indication of the production of antiferromagnetism in such a substance. The initial slopes are about twice as large as expected, which may be due to the contribution of the paramagnetic impurities to the local frequency. The plateau is reached at $R \simeq 0{\cdot}5$, with values of $\chi_\perp$ about 30 to 40 per cent larger than expected. This discrepancy may be due to inaccuracies of the Weiss-field theory, or to faulty

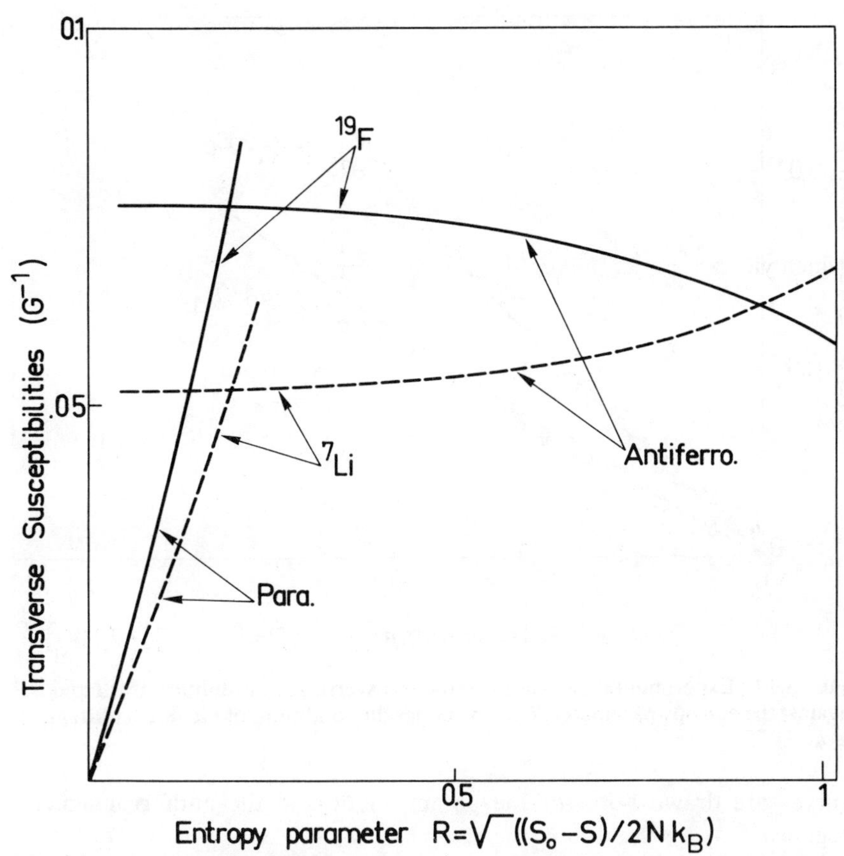

FIG. 8.13. Theoretical variation of the transverse susceptibilities of ^{7}Li and ^{19}F against the entropy parameter R in a platelet of LiF, after ADRF with $H_0 \| [100]$ and $T<0$ (expected antiferromagnetic ordering III), according to 1st order high temperature and Weiss field approximation. (After Cox *et al.* 1975.)

calibrations or to both. The ratio of the susceptibilities,

$$\chi_\perp(\mathrm{F})/\chi_\perp(\mathrm{Li})_{\mathrm{exp}} \simeq 1{\cdot}63,$$

is however reasonably close to the theoretical value 1.46.

Figure 8.15 reproduces the measurements of $\chi_\perp(^7\mathrm{Li})$ and $\chi_\perp(^1\mathrm{H})$ in lithium hydride (Roinel and Bouffard, private communication). The sample is a thin parallelepiped parallel to the external field. The ADRF is performed with $H_0 \| [001]$ and $T<0$, and the antiferromagnetic structure is structure III of Fig. 8.3.

The susceptibilities, obtained by a Kramers–Kronig transform of the absorption signals of ^{1}H and ^{7}Li, are plotted against the dipolar energy. The energy is measured by the method of chapter 5, section **C**(c)2. Theoretical

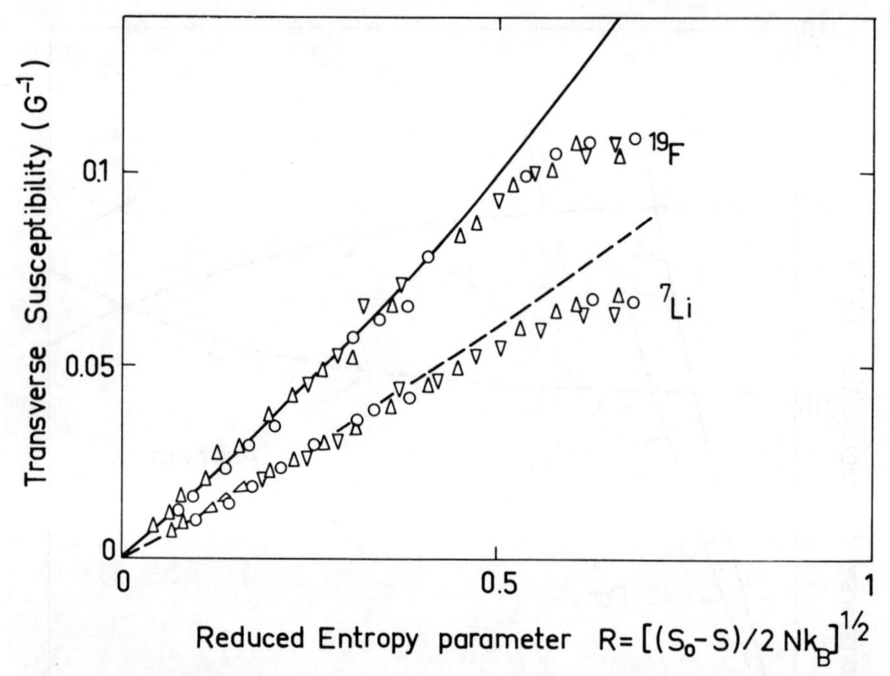

FIG. 8.14. Experimental variations of the transverse susceptibilities of ^{7}Li and ^{19}F against the entropy parameter R, in LiF under the conditions of Fig. 8.13. (After Cox *et al.* 1975.)

curves are drawn both for the paramagnetic and the antiferromagnetic regions.

As with LiF, the initial slopes are lower than expected, because of the influence of the paramagnetic impurities. Within experimental uncertainty, the quasi-plateaus are in fair agreement with the Weiss-field theory, and exhibit the small variations predicted by this theory. The turnover from high temperature to Weiss-field behaviour coincides with the transition to antiferromagnetism, as observed by neutron diffraction (section **D**(f)).

(b) *Longitudinal susceptibility*

We consider the case of a system with one spin species only. In zero effective field, the sublattice polarizations are opposite and the bulk polarization of the sample is zero. In the presence of a small longitudinal field, the magnitudes of the sublattice polarizations cease to be equal, and there appears a bulk polarization proportional to the field. The longitudinal susceptibility is defined through:

$$\chi_{\parallel} = \tfrac{1}{2}(p_z^{\mathrm{A}} + p_z^{\mathrm{B}})/\Delta, \tag{8.103}$$

where Δ is the effective frequency.

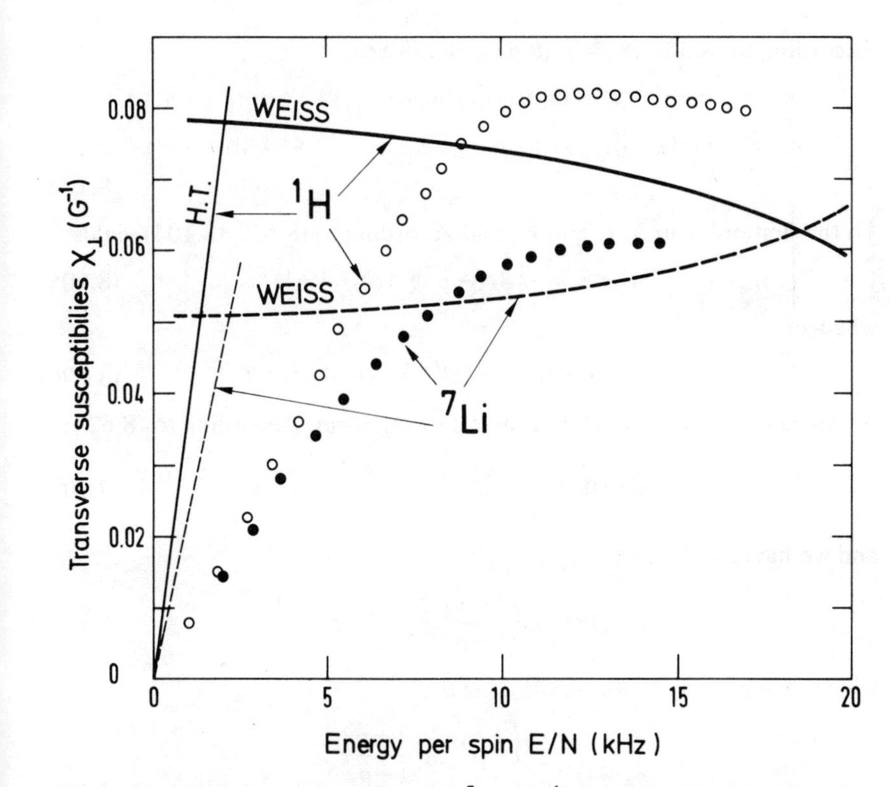

FIG. 8.15. Transverse susceptibilities of ^{7}Li and ^{1}H as a function of energy in a platelet of LiF, after ADRF with $H_0 \| [100]$ and $T < 0$ (antiferromagnetic ordering III), together with 1st order high temperature and Weiss field predictions. The susceptibilities are obtained by the Kramers–Kronig transforms of the absorption signals, and the dipolar energy by the saturation method. (After Roinel and Bouffard, private communication.)

The property which, together with the constancy of $\chi_\perp$, is characteristic of antiferromagnets is that their longitudinal susceptibility decreases below the transition and vanishes at zero temperature. The easiest way to see this is through the Weiss-field approximation. At zero temperature, the sublattice polarizations are unity and their orientation is determined by that of the field they experience. A small longitudinal field superimposed on the dipolar field changes slightly the amplitude but not the orientation of the field in each sublattice. At zero temperature there will be no change in the sublattice polarizations and the bulk polarization will remain zero, whence $\chi_\| = 0$.

When the temperature is not zero, let us write:

$$p_z^{\mathrm{A}} = p + \varepsilon \quad \text{and} \quad p_z^{\mathrm{B}} = -p + \varepsilon'.$$

According to (8.54), (8.54′), (8.55), and (8.56):

$$\begin{aligned} p+\varepsilon &= \tanh\{-\tfrac{1}{2}\beta[A(\mathbf{k}_0)p+\tfrac{1}{2}A(\mathbf{k}_0)(\varepsilon-\varepsilon')+\tfrac{1}{2}A(0)(\varepsilon+\varepsilon')+\Delta]\} \\ -p+\varepsilon' &= \tanh\{-\tfrac{1}{2}\beta[-A(\mathbf{k}_0)p-\tfrac{1}{2}A(\mathbf{k}_0)(\varepsilon-\varepsilon')+\tfrac{1}{2}A(0)(\varepsilon+\varepsilon')+\Delta]\}. \end{aligned} \tag{8.104}$$

To the first order in Δ, ε, and ε', and according to (8.62), (8.104) yields:

$$\varepsilon=\varepsilon'=-\tfrac{1}{2}\beta(1-p^2)[A(0)\varepsilon+\Delta], \tag{8.105}$$

whence:

$$\chi_{\|}=\varepsilon/\Delta=-\tfrac{1}{2}\beta(1-p^2)/[1+\tfrac{1}{2}\beta(1-p^2)A(0)], \tag{8.106}$$

which can also be expressed in the following form. According to (8.62′):

$$-\tfrac{1}{2}\beta A(\mathbf{k}_0)p=\tanh^{-1}(p)=\frac{1}{2}\ln\left(\frac{1+p}{1-p}\right), \tag{8.107}$$

and we have:

$$\beta=-\ln\left(\frac{1+p}{1-p}\right)\Big/[A(\mathbf{k}_0)p], \tag{8.108}$$

which, when carried into (8.106), yields:

$$\chi_{\|}=\frac{(1-p^2)\ln\left(\dfrac{1+p}{1-p}\right)}{2A(\mathbf{k}_0)p-(1-p^2)\ln\left(\dfrac{1+p}{1-p}\right)A(0)}. \tag{8.109}$$

A calculation of $\chi_{\|}$ in the paramagnetic state similar to that of $\chi_{\perp}$ in (8.101) yields:

$$\chi_{\|\text{para}}=\frac{\mp p_i}{D\pm A(0)p_i}, \tag{8.109′}$$

where the sign is opposite to that of the temperature ($-$ for $T>0$ and $+$ for $T<0$). Again, its matching to the antiferromagnetic value (8.109) at the transition, that is for $p=0$, yields for the critical initial polarization the value (8.66). It is easily verified that $\chi_{\|}$ is maximum when $p=0$, that is at the transition, and decreases monotonically to zero when p increases up to 1.

The variation of $\chi_{\|}$ has been subjected to two experimental investigations in spherical samples of CaF_2 ($A(0)=0$) with $H_0\|[001]$ and $T<0$ (corresponding to antiferromagnetic structure III).

The first study was simply aimed at proving that $\chi_{\|}$ does indeed decrease below the transition. Its principle is to detect the bulk magnetization of the CaF_2 sphere through the dipolar field it creates in its neighbourhood, by observing a shift in the NMR resonance frequency of liquid ^{3}He surrounding the sample (Chapellier, 1971). The experimental procedure is schematically as follows. After ADRF of the fluorine spins, the r.f. field is subjected to a small and slow frequency modulation around the ^{19}F Larmor frequency. This is equivalent to an adiabatic modulation of the longitudinal effective field seen by the ^{19}F spins without actually changing the external field seen by the ^{3}He. It produces a modulation of the magnetization of the CaF_2 sample proportional to $\chi_\parallel$, which in turn produces a modulation of the magnetic field seen by the spins of ^{3}He and of their resonance frequency. This last frequency modulation is detected by recording the modulation of the lock-in signal of ^{3}He in a coil located just below the CaF_2 sphere. In the course of time, the ^{19}F system warms up under the effect of spin–lattice relaxation, and one expects that the amplitude of modulation of the helium signal will first increase and then decrease, when the system becomes paramagnetic. The results for a typical run are shown in Fig. 8.16. The small

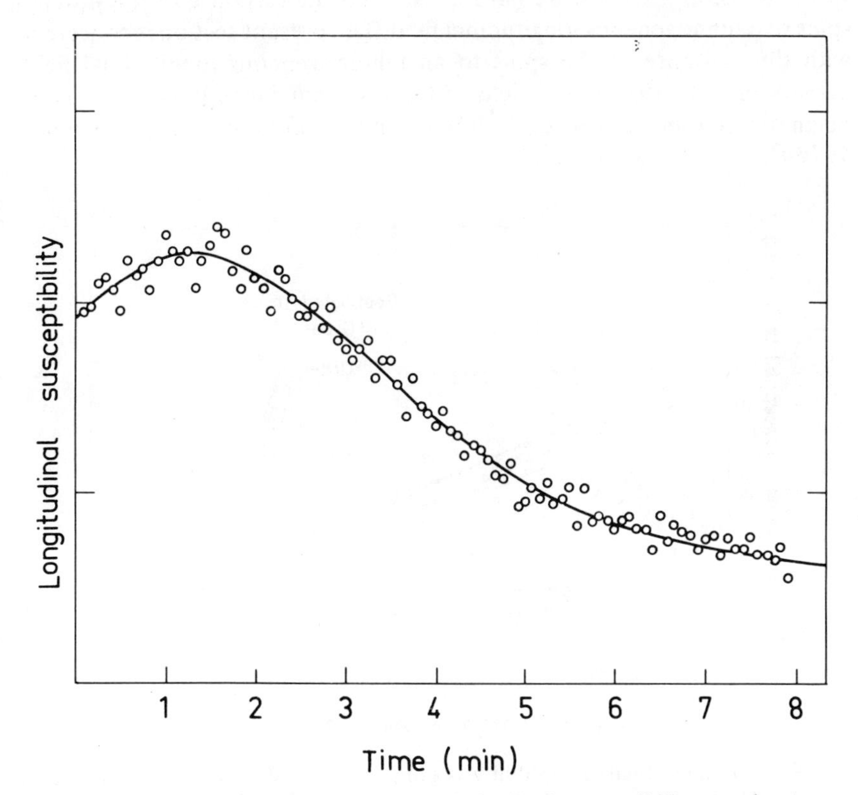

FIG. 8.16. Variation with time of the longitudinal susceptibility, as the system warms up, in structure III of CaF_2. The ^{19}F susceptibility is monitored by the resonance frequency shift of a ^{3}He probe. (After Chapellier, 1971.)

initial increase of the modulation amplitude has been repeatedly observed in all runs.

The second study, described in Section **D**(c)2 of chapter 5, consisted in observing the absorption signal following a succession of sudden field jumps back and forth from resonance (Jacquinot, Chapellier, and Goldman, 1974). The values of $\chi_\parallel$ are calibrated against those of $\chi_\perp$ at high temperature, where they tend toward equal values, and the latter are calibrated assuming that $\chi_\perp$ in the antiferromagnetic plateau is equal to the Weiss-field value (8.96).

The results are shown in Fig. 8.17, together with various predictions: 3rd order expansion in β, Weiss-field, 1st and 2nd order restricted trace, and RPA (described in section **H**(b)). The agreement is qualitatively correct. The decrease of $\chi_\parallel$ at high energy is small, but as expected.

(c) *Non-uniform longitudinal susceptibility of the* ^{19}F *spins in* CaF_2

In the preceding section we have considered the response of the ordered spins to a homogeneous longitudinal field. The present section is concerned with the response of the spins to an inhomogeneous longitudinal field, namely in CaF_2 the dipolar field of the low-abundance (~0·13 per cent) magnetic isotope of calcium, ^{43}Ca of spin $\frac{7}{2}$ (Goldman and Jacquinot, 1976a).

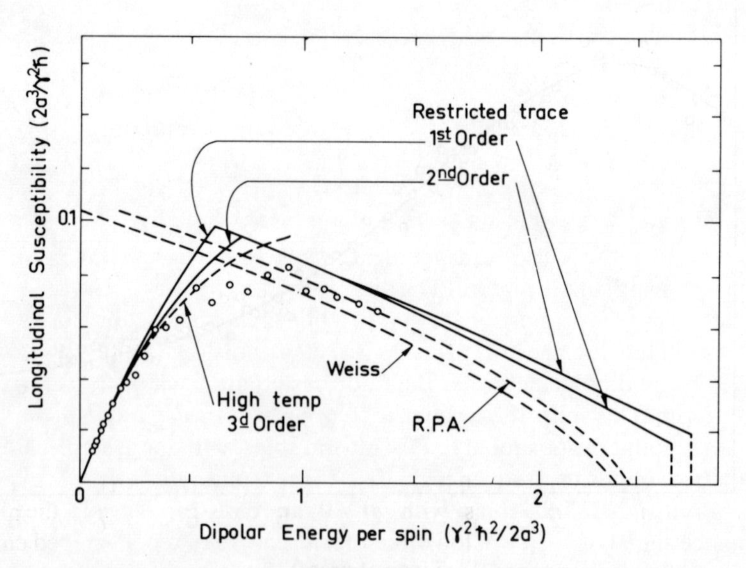

FIG. 8.17. Longitudinal susceptibility as a function of dipolar energy in structure III of CaF_2. The values of $\chi_\parallel$ are calibrated against those of $\chi_\perp$ (Fig. 8.11). Theoretical curves for high-temperature, Weiss field, RPA and restricted-trace approximations. (After Goldman and Sarma, 1975.)

It was stated in section **B**(c) that, because of their low abundance and their small gyromagnetic ratio, the ^{43}Ca spins have a very small influence on the ordering of the ^{19}F spins. This small influence can however be detected by observing the resonance signal of the ^{43}Ca spins themselves. Its mechanism is the following. In the dipolar field of the ^{43}Ca spins the fluorine spins acquire polarizations departing slightly from their undisturbed values. These departures are non-uniform. They affect the shape of the ^{43}Ca resonance signal in a way which depends on the non-uniform longitudinal susceptibility of the ^{19}F spins to the dipolar field of the ^{43}Ca spins.

As a consequence of the low abundance and small magnetic moment of the ^{43}Ca spins, the effect of each one can be treated independently from that of the others, for the following reasons.

(i) The heat capacity of the ^{43}Ca–^{43}Ca dipolar interactions $\mathcal{H}'_{SS}$ is much less than that of $\mathcal{H}'_{IS}$ and $\mathcal{H}'_{II}$ (S and I refer to ^{43}Ca and ^{19}F, respectively).

(ii) The mean time $\tau(^{43}\text{Ca})$ between successive flip-flops, for two ^{43}Ca spins at their average distance is very long, of the order of one hour, whereas internal equilibrium within the ^{19}F spin system takes place within a time $T_2(^{19}\text{F})$ which is a small fraction of a millisecond. The condition $\tau(^{43}\text{Ca}) \gg T_2(^{19}\text{F})$ is thus amply satisfied and one can assume that each ^{43}Ca spin is in a well-defined and time-independent state $S_z = m$.

(iii) The dipolar field of a ^{43}Ca spin falls down to a negligible value at distances much less than the average distance between neighbouring ^{43}Ca spins. There are then negligibly few ^{19}F spins influenced by more than one ^{43}Ca spin.

The Hamiltonian of the system is to a good approximation:

$$\mathcal{H}'_{\mathrm{D}} = \mathcal{H}'_{II} + \mathcal{H}'_{IS} \tag{8.110}$$

where $\mathcal{H}'_{II}$ and $\mathcal{H}'_{IS}$ are given by (5.43) and (5.45), respectively.

The shape of the ^{43}Ca resonance signal can be qualitatively accounted for by the following simple picture. The dipolar field created by a calcium spin in the state $S_z = m$ is proportional to m. In a linear approximation the changes it induces in the fluorine polarizations is also proportional to m, and so is the change in the dipolar field that the fluorines create back at the site of the calcium spin. This shifts the resonance frequency of the ^{43}Ca spin in the state m by an amount proportional to m. Calcium spins with opposite values of m experience opposite frequency shifts. Since spins with $m<0$ give an absorption signal and spins with $m>0$ an emission signal, the total resonance signal of ^{43}Ca will look like the derivative of a bell-shaped curve, as expected for spins in a demagnetized state.

As for most NMR signals in solids, it is impossible to give a detailed description of the shape of the calcium resonance signal. A simple and well-defined information is however obtained from its first moment. Since the coupling $\mathcal{H}'_{SS}$ is neglected in the Hamiltonian (8.110), the first moment $\mathcal{M}_1^S$ yields, according to (5.106), the ^{43}Ca–^{19}F dipolar energy $\langle \mathcal{H}'_{IS} \rangle$.

This energy will now be calculated as a function of temperature in the Weiss-field approximation. In accordance with the approximation of independent calcium spins, we first consider the system consisting of all spins I plus one spin S in a state $S_z = m$.

The polarization of a spin I_i is, according to the local Weiss-field eqn (8.5), related to its local Weiss frequency ω_i through:

$$p_i = \tanh(-\tfrac{1}{2}\beta\omega_i). \tag{8.111}$$

If the spin S were absent, the undisturbed antiferromagnetic structure would correspond to:

$$p_i = p_i^0 = \pm p,$$

and, according to (8.9) and (8.16):

$$\omega_i = \omega_i^0 = A(\mathbf{k}_0)p_i^0.$$

In the presence of the spins S we have:

$$p_i = p_i^0 + \varepsilon_i,$$

$$\omega_i = \omega_i^0 + \delta\omega_i = \omega_i^0 + \sum_j A_{ij}\varepsilon_j + 2mB_{i\mu}. \tag{8.112}$$

We assume that $\beta\delta\omega_i \ll 1$, i.e. that we can expand the right-hand side of (8.111) to the first order in $\beta\delta\omega_i$. This yields:

$$\varepsilon_i = -\tfrac{1}{2}\beta(1 - p_1^{02})\Big(\sum_j A_{ij}\varepsilon_j + 2mB_{i\mu}\Big). \tag{8.113}$$

For the antiferromagnet in zero field $(p_i^0)^2$ is the same for all the spins I_i and can be rewritten as p^2. Equation (8.113) expresses the fact that the change ε_i in the polarization of the spin I_i depends not only on the dipolar field exerted on it by the spin S, but also on the polarization changes induced in *all* the spins I_j. The response of the spin I_i to the non-uniform field produced by a spin S is non-local.

Through a space Fourier transform of (8.113) we obtain:

$$\varepsilon(\mathbf{k}) = -\tfrac{1}{2}\beta(1-p^2)\left[A(\mathbf{k})\varepsilon(\mathbf{k}) + \frac{2}{N_I^{1/2}}\,mB(\mathbf{k})\right], \tag{8.114}$$

where N_I is the number of spins I and $\mathbf{k}$ belongs to the first Brillouin zone of the fluorine lattice.

Using the Weiss-field value of the critical inverse temperature:

$$\beta_c = -2/A(\mathbf{k}_0),$$

we obtain:

$$\varepsilon(\mathbf{k}) = \frac{2mB(\mathbf{k})}{N_I^{1/2}A(\mathbf{k}_0)} \cdot \frac{(\beta/\beta_c)(1-p^2)}{1-(\beta/\beta_c)(1-p^2)A(\mathbf{k})/A(\mathbf{k}_0)}. \tag{8.115}$$

According to (8.115), $\varepsilon(\mathbf{k})$ is proportional to the Fourier component with wave vector $\mathbf{k}$ of the calcium dipolar field: $2mB(\mathbf{k})$, and we can write:

$$\varepsilon(\mathbf{k}) = \chi_{\parallel}(\mathbf{k}) \,.\, 2mB(\mathbf{k}). \tag{8.115$'$}$$

In contrast to eqn (8.106) for the uniform susceptibility, (8.115) is valid above as well as below the transition, since the only assumption made with respect to the polarizations p_i^0 of the spins I was a unique value for $(p_i^0)^2$ which is true for both the paramagnetic and the antiferromagnetic phases. It is easily shown that $(\beta/\beta_c)(1-p^2)$ is maximum at the transition, where it is equal to 1, so that when $A(\mathbf{k})/A(\mathbf{k}_0)$ is close to 1, $\chi_{\parallel}(\mathbf{k})$ is sharply peaked at the transition.

The effect of the spins S being treated independently from each other is that the energy $\langle \mathcal{H}'_{IS} \rangle$ is equal to:

$$\begin{aligned}\langle \mathcal{H}'_{IS} \rangle &= 2\sum_{i\mu} B_{i\mu} \langle I_z^i S_z^\mu \rangle \\ &= \sum_m m\mathcal{P}_m \sum_i B_{i\mu} p_i(\beta, m),\end{aligned} \tag{8.116}$$

where $\mathcal{P}_m$ is the population of the spins S with $S_z = m$. In CaF_2, the spins $S(^{43}Ca)$ are located at the centres of cubes of spins $I(^{19}F)$. Each site of ^{43}Ca is a centre of symmetry not only for the crystalline lattice but also for the unperturbed antiferromagnetic structure of the ^{19}F spins, and the unperturbed Weiss field they produce at this site vanishes:

$$\sum_i B_{i\mu} p_i^0 = 0.$$

The energy $\langle \mathcal{H}'_{IS} \rangle$ is then:

$$\begin{aligned}\langle \mathcal{H}'_{IS} \rangle &= \sum_m m\mathcal{P}_m \sum_i B_{i\mu}\varepsilon_i \\ &= N_I^{-1/2} \sum_m m\mathcal{P}_m \sum_{\mathbf{k}} B(-\mathbf{k})\varepsilon(\mathbf{k}) \\ &= 2N_I^{-1/2} \sum_m m^2\mathcal{P}_m \sum_{\mathbf{k}} B(\mathbf{k})^2 \chi_{\parallel}(\mathbf{k})\end{aligned}$$

where we have used (8.115$'$) and the fact that $B(\mathbf{k}) = B(-\mathbf{k})$.

We obtain finally, according to (8.115):

$$\langle \mathcal{H}'_{IS} \rangle = 2\frac{N_S}{N_I}\langle S_z^{\mu 2}\rangle \beta(1-p^2)\sum_{\mathbf{k}} \frac{B(\mathbf{k})^2}{\beta_c A(\mathbf{k}_0) - \beta(1-p^2)A(\mathbf{k})}. \quad (8.117)$$

If we assume that all populations $\mathcal{P}_m$ are equal, which can be achieved if necessary by saturating the spins S before the ADRF of the spins I, these populations will remain equal thereafter. The flip-flop processes between spins S which might conceivably destroy this equality are exceedingly slow ($\tau(^{43}\mathrm{Ca}) \sim 1$ hour). In that case, $\langle S_z^{\mu 2}\rangle = \frac{1}{3}S(S+1)$ in (8.117). We shall consider in the next section the possibility of a non-vanishing quadrupole alignment $\langle S_z^{\mu 2} - \frac{1}{3}S(S+1)\rangle$ of rare spins S.

In the sum on the right-hand side of (8.117), those terms for which $A(\mathbf{k})/A(\mathbf{k}_0)$ is close to unity will cause the energy $\langle \mathcal{H}'_{IS}\rangle$ to be sharply peaked at the transition.

The experiment consists in measuring the ^{43}Ca–^{19}F dipolar energy as a function of the F–F dipolar energy. Since, as will be seen shortly, $\langle \mathcal{H}'_{IS}\rangle \ll \langle \mathcal{H}'_{II}\rangle$, the latter is obtained from the first moment $\mathcal{M}_1^I$ of the resonance signal of the spins I.

In a rough makeshift approximation, $\langle \mathcal{H}'_{II}\rangle$ is calculated as a function of β by the 1st order restricted-trace approximation. The resulting curve for structure III ($H_0 \| [001]$, $T<0$) is shown in Fig. 8.18, together with high temperature and Weiss-field predictions, and the combined experimental results of several runs.

The expected maximum of $\langle \mathcal{H}'_{IS}\rangle$ is clearly exhibited, and its overall variation is as predicted. None of the approximations however is very good, possibly because the non-uniform fluorine susceptibility depends very much on short-range correlations. There may be another cause of error: although the gyromagnetic ratio of the spins S is much smaller than that of the spins I $(\gamma_S/\gamma_I \simeq \frac{1}{14})$, the ratio of their magnetic moments:

$$\frac{\mu_S}{\mu_I} = \frac{S\gamma_S}{I\gamma_I} = 7\frac{\gamma_S}{\gamma_I} \simeq 0{\cdot}5,$$

is not very small, and the linear expansion of (8.111) as a function of $\delta\omega_i$ is of questionable validity.

The combination of Weiss-field and high temperature approximations is completely inadequate. The restricted-trace prediction, although qualitatively correct, is still far from satisfactory.

The energy per spin $\langle \mathcal{H}'_{II}\rangle/N_I$ at which $\langle \mathcal{H}'_{IS}\rangle$ is maximum, identified with the transition energy, is about 2·5 kHz, whereas the values predicted by the restricted-trace method are 1·62 kHz and 1·93 kHz for the 1st order and 2nd order approximations, respectively.

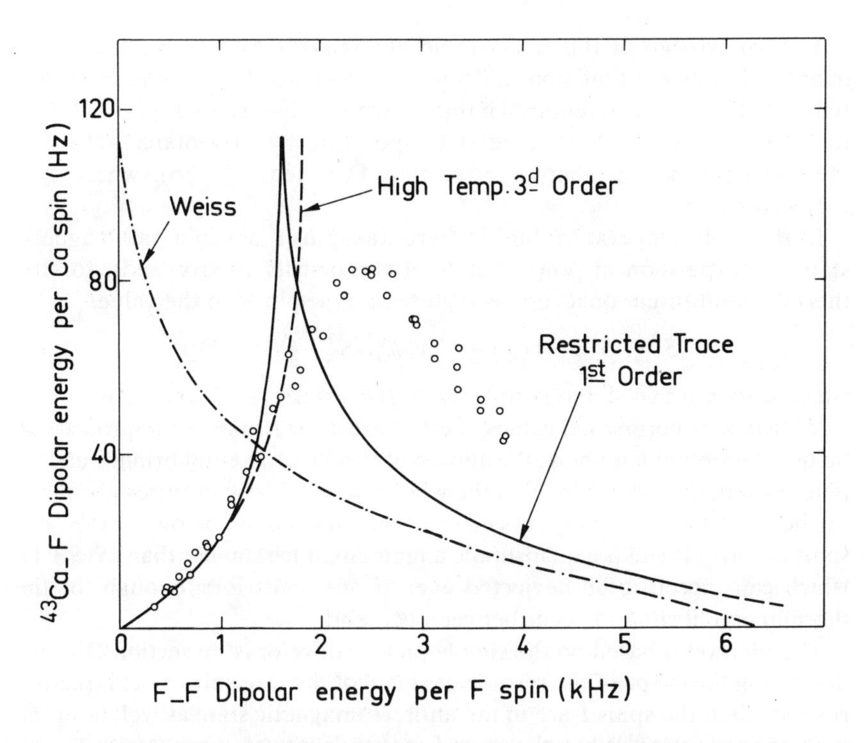

FIG. 8.18. ^{19}F–^{43}Ca energy against ^{19}F–^{19}F energy in CaF_2 demagnetized with $H_0 \| [100]$ and $T<0$ (antiferromagnetic structure III). Theoretical curves for 3rd order high temperature, Weiss field and 1st order restricted-trace approximations. (After Goldman and Jacquinot, 1976*a*.)

1. *Quadrupole alignment of the rare spins*

In interpreting the results of the experiment described in the last section it was assumed that the populations $\mathcal{P}_m$ of the rare spins remained equal, so that in (8.117), $\langle S_z^{\mu 2} \rangle$ was equal to $\frac{1}{3}S(S+1)$. However, this assumption does not correspond to thermal equilibrium between $\mathcal{H}'_{II}$ and $\mathcal{H}'_{IS}$. We shall see presently that thermal equilibrium between these terms leads to a non-vanishing quadrupole alignment of the spins S. The time required for reaching this equilibrium is of the order of the flip-flop time $\tau(S)$ between the rare spins. In CaF_2, this time was far longer than the time elapsed between the ADRF of the spins I and the observation of the resonance of the spins S, and the assumption of equal populations $\mathcal{P}_m$ (and therefore of vanishing quadrupole alignment) was justified. Under experimental conditions different from those of the experiment with CaF_2 described in section **D**(c), the quadrupole alignment of the rare spins S might reach its thermal equilibrium value.

The knowledge of this quadrupole alignment may be necessary for a quantitative study of the non-uniform susceptibility of the abundant spins I (eqn (8.117)). It is also required if the absorption signal of the spins S is used for a measurement of the dipolar temperature by the cotanh transform method described in chapter 5, section **C**(g) (eqn (5.136), where I is replaced by S).

In the high temperature limit, where the spins I are in a paramagnetic state, an expansion in powers of β of the density matrix leads, for the thermal equilibrium quadrupole alignment of a spin S, to the value:

$$\langle S_z^2 - \tfrac{1}{3}S(S+1)\rangle = \tfrac{1}{2}\beta^2 \operatorname{Tr}\langle \mathcal{H}_{IS}'^2[S_z^2 - \tfrac{1}{3}S(S+1)]\rangle,$$

which does not vanish but is small since $\beta\mathcal{H}'_{II} \gg \beta\mathcal{H}'_{IS}$ is itself small.

When the experimental values of $\langle \mathcal{H}'_{IS}\rangle$ are known, another approach can be used to obtain a numerical estimate of the thermal equilibrium quadrupole alignment of the spins S in the whole range of temperatures produced by the ADRF of the spins I. As an example, this is done below for the ^{43}Ca spins in CaF_2. It yields a quadrupole alignment much smaller than $\frac{1}{3}S(S+1)$, which can therefore be neglected even if one waits long enough for the thermal equilibrium to occur between $\mathcal{H}'_{II}$ and $\mathcal{H}'_{IS}$.

The method is based on the simple picture developed in section **D**(c) for describing the shape of the resonance signal of the rare spins S. This picture is valid when the spins I are in the antiferromagnetic state as well as in the paramagnetic state. When the spins I in the vicinity of a spin S with $S_z = m$ are at internal equilibrium, they produce on that spin S a dipolar field proportional to m, say Ωm, where Ω depends on the inverse temperature β. The energy of the spins S in this field is:

$$E_m = \Omega m \times m = \Omega m^2. \tag{8.118}$$

This energy is the same as if the spins S experienced a fictitious quadrupole interaction:

$$\mathcal{H}_Q = \Omega S_z^2. \tag{8.119}$$

At thermal equilibrium with respect to this interaction, the spins S acquire a non-vanishing quadrupole alignment, which corresponds to unequal populations $\mathcal{P}_m$.

The dipolar fields experienced by the spins S are maximum at the transition. In CaF_2, the maximum I–S dipolar energy per calcium spin is, according to Fig. 8.18, of the order of:

$$E \simeq 100 \text{ Hz}.$$

The populations $\mathcal{P}_m$ being equal in this experiment, this energy is, according to (8.118) and (8.119):

$$E = \Omega\langle S_z^{\mu 2}\rangle = \tfrac{1}{3}S(S+1)\Omega = \tfrac{21}{4}\Omega \simeq 5\ \Omega,$$

since for ^{43}Ca $S=7/2$, whence:

$$\Omega \simeq 20\ \text{Hz}.$$

The critical temperature for structure III of CaF_2 is, according to (8.46), of the order of:

$$T_c \sim -0{\cdot}5\ \mu\text{K} \simeq -10\ \text{kHz},$$

so that $\beta_c\Omega \sim 2\,.\,10^{-3} \ll 1$. We can then use the high temperature approximation for calculating the thermal equilibrium quadrupole alignment of a spin S:

$$\begin{aligned}\langle S_z^2 - \tfrac{1}{3}S(S+1)\rangle_{\text{eq}} &\simeq -\beta\Omega\,\text{Tr}\{S_z^2[S_z^2 - \tfrac{1}{3}S(S+1)]\}\\ &\simeq 4{\cdot}2\times 10^{-2} \quad \text{for } S=\tfrac{7}{2}\\ &\simeq \tfrac{1}{3}S(S+1)\times 8\,.\,10^{-3},\end{aligned}$$

whence, in eqn (5.136) and at thermal equilibrium for the spins S:

$$S(S+1) - (S_z^{\mu 2}) \simeq \tfrac{2}{3}S(S+1)\times(1-4\times 10^{-3}),$$

which shows that, for all practical purposes, the quadrupole alignment of the rare spins can indeed be neglected.

(*d*) *Spin–lattice relaxation*

Once the nuclear spin system has been cooled by adiabatic demagnetization in the rotating frame, it warms up toward the lattice temperature under the effect of spin–lattice relaxation. In solids, this relaxation is known to originate from the coupling of the nuclear spins with paramagnetic impurities. It depends on their nature, their concentration and their coupling to the lattice.

The problem investigated in this section is the following: in the presence of given impurities, how does the spin–lattice relaxation of the nuclear spin system depend on its internal state, paramagnetic or antiferromagnetic (Goldman and Jacquinot, 1976*b*).

In the high temperature limit, one can define unambiguously a spin–lattice relaxation time T_1. Most if not all measurable quantities are proportional to the inverse temperature β, which tends exponentially toward the lattice inverse temperature β_L. At low spin temperatures, and in particular in an ordered state, the various physical parameters are not proportional to each other and none is expected *a priori* to vary exponentially with time. It is then necessary to specify the physical quantity whose time-variation is being studied. We are concerned in this section with the variation under the effect of spin–lattice relaxation of the dipolar energy of the nuclear spin system. We limit ourselves to the case of CaF_2, a system with one spin species (^{19}F) whose dipolar energy is easily obtained from the first moment of the absorption signal (chapter 5, section **C**(*d*)).

Experimentally, after the ADRF, the fluorine absorption signal observed at a non-saturating r.f. level is recorded at regular time intervals, and its first moment $\mathcal{M}_1$ is computed.

As in standard treatments of spin–lattice relaxation (Abragam, 1961, chapter IX; Goldman, 1970, chapter 3; chapter 6 of this book), the spin orientations of the paramagnetic impurities are treated as stochastic variables which vary randomly around their average value under the effect of their own relaxation (let it be electronic spin–lattice relaxation or flip-flop between impurities as in chapter 6, as the case may be). The dipolar fields created at the nuclear sites by these impurities have the same random variation. The perturbation responsible for the relaxation of the nuclear spins has then the form of a Zeeman coupling with these random fields. At low lattice temperature, the correlation time τ_c of the field fluctuation is much longer than the nuclear Larmor period: $\omega_0\tau_c \gg 1$, and only the longitudinal components of the local fields are effective for relaxing the nuclear dipolar reservoir.

The coupling responsible for the relaxation is written in the form:

$$\mathcal{H}_1(t) = VF(t), \tag{8.120}$$

where:

$$V = \sum_i \delta_i I_z^i, \tag{8.121}$$

and $F(t)$ is a random function satisfying:

$$\langle F(t)\rangle_{av} = 0; \qquad \langle F(0)F(t)\rangle_{av} = \exp(-|t|/\tau_c). \tag{8.122}$$

Equation (8.120) describes the nuclear coupling with one impurity. The various impurities contribute additively to the nuclear relaxation.

The master equation for the rate of change of the dipolar energy is of the form (chapter 1, section **D**):

$$\frac{d}{dt}\langle \mathcal{H}'_D\rangle = -\int_0^\infty \exp(-t/\tau_c)\langle[[\mathcal{H}'_D, V(t)], V]\rangle\, dt, \tag{8.123}$$

where:

$$V(t) = \exp(i\mathcal{H}'_D t)\, V \exp(-i\mathcal{H}'_D t). \tag{8.124}$$

Let T_2 be the decay time of the expectation value on the right-hand side of (8.123), roughly comparable with the inverse of the nuclear linewidth. We consider two limiting cases: $\tau_c \ll T_2$ and $\tau_c \gg T_2$.

1. *Short correlation time:* $\tau_c \ll T_2$

In the integrand of (8.123) the exponential decays much faster than the trace, whose decay can therefore be neglected. This yields:

$$\frac{d}{dt}\langle \mathcal{H}'_D\rangle = -\tau_c\langle[[\mathcal{H}'_D, V], V]\rangle, \tag{8.124'}$$

or else, according to (5.43) and (8.121):

$$\frac{\mathrm{d}}{\mathrm{d}t}\langle \mathcal{H}'_{\mathrm{D}}\rangle = \tfrac{1}{4}\tau_{\mathrm{c}} \sum_{i,j} A_{ij}(\delta_i - \delta_j)^2 \langle I^i_+ I^i_- + I^i_- I^i_+ \rangle. \tag{8.125}$$

The time-derivative of the energy is proportional to the correlations of the transverse spin components. This enables one to guess the qualitative features of relaxation. At high nuclear spin temperature, that is at low dipolar energy, both the energy and the transverse correlations are proportional to the inverse spin temperature β, and the relaxation of $\langle \mathcal{H}'_{\mathrm{D}}\rangle$ is exponential. Below the transition, on the other hand, the energy increases very steeply, because of the energy associated with long-range order, whereas the transverse correlations which are short-ranged are expected to increase but slowly and even to decrease when the sublattice polarization becomes large.

Consequently, the relaxation rate defined as $-(\mathrm{d}/\mathrm{d}t\langle \mathcal{H}'_{\mathrm{D}}\rangle)/\langle \mathcal{H}'_{\mathrm{D}}\rangle$ is expected to remain approximately constant in the paramagnetic phase and to be a decreasing function of β in the ordered phase.

In a crude estimate, the 1st order restricted-trace approximation was used to compute $\langle \mathcal{H}'_{\mathrm{D}}\rangle$ and the transverse correlations. The corresponding predictions, as deduced from (8.125) are shown in Figs 8.19 and 8.20. They will be discussed together with the experimental results.

2. *Long correlation time:* $\tau_{\mathrm{c}} \gg T_2$

The result yielded by (8.123) in that case can most simply be obtained by the following physical argument. If the coupling (8.120) were time-independent (i.e. $F(t) = F(0) = 1$) the terms $\mathcal{H}'_{\mathrm{D}}$ and V would merge into a single thermal reservoir, and it would take a time of the order of T_2 for this reservoir to reach internal thermal equilibrium. If $F(t)$ varies slowly ($\tau_{\mathrm{c}} \gg T_2$) both terms are practically at thermal equilibrium at all times, and the relaxation rate of the reservoir is a weighted average of the partial relaxation rates of its components. The part V is the only one to be in contact with the lattice, and its relaxation rate is τ_{c}^{-1}, whence:

$$\frac{\mathrm{d}}{\mathrm{d}t}\langle \mathcal{H}'_{\mathrm{D}} + V\rangle = -\frac{1}{\tau_{\mathrm{c}}}\langle V\rangle,$$

or, assuming $\langle \mathcal{H}'_{\mathrm{D}}\rangle \gg \langle V\rangle$:

$$\frac{\mathrm{d}}{\mathrm{d}t}\langle \mathcal{H}'_{\mathrm{D}}\rangle = -\frac{1}{\tau_{\mathrm{c}}}\langle V\rangle. \tag{8.126}$$

The term V is the magnetic coupling of the nuclear spins with the dipolar field of the impurities.

The behaviour of $\langle V\rangle$ with the nuclear spin temperature does not differ qualitatively from the ^{43}Ca–^{19}F energy $\langle \mathcal{H}'_{IS}\rangle$ discussed in section **D**(c). Both

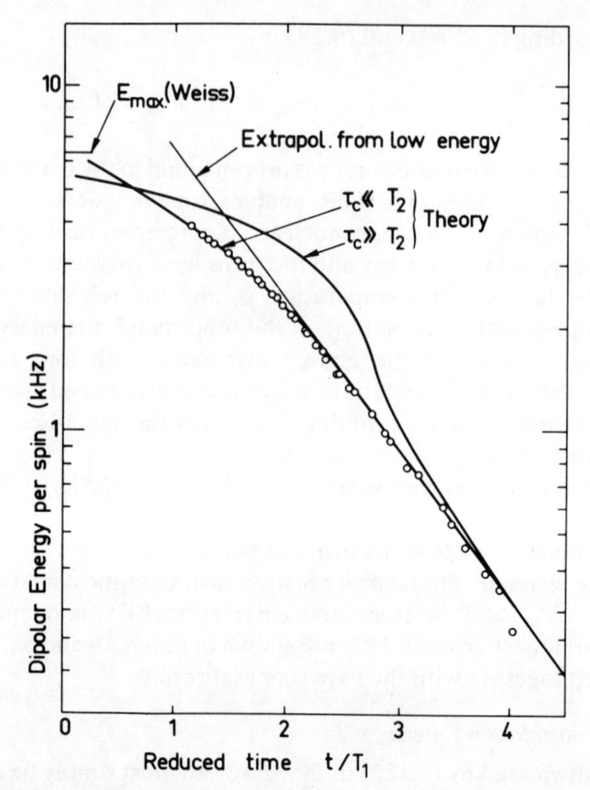

FIG. 8.19. Reduced time variation of dipolar energy in CaF_2 (antiferromagnetic structure III), under the effect of relaxation. Theoretical curves for $\tau_c \ll T_2$ and $\tau_c \gg T_2$, and experimental results. The value of T_1 is adjusted to the low-energy exponential decay. (After Goldman and Jacquinot, 1976*b*.)

measure the polarizability of the spin system I by inhomogeneous dipolar fields of impurities. In the paramagnetic phase $\langle V \rangle$ increases faster than $\langle \mathscr{H}'_D \rangle$, it goes through a maximum at the transition and then decreases. One then expects that the relaxation rate of the energy $\langle \mathscr{H}'_D \rangle$ will exhibit a maximum at the transition. The approximate calculation of the expectation values in (8.126) is described in Goldman and Jacquinot (1976*b*). The corresponding results are shown in Figs 8.19 and 8.20.

The experimental variation of the energy as a function of reduced time is shown in Fig. 8.19. These data are used to compute the points of Fig. 8.20. The only adjustable parameter is the relaxation time T_1 in the limit of small energy.

The results are consistent with a relaxation corresponding to $\tau_c \ll T_2$. The striking agreement with the theoretical curves pertaining to this case is accidental, since the transition energy obtained in section **D**(*c*) is higher than

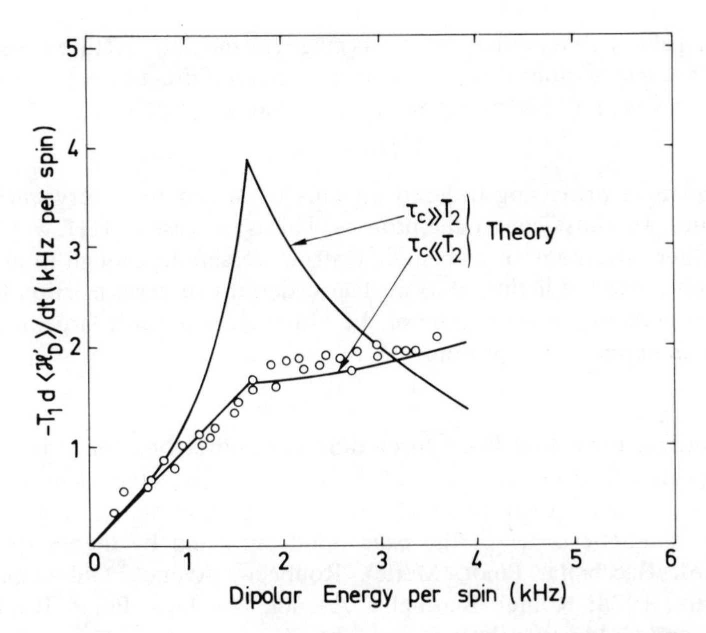

FIG. 8.20. Variation against dipolar energy of the reduced time-derivative of this energy in CaF_2. The points are deduced from those of Fig. 8.19, and the curves correspond to the same approximations in both figures. (After Goldman and Jacquinot, 1976*b*.)

that predicted by the 1st order restricted-trace approximation. The dipolar energy depends on both longitudinal and transverse spin correlations whereas, according to (8.125), its decay rate depends on transverse correlations only. There is already a slowing down of the relaxation rate below the transition. This is indicative of the fact that in the paramagnetic phase at the approach of the transition, the longitudinal correlations increase faster than the transverse correlations, which is physically reasonable.

(*e*) *Neutron diffraction in lithium hydride*

As explained in chapter 7, neutron diffraction on nuclear magnetically ordered substances is rendered possible by the fact that the nuclear interactions between neutrons and nuclei result in a spin-dependent neutron–nucleus scattering amplitude much larger than the nuclear magnetic scattering amplitude. In the terminology of chapter 7, this is equivalent to saying that the nuclear pseudomagnetic moment μ^* is much larger than the true magnetic moment μ.

According to Table 7.1, the pseudomagnetic moment of ^{19}F,

$$\mu^*(^{19}F) \simeq -0{\cdot}018\mu_B,$$

although about 12 times larger than its magnetic moment, is still too small to allow the observation of superstructure neutron diffraction lines in CaF_2.

The largest of all known moments μ^* is that of the proton,

$$\mu^*(^1H) \simeq 5{\cdot}4\mu_B,$$

and the most promising ordered systems on which to observe neutron diffraction are those containing protons. This is the case of LiH, which has the further advantage of a crystalline structure simple enough to allow a reasonably safe prediction of its nuclear ordered structures (section **B**(*d*)).

The pseudomagnetic moment of ^{7}Li, although noticeably large, is much smaller than that of the proton:

$$\mu^*(^7Li) \simeq -0{\cdot}63\mu_B.$$

According to section **B**(*d*), three different antiferromagnetic orderings are expected to occur in LiH. All three have been detected by neutron diffraction, which provides a direct verification of the existence of long range nuclear magnetic order of the exact kind predicted by theory (Roinel, Bouffard, Bacchella, Pinot, Meriel, Roubeau, Avenel, Goldman, and Abragam, 1978; Roinel, Bacchella, Avenel, Bouffard, Pinot, Roubeau, Meriel, and Goldman, 1980).

With the experimental arrangement used for neutron diffraction studies, one can observe the neutrons in a plane perpendicular to the external field, so that superstructure neutron diffraction lines can be observed only when the antiferromagnetic diffracting planes are parallel to the external field. They are shown as hatched planes for the structures I, II, and III in Fig. 8.3.

With $H_0 \| [001]$, the observed antiferromagnetic neutron reflection is the reflection 110, forbidden in the absence of ordering. It is the same reflection for both signs of the spin temperature: since $\mu^*(^1H) \gg \mu^*(^7Li)$, the neutrons are essentially sensitive to the spin orientations of the proton spins, which are the same in structures I and III. These structures differ only by the relative orientations of the proton and lithium spins: at $T>0$, they are parallel in each diffracting plane and antiparallel at $T<0$. The proof that these structures are different, although they show up by the same neutron reflection, is inferred from the difference of their transition entropies, much lower at $T>0$ than at $T<0$, as for the corresponding structures in CaF_2: the minimum initial proton polarizations experimentally necessary for observing a superstructure neutron line after ADRF are approximately:

$$p_0(^1H) \simeq 0{\cdot}60 \quad \text{at } T<0;$$
$$p_0(^1H) \simeq 0{\cdot}95 \quad \text{at } T>0.$$

Figure 8.21 shows a rocking curve of the 110 antiferromagnetic reflection observed at negative spin temperature, as compared to the rocking curve of the 220 crystalline reflection.

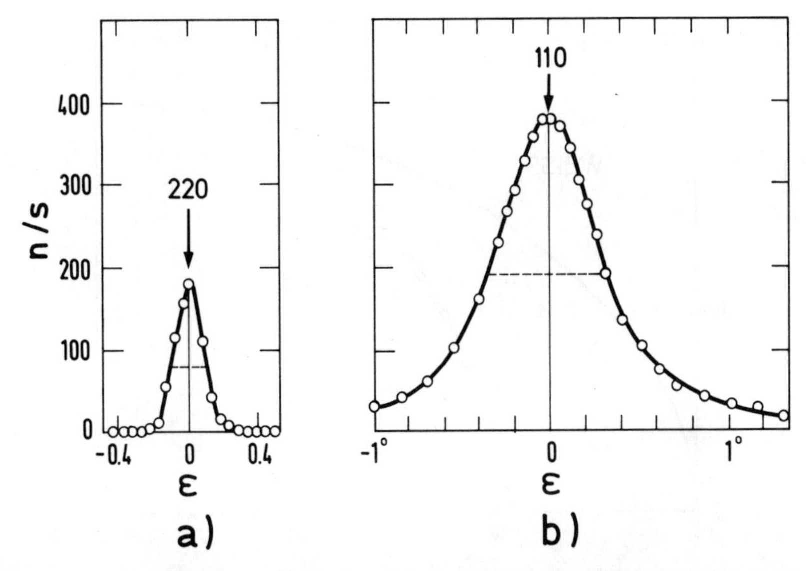

FIG. 8.21. Neutron diffraction rocking curves in LiH with $H_0 \| [001]$. The crystalline reflection 220 is observed with the unpolarized sample. The antiferromagnetic reflection 110 is observed after ADRF at $T<0$ of the polarized spins. (After Roinel, Bouffard, Bacchella, Pinot, Meriel, Roubeau, Avenel, Goldman, and Abragam, 1978.)

With $H_0 \| [011]$, the antiferromagnetic structure II at $T>0$ is detected by the observation of the neutron reflection 100.

Several neutron diffraction studies were performed on structure III of LiH, with $H_0 \| [001]$ and $T<0$. The sublattice polarization of the proton spins is obtained by comparing the area of the rocking curve of the antiferromagnetic reflection 110 with that of the crystalline reflection 220. The small contribution of the ^{7}Li spins to the intensity of the 110 line (~20 per cent) is estimated by assuming that the proton and lithium sublattice polarizations are in the ratio predicted by the Weiss-field theory. The largest measured proton sublattice polarization is close to 80 per cent.

The variation of the proton sublattice polarization is plotted in Figs 8.22 and 8.23 against the first moment $\mathcal{M}_1(^7\text{Li})$ of lithium and against the entropy as deduced from the initial spin polarizations. The difference between the observed variations and the predictions of the Weiss-field theory originate in the contributions of short-range spin correlations to $\mathcal{M}_1$ and S, which are ignored by this theory.

The measurement of the dipolar spin temperature as a function of the first moment $\mathcal{M}_1(^7\text{Li})$, by the cotanh transform method, was described in chapter 5, section **C**(*f*) (Fig. 5.10). By combining these results with those of Fig. 8.22, one obtains the variation of the proton sublattice polarization as a

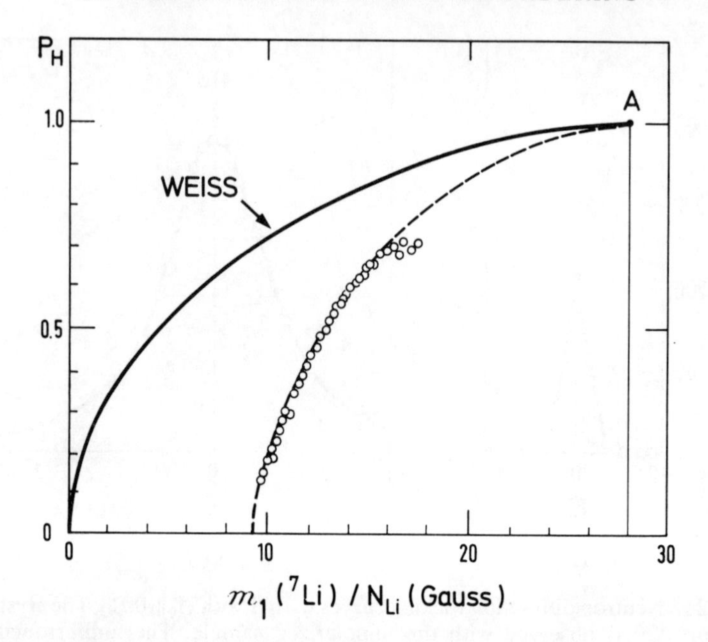

FIG. 8.22. Antiferromagnetic structure III in LiH. Proton sublattice polarization versus first moment $\mathcal{M}_1$ of the ^{7}Li absorption signal, together with Weiss field prediction. (After Goldman, Roinel, Meriel, Bouffard, Bacchella and Pinot, private communication.)

function of temperature, which is shown in the plot of Fig. 8.24. The polarization $p(\mathrm{H})$ is plotted against the reduced temperature T/T_c. The critical temperature T_c is adjusted so as to give the best fit with the Weiss-field prediction. Within this adjustment, the overall agreement is very satisfactory. The departure from Weiss-field behaviour observed in the vicinity of the transition can be accounted for by a small inhomogeneity of the spin temperature (of the order of 10 per cent) within the sample. Such a temperature inhomogeneity may result either from an inhomogeneity of the initial polarization or an inhomogeneity of the dipolar spin–lattice relaxation rate. It precludes for the time being the possibility of investigating in detail the critical behaviour of the sublattice polarization in the vicinity of T_c. (Meriel, Roinel, Goldman, Bacchella, Bouffard and Pinot, private communication.)

The adjusted value of the critical temperature, with a calibration uncertainty of about 10 per cent

$$T_c \exp \simeq -1{\cdot}16\ \mu\mathrm{K},$$

is not too far from the Weiss-field value,

$$T_c \,\mathrm{th} \simeq -1{\cdot}47\ \mu\mathrm{K}.$$

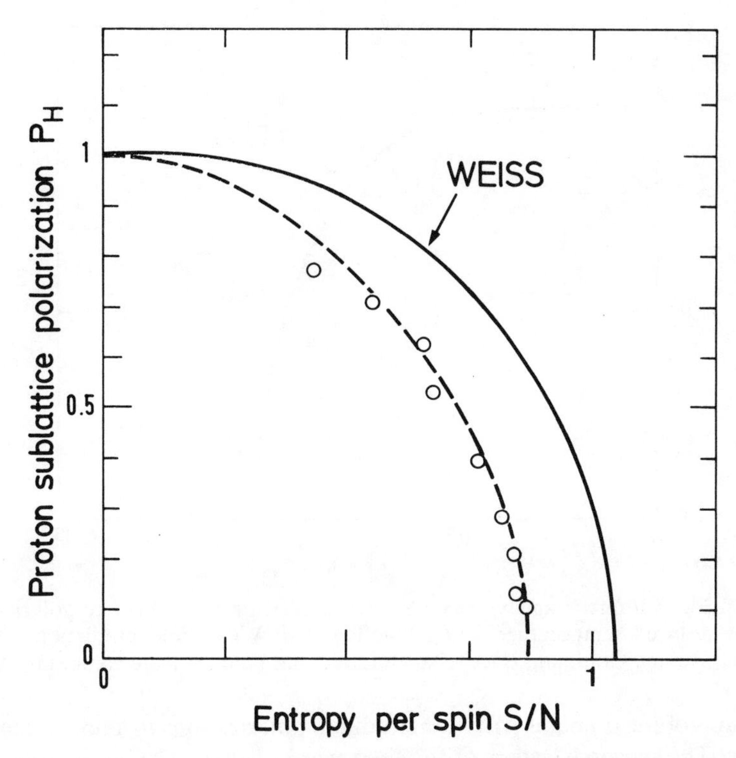

FIG. 8.23. Antiferromagnetic structure III in LiH. Proton sublattice polarization versus entropy, together with Weiss field prediction. (After Goldman, 1979.)

E. Ferromagnetism

We describe in this section an investigation of ferromagnets with domains.

(a) Characteristic properties of ferromagnets with domains

The Weiss-field theory is here applied to the example of the longitudinal ferromagnet at $T<0$ described in section **B**(c) for CaF_2, that is for a simple cubic lattice of identical spins $\frac{1}{2}$. We suppose for simplicity (and also in accordance with the experimental conditions) that the sample is spherical, so that $A(0)=0$.

In zero effective field, the structure IV described in section **B**(c) consists of thin slices perpendicular to the external field with opposite longitudinal polarizations in adjacent domains. The relative volumes of the two types of domains are equal, so that the bulk polarization vanishes.

When an external longitudinal field is applied, it is a characteristic property of ferromagnets that the relative size of the domains will vary through the motion of the domain walls. Let x and $(1-x)$ be the relative

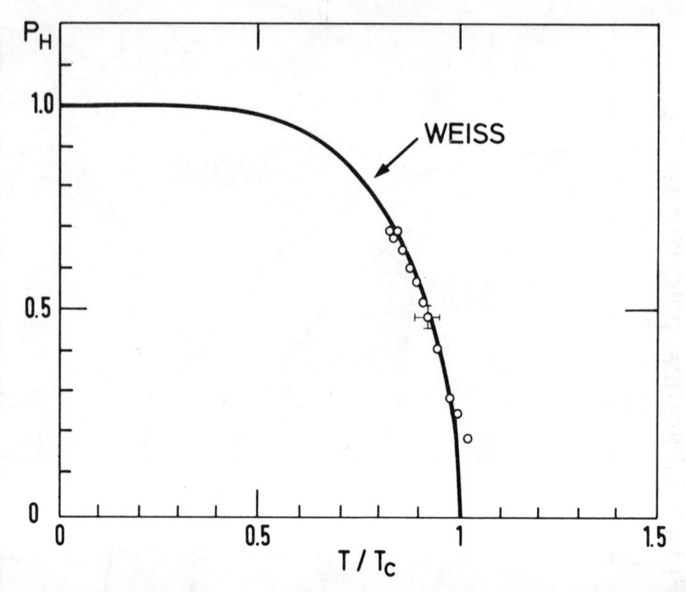

FIG. 8.24. Antiferromagnetic structure III in LiH. Proton sublattice polarization versus reduced temperature T/T_c, together with Weiss field prediction. (After Meriel, Roinel, Goldman, Bacchella, Bouffard and Pinot, private communication.)

volumes of the domains with (longitudinal) polarizations p_A and p_B, respectively. The successive steps of the theory are:

(i) calculation of the total frequencies experienced by the spins in the two types of domains;

(ii) calculation of the energy and of the entropy;

(iii) self-consistent determination of p_A, p_B and x by the condition that $-\beta F = S - \beta E$ is maximum. We consider here that the temperature of the system is fixed, which in the present case is more convenient than to fix initially the entropy or the energy. If necessary, it is possible to express afterwards the temperature as a function of energy or entropy.

We calculate first the Weiss frequencies ω_A and ω_B experienced by the spins I in the domains $\mathscr{A}$ and $\mathscr{B}$. We use two different methods which illustrate two different approaches.

In the first, closely related to the methods of section **B**(a)1 we start from the last relation (8.4) and use (8.13) and (8.14):

$$\begin{aligned}\omega_i &= \sum_j A_{ij} p_j = N^{-1/2} \sum_{j,\mathbf{k}} A_{ij} p(\mathbf{k}) \exp(-i\mathbf{k}\,.\,\mathbf{r}_j)\\ &= N^{-1/2} \sum_{\mathbf{k}} p(\mathbf{k}) \exp(-i\mathbf{k}\,.\,\mathbf{r}_i) \sum_j A_{ij} \exp[i\mathbf{k}\,.\,(\mathbf{r}_i - \mathbf{r}_j)]\\ &= N^{-1/2} \sum_{\mathbf{k}} A(\mathbf{k}) p(\mathbf{k}) \exp(-i\mathbf{k}\,.\,\mathbf{r}_i). \end{aligned} \tag{8.127}$$

In a longitudinal ferromagnet with domains,

$$p_i = p(\mathbf{r}_i) = \begin{cases} p_A & \mathbf{r}_i \in \mathcal{A} \\ p_B & \mathbf{r}_i \in \mathcal{B} \end{cases}.$$

In the non-uniform polarization $p_i = p(\mathbf{r}_i)$ we can separate a uniform average part:

$$\bar{p}_i = p_{av} = xp_A + (1-x)p_B, \tag{8.128}$$

and an oscillating part p_i' with vanishing average value:

$$\begin{aligned} p_i' &= (1-x)(p_A - p_B) \quad \mathbf{r}_i \in \mathcal{A} \\ &= -x(p_A - p_B) \qquad \mathbf{r}_i \in \mathcal{B}. \end{aligned} \tag{8.128'}$$

It is easy to verify from (8.128) and (8.128′) that:

$$\begin{aligned} p_i = p_i' + p_{av} &= p_A \quad \mathbf{r}_i \in \mathcal{A} \\ &= p_B \quad \mathbf{r}_i \in \mathcal{B} \end{aligned}$$

and that:

$$\overline{p_i'} = xp'(\mathcal{A}) + (1-x)p'(\mathcal{B}) = 0.$$

If the size of the domains is both large compared with the lattice spacing and small compared with the size R of the sample the Fourier expansion of $p(\mathbf{r}_i)$ will contain only the component $\mathbf{k} = 0$ relative to p_{av}, and wave vectors $\mathbf{k}$ with $\mathbf{k} \| \mathbf{H}_0$ and $ka \ll 1 \ll kR$ in the expansion of p_i'.

For these wave vectors, according to (8.40):

$$A(\mathbf{k}) = \frac{8\pi}{3}\gamma^2\hbar/(2a^3) = q, \tag{8.129}$$

is independent of $|\mathbf{k}|$.

From (8.127) and (8.129) we then get: $\omega_i = qp_i' + A(0)p_{av} = qp_i'$, or:

$$\omega_A = q(1-x)(p_A - p_B); \tag{8.130a}$$

$$\omega_B = -qx(p_A - p_B). \tag{8.130b}$$

The limitation of the wave vectors k in the summation (8.127) by the condition $ka \ll 1$ implies that both on the surface of the sample and between domains, the change in polarization from p_A to p_B, or from $p_{A,B}$ to zero must take place over more than one lattice spacing.

The same result is obtained by a standard magnetostatic calculation. The dipolar frequency experienced by a spin I_i in a given domain $\mathcal{A}$ is equal to the frequency originating from the other spins in the same domain, plus that produced by the spins in the rest of the sample. The first is given by $-\gamma H_{WA}$, where H_{WA} is the Weiss field seen by a spin in a cubic environment inside a flat disc. Its well-known value is:

$$H_{\mathrm{WA}} = -\frac{8\pi}{3} M_{\mathrm{A}},$$

where:

$$\begin{aligned} M_{\mathrm{A}} &= \tfrac{1}{2}\gamma\hbar n p_{\mathrm{A}} \\ &= \gamma\hbar p_{\mathrm{A}}/(2a^3), \end{aligned}$$

in a simple cubic lattice, is the nuclear magnetization inside the disc. Its contribution $-\gamma H_{\mathrm{WA}}$ to ω_{A} is $q p_{\mathrm{A}}$, where q is the same as in (8.129). The contribution of the rest of the sample is the same as if this external region had a homogeneous polarization equal to p_{av}. We have then:

$$\omega_{\mathrm{A}} = q p_{\mathrm{A}} + r p_{\mathrm{av}}. \tag{8.131}$$

The constant r is obtained by considering the particular case when $p_{\mathrm{A}} = p_{\mathrm{B}} = p$. Equation (8.131) yields then:

$$\omega_{\mathrm{A}} = (q + r)p.$$

Since in an homogeneously polarized sphere the dipolar field vanishes, one obtains:

$$r = -q.$$

By carrying this value into (8.131) and (8.128), one recovers eqn (8.130*a*). Equation (8.130*b*) is obtained in a similar fashion.

The external frequency being Δ, the total frequencies in the domains $\mathscr{A}$ and $\mathscr{B}$ are:

$$\omega_{\mathrm{A}}^{\mathrm{T}} = \Delta + q(1-x)(p_{\mathrm{A}} - p_{\mathrm{B}}); \tag{8.132a}$$

$$\omega_{\mathrm{B}}^{\mathrm{T}} = \Delta - qx(p_{\mathrm{A}} - p_{\mathrm{B}}). \tag{8.132b}$$

The energy and entropy per spin are equal respectively to:

$$\begin{aligned} E/N &= \tfrac{1}{2}\Delta\{x p_{\mathrm{A}} + (1-x) p_{\mathrm{B}}\} + \tfrac{1}{4}\{x\omega_{\mathrm{A}} p_{\mathrm{A}} + (1-x)\omega_{\mathrm{B}} p_{\mathrm{B}}\} \\ &= \tfrac{1}{2}\Delta\{x p_{\mathrm{A}} + (1-x) p_{\mathrm{B}}\} + \tfrac{1}{4} q x (1-x)(p_{\mathrm{A}} - p_{\mathrm{B}})^2; \end{aligned} \tag{8.133}$$

$$S/N = x s_{\mathrm{A}} + (1-x) s_{\mathrm{B}}; \tag{8.134}$$

where $s_{\mathrm{A}} = s(p_{\mathrm{A}})$ is of the form (8.59).

The conditions:

$$\frac{\partial}{\partial p_{\mathrm{A}}}(-\beta F) = \frac{\partial}{\partial p_{\mathrm{B}}}(-\beta F) = 0$$

yield the Weiss-field equations:

$$u_{\mathrm{A}} = \tanh^{-1}(p_{\mathrm{A}}) = -\tfrac{1}{2}\beta\omega_{\mathrm{A}}^{\mathrm{T}};$$
$$u_{\mathrm{B}} = \tanh^{-1}(p_{\mathrm{B}}) = -\tfrac{1}{2}\beta\omega_{\mathrm{B}}^{\mathrm{T}}. \tag{8.135}$$

The condition $\partial/\partial x(-\beta F) = 0$ yields:

$$s_{\mathrm{A}} - s_{\mathrm{B}} = \tfrac{1}{2}\beta(p_{\mathrm{A}} - p_{\mathrm{B}})\{\Delta + \tfrac{1}{2}q(1-2x)(p_{\mathrm{A}} - p_{\mathrm{B}})\} \tag{8.136}$$

or else, according to (8.132) and (8.135):

$$s_{\mathrm{A}} - s_{\mathrm{B}} = -\tfrac{1}{2}(u_{\mathrm{A}} + u_{\mathrm{B}})(p_{\mathrm{A}} - p_{\mathrm{B}}). \tag{8.137}$$

Equation (8.137) is obviously satisfied when:

$$p_{\mathrm{A}} = \pm p_{\mathrm{B}}. \tag{8.138}$$

It can be shown that it has no other solution.

The two possible structures are locally stable with respect to small changes of p_{A}, p_{B}, or x. The more stable of the two is found by comparing either their free energies at constant temperature, or else their energies at constant entropy, which will eventually prove simpler.

When $p_{\mathrm{A}} = p_{\mathrm{B}}$, the polarization is homogeneous in the whole sample and the quantity x is meaningless. We begin by considering the ferromagnetic phase where according to (8.138) $p_{\mathrm{A}} = -p_{\mathrm{B}} = p$. The polarizations of the two types of domains are opposite, irrespective of the value of x.

Since the entropy is an even function of polarization, (8.134) yields:

$$S/N = s(p),$$

which means that during an adiabatic variation of the external field the polarizations of the domains (when they exist) remain constant. Their absolute value is equal to the initial polarization p_{i}. Equation (8.136) becomes:

$$0 = \beta p\{\Delta + q(1-2x)p\}, \tag{8.139}$$

that is:

$$x = \tfrac{1}{2}(1 + \Delta/qp). \tag{8.140}$$

The relative domain sizes vary linearly with the effective frequency. Since $|x| \leqslant 1$, the ferromagnetic structure with domains can possibly exist only when:

$$|\Delta| \leqslant \Delta_{\mathrm{c}} = qp. \tag{8.141}$$

At constant entropy, the ferromagnet is stable with respect to the paramagnetic phase ($p_{\mathrm{A}} = p_{\mathrm{B}}$) if $E_{\mathrm{ferro}} - E_{\mathrm{para}} > 0$. According to (8.133) and

(8.140), we have:

$$E_{\text{ferro}} - E_{\text{para}} = (\Delta - qp)^2/4q \geqslant 0,$$

so that for $|\Delta| \leqslant \Delta_c$, it is the ferromagnet with domains which is stable. Δ_c is then the critical field of transition between paramagnetism and ferromagnetism with domains.

The total frequencies in the domains are, according to (8.132) and (8.140):

$$\omega_A^T = qp \quad \text{and} \quad \omega_B^T = -qp.$$

When p is constant, (8.135) yield $\beta = \text{const}$: the isentropic magnetization is also isothermal within the range of existence of the domains.

The bulk polarization (8.128) is:

$$p_{av} = xp - (1-x)p = (2x-1)p,$$

that is, according to (8.140):

$$p_{av} = \Delta/q. \tag{8.142}$$

It varies linearly with Δ in the whole range of existence of the domains. The parallel susceptibility is equal to:

$$\chi_{\parallel} = p_{av}/\Delta = 1/q. \tag{8.143}$$

It is independent of p, that is of temperature.

As for the transverse susceptibility $\chi_{\perp}$, its calculation is analogous to that of $\chi_{\perp}$ in an antiferromagnet (section **D**(a)1): in the presence of a small transverse frequency ω_1, the polarizations $\mathbf{p}_A$ and $\mathbf{p}_B$ of the domains tilt by a small angle θ, so as to become aligned with the total frequencies experienced by the various spins. The result is:

$$\chi_{\perp} = \chi_{\parallel} = 1/q, \tag{8.144}$$

i.e. the susceptibility is isotropic. This result is valid only when $A(0) = 0$ (i.e. in a sphere of CaF_2). In a system where $A(0) \neq 0$, it can be shown that the susceptibilities have the following unequal values:

$$\chi_{\parallel} = 1/[q - A(0)]; \tag{8.145a}$$

$$\chi_{\perp} = 1/[q + \tfrac{1}{2}A(0)]. \tag{8.145b}$$

In a system with one spin species only, and within the Weiss-field approximation, the constancy of $\chi_{\parallel}$ is a characteristic difference between ferromagnets with domains, and antiferromagnets where $\chi_{\parallel}$ decreases with decreasing temperature.

(*b*) *Resonance of* ^{43}Ca *in ferromagnetic* CaF_2

The ferromagnetic structure under investigation is structure IV of section **B**(*c*) (Fig. 8.1). It is produced by ADRF at negative temperature with the external field oriented with respect to the crystalline axes sufficiently far away from the direction [001]. In most of the studies, the field is oriented along [111].

The ^{43}Ca resonance is observed at a high polarization, obtained as follows. The solid effect, used for polarizing the ^{19}F spins is ineffective for polarizing the ^{43}Ca. An obvious limitation, among others, to the polarization process is the very slow spin diffusion within the dilute ^{43}Ca spin system, which limits severely the rate at which the polarization of the ^{43}Ca spins close to the paramagnetic impurities can be carried to the distant ones. The polarization of the calcium is obtained by the method described in chapter 5, section **C**(*g*): after the ADRF on the fluorine spins, an r.f. field is applied at the distance Δ from the Larmor frequency of the calcium spins. The latter become polarized as a result of the thermal coupling of their effective Zeeman interaction with the fluorine dipole–dipole reservoir. Polarizations as high as 80 per cent are readily produced. One then uses a non-saturating field to observe the ^{43}Ca signal during linear sweeps of the external field.

1. *Evidence for the production of ferromagnetism*

The observation of the resonance signal of the ^{43}Ca spins provides the most direct evidence that the ordered phase of the demagnetized fluorine spins is a ferromagnet with domains.

At low entropy of the fluorine system, the ^{43}Ca resonance signal is split into two well-separated lines, which proves that there are two types of calcium spins experiencing different and well-defined dipolar fields superimposed on the external field. This observation rules out the possibility of a two-sublattice antiferromagnetic structure for the fluorine spins, a structure for which the dipolar field at the calcium sites would be zero. In the antiferromagnetic state investigated in section **D**, the calcium signal is indeed observed to consist of a single line. The observation of two lines in a domain ferromagnet follows immediately from the existence of macroscopic domains with opposite fluorine polarizations, since calcium located in different domains will experience different non-zero dipolar fields produced by the fluorine spins.

Examples of ^{43}Ca resonance signals are shown in Figs 8.25 and 8.26. They were observed after ADRF of the fluorine spins at $T<0$, with H_0 parallel to [111] and [001], respectively. Also shown are the fluorine resonance signals observed in both cases. They look very much alike, and their observation does not allow us to discriminate between the two differently ordered structures.

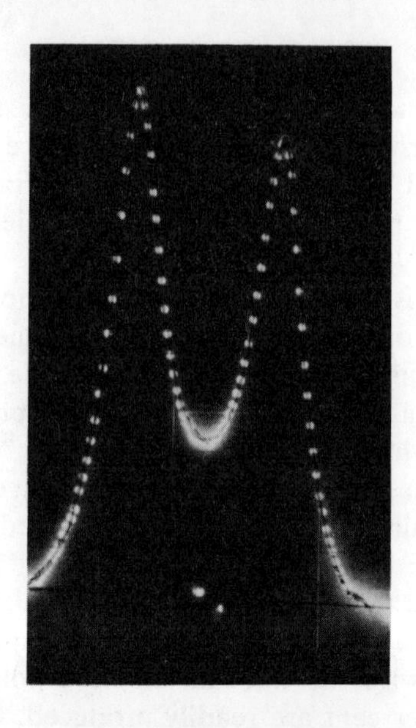

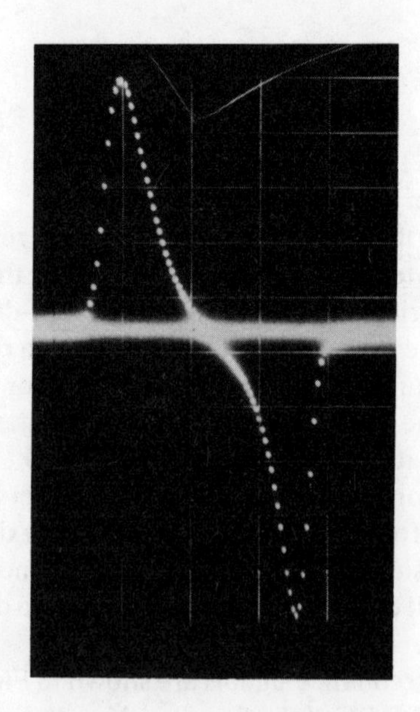

FIG. 8.25. ^{43}Ca and ^{19}F resonance signals (top and bottom, respectively) in CaF_2, in the ferromagnet with domains IV produced by ADRF with $H_0\|[111]$ and $T<0$. (After Goldman, 1977.)

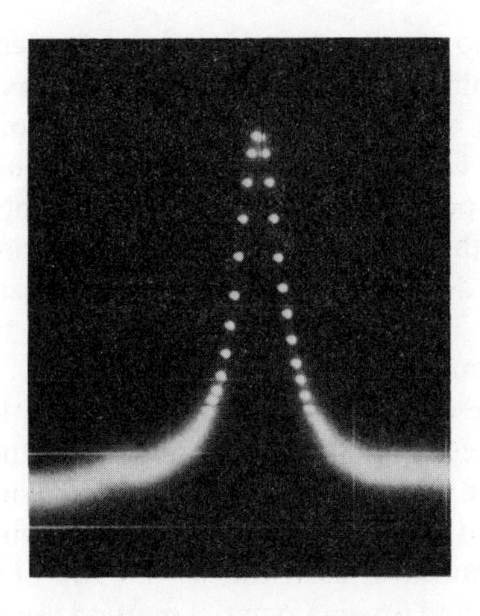

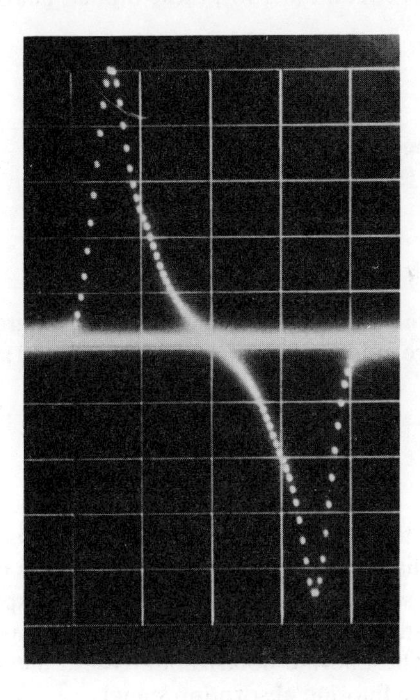

FIG. 8.26. ^{43}Ca and ^{19}F resonance signals (top and bottom, respectively) in CaF_2, in the antiferromagnetic structure III produced by ADRF with $H_0 \| [001]$ and $T < 0$. (After Goldman, 1977.)

Another observation is that the relative intensities of the calcium lines (normalized to equality in zero effective field) vary monotonically with effective field during an adiabatic remagnetization of the fluorine system. The intensity of each line being proportional to the number of calcium spins present in the corresponding type of domains, this observation shows that the relative sizes of the different domains vary continuously with external field, i.e. that these domains are of macroscopic size with respect to the interatomic spacing. This completes the proof that the fluorine state is a ferromagnet with domains.

Quantitatively, the dipolar fields experienced by the ^{43}Ca spins in each type of domains are equal to those experienced by the fluorine spins. This follows from the fact that because of the cubic environment of ^{43}Ca, the dipolar field produced at it by the parallel fluorine spins in a sphere centred on a calcium spin vanishes, as it did at the site of a fluorine spin. The calculation by the second (magnetostatic) method used in section **E**(a) is the same whether computing the dipolar field at a calcium or at a fluorine site, and it yields the same result.

According to (8.130) and (8.138), these dipolar fields are:

$$\begin{aligned} h_{\mathrm{A}} &= -2(q/\gamma_I)p(1-x); \\ h_{\mathrm{B}} &= 2(q/\gamma_I)px; \end{aligned} \tag{8.146}$$

where q is the same as in (8.129).

Using the value (8.140) for x, one finds:

$$\begin{aligned} h_{\mathrm{A}} &= (-qp+\Delta)/\gamma_I; \\ h_{\mathrm{B}} &= (qp+\Delta)/\gamma_I. \end{aligned} \tag{8.147}$$

The difference of the dipolar fields, and the centre of gravity of the calcium resonance signal are independent of Δ. We have indeed according to (8.147) and (8.146):

$$h_{\mathrm{A}} - h_{\mathrm{B}} = -2qp/\gamma_I; \tag{8.148}$$

$$h_0 = xh_{\mathrm{A}} + (1-x)h_{\mathrm{B}} = 0. \tag{8.148$'$}$$

Experimentally, the ADRF to zero field of the ^{19}F system is followed by a series of partial adiabatic remagnetizations to various effective fields. The calcium signal is observed after each step. The whole experiment takes place in a time short compared with $T_{1\mathrm{D}}$, so that the entropy remains approximately constant. The fluorine polarization is also measured at each step, from the area of the fluorine absorption signal.

The ^{43}Ca resonance intensities and frequency shifts, and the ^{19}F polarization, are shown in Figs 8.27, 8.28, and 8.29, respectively. The results are compared with the Weiss-field predictions, using the theoretical value

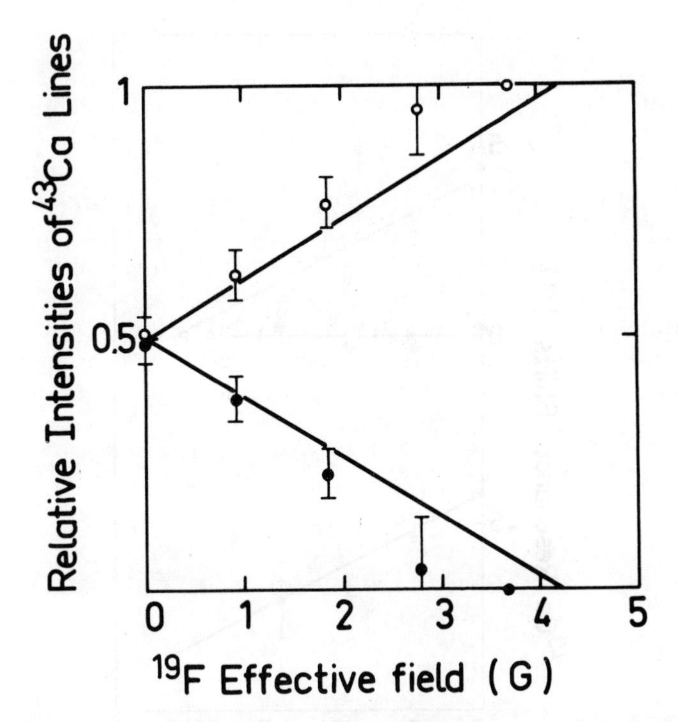

FIG. 8.27. Relative intensities of ^{43}Ca resonance lines against effective field in CaF_2 ferromagnetic with domains. The solid lines are the Weiss field predictions for $p = 0{\cdot}7$. (After Goldman, 1977.)

$q/\gamma_I \simeq 5{\cdot}5$ G, and the best-fit polarization $p = 0{\cdot}7$. The results exhibit a reasonable, qualitative fit to these predictions.

It must be stressed that the vanishing of the dipolar field at the calcium sites in an antiferromagnetic fluorine phase is the accidental consequence of the crystalline position of calcium in CaF_2. In lithium hydride, for instance, the rare spins ^{6}Li replace substitutionally the ^{7}Li abundant spins and in an antiferromagnetic state, ^{6}Li spins in different sublattices experience different dipolar fields, which shows up in a splitting of the ^{6}Li resonance line (Roinel, Bouffard, and Roubeau, 1978), and the distinction between ferromagnetism with domains and antiferromagnetism cannot be obtained by the same method as in CaF_2.

2. *Dipolar field as a function of energy*

The difference δh between the dipolar fields of domains with opposite fluorine polarizations is equal to the splitting between the calcium resonance lines (eqn (8.148)). It was measured as a function of the dipolar energy, as deduced from the moment $\mathcal{M}_1$ of the fluorine signal. The fluorine system was

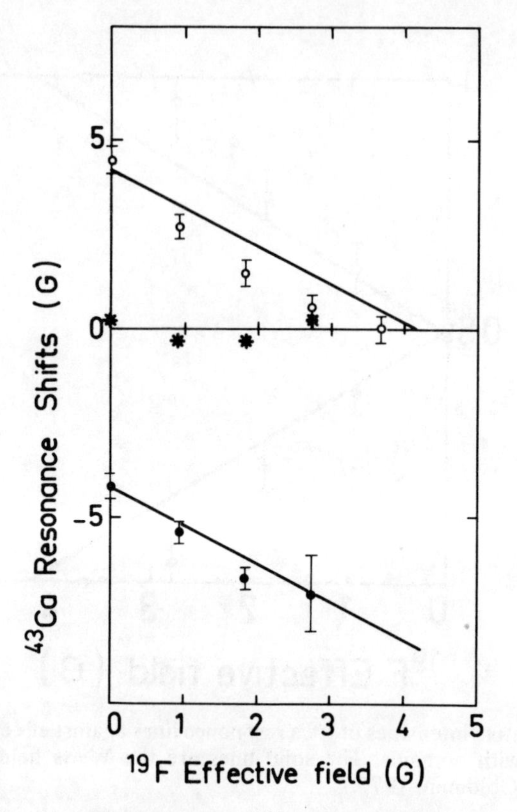

FIG. 8.28. Resonance frequencies of the ^{43}Ca resonance lines against effective field in CaF_2 ferromagnetic with domains. The stars are the centres of gravity of each pair of lines. Same theoretical approximation as in Fig. 8.27. (After Goldman, 1977.)

demagnetized to zero field, and its energy varied either under the effect of spin–lattice relaxation or by successions of slightly saturating passages. The field was along the direction [111].

According to the Weiss-field theory, the energy per spin is equal to:

$$E/N = \tfrac{1}{4}qp^2, \tag{8.149}$$

whence, according to (8.148):

$$\delta h = (4/\gamma_I)(qE/N)^{1/2}. \tag{8.150}$$

The results are shown in Fig. 8.30, together with the Weiss-field and the 1st order restricted-trace approximation. No attempt was made to take into account the effect of the distribution of dipolar fields from the impurities.

Within experimental uncertainty, the overall variation is qualitatively as expected.

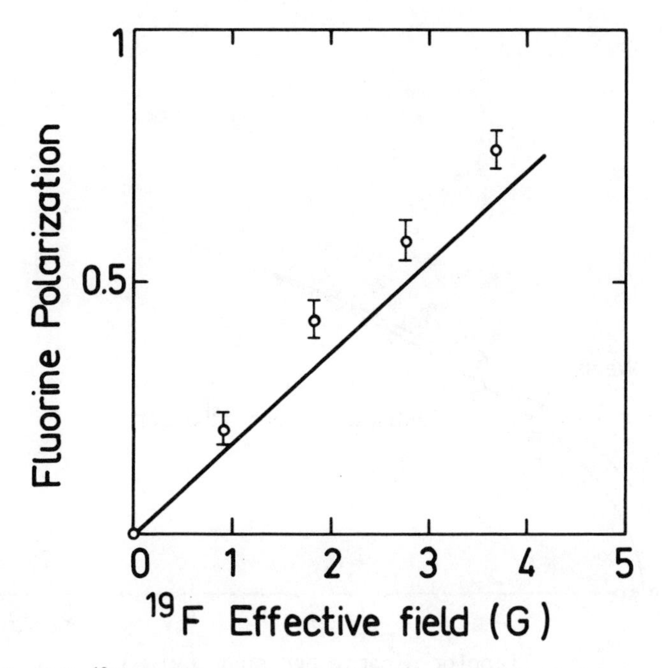

FIG. 8.29. Bulk ^{19}F polarization against effective field in CaF_2 ferromagnet with domains. Same theoretical approximation as in Fig. 8.27. (After Goldman, 1977.)

(c) *Ferromagnetic–antiferromagnetic transition in* CaF_2 *in zero field*

The regions of existence of the ferromagnetic and antiferromagnetic phases, as a function of the orientation of the external field with respect to the crystalline axes, are determined from the shape of the calcium resonance signal which consists respectively of two lines and one line.

After ADRF at negative temperature and polarization of the calcium spins, the sample was rotated around an axis perpendicular to H_0 and parallel to a binary crystalline axis, so that the field orientation swept a (110) plane from the [110] axis where the system is ferromagnetic to the [001] axis where it is antiferromagnetic. Let θ be the angle between H_0 and the axis [001]. The calcium signal consists of two lines (ferromagnetic structure IV) when θ varies from $\pi/2$ ($H_0 \| [110]$) to a critical angle comprised between 19° and 22°, where this signal undergoes an abrupt change. Instead of two lines symmetrically shifted with respect to the Larmor frequency, characteristic of a ferromagnet, one observes a strong central line with, however, two symmetrical satellites. This corresponds to a partial transformation of the ferromagnet IV to the antiferromagnet III. The satellites disappear at smaller values of θ, that is closer to [001]. The transformation is reversible when rotating the crystal back.

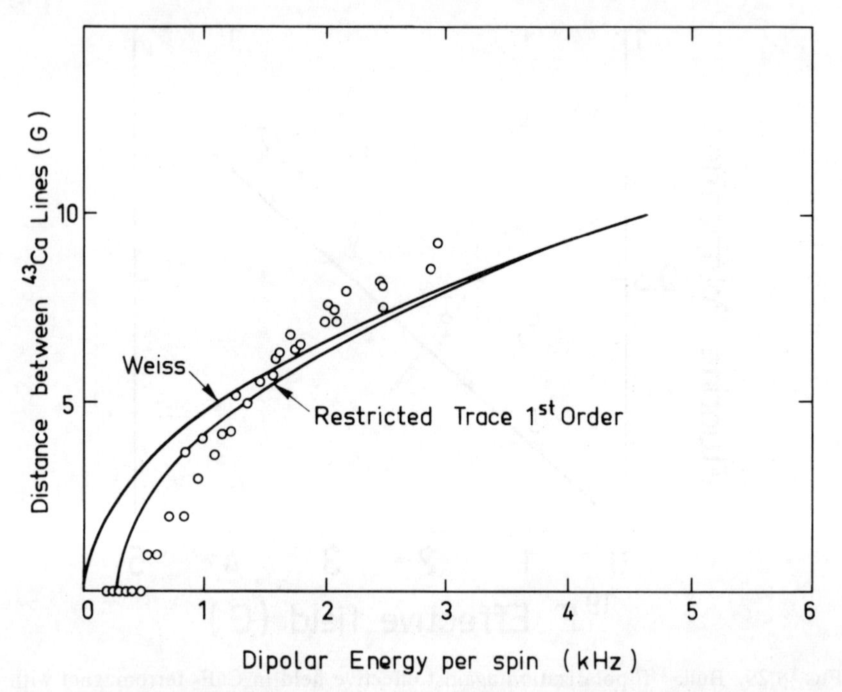

FIG. 8.30. Field separation between ^{43}Ca resonance lines against dipolar energy in CaF_2 ferromagnetic with domains. The curves correspond to the Weiss field and 1st order restricted-trace approximations. (After Goldman, 1977.)

In the Weiss-field approximation, for one spin species and in zero field, the energy per spin is $E = \frac{1}{4}qp^2$ and the entropy $S(p)$ given by (8.59) is a universal function of E/q. Two curves $S(E)$ relative to two structures with different values of q will not cross. When the sample is rotated from say [110] to [001], q_{ferro} is independent of θ but q_{AF} is not, and exceeds q_{ferro} for $\theta \leq \theta_0 = 17{\cdot}5°$. Whatever the entropy, the Weiss-field theory predicts ferromagnetism for $\theta > \theta_0$ and antiferromagnetism for $\theta < \theta_0$.

On the other hand, the observation of two satellite lines of ^{43}Ca for θ in the vicinity of 19–22° can be interpreted as a coexistence of a ferromagnetic and an antiferromagnetic phase, akin to a first-order transition.

Physically, it is well known that for a first-order transition there is a temperature where the two phases can coexist in different proportions. This implies that the curves $S(E)$ for the two phases must cross. Two such curves are shown in Fig. 8.31. The inverse temperature at the first-order transition is equal to the slope $\beta = \mathrm{d}S/\mathrm{d}E$ of the straight line AB, common tangent to the two entropy curves. The two phases coexist between A and B.

The corrections to the Weiss-field curves $S(E)$ for the structures III and IV under study, which would result in their crossing, arise from short-range

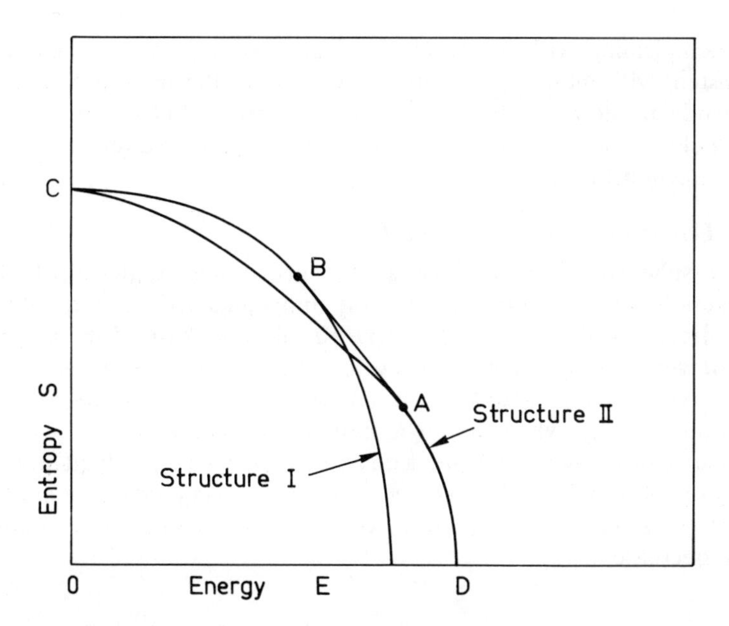

FIG. 8.31. Schematic energy–entropy diagram for two different structures. The stable states are: along CB: structure I; along AD: Structure II: along AB: both structures I and II in various proportions.

correlation effects. These corrections were calculated by the 1st order restricted-trace approximation (Goldman, 1977). The angle of transition is predicted to vary from about 20·5° to 22°, depending on entropy. For a given entropy, there is an angular interval where both phases can coexist. This angle interval is small, at most about 0·1°. For a given angle the theory predicts that the antiferromagnet is the more stable phase at the lower entropy, so that by warming up, the system is expected to undergo a transition from antiferromagnetism to ferromagnetism with domains before becoming paramagnetic. This prediction of the 1st order restricted trace approximation is disproved by experiment, which exhibits the opposite behaviour: at low entropy, the calcium signal consists of two lines and, as the system warms up, there appears a central line which grows at the expense of the initial lines.

(*d*) *Transverse susceptibilities*

Earlier in this book, the transverse susceptibility was defined by the linear response of the system to a small probing field H_1 applied along an axis $0x$ at right angle to the large d.c. field H_0:

$$\chi_{\perp} = p_x/\omega_1.$$

When dealing with longitudinal ordered structures, this definition is consistent with the usage current in magnetism: the small probing field is perpendicular to the spin orientations in the ordered phase which exists in zero field. It may not be so when studying a transverse ferromagnet (see below section **E**(d)1).

1. *Ferromagnetic calcium fluoride*

In a sphere of CaF_2, the Weiss-field theory predicts that in the ferromagnetic state, $\chi_\perp$ must be constant and equal to $1/q$ (eqn (8.144)). Measurements of $\chi_\perp$ by the Kramers–Kronig transform of the absorption signal, were performed at various orientations of the field H_0 in the (110) plane. As an example, Fig. 8.32 shows the experimental variation of $\chi_\perp$ as a function of $\langle \mathscr{H}'_D \rangle$, when the angle between $\mathbf{H}_0$ and [001] is $\theta = 65°$. The transverse susceptibility reaches a quasi-plateau, but with a slight monotonic increase. This increase is observed in the ferromagnetic phase at all field orientations, whereas in the antiferromagnetic phase one always observes a small decrease in $\chi_\perp$.

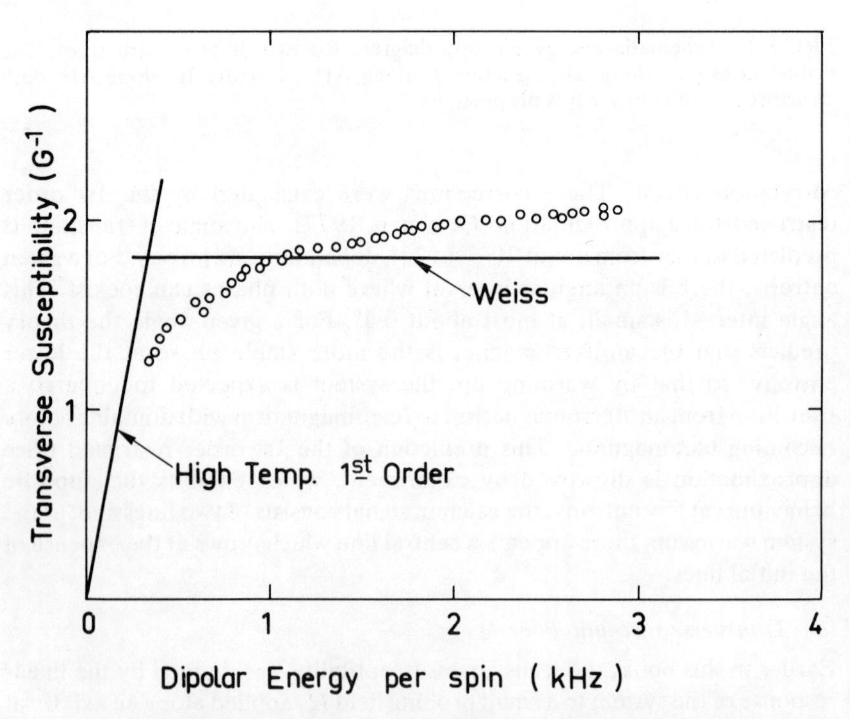

FIG. 8.32. Transverse susceptibility against dipolar energy in CaF_2 ferromagnetic with domains. The field H_0 is oriented in the (001) crystalline plane at an angle $\theta = 65°$ from the [001] axis. (After Goldman, 1977.)

The absolute value of $\chi_\perp$ is larger than expected from the Weiss field theory, as was already observed in the antiferromagnetic phase (section **D**(a)1). There is some arbitrariness in comparing the values of $\chi_\perp$ for different field orientations, since its variation with energy is not always the same. Figure 8.33 is a plot of $\chi_\perp$ as a function of the angle θ between $\mathbf{H}_0$ and [001], at constant energy $E/N = 2{\cdot}6$ kHz. Except for an excess of 10–20 per cent, it is not inconsistent with the predictions of the Weiss field theory. The experimental uncertainties preclude more detailed interpretations.

2. $Ca(OH)_2$ *at negative temperature*

The sample is in the form of a thin hexagon, and the field H_0 is parallel to the c-axis (direction $0Z$ in section **B**(e)). The ordered structure predicted in section **B**(e) at negative spin temperature is structure I: longitudinal ferro-

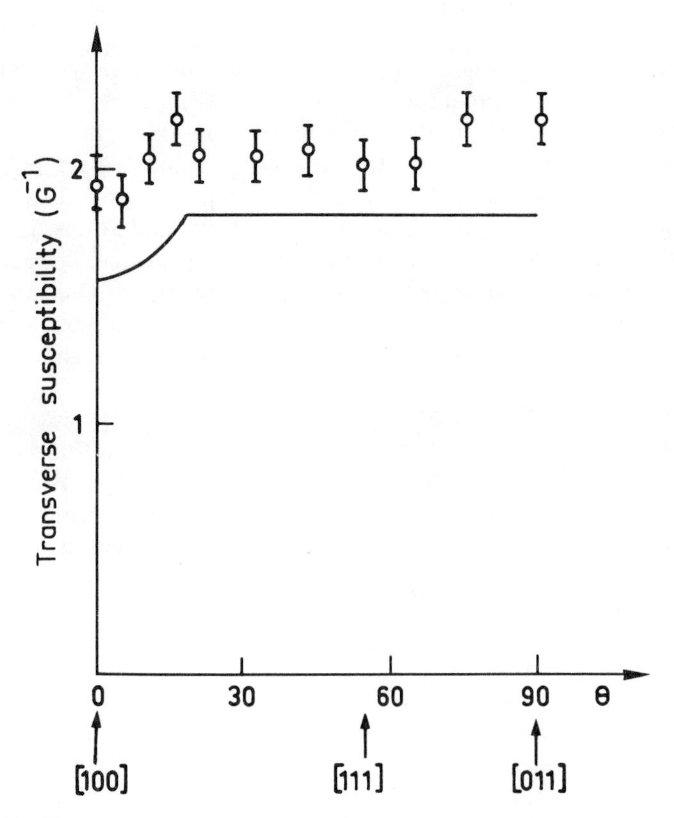

FIG. 8.33. Transverse susceptibility in CaF_2 demagnetized at $T<0$, against the angle between the external field H_0 and the crystalline axis [001], in the (011) plane, at constant energy $E/N = 2{\cdot}6$ kHz. The solid curve is the Weiss field prediction. (After Goldman, 1977.)

magnet analogous to structure IV in CaF_2, which is however nearly degenerate with a longitudinal antiferromagnetic structure.

The transverse susceptibility $\chi_\perp$ in the ferromagnetic state is of the form (8.145*b*):

$$\chi_{\perp\text{ferro}} = \frac{1}{q_{\mathrm{I}} + \frac{1}{2}A(0)}. \tag{8.151}$$

The Weiss field factor $q_I = \frac{1}{2}\lambda_I$ is positive, and the value of $A(0)$ in the thin hexagonal sample is also positive and equal to:

$$A(0) \simeq 0{\cdot}9 q_{\mathrm{I}}. \tag{8.152}$$

In the paramagnetic state, $\chi_\perp$ is, to the first order in β, of the form (8.101) for $T<0$:

$$\chi_{\perp\text{para}} = \frac{p_{\mathrm{i}}}{D + \frac{1}{2}A(0)p_{\mathrm{i}}}. \tag{8.153}$$

Since $A(0)>0$, the variation of $\chi_{\perp\text{para}}$ as a function of p_{i} is slower than linear. $\chi_{\perp\text{para}}$ and $\chi_{\perp\text{ferro}}$ are equal for $p_{\mathrm{i}} = p_0 = D/q_I \simeq 0{\cdot}20$.

The transverse susceptibility is measured by observing the centre of the fast passage signal. The results are shown in Fig. 8.34, together with the predictions of the Weiss-field theory and the 1st and 3rd order expansions in β for the paramagnetic phase (Marks, Wenkebach, and Poulis, 1979). The value of $\chi_\perp$ in the plateau is adjusted to the Weiss-field value. The agreement with the 3rd order paramagnetic expansion is then good. The existence of the plateau provides convincing evidence for the production of an ordering, but no clue as to whether it is ferromagnetic or antiferromagnetic.

3. $Ca(OH)_2$ *at positive temperature*

The sample and the field orientation $H_0 \| 0Z$ are the same as in the preceding case. The ordered structure predicted in section **B**(*e*) is transverse. It may be either a ferromagnetic structure with thin domains perpendicular to $0Z$, or a helical structure with $\mathbf{k} \| 0Z$. The Weiss-field factor is:

$$q_{\mathrm{II}} = \tfrac{1}{2}\lambda_{\mathrm{II}} = -\tfrac{1}{2}q_{\mathrm{I}}. \tag{8.154}$$

By analogy with negative temperatures, these structures are nearly degenerate with a transverse antiferromagnet where alternating proton planes would carry opposite magnetizations in the $0xy$ plane. It is worth recalling once more that all transverse structures are stationary in the frame rotating at the Larmor frequency and appear as precessing at that frequency in the laboratory frame.

In the paramagnetic state, $\chi_\perp$ is of the form (8.101) for $T>0$:

$$\chi_{\perp\text{para}} = \frac{-p_i}{D - \frac{1}{2}A(0)p_i}, \tag{8.155}$$

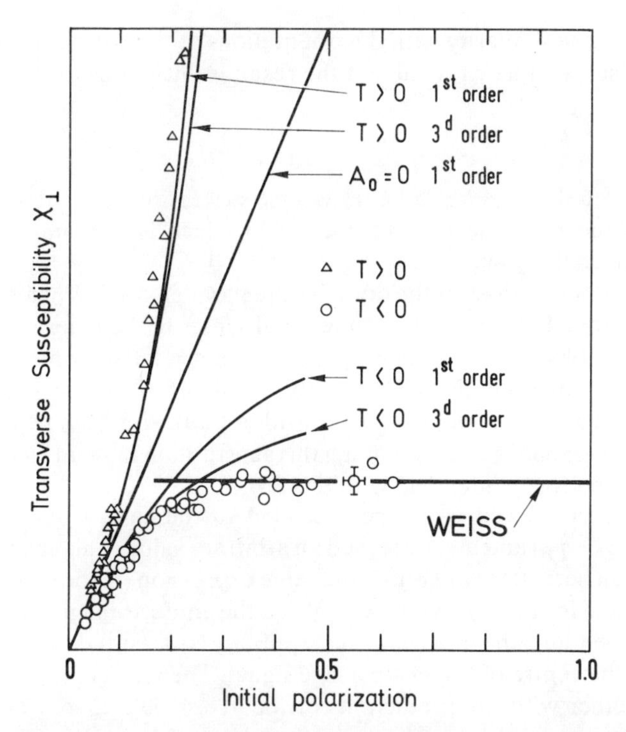

FIG. 8.34. Proton transverse susceptibility against initial polarization, after ADRF of a platelet of $Ca(OH)_2$ perpendicular to the c-axis. The field H_0 is parallel to the c-axis. The theoretical curves correspond to high-temperature approximations and at $T<0$, to the Weiss field approximation for the predicted longitudinal ferromagnet with domains. (After Marks, Wenkebach, and Poulis, 1979.)

and, since $D \simeq 0{\cdot}22A(0)$, the increase of $|\chi_{\perp\text{para}}|$ as a function of p_i is much faster than linear.

In order to calculate $\chi_\perp$ in the ordered phase, one must make an assumption on the orientations of the transverse polarizations with respect to ω_1. That such an assumption is necessary stems from the fact that in the absence of a transverse field it is only the relative orientations of the spin polarizations which are specified by the theory for a transverse structure. The orientation of the pattern of correlated transverse polarizations relative to any given direction in the rotating frame is undetermined. We shall first assume that the ordered state is a transverse ferromagnet with domains, and consider two extreme cases.

(i) The polarizations of the domains are nearly perpendicular to the direction $0x$ of ω_1, i.e.:

$$p_x^{\mathrm{A}} = p_x^{\mathrm{B}}; \qquad p_y^{\mathrm{A}} = -p_y^{\mathrm{B}}.$$

There is a close similarity with the orientations of the polarizations at $T<0$ in the presence of an r.f. field and the result is analogous to (8.151):

$$\chi_{\perp \mathrm{ferro}} = \frac{1}{q_{\mathrm{II}} + \frac{1}{2}A(0)}. \tag{8.156}$$

The susceptibility is indeed a transverse susceptibility in the sense that it measures the response to a probing field which is *perpendicular* to the domain magnetizations.

(ii) The polarizations of the domains are parallel to ω_1. What is measured then is a parallel susceptibility in the usual sense. In the external frequency ω_1, the two types of domains acquire different relative volumes, and the form of the susceptibility is analogous to (8.145a). There is however a difference: the Weiss field associated with a *transverse* bulk polarization is $-\frac{1}{2}A(0)$ rather than $A(0)$. By making this substitution in (8.145a), we obtain an expression for $\chi_\perp$ identical with (8.156).

The calculation can easily be extended to the case when the angle θ between $(\mathbf{p}_\mathrm{A} - \mathbf{p}_\mathrm{B})$ and $\boldsymbol{\omega}_1$ is assigned an arbitrary value. This calculation will not be given here. Its result is that the same expression (8.156) is valid for $\chi_\perp$. It is also valid for the transverse helical and the antiferromagnetic structures.

Experimentally, the transverse susceptibility is measured by observing the height of the centre of the fast passage signal. The results are shown in Fig. 8.35, together with the predictions of the Weiss field theory and the 3rd order expansion in β for the paramagnetic phase (Marks, Wenckebach, and Poulis, private communication). The value of $\chi_\perp$ is calibrated against the results obtained at negative temperature.

At low initial polarizations, the results are in fair agreement with the high temperature predictions, but at large initial polarization, $\chi_\perp$ continues to grow to values much larger than the Weiss-field prediction.

These results do suggest the actual occurrence of an ordering: The value of $\chi_\perp$ at $p_i = 0{\cdot}815$ is the same as that predicted by the high-temperature expansion for $p_i = 0{\cdot}575$ and $\beta D \simeq 0{\cdot}8$. If the system were still disordered, the 3rd order expansion in β should still not be too bad for such a value of βD, and it would be difficult to account for a short-range dipolar entropy so much lower than that predicted by this expansion. Further conclusions on the basis of the available experimental evidence would have a speculative character.

An unambiguous answer as to the nature of the structure produced could be provided by a neutron diffraction experiment.

(*e*) *Neutron diffraction in* LiH

Nuclear ferromagnetism was produced in a sample of LiH with its axis [110] oriented along the field H_0, by an ADRF at negative temperature. The domain structure is that of structure IV: thin slices perpendicular to H_0 (Fig. 8.3).

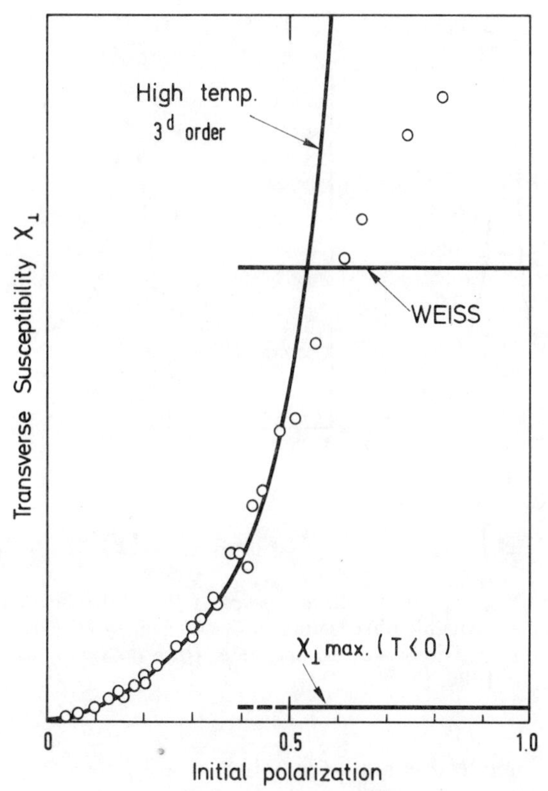

FIG. 8.35. Proton transverse susceptibility against initial polarization after ADRF at $T > 0$ of $Ca(OH)_2$. Same sample and field orientation as in Fig. 8.34. The solid curves correspond to the 3rd order high-temperature prediction for the paramagnetic phase, and to the Weiss field prediction for the expected transverse ferromagnet with domains. (After Marks, Wenckebach, and Poulis, private communication.)

In a neutron diffraction experiment, the occurrence of ferromagnetism can only show up by observing a crystalline neutron reflection. In the present case, it is the reflection 200. Figure 8.36 shows a 200 rocking curve observed (a) after dynamic polarization of the spins and (b) after their ADRF at $T < 0$ (Roinel *et al.* 1980). The production of ferromagnetic domains with quasi-macroscopic sizes by the ADRF shows up in Fig. 8.36(b) as a broad line superimposed on the narrow crystalline line originating from the spin-independent neutron scattering amplitudes. The comparison of the areas of the rocking curves (a) and (b) is complicated by the phenomenon of extinction of the diffracted neutron beam. Its discussion is outside the scope of this book. The analysis of the ferromagnetic neutron diffraction pattern, and of the information it yields on the size and shape of the domains, is deferred to section **G**(b).

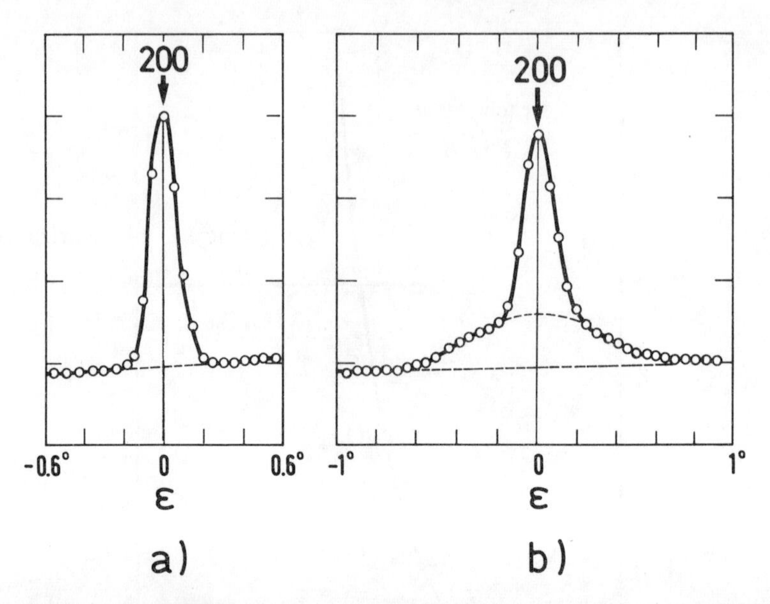

FIG. 8.36. Neutron diffraction rocking curves of the 200 reflection in LiH with $H_0 \| [011]$. (a) After dynamic polarization of the nuclear spins. (b) After ADRF at negative temperature. The broad line originates from the ferromagnetic domains. (After Roinel *et al.* 1980.)

F. Antiferromagnetism in non-zero field

The description of antiferromagnetism given in section **D** was restricted to zero external field or to small probing fields. The present section is devoted to the study of the field–entropy phase diagram of antiferromagnets. The field is longitudinal. Although it is necessary to use an r.f. field to ensure thermal equilibrium in the rotating frame, we neglect the contribution to the energy of the Zeeman coupling with this r.f. field and the transverse components of the spin polarizations. We use the Weiss-field approximation to discuss the qualitative features of the phase diagram. We treat as an example the case of the antiferromagnetic structure III in CaF_2 (section **B**(*c*): $H_0 \| [001]$, $T < 0$) in a spherical sample, for which $A(0) = 0$.

In the following we shall have to discuss the nature of the various transitions that occur in the spin system. It may be well at this stage to define clearly what is meant by first order and second order transitions. Let us consider a transition between two different phases and suppose that the values of the external parameters are infinitely close to those at which the transition takes place. If under these conditions the two phases are infinitely close to each other, the transition is of second order. If the two phases are distinctly different, the transition is of first order.

(a) Second order transition

In a simple-minded approach, we suppose that the system can be either antiferromagnetic or paramagnetic, and we compare at constant entropy the energies of the paramagnetic and the antiferromagnetic structures as a function of the external field. The entropy is fixed by the initial polarization p.

For a paramagnetic structure in an external frequency Δ, the polarization of each spin is equal to p and the energy is equal to:

$$E_{\text{para}} = \tfrac{1}{2}N\Delta p. \tag{8.157}$$

The steps in the calculation for the antiferromagnetic structure are outlined below.

(i) Let p_A and p_B be the sublattice polarizations. Each sublattice contains $N/2$ spins. The entropy is the same as in the paramagnetic state:

$$s(p_A) + s(p_B) = 2s(p). \tag{8.158}$$

When p_A has a given positive value, with say $p_A \geqslant p$, we can calculate p_B so as to satisfy eqn (8.158) (Fig. 8.37). At this stage both values $p_B = \pm|p_B|$ are acceptable. As a consequence of the curvature of the function $s(p)$ (Fig. 8.37) we always have:

$$0 \leqslant p_A + p_B \leqslant 2p. \tag{8.159}$$

(ii) For each couple of values of p_A and p_B chosen in accordance with (8.158), we can write down the value of Δ consistent with the Weiss-field eqns (8.56). With the notations:

$$u_A = \tanh^{-1}(p_A); \qquad u_B = \tanh^{-1}(p_B),$$

eqns (8.54), (8.54′), (8.55), and (8.56) yield:

$$\frac{u_A}{u_B} = \frac{-\frac{1}{2}\beta\omega_A^T}{-\frac{1}{2}\beta\omega_B^T} = \frac{\omega_A^T}{\omega_B^T} = \frac{\Delta + \frac{1}{2}A(\mathbf{k}_0)(p_A - p_B)}{\Delta - \frac{1}{2}A(\mathbf{k}_0)(p_A - p_B)}, \tag{8.160}$$

whence:

$$\Delta = \tfrac{1}{2}A(\mathbf{k}_0)(p_A - p_B)(u_A + u_B)/(u_A - u_B), \tag{8.161}$$

which is positive as a consequence of the choice $p_A \geqslant p > 0$. We can extract from (8.158) and (8.161) the values of p_A and p_B as functions $p_A = f_A(p, \Delta)$, $p_B = f_B(p, \Delta)$ and plot them against Δ for a given value of p, that is of the entropy. As an example, Fig. 8.38 shows the isentropic variation of $p_A(\Delta)$ and $p_B(\Delta)$ for a polarization $p = 0{\cdot}35$.

(iii) We compare the energy (8.157) of the paramagnetic phase with that of the antiferromagnetic phase:

$$\begin{aligned} E_{AF} &= \tfrac{1}{4}N\Delta(p_A + p_B) + \tfrac{1}{8}N(\omega_A p_A + \omega_B p_B) \\ &= \tfrac{1}{4}N\Delta(p_A + p_B) + \tfrac{1}{16}NA(\mathbf{k}_0)(p_A - p_B)^2. \end{aligned} \tag{8.162}$$

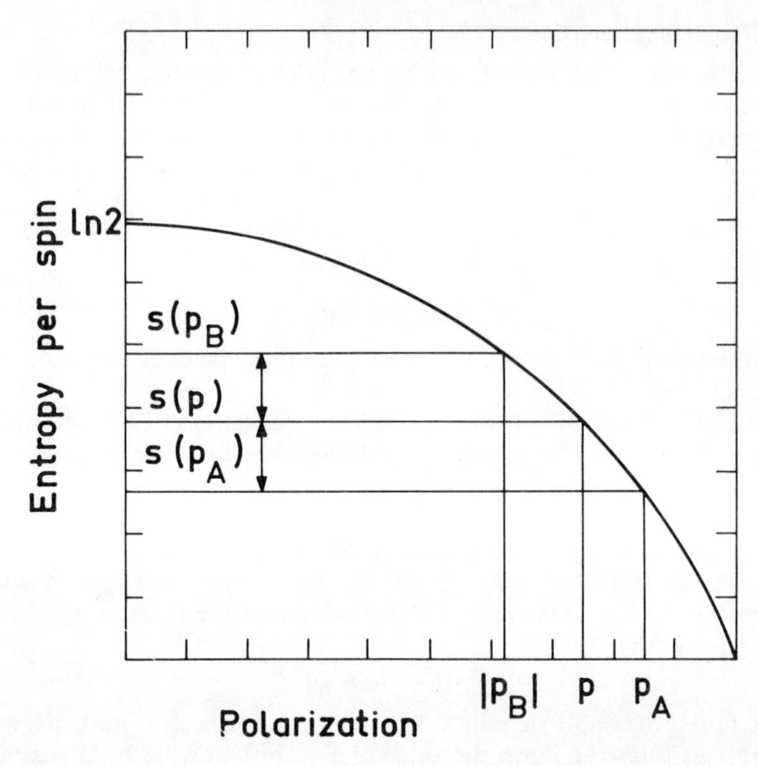

FIG. 8.37. Antiferromagnetism in non-zero field in CaF_2. Selection of a couple of values of sublattice polarizations p_A and $|p_B|$ corresponding within the Weiss field approximation to the same entropy as for the initial polarization p. (After Goldman, Chapellier, Vu Hoang Chau, and Abragam, 1974.)

The temperature being negative, the stable structure is that whose energy is the highest.

Let us compare the variations of the energies of the two phases corresponding to an isentropic increase by $d\Delta$ of the external frequency Δ. According to (8.157) and (8.162) we have:

$$dE_{para} = \tfrac{1}{2}Np\,d\Delta; \tag{8.163}$$

$$dE_{AF} = \tfrac{1}{4}N(p_A + p_B)\,d\Delta + \tfrac{1}{4}N\{\Delta(dp_A + dp_B) + \tfrac{1}{2}A(\mathbf{k}_0)(p_A - p_B)(dp_A - dp_B)\}; \tag{8.163'}$$

whereas, according to (8.158), dp_A and dp_B are related through:

$$\begin{aligned} 0 &= \frac{ds(p_A)}{dp_A}\,dp_A + \frac{ds(p_B)}{dp_B}\,dp_B \\ &= -u_A\,dp_A - u_B\,dp_B. \end{aligned} \tag{8.164}$$

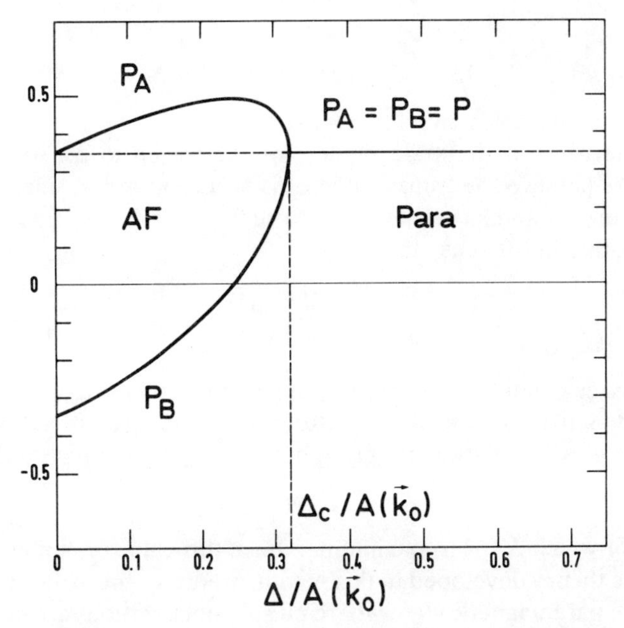

FIG. 8.38. Weiss-field prediction for the isentropic variation of the sublattice polarizations against reduced effective frequency in antiferromagnetic CaF_2. The entropy corresponds to the initial polarization $p = 0{\cdot}35$. The transition to paramagnetism is of second order.

It is easily found that, as a consequence of (8.161) and (8.164), the last term on the right-hand side of (8.163′) vanishes, and we obtain:

$$dE_{AF} = \tfrac{1}{4}N(p_A + p_B)\,d\Delta \tag{8.163''}$$

whence, according to (8.163) and (8.159):

$$0 \leqslant \frac{dE_{AF}}{d\Delta} \leqslant \frac{dE_{para}}{d\Delta}. \tag{8.165}$$

Provided that at constant p, the polarizations p_A and p_B are uniform functions of Δ, as in Fig. 8.38, $E_{AF}(\Delta)$ and $E_{para}(\Delta)$ are well-defined functions of Δ which, according to (8.165), can be equal at one value of Δ only. It is obviously the value Δ_c at which $p_A = p_B = p$. From (8.161), by making $p_A \to p_B$ we get:

$$\Delta_c(p) = \tfrac{1}{2}A(\mathbf{k}_0)\frac{dp}{du} \times 2u = A(\mathbf{k}_0)(1-p^2)\tanh^{-1}(p). \tag{8.166}$$

Since:

$$E_{AF}(0) = \tfrac{1}{4}NA(\mathbf{k}_0)p^2 > E_{para}(0) = 0,$$

one has:

$$E_{\mathrm{AF}}(\Delta) \geqslant E_{\mathrm{para}}(\Delta) \quad \text{for } 0 \leqslant \Delta \leqslant \Delta_c,$$

where the equality holds only for $\Delta = \Delta_c$.

The antiferromagnetic phase is therefore stable up to Δ_c, which is the transition frequency. The transition at Δ_c is of second order, since immediately below Δ_c, p_{A} and p_{B} are infinitely close to p. For $\Delta = \Delta_c$, E_{AF} and E_{para} match smoothly in the sense that:

$$\left.\frac{\mathrm{d}E_{\mathrm{AF}}}{\mathrm{d}\Delta}\right|_{\Delta_c} = \left.\frac{\mathrm{d}E_{\mathrm{para}}}{\mathrm{d}\Delta}\right|_{\Delta_c}$$

The above calculation assumes explicitly that p_{A} and p_{B}, as deduced from (8.158) and (8.161), are uniform functions of Δ. The result of a straightforward but lengthy calculation not given here is that this happens only when:

$$2p \tanh^{-1} p \leqslant 1, \quad \text{or} \quad p \leqslant 0{\cdot}647,$$

the value for which $\Delta_c(p)$ is maximum. Above this value of p, that is at low entropy, the theory developed in this section predicts a first order transition between the paramagnetic state where all spin polarizations are equal to p, and an antiferromagnetic state with $p_{\mathrm{A}} \neq p_{\mathrm{B}} \neq p$.

We show next that this transition is more complicated than predicted by the simple-minded approach used here.

(*b*) *The mixed state*

1. *Qualitative discussion*

In the preceding section we have made the assumption that the system was either entirely antiferromagnetic or entirely paramagnetic and we have systematically used a variational approach to determine the stable structure which at constant entropy would maximize the energy (minimize at $T > 0$).

As in all variational problems, extending the range of trial functions (structures in the present case), may lead to a higher value of the energy and to the discovery of a more stable structure. The choice we take, which will be substantiated in section **F**(*d*), is to consider the possibility of coexistence of both paramagnetic and antiferromagnetic regions in the sample.

In order to understand why a uniform isentropic change in the microscopic structure of the whole sample is unlikely to occur at a well-defined field at low entropy, it is instructive to consider the limiting case of exceedingly low negative temperatures. At such temperatures, in zero field the sublattice polarizations p_{A} and p_{B} are very nearly ± 1 and will not be modified even by a sizeable external frequency until it becomes comparable to the Weiss frequency $A(\mathbf{k}_0)p \simeq A(\mathbf{k}_0)$. (This is of course in agreement with the vanishing of the longitudinal susceptibility of antiferromagnets at zero temperature but it holds even for sizeable external frequencies.) Until then the Zeeman energy of the antiferromagnet remains zero and its total energy remains the same as in zero field:

$$E = \tfrac{1}{4}NA(\mathbf{k}_0)p^2 \simeq \tfrac{1}{4}NA(\mathbf{k}_0).$$

To get a contribution to the Zeeman energy we replace tentatively a very small fraction ε of the antiferromagnet, in the shape of a slice very thin but macroscopic, perpendicular to the field H_0, by an equal volume of paramagnet with the same absolute value of the polarization $p \simeq 1$. In the Weiss-field approximation this does not change the entropy of the sample.

The resulting change in the energy of the system will be:

(a) A decrease $\tfrac{1}{4}N\varepsilon A(\mathbf{k}_0)p^2 \simeq \tfrac{1}{4}N\varepsilon A(\mathbf{k}_0)$ in the energy of the antiferromagnet through suppression of an antiferromagnetic slice.

(b) An increase $\tfrac{1}{4}N\varepsilon qp^2 \simeq \tfrac{1}{4}N\varepsilon q$, corresponding to the Weiss-field energy of the paramagnetic slice, in which the Weiss frequency is qp. This increase of energy is maximum when q is maximum, that is when the orientation of the slice is perpendicular to H_0, in accordance with the choice made. The value of q is then $A(\mathbf{k})_{\max}$ (eqn (8.49)). It is smaller than $A(\mathbf{k}_0) = A(\mathbf{k}_{\mathrm{III}})$ (eqn (8.46)) which, as seen in section **B**(a)1, is a prerequisite for the antiferromagnetic structure to be the stable one in zero field, and the present increase of energy is smaller than the decrease of antiferromagnetic energy. Actually the ratio is:

$$q/A(\mathbf{k}_0) = \frac{8\pi}{3 \times 9{\cdot}687} = 0{\cdot}865.$$

(c) An increase $\tfrac{1}{2}N\varepsilon\,\Delta p \simeq \tfrac{1}{2}N\varepsilon\,\Delta$ representing the Zeeman energy of the paramagnetic slice.

In spite of the long range of the dipolar interactions there is no contribution from magnetic interactions between the paramagnetic slice and the antiferromagnet: the field produced by the extremely thin paramagnetic slab outside of its boundaries is vanishingly small, according to standard magnetostatics, and it does not affect the antiferromagnetic sublattice polarizations which remain opposite. Under these conditions the antiferromagnet behaves outside of its boundaries as a completely non-magnetic material which produces no dipolar field in the paramagnetic slab.

The change in energy due to the introduction of the paramagnetic slice is:

$$\delta E = \tfrac{1}{4}N\varepsilon p\{2\Delta - p[A(\mathbf{k}_0) - q]\}.$$

The introduction of the paramagnetic slice stabilizes the system as soon as this change in energy is positive, that is as soon as Δ exceeds the threshold value:

$$\begin{aligned}\Delta_{\mathrm{c}}^{(1)} &= \tfrac{1}{2}A(\mathbf{k}_0)[1 - q/A(\mathbf{k}_0)]p \\ &= 0{\cdot}068A(\mathbf{k}_0)p \simeq 0{\cdot}068A(\mathbf{k}_0). \qquad (8.167)\end{aligned}$$

A second threshold frequency $\Delta_c^{(2)}$ can be defined as the onset of a heterogeneous phase as Δ is reduced from a very large value. Starting from a homogeneously polarized sample of polarization $p=1$ and of spherical shape, we introduce now a thin *antiferromagnetic* slice with sublattice polarizations $\pm p=\pm 1$. The slice is perpendicular to H_0 and its relative volume is ε.

The change in energy of the system is made of:

(a) The antiferromagnetic energy

$$\tfrac{1}{4}N\varepsilon A(\mathbf{k}_0)p^2.$$

(b) A change $\frac{1}{4}N\varepsilon qp^2$ due to the removal of the paramagnetic slice, which can be described as an *addition* of a paramagnetic slice of opposite polarization. This value $\frac{1}{4}N\varepsilon qp^2$ holds only for a spherical sample.

(c) A reduction $-\frac{1}{2}N\varepsilon\,\Delta p$ in the Zeeman energy of the sample.

Altogether:

$$\delta E=\tfrac{1}{4}N\varepsilon p\{[A(\mathbf{k}_0)+q]p-2\Delta\},$$

which is positive when Δ is smaller than:

$$\begin{aligned}\Delta_c^{(2)}&=\tfrac{1}{2}A(\mathbf{k}_0)[1+q/A(\mathbf{k}_0)]p\\&=0{\cdot}932A(\mathbf{k}_0)p=0{\cdot}932A(\mathbf{k}_0).\end{aligned}\tag{8.168}$$

We must check however that the assumption of an antiferromagnet with zero net polarization in the applied frequency $\Delta_c^{(2)}$ is consistent. It is so actually: to the frequency $\Delta_c^{(2)}$ we must add the frequency produced in the antiferromagnetic slice by the surrounding paramagnet, which is equal to $-q$. The total uniform field seen by the antiferromagnetic slice is:

$$\Delta_c^{(2)}-q=\tfrac{1}{2}[A(\mathbf{k}_0)-q]<A(\mathbf{k}_0).$$

At zero temperature, it does not affect the polarizations ± 1 of the sublattices and there is no contribution to the energy from the magnetic interaction between the antiferromagnetic slice and the surrounding paramagnet.

2. *Quantitative description*

The previous discussion was for the sake of simplicity restricted to zero temperature and incipient paramagnetism or antiferromagnetism. It made it plausible that during a remagnetization the transition from an antiferromagnetic to a paramagnetic phase passes through the formation of a heterogeneous phase where both paramagnetic and antiferromagnetic regions are present.

Lifting the simplifying restrictions of the last section we assume now that the remagnetization of the antiferromagnetic structure III ($H_0\|[001]$; $T<0$) passes through the formation of a so-called mixed state. This state involves the coexistence of paramagnetic and antiferromagnetic slices or domains, perpendicular to the [001] axis and spread uniformly on a macroscopic scale through the sample.

The relative volume of the paramagnetic domains is x and their spin polarization is p_C. The relative volume of the antiferromagnetic domains is $(1-x)$ and their sublattice polarizations p_A and p_B need not be opposite anymore.

In contrast with the limiting cases described in the last section, the interactions between the various domains do contribute to the energy of the system. The calculation of the total frequencies experienced by the various spins is analogous to that leading from (8.127) to (8.132). The most convenient procedure is to use the model of section **B**(c): one computes the dipolar frequencies experienced by a spin I_i by considering first the contribution from the other spins in the same domain (the so-called internal frequency), and then treating the rest of the sample as if it had a homogeneous polarization. This last contribution combines with the applied frequency to form the so-called external frequency.

The results for a spherical sample are the following:

$$\begin{aligned}\omega_A^T &= \Delta+\omega_A = \Delta - xq[p_C - \tfrac{1}{2}(p_A+p_B)] + \tfrac{1}{2}A(\mathbf{k}_0)(p_A-p_B);\\ \omega_B^T &= \Delta+\omega_B = \Delta - xq[p_C - \tfrac{1}{2}(p_A+p_B)] - \tfrac{1}{2}A(\mathbf{k}_0)(p_A-p_B); \quad (8.169)\\ \omega_C^T &= \Delta+\omega_c = \Delta + (1-x)q[p_C - \tfrac{1}{2}(p_A+p_B)];\end{aligned}$$

where $q = A(\mathbf{k})_{\max}$ (eqn (8.49)) is the Weiss-field factor associated with the bulk polarization of a domain.

The internal and external contributions to these frequencies are:

$$\begin{aligned}\omega_{A\,\mathrm{int}} &= \tfrac{1}{2}A(\mathbf{k}_0)(p_A-p_B) + \tfrac{1}{2}q(p_A+p_B)\\ \omega_{B\,\mathrm{int}} &= -\tfrac{1}{2}A(\mathbf{k}_0)(p_A-p_B) + \tfrac{1}{2}q(p_A+p_B) \qquad (8.170)\\ \omega_{C\,\mathrm{int}} &= qp_C;\end{aligned}$$

$$\omega_{A,B,C\,\mathrm{ext}}^T = \Delta - q[xp_C + \tfrac{1}{2}(1-x)(p_A+p_B)].$$

The energy and the entropy per spin are respectively equal to:

$$E/N = x[\tfrac{1}{2}\Delta p_C + \tfrac{1}{4}\omega_C p_C] + \tfrac{1}{2}(1-x)[\tfrac{1}{2}\Delta(p_A+p_B) + \tfrac{1}{4}(\omega_A p_A + \omega_B p_B)]; \quad (8.171)$$

$$S/N = xs(p_C) + \tfrac{1}{2}(1-x)[s(p_A)+s(p_B)]. \quad (8.172)$$

The calculation procedure which turns out to be the most convenient in the present case is to consider fixed values of Δ and β, and to determine p_A, p_B, p_C and x by the condition that they maximize $-\beta F = S - \beta E$, i.e. by setting equal to zero the partial derivatives of $(-\beta F)$ with respect to these four quantities. We obtain in that way a system of four equations which must be solved numerically. For instance the three conditions:

$$\frac{\partial(-\beta F)}{\partial p_\lambda}=0 \qquad (\lambda = \text{A, B, C}),$$

yield the Weiss-field equations:

$$p_\lambda = \tanh(-\tfrac{1}{2}\beta\omega_\lambda^{\text{T}}). \tag{8.173}$$

Since the p_λ and x have been obtained by a variational method, which is consistent only if the stable state of the system *is* a mixed state, it is furthermore necessary to ensure that the value obtained for $(-\beta F)$ is larger than for a purely paramagnetic or purely antiferromagnetic state with the same values of Δ and β. The cases when there is no mixed state and when the transition is of second order show up as special cases in this general theory.

A remarkable result of the theory is that at constant temperature, in the range of coexistence of the two phases, the polarizations p_{A}, p_{B}, and p_{C} remain constant. This means, according to (8.173) that the total frequencies are also constant, whence according to (8.169) that x depends linearly on Δ:

$$x = (\Delta - \Delta_{\text{c}}^{(1)})/(\Delta_{\text{c}}^{(2)} - \Delta_{\text{c}}^{(1)}) \tag{8.174}$$

The critical fields $\Delta_{\text{c}}^{(1)}$ and $\Delta_{\text{c}}^{(2)}$ correspond to $x = 0$ and 1, respectively. According to (8.169), they are related through:

$$\Delta_{\text{c}}^{(2)} - \Delta_{\text{c}}^{(1)} = q[p_{\text{C}} - \tfrac{1}{2}(p_{\text{A}} + p_{\text{B}})]. \tag{8.175}$$

According to (8.170), in the range of coexistence of the two phases, the internal *and* the external components of the frequencies remain separately constant, at constant β. The various frequencies are:

$$\begin{aligned}\omega_{\text{A,B}}^{\text{T}} &= \Delta_{\text{c}}^{(1)} \pm \tfrac{1}{2}A(\mathbf{k}_0)(p_{\text{A}} - p_{\text{B}});\\ \omega_{\text{c}}^{\text{T}} &= \Delta_{\text{c}}^{(2)};\\ \omega_{\text{A,B,C ext}}^{\text{T}} &= \Delta_{\text{c}}^{(2)} - q p_{\text{C}} = \Delta_{\text{c}}^{(1)} - \tfrac{1}{2}q(p_{\text{A}} + p_{\text{B}}).\end{aligned} \tag{8.176}$$

We have treated so far the case when $\Delta > 0$, a consequence of the choice $p_{\text{A}} \geqslant p > 0$. It is almost self-evident that the solution is symmetrical with respect to Δ: at constant β, the stable states for opposite values of Δ differ only by a reversal of all polarizations.

As will appear in section **F**(c), the actual experiment is performed at constant entropy rather than at constant temperature. Having calculated the polarizations $p_{\text{A,B,C}}(\beta)$, the critical fields $\Delta_{\text{c}}^{(1)}(\beta)$ and $\Delta_{\text{c}}^{(2)}(\beta)$, and the relative abundance $x(\beta, \Delta)$ given by (8.174), it is possible to calculate $S(\beta, \Delta)$ by eqn (8.172), and to extract from it the inverse temperature $\beta = \beta(S, \Delta)$.

Fixing the entropy S, it is then possible to express $p_{A,B,C}$ and x as functions of S and Δ, and to compare these quantities with the experimental quantities x, p_A, p_B, p_C measured while varying Δ at constant entropy.

This problem did not arise for the ferromagnetism with domains of section **E**(a), where below the critical frequency Δ_c, the variation of the state of the system with Δ turns out to be *both* isentropic and isothermal.

The theoretical fields, $\Delta_c^{(1)}$ and $\Delta_c^{(2)}$ when there is a mixed phase, and Δ_c when the transition is of second order, are plotted against the initial polarization (that is as a function of the entropy) in Fig. 8.39.

As always, the Weiss-field approximation is inadequate at low initial polarization. Also plotted in Fig. 8.39 are the predictions of the 1st order

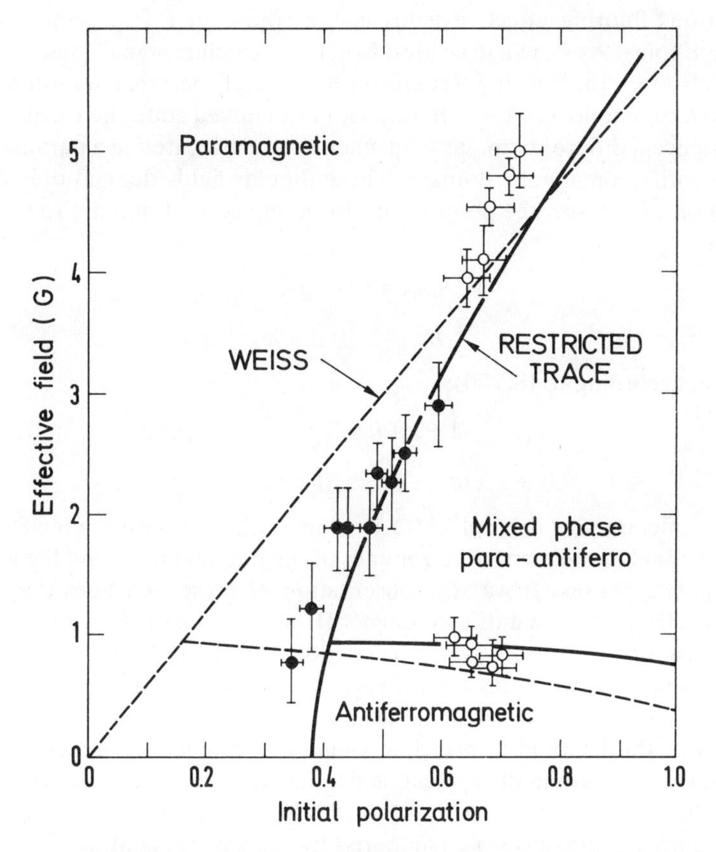

FIG. 8.39. Field-initial polarization phase diagram for the antiferromagnet III in CaF_2. Solid circles: transition field measured from the ^{19}F fast-passage signal. Open circles: transition field measured from the ^{43}Ca resonance signal. Broken curves: Weiss field approximation. Solid curves: 1st order restricted-trace approximation. (After Urbina and Jacquinot, to be published.)

restricted-trace approximation. The calculation procedure is very similar to that outlined above, the only difference being that one uses the restricted-trace expressions for the energy and the entropy in place of (8.171) and (8.172).

The range of entropy (initial polarization), where the transition is expected to be of second order is much narrower than predicted in section **F**(*a*), both because of the inadequacy of the Weiss field theory at low initial polarization, and of the occurrence of the mixed state.

(*c*) *Experimental results*

Two experimental methods can be used for measuring the critical fields. The first method consists in observing the absorption signal of the rare spins ^{43}Ca at various fluorine effective fields in the course of a fast passage of the fluorine spins. As shown in section **E**(*b*)1, the calcium signal consists of one line only when the fluorine system is either entirely paramagnetic or entirely antiferromagnetic. On the other hand, in the mixed state the calcium spins experience a different dipolar field when they are located in a paramagnetic or an antiferromagnetic domain. These dipolar fields depend only on the variation of the average polarization from one type of domain to the next, that is:

$$h_{\text{para}} = -\omega_{\text{C}}/\gamma_{\text{I}};$$
$$h_{\text{AF}} = -\tfrac{1}{2}(\omega_{\text{A}} + \omega_{\text{B}})/\gamma_{\text{I}};$$

or else, according to (8.170):

$$\begin{aligned} h_{\text{para}} &= -(1-x)(q/\gamma_{\text{I}})[p_{\text{C}} - \tfrac{1}{2}(p_{\text{A}} + p_{\text{B}})]; \\ h_{\text{AF}} &= x(q/\gamma_{\text{I}})[p_{\text{C}} - \tfrac{1}{2}(p_{\text{A}} + p_{\text{B}})]. \end{aligned} \tag{8.177}$$

The calcium signal is split into two lines whose relative intensities are proportional to the respective volumes of the paramagnetic and the antiferromagnetic phases. It was the observation of these two lines during the remagnetization of an antiferromagnet III which gave the first inkling of the existence of the mixed state. The two critical fields are obtained by extrapolation, as the fields at which the intensity of one of the lines vanishes (Urbina and Jacquinot, to be published).

This method should in principle yield very detailed information on the mixed state including the critical fields, the relative proportions of the two types of domains, and the bulk polarizations therein. In practice, these measurements are seriously hampered by the weakness of the ^{43}Ca signal and the brevity of the lifetime of the ordered state. Furthermore, in order to determine the relative intensities of the two lines, their splitting $|h_{\text{AF}} - h_{\text{para}}|$ must be larger than their width, which limits the measurements to low entropies.

In the second method one observes the shape of the fluorine fast passage dispersion signal. We have discussed in section **D**(a)1 the amplitude of the fast passage signal at exact resonance, which undergoes a qualitative change below the transition from the paramagnetic to the antiferromagnetic state. The same is true for the shape of the entire fast passage signal. The principle of the calculation of the transverse polarization within the Weiss-field approximation is always the same: in the presence of a small transverse frequency, the polarization of each spin—be it in a paramagnetic or an antiferromagnetic domain—is tilted away from the z-direction so as to become aligned with the total dipolar plus applied frequency it experiences. The result is that in the whole longitudinal effective field interval of existence of either the mixed or the antiferromagnetic phases, the transverse susceptibility $\chi_{\perp}$ undergoes only a small variation (Goldman, 1977). This is in accordance with observation: whereas in the paramagnetic state the fast-passage signal has the usual Lorentzian-like shape, it exhibits a flat top in the ordered state (Chapellier *et al.* 1969).

As an example, Fig. 8.40 shows the fast passage signals observed with initial polarizations $p_{\mathrm{i}}=0{\cdot}28$ and $0{\cdot}59$, respectively, that is at entropies above and below the transition. Also shown is the signal predicted by the Weiss-field theory for $p_{\mathrm{i}}=0{\cdot}59$. The restricted-trace approximation yields a very similar curve. According to theoretical expectation, the effective field at which the experimental flat top begins is identified with the upper critical field.

The fast passage method, the first to be used, gives a measure of the upper critical field but tells nothing about the structure or even the existence of the mixed state.

The results obtained by the two methods are plotted in Fig. 8.39 against the initial polarization. They provide convincing evidence for the actual existence of the mixed state expected on theoretical grounds. Within experimental uncertainty, the critical fields are in reasonable agreement with the prediction of the 1st order restricted-trace approximation, a prediction which involves no adjustable parameter.

(d) *Comparison with other antiferromagnetic systems*

The occurrence of a first order transition during the magnetization of an antiferromagnet is well known in electronic antiferromagnetism: it is the transition to the so-called spin-flop phase. It is instructive to outline both the analogies and the differences between the behaviour of the electronic and the nuclear spins.

We consider an idealized system where the spins of the electronic antiferromagnet experience isotropic and short range Heisenberg interactions, but no long range interactions. These interactions are usually supplemented with a much weaker anisotropy energy which lifts the degeneracy connected

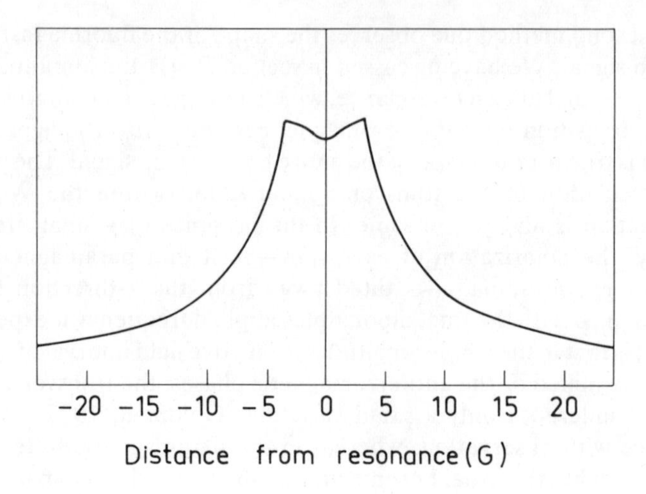

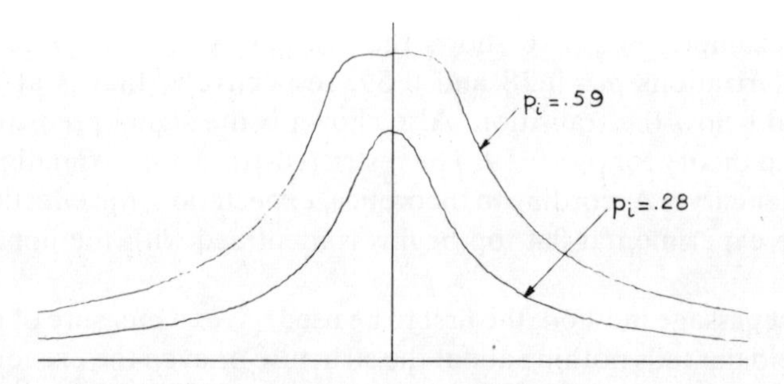

FIG. 8.40. Fast passage signal of ^{19}F in a sphere of CaF_2, at $T<0$ and with $H_0 \| [100]$. Bottom: experimental signals with initial polarizations $p_i = 0{\cdot}28$ and $0{\cdot}59$. Top: Weiss field prediction for the fast passage signal with $p_i = 0{\cdot}59$. (After Goldman, 1977.)

with the isotropy of the Heisenberg interactions and selects a preferential orientation for the sublattice polarizations along a direction $0z$, resulting in an energy lower than if the sublattice polarizations were in the $0xy$ plane. The strong coupling of the electronic spins with the lattice precludes the observation of negative temperatures and in contrast to nuclear magnetism the stable structures are always those of lowest energy.

At very low temperature, the parallel susceptibility of the antiferromagnet is vanishingly small, so that if we assume that the sublattice polarizations remain parallel to the anisotropy axis, a moderate field parallel to this axis would not affect the antiferromagnetic structure. Its bulk magnetization would remain zero and there would be no Zeeman contribution to the energy of the system. If on the other hand we consider an antiferromagnetic structure differing from the previous one by a 90° rotation of the spin polarizations, so as to bring them in the $0xy$ plane, the Heisenberg interaction energy is the same as before, because this interaction is isotropic, but the anisotropy energy vanishes, and this structure is in zero field less stable than the longitudinal antiferromagnet. However, the susceptibility χ of this structure to a field parallel to the anisotropy axis is large: the spin polarizations can rotate so as to become parallel to the total field they experience. This is the so-called spin-flop phase, in which the interactions of the spins with the external field as well as with their internal Weiss-field both yield sizeable contributions to the energy of the system. Above a threshold external field $H_c^{(1)}$, the energy of the spin-flop phase (including a Zeeman contribution) is lower than that of the longitudinal antiferromagnetic phase, and the system undergoes a first-order transition to the spin-flop phase.

The critical field $H_c^{(1)}$ depends only on the temperature T. At constant temperature T and field $H = H_c^{(1)}(T)$, both phases can coexist in any proportion, and the properties of each phase remain constant. This is a characteristic feature of first-order transitions in systems with short-range interactions, where the two phases do not influence each other: in a first-order transition at given temperature, the two phases can coexist in any proportion with constant properties, for uniquely defined values of the other external parameters. Slightly above $H_c^{(1)}$ the whole system is in the spin-flop phase. When H increases, the angle between the sublattice polarizations decreases until it vanishes at a critical field $H_c^{(2)}$, which corresponds to a second-order transition to the paramagnetic phase where the polarizations of all spins are equal.

This is what happens if the sample is kept at a constant temperature by thermal contact with an external source, a frequent procedure in electronic magnetism. It is also possible to perform an adiabatic magnetization of a system of thermally isolated electronic spins. The entropy of the system remains constant, but its temperature changes. In the course of the phase transition, the state of the system is represented in the H–T phase diagram by a point which moves along the $H_c^{(1)}(T)$ curve, from $H_c^{(1)}(T_i)$ where the spin-flop phase begins to appear, to $H_c^{(1)}(T_f)$ where the whole system is in the spin-flop phase. At each field, the relative proportions of the two phases are uniquely determined by the value of the entropy. Further increase of the field above $H_c^{(1)}(T_f)$ brings the system into the paramagnetic state as before in a second-order transition.

We have described in the three previous sections the behaviour of a nuclear antiferromagnet undergoing magnetization and found it very different from an electronic spin-flop transition. However, the philosophy is the same: stabilize the system by allowing it to include, besides the low-susceptibility phase which is stable in low field, regions corresponding to a different structure with a larger susceptibility. The qualitative difference in behaviour stems from a qualitative difference in the nature of the spin–spin interactions.

The absence of a spin-flop transition in nuclear magnetic ordering in high field resides in the very strong anisotropy of the truncated dipolar interactions. Whereas in an electronic antiferromagnet the rotation of the sublattice polarizations by 90° costs only the small anisotropy energy, for truncated dipolar interactions it would cause the Weiss-field (and therefore the spin–spin energy) to be multiplied by $-\frac{1}{2}$. The nature of the mixed state which is produced is determined by the permanent structures with large susceptibility whose spin–spin energies are close to that of the antiferromagnetic structure. In CaF_2 with $H_0\|[001]$ and $T<0$, there is just one such structure: the ferromagnetic structure IV with thin domains perpendicular to H_0.

The mixed state described in the preceding sections can be viewed as a compromise between the antiferromagnetic structure III and the ferromagnetic structure IV: it contains antiferromagnetic domains, and its domain-ferromagnetic character shows up in the square-wave modulation of the bulk polarization from the paramagnetic to the antiferromagnetic domains, whose shapes and orientations are the same as in the ferromagnet IV.

There is a further important difference between an electronic and a nuclear antiferromagnet, originating in the long range of the dipolar interactions, which are the *only* interactions producing nuclear magnetic ordering, whereas in electronic spin systems they are negligible compared with the short-range Heisenberg interactions. According to eqns (8.176), the mixed state retains a characteristic property of first-order transitions: at constant temperature, and whatever their relative concentration, the two phases coexist with constant properties in a *fixed* external field in each phase. However, by contrast with the electronic case, the external field in a given domain does *not* reduce to the applied field, because of the long-range dipolar contribution from the spins outside this domain. This dipolar contribution to the external fields depends on the relative volume of the two phases. This is why, when this relative volume is varied, the only way of keeping the external fields constant is to vary the applied field. At constant temperature, what was a *single* transition field in an electronic antiferromagnet is replaced in a nuclear antiferromagnet by a transition field *interval*, between $-\Delta_c^{(1)}/\gamma$ and $-\Delta_c^{(2)}/\gamma$. With our definition of the external fields in the mixed state, they are the same in the two phases.

The description of the transition from nuclear antiferromagnetism to paramagnetism was limited to the only case where an experimental investigation has been performed: CaF_2 with $H_0\|[001]$ and $T<0$. Under different conditions of either field orientation, sign of the spin temperature, or nuclear spin system, there may be other structures than a domain ferromagnet, having both a large susceptibility and a dipolar energy close to that of the antiferromagnetic structure stable in zero field. The transition from antiferromagnetism to paramagnetism in the course of a nuclear magnetization would then be expected to take place through the formation of a mixed state whose nature would be different from that investigated in the preceding sections. In the absence of any experimental study, the theoretical analysis of such mixed states can only have a speculative character and it will not be given here.

G. Investigation of the ordered domains

We have stated in sections **B** and **E** that, except for very special sample shapes, nuclear ferromagnetism occurs with domains whose shape and orientation are well-defined, but nothing was said of their actual size.

The occurrence of domains is a necessity in ferromagnetic structures because of the long range of the truncated dipole–dipole interactions. As was seen in section **B**(*c*), the Weiss field experienced by a given spin in a ferromagnetic domain does not originate from the spins in its vicinity (at least in cubic symmetry), but from distant spins, and this Weiss field depends very much on the shape and the orientation of the domains.

The situation is different in an antiferromagnet, where successive nuclear planes carry opposite magnetizations in zero external field. Let us consider in an antiferromagnetic system a region of macroscopic size l at a distance $r \gg l$ from a given spin I_i. The truncated dipolar interactions of the spin I_i with the various spins of that region, proportional to $(1-3\cos^2\theta_{ij})r_{ij}^{-3}$, are nearly all the same, and since the bulk polarization of the region is zero, the Weiss field it produces at the site of the spin I_i is vanishingly small. The Weiss field experienced by each spin is therefore essentially due to the spins in its neighbourhood, despite the long range of the dipole–dipole interactions.

The antiferromagnetic structures predicted by the methods of section **B** correspond to a regular succession of up and down nuclear planes extending over the whole sample. However, antiferromagnetic domains are likely to exist for the following reason. We have seen in section **F**(*b*) that during an adiabatic demagnetization at low entropy, small antiferromagnetic slices appear at the external frequency $\Delta_c^{(2)}$, and grow in size when Δ is further decreased to the frequency $\Delta_c^{(1)}$, where the various slices join together and the system becomes entirely antiferromagnetic. At the junction between two slices, there is a probability $\frac{1}{2}$ of obtaining a matching of the two antiferromagnetic arrangements, that is the right succession of planes with polarizations up, down, up, down . . . etc. There is also a probability $\frac{1}{2}$ to have a mismatch, that is a succession of nuclear planes with polarizations up, down, down, up, down . . . or up, down, up, up, down, up . . . . The possibility of a mismatch gives rise to antiferromagnetic domains, the polarization of an individual spin I_i being of the form:

$$p_i = p \exp(i\mathbf{k}_0 \cdot \mathbf{r}_i) g_i, \tag{8.178}$$

where g_i is equal to either $+1$ or -1, depending on the domain where the spin I_i sits.

Since a given domain produces no Weiss field in the other domains, there is no fundamental reason why the antiferromagnetic domains should have any particular shape.

The purpose of this section is to describe several experimental properties of the domains, as obtained by neutron diffraction in lithium hydride and by NMR in calcium fluoride.

This description is preceded by a theoretical discussion of the size of the ferromagnetic domains in a perfect crystal.

(*a*) *An estimate of the thickness of ferromagnetic domains*

This estimate of the ferromagnetic domain thickness in a perfect crystal is restricted to the ferromagnetic structure IV (section **E**). The domains are in the form of thin slices perpendicular to the direction of the external field H_0.

We use a very crude model, whose aim is to provide only an order of magnitude for the domain thickness. An elaborate treatment would be pointless because, as will be seen below, the domain thickness turns out to be much less than predicted. Still, the present estimate is not useless: its failure to account for the experimental results has revealed that the domain formation is triggered by a specific mechanism not included in the model. A possible candidate for such a mechanism will be described in the next section.

We assume, in accordance with the prediction of section **B**(*c*), that the ferromagnetic slices extend through the whole sample. The shape of a domain is approximated by a disk of radius R and thickness d. The polarization has a value p in the bulk of the domain and a smaller value in the domain wall and close to the edge of the disc, because of surface effects. In an idealized description, the polarization is taken to be zero in a plane sheet of thickness ε (the wall) at the junction between two domains, and in a ring of thickness αd, where α is a coefficient of order unity. Such an idealized domain is shown in Fig. 8.41. Since only the polarized portions of the domain contribute to the energy and the negentropy of the system, the equilibrium configuration is that for which the ratio of polarized to total volume is maximum:

$$\rho = \frac{V_{\text{pol}}}{V_{\text{tot}}} = \frac{\pi(R-\alpha d)^2(d-\varepsilon)}{\pi R^2 d},$$

that is, for $d/R,\ \varepsilon/d \ll 1$:

$$\rho \simeq 1 - 2\alpha d/R - \varepsilon/d. \tag{8.179}$$

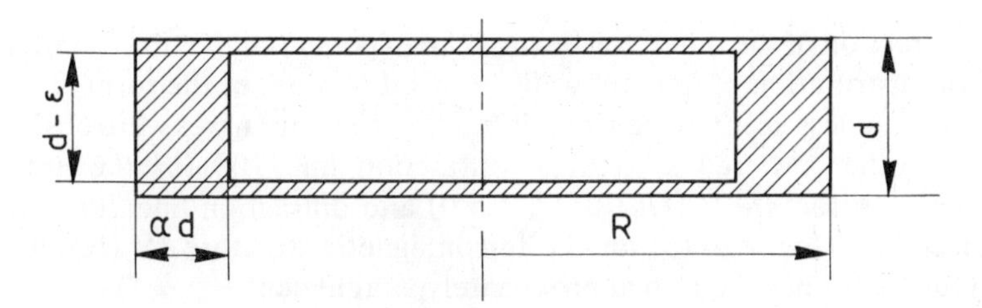

FIG. 8.41. Model of a ferromagnetic domain. The polarization is zero in the dashed part and homogeneous in the central part.

This ratio is maximum for:

$$d^2 = \varepsilon R/(2\alpha). \tag{8.180}$$

The domain thickness depends on the wall thickness ε, whose calculation is difficult. According to an approximate theory, not reproduced here (Goldman, 1977), ε which is a decreasing function of the domain polarization p, is comparable with the interatomic spacing a in a large range of values of p. Since according to (8.180) $d \propto \varepsilon^{1/2}$, the domain thickness is not very sensitive to the exact value of ε. The very existence of two well-separated resonance lines of ^{43}Ca (Fig. 8.25) indicates at least that the wall thickness ε is much smaller than the domain thickness d. With a sample dimension R in the millimeter range, ε of the order of a and α of the order of unity, (8.180) yields for d a value of the order of a few thousand Ångstrom units.

(*b*) *Neutron diffraction and domains in* LiH

We have seen in section **F**(*b*) of chapter 7 that the neutron diffraction pattern of a magnetically ordered nuclear system could be used to obtain the correlation function of the distribution of domain dimensions in various directions. The diffraction rocking curve yields the domain correlation function in the direction of the diffracted beam, that is perpendicular to the external field $\mathbf{H}_0$. It is only through the use of a 2-dimensional neutron detector that one can obtain the domain correlation function in other directions.

The neutron diffraction rocking curve for the reflection 110, corresponding to the antiferromagnetic structure III ($H_0 \| [001]$, $T<0$) is shown in Fig. 8.21(b). It is very nearly Lorentzian and much broader than the Gaussian-like crystalline diffraction line 220 shown in Fig. 8.21(a). The finite width of the line 220 is due to the imperfection of the incident neutron beam: distribution of wavelength and angular spread, and to the crystalline defects of the sample. When the experimental rocking curve 110 is corrected for these spurious broadening effects, the resulting curve is still very nearly Lorentzian. Its Fourier transform, which is the domain correlation function in the direction of the diffracted beam, is an exponential. It corresponds to a Poisson distribution of domain walls in this direction, of the form (7.124).

Somewhat less accurate neutron diffraction rocking curves were observed for two other ordered structures: diffraction line 100 for the antiferromagnetic structure II ($H_0 \| [011]$, $T > 0$) and diffraction line 200 (broad component of Fig. 8.36(b) for the ferromagnetic structure IV ($H_0 \| [011]$, $T < 0$). These lines are also approximately Lorentzian.

The average transverse domain dimensions deduced from the rocking curves for these three structures are the following:

$$\text{AF II: } l_0 \simeq 160 \text{ to } 200 \text{ Å;}$$

$$\text{AF III: } l_0 \simeq 160 \text{ to } 200 \text{ Å;}$$

$$\text{Ferro IV: } l_0 \simeq 100 \text{ to } 160 \text{ Å.}$$

The observation of the 2-dimensional distribution of diffracted neutrons, in the collecting plane of a multidetector, was limited to the antiferromagnet structure III and the ferromagnet IV, with the crystal oriented at the exact Bragg angle (reflections 110 for AF III, and 200 for Ferro IV).

The results obtained for the ferromagnet IV are shown in Fig. 8.42. The field H_0 is parallel to [011] and one observes the neutron reflection 200. In this figure are drawn the lines of equal neutron intensity (a) after the dynamic polarization, that is with all spin polarizations parallel to H_0, (b) after ADRF of the polarized spins at negative temperature. The enormous elongation of the neutron blot in the direction of H_0 shows that the average domain size is much less in the direction of H_0 than perpendicular to it. This is a direct proof that the shapes of the ferromagnetic domains, as well as their orientation with respect to the field, are indeed as expected theoretically.

In the antiferromagnetic state III, the 110 neutron reflection pattern also exhibits a vertical elongation, somewhat less pronounced than for the ferromagnet. The antiferromagnetic domains are therefore also anisotropic, much thinner in the direction of the field than across it. Whereas the domain anisotropy is an intrinsic feature of the ferromagnet IV, it does not correspond to any fundamental necessity in the antiferromagnet and is not understood at present.

In the absence of a complete 2-dimensional neutron pattern for all crystal orientations, one can obtain but a crude estimate of the average domain thickness in the direction of the external field. This thickness is of the order of 20 to 40 Å.

All domain dimensions are thus very much smaller than predicted for the ferromagnet in section **G**(a).

On the other hand, the domain thickness is comparable with the average distance between the paramagnetic F-centres present in the sample, which

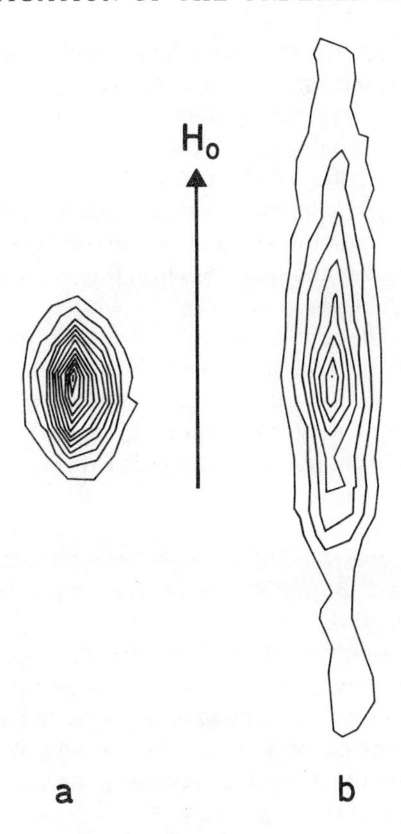

FIG. 8.42. Lines of equal intensity of the neutron beam for the reflection 200 at the centre of the rocking curve, in LiH with $H_0 \| [001]$. (a) Before demagnetization. No dipolar order. The intensity is enhanced by the large nuclear polarization. (b) After ADRF at $T < 0$. Ferromagnetic structure IV. (After Roinel *et al.* 1980.)

is about 40 Å. This is an indication that paramagnetic impurities might have an important influence in the initiation of the domains in the course of the ADRF.

This influence can tentatively be envisaged as follows. In high field and low lattice temperature, the electronic polarization is nearly unity and the magnetization of all electronic spins is parallel to H_0. Each electronic spin produces a steady dipolar field on the neighbouring nuclei. This field is either parallel or antiparallel to the external field, depending on the position of the nuclear spins with respect to the electronic spin, and adds up to the external field. In the course of an ADRF, the total effective field external plus electronic dipolar reaches the critical value in small regions near each impurity earlier than in the bulk of the sample. This produces a local rearrangement of the nuclear polarizations in these regions akin to the ordered structure. When the external effective field is lowered further, the local ordering spreads out through co-operative nuclear spin–spin processes, and gives rise to macroscopic ordered domains.

In the ferromagnetic phase, the initial domains, which are thin slices of polarization opposite to that of the rest of the sample, grow in thickness when the field is lowered so as to fill half the volume of the sample in zero effective field. The average thickness of the ferromagnetic domains in zero effective field is then expected to be equal to the average distance between the initial incipient domains.

In the antiferromagnetic phase, the initial antiferromagnetic domains grow in thickness until they fill the *whole* volume of the sample in zero field. Since there is a probability $\frac{1}{2}$ that at the junction between two adjacent antiferromagnetic slices they merge into a single domain, the average thickness of the antiferromagnetic domains is expected to be twice as large as the average distance between the initial domains at the onset of antiferromagnetism.

The 2-dimensional neutron diffraction patterns do indeed suggest that the longitudinal thickness of the antiferromagnetic domains is larger than that of the ferromagnetic domains.

The transverse dimensions of the domains are much larger than the average distance between impurities, which suggests that not all local orderings around the impurities expand so as to give rise to domains, but that some of them must shrink back. The physical choice of the 'successful' impurities that produce domains may depend on the statistical fluctuation in the spatial distribution of the impurities.

Although plausible, these considerations can only be considered as speculative in the absence of conclusive experimental tests.

(*c*) *NMR investigation of ferromagnetic domains in* CaF_2

Several properties of the ferromagnetic domains in structure III of CaF_2 were investigated through the observation of the resonance signal of the rare spins ^{43}Ca. We describe in turn the study of the reproducibility of the domain positions, and an attempt to estimate their thickness.

1. *Reproducibility of the domains*

When one produces a ferromagnet with domains in CaF_2, by an ADRF of the fluorine spins at negative spin temperature with say $H_0 \| [111]$, the positions of the individual domains are nearly reproducible from one experiment to the other. This is shown by the following experiment.

(i) After production of the ferromagnet by ADRF and polarization of the ^{43}Ca spins as explained in section **E**(*b*), one can materialize the positions of the domains by imparting different polarizations to the ^{43}Ca spins located in different domains. For instance, since the two calcium resonance lines originate from ^{43}Ca spins in different types of domains, the saturation by r.f. irradiation of one of the lines will decrease the calcium polarization to zero in the corresponding type of domain while leaving unaffected the calcium polarization in the other type of domain. Alternatively, one can irradiate the calcium spins at their Larmor frequency, i.e. midway between the two resonance lines: ^{43}Ca spins in domains of different type experience opposite fluorine dipolar fields and, through thermal mixing with the fluorine system, acquire opposite polarizations. The differential polarization of the calcium spins in the different fluorine domains constitutes an imprint of the positions of these domains. This differential polarization is monitored by the sizes of the two ^{43}Ca signals observed with a small non-saturating r.f. field. We suppose for the sake of definiteness that the calcium polarization has been saturated in one type of domains.

(ii) The fluorine spins are remagnetized to high effective field. All fluorine polarizations become equal, the domains disappear and the two ^{43}Ca lines merge together. The calcium imprint of the former fluorine domains is however still present as long as spin diffusion among the ^{43}Ca spins has not homogenized their polarization. We suppose for the moment that calcium spin diffusion is negligible during the whole experiment.

(iii) The fluorine system is demagnetized back to zero effective field. The calcium line which had been saturated in the step (i), and had then a vanishing intensity, is now observed, but with a far smaller intensity than the other line, of the order of 30 per cent of the total intensity of the two lines. This proves that the new fluorine domains occupy the same position, to within about 30 per cent, and have the same orientation of their polarization, as the initial ones. If indeed the new domains were randomly located with respect to the initial ones, each type of domains would contain on the average equal proportions of polarized calcium spins, and the two calcium lines would have equal intensities.

This reproducibility of the domains was quite unexpected. We describe below several of its experimental features.

(a) The size of the domains does not vary when the fluorine domain polarization decreases through spin–lattice relaxation. When after the fluorine ADRF, the system is prepared with opposite calcium polarizations in the two types of domains, the calcium resonance signal has the shape of the derivative of a bell-shaped curve: calcium spins polarized up give an absorption line whereas calcium spins polarized down give an emission line, and these lines are shifted in frequency by an amount proportional to the fluorine domain polarization. When this polarization decreases through spin–lattice relaxations, the shift between the two calcium lines becomes smaller, and since the lines overlap, the calcium resonance signal decreases. The observed decrease of the calcium signal is consistent with this picture. If on the other hand the size of the domains changed in the course of the relaxation, individual domains would spill over regions with opposite calcium polarizations, and the calcium signal would decrease much faster than observed.

The constancy of the domain size is in contradiction with eqn (8.180) of section **G**(a), for the wall thickness ε is expected to decrease with the domain polarization p (Goldman, 1977).

(b) The ferromagnetic domains are pinned to their position. This is shown as follows. After production of the ferromagnet with $H_0 \| [111]$, and materialization of the domain positions with opposite calcium polarizations, the sample is rotated by small successive angular steps around an axis perpendicular to H_0. In a large range of field orientations around the direction [111], the stable state of the system is still a ferromagnetic state with thin domains perpendicular to H_0. During the rotation the calcium imprint of the initial domains remains fixed with respect to the crystal, i.e. the slices with opposite calcium polarizations become tilted with respect to H_0. If the fluorine ferromagnetic domains evolved so as to be at equilibrium at all times i.e. perpendicular to H_0, they would be tilted with respect to the calcium imprint. Each type of fluorine domain would contain calcium spins with opposite polarizations, and the calcium signal would decrease continuously with rotation angle.

Experimentally, the calcium signal remains unaffected when the rotation angle is less than about 10°. At $\theta_0 \sim 10°$, it disappears suddenly, which shows that the fluorine domains undergo an abrupt reorientation. If the crystal is then rotated back to its initial orientation, the calcium signal reappears with an intensity decreased by about 30 per cent: the fluorine domains have come back to their initial positions.

It is possible from this experiment to obtain an estimate of the energy of pinning of the domains to their position: the tilted fluorine domains will reorient perpendicularly to H_0 when the corresponding gain in dipolar energy is equal to the pinning energy.

When the ferromagnetic slices are tilted by an angle θ with respect to H_0, the Weiss field energy of the system is:

$$E(\theta) = \tfrac{1}{4}Nqp^2 \,.\, (3\cos^2\theta - 1)/2. \qquad (8.181)$$

If θ_0 is the rotation angle at which the domains reorient, the pinning energy is of the order:

$$\varepsilon_{\text{pin}} = [E(0) - E(\theta_0)] = \tfrac{3}{8}Nqp^2 \sin^2\theta_0. \qquad (8.182)$$

For $p = 0{\cdot}7$ and $\theta_0 = 10°$, this yields in CaF_2:

$$\frac{1}{N}\varepsilon_{\text{pin}} \simeq 0{\cdot}12 \text{ kHz}.$$

(c) The positions of the domains after the ADRF do not depend on the initial orientation of the fluorine polarization. This is shown as follows. Let us suppose that the first ADRF is performed with fluorine polarizations initially up, and let us saturate one of the calcium lines. The reproducibility of the domain positions is monitored by observing this signal after a remagnetization–demagnetization process. This can be achieved in two different ways, either by sweeping the external field back to its original value, or by sweeping it further away from the fluorine resonance value. In the first case, the remagnetization brings the fluorine polarizations up, and in the succeeding demagnetization, the ferromagnetic transition corresponds to the appearance of small domains polarized down in a sea of fluorine spins polarized up.

In the second case, the opposite occurs: the fluorine spins are polarized down by the remagnetization, and in the succeeding demagnetization small domains polarized up appear in a sea of spins polarized down.

Experimentally, the final calcium signal is the same in both cases: neither the position, nor the orientation of the polarization of the various domains depend on the way the second remagnetization–demagnetization is performed.

(d) The reproducibility of the domains is not affected by the dipolar field of the ^{43}Ca spins. The fluorine–calcium dipolar energy is very small, but since after fluorine remagnetization the ^{43}Ca polarization imprint is the only remnant of the position of the initial domains, it is conceivable that this energy might trigger the domain formation in the second demagnetization. This possibility is ruled out by comparing the results of two experiments A and B, which differ through the imprints of the domains produced after the first demagnetization. In experiment A, the calcium polarizations are respectively up and zero in the domains with fluorine polarizations up and down. They are zero and up in experiment B. The dipolar interaction energy between the ^{43}Ca and the ^{19}F spins has opposite values in the two experiments. If this interaction were responsible for the nucleation of the domains after the remagnetization–demagnetization sequence, the polarizations of the new domains would be oriented so as to maximize the ^{43}Ca–^{19}F energy (remember $T<0$), that is, the initial up and down fluorine domains would remain up and down in one of the two experiments, and become down and up in the other one, with a corresponding change in the ^{43}Ca resonance signal. This is not observed: the initial up and down domains remain up and down in the two experiments A and B.

2. *Thickness of the domains*

In contrast with LiH, where neutron diffraction gives direct access to the size of the domains, nuclear magnetic resonance, the only experimental tool for investigating magnetic ordering in CaF_2, can only give indirect information on the thickness of the ferromagnetic domains.

This investigation, performed before the neutron diffraction measurements in LiH, was initially undertaken in the belief that the domain thickness estimate given in section **G**(a) was realistic, i.e. that this thickness was a few thousand Ångstrom units.

The first attempt to measure domain thickness of that order of magnitude relied on fluorine spin diffusion. The phenomenon of spin diffusion is described in Abragam (1961), chapter V: flip–flop transitions between like spins tend to homogenize their polarization throughout the sample; the evolution toward homogeneity of an initially inhomogeneous polarization is governed by a diffusion equation, with a diffusion constant D whose value is approximately known in systems such as ^{19}F in CaF_2.

This simple picture, which is valid in the paramagnetic state, must be analysed more closely when the system is in an ordered state. In a ferromagnet with domains the nuclear magnetization is highly inhomogeneous, and if such an ordered state is to be stable at all, there must be some mechanism inhibiting spin diffusion among the fluorine spins. This mechanism is best understood by considering first the simpler case of spin diffusion in a very inhomogeneous magnetic field such as exists for instance in type II superconductors (Genack and Redfield, 1975): a flip–flop between two spins experiencing different fields does not conserve the Zeeman energy, and can take place only if the Zeeman energy imbalance is compensated by a change of local dipole–dipole energy. The change of energy of the dipolar interactions is accompanied by a variation of their entropy. Eventually, the decrease of short-range dipolar entropy limits the continuation of flip–flop processes and the equilibrium state in the inhomogeneous field does *not* correspond in general to a homogeneous polarization. The situation is very similar in a ferromagnet with domains, except that the inhomogeneous field is the dipolar Weiss field created by the spins themselves. The variation of the domain polarizations can only result from flip–flops between spins belonging to different domains. Such flip–flops do not conserve the long-range dipolar energy. They create therefore short-range dipolar energy, and the flip–flops are limited by the decrease of the short-range entropy. One is then naturally led to distinguish between long-range order and short-range order. In a ferromagnet with domains at internal equilibrium, there is a well-defined ratio between long-range and short-range entropies (or energies).

It is clear at this point that if one were able to destroy continuously the short-range order, nothing would hamper flip–flops and the fluorine polarization would decrease to zero through spin diffusion.

Starting from a square wave shape of the polarization,

$$p(z, 0) = \frac{4}{\pi} p_0 \sum_{n \text{ odd}} \frac{1}{n} \sin\left(\frac{n\pi z}{d}\right), \tag{8.183}$$

which corresponds to domains of thickness d, we would have at time t:

$$p(z, t) = \frac{4}{\pi} p_0 \sum_{n \text{ odd}} \frac{1}{n} \sin\left(\frac{n\pi z}{d}\right) \exp\left(-\frac{D\pi^2 n^2}{d^2} t\right). \tag{8.184}$$

The harmonic $n = 1$ becomes rapidly the leading term. The first moment of the fluorine signal, which is proportional to $\langle p(z, t)^2 \rangle_{\text{av}}$ should then decay exponentially with a time constant:

$$\tau = d^2/(2\pi^2 D), \tag{8.185}$$

whence:

$$d = \pi(2D\tau)^{1/2}. \tag{8.186}$$

In CaF_2, the diffusion constant D is of the order of 10^{-12} cm^2 sec. (Roinel and Winter, 1970). It is then sufficient to measure the time constant τ to obtain an estimate of d. For $d = 5000$ Å, (8.185) predicts:

$$\tau \sim 125 \text{ sec.}$$

The method for saturating the local dipolar order consists in applying a strong r.f. field whose frequency is modulated by a few kHz around the ^{43}Ca resonance value (Goldman (1970), chapter 5). The effective calcium Zeeman interaction in the rotating frame is thermally coupled via $\mathcal{H}'_{IS}$ to $\mathcal{H}'_{II}$. The non-adiabatic modulation of $\langle \mathcal{H}'_{IS} \rangle$ results in a heating of $\langle \mathcal{H}'_{II} \rangle$. The heating process can be illustrated by the following picture. The fluorine short-range order can only be destroyed by a time dependent magnetic field that is inhomogeneous on an atomic scale and has a frequency 'on speaking terms' with the fluorine dipolar spectrum. Such a field is precisely produced by the ^{43}Ca magnetic moment whose z-component, under the action of the modulated r.f. field H_1, oscillates at the rate of the modulation. When the fluorine system is paramagnetic, the decay time of the fluorine dipolar energy is equal to the dipolar saturation time. When the fluorine system is ferromagnetic, the decay of the fluorine dipolar energy proceeds through saturation of the local order and spin diffusion. Its decay rate is governed by the slower of the two processes, which was anticipated to be the spin diffusion. The expected result of this study was that after an ADRF with a large initial polarization (yielding a ferromagnetic order) the decay rate of the first moment of the fluorine signal would be much slower than when the ADRF is performed with a small initial polarization (yielding a paramagnetic state). These decay rates were observed to remain equal down to 1 s, the shortest saturation time that could be obtained in practice. This result implied that the spin-diffusion time constant τ was shorter than 1 s, i.e. according to (8.186):

$$d \leqslant 450 \text{ Å}.$$

This was the first inkling that the domain thickness was not determined by surface effects, as assumed in section **G**(a).

An experimental estimate of the domain thickness could however be obtained by using a much slower spin-diffusion: that of the ^{43}Ca spins. In the experiments described in section **G**(c)1, which revealed a 'memory' of the calcium signal of the order of 60 to 70 per cent, the remagnetization–demagnetization following the calcium imprint of the ferromagnetic domains took place within a few minutes, and the calcium signal 'memory' was entirely attributed to the reproducibility of the domain positions.

If one waits a long time in the remagnetized state before performing the second ADRF, the calcium signal 'memory' is smaller than before. Its variation as a function of the time elapsed between remagnetization and second demagnetization is plotted in Fig. 8.43. This 'memory' decays with a time constant of the order of 20 h, very different from the fluorine spin diffusion time across a domain (less than 1 s), the dipolar spin–lattice relaxation time (1 to 3 h) and the fluorine Zeeman relaxation time (estimated by extrapolation from higher lattice temperatures to be at least a few hundred hours). The decay of the calcium signal 'memory' is attributed to the spin diffusion among the calcium spins, which tends to homogenize their polarization in space and to wipe out the imprints of the domains. The calcium signal memory being proportional to the difference of average calcium polarizations in the two types of domains, its time variation can be computed according to (8.184) (solid line in Fig. 8.43). The time constant τ of the long-time exponential decay is related to the domain thickness d through:

$$d = \pi\sqrt{(D_S\tau)}. \tag{8.187}$$

The spin diffusion coefficient D_S of ^{43}Ca in CaF_2 can be estimated by an approximate theory (Goldman and Jacquinot, to be published), with the result:

$$D_S \sim 1{\cdot}8 \times 10^{-18} \text{ cm}^2 \text{ s},$$

which, from $\tau \simeq 20$ h, yields for d the value:

$$d \sim 100 \text{ Å}.$$

This estimate is very rough because of the theoretical uncertainty in D_S, the experimental uncertainty in the decay time τ, and also because the value obtained for d is not much larger than the average distance between ^{43}Ca

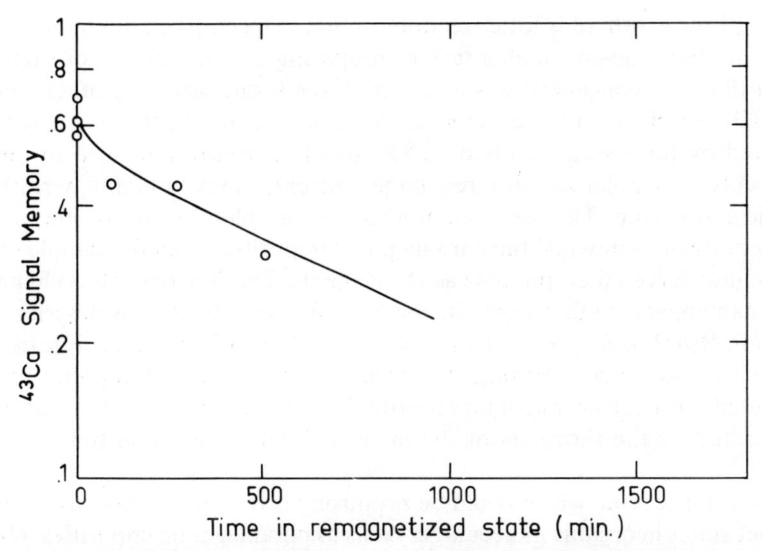

FIG. 8.43. Ferromagnetic structure IV in CaF_2. Variation of the 'memory' of the ^{43}Ca resonance signal as a function of the time spent in the remagnetized state. The theoretical solid curve is for a time constant $\tau = 1200$ min. (After Goldman, 1977.)

spins (~30 Å), and it is doubtful whether a diffusion equation is valid over such small distances.

The estimated thickness d of the ferromagnetic domains is comparable with the average distance between the paramagnetic Tm^{2+} ions present in the sample, of the order of 130 Å. This is in agreement with the result obtained by neutron diffraction in LiH and adds some plausibility to the hypothesis that the ordered domains are nucleated by paramagnetic impurities. As in LiH, the transverse dimensions of the domains, which must be much larger than their thickness, are therefore expected to be much larger than the average distance between impurities.

If the domains are indeed nucleated by the paramagnetic impurities, as seems plausible, it is not clear why, when randomly distributed impurities give rise to domains whose transverse dimensions are much larger than the inter-impurity spacing, the position Δ of these domains should be so precisely reproducible as observed.

(*d*) *Control of paramagnetic impurities*

The experimental results described in section **G**(*c*) strongly suggest that the paramagnetic impurities introduced into the sample for DNP purposes are responsible for the formation and the size of domains. A systematic investigation of this problem would require the possibility of changing at will the concentration of impurities within large limits. Unfortunately, in the present state of the art there is little freedom in that respect. Even discounting the difficult and time-consuming task of preparing a large number of samples with different concentrations and testing them one after the other, there remains the fact that the range of acceptable concentrations is strongly limited by the requirements of DNP: small concentrations lead to unacceptably slow polarization rates, large concentrations to highly perturbed nuclear ordering. The ideal solution of this problem would reside in the possibility of 'removing' the paramagnetic impurities from the sample once they have served their purpose as DNP agents. This is actually less chimerical than appears at first sight. One such 'removal' scheme was described in section **H**(*b*)2 of chapter 6: the transfer of Tm^{2+} ions from a magnetic triplet state into a non-magnetic singlet state had proved feasible. Its application to the study of nuclear magnetic ordering in CaF_2 in low fields was however frustrated by the shortness of the nuclear dipolar relaxation time in zero field.

Another scheme which could be promising is the use of optically excited triplet states in organic molecular crystals as paramagnetic impurities. Once the microwave DNP process is terminated the triplet states are allowed to decay totally or partially into the non-magnetic singlet ground state by suppressing or reducing the exciting light.

In a preliminary experiment a proton polarization enhanced by a factor 10 to a value of $\frac{1}{4}$ per cent has already been obtained in a crystal of paradibromobenzene doped with 1 per cent of paradichlorobenzene (Deimling, Brunner, Dinse, Hausser, and Colpa, 1980). It is clear that there is still a long way to go before polarizations suitable for producing nuclear long range order are reached. Another difficulty is the complexity of the structure of these crystals, far greater than that of CaF_2 and LiH, studied so far. However now that the observability of nuclear long range order by neutron diffraction has been firmly established, this tool could be used with confidence to disentangle more complicated magnetic structures.

Besides the systematic study of domains, the possibility of partial or total removal of impurities would have other advantages. It might lengthen the nuclear dipolar relaxation time allowing more perfect ADRF and even more important, longer lifetimes of the ordered states. It might reduce the inhomogeneity of the order parameter over the sample, and perhaps allow studies of critical behaviour frustrated by this inhomogeneity.

Above all it could change the whole approach to the problem of nuclear magnetic ordering. Instead of considering the paramagnetic impurities as a nuisance that one must bear, their role in nuclear magnetic ordering could be turned into an object of study in itself. Their influence on various aspects of nuclear order such as a change in the critical temperature or in the phase diagram would be of great interest. Time will tell whether these prospects can materialize into realities.

H. Some further aspects of the theory of nuclear antiferromagnetism

The approximate theoretical methods described in section **C** (Weiss field and restricted-trace approximations) proved qualitatively successful for predicting the overall properties of the ordered phases. These methods are however very primitive insofar as they either neglect completely, or give but a crude account of the short-range correlation effects. From a fundamental point of view, it is instructive to obtain a deeper description of reality in the ordered state, if only in a limited temperature range.

In the theory of magnetism, the theoretical effort has focussed on two temperature ranges: the vicinity of the transition and the low temperature region.

The transition to ordering has been tackled by methods of extrapolation of the high-temperature expansion of the partition function, and more recently by the renormalization group theory. Both methods are much more difficult with long-range dipolar interactions than with short-range interactions, and have not been applied to nuclear magnetic ordering.

At very low temperature, two methods are known to be much better than the Weiss field approximation: the spin wave and the random phase approximations. These two theories have played such an important part in magnetism that it would be hardly justified not to mention them in a chapter devoted to magnetic ordering. The spin wave approximation is useful in providing an improved description of the influence of quantum effects on the actual ordered state. The random phase approximation is useful for estimating the validity of the Weiss field theory at low temperature.

Finally, we describe the results of a Monte Carlo analysis of the ordering in a model system of classical magnets subjected to truncated dipolar interactions.

The description of these theories will be very brief and limited to the case of nuclear antiferromagnetism.

(*a*) *The spin wave approximation*

This method has been used for a long time and is described in many textbooks (e.g. Kittel, 1963; Anderson, 1963; Mattis, 1965). Its application to nuclear dipolar antiferromagnetism is described in Vu Hoang Chau (1970).*

The principle of this theory will be recalled for the example of truncated dipolar antiferromagnetism in a simple cubic system of spins $\frac{1}{2}$(^{19}F in CaF_2), using the simplest of a number of approaches.

To start with, we take it for granted that the ordered state is an antiferromagnetic state predicted in section **B**(*c*). We depart from the notations used so far, and we call I^i and S^μ the spins belonging to the sublattices $\mathscr{A}$ and

* Vu Hoang Chau (1970). Thesis, Orsay. CNRS no. 708.

$\mathscr{B}$, respectively. Each sublattice contains N spins (the total number of spins is therefore $2N$). The Hamiltonian is written in the form:

$$\begin{aligned}\mathscr{H}'_{\mathrm{D}} = \tfrac{1}{2}\sum_{i,j} B_{ij}\{2I^i_z I^j_z - \tfrac{1}{2}(I^i_+ I^j_- + I^i_- I^j_+)\} \\ + \tfrac{1}{2}\sum_{\mu,\nu} B_{\mu\nu}\{2S^\mu_z S^\nu_z - \tfrac{1}{2}(S^\mu_+ S^\nu_- + S^\mu_- S^\nu_+)\} \\ + \sum_{i,\mu} C_{i\mu}\{2I^i_z S^\mu_z - \tfrac{1}{2}(I^i_+ S^\mu_- + I^i_- S^\mu_+)\}.\end{aligned} \tag{8.188}$$

At zero temperature, the state predicted by the local Weiss field approximation corresponds to:

$$\langle I^i_z\rangle = \tfrac{1}{2}, \qquad \langle S^\mu_z\rangle = -\tfrac{1}{2}, \tag{8.189}$$

whence:

$$\langle I_z\rangle = \sum_i \langle I^i_z\rangle = \frac{N}{2}, \qquad \langle S_z\rangle = \sum_\mu \langle S^\mu_z\rangle = -\frac{N}{2}.$$

This cannot be the exact antiferromagnetic ground state, since it is not an eigenstate of the Hamiltonian: flip–flop terms of the form $I^i_- S^\mu_+$ mix it with states corresponding to $\langle I_z\rangle = \frac{1}{2}(N-1)$ and $\langle S_z\rangle = -\frac{1}{2}(N-1)$.

As a consequence of this mixing, the sublattice polarizations in the actual ground state depart from unity, i.e. we have in place of (8.189):

$$\langle I^i_z\rangle = \tfrac{1}{2}(1-\varepsilon), \qquad \langle S^\mu_z\rangle = -\tfrac{1}{2}(1-\varepsilon).$$

This is a characteristic feature of antiferromagnets. ε is the so-called spin deviation at zero temperature.

Our definition of the ground state needs to be sharpened. The ground manifold of the Hamiltonian (8.188) is invariant with respect to a 180° rotation of all the spins around an axis at right angle to $0z$, and their expectation values vanish inside this manifold. We make the usual assumption that in the ordered phase the symmetry relative to that rotation has been broken and define the ground state as the state of lowest (or highest) energy consistent with this broken symmetry.

In the spin wave theory, one obtains the ground state by a perturbation correction of the state described by (8.189), as well as the energies of the first few excited states. The main steps of the theory are the following.

1. *Elementary excitations*

An operator $\alpha^\dagger$ for the creation of an elementary excitation of energy ω is such that its evolution equation in the Heisenberg representation is:

$$-\mathrm{i}\frac{\mathrm{d}}{\mathrm{d}t}\alpha^\dagger = [\mathscr{H}, \alpha^\dagger] = \omega\alpha^\dagger. \tag{8.190}$$

If $|\psi\rangle$ is an eigenstate of $\mathscr{H}$ with eigenvalue E:

$$\mathscr{H}|\psi\rangle = E|\psi\rangle,$$

we have indeed, according to (8.190):

$$\mathscr{H}\alpha^\dagger|\psi\rangle = \{[\mathscr{H}, \alpha^\dagger] + \alpha^\dagger \mathscr{H}\}|\psi\rangle$$
$$= (E+\omega)\alpha^\dagger|\psi\rangle,$$

i.e. $\alpha^\dagger|\psi\rangle$ is an eigenstate of $\mathscr{H}$ with energy $E+\omega$: $\alpha^\dagger$ has created an excitation of energy ω.

When acting on the state described by (8.189) considered as 'the vacuum', the operators I^i_- and S^μ_+ act as creation operators of spin reversals, and it is a natural starting point to consider their equation of motion.

We have:

$$-\mathrm{i}\frac{\mathrm{d}}{\mathrm{d}t}I^i_- = [\mathscr{H}'_\mathrm{D}, I^i_-]$$
$$= -\sum_j B_{ij}(2I^j_z I^i_- + I^i_z I^j_-) - \sum_\mu C_{i\mu}(2S^\mu_z I^i_- + I^i_z S^\mu_-). \qquad (8.191)$$

This equation is linearized by replacing the operators I^i_z and S^μ_z by their expectation value (8.189) in the 'vacuum'. In the same approximation:

$$[I^i_+, I^i_-] = 2I^i_z \simeq 1,$$

and:

$$[S^\mu_-, S^\mu_+] = -2S^\mu_z \simeq 1, \qquad (8.192)$$

i.e. I^i_- and S^μ_+ behave as boson creation operators. The analogy is not perfect since for spins $\frac{1}{2}$: $I^{i2}_- = S^{\mu 2}_+ = 0$. However at low temperature it is a good approximation to use boson operators, because spurious states corresponding to more than one boson at a given site have a negligible weight in determining the properties of the system.

We use the following Fourier transforms:

$$c^\dagger_\mathbf{k} = \frac{1}{\sqrt{N}}\sum_i I^i_- \exp(-\mathrm{i}\mathbf{k}\,.\,\mathbf{r}_i); \qquad (8.193a)$$

$$d^\dagger_\mathbf{k} = \frac{1}{\sqrt{N}}\sum_\mu S^\mu_+ \exp(-\mathrm{i}\mathbf{k}\,.\,\mathbf{r}_\mu); \qquad (8.193b)$$

$$c_\mathbf{k} = (c^\dagger_\mathbf{k})^\dagger; \qquad d_\mathbf{k} = (d^\dagger_\mathbf{k})^\dagger.$$

$$B(\mathbf{k}) = \sum_i B_{ij} \exp[\mathrm{i}\mathbf{k}\,.\,(\mathbf{r}_i - \mathbf{r}_j)]$$
$$= \sum_\mu B_{\mu\nu} \exp[\mathrm{i}\mathbf{k}\,.\,(\mathbf{r}_\mu - \mathbf{r}_\nu)]; \qquad (8.194a)$$

$$C(\mathbf{k}) = \sum_i C_{i\mu} \exp[\mathrm{i}\mathbf{k}\,.\,(\mathbf{r}_i - \mathbf{r}_\mu)]$$
$$= \sum_\mu C_{i\mu} \exp[\mathrm{i}\mathbf{k}\,.\,(\mathbf{r}_i - \mathbf{r}_\mu)], \qquad (8.194b)$$

where **k** belongs to the first Brillouin zone of an antiferromagnetic sublattice. Its size is half of that of the Brillouin zone of the crystalline lattice.

According to (8.192), the operators defined by (8.193) are boson creation operators:

$$[c_{\mathbf{k}}, c^{\dagger}_{\mathbf{k}'}] = [d_{\mathbf{k}}, d^{\dagger}_{\mathbf{k}'}] = \delta(\mathbf{k}-\mathbf{k}'). \tag{8.195}$$

The Fourier transforms (8.194) are related to the Fourier transforms $A(\mathbf{k})$ (eqn (8.14)) as follows. Let $\mathbf{k}_0$ be the wave vector corresponding to the antiferromagnetic structure:

$$\exp[\mathrm{i}\mathbf{k}_0 \,.\, (\mathbf{r}_i - \mathbf{r}_j)] = \exp[\mathrm{i}\mathbf{k}_0 \,.\, (\mathbf{r}_\mu - \mathbf{r}_\nu)] = 1;$$

$$\exp[\mathrm{i}\mathbf{k}_0 \,.\, (\mathbf{r}_i - \mathbf{r}_\mu)] = -1.$$

We have:

$$A(\mathbf{k}) = B(\mathbf{k}) + C(\mathbf{k});$$

$$A(\mathbf{k}+\mathbf{k}_0) = B(\mathbf{k}) - C(\mathbf{k});$$

whence:

$$\begin{aligned} B(\mathbf{k}) &= \tfrac{1}{2}[A(\mathbf{k}) + A(\mathbf{k}+\mathbf{k}_0)]; \\ C(\mathbf{k}) &= \tfrac{1}{2}[A(\mathbf{k}) - A(\mathbf{k}+\mathbf{k}_0)]. \end{aligned} \tag{8.196}$$

By carrying these Fourier transforms into (8.191) and the analogous equation of motion of S^{μ}_{-}, we obtain after a little algebra:

$$\begin{aligned} -\mathrm{i}\frac{\mathrm{d}}{\mathrm{d}t}c^{\dagger}_{\mathbf{k}} &= -[B(0) - C(0) + \tfrac{1}{2}B(\mathbf{k})]c^{\dagger}_{\mathbf{k}} - \tfrac{1}{2}C(\mathbf{k})d_{\mathbf{k}}, \\ -\mathrm{i}\frac{\mathrm{d}}{\mathrm{d}t}d_{\mathbf{k}} &= [B(0) - C(0) + \tfrac{1}{2}B(\mathbf{k})]d_{\mathbf{k}} + \tfrac{1}{2}C(\mathbf{k})c^{\dagger}_{\mathbf{k}}, \end{aligned} \tag{8.197}$$

and similar coupled equations for the evolutions of $c_{\mathbf{k}}$ and $d^{\dagger}_{\mathbf{k}}$.

Equations (8.197) are diagonalized by the so-called Bogoliubov transformation:

$$\begin{aligned} \alpha^{\dagger}_{\mathbf{k}} &= V(\mathbf{k})c^{\dagger}_{\mathbf{k}} - U(\mathbf{k})d_{\mathbf{k}}, \\ \beta_{\mathbf{k}} &= V(\mathbf{k})d_{\mathbf{k}} - U(\mathbf{k})c^{\dagger}_{\mathbf{k}}, \end{aligned} \tag{8.198}$$

where $V(\mathbf{k})$ and $U(\mathbf{k})$ depend on $B(0)$, $B(\mathbf{k})$, $C(0)$ and $C(\mathbf{k})$. The operators $\alpha^{\dagger}_{\mathbf{k}}$ and $\beta^{\dagger}_{\mathbf{k}}$ are boson creation operators if we choose:

$$V(\mathbf{k})^2 - U(\mathbf{k})^2 = 1. \tag{8.199}$$

Equations (8.197) become:

$$\begin{aligned} -\mathrm{i}\frac{\mathrm{d}}{\mathrm{d}t}\alpha^{\dagger}_{\mathbf{k}} &= \omega(\mathbf{k})\alpha^{\dagger}_{\mathbf{k}}, \\ -\mathrm{i}\frac{\mathrm{d}}{\mathrm{d}t}\beta_{\mathbf{k}} &= -\omega(\mathbf{k})\beta_{\mathbf{k}}, \end{aligned} \tag{8.200}$$

whereas the evolution equations of $c_{\mathbf{k}}$ and $d_{\mathbf{k}}^{\dagger}$ yield likewise:

$$-\mathrm{i}\frac{\mathrm{d}}{\mathrm{d}t}\alpha_{\mathbf{k}}=-\omega(\mathbf{k})\alpha_{\mathbf{k}},$$

$$-\mathrm{i}\frac{\mathrm{d}}{\mathrm{d}t}\beta_{\mathbf{k}}^{\dagger}=\omega(\mathbf{k})\beta_{\mathbf{k}}^{\dagger}, \tag{8.200'}$$

where $\alpha_{\mathbf{k}}=(\alpha_{\mathbf{k}}^{\dagger})^{\dagger}$, $\beta_{\mathbf{k}}^{\dagger}=(\beta_{\mathbf{k}})^{\dagger}$.

According to (8.198) and (8.199), the $\alpha_{\mathbf{k}}^{\dagger}$ and $\beta_{\mathbf{k}}^{\dagger}$ are boson creation operators. The elementary excitations they create are the spin waves.

The spin wave energy $\omega(\mathbf{k})$ is found after a little algebra to be equal to:

$$\begin{aligned}\omega(\mathbf{k})&=\pm\{[B(0)-C(0)+\tfrac{1}{2}B(\mathbf{k})]^2-\tfrac{1}{4}C(\mathbf{k})^2\}^{1/2}\\&=\pm\{[A(\mathbf{k}_0)+\tfrac{1}{2}A(\mathbf{k})][A(\mathbf{k}_0)+\tfrac{1}{2}A(\mathbf{k}+\mathbf{k}_0)]\}^{1/2},\end{aligned} \tag{8.201}$$

where the sign is opposite to that of:

$$B(0)-C(0)+\tfrac{1}{2}B(\mathbf{k})=A(\mathbf{k}_0)+\tfrac{1}{2}\{A(\mathbf{k})+A(\mathbf{k}+\mathbf{k}_0)\}.$$

A condition for consistency is that the spin wave energies be real.

The spin wave populations are given by the Bose–Einstein formula:

$$\langle\alpha_{\mathbf{k}}^{\dagger}\alpha_{\mathbf{k}}\rangle=\langle\beta_{\mathbf{k}}^{\dagger}\beta_{\mathbf{k}}\rangle=\{\exp[\beta\omega(\mathbf{k})]-1\}^{-1}, \tag{8.202}$$

whereas the expectation values of all other bilinear products of these operators vanish.

In order for the structure to be stable, i.e. that the spin wave populations remain finite and non-negative, it is necessary that $\beta\omega(\mathbf{k})$ be positive, that is:

$$\text{at } T>0\text{: } \omega(\mathbf{k})>0$$

$$\text{at } T<0\text{: } \omega(\mathbf{k})<0.$$

At zero temperature ($\beta=\infty$), (8.202) yield:

$$\langle\alpha_{\mathbf{k}}^{\dagger}\alpha_{\mathbf{k}}\rangle=\langle\beta_{\mathbf{k}}^{\dagger}\beta_{\mathbf{k}}\rangle=0, \tag{8.202'}$$

that is:

$$\langle\alpha_{\mathbf{k}}\alpha_{\mathbf{k}}^{\dagger}\rangle=\langle\beta_{\mathbf{k}}\beta_{\mathbf{k}}^{\dagger}\rangle=1. \tag{8.202''}$$

We have already used the formalism of operators of creation and destruction of elementary excitations in the theory of superfluid ^{3}He (chapter 4). These excitations were fermions, whereas spin wave excitations are bosons.

2. *Sublattice polarization*

From the relations:

$$I_-^iI_+^i=\tfrac{1}{2}-I_z^i,$$

$$S_+^{\mu}S_-^{\mu}=\tfrac{1}{2}+I_z^i,$$

the sublattice polarizations are:

$$p_{\mathrm{A}} = \frac{2}{N}\sum_i I_z^i = 1 - \frac{2}{N}\sum_i \langle I_-^i I_+^i \rangle$$

$$= 1 - \frac{2}{N}\sum_{\mathbf{k}} \langle c_{\mathbf{k}}^\dagger c_{\mathbf{k}} \rangle \tag{8.203}$$

and likewise:

$$p_{\mathrm{B}} = -1 + \frac{2}{N}\sum_{\mathbf{k}} \langle d_{\mathbf{k}}^\dagger d_{\mathbf{k}} \rangle. \tag{8.203'}$$

In order to calculate the expectation values on the right-hand sides of (8.203) and (8.203′), the operators $c_{\mathbf{k}}$, $c_{\mathbf{k}}^\dagger$, $d_{\mathbf{k}}$ and $d_{\mathbf{k}}^\dagger$ are written as functions of $\alpha_{\mathbf{k}}$, $\alpha_{\mathbf{k}}^\dagger$, $\beta_{\mathbf{k}}$, $\beta_{\mathbf{k}}^\dagger$ by inverting (8.198). One then uses eqns (8.202). The result is:

$$\langle c_{\mathbf{k}}^\dagger c_{\mathbf{k}} \rangle = V(\mathbf{k})^2 \langle \alpha_{\mathbf{k}}^\dagger \alpha_{\mathbf{k}} \rangle + U(\mathbf{k})^2 [1 + \langle \beta_{\mathbf{k}}^\dagger \beta_{\mathbf{k}} \rangle];$$

$$\langle d_{\mathbf{k}}^\dagger d_{\mathbf{k}} \rangle = V(\mathbf{k})^2 \langle \beta_{\mathbf{k}}^\dagger \beta_{\mathbf{k}} \rangle + U(\mathbf{k})^2 [1 + \langle \alpha_{\mathbf{k}}^\dagger \alpha_{\mathbf{k}} \rangle]. \tag{8.204}$$

According to (8.202), (8.203), and (8.203′), we have:

$$\langle d_{\mathbf{k}}^\dagger d_{\mathbf{k}} \rangle = \langle c_{\mathbf{k}}^\dagger c_{\mathbf{k}} \rangle,$$

and $p_{\mathrm{B}} = -p_{\mathrm{A}}$ as it should.

The expectation values (8.204) do not vanish in the ground state, at zero temperature. According to (8.202′):

$$\langle c_{\mathbf{k}}^\dagger c_{\mathbf{k}} \rangle(0) = \langle d_{\mathbf{k}}^\dagger d_{\mathbf{k}} \rangle(0) = U(\mathbf{k})^2,$$

so that, as anticipated, the sublattice polarizations are not exactly equal to ± 1.

The spin wave approximation is valid only if the ground state it predicts does not differ too much from the ideal antiferromagnetic state described by (8.189), that is if the ground state spin deviation ε is small.

3. *Energy*

In the Hamiltonian (8.188) the energy of the flip–flop terms is obtained in a straightforward manner. It is easy to show through (8.193) and (8.194) that a term such as:

$$\sum_{i,j} B_{ij} I_-^i I_+^j = \sum_{\mathbf{k}} B(\mathbf{k}) c_{\mathbf{k}}^\dagger c_{\mathbf{k}}$$

and $\langle c_{\mathbf{k}}^\dagger c_{\mathbf{k}} \rangle$ is given by (8.204).

A term such as $I_z^i I_z^j$ is written in the form:

$$I_z^i I_z^j = (\tfrac{1}{2} - I_-^i I_+^i)(\tfrac{1}{2} - I_-^j I_+^j).$$

At low temperature, since $\langle I_z^i \rangle \# \frac{1}{2}$, we neglect the expectation value of the product of four operators on the right-hand side. We obtain:

$$\sum_{i,j} B_{ij}\langle I_z^i I_z^j \rangle \simeq \sum_{i,j} B_{ij}(\tfrac{1}{4} - \tfrac{1}{2}\langle I_-^i I_+^i \rangle - \tfrac{1}{2}\langle I_-^j I_+^i \rangle)$$

$$= \tfrac{1}{4}NB(0) - \sum_{\mathbf{k}} B(\mathbf{k})\langle c_{\mathbf{k}}^{\dagger} c_{\mathbf{k}} \rangle.$$

The other terms of $\mathcal{H}_{\mathrm{D}}'$ are calculated in a similar manner.

Spin wave calculations have been performed for the three antiferromagnetic structures of section **B**(c). The spin wave energies are all real and have the proper sign: positive at $T > 0$ and negative at $T < 0$. In Table 8.2 are listed the zero point spin deviations and the variations of the ground state energy with respect to the Weiss field energy for these three structures. The spin deviations are small and the zero point energies do not differ much from the Weiss field values.

The spin wave spectrum differs markedly from that of Heisenberg antiferromagnets. Since Heisenberg interactions are rotationally invariant, a uniform rotation of all spin polarizations does not change the energy, i.e. the energy $\omega(0)$ of the uniform excitation $\mathbf{k} = 0$ vanishes. At small $|\mathbf{k}|$, it can be shown that $\omega(\mathbf{k}) \propto |\mathbf{k}|$ (see e.g. Kittel, 1963). According to (8.202) $\langle \alpha_{\mathbf{k}}^{\dagger} \alpha_{\mathbf{k}} \rangle$ diverges at finite temperature when $\omega(\mathbf{k}) \propto |\mathbf{k}| \to 0$. However, the integral overall values of $\mathbf{k}$ yielding the sublattice polarization $p(T)$ does not diverge, and the departure $p(0) - p(T)$ can be expressed in the form of an expansion in powers T^m of the temperature T, where m may be a fraction (Kittel, 1963).

By contrast, since truncated dipolar interactions are highly anisotropic, $\omega(0)$ does not vanish, nor does any $\omega(\mathbf{k})$. According to (8.196) and (8.201) we have, for a sphere:

$$\omega(0) = -\sqrt{(3/2)}A(\mathbf{k}_0). \tag{8.205}$$

As an example, in the antiferromagnetic structure III ($H_0 \| [001]$, $T < 0$) one has:

$$\omega(0) = -11{\cdot}864\gamma^2\hbar/(2a^3),$$

Table 8.2

Structure	Spin deviation (per cent)	$(E_{\mathrm{SW}}^0 - E_{\mathrm{W}}^0)/E_{\mathrm{SW}}^0$ (per cent)
I	5·4	7·5
II	4·2	23
III	1·2	2·9

and $\omega(\mathbf{k})$ extends from

$$-8{\cdot}241\gamma^2\hbar/(2a^3) \text{ for } \mathbf{k}=(\pi/a, \pi/a, 0)$$

$$\text{to } -14{\cdot}204\gamma^2\hbar/(2a^3) \text{ for } \mathbf{k}=(0, 0, \pi/2a).$$

According to (8.202), the low temperature sublattice polarization departure $p(0)-p(T)$ is approximately an exponential function of $1/T$, and cannot be expanded in powers of T. This behaviour is similar to that predicted by the Weiss field theory. Together with the small value of the spin deviation and the small spin wave correction to the zero point energy, it shows that the Weiss field theory gives a reasonably good description of the antiferromagnetic ground state.

(b) The random phase approximation (RPA)

The most usual formulation of RPA is through the Green function formalism (Zubarev, 1960; Tyablikov, 1967). Our aim is to give only a brief report of this approximation, which can be done through a much simpler method, similar to that used for the spin wave theory, namely the linearized equations of motion (8.191). (Brout, 1965*a*, chapter 5.)

The main difference with spin wave theory is that, instead of taking the values (8.189) of $\langle I_z^i\rangle$ and $\langle S_z^\mu\rangle$, one uses:

$$\langle I_z^i\rangle = \tfrac{1}{2}p, \qquad \langle S_z^\mu\rangle = -\tfrac{1}{2}p, \tag{8.206}$$

where p is the, so far unknown, sublattice polarization, whose value is calculated later by a self-consistent procedure.

The popularity of RPA stemmed from the fact that, because of its use of the renormalized conditions (8.206), it was expected to be approximately valid over a large range of values of p and not only for $p \simeq 1$ as the spin wave theory.

The solutions of the linearized equations of motion (8.191) are similar to those of the spin wave theory. The elementary excitation creation and annihilation operators $\alpha_\mathbf{k}^\dagger$, $\alpha_\mathbf{k}$, ... are of the same form as (8.198). There are two differences with the spin wave theory:

(i) The energies of the elementary excitations are renormalized:

$$\Omega(\mathbf{k}) = p\omega(\mathbf{k}), \tag{8.207}$$

where $\omega(\mathbf{k})$ is the spin wave excitation energy (8.201).

(ii) The elementary excitation operators are *not* boson operators: one has as a consequence of (8.206):

$$[\alpha_\mathbf{k}, \alpha_\mathbf{k}^\dagger] = [\beta_\mathbf{k}, \beta_\mathbf{k}^\dagger] = p. \tag{8.208}$$

It can be shown that as a consequence of (8.207) and (8.208) the expression (8.202) for $\langle\alpha_\mathbf{k}^\dagger\alpha_\mathbf{k}\rangle$ given by the spin wave theory is replaced in

RPA by:

$$\langle\alpha_{\mathbf{k}}^{\dagger}\alpha_{\mathbf{k}}\rangle = \langle\beta_{\mathbf{k}}^{\dagger}\beta_{\mathbf{k}}\rangle$$
$$= p\{\exp[\beta p\omega(\mathbf{k})]-1\}^{-1}. \qquad (8.209)$$

We omit the proof, based on the fluctuation–dissipation theorem (see e.g. Brout, 1965*a*, p. 110).

The sublattice polarization $p = p_A = -p_B$ is calculated through (8.203) and (8.204), where $\langle\alpha_{\mathbf{k}}^{\dagger}\alpha_{\mathbf{k}}\rangle$ and $\langle\beta_{\mathbf{k}}^{\dagger}\beta_{\mathbf{k}}\rangle$ are replaced by their RPA value (8.209) which depends on p. One obtains therefore a self-consistent equation for p, which is solved numerically.

The energy $\langle\mathcal{H}'_D\rangle$ consists of two parts, corresponding respectively to flip–flop terms and to longitudinal correlation terms. The former cause no problem, since they can be written as a function of $\langle\alpha_{\mathbf{k}}^{\dagger}\alpha_{\mathbf{k}}\rangle$ and $\langle\beta_{\mathbf{k}}^{\dagger}\beta_{\mathbf{k}}\rangle$. The calculation of, say, $\langle I_z^i I_z^j\rangle_0$ is more complicated, and it involves the use of the Green function formalism (Vu Hoang Chau, 1970).* This calculation will not be described here. Its result is that the longitudinal contribution to the dipolar energy is nearly the same as if we had used the approximation:

$$\langle I_z^i I_z^j\rangle_0 \# \langle I_z^i\rangle_0\langle I_z^j\rangle_0 = \tfrac{1}{4}p^2 \text{ etc.} \ldots. \qquad (8.210)$$

The entropy is obtained by a numerical computation, according to the thermodynamic relation:

$$dS = \beta\, d\langle\mathcal{H}'_d\rangle. \qquad (8.211)$$

RPA calculations were performed for the antiferromagnetic structures I, II, and III of CaF_2 up to the transition. As an example, the critical values for structure III are:

$$T_c = -0{\cdot}59\ \mu K,$$
$$S_c/N = 0{\cdot}686 k_B,$$

to be compared with the values obtained in section **C** with the Weiss field-high temperature approximation:

$$T_c = -0{\cdot}613\ \mu K,$$
$$S_c/N = 0{\cdot}638 k_B.$$

However close, the critical entropies correspond to markedly different critical initial polarizations: $p_0 = 12{\cdot}2$ per cent in RPA, as compared with $p_0 = 32{\cdot}6$ per cent in the Weiss field plus high temperature approximation used in section **C**(*a*)2.

The RPA value of p_0 is much less than the experimental values, and RPA is therefore a bad approximation close to the transition.

* Vu Hoang Chau (1970). Thesis, Orsay. CNRS no. 708.

On the basis of the extensive use of RPA in magnetism, it is believed that this approximation should be reasonably good well below the transition.

For the antiferromagnetic structure III of CaF_2, the predictions of RPA are close to those of Weiss field theory. An example of this is the similarity of the predictions of the two theories for the variation of $\chi_{\|}$ as a function of the dipolar energy, shown in Fig. 8.17. In so far as RPA is a good approximation outside the transition region, as borne out by the results of Fig. 8.17, so should be the Weiss field theory. From the theoretical point of view, the origin of the relative concordance between the two theories lies in the small relative spread of the elementary excitation spectrum around the value for $\mathbf{k}=0$, for this structure:

$$\left|\frac{\Omega(\mathbf{k})-\Omega(0)}{\Omega(0)}\right| \leq 0{\cdot}305.$$

This implies that short-range correlations, neglected in the Weiss field theory, are not too important. The calculation leading to this conclusion is a little lengthy and will not be given here.

(*c*) *Monte Carlo calculations*

We mention briefly the main results of a Monte Carlo analysis of magnetic ordering in systems of classical magnets subjected to truncated dipole–dipole interactions (Mouritsen and Knak Jensen, 1978, 1980). The systems investigated are either a simple cubic lattice of like moments (akin to CaF_2) or a NaCl type lattice of unlike moments (akin to LiH).

For reasons of tractability, the range of the truncated dipolar interactions is limited as a rule to the 26 closest neighbours of each spin (nearest, next-nearest and next-next-nearest neighbours) and occasionally to the 124 closest neighbours. The sample used for the computation includes between 1000 and 8000 spins.

The principle of the calculation is to construct a canonical ensemble of spin configurations corresponding to a given temperature T. Each configuration differs from the preceding one by the random variation of the orientation of one spin. The size of the ensemble is determined by the condition that the ensemble averages of several physical quantities reach reasonably stable values.

For $H_0 \| [001]$, $T<0$ and $T>0$, and for $H_0 \| [011]$, $T<0$, the same antiferromagnetic structures are predicted as in section **B**. However, the Monte Carlo calculation does not predict the ferromagnetic structures IV and V, since it neglects the dipolar interactions between distant spins which are responsible for the Weiss field experienced by each spin in these structures.

The predicted values of the transition temperature, energy and entropy, derived for classical magnets, i.e. infinite spins, can hardly be compared with those of systems of quantum spins $\frac{1}{2}$ or $\frac{3}{2}$.

The critical exponents for these systems are found to be very close to those of a three-dimensional Ising ferromagnet. It is difficult to ascertain whether this is not due to the limitation of the range of the interactions. The measurement of critical exponents is unfortunately out of range of experiment in the present state of the art.

I. Magnetic ordering in solid ^{3}He

This section departs from the rest of this chapter, which is devoted to nuclear dipolar magnetic ordering, whereas the spin–spin interactions in solid helium 3 are essentially exchange interactions. This complement to chapter 3 has been deferred to the present chapter in order to benefit from the background on magnetic ordering described in the preceding sections.

We describe in turn the properties of ordering in solid ^{3}He which contradict the predictions based on a Heisenberg Hamiltonian consisting of two-spin exchange between nearest neighbours, the nature of the ordered phase in zero field as deduced from NMR experiments, and the interpretation of the low temperature properties of solid helium 3 with á Hamiltonian consisting of 3 and 4-spin exchange interactions. The discussion is limited throughout to solid ^{3}He on the melting curve.

(*a*) *Ordering properties of solid* ^{3}He

We consider first the ordering that would be expected with 2-spin exchange interactions between nearest neighbours.

The exchange coupling J being negative, as deduced from the negative value of the Curie–Weiss temperature, the expected structure in zero field is an antiferromagnet where the sublattices are the imbricated simple cubic lattices of the b.c.c. structure. The transition to antiferromagnetism is of second order. Within the Weiss field theory, the critical temperature T_c is equal to $-\theta \simeq 3$ mK, whereas according to more refined theories and to the experimental results obtained with well-behaved electronic antiferromagnets, it should be of the order of:

$$T_c \simeq -0{\cdot}7\theta \simeq 2 \text{ mK}.$$

Anisotropy in the system can only arise from dipole–dipole interactions. In the expected structure, the Weiss field value of the dipolar energy vanishes. Its non-vanishing value, which arises from higher order effects, must therefore be very small. As a consequence of the very small anisotropy energy, the transition field to the spin–flop phase must be very small.

The Weiss field prediction of the variation with external field of the critical temperature of the second order spin–flop–paramagnetic transition is shown in Fig. 8.44.

Experimentally, in zero field one observes a *first-order* transition at a temperature $T_c \simeq 1{\cdot}03$ mK, as evidenced by a sharp drop in the entropy ($\Delta S/N \simeq 0{\cdot}443\ k_B \ln 2$), and a sharp drop of the susceptibility χ to a temperature-independent value:

$$\chi \simeq -C/(2\theta), \tag{8.212}$$

where C is the Curie constant (Prewitt and Goodkind, 1977). The first-order nature of the transition is confirmed by the NMR measurements described in the next section. In the presence of a magnetic field, this 1st order transition persists up to $H \simeq 4{\cdot}1$ kG, with a slight decrease of the critical temperature.

At higher values of the field one observes a second-order transition to a phase with a large magnetization increasing slightly with field: well below the transition this magnetization varies from $M \simeq 0{\cdot}6M_0$ at $H = 4{\cdot}1$ kG, to $M \simeq 0{\cdot}7M_0$ at $H \simeq 70$ kG, where M_0 is the saturation magnetization. The

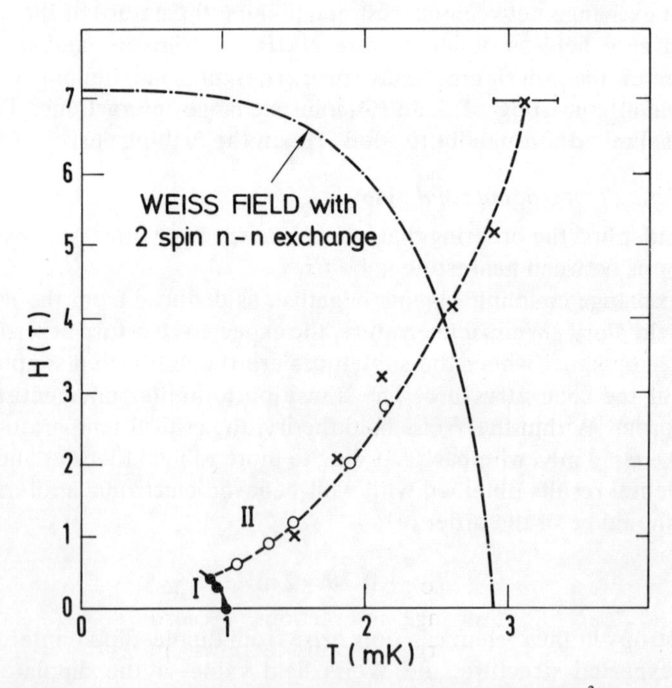

FIG. 8.44. Field-temperature phase diagram of ordering in solid ^{3}He. Solid curve: 1st order transition. Broken curve: 2nd order transition. Broken-dotted curve: Weiss field prediction for spin–flop–paramagnetic transition, for 2-spin nearest-neighbour exchange interactions. (After Roger (1980). Thesis, Orsay. No. 2297.)

critical transition temperature, for this phase, is up to 70 kG, a steadily increasing function of the field. The experimental H–T phase diagram is shown in Fig. 8.44. It is in violent disagreement with the Weiss field predictions for a Heisenberg Hamiltonian.

(b) NMR study of the ordered phase in low field

The nuclear magnetic resonance of ^{3}He in the ordered state has been observed in a field interval ranging from ~50 to 500 G (Osheroff *et al.* 1980). The experiments were performed with single crystals. Above 300 G, one observes three different resonance frequencies, which lead to the assumption that the anisotropy axis is along a fourfold crystalline axis, and that the three resonance lines correspond to three types of domains with their anisotropy axes parallel to either one of the three crystalline axes A_4.

The NMR results are in excellent agreement with those predicted for an antiferromagnetic structure with so-called planar anisotropy, that is when the anisotropic dipolar energy is minimum and degenerate for sublattice polarizations perpendicular to the anisotropy axis.

First we derive the resonance frequencies for that case before describing the experimental results and the structure which fits them.

Let $\mathscr{A}$ and $\mathscr{B}$ be the sublattices, and $\mathbf{p}_A$ and $\mathbf{p}_B$ the sublattice polarizations. The Weiss fields experienced by the various spins consist of two parts: an isotropic exchange field and a dipolar field axially symmetric around the direction 0z. The latter can be split into an isotropic field plus a field parallel to $0z$. The dipolar frequency is, as seen in previous sections, of the order of 1 μK. It is several orders of magnitude smaller than the exchange frequency, which is of the order of θ, that is in the mK range. We shall call for brevity exchange field the isotropic part of the Weiss field (which does in fact contain a small dipolar contribution), and dipolar field the longitudinal component of this field along the anisotropy axis of unit vector $\hat{\mathbf{z}}$.

In usual magnetic systems, the interactions are bilinear with respect to the individual spin operators. In the Weiss field approximation, where each spin operator is replaced by its expectation value, the Weiss frequency experienced by a given spin,

$$\boldsymbol{\omega}_i = \partial E/\partial\langle \mathbf{I}_i\rangle, \tag{8.213}$$

is a linear function of the various spin polarizations. The situation is different in solid ^{3}He: as shown in section $\mathbf{G}(b)$ of chapter 3, the Hamiltonian is likely to contain four-spin exchange interactions. According to (3.195) and (3.196), the Weiss field energy contains therefore products of four spin polarizations, and the Weiss fields contain products of three spin polarizations.

We can nevertheless derive some general properties of the Weiss fields. If, as we assume, the stable structure in zero field is a two-sublattice antiferromagnet, the exchange Weiss fields experienced by spins in sublattices with opposite polarizations are opposite. On the other hand, if the sublattice polarizations depart from opposite orientations, the extra exchange Weiss field originating in the bulk polarization will be the same in the two sublattices.

We can then write quite generally the Weiss frequencies in the form:

$$\boldsymbol{\omega}_{\mathrm{A}} = -q(\mathbf{p}_{\mathrm{A}} - \mathbf{p}_{\mathrm{B}}) + r(\mathbf{p}_{\mathrm{A}} + \mathbf{p}_{\mathrm{B}}) + [s(p_{\mathrm{A}z} - p_{\mathrm{B}z}) + t(p_{\mathrm{A}z} + p_{\mathrm{B}z}]\hat{\mathbf{z}}; \quad (8.214)$$

$$\boldsymbol{\omega}_{\mathrm{B}}(\mathbf{p}_{\mathrm{A}}, \mathbf{p}_{\mathrm{B}}) = \boldsymbol{\omega}_{\mathrm{A}}(\mathbf{p}_{\mathrm{B}}, \mathbf{p}_{\mathrm{A}}). \quad (8.214')$$

The exchange Weiss field coefficients q and r may depend (symmetrically) on the polarizations $\mathbf{p}_{\mathrm{A}}$ and $\mathbf{p}_{\mathrm{B}}$. The dipolar Weiss field coefficients s and t, which originate from bilinear interactions, are independent of the polarizations. If the sample is not an ellipsoid, t (but not s) may vary through the sample, and the Weiss fields may not be the same for all spins of a given sublattice. We will see below that this has no practical consequences, and we assume for the time being that t is a constant.

At thermal equilibrium in zero field, $\mathbf{p}_{\mathrm{A}} = -\mathbf{p}_{\mathrm{B}}$, and $|\mathbf{p}_{\mathrm{A}}| = |\mathbf{p}_{\mathrm{B}}| = p$. The isotropic exchange energy is:

$$E_{\mathrm{e}} = -\tfrac{1}{2}Nqp^2, \quad (8.215)$$

and the anisotropic dipolar energy is:

$$E_{\mathrm{d}} = \tfrac{1}{4}Ns\{(\mathbf{p}_{\mathrm{A}} - \mathbf{p}_{\mathrm{B}}) \cdot \hat{\mathbf{z}}\}^2 = \tfrac{1}{2}Nsp_z^2. \quad (8.216)$$

Since the stable structure corresponds to positive temperature, one has $q > 0$, and since the anisotropy is planar, the anisotropy energy is maximum when p_{A} and p_{B} are aligned with $0z$, i.e. one has $s > 0$.

The exchange field factors q and r can be related to the susceptibility as follows. In the presence of an external field much smaller than the exchange field but much larger than the dipolar field, the equilibrium magnetization is practically the same as if the dipolar field were absent, and the susceptibility is very nearly isotropic. The equilibrium configuration in an external field is of the spin–flop type:

$$p_{\mathrm{A}Z} = p_{\mathrm{B}Z}; \qquad p_{\mathrm{A}X} = -p_{\mathrm{B}X},$$

where $0Z$ is the direction of $\mathbf{H}$ and $0X$ is perpendicular to it. By writing that $\mathbf{p}_{\mathrm{A}}$ is parallel to $\boldsymbol{\omega}_{\mathrm{A}}^{\mathrm{T}} = \boldsymbol{\omega}_{\mathrm{A}} + \boldsymbol{\Delta}$, where $\boldsymbol{\Delta} = -\gamma\mathbf{H}$, we obtain:

$$\frac{\Delta + 2rp_{\mathrm{A}Z}}{p_{\mathrm{A}Z}} = -\frac{2qp_{\mathrm{A}X}}{p_{\mathrm{A}X}},$$

whence:

$$p_{\mathrm{A}Z} = -\Delta/2(q+r) \quad (8.217)$$

or else:

$$M_Z = \tfrac{1}{2}N\gamma\hbar p_{AZ} = -\tfrac{1}{4}N\gamma\hbar\Delta/(q+r) = \tfrac{1}{4}N\gamma^2\hbar H/(q+r),$$

whence:

$$\chi = M_Z/H = \tfrac{1}{4}N\gamma^2\hbar/(q+r). \tag{8.218}$$

Experimentally, χ is temperature-independent in the order phase (eqn (8.212)), so that $(q+r)$ does not depend on the sublattice polarizations.

Now, let us specify that the external field is in the plane $0yz$, at an angle θ from the anisotropy axis $0z$ (Fig. 8.45). Since this field is much smaller than the exchange field, at equilibrium one has $|\mathbf{p}_A+\mathbf{p}_B| \ll 2p$ and $|\mathbf{p}_A-\mathbf{p}_B| \simeq 2p$, and the dipolar energy (8.216) is minimum when $\mathbf{p}_A-\mathbf{p}_B$ is perpendicular to $0Z$ and $0z$, that is along $0x$.

In the Weiss field approximation, the equations of motion of the polarizations are:

$$\begin{aligned}\frac{d}{dt}\mathbf{p}_A &= \boldsymbol{\omega}_A^T \wedge \mathbf{p}_A = (\boldsymbol{\omega}_A + \boldsymbol{\Delta}) \wedge \mathbf{p}_A,\\ \frac{d}{dt}\mathbf{p}_B &= \boldsymbol{\omega}_B \wedge \mathbf{p}_B = (\boldsymbol{\omega}_B + \boldsymbol{\Delta}) \wedge \mathbf{p}_B,\end{aligned} \tag{8.219}$$

whence, according to (8.214) and (8.214′) and after a little algebra:

$$\frac{d}{dt}(\mathbf{p}_A+\mathbf{p}_B) = \boldsymbol{\Delta} \wedge (\mathbf{p}_A+\mathbf{p}_B) + t(p_{Az}+p_{Bz})\hat{\mathbf{z}} \wedge (\mathbf{p}_A+\mathbf{p}_B) + s(p_{Az}-p_{Bz})\hat{\mathbf{z}} \wedge (\mathbf{p}_A-\mathbf{p}_B), \tag{8.220}$$

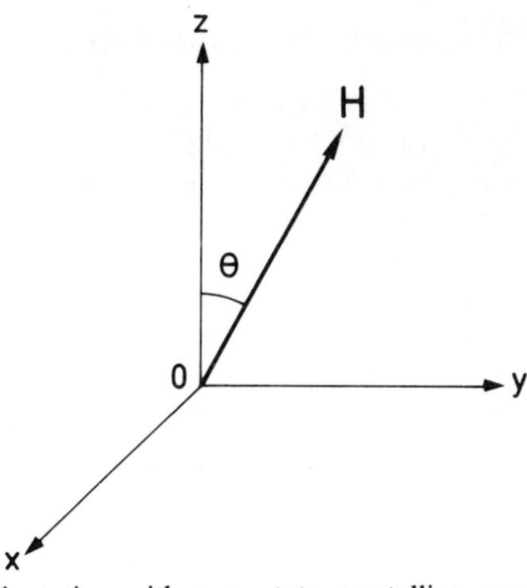

FIG. 8.45. Field orientation with respect to crystalline axes in solid ^{3}He. The anisotropy axis of the dipolar energy is along $0z$.

and:

$$\frac{\mathrm{d}}{\mathrm{d}t}(\mathbf{p}_\mathrm{A}-\mathbf{p}_\mathrm{B})=\mathbf{\Delta}\wedge(\mathbf{p}_\mathrm{A}-\mathbf{p}_\mathrm{B})-(q+r)(\mathbf{p}_\mathrm{A}-\mathbf{p}_\mathrm{B})\wedge(\mathbf{p}_\mathrm{A}+\mathbf{p}_\mathrm{B})$$
$$+s(p_{\mathrm{A}z}-p_{\mathrm{B}z})\hat{\mathbf{z}}\wedge(\mathbf{p}_\mathrm{A}+\mathbf{p}_\mathrm{B})+t(p_{\mathrm{A}z}+p_{\mathrm{B}z})\hat{\mathbf{z}}\wedge(\mathbf{p}_\mathrm{A}-\mathbf{p}_\mathrm{B}). \tag{8.221}$$

Since $s, t \ll q+r$, we can discard the last two terms of the right-hand side of (8.221). We consider only motions of small amplitude of the spins around their equilibrium positions. Under these conditions, the second term on the right-hand side of (8.220) is of second order and it can be discarded.

We introduce now the following notations:

$$\mathbf{S}=\sum_i \mathbf{I}_i=\tfrac{1}{4}N(\mathbf{p}_\mathrm{A}+\mathbf{p}_\mathrm{B});$$

$$\mathbf{d}=(\mathbf{p}_\mathrm{A}-\mathbf{p}_\mathrm{B})/2p, \qquad |\mathbf{d}|\simeq 1;$$

$$\mathbf{l}=\hat{\mathbf{z}};$$

$$\lambda = Nsp^2.$$

To within a term proportional to $|\mathbf{p}_\mathrm{A}+\mathbf{p}_\mathrm{B}|^2$, which is negligible, the dipolar energy E_d is still given by (8.216) or else, with the new notation:

$$E_\mathrm{d}=\tfrac{1}{2}\lambda(\mathbf{d}\,.\,\mathbf{l})^2. \tag{8.222}$$

Equations (8.220), (8.221), and (8.218) yield then:

$$\dot{\mathbf{d}}=\gamma\mathbf{d}\wedge\left(\mathbf{H}-\frac{\gamma\hbar}{\chi}\mathbf{S}\right); \tag{8.223}$$

$$\dot{\mathbf{S}}=\gamma\mathbf{S}\wedge\mathbf{H}-\lambda(\mathbf{d}\,.\,\mathbf{l})(\mathbf{d}\wedge\mathbf{l}). \tag{8.224}$$

These equations are identical with the Leggett equations (4.193*a*) and (4.193*b*) for the superfluid phase A of liquid ^{3}He. They could have been derived by a hydrodynamic approach analogous to that used in chapter 4. Since the motions of the vectors **S** and **d** are much slower than the exchange frequency, the state of the system can be characterized by these two macroscopic variables. The present derivation has the advantage of short-circuiting the lengthy developments of chapter 4 on superfluid ^{3}He where the Leggett equations are derived.

These equations are solved for small motions as follows. We introduce the notations:

$$\mathbf{u}=\mathbf{S}-(\chi/\gamma\hbar)\mathbf{H},$$

and:

$$\gamma^2\hbar/\chi=\alpha.$$

Equations (8.223) and (8.224) become:

$$\dot{\mathbf{d}} = \alpha \mathbf{u} \wedge \mathbf{d}$$

$$\dot{\mathbf{u}} = \mathbf{\Delta} \wedge \mathbf{u} - \lambda\, d_z (\mathbf{d} \wedge \mathbf{l})$$

with $d_x \simeq 1$, and to the first order in d_y and d_z, we obtain:

$$\begin{aligned} \dot{d}_y &= \alpha u_z \\ \dot{d}_z &= -\alpha u_y \end{aligned} \tag{8.225}$$

$$\begin{aligned} \dot{u}_x &= s\Delta u_z - c\Delta u_y \\ \dot{u}_y &= c\Delta u_x + \lambda d_z \\ \dot{u}_z &= -s\Delta u_x \end{aligned} \tag{8.226}$$

where $c = \cos\theta$, $s = \sin\theta$.

A second differentiation yields:

$$\begin{aligned} \ddot{u}_y &= -(c^2\Delta^2 + \alpha\lambda) u_y + cs\Delta^2 u_z; \\ \ddot{u}_z &= -s^2\Delta^2 u_z + cs\Delta^2 u_y. \end{aligned} \tag{8.227}$$

A periodic motion of frequency Ω corresponds to:

$$\ddot{u}_y = -\Omega^2 u_y, \qquad \ddot{u}_z = -\Omega^2 u_z,$$

whence, according to (8.227):

$$\Omega^4 - (\Delta^2 + \alpha\lambda)\Omega^2 + s^2\Delta^2\alpha\lambda = 0,$$

or:

$$\Omega^2 = \tfrac{1}{2}\{\Delta^2 + \Omega_0^2 \pm [(\Delta^2 - \Omega_0^2)^2 + 4\Delta^2\Omega_0^2 \cos^2\theta]^{1/2}\}, \tag{8.228}$$

where Δ is the Larmor frequency and:

$$\begin{aligned} \Omega_0^2 &= \alpha\lambda = \gamma^2\hbar\lambda/\chi \\ &= N\frac{\gamma^2\hbar}{\chi} sp^2. \end{aligned} \tag{8.229}$$

The resonance frequencies observed with one sample at 0·487 mK are shown in Fig. 8.46 together with three sets of best-fit curves (8.228). The frequency $\Omega_0/2\pi$ is 777·7 kHz, and the values of $\cos^2\theta$ are 0·0271, 0·976, and 0·0039. They add up to 1·007 which is sufficiently close to 1 to prove that the anisotropy axes in the three domains are perpendicular to one another as expected if they are A_4 axes.

It can be checked that with an axial asymmetry ($\lambda < 0$), the variation of the resonance frequencies as a function of the field amplitude and orientation would be very different from (8.228).

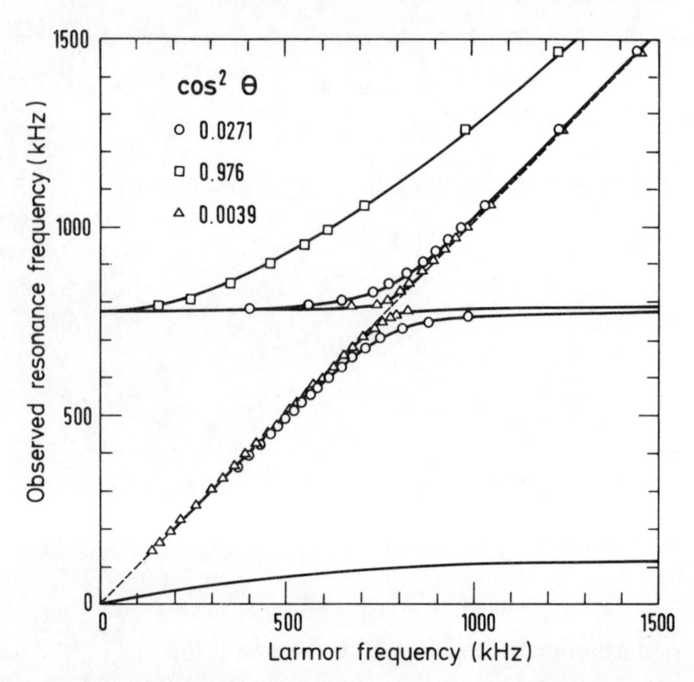

FIG. 8.46. Resonance frequencies of a magnetically ordered single crystal of ^{3}He against Larmor frequency in the external field. The solid curves correspond to eqn (8.228) with values of $\cos^2\theta$ written in the figure. (After Osheroff *et al.* 1980.)

The antiferromagnetic resonance frequency $\Omega_0/2\pi$ extrapolates to 825 kHz for $T = 0$ and decreases down to 524 kHz at T_c, where it vanishes suddenly, which is the clearest proof that the transition is of first order.

If one looks for an antiferromagnetic structure with a well defined value of $\mathbf{k}_0$, i.e.:

$$p_x^i = \pm p = a\,\exp(i\mathbf{k}_0\,.\,\mathbf{r}_i) + a^*\,\exp(-i\mathbf{k}_0\,.\,\mathbf{r}_i), \tag{8.230}$$

there are only two possibilities, as discussed in section $\mathbf{B}(a)1$:

either $\mathbf{k}_0 \| 0z$ is at the boundary of the Brillouin zone, which corresponds to alternating planes of opposite magnetizations,

or $2\mathbf{k}_0$ is at the boundary of the Brillouin zone, which corresponds to a succession of pairs of planes with alternating opposite magnetizations.

As stated in section $\mathbf{I}(a)$, for the first structure the dipolar energy is very small and could not account for the large value of Ω_0.

On the other hand, the second structure yields a reasonable account for these values. The value of χ is measured, and s is calculated along the lines of section $\mathbf{B}(a)$. The only unknown in (8.229) is the sublattice polarization p. Osheroff *et al.* (1980) assume that the zero point spin deviation $(1-p)$ is the same as that calculated for normal Heisenberg antiferromagnets, namely 0·15. The calculated value at $T=0$ is:

$$\Omega_0/2\pi = 880\ \text{kHz},$$

which is in excellent agreement with the experimental value of 825 kHz, if one considers the uncertainties in the theory, the zero point sublattice polarization, and the influence of the zero point motion of the spins on the dipolar interactions.

It is therefore very likely that the low field ordered structure is indeed this antiferromagnetic structure, called by the authors uudd (for up, up, down, down). This structure is shown in Fig. 8.47.

(c) *Multiple-spin exchange model: a summary*

On the basis of the physical arguments developed in section **G**(b) of chapter 3, one can attempt to account for the low temperature properties of solid helium 3, both above and below the ordering, by a Hamiltonian including 3 and 4-spin exchange interactions.

It results from extensive local Weiss field calculations that, in order to account for the uudd phase suggested by the NMR results, the Hamiltonian must contain both 3-spin and planar 4-spin exchange interactions (Roger, 1980).*

The simplest attempt is to use a Hamiltonian containing only these two interaction terms:

$$\mathcal{H}_e = J_t \sum_{i<j<k}^{(1)} \mathcal{P}_{ijk} - K_p \sum_{i<j<k<l}^{(2)} \mathcal{P}_{ijkl}, \qquad (8.231)$$

where the first sum is restricted to the triangular cycles of Fig. 3.19, and the second to the planar cycles of Fig. 3.20.

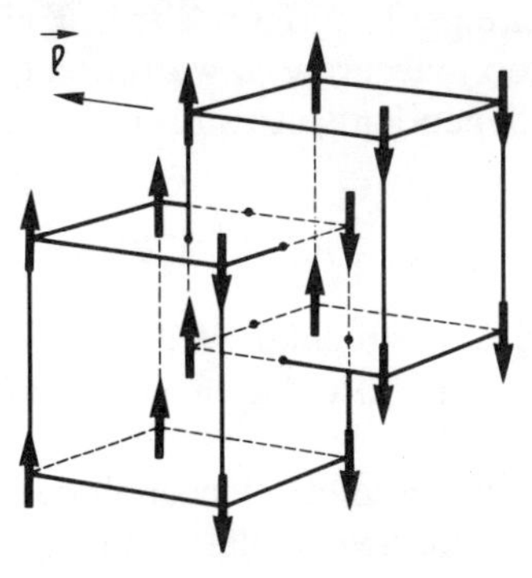

FIG. 8.47. uudd low field ordering in solid ^{3}He, as inferred from the NMR measurements.

* M. Roger (1980). Thesis no 2297, Orsay, France.

It is found that in order for the uudd phase to be the stable phase in low field, one must have:

$$4|J_t| > |K_p|.$$

The transition to this phase is of 1st order.

The two parameters J_t and K_p of the model Hamiltonian (8.231) can be adjusted to the low-field experimental values of the critical temperature T_c, the Curie–Weiss temperature θ and the coefficient $\tilde{e}_2$ of the specific heat. The best fit values:

$$J_t = -0{\cdot}1 \text{ mK},$$

$$\mathrm{K_p} = -0{\cdot}355 \text{ mK},$$

correspond to:

$$T_c = 1{\cdot}06 \text{ mK},$$

$$\theta = -2{\cdot}79 \text{ mK},$$

$$\tilde{e}_2 = 7{\cdot}15 \text{ mK}^2,$$

to be compared with the experimental values:

$$T_c = 1{\cdot}03 \text{ mK},$$

$$\theta = -3 \pm 0{\cdot}4 \text{ mK},$$

$$\tilde{e}_2 = 6{\cdot}95 \text{ mK}^2.$$

With these adjusted values, the Hamiltonian (8.231) is able to account, at least qualitatively, for *all* the observed properties of solid helium 3.

In the paramagnetic region, the coefficient B of eqn (3.192) is negative, and the coefficient $\tilde{e}_3$ of eqn (3.193) is positive. The experimental values of these coefficients are too inaccurate to warrant a quantitative comparison.

The temperature-independent susceptibility in the uudd phase is found equal to:

$$\chi \simeq -C/(1{\cdot}91\theta),$$

as compared with the experimental value $-C/(2\theta)$. According to eqn (8.229), the theory accounts also for the value of the antiferromagnetic resonance frequency.

In high field, the theory predicts a second-order transition to a canted antiferromagnetic phase: the two sublattices $\mathscr{A}$ and $\mathscr{B}$ are the imbricated simple cubic lattices of the b.c.c. crystalline lattice. The sublattice polarizations make equal angles with the external field, i.e.:

$$p_{\mathrm{A}z} = p_{\mathrm{B}z}, \qquad p_{\mathrm{A}x} = -p_{\mathrm{B}x}.$$

The theory gives an excellent account of the large magnetization and of its small dependence on the field, as well as of the field dependence of the transition temperature.

However, the field of transition between the uudd phase and the canted antiferromagnetic phase is calculated to be 12·1 kG, which is much larger than the experimental value of 4·1 kG. This faulty prediction might be due to the crudeness of the Weiss field approximation.

Finally, a spin wave calculation has been used to compute the low temperature specific heat in the uudd phase. The agreement with the experimental result is satisfactory.

The Hamiltonian (8.231) is certainly approximate, and the true Hamiltonian should also contain 2-spin and folded 4-spin exchange interactions. However, any attempt to improve the agreement between theory and experiment by increasing the number of adjustable parameters is pointless in the absence of a theory of magnetic ordering much more reliable than the existing approximate theories.

The ability of the model Hamiltonian (8.231) to describe reasonably well the many available experimental properties with only two adjustable parameters points strongly toward the physical reality of 3 and 4-spin exchange interactions in solid helium 3. In any case, it is a strong incentive for more theoretical and experimental work on the fascinating subject of solid ^{3}He at very low temperatures.

REFERENCES

Abragam, A. (1960). *C.r. hebd. Séanc. Acad. Sci., Paris* **251**, 225.

—— (1961). *The principles of nuclear magnetism.* Clarendon Press, Oxford.

—— (1962). *C.r. hebd. Séanc. Acad. Sci., Paris* **254**, 1267.

—— (1963). *Cryogenics* **3**, 42.

—— (1965). *Cryogenics* **5**, 107.

—— (1973). *Trends in Physics.* European Physical Society, Geneva, Switzerland (p. 177).

—— (1976). *C.r. hebd. Séanc. Acad. Sci., Paris* **282**, B 247.

—— Bleaney, B. (1970). *Electron paramagnetic resonance of transition ions.* Clarendon Press, Oxford.

—— Borghini, M. (1964). *Progr. low Temp. Phys.* (ed. C. J. Gorter) **4**, 384. North-Holland, Amsterdam.

—— Chapellier, M. (1964). *Phys. Lett.* **11**, 207.

—— Goldman, M. (1978). *Rep. Prog. Phys.* **41**, 395.

—— Proctor, W. G. (1958). *C.r. hebd. Séanc. Acad. Sci. Paris* **246**, 2253.

—— Winter, J. (1959). *C.r. hebd. Séanc. Acad. Sci., Paris* **249**, 1633.

—— Bouffard, V., and Roinel, Y. (1976). *J. magn. Reson.* **22**, 53.

—— —— —— (1980). *C.r. hebd. Séanc. Acad. Sci., Paris.* **290**, B 203.

—— Chapellier, M., Jacquinot, J. F., and Goldman, M. (1973). *J. magn. Reson.* **10**, 322.

—— Jacquinot, J. F., Chapellier, M., and Goldman, M. (1972). *J. Phys. C.* **5**, 2629.

—— Chapellier, M., Goldman, M., Jacquinot, J. F., and Vu Hoang Chau (1971). *Proc. 2nd Int. Conf. Polar. Targets* (Ed. G. Shapiro). Lawrence Berkeley Lab., Berkeley (p. 247).

—— Bacchella, G. L., Glättli, H., Meriel, P., Piesvaux, J., and Pinot, M. (1972). *C.r. hebd. Séanc. Acad. Sci., Paris* **274**, B 423.

—— —— —— —— —— —— (1976). *Physica, 's Grav.* **81B**, 245.

—— —— —— —— Pinot, M., and Piesvaux, J. (1973). *Phys. Rev. Lett.* **31**, 776.

—— —— Long, C., Meriel, P., Piesvaux, J., and Pinot, M. (1972). *Phys. Rev. Lett.* **28**, 805.

—— Borghini, M., Catillon, P., Coustham, J., Roubeau, P., and Thirion, J. (1962). *Phys. Lett.* **2**, 310.

—— Bacchella, G. L., Bouffard, V., Goldman, M., Meriel, P., Pinot, M., Roinel, Y., and Roubeau, P. (1978). *C.r. hebd. Séanc. Acad. Sci., Paris* **286**, B 311.

Ahonen, A. I., Krusius, M., and Paalanen, M. A. (1975). *Low. Temp. Phys. LT14* (ed. M. Krusius and M. Vuorio) **1**, 107. North-Holland, Amsterdam, and American Elsevier, New York.

—— Alvesalo, T. A., Haikala, M. T., Krusius, M., and Paalanen, M. A. (1975). *Phys. Lett.* **51A**, 279.

Akopyan, G. G., Alfimenkov, V. P., Lason, L., Ovchinnikov, O. N., and Shapiro, E. I. (1975). *Zh. éksp. teor. Fiz.* **69**, 777. [Translat.: *Soviet Phys. JETP* (1976) **42**, 397.]

Ambegaokar, V. P., De Gennes, P. G., and Rainer, D. (1974). *Phys. Rev., Sec. A* **9**, 2676.
Anderson, P. W. (1963). *Concepts in solids*. Benjamin, New York. [Ch. 3].
Anderson, P. W. and Brinkman, W. F. (1978). *Theory of anisotropic superfluidity in ^{3}He*, in: *The physics of liquid and solid helium* (Ed. K. H. Bennemann and J. B. Ketterson). Part II, p. 177. Wiley and Sons, New York.
Andrew, E. R. (1971). *Prog. nucl. magn. reson. Spectrosc.* **8**, 1.
—— Firth, M., Jasinski, A., and Randall, P. J. (1970). *Phys. Lett.* **31A**, 446.
Atsarkin, V. A. (1970). *Zh. éksp. teor. Fiz.* **58**, 1884. [Translat.: *Soviet Phys. JETP* **31**, 1012.]
Aue, W. P., Bartholdi, E. and Ernst, R. R. (1976). *J. chem. Phys.* **64**, 2229.
Balian, R. and Werthamer, N. R. (1963) *Phys. Rev.* **131**, 1553.
Barnaal, D. and Lowe, I. J. (1963). *Phys. Rev. Lett.* **11**, 258.
Baryshevskii, V. G. and Podgoretskii, M. I. (1964). *Zh. éksp. teor. Fiz.* **47**, 1050. [Translat.: *Soviet Phys. JETP* **20**, 704.]
Bernier, M. and Delrieu, J. M. (1976). In: *Magnetic resonance and related phenomena. Proc. XIXth Congress Ampère, Heidelberg* (ed. H. Brunner, K. H. Hausser, and D. Schweitzer) Groupement Ampère, Heidelberg and Geneva.
Bernier, M. and Landesman, A. (1971). *J. Phys (Fr.)* **32**, C5a-213.
Blatt, J. M. and Weisskopf, V. F. (1952). *Theoretical nuclear physics*. Wiley and Sons, New York.
Bloembergen, N. (1949). *Physica, 's Grav.* **15**, 386.
Borghini, M. (1968). *Phys. Rev. Lett.* **20**, 419.
—— De Boer, W., and Morimoto, K. (1974). *Phys. Lett.* **48A**, 244.
Borisov, N. S., Glonti, L. N., Kazarinov, M. Yu., Kazarinov, Yu. M., Kiselev, Yu. F., Kiselev, V. S., Matafonov, V. N., Macharashvili, G. G., Neganov, B. S., Strakhota, I., Trofimov, V. N., Usov, Yu. A., and Khachaturov, B. A. (1977). *Zh. éksp. teor. Fiz.* **72**, 405. [Translat.: *Soviet Phys. JETP* **45**, 212.]
Bragg, W. and Williams, E. (1934). *Proc. Roy. Soc.* **145**, 609.
Brinkman, W. F. (1974). *Phys. Lett.* **49A**, 411.
—— Smith, S. (1975*a*). *Phys. Lett.* **51A**, 449.
—— —— (1975*b*). *Phys. Lett.* **53A**, 43.
—— —— Osheroff, D. D., and Blount, E. I. (1974). *Phys. Rev. Lett.* **33**, 624.
Brout, R. (1965*a*) *Phase transitions*. Benjamin, New York.
—— (1965*b*). *Statistical mechanics of ferromagnetism*. In *Magnetism*, Vol II A, p. 43 (ed. G. T. Rado and H. Suhl). Academic Press, New York.
Buishvili, L. L. (1965). *Zh. éksp. teor. Fiz.* **49**, 1886. [Translat.: (1966) *Soviet Phys. JETP* **22**, 1277.]
Button-Shafer, J. (1979). In: *High energy phys. with polarized beams and polarized targets*, p. 41 (Ed. G. H. Thomas.) American Institute of Physics, New York.
—— Lichti, R. I., and Potter, W. H. (1977). *Phys. Rev. Lett.* **39**, 677.
Chang, J. J., Griffin, R. G., and Pines, A. (1975). *J. chem. Phys.* **62**, 4923.
Chapellier, M. (1971). *Proc. 12th Int. Conf. on Low Temp. Phys.* (p. 637) (ed. Eizo Kanda). Academic Press of Japan.
—— Goldman, M., Vu Hoang Chau, and Abragam, A. (1969). *C.r. hebd. Séanc. Acad. Sci., Paris* **268B**, 1530.
Cohen, M. H. and Keffer, F. (1955). *Phys. Rev.* **99**, 1128.
Cohen-Tannoudji, C., Diu, B., and Laloe, F. (1973). *Mécanique quantique*, p. 944. Hermann, Paris.
Cooper, L. N. (1956). *Phys. Rev.* **104**, 1189.

Cox, S. F. J., Bouffard, V., and Goldman, M. (1973). *J. Phys. C* **6**, L100.
——————(1975). *J. Phys. C* **8**, 3664.
—— Read, S. F. J., and Wenckebach, W. Th. (1977). *J. Phys. C* **10**, 2917.
De Boer, W. (1976). *J. low temp. Phys.* **22**, 185.
—— and Niinikoski, T. O. (1974). *Nucl. Instrum. Meth.* **114**, 495.
De Boer, W., Borghini, M., Morimoto, K., and Niinikoski, T. O. (1974) *J. Low Temp. Phys.* **15**, 249.
Delrieu, J. M. and Roger, M. (1978). *J. Phys. (Fr.)* **39**, C6-123.
Demco, D. E., Tegenfeld, J., and Waugh, J. S. (1975). *Phys. Rev. B* **11**, 4133.
Deimling, M., Brunner, H., Dinse, K. P., Hausser, K. H., and Colpa, J. P. (1980). *J. Magn. Res.* **39**, 185.
Deville, G. (1976). *J. Phys. (Fr.)* **37**, 781.
—— Bernier, M., and Delrieu, J. M. (1979). *Phys. Rev. B* **19**, 5666.
Dirac, P. A. M. (1930). *The principles of quantum mechanics.* Clarendon Press, Oxford.
Engelsberg, S., Brinkman, W. F., and Anderson, P. W. (1974). *Phys. Rev. A* **9**, 2592.
Englert, F. (1963). *Phys. Rev.* **129**, 567.
Faughnan, B. W. and Strandberg, M. W. P. (1961). *J. Phys. Chem. Solids* **19**, 155.
Garroway, A. N., Mansfield, P., and Stalker, D. C. (1975). *Phys. Rev. B* **11**, 121.
Genack, A. Z. and Redfield, A. G. (1975). *Phys. Rev. B* **12**, 78.
Gillet, V. and Normand, J. M. (1971). *Nucl. Phys. A* **176**, 225.
Glättli, H., Abragam, A., Bacchella, G. L., Fourmond, M., Mériel, P., Piesvaux, J., and Pinot, M. (1978). *Phys. Rev. Lett.* **40**, 748.
—— Bacchella, G. L., Fourmond, M., Malinovski, A., Mériel, P., Pinot, M., Roubeau, P., and Abragam, A. (1979). *J. Phys. (Fr.)* **40**, 629.
Glauber, R. J. (1962). In *Lectures in theoretical physics* Vol. IV, p. 571 (ed. W. E. Brittin, B. W. Downs, and J. Downs). Interscience, New York.
Goldman, M. (1970). *Spin temperature and nuclear magnetic resonance in solids.* Clarendon Press, Oxford.
—— (1975). *J. magn. Reson.* **17**, 393.
—— (1977). *Phys. Rep.* **32G**, 1.
—— (1979). *J. Magn. magnetic Mat.* **14**, 105.
—— (1980). *J. Phys. (Fr.)* **41**, 885.
—— Jacquinot, J. F., (1976*a*). *Phys. Rev. Lett.* **36**, 330.
————, (1976*b*). *J. Phys. (Fr.)* **37**, 617.
———— (1981). *To be published.*
—— Sarma, G. (1975). *J. Phys. (Fr.)* **36**, 1353.
—— Cox, S. F. J., and Bouffard, V. (1974). *J. Phys. C* **7**, 2940.
—— Roinel, Y., and Bouffard, V. *To be published.*
—— Chapellier, M., Vu Hoang Chau, and Abragam, A. (1974). *Phys. Rev. B* **10**, 226.
—— Jacquinot, J. F., Chapellier, M., and Vu Hoang Chau (1975). *J. Magn. Reson.* **18**, 22.
Gully, W. J., Gould, C. M., Richardson, R. C., and Lee, D. M. (1976). *J. low temp. Phys.* **24**, 563.
Gunter, T. E. and Jeffries, C. D. (1966). *Bull. Am. Phys. Soc.* **11**, 906.
———— (1967). *Phys. Rev.* **159**, 290.
Haeberlen, U. (1976). *High resolution NMR in solids. Selective averaging. Supplement 1 of Advances in magnetic resonance* (ed. J. S. Waugh). Academic Press, New York.

—— Waugh, J. S. (1968). *Phys. Rev.* **175**, 453.
—— Kohlschutter, U., Kempf, J., Spiess, H. W., and Zimmermann, H. (1974). *Chem. Phys.* **3**, 248.
Halperin, W. P., Rasmussen, F. B., Archie, C. N., and Richardson, R. C. (1978). *J. low temp. Phys.* **31**, 61.
—— Archie, C. N., Rasmussen, F. B., Buhrman, R. A., and Richardson, R. C. (1974). *Phys. Rev. Lett.* **32**, 927.
Hansch, T. W., and Schawlow, A. L. (1975). *Opt. Commun.* **13**, 68.
Hartmann, S. R. and Hahn, E. L. (1962). *Phys. Rev.* **128**, 2042.
Harris, A. B. (1971). *Solid State Commun.* **9**, 2255.
Herpin, A. and Mériel, P. (1973). *J. Phys. (Fr.)* **34**, 423.
Herring, C. (1966). *Exchange interactions among itinerant electrons.* In: *Magnetism* Vol. IV (ed. G. T. Rado and H. Suhl). Academic Press, New York.
Hester, R. K., Ackerman, J. L., Cross, V. R. and Waugh, J. S. (1975). *Phys. Rev. Lett.* **34**, 993.
—— —— Neff, B. L., and Waugh, J. S. (1976). *Phys. Rev. Lett.* **36**, 1081.
—— Cross, V. R., Ackerman, J. L., and Waugh, J. S. (1975). *J. Chem. Phys.* **63**, 3606.
Hetherington, J. H. (1968). *Phys. Rev.* **176**, 231.
—— Willard, F. D. C. (1975). *Phys. Rev. Lett.* **35**, 1442.
Hirschfelder, J. O., Curtiss, C. F., and Bird, R. B. (1954). *Molecular theory of gases and liquids* (p. 1040). John Wiley, New York.
Horwitz, G. and Callen, H. (1961). *Phys. Rev.* **124**, 1757.
Hwang, C. F. and Hill, D. A. (1967). *Phys. Rev. Lett.* **19**, 1011.
Ito, Y. and Shull, C. G. (1969). *Phys. Rev.* **185**, 961.
Ivanov, Yu. N., Provotorov, B. N. and Fel'dman, E. B. (1978). *Pis'ma Zh. éksp. teor. Fiz.* **27**, 164. [Translat.: *JETP Lett.* (1978) **27**, 153.]
Jacquinot, J. F., Chapellier, M., and Goldman, M. (1974). *Phys. Lett.* **48A**, 303.
—— Lounasmaa, O. V., and Urbina, C. (1978). *Physica, 's Grav.* **95B**, 76.
—— Wenckebach, W. Th., Goldman, M., and Abragam, A. (1974). *Phys. Rev. Lett.* **32**, 1096.
—— —— Chapellier, M., Goldman, M., and Abragam, A. (1974). *C.r. hebd. Séanc. Acad. Sci., Paris* **278**, B 93.
Jeener, J. and Broekaert, P. (1967). *Phys. Rev.* **157**, 232.
Jeffries, C. D. (1963). *Cryogenics* **3**, 41.
—— (1972). In *Electron paramagnetic resonance*, p. 244 (ed. S. Geshwind). Plenum Press, New York.
Kessemeier, H. and Norberg, R. E. (1967). *Phys. Rev.* **155**, 321.
King, A. R., Wolfe, J. P., and Nallard, R. L. (1972). *Phys. Rev. Lett.* **28**, 109.
Kirkwood, J. G. (1938). *J. Chem. Phys.* **6**, 70.
Kittel, C. (1948). *Phys. Rev.* **73**, 155.
—— (1963). *Quantum Theory of Solids*, Ch. 4. Wiley and Sons, New York.
Koester, L., and Steyerl, A. (1977). *Neutron physics.* Springer-Verlag, Berlin.
Kozhushner, M. A., and Provotorov, B. N. (1964). *Proc. Krasnoyarks Conf.* In: Fiz. Tverdogo Tela **6**, 1472. [Translat.: *Soviet Phys. Solid State* **6**, 1152.]
Kubo, R. (1957). *J. Phys. Soc. Japan* **12**, 570.
Lado, F., Memory, J. D., and Parker, G. W. (1971). *Phys. Rev. Sec. B.* **4**, 1406.
Landau, D. L. (1956). *Zh. éksp. teor. Fiz.* **30**, 1058. [Translat.: *Soviet Phys. JETP* (1957) **3**, 920.]
—— Lifschitz, E. M. (1960). *Electrodynamics of continuous media.* Pergamon, New York.

Landesman, A. (1973). *Ann. Phys. (Fr.)* **8**, 53.
—— (1974). *J. low temp. Phys.* **17**, 365.
—— (1975). *Ann. Phys. (Fr.)* **9**, 69.
—— (1978). *J. Phys. (Fr.)* **39**, C6-1305.
Lang, D. V. and Moran, P. R. (1970). *Phys. Rev. Sec. B.* **1**, 53.
Lee, M. and Goldburg, W. I. (1965). *Phys. Rev.* **140**, A 1261.
Leggett, A. J. (1972). *Phys. Rev. Lett.* **29**, 1227.
Leggett, A. J. (1973). *Phys. Rev. Lett.* **31**, 352.
—— (1974). *Ann. of Physics* **85**, 11.
—— (1975). *Rev. Mod. Phys.* **47**, 331.
—— Takagi, S. (1975). *Phys. Rev. Lett.* **34**, 1424.
—— —— (1977). *Ann. of Physics* **106**, 79.
Lushchikov, V. I., Taran, Yu. V., and Shapiro, F. L. (1969). *Jadernaja Fiz.* **10**, 1178. [Translat.: *Sov. J. Nucl. Phys.* (1970) **10**, 669.]
McArthur, D. A., Hahn, E. L., and Walstedt, R. E. (1969). *Phys. Rev.* **188**, 609.
Malinovski, A., Coustham, J., and Glättli, H. *To be published.*
Mansfield, P. (1971). *J. Phys. C* **4**, 1444.
Marks, J., Wenckebach, W. Th., and Poulis, N. J. (1979). *Physica, 's Grav.* **96B**, 337.
Marks, J., Wenckebach, W. Th., and Poulis, N. J. *To be published.*
Mattis, D. C. (1965). *The theory of magnetism*, Ch. 6. Harper and Row, New York.
Mehring, M. (1976). *High resolution NMR spectroscopy in solids.* Springer-Verlag, Berlin.
—— Waugh, J. S. (1972). *Phys. Rev. sec. B.* **5**, 3459.
—— Griffin, R. G., and Waugh, J. S. (1971). *J. Chem. Phys.* **55**, 746.
—— Raber, H., and Sinning, G. (1974). *Proc. 18th Ampère Congress, Nottingham* p. 35 (ed. P. S. Allen, E. R. Andrew and C. A. Bates), The University Nottingham.
—— Sinning, G., and Pines, A. (1976). *Zeitschrift Phys.* **B24**, 53.
Melikiya, M. G. (1968). *Fizika tverd. Tela* **10**, 858. [Translat.: *Soviet Phys. solid St.* **10**, 673.]
Mori, H. (1965). *Prog. theor. Phys. Kyoto* **33**, 423.
Mouritsen, O. G. and Knak Jensen, S. J. (1978). *Phys. Rev. Sec. B.* **18**, 465.
—— —— (1980). *Phys. Rev. Sec. B.* **22**, 1127.
Niinikoski, T. O. and Udo, F. (1976). *Nucl. Instrum. Meth.* **134**, 219.
Odehnal, M. (1967). *C.r. hebd. Séanc. Acad. Sci., Paris* **264**, 334.
—— (1975). *Phys. Lett.* **53A**, 9.
Osheroff, D. D. (1975). *Low temp. Phys. LT14* Vol. 1, p. 100 (ed. M. Krusius and M. Vuorio) North-Holland, Amsterdam and American Elsevier, New York.
—— Corruccini, L. R. (1975). *Phys. Lett.* **51A**, 447.
—— Cross, M. C., and Fisher, D. S. (1980). *Phys. Rev. Lett.* **44**, 792.
—— Gully, W. J., Richardson, R. C., and Lee, D. M. (1972). *Phys. Rev. Lett.* **29**, 920.
—— Richardson, R. C., and Lee, D. M. (1972). *Phys. Rev. Lett.* **28**, 885.
Pines, A. and Shattuck, T. W. (1974). *J. Chem. Phys.* **61**, 1255.
—— Waugh, J. S. (1972). *J. Magn. Reson.* **8**, 354.
—— —— (1974). *Phys. Lett.* **47A**, 337.
—— Gibby, M. G., and Waugh, J. S. (1972*a*). *Chem. Phys. Lett.* **15**, 373.
—— —— —— (1972*b*). *J. Chem. Phys.* **56**, 1776.
—— —— —— (1973). *J. Chem. Phys.* **59**, 569.
Prewitt, T. C. and Goodkind, J. M. (1977). *Phys. Rev. Lett.* **39**, 1283.
Provotorov, B. N. (1961). *Zh. éksp. teor. Fiz.* **41**, 1582. [Translat.: *Soviet Phys. JETP* (1962) **14**, 1126.]
Ramsey, N. F. (1949). *Phys. Rev.* **76**, 996.

Redfield, A. G. (1955). *Phys. Rev.* **98**, 1787.
—— Yu, W. N. (1968). *Phys. Rev.* **169**, 443.
—— —— (1969). *Phys. Rev.* **177**, 1018.
Rhim, W. K., Burum, D. P., and Elleman, D. D. (1976). *Phys. Rev. Lett.* **37**, 1764.
—— Elleman, D. D., and Vaughan, R. W. (1973*a*). *J. Chem. Phys.* **58**, 1772.
—— —— —— (1973*b*). *J. Chem. Phys.* **59**, 3740.
—— Pines, A., and Waugh, J. S. (1971). *Phys. Rev. Sec. B.* **3**, 684.
Richardson, R. C., Landesman, A., Hunt, E. and Meyer, H. (1966). *Phys. Rev.* **146**, 244.
Robinson, F. N. H. (1963). *Phys. Lett.* **4**, 180.
Rodak, M. I. (1973). *Fiz. tverd. Tela* **15**, 404. [Translat.: *Soviet Phys. solid St.* **15**, 290.]
Roger, M., Delrieu, J. M., and Hetherington, J. H. (1980). *J. Phys. (Fr.)* **41**, C7–241.
Roinel, Y. and Bouffard, V. (1975). *J. Magn. Reson.* **18**, 304.
—— —— (1976). In *Magnetic resonance and related phenomena* p. 615 (Ed. K. H. Hausser and D. Schweitzer) Groupement Ampère, Heidelberg, Geneva.
—— —— (1977). *C.r. hebd. Séanc. Acad. Sci., Paris* **284**, B29.
—— Winter, J. M. (1970). *J. Phys. (Fr.)* **31**, 351.
—— Bouffard, V., and Roubeau, P. (1978). *J. Phys. (Fr.)* **39**, 1097.
—— Bacchella, G. L., Avenel, O., Bouffard, V., Pinot, M., Roubeau, P., Mériel, P., and Goldman, M. (1980). *J. Phys. (Fr.) Lett.* **41**, L-123.
Roinel, Y., Bouffard, V., Bacchella, G. L., Pinot, M., Mériel, P., Roubeau, P., Avenel, O., Goldman, M., and Abragam, A. (1978). *Phys. Rev. Lett.* **41**, 1572.
Rybaczewski, E. F., Neff, B. L., Waugh, J. S. and Sherfinski, J. S. (1977). *J. Chem. Phys.* **67**, 1231.
Schaefer, J. and Stejskal, E. O. (1976). *J. Am. chem. Soc.* **98**, 1031.
Schmugge, T. J. and Jeffries, C. D. (1065). *Phys. Rev.* **138**, A1785.
Shull, C. G. and Ferrier, R. P. (1963). *Phys. Rev. Lett.* **10**, 295.
Smart, J. S. (1966). *Effective field theories of magnetism.* Saunders, Philadelphia.
Solomon, I. (1963). In *Magnetic and electric resonance and relaxation*, p. 25 (ed. J. Smidt) North-Holland, Amsterdam.
Squires, G. L. (1978). *Thermal neutron scattering.* Cambridge University Press, Cambridge.
Stinchcombe, R., Horwitz, G., Englert, F., and Brout, R. (1963). *Phys. Rev.* **130**, 155.
Stokes, H. T. and Ailion, D. C. (1977). *Phys. Rev. Sec. B.* **15**, 1271.
Strombotne, R. L. and Hahn, E. L. (1964). *Phys. Rev.* **133**, A1616.
Sullivan, N. S. and Chapellier, M. (1974). *J. Phys. C* **7**, L195.
—— Deville, G., and Landesman, A. (1975). *Phys. Rev. Sec. B.* **11**, 1858.
Thouless, D. J. (1965). *Proc. Phys. Soc.* **86**, 893.
Trickey, S. B., Kirk, W. P., and Adams, E. D. (1972). *Rev. Mod. Phys.* **44**, 668.
Tyablikov, S. V. (1967). *Methods in the quantum theory of magnetism.* Plenum Press, New York.
Urbina, C. and Jacquniot, J. F. (1980). *Physica, 's Grav.* **100B**, 333.
Urbina, C. and Jacquinot, J. F. *To be published.*
Uttley, C. A. and Diment, K. M. (1968). *Report No. AERE-PR/NP14*; *(1969). Reports No. AERE-PR/NP15 and AERE-PR/NP16.* (Atomic Energy Research Establishment, Harwell, England. Copies available on request.)
Van Den Heuvel, G. M., Swanenburg, T. J. B., and Poulis, N. J. (1971). *Physica, 's Grav.* **56**, 356.

Vermeulen, J. (1972). *Proc. 2nd Int. Conf. on Polarized Targets, Berkeley 1971, LB500, UC-34 Physics,* p. 69 (ed. G. Shapiro). National Technical Information Service, Springfield, Virginia.
Villain, J. (1959). *Physics Chem. Solids* **11**, 303.
Vu Hoang Chau, Chapellier, M., and Goldman, M. (1969). *J. Phys.* (*Fr.*) **30**, 427.
Waugh, J. S., Huber, L. M., and Haeberlen, U. (1968). *Phys. Rev. Lett.* **20**, 180.
Webb, R. A., Kleinberg, R. L., and Wheatley, J. C. (1974). *Phys. Lett.* **48A**, 421.
—— Sager, R. E., and Wheatley, J. C. (1975). *Phys. Rev. Lett.* **35**, 1164.
Wenckebach, W. Th., Van Den Heuvel, G. M., Hoogstraate, H., Swanenburg, T. J. B., and Poulis, N. J. (1969). *Phys. Rev. Lett.* **22**, 581.
Wheatley, J. C. (1975). *Rev. Mod. Phys.* **47**, 415.
Wineland, D. J., Drullinger, R. E., and Walls, F. L. (1978). *Phys. Rev. Lett.* **40**, 1639.
Zubarev, D. N. (1960). *Usp. fiz. Nauk* **71**, 71. [Translat.: *Soviet Phys. Usp.* **3**, 320]

NAME INDEX

Abragam, A. 1, 11, 15, 26, 29, 49, 51, 52, 53, 55, 56, 63, 94, 108, 123, 125, 166, 288, 289, 290, 291, 292, 294, 296, 303, 305, 310, 339, 341, 342, 346, 348, 359, 360, 371, 376, 390, 391, 392, 394, 400, 401, 402, 403, 404, 406, 408, 419, 434, 438, 439, 441, 442, 446, 447, 451, 470, 472, 477, 489, 492, 512, 514, 515, 532, 536, 537, 562
Ackerman, J. L. 89, 90, 92
Adams, E. D. 116
Ahonen, A. I. 223, 246, 247
Ailion, D. C. 38
Akopyan, G. C. 443
Alfimenkov, V. P. 443
Ambegaokar, V. P. 256
Anderson, P. W. 187, 234, 589
Andrew, E. R. 58, 60
Archie, C. N. 108
Atsarkin, V. A. 365, 369
Aue, W. P. 91
Avenel, D. 536, 537

Bacchella, G. L. 348, 360, 434, 441, 442, 446, 447, 536, 537
Balian, R. 187, 196, 202
Bardeen, J. 194
Barnaal, D. 63
Bartholdi, E. 91
Baryshevskii, V. G. 452
Bernier, M. 120, 162, 163, 164, 166
Bird, R. B. 129
Blatt, J. M. 422
Bleaney, B. 341, 342
Bloembergen, N. 134
Blount, E. I. 257
Borghini, M. 376, 379, 393, 398, 406
Borisov, N. S. 407
Bouffard, V. 283, 284, 285, 286, 287, 298, 311, 346, 370, 371, 395, 397, 401, 403, 404, 536, 537
Bragg, W. 501
Brinkman, W. F. 187, 234, 248, 251, 257, 261, 272
Broekaert, P. 44, 45
Brout, R. 306, 315, 505, 596
Brunner, H. 588
Buishvili, L. L. 376
Buhrman, R. A. 108
Burum, D. P. 99
Button-Shafer, J. 410

Callen, H. 315
Catillon, P. 406
Chang, J. J. 89
Chapellier, M. 138, 140, 288, 289, 290, 291, 292, 294, 295, 296, 327, 337, 346, 360, 400, 419, 477, 489, 492, 512, 514, 515, 523, 524, 562, 571
Cohen, M. H. 167, 168, 327, 487, 490
Cohen-Tannoudji, C. 423
Colpa, J. P. 588
Cooper, L. N. 193, 194
Corruccini, L. R. 255
Coustham, J. 406, 447, 450
Cox, S. F. J. 355, 370, 371, 494, 519, 520
Cross, M. C. 108
Cross, V. R. 90, 92
Curtiss, C. F. 129

De Boer, W. 393, 394, 395, 396, 398, 399, 401
De Gennes, P. G. 256
Deimling, M. 588
Delrieu, J. M. 114, 166, 180, 187
Demco, D. E. 39, 40, 41
Deville, G. 127, 141, 142, 143, 155, 166, 175, 176, 178, 179
Diment, K. M. 456
Dinse, K. P. 588
Dirac, P. A. M. 113, 114
Diu, B. 423
Drullinger, R. E. 417

Elleman, D. D. 75, 99
Engelsberg, S. 234
Englert, F. 315
Ernst, R. R. 91

Faughnan, B. W. 346
Fel'dman, E. B. 105
Ferrier, R. P. 442
Firth, M. 60
Fisher, D. S. 108
Fourmond, M. 447

Garroway, A. N. 83
Gibby, M. G. 86
Gillet, V. 441
Glattli, H. 348, 434, 446, 447, 450, 456
Glauber, R. J. 452
Glonti, L. N. 407
Goldburg, W. I. 63

Goldman, M. 1, 11, 22, 24, 35, 36, 49, 51, 63, 101, 156, 288, 289, 290, 291, 292, 294, 296, 298, 299, 302, 303, 306, 310, 311, 327, 335, 336, 337, 338, 346, 360, 364, 366, 370, 371, 419, 465, 471, 477, 489, 492, 509, 512, 514, 517, 517, 524, 529, 531, 532, 534, 535, 536, 537, 539, 540, 546, 547, 549, 550, 551, 552, 554, 555, 562, 571, 572, 582, 585, 586, 587
Goodkind, J. M. 179, 600
Gould, C. M. 245
Griffin, R. G. 85, 89
Gully, W. J. 244, 245
Gunter, T. E. 351

Haeberlen, U. 54, 64, 73, 83, 84
Hahn, E. L. 24, 32, 45
Halperin, W. P. 108
Hänsch, T. W. 417
Harris, A. B. 127
Hartmann, S. R. 32
Hausser, K. H. 588
Herpin, A. 443
Herring, C. 115
Hester, R. K. 89, 90, 92
Hetherington, J. H. 120, 180
Hill, D. A. 375
Hirschfelder, J. O. 129
Hoogstraate, H. 369
Horwitz, G. 315
Huber, L. M. 73
Hunt, E. 141
Hwang, C. F. 375

Ito, Y. 442, 443
Ivanov, Yu. N. 105

Jacquinot, J. F. 288, 289, 290, 291, 292, 310, 312, 327, 337, 346, 362, 412, 413, 415, 415, 416, 515, 524, 529, 531, 534, 535, 586
Jasinski, A. 60
Jeener, J. 44, 45
Jeffries, C. D. 351, 359, 408

Kazarinov, Yu. M. 407
Kazarinov, M. Yu. 407
Keffer, F. 167, 168, 327, 487, 490
Kempf, J. 84
Kessemeier, H. 59
Khachaturov, B. A. 407
King, A. R. 354
Kirk, W. P. 116
Kirkwood, J. G. 505, 507
Kittel, C. 282, 589, 595
Kleinberg, R. L. 254
Knak Jensen, S. J. 598
Koester, L. 441
Kohlshutter, U. 84
Kozhushner, M. A. 376
Krusius, M. 223, 246, 247
Kubo, R. 302

Lado, F. 28
Laloe, F. 423
Landau, D. L. 189, 282
Landesman, A. 115, 116, 119, 120, 139, 141, 144, 153, 154, 155, 160, 162, 163, 164, 165, 180, 182
Lason, L. 443
Lee, D. M. 187, 194, 244, 245
Lee, M. 63
Leggett, A. J. 186, 196, 218, 223, 225, 234, 244, 245, 258, 263, 264, 266
Lichti, R. L. 410
Lifschitz, E. M. 282
Long, C. 360, 441, 442, 446
Lounasmaa, O. V. 362
Lowe, I. J. 63
Lushchikov, V. I. 458

McArthur, D. A. 32, 37
Malinovski, A. 447
Mansfield, P. 75, 83
Marks, J. 359, 360, 397, 495, 496, 556, 557, 558, 559
Mattis, D. C. 589
Mehring, M. 16, 31, 38, 54, 73, 83, 85, 93, 95
Melikiya, M. G. 366
Memory, J. D. 28
Meriel, P. 348, 360, 434, 441, 442, 443, 446, 447, 536, 537
Meyer, H. 141
Mori, H. 16
Morimoto, K. 393
Mouritsen, O. G. 598

Nallard, R. L. 354
Neff, B. L. 89, 92
Neganov, B. S. 407
Niinikoski, T. O. 393, 406, 407
Norberg, R. E. 59
Normand, J. M. 441

Osheroff, D. D. 108, 187, 194, 244, 255, 256, 257, 601, 606
Ovchinnikov, O. N. 443

Paalanen, M. A. 223, 246, 247
Parker, G. W. 28
Piesvaux, J. 348, 360, 434, 441, 442, 446
Pines, A. 38, 76, 86, 89, 93, 97, 99

Pinot, M. 348, 360, 434, 441, 442, 446, 447, 536, 537
Podgoretskii, M. I. 452
Potter, W. H. 410
Poulis, N. J. 359, 369, 556, 557
Prewit, T. C. 179, 600
Proctor, W. G. 339
Provotorov, B. N. 8, 105, 376, 376

Raber, H. 38
Rainer, D. 256
Ramsey, N. F. 444
Randall, P. J. 60
Rasmussen, F. B. 108
Read, S. F. J. 355, 362
Redfield, A. G. 63, 137, 149, 375, 376
Rhim, W. K. 75, 82, 83, 84, 99, 102, 103, 104
Richardson, R. C. 108, 141, 187, 194, 244, 245
Rodak, M. I. 364
Roger, M. 114, 180, 185, 600, 607
Roinel, Y. 283, 284, 285, 286, 287, 298, 311, 346, 371, 395, 397, 401, 403, 404, 519, 521, 536, 537, 538, 560, 579
Roubeau, P. 401, 406, 447, 536, 537
Rybaczewski, E. F. 92, 93

Sager, R. E. 270, 271
Sarma, G. 515, 516, 524
Schawlow, A. L. 417
Schmugge, T. J. 359
Schrieffer 194
Schull, C. G. 442, 443
Shaeffer, J. 89
Shapiro, E. I. 443
Shapiro, F. L. 458
Shattuck, T. W. 38
Sherfinski, J. S. 92
Sinning, G. 38, 93
Smart, J. S. 281
Smith, S. 248, 251, 257, 272
Solomon, I. 375, 376
Spiess, H. W. 84
Squires, G. L. 428
Stalker, D. C. 83
Stejskal, E. O. 89
Steyerl, A. 441
Stinchcombe, R. 315, 327
Stokes, H. T. 38
Strakhota, I. 407
Strandberg, M. W. P. 346
Strombotne, R. L. 24
Sullivan, N. S. 138, 140, 155, 157, 158
Swanenburg, T. J. B. 369

Takagi, S. 223, 263, 264, 266
Taran, Yu. V. 458
Tegenfeld, J. 39
Thirion, J. 406
Thouless, D. J. 180
Trickey, S. B. 116
Trofimov, V. N. 407
Tyablikov, S. V. 596

Udo, F. 406, 407
Urbina, C. 362, 411, 412, 413, 415, 416, 569, 570
Usov, Yu. A. 407
Uttley, C. A. 456

Van Den Heuvel, G. M. 369
Vaughan, R. W. 75
Vermeulen, J. 406, 407
Villain, J. 477
Vu Hoang Chau, 327, 360, 419, 477, 489, 492, 512, 514, 562, 589, 597

Walls, F. L. 417
Walstedt, R. E. 32
Waugh, J. S. 39, 64, 73, 76, 85, 86, 89, 90, 92, 97, 99
Webb, R. A. 254, 270, 271
Weiss, Pierre, 438
Weisskopf, V. F. 422
Wenckebach, W. Th. 310, 355, 359, 369, 515, 556, 557
Werthamer, N. R. 187, 196, 202
Wheatley, J. C. 193, 254, 270, 271
Willard, F. D. C. 180
Williams, E. 501
Wineland, D. J. 417
Winter, J. 95
Wolfe, J. P. 354

Yu, W. N. 137, 149

Zimmermann, H. 84
Zubarev, D. N. 596

SUBJECT INDEX

ABM phase 213, 238–40, 265
  degeneracy of 232
  equations of motion 240
  equilibrium position in 232–3
  free precession 248–51, 255
  identified with ^{3}He A phase 223, 244
  large amplitude motions 247–54
  Leggett's equations 238–40
  longitudinal ringing 247–8, 254–5
  small amplitude motions 242–3
  susceptibility 221
  walls 256
absorption signal (*see also* Zeeman resonance signal)
  area of 293
  cotanh transform 306–13
  first moment 299–301
  linear passages, slightly saturating 294–9
  qualitative shape 301–2
  rate of change of polarization 293–4
  saturation of dipolar energy 295–9
  saturation of polarization 294–5
adiabatic 'fast passage' 8
adiabatic line width 127–31
adiabatic susceptibility 223–6, 264
ADRF (adiabatic demagnetization in rotating frame) 8, 32, 44, 97, 98, 104, 301, 466–76, 504, 530, 531, 579, 585–6
ammonium hydrogen malonate
  measurement of dipolar interaction 92–3
Anderson–Brinkman–Morel phase, *see* ABM phase
antiferromagnetic domains 567–71, 575–6
antiferromagnetic exchange in ^{3}He 115
antiferromagnetic ordering in ^{3}He 181
antiferromagnetic stuctures
  calcium fluoride, *see under* calcium fluoride
  calcium hydroxide 497
  lithium hydride 536–40
antiferromagnetic substances
  neutron scattering 459–62
antiferromagnetism
  electronic and nuclear spins 571–5
  elementary excitations 590–3
  Monte Carlo calculations 598–9
  random phase approximation (RPA) 596–8
  spin wave
    approximation 589–96
    spectrum 595
    sublattice polarization 593–4
antiferromagnetism in non-zero field 560–75
  experimental results 570–1
  mixed state 564–70
  second order transition 561–4
  Weiss-field approximation 565, 569–70
  Weiss-field predictions 561
antiferromagnetism in zero field 510–39
  longitudinal susceptibility 520–31; non-uniform 524–31
  neutron diffraction in lithium hydride 535–9
  quadrupole alignment of rare spins 529–31
  spin–lattice relaxation 531–5
  transverse susceptibility 511–20

Balian–Werthamer phase (BW), *see* BW phase
BCS equations
  A_1 phase 213–14
  Anderson–Brinkman–Morel (ABM) state 213
  Balian–Werthamer (BW) state 212–13
  below critical temperature 208–10
  critical temperature 206–8
  for unitary states 206
  free energy 210–12
  generalized 206
  Ginsberg–Landau approximation 210–12
  p-wave solutions 212
  planar solution 213
  polar solution 213
  solutions 206–14
BCS theory 193–5
  generalized 196
benzene, measurement of dipolar interaction 92
bootstrap effect 411–15
Borghini's relation 381
Born approximation 421, 422, 425
Bragg peaks 428–9, 460, 467
Bragg scattering 426–33, 455, 467
  on polarized targets 441–3
broadening, inhomogeneous, *see* inhomogeneous broadening
BW phase 212–13, 255, 265, 272
  degeneracy of 232
  equations of motion 240–2
  equilibrium position in 233–4

free precession 251–4, 255
identified with ^{3}He B phase 223
in restricted geometry 258–61
Leggett equations 238–40
motion of axis of rotation 244
small amplitude motions 243–4
susceptibility 220–1
wall-pinned mode 261–3
walls 257, 258–61

calcium fluoride
antiferromagnetic in non-zero field 560–75
cooling of electron spins by nuclei 411–15
structure I (antiferromagnetic) 488
structure II (antiferromagnetic) 488–9
structure III (antiferromagnetic) 489, 509–10, 514–35, 547, 577–8
longitudinal susceptibility 524–31
transverse susceptibility 511–16
structure IV (ferromagnetic with domains) 489, 545–55, 578–80
NMR investigation of domains 580–7
transverse susceptibility 554–5
calcium hydroxide
ordered structures 495–8
transverse susceptibility
at negative temperature 555–6
at positive temperature 556–8
Cayley trees, vertex decoration by 327
chemical shift 53, 56, 61–3, 75–6, 82, 85
anistropic 62–3
heteronuclear narrowing 86–9
commutation relations in Legget's equations 236–7
Cooper pairs 193–5, 196, 218, 230
influence of wall on 256
susceptibility of a single pair $d(n)$ 214–20
cotanh transform of absorption signal 306–13
coupling, *see* spin coupling
critical temperature 206–8
Curie–Weiss law 179, 181
CW spin rotation 58–64

Debye approximation for phonon spectrum 162
Debye frequency 110, 116
Debye temperature 110
defects–phonons coupling 161–4
relaxation 161–4
deuteron polarization 396, 398
diagrammatic method for spin operators 315–27
contractions and diagrams 318–22
general case 322–7
longitudinal case 316–22
semi-invariants 317–18, 323

differential cross effect 375
diffusion, spin, *see* spin diffusion
dipolar cooling 44
dipolar energy 328–9
and Zeeman energy 6, 7
variation with polarization in high field 332–6
dipolar interactions
heteronuclear narrowing 89–93
quenching by effective field 60–1
dipolar order in high field 43–5
dipolar relaxation rate 372–3
dipolar temperature
cotanh transform of absorption signal 306–13
DNP (dynamic nuclear polarization) 472, 473
and cooling by laser irradiation 417–18
applications 402–7
electronic spin-lattice relaxation 344–9
electronic spin–spin interactions 362–5
electronic spin–spin temperature 362–5
historical background 373–7
hyperfine structure 342–4
in measurement of pseudomagnetic moments 449–50
Nedor method 402–5
nuclear hyperfine relaxation 346–9
solid effect, *see* solid effect
spin Hamiltonian, 342–4
domains
ferromagnetic (*see also* ferromagnets with domains):
reproducibility 580–3
size 541, 543
thickness 576–7, 583–7
antiferromagnetic 575–6
dynamic nuclear polarization, *see* DNP

echoes, spin, *see* spin echoes
electronic spin–spin reservoir 365–8, 373–402
coupling to nuclear Zeeman reservoir 365–8, 408–18
experimental results 393–402
high temperature case 377–9
low temperature case 379–88
numerical results 388–93
electronic spin–spin temperature 362–5
electronic Zeeman inverse temperature 377
entropy (expansion of) 331–2

EPR (or ESR) line 363–5, 367, 369, 379, 382, 386, 387–8, 388–95, 404–5
singlet, 415–16
triplet, 411–15
exchange, spin
four-spin 184–5
Hamiltonian 112–14
in solid ^{3}He 111–12, 112–16, 138–44
physical origin of multiple spin 182–5
three-spin 184
exchange integral 114–16
exchange–defects coupling 156–61
impurities 159–61
vacancies 156–9

Fermi fluids 196, 264, 265
Landau theory of 188–93
molecular fields 190–1
Fermi potential, fictitious 422, 433
ferromagnets with domains (*see also under* domains) 539–60
bulk polarization 544
calcium fluoride, *see under* calcium fluoride
calcium hydroxide 495–7, 498, 555–8
characteristic properties 539–44
control of paramagnetic impurities 587–8
dipolar field as function of energy 549
evidence for production of 545–9
ferromagnetic–antiferromagnetic transition in zero field 551–3
lithium hydride 558–60
parallel susceptibility 544
resonance of ^{43}Ca 545–51
transverse susceptibilities 544, 553–8
Weiss-field theory 539–44
fictitious Fermi potential 422, 433
FID, *see* free induction decay
four-spin exchange 184–5
free energy 330–1
free induction decay (FID) 10, 98
free precession 87, 90
ABM phase 248–51
BW phase 251–4
free precession signal 9–10, 63, 66

'Gaussian memory shape' 28–32
Gaussian memory function 131, 152
Ginzburg–Landau approximation 210–12
Grüneisen constant 116

heading distance 257–8
^{3}He, normal liquid 110, 263
and Landau interactions 188–93
numerical data 193
^{3}He, solid 108–85
antiferromagnetic exchange 115
antiferromagnetic structure 606–7
defects–phonons coupling 161–4
exchange, 111–12, 112–16
effects of 122–56
experimental results 138–44
Hamiltonian 112–14
exchange–defects coupling 156–61
vacancies 156–9
impurities 159–61
exchange-integral 114–16
field-temperature phase diagram 600
impurities in 111, 121–2
inadequacy of Heisenberg Hamiltonian in paramagnetic state 180–2
Leggett equations 604–5
low temperature properties 180–5
magnetic ordering in 599–609
motions in 110–22
multiple echoes in 166–79
effects of relaxation 172–3
theory 167–72
multiple-spin exchange model 607–9
NMR study of ordered phase in low field 601–7
ordering properties 599–601
paramagnetic phases 109–10
phase diagram 109
phonons in 112, 122
spin diffusion, *see* spin diffusion
spin–lattice relaxation theory 123–7
thermal dilatation 118
tunnelling frequency 118–20
vacancies in 111, 116–21
effects of 144–56
formation energy 117–18
vacancy waves 121
Weiss fields 601–4
Zeeman relaxation time 138–44
^{3}He, superfluid 186–272
A phase (*see also* ABM phase) 222–3, 244, 245, 246, 254–5
adiabatic susceptibility 264
B phase (*see also* BW phase) 223, 244–5, 247, 255, 270, 271
Cooper pairs and BCS theory 193–5
diagonalization of energy matrix 202–4
energy gap in 200–6
equations of motion 266–7
healing distance 257
large amplitude motions 247–54
linearization of interaction Hamiltonian 200–2
local field 226
magnetic dipole–dipole interactions 226–32
magnetic resonance 232–55
experimental results 244–7
magnetic susceptibility 214–26

phase diagrams 187, 188
relaxation, *see* relaxation
self-consistency equation 204–6
walls 255–63
^{4}He,
impurities in ^{3}He 121–2
vacancy waves 120–1
heteronuclear narrowing, 84–93
chemical shift measurement 86–9
measurement of dipolar interactions 89–93
heteronuclear spin manipulation (*see also* heteronuclear narrowing) 93–5
holmium ethyl sulphate 348, 371
homonuclear narrowing 58–84
effective periodic Hamiltonian 66–9
large effective field in rotating frame 58–64
limitations due to pulse imperfections 82–4
Magnus expansion 69–73, 77, 78
multipulse cycles 73–6
multipulse methods 64–84
off-resonance irradiation 76–7
pulses of finite width 77–82

indirect spin couplings 26, 56–7
inhomogeneous broadening 363–4
experimental results 395
first type 379–83
second type 383–8
'interaction representation' Hamiltonian 71
inverse dipolar temperature 7
inverse lattice temperature 376
inverse spin temperature 8–9, 101
inverse Zeeman temperature 7

Kramers centres 341–2, 343

L–T equations, *see* Leggett–Takagi equations
Landau corrections to susceptibility 221–2
Landau theory of Fermi fluids 188–93
free energy 191
static magnetic susceptibility 192–3
lanthanum magnesium double nitrate, *see* LMN
laser irradiation, cooling by 417–18
lattice correlation function 50–1
lattice correlation time 51
Leggett's equations of motion 238–40, 604–5
Leggett–Takagi (L–T) equations 267, 268, 269
line broadening, *see* inhomogeneous broadening line narrowing
by vacancies 147–52
homonuclear, *see* homonuclear narrowing
line width
adiabatic 127–31
correction functions 31
linear response 302–6
Liouville formalism 15–25, 61
memory functions, *see* memory functions
operators of interest 17–19
lithium fluoride
transverse susceptibility 516–20
lithium hydride
antiferromagnetic structure 538, 539, 540
neutron diffraction 535–9, 558–60
and domains 577–80
transverse susceptibility 516–20
LMN (lanthanum magnesium double nitrate) 351, 359, 369, 396–7, 400, 409, 410, 411, 434, 439
local field and superfluid ^{3}He 226
longitudinal susceptibility, 337, 338, 520–31
antiferromagnets 522–31, 564–5
non-uniform 524–31
Luttinger and Tisza's method 485–7, 493

'magic angle' mechanical spinning 89
magic axis 64
magic sandwich 47–9
magnetic susceptibility 214–26
ABM phase 221
adiabatic 223–6
BW phase 220–1
experimental results 222–3
Landau corrections 221–2
longitudinal, *see* longitudinal susceptibility
transverse, *see* transverse susceptibility
Magnus expansion 69–73, 78, 97
Magnus Hamiltonian 96, 97, 105, 106, 107
Magnus paradox 103–7
memory functions 19, 131, 150–1
and resonance line shape 25–32
Gaussian approximation 28–32, 39–41
model system and Hamiltonian 95–6, 99–102, 105
Monte Carlo calculations 598–9
MREV sequence 75–7
advantage over WHH-4 83
off-resonance irradiation 76–7
scaling factor 82
multiple spin echoes, *see* spin echoes, multiple
multiple spin exchange, *see under* exchange, spin
multipulse cycles 73–6
multipulse methods 64–84
spin temperature experiments 95–103

Nedor method 402–5
neutron(s)
  absorption 453–4
  refractive index 452–3
  spin-dependence of absorption cross-sections 454–6
  transmission 453–4
  wave-like description 452–9
neutron scattering
  absorption cross-section 423
  antiferromagnetic nuclear states 459–62
  Bragg scattering
    nuclei with spin 430–3
    nuclei without spin 427–30
  by isolated nucleus
    with spin 424–6
    without spin 420–4
  by macroscopic target 426–33
  correlation function 459–62
    mapping 462–6
  domains 459–69, 462–6, 577–80
  scattering amplitude 420–2, 426, 440–1
  short-range spin correlations in paramagnetic dipolar state 466–9
NMR rigorous results 293–313
NMR signals 9–12
  absorption and dispersion at low r.f. level 10–12
  free precession 9–10, 63, 66
  formal expression of 274–6
  from organic substances 84
  Zeeman, *see* Zeeman resonance signal
NMR spectrum
  solid samples compared with liquid 53–4
non-Kramers centres 341–2
non-Zeeman relaxation mechanism 156–66
nuclear direct process 348
nuclear magnetic ordering
  antiferromagnetism in zero field 510–39
  approximate theories 500–10
  domains 575–88
  ferromagnetism 539–60
  general character of structures 498–500
  general features 473–6
  helical structures 477, 482
  Luttinger and Tisza's method 485–7
  magnetic measurements 474
  negative temperature structures 489–93
  neutron diffraction method 474
  permanent structures 481–2, 486
  positive temperature structures 488–9
  prediction of ordered structures 476–500
  principle of production 470–3
  restricted-trace approximation 505–10
  stable structures
    in calcium fluoride 487–8
    in calcium hydroxide 495–8
    in lithium fluoride and hydride 493–5
  Villain's method 477–85
  Weiss-field approximation 500–5
    equations 500–3
nuclear pseudomagnetic field, *see* pseudomagnetic nuclear field
nuclear pseudomagnetic moments (*see also* pseusomagnetic nuclear field)
  DNP method 449–50
  experimental determination 443–50
  static polarization method 447–9
  systemactic measurements 440–51
nuclear relaxation
  by isolated paramagnetic impurities 349–53
    isolated electron-nucleus pair 349–53
  non-interacting centres 354–5
    random field approach 349–51
    scrambled states approach 351–3
  dipolar 371–3
  Zeeman 368–71
nuclear Zeeman energy
  and dipolar energy 6, 7
  and spin–spin energy 39, 376
    experimental results 393
nuclear Zeeman reservoir
  coupling with electronic spin–spin reservoir 365–8, 408–18
off-resonance irradiation 76–7

Orbach process 348
organic substances
  NMR signals 84

paramagnetic centres
  structure and relaxation 341–9
  types 341
paramagnetic dipole state
  short-range spin correlations 466–9
paramagnetic impurities
  and ferromagnetic domains 587–8
  nuclear relaxation by, *see under* nuclear relaxation
paramagnetic state
  inadequacy of Heisenberg Hamiltonian 180–2
  non-linear domain 313–38
  transverse susceptibility 518–20
phase refocusing 45–9
phonon bottlenecks 358–9, 360–1, 369, 404
  coefficient 345
phonon spectrum, Debye approximation for 162
phonons
  collisions with vacancy waves 146

coupling with defects, *see* defects–phonons coupling
coupling with vacancy waves 164–6
in solid ^{3}He 112, 122
polarization, spin, *see* spin polarization
proton polarization 396, 398, 410
Provotorov equations 8, 12–15, 35, 36, 132, 378
criterion for validity of 14
generalized 19–25, 124, 366
long term trend toward thermal equilibrium 19–22
short term transient oscillations 22–5
spin–lattice relaxation 49–52
pseudomagnetic moments, *see* nuclear pseudomagnetic moments
pseudomagnetic nuclear field 433–40
and Weiss field 438
experimental proof 434–8
rotating 435–6
pseudomagnetic precession 443–50, 450–1, 456–9

quadrupole alignment of rare spins 529–31

Raman process 348
random phase approximation (RPA) 596–8
rare isotope spins
heteronuclear narrowing 84–93
Rayleigh formula 163
relaxation
dipolar rate 372–3
equations of motion 266–7
non-Zeeman mechanism 156–66
nuclear, *see* nuclear relaxation
of wall-pinned mode 268–9
proton, through ^{165}Ho spins 371
small amplitude oscillations 269–72
spin–lattice, *see* spin–lattice relaxation
superfluid ^{3}He 263–72
theory of in superfluid ^{3}He 263–8
transverse 132–4
resonance line shape 25–32
resonance signals, *see* NMR signals
rotating crystal 408–11

scattering amplitude, neutron 420–2, 426
spin-independent 440–1
silver fluoride
heteronuclear spin manipulation 93–5
slow neutron scattering, *see* neutron scattering
solid effect, DNP 339–41, 351, 436
experimental results 359–62
rate equations 355–9
well-resolved 355–62, 375, 395–6

spin coupling
indirect, *see* indirect spin coupling
unlike systems, 32–41
experimental results 36–41
spin diffusion 134–7
in measurement of domain thickness 584
vacancies and 146–7
experimental results 153–4
spin echoes 45–6, 47–9, 166–79
multiple
general case 172–9
sample shape and field gradient 173–4
theory of 167–72
spin exchange, *see* exchange, spin
spin Hamiltonians 54–7
chemical shift 56
indirect couplings 56–7
quadrupole interaction 55–6
spin manipulation, coherent
fast rotation of sample 57–8
homonuclear narrowing, *see* homonuclear narrowing
spin operators, *see* diagrammatic method for spin operators
spin polarization 87, 90, 329–30
spin precession 59, 63
spin rotation
methods of producing 54
spin system
thermodynamic properties 327–32
spin temperature 362–5
and Weiss field 503–5
concept of 4–9
validity 41–9
dipole–dipole energy 328–9
free energy 330–1
Hamiltonian 2–4
high fields 6–9
dipolar order in 43–5
with r.f. 7–9
high temperature approximation 5, 51–2
low fields 5–6
magic sandwich 47–9
multipulse experiments 95–103
non-linear effects 313–38
experimental illustration 332–8
phase refocussing 45–9
Provotorov equations, *see* Provotorov equations
random phase assumption 41–3
susceptibilities in zero effective field 336–8
theory of 2–15
variation of dipolar energy with polarization in high field 332–6
spin wave approximation 589–96
spin–lattice equation 50

spin–lattice relaxation 49–52, 89
  antiferromagnetism in zero field 531–5
  'extreme narrowing' 51
  long correlation time in ordered state 533–5
  pseudomagnetic precession method 450–1
  short correlation time in ordered state 532–3
  solid ^{3}He 123–7
spin–spin energy
  and nuclear Zeeman energy 369, 376
    experimental results 393
static polarization for pseudomagnetic moments 447–9
sublattice polarization 593–4
superfluid ^{3}He, *see* ^{3}He, superfluid
superfluid phases
  ABM, *see* ABM phase
  adiabatic approximation 234–5
  BW, *see* BW phase
  commutation relations 236–7
  equilibrium positions 232–4
  Leggett's equations of motion 238–40
  magnetic resonance 232–55
superstructure peaks 460
susceptibility, magnetic, *see* magnetic susceptibility

temperature
  in statistical mechanics 42
  spin, *see* spin temperature
thermodynamic properties of spin system 327–32
three-spin exchange 184
transverse susceptibility 302–6, 332, 511–20
  calcium fluoride 511–16
  ferromagnets 544, 553–8
  in zero effective field 336–8
  lithium fluoride and hydride 516–20
tunnelling frequency 118–20, 150, 153
tunnelling Hamiltonian 147, 148
two-coils method for pseudomagnetic precession 443–50

vacancies in solid ^{3}He, 111, 116–21
  and spin diffusion 146–7, 153–4
  effects of 144–56
  exponential Arrhenius law for motion of 145
  formation energy 117–18
  line narrowing by 147–53
  relaxation by 147–52
    experimental results 153
vacancies–phonon relaxation rate 159
vacancy activation energy 158
vacancy waves
  collisions with phonons 146
  coupling with phonons 164–6
  solid ^{3}He 121
  solid ^{4}He 120–1
vertex decoration by Cayley trees 327
Villain's method 477–85, 493
  general form of solutions 480–2
  systems with two spin species 483–5

wall-pinned mode 261–3, 271, 272
  relaxation of 268–9
walls in superfluids 255–63
  influence on Cooper pair 256
Weiss-field equations 500–3
Weiss-field theory 282, 438, 500–5, 513, 514–16, 517–20
  and spin temperature 503–5
  antiferromagnets 561, 565, 569–70
  ferromagnets 539–44, 550, 552
  solid ^{3}He 599–600, 601–4
WHH-4 sequence 73–5, 91, 92
  finite-width pulses 79–82
  MREV advantage over 83
  off-resonance irradiation 76–7
  scaling factor 82

Yosida function 216
yttrium ethyl sulphate 354

Zeeman deuteron reservoir 399
Zeeman energy
  and dipolar energy 6, 7
Zeeman relaxation 138–44
  due to vacancies: experimental results 154–6
  theory of 147–53
Zeeman resonance signal (*see also* absorption signal) 276–92
  moments of, 279–81
    first 281–8
    second 288–91
    higher 291–2
  one spin species 281–5
  several spin species 284–8